BASIC CURVES

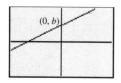

$$y = mx + b$$

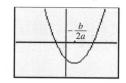

$$y = ax^2 + bx + c \quad (a > 0)$$

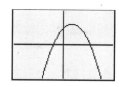

$$y = ax^2 + bx + c \quad (a < 0)$$

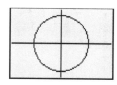

$$x^2 + y^2 = a^2$$

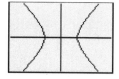

$$y^2 = 4px \quad (p > 0)$$

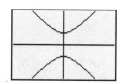

$$x^2 = 4py \quad (p > 0)$$

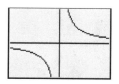

$$\frac{x^2}{a^2} + \frac{y^2}{b^2} = 1$$

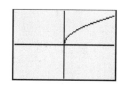

$$\frac{y^2}{a^2} + \frac{x^2}{b^2} = 1$$

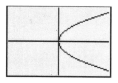

$$\frac{x^2}{a^2} - \frac{y^2}{b^2} = 1$$

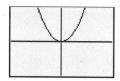

$$\frac{y^2}{a^2} - \frac{x^2}{b^2} = 1$$

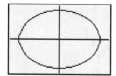

$$xy = a \quad (a > 0)$$

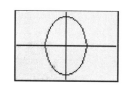

$$y = a\sqrt{x} \quad (a > 0)$$

$$y = ax^3 \quad (a > 0)$$

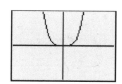

$$y = ax^4 \quad (a > 0)$$

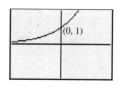

$$y = b^x \quad (b > 1)$$

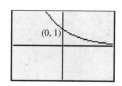

$$y = b^{-x} \quad (b > 1)$$

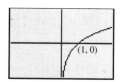

$$y = \log_b x$$

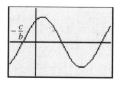

$$y = a \sin(bx + c)$$
$$(a > 0, c > 0)$$

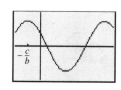

$$y = a \cos(bx + c)$$
$$(a > 0, c > 0)$$

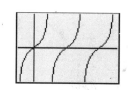

$$y = a \tan x \quad (a > 0)$$

TECHNICAL CALCULUS
WITH ANALYTIC GEOMETRY

OTHER ADDISON-WESLEY TITLES OF RELATED INTEREST

- *Basic Technical Mathematics,* Seventh Edition, by Allyn J. Washington
- *Basic Technical Mathematics with Calculus,* Seventh Edition, by Allyn J. Washington
- *Basic Technical Mathematics with Calculus, Metric Version,* Seventh Edition, by Allyn J. Washington
- *Introduction to Technical Mathematics,* Fourth Edition, by Allyn J. Washington and Mario F. Triola

TECHNICAL CALCULUS
WITH ANALYTIC GEOMETRY

— FOURTH EDITION —

ALLYN J. WASHINGTON
Dutchess Community College

Addison
Wesley

An imprint of Addison Wesley Longman, Inc.

Boston, Massachusetts • Menlo Park, California • New York • Harlow, England
Don Mills, Ontario • Sydney • Mexico City • Madrid • Amsterdam

Publisher: Jason A. Jordan

Acquisitions Editor: Maureen O'Connor

Editorial Assistant: Melissa Wright

Associate Editor: Suzanne Alley

Managing Editor: Ron Hampton

Text Design and Cover Art Direction: Dennis Schaefer

Production Services: Greg Hubit Bookworks

Associate Production Supervisor: Sheila Spinney

Prepress Services Buyer: Caroline Fell

Marketing Manager: Dona Kenly

Manufacturing Buyer: Evelyn Beaton

Composition: The Beacon Group

ABOUT THE COVER

Millennial Cityscape is a three-dimensional rendering of a computer motherboard created using KPT Bryce software with enhanced textures added in Adobe Photoshop.

LIBRARY OF CONGRESS CATALOGING-IN-PUBLICATION DATA

Washington, Allyn J.
    Technical calculus with analytic geometry / Allyn J. Washington.—4th ed.
        p.   cm.
    Includes indexes.
    ISBN 0-201-71112-5
    1. Calculus.   2. Geometry, Analytic.   I. Title.
QA303.W38   2002
515'.15—dc21   00-066376

    6 7 8 9 10—CRW—05

CONTENTS

SCOPE OF THE BOOK

This book is intended primarily for students taking technical programs at two-year and four-year colleges and technical institutes. The text emphasizes elementary topics in calculus and certain selected topics from the more advanced areas. They have been selected because of their application in the technologies, especially electrical and mechanical technologies. The topics are developed in a nonrigorous and intuitive manner, with stress being placed on the interpretation and application of the material being presented.

It is assumed that students using this text will have had courses covering basic algebra and trigonometry. However, several sections on algebra and trigonometry are included for review and reference purposes.

The general topics covered in the text are basic analytic geometry, differentiation and integration of algebraic and elementary transcendental functions, an introduction to partial derivatives and double integrals, expansion of functions in series, and differential equations.

Analytic geometry is developed primarily for use in the calculus, although numerous direct applications of curves, such as the conic sections, are discussed. In the chapter dealing with partial derivatives and double integrals, a section is devoted to solid analytic geometry. The chapter on polar coordinates demonstrates the use of different coordinate systems, including cylindrical coordinates. Also, the graphs of the trigonometric, inverse trigonometric, exponential, and logarithmic functions are reviewed.

The calculus will give the student a mathematical understanding of many topics that arise in other courses. The standard methods of differentiation and integration are covered, along with certain numerical methods, applicable for use on a calculator or a computer.

Partial derivatives and double integrals are included for use in courses in which these topics are deemed appropriate. Applications are shown for such fields as electronics, thermodynamics, and mechanics.

The chapter on expansion of functions in series will give students an understanding of various types of series and operations using them. The sections on Fourier series are appropriate for curricula in electrical technology.

Differential equations provide one of the richest areas of mathematical applications in technical areas, especially in electrical and mechanical technologies. Here a mathematical basis can be established for many concepts and formulas that students have had to accept without basis.

Numerical methods of solving differential equations are appropriate for use with calculators and computers, and familiarity with them will make the technician more effective in modern industry. Laplace transforms are introduced to show methods and terminology for electrical and mechanical students.

NEW FEATURES

This fourth edition of *Technical Calculus with Analytic Geometry* includes all the basic features of the first three editions. However, most sections have been rewritten to some degree to include additional or revised explanatory material, examples, and exercises. Some sections have been extensively revised. Specifically, among the new features are the following:

THE GRAPHING CALCULATOR

The graphing calculator is used throughout the text in examples and exercises to help develop and reinforce the coverage of many topics. There are over 60 graphing calculator screens displayed with the text material.

Appendix C presents a discussion of the use of a graphing calculator. Also in Appendix C, graphing calculator programs show how the calculator's use can be expanded even more.

REVISED COVERAGE

Chapter 1 is now devoted to an introduction to functions and graphs. Among the topics included are domain and range and an introduction to the use of a graphing calculator. Coverage of analytic geometry is now in Chapter 2.

Linear approximations are now included with differentials and are now in the final section of Chapter 4 on Applications of the Derivative. Cylindrical coordinates are now included with polar coordinates in Chapter 12. The coverage of the Fourier series has been expanded to two sections to include additional coverage, including half-range expansions. There is additional coverage of higher-order differential equations in Chapter 15. The Euler and Runge-Kutta numerical methods of solving differential equations are included in Chapter 16. In Appendix A there are sections on Rotation of Axes and Regression. This material on regression replaces the chapter on empirical curve fitting in the third edition.

PAGE LAYOUT

Special attention has been given to the page layout. Nearly all examples are started and completed on the same page (there are only three exceptions, and each of these is presented on facing pages). Also, all figures are shown immediately adjacent to the material in which they are discussed.

SPECIAL CAUTION AND NOTE INDICATORS

CAUTION ▶

NOTE ▶

Two special margin indicators (as shown at the left) are used. The caution indicator is used to identify points with which students commonly make errors or which they tend to have more difficulty in handling. The note indicator is used to point out text material that is of particular importance in developing or understanding the topic under discussion.

WRITING EXERCISES

There is one specific writing exercise included at the end of each chapter. These exercises give the student practice in writing explanations of problem solutions. Also, there are over 110 additional exercises throughout the book (at least five in each chapter) that require at least a sentence or two of explanation along with the answer. These are noted by the symbol Ⓦ placed to the left of the exercise number. A special *Index of Writing Exercises* is included in the back of the book.

PROBLEM-SOLVING TECHNIQUES

Techniques and procedures that summarize the approaches to solving many types of problems have been outlined in color-shaded boxes.

SUBHEADS AND KEY TERMS

Many sections include subheads to indicate where the discussion of a new topic starts within the section, and other key terms are noted in the margin for emphasis and easy reference. Also, where appropriate, chapter equations from earlier material are shown in the margin.

CHAPTER INTRODUCTIONS

Each chapter is introduced by identifying the topics to be developed and some of the important areas of technical applications. A particular type of application is illustrated through a drawing at the beginning of each chapter, and a problem related to this application is solved in an example later in the chapter.

CHAPTER EQUATIONS, REVIEW EXERCISES, AND PRACTICE TEST

At the end of each chapter, all important equations are listed together for easy reference. Each chapter is also followed by a set of review exercises that covers all the material of the chapter. Following the chapter equations and review exercises is a chapter practice test that students can use to check their understanding of the material. Solutions to all practice test problems are given in the back of the book.

EXERCISES

There are about 4400 exercises (an increase of about 20%), including about 2000 that are new to this edition. Technical applications are illustrated in about 1000 of these exercises, including over 650 that are new to this edition.

EXAMPLES

Over 620 worked examples are included in this edition. Of these, over 120 illustrate technical applications.

FIGURES

There are about 620 figures in this edition, an increase of about 200 from the third edition. These new figures assist the students in the examples and exercises.

ADDITIONAL FEATURES

APPLICATIONS

Examples and exercises illustrate the application of mathematics to all fields of technology. Many relate to modern technology such as computers, electronics, solar energy, lasers, fiber optics, holography, and space technology. A special *Index of Applications* for the exercises is included in the back of the book.

SPECIAL EXPLANATORY COMMENTS

Special explanatory comments in color have been used in many examples to emphasize and clarify certain important points. Arrows are often used to indicate clearly the part of the example to which reference is made.

IMPORTANT FORMULAS

Throughout the book, important formulas are set off and displayed so that they can be easily located and used.

ANSWERS TO EXERCISES

The answers to all the odd-numbered exercises (except the end-of-chapter writing exercises) are given at the back of the book. The *Student's Solutions Manual* contains solutions for all of the odd-numbered exercises and the *Instructor's Solutions Manual* contains solutions for all of the exercises. The answers to all exercises are given in the *Answer Book*.

FLEXIBILITY OF MATERIAL COVERAGE

The order of material coverage can be changed in many places, and certain sections may be omitted without loss of continuity of coverage. Users of earlier editions have indicated the successful use of several variations in coverage. Any changes will depend on the type of course and completeness required.

SUPPLEMENTS

Supplements to this text include an *Instructor's Solutions Manual* with detailed solutions to every exercise, as well as a *Student's Solutions Manual* with detailed solutions for every odd-numbered exercise (except the end-of-chapter writing exercises). An *Answer Book* for the instructors includes answers to all exercises.

Instructors may obtain copies of any of these supplements by contacting their Addison-Wesley sales consultant.

QUESTIONS/COMMENTS/INFORMATION

We welcome your comments about this text. You may write to the publisher or author at:

> Addison-Wesley
> Mathematics Marketing
> 75 Arlington Street, Suite 300
> Boston, MA 02116

You may also contact us via e-mail at math@awl.com to leave comments, suggestions, or questions for the publisher or the author.

ACKNOWLEDGMENTS

The author gratefully acknowledges the contributions of the following reviewers. Their detailed comments and many suggestions were of great assistance in preparing this fourth edition.

Nan Byars   James Runyon
University of North Carolina   *Rochester Institute of Technology*

Carole Goodson
University of Houston

Sue Schroeder
University of Houston

Deborah Hochstein
University of Memphis

Irving Tang
Oklahoma State University

George Poole
East Tennessee State University

Jerry Wilkerson
Missouri Western State College

Scott Randby
Rochester Institute of Technology

The author also wishes to acknowledge the comments of the following reviewers of the calculus chapters of *Basic Technical Mathematics with Calculus,* Seventh Edition, for their comments and suggestions that were of assistance in preparing many of the chapters of this text.

Ginny Anson
Northeastern Iowa Community College

Wayne Braith
St. Cloud State University

Michael Chen
*British Columbia Institute of
Technology*

Marcel Maupin
*Oklahoma State University
Oklahoma City*

Carol A. McVey
Florence Darlington Technical College

E. Kurt Mobley
Augusta Technical Institute

Robert Opel
Waukesha County Technical College

Derek Randall
Camosun College

Jack Sontrop
Loyalist College

Special thanks go to Bob Martin of Tarrant County Community College for preparing the *Answer Book,* the *Student's Solutions Manual,* and the *Instructor's Solutions Manual.* My thanks and gratitude also go to Jim Bryant who drew all of the chapter-opener drawings.

I also wish to express my appreciation and thanks to James Runyon of the Rochester Institute of Technology and Sue Schroeder of the University of Houston for their assistance in the tedious task of checking the exercises and answers.

I gratefully acknowledge the cooperation and support of my editors, Jennifer Crum and Maureen O'Connor. The very valuable assistance of the production editor, Greg Hubit, was also greatly appreciated.

Also of great assistance during the production of this edition were Suzanne Alley, Ruth Berry, Ron Hampton, Sheila Spinney, Dona Kenly, and Melissa Wright of the Addison-Wesley staff. Finally, special mention is due my wife, Millie, for her patience and support all through the preparation of this edition and all earlier editions.

A. J. W.

1

FUNCTIONS AND GRAPHS

In technology and science, as well as in everyday life, we often find that one quantity depends on one or more other quantities. Plant growth depends on sunlight and rainfall; traffic flow depends on roadway design; the sales tax on an item depends on the cost of the item; the time required to access an Internet site depends on the speed at which a computer processes data. These are but a few of the innumerable possible examples.

Determining how one quantity depends on other quantities is one of the primary goals of science. A rule that relates such quantities is of great importance and usefulness in science and technology. In mathematics such a rule is called a *function,* and we start this chapter with a discussion of functions. We also will develop basic methods of drawing and using *graphs,* which give us a way of visualizing functions.

An important type of problem that arises involves the rate of change of one quantity with respect to another. Examples of rates of change are velocity (the rate of change of distance with respect to time), the rate of change of the length of a metal rod with respect to the temperature, the rate of change of light intensity with respect to the distance from the source, the rate of change of electric current with respect to time, and many other similar physical situations. The methods of **differential calculus,** which we shall start developing in Chapter 3, will enable us to solve problems involving these quantities.

Another principal type of problem that calculus enables us to solve is that of finding a function when its rate of change is known. This is **integral calculus,** a study of which starts in Chapter 5. One of the principal applications comes from electricity, where current is the time rate of change of electric charge. Integral calculus also leads to the solutions of a great many other apparently unrelated problems, which include plane areas, volumes, and the physical concepts of work and force due to liquid pressure.

Before starting a study of the calculus, in Chapter 2 we will first develop some of the basic concepts of **analytic geometry.** This branch of mathematics deals with the relationship between algebra and geometry. The definitions and methods developed in Chapters 1 and 2, along with certain basic curves and their characteristics, will enable us to develop the calculus more readily. Also, many technical applications of functions and analytic geometry are illustrated.

The electric power produced in a circuit depends on the resistance R in the circuit. In Section 1-4 we draw a graph of
$$P = \frac{100R}{(0.50 + R)^2}$$ to see this type of relationship.

1-1 INTRODUCTION TO FUNCTIONS

If we were to perform an experiment to determine whether or not a relationship exists between the distance an object falls from rest and the time it falls, observation of the results would indicate (approximately, at least) that $s = 16t^2$, where s is the distance in feet and t is the time in seconds. Here we note that t may take on essentially any positive value and that the value of s will vary, depending on the value of t.

In general, *those literal numbers (numbers represented by letters) that may vary are called* **variables,** and *those that are held fixed are called* **constants.** Generally letters toward the end of the alphabet are used to denote variables, and letters near the beginning of the alphabet denote constants, although there are occasional exceptions to this usage.

■**EXAMPLE 1**   In physics it is shown that the equation relating the distance s an object falls in t seconds, when the acceleration due to gravity is g, is given by $s = \frac{1}{2}gt^2$. Here, s and t are variables, and g is a constant.

In the equation for the area of a circle, $A = \pi r^2$, A and r are variables, and π is a constant.   ■

The equations in Example 1 illustrate how one quantity may be shown to be related to another quantity. Such relationships lead to one of the most important and basic concepts in mathematics, that of a *function*.

DEFINITION OF A FUNCTION

> *Whenever a relationship exists between two variables such that for every value of the first, there is one and only one corresponding value of the second, we say that the second variable is a* **function** *of the first variable.*
>
> *The first variable is called the* **independent variable,** *and the second variable is called the* **dependent variable.**

The first variable is termed *independent* since permissible values can be assigned to it arbitrarily, and the second variable is termed *dependent* since its value is determined by the choice of the independent variable. *For our work, values of the independent variable and dependent variable will be real numbers.* Therefore, there may be restrictions on their possible values. This is discussed in the following section.

■**EXAMPLE 2**   In the equation $y = 2x$, we see that y is a function of x, since for each value of x there is only one value of y. For example, if we substitute $x = 3$, we get $y = 6$ and no other value. By arbitrarily assigning values to x and then substituting, we see that the values of y we obtain *depend* on the values chosen for x. Therefore, x is the independent variable and y is the dependent variable.   ■

■**EXAMPLE 3**   The power P developed in a certain resistor by a current I is given by $P = 4I^2$. Here P is a function of I. The dependent variable is P, and the independent variable is I.

If the equation relating the power and current is written as $I = \frac{1}{2}\sqrt{P}$; that is, if the current is expressed in terms of the power, I is a function of P. In this case, I is the dependent variable and P is the independent variable.   ■

There are many ways to express functions. Formulas, tables, charts, and graphs can also define functions. Functions will be of importance throughout the book, and we will use a number of different types of functions in later chapters.

Functional Notation

For convenience of notation, the phrase

"function of x" is written as f(x).

Therefore, "*y* is a function of *x*" may be written as $y = f(x)$. Here, *f* represents the function and does not represent a variable. (Although the function is *f*, it is common usage to refer to the function as $f(x)$ in order to show the variable on which *f* depends). Note carefully that $f(x)$ *does not mean f times x.*

■ EXAMPLE 4 If $y = 6x^3 - 5x$, we say that *y* is a function of *x*. This function is $6x^3 - 5x$. It is also common to write such a function as $f(x) = 6x^3 - 5x$. However, *y* and $f(x)$ represent the same expression, $6x^3 - 5x$. Using *y*, the quantities are shown, and using $f(x)$, the functional dependence is shown. ----------■

One of the most important uses of functional notation is to designate the value of the function for a particular value of the independent variable. That is,

the value of the function f(x) when x = a is written as f(a).

This is illustrated in the next example.

■ EXAMPLE 5 For a function $f(x)$, the value of $f(x)$ for $x = 2$ may be expressed as $f(2)$. Thus, substituting 2 for *x* in $f(x) = 3x - 7$, we have

$$f(2) = 3(2) - 7 = -1 \qquad \text{substitute 2 for } x$$

The value of $f(x)$ for $x = -1.4$ is

$$f(-1.4) = 3(-1.4) - 7 = -11.2 \qquad \text{substitute } -1.4 \text{ for } x \quad ----■$$

We must note that *whatever number a represents, to find f(a), we substitute a for x in f(x).* This is true even if *a* is a literal number, as in the following examples.

■ EXAMPLE 6 The electric resistance *R* of a particular resistor as a function of the temperature *T* is given by $R = 10.0 + 0.10T + 0.001T^2$. If a given temperature *T* is increased by 10°C, what is the value of *R* for the increased temperature as a function of the temperature *T*?

We are to determine *R* for a temperature of $T + 10$. Since

$$f(T) = 10.0 + 0.10T + 0.001T^2$$

then

$$f(T + 10) = 10.0 + 0.10(T + 10) + 0.001(T + 10)^2 \qquad \text{substitute } T + 10 \text{ for } T$$
$$= 10.0 + 0.10T + 1.0 + 0.001T^2 + 0.02T + 0.1$$
$$= 11.1 + 0.12T + 0.001T^2 \qquad ----■$$

All calculator screens shown with text material are for a TI-83. They are intended only as an illustration of a calculator screen for the particular operation. Screens for other models may differ.

CAUTION ▶

A graphing calculator can be used to evaluate a function in several ways. One is to directly substitute the value into the function as shown in Fig. 1-1(a). A second is to enter the function as Y_1 and evaluate as shown in Fig. 1-1(b). A third way, which is very useful when many values are to be used, is to enter the function as Y_1 and use the *table* feature as shown in Fig. 1-1(c).

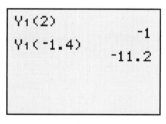

(a)

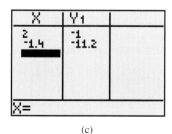

(b)

(c)

Fig. 1-1

At times we need to define more than one function of the independent variable. We then use different symbols to denote the different functions. For example, $f(x)$ and $g(x)$ may represent different functions, such as $f(x) = 5x - 3$ and $g(x) = ax^2 + x$. Special functions are represented by particular symbols. For example, in Chapter 7 when we discuss trigonometric functions, we come across the "sine of the angle θ," where the sine is a function of θ. This is designated by $\sin \theta$.

■ EXAMPLE 7   For $f(x) = 5x - 3$ and $g(x) = ax^2 + x$, we have

$$f(-4) = 5(-4) - 3 = -23 \qquad \text{substitute } -4 \text{ for } x \text{ in } f(x)$$

$$g(-4) = a(-4)^2 + (-4) = 16a - 4 \qquad \text{substitute } -4 \text{ for } x \text{ in } g(x) \qquad ■$$

A function may be looked upon as a set of instructions. These instructions tell us how to obtain the value of the dependent variable for a particular value of the independent variable, even if the instructions are expressed in literal symbols.

■ EXAMPLE 8   The function $f(x) = x^2 - 3x$ tells us to "square the value of the independent variable, multiply the value of the independent variable by 3, and subtract the second result from the first." An analogy would be a computer that was programmed so that when a number was entered into the program it would square the number, then multiply the number by 3, and finally subtract the second result from the first.

The functions $f(t) = t^2 - 3t$ and $f(n) = n^2 - 3n$ are the same as the function $f(x) = x^2 - 3x$, since the operations performed on the independent variable are the same. ***Although different literal symbols appear, this does not change the function.***   ■

NOTE ▶

The Real Number System

The numbers that we use, such as in evaluating functions, may be represented by points on a line. We draw a horizontal line and designate some point on it by O, which we call the **origin** (see Fig. 1-2). Starting from the origin, equal intervals are marked off along the line, and the **positive integers** *(the numbers $1, 2, 3, \ldots$, which represent whole quantities)* are placed at these positions to the right of the origin. **Zero** *(which is an integer, but is neither positive nor negative)* is placed at the origin, and the **negative integers** are placed to the left of the origin

Negative numbers were not widely accepted by mathematicians until late in the sixteenth century.

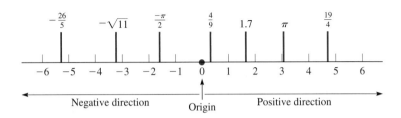

Fig. 1-2

The points between the integers represent either **rational numbers**—*numbers that can be represented by the division of one integer by another (not zero)*—or **irrational numbers** (*those numbers that cannot be represented by the division of one integer by another;* π *and* $\sqrt{2}$ *are such numbers*). In this way all **real numbers,** *which include all the rational numbers (the integers are rational) and irrational numbers,* may be represented as points on the line. We shall use real numbers throughout this text unless otherwise noted. We will occasionally refer to and use **imaginary numbers,** *which is the name given to square roots of negative numbers.*

Irrational numbers were discussed by the Greek mathematician Pythagoras in about 540 B.C.

■**EXAMPLE 9** The number 7 is an integer (since $7 = 7/1$), rational, and real (the real numbers include all the rational numbers); 3π is irrational and real; $\sqrt{5}$ is irrational and real; $\frac{1}{8}$ is rational and real; $7\sqrt{-1}$ is imaginary; $\frac{6}{3}$ is rational and real (it is an integer when we use the symbol "2" to represent it); $\pi/6$ is irrational and real.

Imaginary numbers were so named because the French mathematician René Descartes (1596–1650) referred to them as "imaginaries." Most of the development of them occurred in the eighteenth century.

We next define another important concept. *The* **absolute value** *of a nonnegative number is the number itself, and the absolute value of a negative number is the corresponding positive number.* On the number line we may interpret the absolute value of a number as the distance between the origin and the number. The absolute value is denoted by writing the number between vertical lines, as shown in the following example.

■**EXAMPLE 10** The absolute value of 6 is 6, and the absolute value of -7 is 7. We write these as $|6| = 6$ and $|-7| = 7$. See Fig. 1-3.

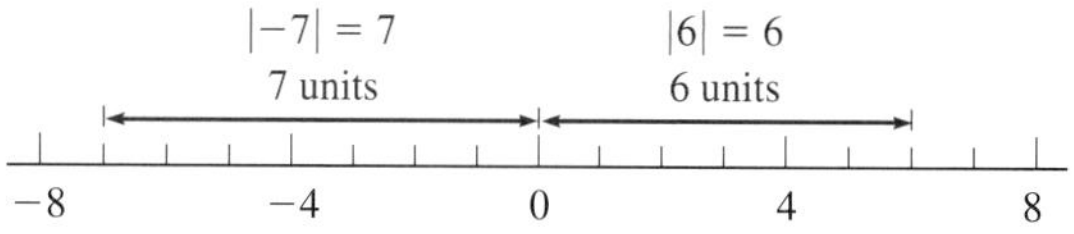

$$|-7| = 7 \qquad |6| = 6$$
$$7\ \text{units} \qquad 6\ \text{units}$$

Fig. 1-3
$$-8 \qquad -4 \qquad 0 \qquad 4 \qquad 8$$

Other examples are: $\left|\dfrac{7}{5}\right| = \dfrac{7}{5}, \quad |-\sqrt{2}| = \sqrt{2}, \quad |0| = 0, \quad -|-9| = -9$ ■

On the number line, *if a first number is to the right of a second number, then the first number is said to be* **greater than** *the second. If the first number is to the left of the second, it is* **less than** *the second number.* The symbol $>$ is used to designate "is greater than," and the symbol $<$ is used to designate "is less than." These are called **signs of inequality.** See Fig. 1-4.

The symbols $=$, $<$, and $>$ were introduced by English mathematicians in the late 1500s.

■**EXAMPLE 11**

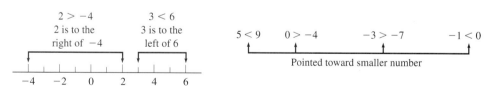

$$2 > -4 \qquad\qquad 3 < 6$$
2 is to the right of -4 \qquad 3 is to the left of 6

$$-4 \quad -2 \quad 0 \quad 2 \quad 4 \quad 6$$

$$5 < 9 \qquad 0 > -4 \qquad -3 > -7 \qquad -1 < 0$$
Pointed toward smaller number

Fig. 1-4 ■

EXERCISES *1-1*

In Exercises 1–8, determine the appropriate functions.

1. Express the area A of a circle as a function of (a) its radius r and (b) its diameter d.

2. Express the circumference c of a circle as a function of (a) its radius r and (b) its diameter d.

3. Express the volume V of a sphere as a function of its diameter d.

4. Express the edge e of a cube as a function of its surface area A.

5. Express the area A of a rectangle of width 5 as a function of its length l.

6. Express the volume V of a right circular cone of height 8 as a function of the radius r of the base.

7. Express the area A of a square as a function of its side s; express the side s of a square as a function of its area A.

8. Express the perimeter p of a square as a function of its side s; express the side s of a square as a function of its perimeter p.

In Exercises 9–20, evaluate the given functions.

9. $f(x) = 2x + 1$; find $f(1)$ and $f(-1)$.

10. $f(x) = 5x - 9$; find $f(2)$ and $f(-2)$.

11. $f(x) = 5 - 3x$; find $f(-2)$ and $f(0.4)$.

12. $f(T) = 7.2 - 2.5T$; find $f(2.6)$ and $f(-4)$.

13. $\phi(x) = \dfrac{6 - x^2}{2x}$; find $\phi(1)$ and $\phi(-2)$.

14. $H(q) = \dfrac{8}{q} + 2\sqrt{q}$; find $H(4)$ and $H(0.16)$.

15. $g(t) = at^2 - a^2t$; find $g(-\frac{1}{2})$ and $g(a)$.

16. $s(y) = 6\sqrt{y + 1} - 3$; find $s(8)$ and $s(a^2)$.

17. $K(s) = 3s^2 - s + 6$; find $K(-s)$ and $K(2s)$.

18. $T(t) = 5t + 7$; find $T(-2t)$ and $T(t + 1)$.

19. $f(x) = 2x + 4$; find $f(3x) - 3f(x)$.

20. $f(x) = 2x^2 + 1$; find $f(x + 2) - [f(x) + 2]$.

In Exercises 21–36, solve the given problems.

21. Designate each of the following numbers as being an integer, rational, irrational, real, or imaginary. (More than one designation may be correct.) $3, -\pi, -\sqrt{-6}, \frac{\sqrt{7}}{3}$

22. Designate each of the following numbers as being an integer, rational, irrational, real, or imaginary. (More than one designation may be correct.) $\frac{5}{4}, \sqrt{-4}, -\frac{7}{3}, \frac{\pi}{6}$

23. Find the absolute value of each of the following numbers: $3, \frac{7}{2}, -\frac{6}{7}, -\sqrt{3}, \sqrt{5} - 3$.

24. Find the absolute value of each of the following numbers: $-4, \sqrt{2}, -\frac{\pi}{2}, -\frac{19}{4}, 3 - \pi$.

25. Insert the correct sign of inequality ($>$ or $<$) between the given pairs of numbers: (a) $-4 \quad 0$ (b) $1 \quad -\pi$ (c) $-\frac{1}{3} \quad -\frac{1}{2}$

26. Insert the correct sign of inequality ($>$ or $<$) between the given pairs of numbers: (a) $-3 \quad -2$ (b) $-\sqrt{2} \quad -1.42$ (c) $-4 \quad -|-3|$

27. If a and b are positive integers and $b > a$, what type of number is represented by

    (a) $b - a$, (b) $a - b$, (c) $\dfrac{b - a}{b + a}$?

28. If a and b represent positive integers, what kind of number is represented by (a) $a + b$, (b) a/b, (c) $a \times b$?

29. For any positive or negative integer: (a) Is its absolute value always an integer? (b) Is its reciprocal always a rational number?

30. For any positive or negative rational number: (a) Is its absolute value always a rational number? (b) Is its reciprocal always a rational number?

(W) 31. For a number $x < 0$, describe the location on the number line of the number with a value of $|x|$.

(W) 32. Describe the location of a number x on the number line when (a) $|x| < 1$, (b) $|x| > 2$.

33. A demolition ball is used to tear down a building. Its distance s (in m) above the ground as a function of the time t (in s) after it is dropped is $s = 17.5 - 4.9t^2$. Since $s = f(t)$, find $f(1.2)$.

(W) 34. The change C (in in.) in the length of a 100-ft steel bridge girder from its length at 40°F, as a function of the temperature T, is given by $C = 0.014(T - 40)$. Since $C = f(T)$, find $f(15)$ and state the meaning of the number obtained.

35. The stopping distance d (in ft) of a car going v mi/h is given by $d = v + 0.05v^2$. Since $d = f(v)$, find $f(30)$, $f(2v)$, and $f(60)$, using both $f(v)$ and $f(2v)$.

36. The electric power P (in W) dissipated in a resistor of resistance R (in Ω) is given by the function $P = \dfrac{200R}{(100 + R)^2}$. Since $P = f(R)$, find $f(R + 10)$.

1-2 ALGEBRAIC FUNCTIONS

The functions studied in mathematics fall into two general classes: **algebraic functions** and **transcendental functions.** In this chapter we shall be concerned with algebraic functions. Transcendental functions, such as the trigonometric functions and the logarithmic function, will be taken up in later chapters.

There are various types of algebraic functions. Those which are of primary concern here are as follows.

Polynomials *are of the form*

$$P(x) = a_0 x^n + a_1 x^{n-1} + \cdots + a_{n-1} x + a_n \tag{1-1}$$

where $a_0, a_1, a_2, \ldots, a_n$ *are constants and* n *is a nonnegative integer.*

Rational functions *are of the form*

$$F(x) = \frac{f(x)}{g(x)} \tag{1-2}$$

where $f(x)$ *and* $g(x)$ *are polynomials.* (A polynomial is a rational function since it can be expressed with a denominator of 1, and 1 is a polynomial with $a_0 = 1$ and $n = 0$.)

Roots *of rational functions are of the form*

$$I(x) = \sqrt[n]{F(x)} \tag{1-3}$$

where $F(x)$ *is a rational function.* Combinations of functions are also encountered.

■EXAMPLE 1 The current I in an electric circuit equals the voltage E divided by the resistance R. If the voltage and resistance are functions of the time t, given by

$$E(t) = 2t^2 + t + 5 \quad \text{and} \quad R(t) = 3t + 20$$

we see from Eq. (1-1) that both are polynomials. The current is given by

$$I(t) = \frac{E(t)}{R(t)} = \frac{2t^2 + t + 5}{3t + 20}$$

which is a rational function. ■

■EXAMPLE 2 If $f(x) = 2x^2 - 3$ and $g(x) = \sqrt{x + 2}$, then

$$f(x) + g(x) = 2x^2 - 3 + \sqrt{x + 2}$$

Also, we find

$$f[g(x)] = f(\sqrt{x + 2}) = 2(\sqrt{x + 2})^2 - 3 = 2x + 1$$
$$g[f(x)] = g(2x^2 - 3) = \sqrt{(2x^2 - 3) + 2} = \sqrt{2x^2 - 1}$$

For $f[g(x)]$ we replace x by $\sqrt{x + 2}$ in $f(x)$, and for $g[f(x)]$ we replace x by $2x^2 - 3$ in $g(x)$. ■

EXPONENTS AND RADICALS

Since we will be dealing with exponents and radicals, we shall now review their basic properties. The basic laws of exponents are as follows.

$$a^m \cdot a^n = a^{m+n} \tag{1-4}$$

$$\frac{a^m}{a^n} = a^{m-n} \quad (a \neq 0) \tag{1-5}$$

$$(a^m)^n = a^{mn} \tag{1-6}$$

$$(ab)^n = a^n b^n, \quad \left(\frac{a}{b}\right)^n = \frac{a^n}{b^n} \quad (b \neq 0) \tag{1-7}$$

$$a^0 = 1 \quad (a \neq 0) \tag{1-8}$$

$$a^{-n} = \frac{1}{a^n} \quad (a \neq 0) \tag{1-9}$$

We shall often find it convenient to express radicals in the form of fractional exponents. The relationship between fractional exponents and radicals, along with the basic operations with radicals, are listed below.

$$a^{m/n} = \sqrt[n]{a^m} \quad (m \text{ and } n \text{ integers}, \quad n > 0) \tag{1-10}$$

$$\sqrt[n]{a^n} = a \tag{1-11}$$

$$\sqrt[m]{\sqrt[n]{a}} = \sqrt[mn]{a} \tag{1-12}$$

$$\sqrt[n]{a} \cdot \sqrt[n]{b} = \sqrt[n]{ab} \tag{1-13}$$

$$\frac{\sqrt[n]{a}}{\sqrt[n]{b}} = \sqrt[n]{\frac{a}{b}} \quad (b \neq 0) \tag{1-14}$$

Unless we state otherwise, when we refer to the root of a number, it is the principal root.

NOTE ▶

To have a single defined value for a root and to consider only real number roots, *the* **principal *n*th root** *of a is defined to be positive if a is positive and to be negative if a is negative and n is odd.* (If *a* is negative and *n* is even, the roots are not real.) Therefore, *the above radical forms hold in general if a > 0 and b > 0*, but not necessarily for negative values of *a* and *b*. For example, a^m may be negative in Eq. (1-10) if *n* is odd, but not if *n* is even. The following examples illustrate the use of the above laws of exponents and radicals with functions.

■EXAMPLE 3   For the function $f(x) = 2x^2 - 1$, we have:

(a) $(2x^2 - 1)^2(2x^2 - 1)^5 = (2x^2 - 1)^{2+5}$ using Eq. (1-4)
$$= (2x^2 - 1)^7$$

(b) $[(2x^2 - 1)^3]^4 = (2x^2 - 1)^{3 \cdot 4}$ using Eq. (1-6)
$$= (2x^2 - 1)^{12}$$

(c) $(2x^2 - 1)^{-2} = 1/(2x^2 - 1)^2$ using Eq. (1-9)

(d) $(2x^2 - 1)^{2/3} = \sqrt[3]{(2x^2 - 1)^2}$ using Eq. (1-10)

(e) $\sqrt[4]{\sqrt{2x^2 - 1}} = \sqrt[4 \cdot 2]{2x^2 - 1}$ using Eq. (1-12)
$$= \sqrt[8]{2x^2 - 1}$$

EXAMPLE 4 Simplify the expression for the function

$$f(x) = \frac{(3x^2 - 1)^{1/3}(2x) - (2x^3)(3x^2 - 1)^{-2/3}}{(3x^2 - 1)^{2/3}}$$

Noting the negative exponent in the second term of the numerator, we could write this term as a fraction, combine terms in the numerator, and then indicate the division. It would then be necessary to use the laws of exponents to complete the simplification.

However, we can also simplify the fraction by multiplying both numerator and denominator by $(3x^2 - 1)^{2/3}$. In this way we eliminate not only the negative exponent in the numerator but also the fractional exponents from the terms of the numerator. Then applying the laws of exponents, we proceed with the simplification. Following this second method, the simplification is as follows.

$$f(x) = \frac{(3x^2 - 1)^{1/3}(2x) - (2x^3)(3x^2 - 1)^{-2/3}}{(3x^2 - 1)^{2/3}} \cdot \frac{(3x^2 - 1)^{2/3}}{(3x^2 - 1)^{2/3}}$$

$$= \frac{(3x^2 - 1)^{1/3}(3x^2 - 1)^{2/3}(2x) - (2x^3)(3x^2 - 1)^{-2/3}(3x^2 - 1)^{2/3}}{[(3x^2 - 1)^{2/3}]^2}$$

$$= \frac{(3x^2 - 1)(2x) - (2x^3)(3x^2 - 1)^0}{(3x^2 - 1)^{4/3}} \qquad \text{using Eqs. (1-4) and (1-6)}$$

$$= \frac{6x^3 - 2x - 2x^3}{(3x^2 - 1)^{4/3}} \qquad \text{using Eq. (1-8)}$$

$$= \frac{4x^3 - 2x}{(3x^2 - 1)^{4/3}}$$

Such algebraic simplifications are commonly needed in calculus problems. ∎

Domain and Range

As we mentioned in the previous section, using only real numbers may result in restrictions as to the permissible values of the independent and dependent variables. *The complete set of possible values of the independent variable is called the* **domain** *of the function,* and *the complete set of all possible resulting values of the dependent variable is called the* **range** *of the function.* Therefore, using real numbers in the domain and range of a function,

NOTE ▶

values that lead to division by zero or
to imaginary values may not be included.

Note meaning of symbol ≥.

EXAMPLE 5 The function $f(x) = x^2 + 2$ is defined for all real values of x. This means its domain is written as *all real numbers.* However, since x^2 is never negative, $x^2 + 2$ is never less than 2. We then write the range as *all real numbers* $f(x) \geq 2$, where the symbol $\geq$ means "is greater than or equal to."

The function $f(t) = \dfrac{1}{t + 2}$ is not defined for $t = -2$, for this value would require division by zero. Also, no matter how large t becomes, $f(t)$ will never exactly equal zero. Therefore, the domain of this function is *all real numbers except* -2, and the range is *all real numbers except* 0. ∎

Note meaning of symbol ≤.

EXAMPLE 6   The function $g(s) = \sqrt{3 - s}$ is not defined for real numbers greater than 3, since such values make $3 - s$ negative and would result in imaginary values for $g(s)$. This means that the domain of this function is *all real numbers* $s \leq 3$, where the symbol $\leq$ means "is less than or equal to."

Also, since $\sqrt{3 - s}$ means the principal square root of $3 - s$ (see page 8), we know that $g(s)$ cannot be negative. This tells us that the range of the function is *all real numbers* $g(s) \geq 0$. This means that the values for the range are zero and all positive numbers, no matter how large.

In Examples 5 and 6 we determined the domain of each function by looking for those values of the independent variable that cannot be used. The range of each was found through an inspection of the function. When possible, this is the procedure we use, although it is often necessary to use more advanced methods to find the range of a function. Even in the second illustration of Example 5, we had to take a special look at the function to find the range. For this reason, until we develop other methods, we will look only for the domain of some functions.

Later in this chapter we will see that the graphing calculator is useful in finding the range of a function.

EXAMPLE 7   Find the domain of the function $f(x) = 16\sqrt{x} + \dfrac{1}{x}$.

From the term $16\sqrt{x}$ we see that x must be greater than or equal to zero in order to have real values. The term $\dfrac{1}{x}$ indicates that x cannot be zero, because of division by zero. Thus, putting these together, the domain is *all real numbers* $x > 0$.

As for the range, it is *all real numbers* $f(x) \geq 12$. More advanced methods (as in Section 4-6) are needed to determine this.

We have seen that the domains of some functions are restricted to particular values. It can also happen that the domain of a function is restricted by definition or by practical considerations in an application. Consider the illustrations in the following example.

EXAMPLE 8   A function defined as

$$f(x) = x^2 + 4 \quad \text{for } x > 2$$

has a domain restricted to real numbers greater than 2 by definition. Thus, $f(5) = 29$, but $f(1)$ is not defined, since 1 is not in the domain. Also, the range is all real numbers greater than 8.

The height h (in m) of a certain projectile as a function of the time t (in s) is found using the equation

$$h = 20t - 4.9t^2$$

Generally, negative values of time do not have meaning in such an application. This leads us to state the domain as values of $t \geq 0$. Of course, the projectile will not continue in flight indefinitely, and there is some upper limit on the value of t. These restrictions are not usually stated unless there is a particular reason that affects the solution.

The following example illustrates a function that is defined differently for different intervals of the domain.

For reference, see Appendix B for units of measurement and the symbols used for them.

■**EXAMPLE 9**　In a certain electric circuit, the current i (in mA) is a function of the time t (in s), which means $i = f(t)$. The function is

$$f(t) = \begin{cases} 8 - 2t & \text{for } 0 \le t \le 4 \text{ s} \\ 0 & \text{for } t > 4 \text{ s} \end{cases}$$

Since negative values of t are not usually meaningful, $f(t)$ is not defined for $t < 0$. Find the current for $t = 3$ s, $t = 6$ s, and $t = -1$ s.

We are to find $f(3)$, $f(6)$, and $f(-1)$, and we see that values of this function are determined differently depending on the value of t. Since 3 is between 0 and 4,

$$f(3) = 8 - 2(3) = 2 \quad \text{or} \quad i = 2 \text{ mA}$$

Since 6 is greater than 4, $f(6) = 0$, or $i = 0$ mA. We see that $i = 0$ mA for all values of t that are 4 or greater.

Since $f(t)$ is not defined for $t < 0$, $f(-1)$ is not defined.　------------■

Functions from Verbal Statements

In the previous section we wrote functions in mathematical form from given statements and by using geometric information. It is often necessary to determine a mathematical function from a given statement. In the following examples we set up such functions.

SOLVING A WORD PROBLEM

■**EXAMPLE 10**　The fixed cost for a company to operate a certain plant is $3000 per day. It also costs $4 for each unit produced in the plant. Express the daily cost C of operating the plant as a function of the number n of units produced.

The daily total cost C equals the fixed cost of $3000 plus the cost of producing n units. Since the cost of producing one unit is $4, the cost of producing n units is $4n$. Thus, the total cost C, where $C = f(n)$, is

$$C = 3000 + 4n$$

Here we know that the domain is all values of $n \ge 0$, with some upper limit on n based on the production capacity of the plant.　------------■

SOLVING A WORD PROBLEM

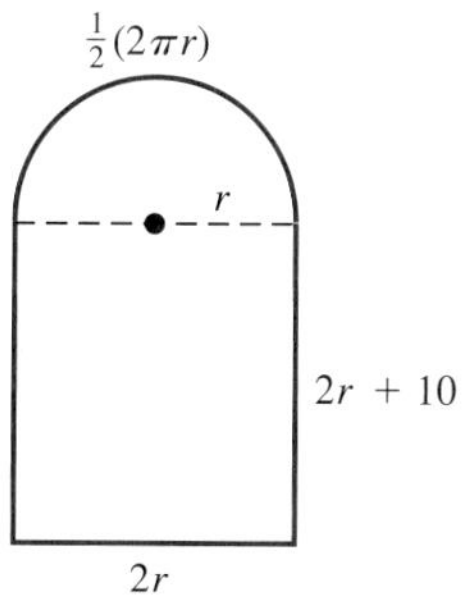

$\frac{1}{2}(2\pi r)$

r

$2r + 10$

$2r$

Fig. 1-5

■**EXAMPLE 11**　An architect designs a window such that it has the shape of a rectangle with a semicircle on top, as shown in Fig. 1-5. The base of the window is 10 cm less than the height of the rectangular part. Express the perimeter p of the window as a function of the radius r of the circular part.

We know that the perimeter is the distance around the window. Since the top part is a semicircle and the circumference of a circle is $2\pi r$, the length of the top circular part is $\frac{1}{2}(2\pi r)$. This also tells us that the dashed line, and therefore the base of the window, is $2r$. Finally, the fact that the base is 10 cm less than the height of the rectangular part tells us that each vertical part has a height of $2r + 10$. This means that the perimeter p, where $p = f(r)$, is

$$p = \tfrac{1}{2}(2\pi r) + 2r + 2(2r + 10)$$
$$= \pi r + 2r + 4r + 20$$
$$= \pi r + 6r + 20$$

We see that the required function is $p = \pi r + 6r + 20$. Since the radius cannot be negative and there would be no window if $r = 0$, the domain of the function is all values $0 < r \le R$, where R is a maximum possible value of r determined by design considerations.　------------■

SOLVING A WORD PROBLEM

■**EXAMPLE 12** A metallurgist melts and mixes m grams of solder that is 40% tin with n grams of another solder that is 20% tin to get a final solder mixture that contains 200 g of tin. Express n as a function of m. See Fig. 1-6.

The statement leads to the following equation.

tin in first solder		tin in second solder		total amount of tin
$0.40m$	$+$	$0.20n$	$=$	200

Since we want $n = f(m)$, we now solve for n.

$$0.20n = 200 - 0.40m$$

$$n = 1000 - 2m$$

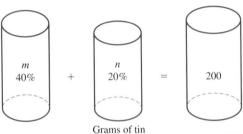

Grams of tin

Fig. 1-6

This is the required function. Since neither m nor n can be negative, the domain is all values $0 \leq m \leq 500$ g, which means that m is greater than or equal to 0 g and less than or equal to 500 g. The range is all values $0 \leq n \leq 1000$ g. ■

In the definition of a function, it was stipulated that any value for the independent variable must yield only a single value of the dependent variable. *If a value of the independent variable yields more than one value of the dependent variable, the relationship is called a* **relation** *instead of a function.* A relation involves two variables related so that the values of the second variable can be determined from values of the first variable. A function is a relation in which each value of the first variable yields only one value of the second. A function is therefore a special type of relation. However, there are relations that are not functions.

RELATION

■**EXAMPLE 13** For $y^2 = 4x^2$, if $x = 2$, then y can be either 4 or -4. Since a value of x yields more than one single value for y, we see that $y^2 = 4x^2$ is a relation, not a function. ■

EXERCISES *1-2*

In Exercises 1–4, use the functions $F(x) = \sqrt{x^2 + 4}$ and $G(x) = x - 1$.

1. Find $F[G(x)]$.   **2.** Find $[G(x)]^3$.

3. Find $G(x)/F(x)$.   **4.** Find $\{G[F(x)]\}^2$.

In Exercises 5–12, perform the indicated operations and simplify.

5. $\sqrt{\sqrt[3]{x^6 + 1}}$

6. $(\sqrt{x} - 1)(\sqrt{x^2 + x + 1})$

7. $\dfrac{(4x - 5)^4}{(4x - 5)^{1/2}}$

8. $\dfrac{3(x^2 + 1)^0}{(x^2 + 4)^{-1/2}}$

9. $(2x + 1)^{1/2} + (x + 3)(2x + 1)^{-1/2}$

10. $(3x - 1)^{-2/3}(1 - x) - (3x - 1)^{1/3}$

11. $\dfrac{(x^2 + 1)^{1/2} - x^2(x^2 + 1)^{-1/2}}{x^2 + 1}$

12. $\dfrac{(1 - 2x^2)^{1/4}(2x) + 4x^3(1 - 2x^2)^{-3/4}}{(1 - 2x^2)^{1/2}}$

In Exercises 13–20, determine the domain and range of the given functions. In Exercises 19 and 20, explain your answers.

13. $f(x) = x + 5$ **14.** $g(u) = 3 - u^2$

15. $G(R) = \dfrac{3.2}{R}$ **16.** $F(r) = \sqrt{r + 4}$

17. $f(s) = \dfrac{2}{s^2}$ **18.** $T(t) = 2t^4 + t^2 - 1$

Ⓦ **19.** $H(h) = 2h + \sqrt{h} + 1$ Ⓦ **20.** $f(x) = \dfrac{6}{\sqrt{2 - x}}$

In Exercises 21–24, determine the domain of the given functions.

21. $Y(y) = \dfrac{y + 1}{\sqrt{y - 2}}$ **22.** $f(n) = \dfrac{n}{6 - 2n}$

23. $f(D) = \dfrac{D}{D - 2} + \dfrac{4}{D + 4} - \dfrac{D - 3}{D - 6}$

24. $g(x) = \dfrac{\sqrt{x - 2}}{x - 3}$

In Exercises 25–28, evaluate the indicated functions.

$$F(t) = 3t - t^2 \quad \text{for } t \le 2 \qquad h(s) = \begin{cases} 2s & \text{for } s < -1 \\ s + 1 & \text{for } s \ge -1 \end{cases}$$

$$f(x) = \begin{cases} x + 1 & \text{for } x < 1 \\ \sqrt{x + 3} & \text{for } x \ge 1 \end{cases} \qquad g(x) = \begin{cases} \dfrac{1}{x} & \text{for } x \ne 0 \\ 0 & \text{for } x = 0 \end{cases}$$

25. Find $F(2)$ and $F(3)$.

26. Find $h(-8)$ and $h\left(-\dfrac{1}{2}\right)$.

27. Find $f(1)$ and $f\left(-\dfrac{1}{4}\right)$.

28. Find $g\left(\dfrac{1}{5}\right)$ and $g(0)$.

In Exercises 29–40, determine the appropriate functions.

29. A motorist travels at 40 mi/h for 2 h and then at 55 mi/h for t hours. Express the distance d traveled as a function of t.

30. Express the cost C of insulating a cylindrical water tank of height 2 m as a function of its radius r, if the cost of insulation is \$3 per square meter.

31. A rocket burns up at the rate of 2 tons/min after falling out of orbit into the atmosphere. If the rocket weighed 5500 tons before reentry, express its weight w as a function of the time t, in minutes, of reentry.

32. A computer part costs \$3 to produce and distribute. Express the profit p made by selling 100 of these parts as a function of the price of c dollars each.

33. Upon ascending, a weather balloon ices up at the rate of 0.5 kg/m after reaching an altitude of 1000 m. If the mass of the balloon below 1000 m is 110 kg, express its mass m as a function of its altitude h if $h > 1000$ m.

34. A chemist adds x liters of a solution that is 50% alcohol to 100 L of a solution that is 70% alcohol. Express the number n of liters of alcohol in the final solution as a function of x.

35. A company installs underground cable at a cost of \$500 for the first 50 ft (or up to 50 ft) and \$5 for each foot thereafter. Express the cost C as a function of the length l of underground cable if $l > 50$ ft.

36. The *mechanical advantage* of an inclined plane is the ratio of the length of the plane to its height. Express the mechanical advantage M of a plane of length 8 m as a function of its height h.

37. The capacities (in L) of two oil-storage tanks are x and y. The tanks are initially full; 1200 L is removed from them by taking 10% of the contents of the first tank and 40% of the contents of the second tank. (a) Express y as a function of x. (b) Find $f(400)$.

38. A spherical buoy 36 in. in diameter is floating in a lake and is more than half above the water. Express the circumference c of the circle of intersection of the buoy and water as a function of the depth d to which the buoy sinks.

39. In studying the electric current that is induced in wire rotating through a magnetic field, a piece of wire 60 cm long is cut into two pieces. One of these is bent into a circle and the other into a square. Express the total area A of the two figures as a function of the perimeter p of the square.

40. The cross section of an air-conditioning duct is in the shape of a square with semicircles on each side. See Fig. 1-7. Express the area A of this cross section as a function of the diameter d (in cm) of the circular part.

Fig. 1-7

In Exercises 41–48, solve the given problems.

41. A computer program displays a circular image of radius 6 in. If the radius is decreased by x in., express the area of the image as a function of x. What are the domain and range of $A = f(x)$?

42. A helicopter 120 m from a person takes off vertically. Express the distance d from the person to the helicopter as a function of the height h of the helicopter. What are the domain and the range of $d = f(h)$? See Fig. 1-8.

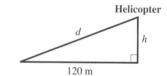

Fig. 1-8

43. A truck travels 300 km in t hours. Express the average speed s of the truck as a function of t. What are the domain and range of $s = f(t)$?

44. A rectangular grazing range with an area of 8 mi^2 is to be fenced. Express the length l of the field as a function of its width w. What are the domain and range of $l = f(w)$?

45. The resonant frequency f (in Hz) in a certain electric circuit as a function of the capacitance is $f = \dfrac{1}{2\pi\sqrt{C}}$. Describe the domain of this function.

46. A jet is traveling directly between Calgary, Alberta, and Portland, Oregon, which are 550 mi apart. If the jet is x mi from Calgary and y mi from Portland, find the domain of $y = f(x)$.

47. Express the mass m of the weather balloon in Exercise 33 as a function of any height h in the same manner as the function in Example 9 (and Exercises 25–28) was expressed.

48. Express the cost C of installing any length l of the underground cable in Exercise 35 in the same manner as the function in Example 9 (and Exercises 25–28) was represented.

1-3 RECTANGULAR COORDINATES

One of the most valuable ways of representing a function is by graphical representation. By using graphs we are able to obtain a "picture" of the function, and by using this picture we can learn a great deal about the function.

To make a graphical representation of a function, we recall from Section 1-1 that numbers can be represented by points on a line. For a function, we have values of the independent variable as well as the corresponding values of the dependent variable. Therefore, it is necessary to use two different lines to represent the values from each of these sets of numbers. We do this by placing the lines perpendicular to each other.

We place one line horizontally and label it the **x-axis.** The values of the independent variable are normally placed on this axis. *The other line is placed vertically and labeled the* **y-axis.** Normally the y-axis is used for values of the dependent variable. *The point of intersection is called the* **origin.** This is the **rectangular coordinate system.**

On the x-axis, positive values are to the right of the origin, and negative values are to the left of the origin. On the y-axis, positive values are above the origin, and negative values are below it. *The four parts into which the plane is divided are called* **quadrants,** which are numbered as in Fig. 1-9.

A point P on the plane is designated by the pair of numbers (x, y), where x is the value of the independent variable and y is the corresponding value of the dependent variable. *The x-value, called the* **abscissa,** *is the perpendicular distance of P from the y-axis. The y-value, called the* **ordinate,** *is the perpendicular distance of P from the x-axis.* The values x and y together, written as (x, y), are the **coordinates** of the point P. Note carefully that *the x-value is always written first and the y-value is written second.*

EXAMPLE 1   Locate the points $A(2, 1)$ and $B(-4, -3)$ on the rectangular coordinate system.

The coordinates $(2, 1)$ for A mean that the point is 2 units to the *right* of the y-axis and 1 unit *above* the y-axis, as shown in Fig. 1-10. The coordinates $(-4, -3)$ for B mean that the point is 4 units to the *left* of the y-axis and 3 units *below* the x-axis, as shown.

The abscissa of point A is 2, and the ordinate of A is 1. For point B, its abscissa is -4, and its ordinate is -3.

EXAMPLE 2   The positions of points $P(4, 5)$, $Q(-2, 3)$, $R(-1, -5)$, $S(4, -2)$, and $T(0, 3)$ are shown in Fig. 1-9 above. We see that this representation allows for *one point for any pair of values* (x, y). Also note that the point $T(0, 3)$ is on the y-axis. Any such point that is on either axis is not *in* any of the four quadrants.

EXAMPLE 3   Three vertices of the rectangle in Fig. 1-11 are $A(-3, -2)$, $B(4, -2)$, and $C(4, 1)$. What is the fourth vertex?

We use the fact that opposite sides of a rectangle are equal and parallel to find the solution. Since both vertices of the base AB of the rectangle have a y-coordinate of -2, the base is parallel to the x-axis. Therefore, the top of the rectangle must also be parallel to the x-axis. Thus, the vertices of the top must both have a y-coordinate of 1, since one of them has a y-coordinate of 1. In the same way, the x-coordinates of the left side must both be -3. Therefore, the fourth vertex is $D(-3, 1)$.

Rectangular (Cartesian) coordinates were developed by the French mathematician Descartes (1596–1650).

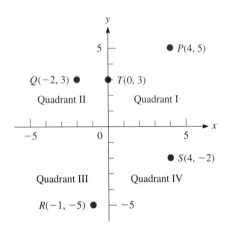

Fig. 1-9

NOTE ▶

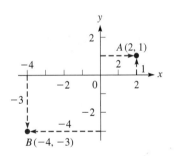

Fig. 1-10

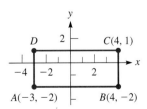

Fig. 1-11

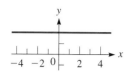

Fig. 1-12

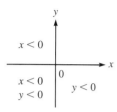

Fig. 1-13

EXAMPLE 4   Where are all points whose ordinates are 2?

Since the ordinate is the *y*-value, we can see that the question could be stated as: "Where are all points for which $y = 2$?" Since all such points are 2 units above the *x*-axis, the answer can be stated as "on a line 2 units above the *x*-axis." See Fig. 1-12. ∎

EXAMPLE 5   Where are all points (x, y) for which $x < 0$ and $y < 0$?

Noting that $x < 0$ means "*x* is less than zero," or "*x* is negative," and that $y < 0$ means the same for *y*, we want to determine where both *x* and *y* are negative. Our answer is "in the third quadrant," since both coordinates are negative for all points in the third quadrant, and this is the only quadrant for which this is true. See Fig. 1-13. ∎

EXERCISES *1-3*

In Exercises 1 and 2, determine (at least approximately) the coordinates of the points specified in Fig. 1-14.

 1. *A, B, C*

 2. *D, E, F*

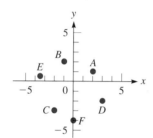

Fig. 1-14

In Exercises 3 and 4, plot the given points.

 3. $A(2, 7)$, $B(-1, -2)$, $C(-4, 2)$

 4. $A(3, \frac{1}{2})$, $B(-6, 0)$, $C(-\frac{5}{2}, -5)$

In Exercises 5–8, plot the given points and then join these points, in the order given, by straight-line segments. Name the geometric figure formed.

 5. $A(-1, 4)$, $B(3, 4)$, $C(1, -2)$

 6. $A(0, 3)$, $B(0, -1)$, $C(4, -1)$

 7. $A(-2, -1)$, $B(3, -1)$, $C(3, 5)$, $D(-2, 5)$

 8. $A(-5, -2)$, $B(4, -2)$, $C(6, 3)$, $D(-3, 3)$

In Exercises 9–12, find the indicated coordinates.

 9. Three vertices of a rectangle are $(5, 2)$, $(-1, 2)$, and $(-1, 4)$. What are the coordinates of the fourth vertex?

 10. Two vertices of an equilateral triangle are $(7, 1)$ and $(2, 1)$. What is the abscissa of the third vertex?

 11. *P* is the point $(3, 2)$. Locate point *Q* such that the *x*-axis is the perpendicular bisector of the line segment joining *P* and *Q*.

 12. *P* is the point $(-4, 1)$. Locate point *Q* such that the line segment joining *P* and *Q* is bisected by the origin.

In Exercises 13–28, answer the given questions.

 13. Where are all the points whose abscissas are 1?

 14. Where are all the points whose ordinates are -3?

 15. Where are all points such that $y = 3$?

 16. Where are all points such that $x = -2$?

 17. Where are all the points whose abscissas equal their ordinates?

 18. Where are all the points whose abscissas equal the negative of their ordinates?

 19. What is the abscissa of all points on the *y*-axis?

 20. What is the ordinate of all points on the *x*-axis?

 21. Where are all the points for which $x > 0$?

 22. Where are all the points for which $y < 0$?

 23. Where are all the points for which $x < -1$?

 24. Where are all the points for which $y > 4$?

 25. In which quadrants is the ratio y/x positive?

 26. In which quadrants is the ratio y/x negative?

 27. Find the distance (a) between $(3, -2)$ and $(-5, -2)$ and (b) between $(3, -2)$ and $(3, 4)$.

 W 28. Find the distance between $(-5, -2)$ and $(3, 4)$. Explain your method. (Hint: See Exercise 27.)

1-4 THE GRAPH OF A FUNCTION

Now that we have introduced the concepts of a function and the rectangular coordinate system, we are in a position to determine the graph of a function. In this way we shall obtain a visual representation of a function.

The graph of a function is the set of all points whose coordinates (x, y) satisfy the functional relationship $y = f(x)$. Since $y = f(x)$, we can write the coordinates of the points on the graph as $(x, f(x))$. Writing the coordinates in this manner tells us exactly how to find them. *We assume a certain value for x and then find the value of f(x). These two numbers are the coordinates of a point.*

Since there is no limit to the possible number of points that can be chosen, we normally select a few values of x, obtain the corresponding values of the function, plot these points, and then join them. Therefore, we use the following basic procedure in plotting a graph.

As to just what values of x to choose and how many to choose, with a little experience you will usually be able to tell if you have enough points to plot an accurate graph.

> **Procedure for Plotting the Graph of a Function**
>
> 1. *Let x take on several values and calculate the corresponding values of y.*
> 2. *Tabulate these values, arranging the table so that **values of x are increasing.***
> 3. *Plot the points and join them from left to right by a **smooth curve** (not short straight-line segments).*

EXAMPLE 1   Graph the function $f(x) = 3x - 5$.

For purposes of graphing, we let $y = f(x)$, or $y = 3x - 5$. We then let x take on various values and determine the corresponding values of y. Note that once we choose a given value of x, we have no choice about the corresponding y-value, as it is determined by evaluating the function. If $x = 0$, we find that $y = -5$. This means that the point $(0, -5)$ is on the graph of the function. Choosing another value of x—for example, 1—we find that $y = -2$. This means that the point $(1, -2)$ is on the graph of the function. Continuing to choose a few other values of x, we tabulate the results, as shown in Fig. 1-15. It is best to arrange the table so that the values of x increase; then there is no doubt how they are to be connected, for they are then connected in the order shown. Finally, we connect the points in Fig. 1-15 and see that the graph of the function is a straight line.

The table of values for a graph can be seen on a graphing calculator using the *table* feature. Below, in Fig. 1-16, is the calculator display of the table for Example 1.

Fig. 1-16

Select values of x and calculate corresponding values of y	Tabulate with values of x increasing	
$f(x) = 3x - 5$	x	y
$f(-1) = 3(-1) - 5 = -8$	-1	-8
$f(0) = 3(0) - 5 = -5$	0	-5
$f(1) = 3(1) - 5 = -2$	1	-2
$f(2) = 3(2) - 5 = 1$	2	1
$f(3) = 3(3) - 5 = 4$	3	4

Plot points and join from left to right with smooth curve

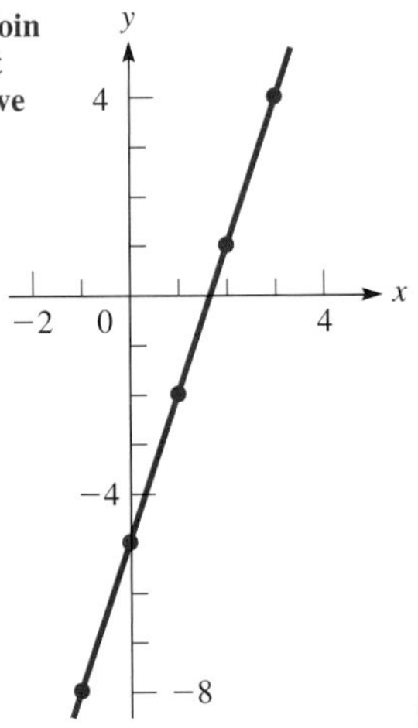

Fig. 1-15

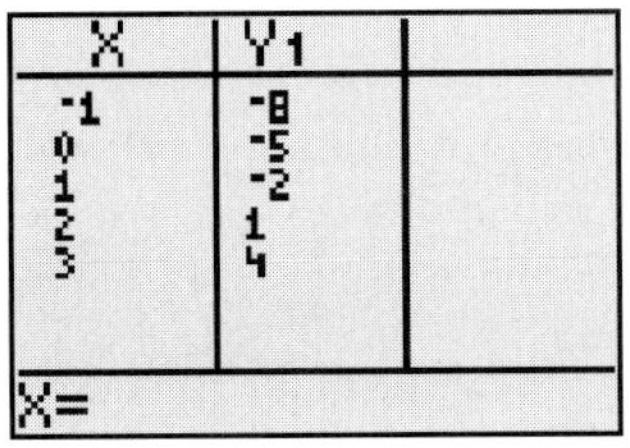

When graphing a function, there are some special points about which we should be careful. These include the following:

RESTRICTIONS ON A GRAPH

1. Since the graphs of most common functions are smooth, any irregularities in the graph should be carefully checked. In these cases it usually helps to take values of x between those values where the question arises.

2. The domain of the function may not include all values of x (remember, *division by zero is not defined, and only real values of the variables are permissible*).

3. In applications, use only values that are meaningful. Negative values for many quantities, such as time, are not generally meaningful.

The following examples illustrate these points.

EXAMPLE 2 Graph the function $y = x - x^2$.

First, we find the values in the table, as shown with Fig. 1-17. We must be careful when dealing with negative values of x. For the value $x = -1$, we have $y = (-1) - (-1)^2 = -1 - 1 = -2$. Once all values in the table have been found and plotted, we note that **$y = 0$ for both $x = 0$ and $x = 1$**. The question arises—*what happens between 0 and 1?* Trying $x = \frac{1}{2}$, we find that $y = \frac{1}{4}$. Using this point completes the information needed for an accurate graph. Note that we do not stop the graph with the last points determined, but indicate that the curve continues by drawing it past these points. ▬

x	y
-2	-6
-1	-2
0	0
1	0
2	-2
3	-6

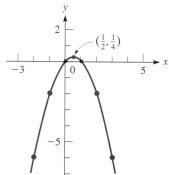

Fig. 1-17

EXAMPLE 3 Graph the function $y = 1 + \dfrac{1}{x}$.

CAUTION ▶

In finding the points on this graph, as shown in Fig. 1-18, we note that y is not defined for $x = 0$, due to division by zero. Thus, $x = 0$ is not in the domain, and ***we must be careful not to have any part of the curve cross the y-axis ($x = 0$)***. Although we cannot let $x = 0$, we can choose other values of x between -1 and 1 that are close to zero. In doing so, we find that as x gets closer to zero, the points get closer and closer to the y-axis, although they do not reach or touch it. In this case the y-axis is called an **asymptote** of the curve. ▬

x	y
-4	$3/4$
-3	$2/3$
-2	$1/2$
-1	0
$-1/2$	-1
$-1/3$	-2
$1/3$	4
$1/2$	3
1	2
2	$3/2$
3	$4/3$
4	$5/4$

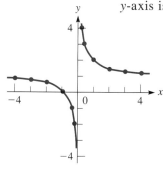

Fig. 1-18

x	y
-1	0
0	1
1	1.4
2	1.7
3	2
4	2.2
5	2.4
6	2.6
7	2.8
8	3

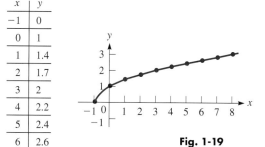

Fig. 1-19

EXAMPLE 4 Graph the function $y = \sqrt{x + 1}$.

NOTE ▶

When finding the points for the graph, we may not let x take on any value less than -1, for ***all such values would lead to imaginary values for y*** and are not in the domain. Also, since we have the positive square root indicated, the range consists of all values of y that are positive or zero ($y \geq 0$). See Fig. 1-19. ▬

See the chapter introduction.

The unit of power, the watt (W), is named for James Watt (1736–1819), a British engineer.

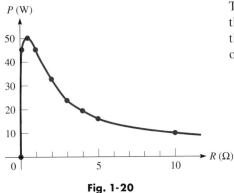

Fig. 1-20

$f(-2) = 2(-2) + 1 = -3$
$f(-1) = 2(-1) + 1 = -1$
$f(0) = 2(0) + 1 = 1$
$f(1) = 2(1) + 1 = 3$
$f(2) = 6 - 2^2 = 2$
$f(3) = 6 - 3^2 = -3$

x	y
-2	-3
-1	-1
0	1
1	3
2	2
3	-3

Fig. 1-21

In Appendix A-2 we see how to find an equation that approximates the function relating the variables in a table of values.

EXAMPLE 5   The electric power P (in W) delivered by a certain battery as a function of the resistance R (in Ω) in the circuit is given by $P = \dfrac{100R}{(0.50 + R)^2}$. Plot P as a function of R.

Since negative values for the resistance have no physical significance, we should not plot any values of P for negative values of R. The following table is obtained.

$R\ (\Omega)$	0	0.25	0.50	1.0	2.0	3.0	4.0	5.0	10.0
$P\ (W)$	0.0	44.4	50.0	44.4	32.0	24.5	19.8	16.5	9.1

The values 0.25 and 0.50 are used for R when it is found that P is less for $R = 2$ than for $R = 1$. The sharp change in direction at $R = 1$ should be checked by using these additional points to better see how the curve changes near $R = 1$ and thereby obtain a smoother curve. See Fig. 1-20.

Also note that the scale on the P-axis is different from that on the R-axis. This reflects the different magnitudes and ranges of values used for each of the variables. Different scales are normally used in such cases.

We can make various conclusions from the graph. For example, we see that the maximum power of 50 W occurs for $R = 0.5\ \Omega$. Also, P decreases as R increases beyond $0.5\ \Omega$. We will consider further the information that can be read from a graph in the next section. ∎

The following example illustrates the graph of a function that is defined differently for different intervals of the domain.

EXAMPLE 6   Graph the function $f(x) = \begin{cases} 2x + 1 & \text{for } x \le 1 \\ 6 - x^2 & \text{for } x > 1 \end{cases}$

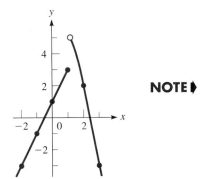

NOTE ▶

First, we let $y = f(x)$ and then tabulate the necessary values. In evaluating $f(x)$ we must be careful to use the proper part of the definition. In order to see where to start the curve for $x > 1$, we evaluate $6 - x^2$ for $x = 1$, but we must realize that **the curve does not include this point** $(1, 5)$ and starts immediately to its right. To show that this is not part of the curve, we draw it as an open circle. See Fig. 1-21. Such a function, with a "break" in it, is called *discontinuous*. ∎

Functions of a particular type have graphs of a specific basic shape, and there are many that have been named. Two examples of this are the *straight line* (Example 1) and the *parabola* (Example 2). In Chapter 2 we will again consider the straight line and parabola, along with other types of curves, with a detailed analysis of some of their properties. Other types of functions and their graphs are found in some of the later chapters.

The use of graphs is extensive in mathematics and in applications. For example, later in this section we show how graphs are used to solve equations. Numerous applications can be noted for most types of curves. For example, the parabola has applications in microwave dish design, suspension bridge design, and the path (trajectory) of a baseball. We will note many more applications of graphs throughout the book.

GRAPH OF A RELATION

In Section 1-1, we defined a function such that it has only one value of the dependent variable for each value of the independent variable. At the end of Section 1-2, we stated that a relation may have more than one value of the dependent variable. The following example illustrates how to use the graph to determine whether or not a relation is a function.

■**EXAMPLE 7**   In Example 13 of Section 1-2, we noted that $y^2 = 4x^2$ is a relation, but not a function. Since $y = 4$ or $y = -4$, normally written as $y = \pm 4$, for $x = 2$, we have two values of y for $x = 2$. Therefore, it is not a function. Making the table as shown at the left, we then draw the graph in Fig. 1-22. When making the table, we also note that there are two values of y for every value of x, except $x = 0$.

If any vertical line that crosses the x-axis in the domain intersects the graph in more than one point, it is the graph of a relation that is not a function. Any such vertical line in any of the previous graphs of this section would intersect the graph in only one point. This shows that they are graphs of functions.   ■

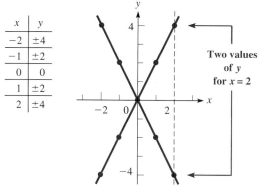

x	y
-2	± 4
-1	± 2
0	0
1	± 2
2	± 4

Two values
of y
for $x = 2$

Fig. 1-22

Graphs on a Graphing Calculator

A graphing calculator can be used for all calculations encountered in this text and in Section 1-1 we noted its use (in the margin) for evaluating a function and finding a table of values for graphing. We will now show that a graphing calculator can display a graph of a function quickly and easily.

A brief discussion of the features of graphing calculators is found in Appendix C. Since any given model uses a specific sequence of keys for a particular operation, the manual should be used for a detailed coverage of its features. For example, on some calculators the *graph* key is used after the function has been entered, whereas on others it puts the calculator into the graphing mode. Following is an example of displaying the graph of a function on a graphing calculator.

See Appendix C for a list of graphing calculator features and examples of calculator use (with page references) that are included in this text.

The specific detail shown in any nongraphic display depends on the model. A graph with the same *window* settings should appear about the same on all models.

■**EXAMPLE 8**   To graph the function $y = 2x + 8$, we first display Y₁= and then enter the $2x + 8$. The display for this is shown in Fig. 1-23(a).

Next we use the *window* (or *range*) feature to set the part of the domain and the range that will be seen in the *viewing window*. For this function we set

Xmin $= -6$, Xmax $= 2$, Xscl $= 1$, Ymin $= -2$, Ymax $= 10$, Yscl $= 1$

in order to get a good view of the graph. The display for the *window* settings is shown in Fig. 1-23(b).

We then display the graph, using the *graph* (or *exe*) key. The display showing the graph of $y = 2x + 8$ is shown in Fig. 1-23(c).

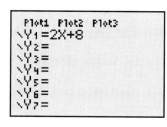

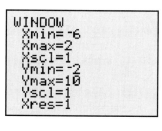

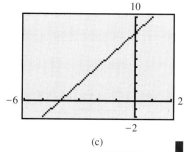

Fig. 1-23   (a)   (b)   (c)   ■

**SOLVING EQUATIONS
GRAPHICALLY**

An equation can be solved by use of a graph. Most of the time the solution will be approximate, but with a graphing calculator it is possible to get a high degree of accuracy in the result. The procedure is as follows:

> ### Procedure for Solving an Equation Graphically
> **1.** *Collect all terms on one side of the equal sign. This gives us the equation $f(x) = 0$.*
> **2.** *Set $y = f(x)$ and graph this function.*
> **3.** *Find the points where the graph crosses the x-axis. These points are called the **x-intercepts** of the graph. At these points, $y = 0$.*
> **4.** *The values of x for which $y = 0$ are the solutions of the equation. (These values are called the **zeros** of the function $f(x)$.)*

■EXAMPLE 9   Using a graphing calculator, solve the equation $x^2 - 2x = 1$.

First, we rewrite the equation as $x^2 - 2x - 1 = 0$ and set $y = x^2 - 2x - 1$. After one or two trial window settings, we have the display shown in Fig. 1-24(a). The scale on each axis is 0.5.

To get more accurate values of x for which $y = 0$, we use the *trace* feature. To find the *y*-value closest to zero, we move the *cursor* (a blinking pixel) with the arrow keys. In this way we have the display shown in Fig. 1-24(b), and we find the solutions to be approximately $x = -0.4$ and $x = 2.4$. We should note that the pixel for which y is closest to zero generally will not have a *y*-coordinate of *exactly zero*.

Much more accurate values can be found by using the *zoom* feature, with which any region of the screen can be greatly magnified. Fig. 1-24(c) shows the screen when *zoom* is used near $x = 2.4$. The cursor shows that the solution is about $x = 2.415$. In the same way, the other solution is about $x = -0.415$.

Checking these solutions *in the original equation,* we get $1.002225 = 1$ for each. This shows that they check, since we know that they are approximate.

Many calculators have a *zero* (or *root*) feature. The equation in Example 9 can be solved using this feature by finding the zeros on the graph of the function $y = x^2 - 2x - 1$.

Many calculators have an *intersect* feature. The equation in Example 9 can be solved using this feature by finding the point where the curves $y = x^2 - 2x$ and $y = 1$ cross.

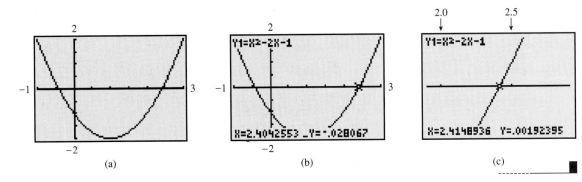

Fig. 1-24   (a)   (b)   (c)

THE RANGE OF A FUNCTION

In Section 1-2 we stated that advanced methods are often necessary to find the range of a function and that we would see that a graphing calculator is useful for this purpose. As with solving equations, we can get very good approximations of the range for most functions by using a graphing calculator.

The method is simply to graph the function and see for what values of y there is a point on the graph. These values of y give us the range of the function.

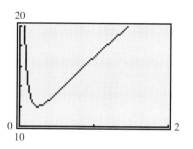

Fig. 1-25

EXAMPLE 10   Graphically find the range of the function $y = 16\sqrt{x} + \dfrac{1}{x}$.

If we start by using the default *window* settings of Xmin $= -10$, Xmax $= 10$, Ymin $= -10$, Ymax $= 10$, we see no part of the graph. Looking back to the function, we note that x cannot be zero because that would require dividing by zero in the $1/x$ term. Also, $x > 0$ since $\sqrt{x}$ is not defined for $x < 0$. Thus, we should try Xmin $= 0$ and Ymin $= 0$.

We also can easily see that $y = 17$ for $x = 1$. This means we should try a value more than 17 for Ymax. Using Ymax $= 20$, after two or three *window* settings, we use the settings shown in Fig. 1-25. The display shows the range to be all real numbers greater than about 12. (Actually, the range is all real numbers $y \geq 12$.)

EXERCISES *1-4*

In Exercises 1–12, plot (by hand) the graphs of the given functions.

1. $y = 2x - 4$
2. $y = 6 - \frac{1}{3}x$

3. $y = 3 - x^2$
4. $y = 2x^2 + 1$

5. $h = 20t - 5t^2$
6. $y = 2 + 3x + x^2$

7. $V = e^3$
8. $y = 3x - x^3$

9. $y = \dfrac{2}{x + 2}$
10. $p = \dfrac{1}{n^2 + 0.5}$

11. $y = \sqrt{4 - x}$
12. $y = \sqrt{x^2 - 16}$

In Exercises 13–24, display the graphs of the given functions on a graphing calculator.

13. $y = 3x + 5$
14. $y = 7 - 2x$

14. $s = -2t^2$
16. $y = \frac{1}{2}x^2 + 2$

17. $y = x^2 - 3x + 1$
18. $i = -2R^3$

19. $y = x^3 - x^2$
20. $y = x^4 - 4x^2$

21. $P = \dfrac{1}{V}$
22. $y = \dfrac{4}{x^2}$

23. $y = \sqrt{x}$
24. $y = \sqrt{16 - x^2}$

In Exercises 25–28, use a graphing calculator to solve the given equations to the nearest 0.1.

25. $x^2 - 41 = 0$
26. $w(w - 4) = 9$

27. $\sqrt{5R + 2} = 3$
28. $x - 2 = \dfrac{1}{x}$

In Exercises 29–36, use a graphing calculator to find the range of each of the given functions.

29. $y = \dfrac{4}{x^2 - 4}$
30. $y = \dfrac{x + 1}{x^2}$

31. $y = \dfrac{x^2}{x + 1}$
32. $y = \dfrac{x}{x^2 - 4}$

33. $Y(y) = \dfrac{y + 1}{\sqrt{y - 2}}$
34. $f(n) = \dfrac{n}{6 - 2n}$

35. $f(D) = \dfrac{D}{D - 2} + \dfrac{4}{D + 4} - \dfrac{D - 3}{D - 6}$
36. $g(x) = \dfrac{\sqrt{x - 2}}{x - 3}$

In Exercises 37–40, plot (by hand) the indicated functions.

37. The consumption of fuel c (in L/h) of a certain engine is determined as a function of the number r of r/min of the engine, to be $c = 0.011r + 4.0$. This formula is valid from 500 r/min to 3000 r/min. Plot c as a function of r. (r is the symbol for revolution.)

38. The rate H (in W) at which heat is developed in the filament of an electric light bulb as a function of the electric current I (in A) is $H = 240I^2$. Plot H as a function of I.

39. The distance p (in m) from a camera with a 50-mm lens to the object being photographed is a function of the magnification m of the camera given by $p = \dfrac{0.05(1 + m)}{m}$. Plot the graph for positive values of m up to 0.50.

40. A measure of the light beam that can be passed through an optic fiber is its numerical aperture N. For a particular optic fiber, N is a function of the index of refraction n of the glass in the fiber, given by $N = \sqrt{n^2 - 1.69}$. Plot the graph for $n \leq 2.00$.

In Exercises 41–44, use a graphing calculator to solve the indicated equations. Assume all data are accurate to two significant digits unless greater accuracy is given.

41. Two cubical coolers together hold 40.0 L (40,000 cm³). If the inside edge of one is 5.00 cm longer than the inside edge of the other, what is the inside edge of each?

42. The length of a rectangular solar panel is 12 cm more than its width. If its area is 520 cm², find its dimensions.

43. The height h (in ft) of a rocket as a function of the time t (in s) of flight is given by $h = 50 + 280t - 16t^2$. Determine when the rocket is at ground level.

44. In finding the illumination at a point x feet from one of two light sources that are 100 ft apart, it is necessary to solve the equation $9x^3 - 2400x^2 + 240{,}000x - 8{,}000{,}000 = 0$. Find x.

In Exercises 45–52, solve the given problems.

45. A land developer is considering several options of dividing a large tract into rectangular building lots, many of which would have perimeters of 200 m. For these, the minimum width would be 30 m, and the maximum width would be 70 m. Express the areas A of these lots as a function of their widths w and plot the graph.

46. A rectangular storage bin is to be made from a rectangular piece of sheet metal 12 in. by 10 in., by cutting out equal corners of side x and bending up the sides. See Fig. 1-26. Find x if the storage bin is to hold 90 in.^{3}.

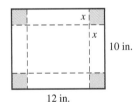

Fig. 1-26

10 in.

12 in.

47. The cutting speed s (in ft/min) of a saw in cutting a particular type of metal piece is given by $s = \sqrt{t - 4t^2}$, where t is the time in seconds. What is the maximum cutting speed in this operation (to two significant digits)? (*Hint*: Find the range.)

W 48. Referring to Exercise 46, explain how to determine the maximum possible capacity for a storage bin constructed in this way. What is the maximum possible capacity (to three significant digits)?

W 49. Plot the graphs of $y = x$ and $y = |x|$ on the same coordinate system. Explain why the graphs differ.

W 50. Plot the graphs of $y = 2 - x$ and $y = |2 - x|$ on the same coordinate system. Explain why the graphs differ.

51. Plot the graph of $f(x) = \begin{cases} 3 - x & \text{for } x < 1 \\ x^2 + 1 & \text{for } x \geq 1 \end{cases}$

52. Plot the graph of $f(x) = \begin{cases} \dfrac{1}{x - 1} & \text{for } x < 0 \\ \sqrt{x + 1} & \text{for } x \geq 0 \end{cases}$

In Exercises 53–56, determine whether or not the indicated graph is that of a function.

53. Fig. 1-27(a)   **54.** Fig. 1-27(b)

55. Fig. 1-27(c)   **56.** Fig. 1-27(d)

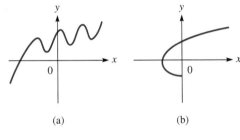

(a)   (b)

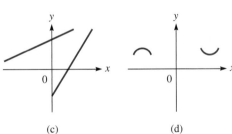

Fig. 1-27   (c)   (d)

CHAPTER EQUATIONS

Polynomial	$P(x) = a_0 x^n + a_1 x^{n-1} + \cdots + a_{n-1} x + a_n$	(1-1)
Rational function	$F(x) = \dfrac{f(x)}{g(x)}$	(1-2)
Root of a rational function	$I(x) = \sqrt[n]{F(x)}$	(1-3)

Laws of exponents

$$a^m \cdot a^n = a^{m+n} \qquad (1\text{-}4)$$

$$\frac{a^m}{a^n} = a^{m-n} \qquad (a \neq 0) \qquad (1\text{-}5)$$

$$(a^m)^n = a^{mn} \qquad (1\text{-}6)$$

$$(ab)^n = a^n b^n, \qquad \left(\frac{a}{b}\right)^n = \frac{a^n}{b^n} \qquad (b \neq 0) \qquad (1\text{-}7)$$

$$a^0 = 1 \qquad (a \neq 0) \qquad (1\text{-}8)$$

$$a^{-n} = \frac{1}{a^n} \qquad (a \neq 0) \qquad (1\text{-}9)$$

Radicals

$$a^{m/n} = \sqrt[n]{a^m} \qquad (m \text{ and } n \text{ integers}, \quad n > 0) \qquad (1\text{-}10)$$

$$\sqrt[n]{a^n} = a \qquad (1\text{-}11)$$

$$\sqrt[m]{\sqrt[n]{a}} = \sqrt[mn]{a} \qquad (1\text{-}12)$$

$$\sqrt[n]{a} \cdot \sqrt[n]{b} = \sqrt[n]{ab} \qquad (1\text{-}13)$$

$$\frac{\sqrt[n]{a}}{\sqrt[n]{b}} = \sqrt[n]{\frac{a}{b}} \qquad (b \neq 0) \qquad (1\text{-}14)$$

REVIEW EXERCISES

In Exercises 1–4, determine the appropriate function.

1. The radius of a circular water wave increases at the rate of 2 m/s. Express the area of the circle as a function of the time t (in s).

2. A conical sheet metal hood is to cover an area 6 m in diameter. Find the total surface area A of the hood as a function of its height h.

3. One computer printer prints at the rate of 2000 lines/min for x minutes, and a second printer prints at the rate of 1800 lines/min for y minutes. Together they print 50,000 lines. Find y as a function of x.

4. Fencing around a rectangular storage depot area costs twice as much along the front as along the other three sides. The back costs $10 per foot. Express the cost C of the fencing as a function of the width w if the length (along the front) is 20 ft longer than the width.

In Exercises 5–12, evaluate the given functions.

5. $f(x) = 7x - 5$; find $f(3)$ and $f(-6)$.

6. $g(I) = 8 - 3I$; find $g(\frac{1}{6})$ and $g(-4)$.

7. $H(h) = \sqrt{1 - 2h}$; find $H(-4)$ and $H(2h)$.

8. $\phi(v) = \frac{3v - 2}{v + 1}$; find $\phi(-2)$ and $\phi(v + 1)$.

9. $f(x) = 3x^2 - 2x + 4$; find $f(x + h) - f(x)$.

10. $F(x) = x^3 + 2x^2 - 3x$; find $F(3 + h) - F(3)$.

11. $f(x) = 3 - 2x$; find $f(2x) - 2f(x)$.

12. $f(x) = 1 - x^2$; find $[f(x)]^2 - f(x^2)$.

In Exercises 13–16, evaluate the given functions. Values of the independent variable are approximate.

13. $f(x) = 8.07 - 2x$; find $f(5.87)$ and $f(-4.29)$.

14. $g(x) = 7x - x^2$; find $g(45.81)$ and $g(-21.85)$.

15. $G(S) = \dfrac{S - 0.087629}{3.0125S}$; find $G(0.17427)$ and $G(0.053206)$.

16. $h(t) = \dfrac{t^2 - 4t}{t^3 + 564}$; find $h(8.91)$ and $h(-4.91)$.

In Exercises 17–20, determine the domain and the range of the given functions.

17. $f(x) = x^4 + 1$

18. $G(z) = \dfrac{4}{z^3}$

19. $g(t) = \dfrac{2}{\sqrt{t + 4}}$

20. $F(y) = 1 - 2\sqrt{y}$

In Exercises 21–32, plot the graphs of the given functions. Check these graphs by using a graphing calculator.

21. $y = 4x + 2$

22. $y = 5x - 10$

23. $y = 4x - x^2$

24. $y = x^2 - 8x - 5$

25. $y = 3 - x - 2x^2$

26. $y = 6 + 4x + x^2$

27. $y = x^3 - 6x$

28. $V = 3 - 0.5s^3$

29. $y = 2 - x^4$

30. $y = x^4 - 4x$

31. $y = \dfrac{x}{x + 1}$

32. $Z = \sqrt{25 - 2R^2}$

In Exercises 33–40, use a graphing calculator to solve the given equations to the nearest 0.1.

33. $7x - 3 = 0$

34. $3x + 11 = 0$

35. $x^2 + 1 = 6x$

36. $3t - 2 = t^2$

37. $x^3 - x^2 = 2 - x$

38. $5 - x^3 = 2x^2$

39. $\dfrac{1}{x} = 2x$

40. $\sqrt{x} = 2x - 1$

In Exercises 41–44, use a graphing calculator to find the range of the given function.

41. $y = x^4 - 5x^2$

42. $y = x\sqrt{4 - x^2}$

43. $A = w + \dfrac{2}{w}$

44. $y = 2x + \dfrac{3}{\sqrt{x}}$

In Exercises 45–52, answer the given questions.

 45. Explain how $A(a, b)$ and $B(b, a)$ may be in different quadrants.

46. Determine the distance from the origin to the point (a, b).

47. Two vertices of an equilateral triangle are $(0, 0)$ and $(2, 0)$. What is the third vertex?

48. The points $(1, 2)$ and $(1, -3)$ are two adjacent vertices of a square. Find the other vertices.

49. An equation used in electronics with a transformer antenna is $I = 12.5\sqrt{1 + 0.5m^2}$. For $I = f(m)$, find $f(0.55)$.

50. The percent p of wood lost in cutting it into boards 1.5 in. thick due to the thickness t (in in.) of the saw blade is $p = \dfrac{100t}{t + 1.5}$. Find p if $t = 0.4$ in. That is, since $p = f(t)$, find $f(0.4)$.

51. The angle A (in degrees) of a robot arm with the horizontal as a function of time t (for 0.0 s to 6.0 s) is given by $A = 8.0 + 12t^2 - 2.0t^3$. What is the greatest value of A to the nearest 0.1°? See Fig. 1-28. (Hint: Find the range.)

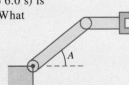

Fig. 1-28

52. The electric power P (in W) produced by a certain battery is $P = \dfrac{24R}{R^2 + 1.40R + 0.49}$, where R is the resistance (in Ω) in the circuit. What is the maximum power produced? (Hint: Find the range.) See Example 5 on page 18.

In Exercises 53–64, display the graphs of the indicated functions on a graphing calculator.

53. When El Niño, a Pacific Ocean current, moves east and warms the water off South America, weather patterns in many parts of the world change significantly. Special buoys along the equator in the Pacific Ocean send data via satellite to monitoring stations. If the temperature T (in °C) at one of these buoys is $T = 28.0 + 0.15t$, where t is the time in weeks between Jan. 1 and Aug. 1 (30 weeks), display the graph of $T = f(t)$.

54. There are 500 L of oil in a tank that has the capacity of 100,000 L. It is filled at the rate of 7000 L/h. Determine the function relating the number of liters N and the time t while the tank is being filled. Display the graph of N as a function of t.

55. For a given temperature, five times the Fahrenheit reading F less nine times the Celsius C reading is 160. Display the graph of $C = f(F)$.

56. A company buys a new copier for $1000 and determines that it costs $10 per day to use it (for paper, toner, etc.). Display the total cost C to operate the copier as a function of the number n of days of use.

57. For a certain laser device, the laser output power P (in mW) is negligible if the drive current i is less than 80 mA. From 80 mA to 140 mA, $P = 1.5 \times 10^{-6}i^3 - 0.77$. Display the graph of $P = f(i)$.

58. It is determined that a reasonable approximation for the cost C (in cents/mi) of operating a certain car at a constant speed v (in mi/h) is given by $C = 0.025v^2 - 1.4v + 35$. Display the C as a function of v for $v = 10$ mi/h to $v = 60$ mi/h.

59. A medical researcher exposed a virus culture to an experimental vaccine. It was observed that the number of live cells N in the culture as a function of the time t (in h) after exposure was given by $N = \dfrac{1000}{\sqrt{t + 1}}$. Display the graph of N as a function of t.

60. The electric field E (in V/m) from a certain electric charge is given by $E = 25/r^2$, where r is the distance (in m) from the charge. Display the graph of $E = f(r)$ for values of r up to 10 cm.

61. The maximum speed v (in mi/h) at which a car can safely travel around a circular turn of radius r (in ft) is given by $r = 0.42v^2$. Display the graph of r as a function of v.

62. The height h (in m) of a certain rocket as a function of the time of flight t (in s) is given by the function $h = 1500t - 4.9t^2$. Display h as a function of t, assuming that the terrain over which it flies is level.

63. The force F (in N) that is exerted by a cam on the arm of a robot is a function of the distance x shown in Fig. 1-29. The function for this force is $F = x^4 - 12x^3 + 46x^2 - 60x + 25$. Display the graph of $F = f(x)$. (Note that x varies from 1 cm to 5 cm.)

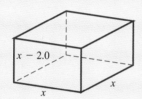

Fig. 1-29

64. A rectangular box is made with a square base and a height that is 2.0 cm less than the length of the side of the base. See Fig. 1-30. Set up the function relating the volume V and the side x of the base and display the function.

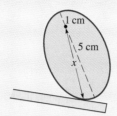

Fig. 1-30

In Exercises 65–72, solve the indicated equations graphically.

65. A person 250 mi from home starts toward home and travels at 60 mi/h for the first 2.0 h and then slows down to 40 mi/h for the rest of the trip. How long does it take the person to be 70 mi from home?

66. One industrial cleaner contains 30% of a certain solvent, and another contains 10% of the solvent. To get a mixture containing 50 gal of the solvent, 120 gal of the first cleaner is used. How much of the second must be used?

67. The solubility s (in kg/m^3 of water) of a certain type of fertilizer is given by $s = 135 + 4.9T + 0.19T^2$, where T is the temperature (in °C). Find T for $s = 500$ kg/m^3.

68. A 2.00-L (2000-cm^3) metal container is to be made in the shape of a right circular cylinder. Express the total area A of metal necessary as a function of the radius r of the base. Then find A for $r = 6.00$ cm, 7.00 cm, and 8.00 cm.

69. In an oil pipeline, the velocity v (in ft/s) of the oil as a function of the distance x (in ft) from the wall of the pipe is given by $v = 7.6x - 2.1x^2$. Find x for $v = 5.6$ ft/s. The diameter of the pipe is 3.50 ft.

70. One ball bearing is 1.00 mm more in radius and has twice the volume of another ball bearing. What is the radius of each?

71. A computer, using data from a refrigeration plant, estimates that in the event of a power failure, the temperature (in °C) in the freezers would be given by $T = \dfrac{4t^2}{t + 2} - 20$, where t is the number of hours after the power failure. How long would it take for the temperature to reach 0°C?

72. Two electrical resistors in parallel (see Fig. 1-31) have a combined resistance R_T given by $R_T = \dfrac{R_1 R_2}{R_1 + R_2}$. If $R_2 = R_1 + 2.0$, express R_T as a function of R_1 and find R_1 if $R_T = 6.0\ \Omega$.

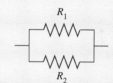

Fig. 1-31

Writing Exercise

73. In one or two paragraphs, explain how you would solve the following problem using a graphing calculator: The inner surface area A of a 250.0-cm^3 cylindrical cup, as a function of the radius r of the base, is $A = \pi r^2 + \dfrac{500.0}{r}$. Find r if $A = 175.0$ cm^2. (What is the answer?) (See if you can derive the formula.) (See Fig. 1-32.)

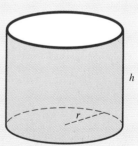

Fig. 1-32

PRACTICE TEST

1. Given $f(x) = 2x - x^2 + \dfrac{8}{x}$, find $f(-4)$ and $f(2.385)$.

2. A rocket has a mass of 2000 Mg at liftoff. If the first-stage engines burn fuel at the rate of 10 Mg/s, find the mass m of the rocket as a function of the time t (in s) while the first-stage engines operate.

3. Plot the graph of the function $f(x) = 4 - 2x$.

4. Use a graphing calculator to solve the equation $2x^2 - 3 = 3x$ to the nearest 0.1.

5. Plot the graph of the function $y = \sqrt{4 + 2x}$.

6. Locate all points (x, y) for which $x < 0$ and $y = 0$.

7. Find the domain and the range of the function $f(x) = \sqrt{6 - x}$.

8. A window has the shape of a semicircle over a square, as shown in Fig. 1-33. Express the area of the window as a function of the radius of the circular part.

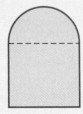

Fig. 1-33

9. The pressure loss P (in lb/in.^{2} per 100 ft) in a fire hose is given by $P = 0.00021Q^2 + 0.013Q$, where Q is the rate of flow (in gal/min). Using a graphing calculator, solve graphically for Q if $P = 1.2$ lb/in.^{2}.

2

PLANE ANALYTIC GEOMETRY

The underlying principle of *analytic geometry* is the relationship of an algebraic equation and the geometric properties of the curve that represents the equation. In this chapter we develop equations for a number of important geometric curves and show how to find the properties of these curves through an analysis of their equations. These curves include the *straight line* and the *conic sections,* which include the *circle, parabola, ellipse,* and *hyperbola.*

We briefly noted the straight line and parabola in Section 1-4, when we also noted that functions of a particular type have graphs of a specific basic shape. In later chapters we will consider curves of other functions—such as the sine curve (in trigonometry), which is a periodic waveform.

The concepts developed in analytic geometry are very useful in the study of calculus. Also, there are a great many technical and scientific applications. These include projectile motion, planetary orbits, and fluid motion. Other important applications range from the design of gears, airplane wings, and automobile headlights to the construction of bridges and nuclear cooling towers.

In Section 2-4 we show an important application of analytic geometry in the design of automobile headlights.

The development of geometry made little progress from the time of the ancient Greeks until the 1600s, when algebra was systematically applied to geometric problems. The French mathematician René Descartes (1596–1650) is recognized as the founder of analytic geometry.

2-1 BASIC DEFINITIONS

As we noted above, analytic geometry deals with the relationship between an algebraic equation and the geometric curve it represents. In this section we develop certain basic concepts that will be needed for future use in establishing the proper relationships between an equation and a curve.

The Distance Formula

The first of these concepts involves the distance between any two points in the coordinate plane. If these points lie on a line parallel to the x-axis, the distance from the first point (x_1, y) to the second point (x_2, y) is $|x_2 - x_1|$. The absolute value is used since we are only interested in the magnitude of the distance. Therefore, we could also denote the distance as $|x_1 - x_2|$. Similarly, the distance between two points (x, y_1) and (x, y_2) that lie on a line parallel to the y-axis is $|y_2 - y_1|$ or $|y_1 - y_2|$.

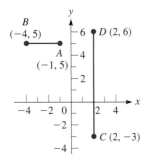

Fig. 2-1

■EXAMPLE 1   The line segment joining $A(-1, 5)$ and $B(-4, 5)$ in Fig. 2-1 is parallel to the x-axis. Therefore, the distance d between these points is

$$d = |-4 - (-1)| = 3 \quad \text{or} \quad d = |-1 - (-4)| = 3$$

Also in Fig. 2-1, the line segment joining $C(2, -3)$ and $D(2, 6)$ is parallel to the y-axis. The distance d between these points is

$$d = |6 - (-3)| = 9 \quad \text{or} \quad d = |-3 - 6| = 9 \quad ■$$

We now wish to find the length of a line segment joining any two points in the plane. If these points are on a line that is not parallel to either axis (see Fig. 2-2), we use the Pythagorean theorem to find the distance between them. By making a right triangle with the line segment joining the points as the hypotenuse and line segments parallel to the axes as legs, we have *the **distance formula**, which gives the distance between any two points in the plane.* This formula is

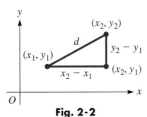

Fig. 2-2

$$\boxed{d = \sqrt{(x_2 - x_1)^2 + (y_2 - y_1)^2}} \tag{2-1}$$

Here we choose the positive square root since we are concerned only with the magnitude of the length of the line segment.

The Pythagorean theorem is named for the Greek mathematician Pythagoras (sixth century B.C.).

■EXAMPLE 2   The distance between $(3, -1)$ and $(-2, -5)$ is given by

$$d = \sqrt{[(-2) - 3]^2 + [(-5) - (-1)]^2}$$
$$= \sqrt{(-5)^2 + (-4)^2} = \sqrt{25 + 16}$$
$$= \sqrt{41} = 6.403$$

See Fig. 2-3.

NOTE▶   *It makes no difference which point is chosen as (x_1, y_1) and which is chosen as (x_2, y_2),* since the differences in the x-coordinates and the y-coordinates are squared. We obtain the same value for the distance when we calculate it as

$$d = \sqrt{[3 - (-2)]^2 + [(-1) - (-5)]^2}$$
$$= \sqrt{5^2 + 4^2} = \sqrt{41} = 6.403 \quad ■$$

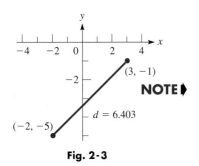

Fig. 2-3

A note regarding the equal sign ($=$) is in order. We will use it for its defined meaning of "equals exactly" and when the result is an approximate number that has been properly rounded off. In Example 2, we have have written $\sqrt{41} = 6.403$, where 6.403 is the rounded-off decimal approximation.

The Slope of a Line

Another important quantity for a line is its *slope,* which gives a measure of the steepness of the line. We now give it a general definition and then develop its meaning in more detail.

The **slope** *of a line is defined as the ratio of the change in y-values (rise) to the corresponding change in the x-values (run) of the coordinates of any two points on the line.* Therefore, the slope, *m,* is given by

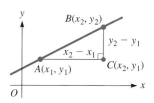

Fig. 2-4

$$m = \frac{y_2 - y_1}{x_2 - x_1} \qquad \text{(2-2)}$$

See Fig. 2-4. When the line is horizontal, $y_2 = y_1$ and $m = 0$. When the line is vertical, $x_2 = x_1$ and the slope is undefined.

EXAMPLE 3 The slope of a line joining $(3, -5)$ and $(-2, -6)$ is

$$m = \frac{-6 - (-5)}{-2 - 3} = \frac{-6 + 5}{-5} = \frac{1}{5}$$

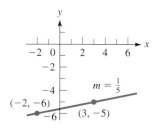

Fig. 2-5

See Fig. 2-5. Again we may interpret either of the points as (x_1, y_1) and the other as (x_2, y_2). We can also obtain the slope of this same line from

$$m = \frac{-5 - (-6)}{3 - (-2)} = \frac{1}{5} \qquad \blacksquare$$

The larger the numerical value of the slope of a line, the more nearly vertical is the line. Also, *a line rising to the right has a positive slope, and a line falling to the right has a negative slope.*

EXAMPLE 4 **(a)** The line in Example 3 has a positive slope, which is numerically small. From Fig. 2-5 it can be seen that the line rises slightly to the right.
(b) The line joining $(3, 4)$ and $(4, -6)$ has a slope of

$$m = \frac{4 - (-6)}{3 - 4} = -10 \qquad \text{or} \qquad m = \frac{-6 - 4}{4 - 3} = -10$$

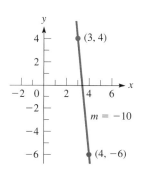

Fig. 2-6

This line falls sharply to the right, as shown in Fig. 2-6. $\blacksquare$

If a given line is extended indefinitely in either direction, it must cross the *x*-axis at some point unless it is parallel to the *x*-axis. *The angle measured from the x-axis in a positive direction to the line is called the* **inclination** *of the line* (see Fig. 2-7). The inclination of a line parallel to the *x*-axis is defined to be zero. *An alternate definition of slope, in terms of the inclination* α, *is*

See Sections 7-1 and 7-2 for a review of the trigonometric functions.

$$m = \tan \alpha \quad (0° \le \alpha < 180°) \qquad \text{(2-3)}$$

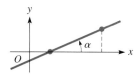

Fig. 2-7

Since the slope can be defined in terms of any two points on the line, we can choose the *x*-intercept and any other point. Therefore, from the definition of the tangent of an angle, we see that Eq. (2-3) is in agreement with Eq. (2-2).

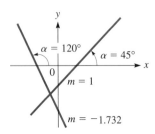

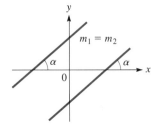

Fig. 2-8

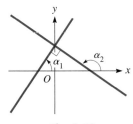

Fig. 2-9

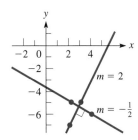

Fig. 2-10

■EXAMPLE 5   **(a)** The slope of a line with an inclination of 45° is

$$m = \tan 45° = 1.000$$

(b) If a line has a slope of -1.732, we know that $\tan \alpha = -1.732$. Since $\tan \alpha$ is negative, α must be a second-quadrant angle (see page 229). Therefore, using a calculator to show that $\tan 60° = 1.732$, we find that $\alpha = 180° - 60° = 120°$. (In using a calculator, the $\tan^{-1}$ function is used; see page 228.) See Fig. 2-8.

We see that if the inclination is an acute angle, the slope is positive, and the line rises to the right. If the inclination is obtuse, the slope is negative, and the line falls to the right. ■

Any two parallel lines crossing the x-axis have the same inclination. Therefore, as shown in Fig. 2-9, the *slopes of parallel lines are equal*. This can be stated as

$$\boxed{m_1 = m_2} \qquad \text{(for } \| \text{ lines)} \qquad \text{(2-4)}$$

If two lines are perpendicular, this means that there must be 90° between their inclinations (Fig. 2-10). The relation between their inclinations is

$$\alpha_2 = \alpha_1 + 90°$$

which can be written as

$$90° - \alpha_2 = -\alpha_1$$

If neither line is vertical (the slope of a vertical line is undefined) and we take the tangent in this last relation, we have

$$\tan(90° - \alpha_2) = \tan(-\alpha_1)$$

or

$$\cot \alpha_2 = -\tan \alpha_1$$

since a function of the complement of an angle equals the cofunction of that angle and since $\tan(-\alpha) = -\tan \alpha$. However, $\cot \alpha = 1/\tan \alpha$, which means that $1/\tan \alpha_2 = -\tan \alpha_1$. Using the inclination definition of slope, we have as *the relation between slopes of perpendicular lines*,

$$\boxed{m_2 = -\frac{1}{m_1} \quad \text{or} \quad m_1 m_2 = -1} \qquad \text{(for } \perp \text{ lines)} \qquad \text{(2-5)}$$

■EXAMPLE 6   The line through $(3, -5)$ and $(2, -7)$ has a slope of

$$m_1 = \frac{-5 + 7}{3 - 2} = 2$$

The line through $(4, -6)$ and $(2, -5)$ has a slope of

$$m_2 = \frac{-6 - (-5)}{4 - 2} = -\frac{1}{2}$$

Since the slopes of the two lines are negative reciprocals, we know that the lines are perpendicular. See Fig. 2-11. ■

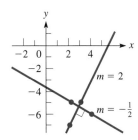

Fig. 2-11

See Appendix C for a graphing calculator program SLOPEDIS. It calculates the slope of a line through two points and the distance between the points.

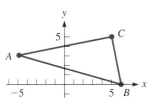

Fig. 2-12

Using the formulas for distance and slope, we can show certain basic geometric relationships. The following examples illustrate the use of the formulas and thereby show the use of algebra in solving problems that are basically geometric. This illustrates the methods of analytic geometry.

■**EXAMPLE 7** Show that the line segments joining $A(-5, 3)$, $B(6, 0)$, and $C(5, 5)$ form a right triangle. See Fig. 2-12.

If these points are vertices of a right triangle, the slopes of two of the sides must be negative reciprocals. This would show perpendicularity. Thus, we find the slopes of the three lines to be

$$m_{AB} = \frac{3 - 0}{-5 - 6} = -\frac{3}{11} \qquad m_{AC} = \frac{3 - 5}{-5 - 5} = \frac{1}{5} \qquad m_{BC} = \frac{0 - 5}{6 - 5} = -5$$

We see that the slopes of AC and BC are negative reciprocals, which means that $AC \perp BC$. From this we can conclude that the triangle is a right triangle. ■

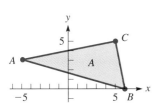

Fig. 2-13

■**EXAMPLE 8** Find the area of the triangle in Example 7. See Fig. 2-13.

Since the right angle is at C, the legs of the triangle are AC and BC. The area is one-half the product of the lengths of the legs of a right triangle. The lengths of the legs are

$$d_{AC} = \sqrt{(-5 - 5)^2 + (3 - 5)^2} = \sqrt{104} = 2\sqrt{26}$$
$$d_{BC} = \sqrt{(6 - 5)^2 + (0 - 5)^2} = \sqrt{26}$$

Therefore, the area is

$$A = \frac{1}{2}(2\sqrt{26})(\sqrt{26}) = 26$$

■

EXERCISES $2\text{-}1$

In Exercises 1–10, find the distance between the given pairs of points.

1. $(3, 8)$ and $(-1, -2)$ **2.** $(-1, 3)$ and $(-8, -4)$

3. $(4, -5)$ and $(4, -8)$ **4.** $(-3, 7)$ and $(2, 10)$

5. $(-12, 20)$ and $(32, -13)$ **6.** $(23, -9)$ and $(-25, 11)$

7. $(\sqrt{32}, -\sqrt{18})$ and $(-\sqrt{50}, \sqrt{8})$

8. $(e, -\pi)$ and $(-2e, -\pi)$

9. $(1.22, -3.45)$ and $(-1.07, -5.16)$

10. $(-5.6, 2.3)$ and $(8.2, -7.5)$

In Exercises 11–20, find the slopes of the lines through the points in Exercises 1–10.

In Exercises 21–24, find the slopes of the lines with the given inclinations.

21. $30°$ **22.** $62.5°$ **23.** $132.7°$ **24.** $135°$

In Exercises 25–28, find the inclinations of the lines with the given slopes.

25. 0.364 **26.** 0.824 **27.** -6.691 **28.** -1.428

In Exercises 29–32, determine whether or not the lines through the two pairs of points are parallel or perpendicular.

29. $(6, -1)$ and $(4, 3)$; $(-5, 2)$ and $(-7, 6)$

30. $(-3, 9)$ and $(4, 4)$; $(9, -1)$ and $(4, -8)$

31. $(-1, -4)$ and $(2, 3)$; $(-5, 2)$ and $(-19, 8)$

32. $(-1, -2)$ and $(3, 6)$; $(2, -6)$ and $(5, 0)$

In Exercises 33–36, determine the value of k.

33. The distance between $(-1, 3)$ and $(11, k)$ is 13.

34. The distance between $(k, 0)$ and $(0, 2k)$ is 10.

35. Points $(6, -1)$, $(3, k)$, and $(-3, -7)$ are on the same line.

36. The points in Exercise 35 are the vertices of a right triangle, with the right angle at $(3, k)$.

In Exercises 37–40, show that the given points are vertices of the given geometric figures.

37. $(2, 3)$, $(4, 9)$, and $(-2, 7)$ are vertices of an isosceles triangle.

38. $(-1, 3)$, $(3, 5)$, and $(5, 1)$ are the vertices of a right triangle.

39. $(-5, -4)$, $(7, 1)$, $(10, 5)$, and $(-2, 0)$ are the vertices of a parallelogram.

40. $(-5, 6)$, $(0, 8)$, $(-3, 1)$, and $(2, 3)$ are the vertices of a square.

In Exercises 41–44, find the indicated areas and perimeters.

41. Find the area of the triangle in Exercise 38.

42. Find the area of the square in Exercise 40.

43. Find the perimeter of the triangle in Exercise 37.

44. Find the perimeter of the parallelogram in Exercise 39.

In Exercises 45–48, use the following definition to find the midpoints between the given points on a straight line.

The *midpoint* between points (x_1, y_1) and (x_2, y_2) on a straight line is the point

$$\left(\frac{x_1 + x_2}{2}, \frac{y_1 + y_2}{2} \right)$$

45. $(-4, 9)$ and $(6, 1)$ **46.** $(-1, 6)$ and $(-13, -8)$

47. $(-12.4, 25.7)$ and $(6.8, -17.3)$

48. $(2.6, 5.3)$ and $(-4.2, -2.7)$

$2\text{-}2$ THE STRAIGHT LINE

In Chapter 1 a number of the graphs that were plotted or displayed on a graphing calculator were straight lines. We now develop methods of finding the equations that are used to represent straight lines. We also show some of the technical and scientific applications of straight lines.

Using the definition of slope, we can derive the general type of equation that represents a straight line. This is another basic method of analytic geometry; that is, equations of a particular form can be shown to represent a particular type of curve. When we recognize the form of the equation, we know the kind of curve it represents. As we have seen, this is of great assistance in sketching the graph.

A straight line can be defined as a *curve* with a constant slope. This means that the value for the slope is the same for any two different points on the line that might be chosen. Thus, considering point (x_1, y_1) on a line to be fixed (Fig. 2-14), and another point $P(x, y)$ that *represents* any other point on the line, we have

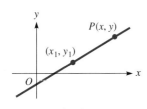

Fig. 2-14

$$m = \frac{y - y_1}{x - x_1}$$

which can be written as

$$\boxed{y - y_1 = m(x - x_1)} \tag{2-6}$$

Equation (2-6) is the **point-slope form** *of the equation of a straight line.* It is useful when we know the slope of a line and some point through which the line passes.

■EXAMPLE 1 Find the equation of the line that passes through $(-4, 1)$ with a slope of -2. See Fig. 2-15.

Substituting in Eq. (2-6), we find that

$$y - 1 = (-2)[x - (-4)]$$

slope

coordinates

which can be simplified to

$$y + 2x + 7 = 0$$

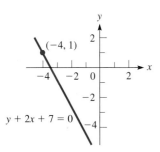

Fig. 2-15

See Appendix C for a graphing calculator program GRAPHLIN. It displays the graph of a line through two given points.

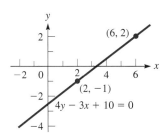

Fig. 2-16

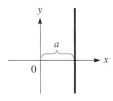

Fig. 2-17

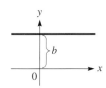

Fig. 2-18

▌EXAMPLE 2   Find the equation of the line through $(2, -1)$ and $(6, 2)$.

We first find the slope of the line through these points:

$$m = \frac{2 + 1}{6 - 2} = \frac{3}{4}$$

Then, by using either of the two known points and Eq. (2-6), we can find the equation of the line:

$$y - (-1) = \frac{3}{4}(x - 2)$$

or

$$4y + 4 = 3x - 6$$

or

$$4y - 3x + 10 = 0$$

This line is shown in Fig. 2-16.   --------▪

Equation (2-6) can be used for any line except for one parallel to the y-axis. Such a line has an undefined slope. However, it does have the property that all points have the same x-coordinate, regardless of the y-coordinate. *We represent a line parallel to the y-axis (see Fig. 2-17) as*

$$\boxed{x = a} \qquad (2\text{-}7)$$

A line parallel to the x-axis has a slope of zero. From Eq. (2-6), we can find its equation to be $y = y_1$. To keep the same form as Eq. (2-7), we normally write this as

$$\boxed{y = b} \qquad (2\text{-}8)$$

See Fig. 2-18.

▌EXAMPLE 3   **(a)** The line $x = 2$ is a line parallel to the y-axis and 2 units to the right of it. This line is shown in Fig. 2-19.

Fig. 2-19

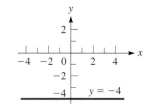

Fig. 2-20

(b) The line $y = -4$ is a line parallel to the x-axis and 4 units below it. This line is shown in Fig. 2-20.   --------▪

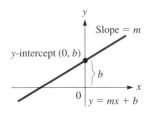

Fig. 2-21

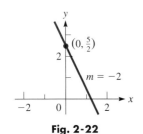

Fig. 2-22

The metric unit for pressure, the pascal, was named for the French mathematician and physicist Blaise Pascal (1623–1662).

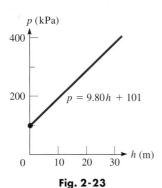

Fig. 2-23

If we choose the special point $(0, b)$, which is the y-intercept of the line, as the point to use in Eq. (2-6), we have $y - b = m(x - 0)$, or

$$y = mx + b \qquad \text{(2-9)}$$

Equation (2-9) is the **slope-intercept form** *of the equation of a straight line.* Its primary usefulness lies in the fact that once we have found the equation of a line and then write it in slope-intercept form, we know that the slope is the coefficient of the x-term and that it crosses the y-axis at the coordinate indicated by the constant term. See Fig. 2-21.

EXAMPLE 4 Find the slope and the y-intercept of the straight line whose equation is $2y + 4x - 5 = 0$.

We write this equation in slope-intercept form:

$$2y = -4x + 5$$

slope ⌐   ⌐ y-coordinate of intercept

$$y = -2x + \frac{5}{2}$$

Since the coefficient of x in this form is -2, the slope is -2. The constant on the right is $5/2$, which means that the y-intercept is $(0, 5/2)$. See Fig. 2-22. ∎

EXAMPLE 5 The pressure p_0 at the surface of a body of water (due to the atmosphere) is 101 kPa. The pressure p at a depth of 10.0 m is 199 kPa. In general, the pressure difference $p - p_0$ varies directly as the depth h. Sketch a graph of p as a function of h.

The solution is as follows:

$$
\begin{array}{ll}
p - p_0 = kh & \text{direct variation} \\
199 - 101 = k(10.0) & \text{substitute given values} \\
k = 9.80 \text{ kPa/m} & \\
p - 101 = 9.80h & \text{substitute in first equation} \\
p = 9.80h + 101 &
\end{array}
$$

We see that this is the equation of a straight line. The slope is 9.80, and the p-intercept is $(0, 101)$. Negative values do not have any physical meaning. The graph is shown in Fig. 2-23. ∎

From Eqs. (2-6) and (2-9) and from the examples of this section, we see that the equation of the straight line has certain characteristics: We have a term in y, a term in x, and a constant term if we simplify as much as possible. *This form is represented by the equation*

$$Ax + By + C = 0 \qquad \text{(2-10)}$$

which is known as the **general form** *of the equation of the straight line.*

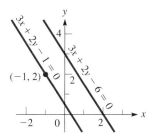

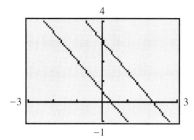

Fig. 2-24

Fig. 2-25

EXAMPLE 6 Find the general form of the equation of the line parallel to the line $3x + 2y - 6 = 0$ and that passes through the point $(-1, 2)$.

Since the line whose equation we want is parallel to the line $3x + 2y - 6 = 0$, it has the same slope. Thus, writing $3x + 2y - 6 = 0$ in slope-intercept form,

$$2y = -3x + 6 \qquad \text{solving for } y$$

$$y = -\frac{3}{2}x + 3$$

Since the slope of $3x + 2y - 6 = 0$ is $-3/2$, the slope of the required line is also $-3/2$. Using $m = -3/2$, the point $(-1, 2)$, and the point-slope form, we have

$$y - 2 = -\frac{3}{2}(x + 1)$$

$$2y - 4 = -3(x + 1)$$

$$3x + 2y - 1 = 0$$

This is the general form of the equation. Both lines are shown in Fig. 2-24.

With $y_1 = -3x/2 + 3$ and $y_2 = -3x/2 + 1/2$, these lines are shown in the calculator display in Fig. 2-25. ------- ∎

In many physical situations, a linear relationship exists between variables. A few examples of this are (1) the distance traveled by an object and the elapsed time, when the velocity is constant, (2) the amount a spring stretches and the force applied, (3) the change in electric resistance and the change in temperature, (4) the force applied to an object and the resulting acceleration, and (5) the pressure at a certain point within a liquid and the depth of the point.

SOLVING A WORD PROBLEM

EXAMPLE 7 For a period of 6.0 s, the velocity v of a car varies linearly with the elapsed time t. If $v = 40$ ft/s when $t = 1.0$ s and $v = 55$ ft/s when $t = 4.0$ s, find the equation relating v and t and graph the function. From the graph, find the initial velocity and the velocity after 6.0 s. What is the meaning of the slope of the line?

With v as the dependent variable and t as the independent variable, the slope is

$$m = \frac{v_2 - v_1}{t_2 - t_1}$$

Using the information given in the statement of the problem, we have

$$m = \frac{55 - 40}{4.0 - 1.0} = 5.0$$

Then, using the point-slope form of the equation of a straight line, we have

$$v - 40 = 5.0(t - 1.0)$$

$$v = 5.0t + 35$$

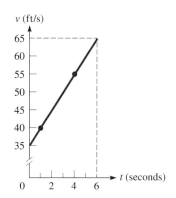

Fig. 2-26

The given values are sufficient to graph the line in Fig. 2-26. There is no need to include negative values of t, since they have no physical meaning. We see that the line crosses the v-axis at 35. This means that the initial velocity (for $t = 0$) is 35 ft/s. Also, when $t = 6.0$ s, we see that $v = 65$ ft/s.

The slope is the ratio of the change in velocity to the change in time. This is the car's *acceleration*. Here the speed of the car increases 5.0 ft/s each second. We can express this acceleration as 5.0 (ft/s)/s = 5.0 ft/s^2. ------- ∎

EXERCISES *2-2*

In Exercises 1–20, find the equation of each of the lines with the given properties. Sketch the graph of each line.

1. Passes through $(-3, 8)$ with a slope of 4.

2. Passes through $(-2, -1)$ with a slope of -2.

3. Passes through $(2, -5)$ and $(4, 2)$.

4. Passes through $(-3, 5)$ and $(-2, 3)$.

5. Passes through $(1, 3)$ and has an inclination of $45°$.

6. Has a y-intercept $(0, -2)$ and an inclination of $120°$.

7. Passes through $(5.3, -2.7)$ and is parallel to the x-axis.

8. Passes through $(-4, -2)$ and is perpendicular to x-axis.

9. Is parallel to the y-axis and is 3 units to the left of it.

10. Is parallel to the x-axis and is 4.1 units below it.

11. Has an x-intercept $(4, 0)$ and a y-intercept of $(0, -6)$.

12. Has an x-intercept of $(-3, 0)$ and a slope of 2.

13. Is perpendicular to a line with a slope of 3 and passes through $(1, -2)$.

14. Is perpendicular to a line with a slope of -4 and has a y-intercept of $(0, 3)$.

15. Is parallel to a line through $(-1, 2)$ and $(3, 1)$ and passes through $(1, 2)$.

16. Is parallel to a line through $(7, -1)$ and $(4, 3)$ and has a y-intercept of $(0, -2)$.

17. Is perpendicular to the line $6.0x - 2.4y - 3.9 = 0$ and passes through $(7.5, -4.7)$.

18. Is parallel to the line $2y - 6x - 5 = 0$ and passes through $(-4, -5)$.

19. Has a slope of -3 and passes through the intersection of the lines $5x - y = 6$ and $x + y = 12$.

20. Passes through the point of intersection of $2x + y - 3 = 0$ and $x - y - 3 = 0$ and through the point $(4, -3)$.

In Exercises 21–28, reduce the equations to slope-intercept form and find the slope and the y-intercept. Sketch each line.

21. $4x - y = 8$

22. $2x - 3y - 6 = 0$

23. $3x + 5y - 10 = 0$

24. $4y = 6x - 9$

25. $3x - 2y - 1 = 0$

26. $4x + 2y - 5 = 0$

27. $11.2x + 1.6 = 3.2y$

28. $11.5x + 4.60y = 5.98$

In Exercises 29–32, determine the value of k.

29. Find k if the lines $4x - ky = 6$ and $6x + 3y + 2 = 0$ are parallel.

30. Find k if the lines given in Exercise 29 are perpendicular.

(W) 31. Find k if the lines $3x - y = 9$ and $kx + 3y = 5$ are perpendicular. Explain how this value is found.

(W) 32. Find k if the lines given in Exercise 31 are parallel. Explain how this value is found.

In Exercises 33–40, determine whether the given lines are parallel, perpendicular, or neither.

33. $3x - 2y + 5 = 0$ and $4y = 6x - 1$

34. $8x - 4y + 1 = 0$ and $4x + 2y - 3 = 0$

35. $6x - 3y - 2 = 0$ and $x + 2y - 4 = 0$

36. $3y - 2x = 4$ and $6x - 9y = 5$

37. $5x + 2y - 3 = 0$ and $10y = 7 - 4x$

38. $48y - 36x = 71$ and $52x = 17 - 39y$

39. $4.5x - 1.8y = 1.7$ and $2.4x + 6.0y = 0.3$

40. $3.5y = 4.3 - 1.5x$ and $3.6x + 8.4y = 1.7$

In Exercises 41–52, some applications involving straight lines are shown.

41. The velocity v of a box sliding down a long ramp is given by $v = v_0 + at$, where v_0 is the initial velocity, a is the acceleration, and t is the time. If $v_0 = 12.2$ ft/s and $v = 35.4$ ft/s when $t = 4.50$ s, find v as a function of t. Sketch the graph.

42. The voltage V across part of an electric circuit is given by $V = E - iR$, where E is a battery voltage, i is the current, and R is the resistance. If $E = 6.00$ V and $V = 4.35$ V for $i = 9.17$ mA, find V as a function of i. Sketch the graph (i and V may be negative).

43. The velocity of sound v increases 0.607 m/s for each increase in temperature T of $1.00°C$. If $v = 343$ m/s for $T = 20.0°C$, express v as a function of T.

44. An acid solution is made from x liters of a 20% solution and y liters of a 30% solution. If the final solution contains 20 L of acid, find the equation relating x and y.

45. One computer printer can print x characters per second, and a second printer can print y characters per second. If the first prints for 50 s and the second for 60 s and they print a total of 12,200 characters, find the equation relating x and y.

(W) 46. A wall is 15 cm thick. At the outside, the temperature is $3°C$, and at the inside, it is $23°C$. If the temperature changes at a constant rate through the wall, write an equation of the temperature T in the wall as a function of the distance x from the outside to the inside of the wall. What is the meaning of the slope of the line?

47. One heating unit uses x gallons of fuel at 72% efficiency, and another heating unit uses y gallons at 90% efficiency. If 135,000 Btu of heat is delivered by these units together, express y as a function of x.

(W) 48. The length of a rectangular solar cell is 10 cm more than the width w. Express the perimeter p of the cell as a function of w. What is the meaning of the slope of the line?

49. A light beam is reflected off the edge of an optic fiber at an angle of 0.0032°. The diameter of the fiber is 48 μm. Find the equation of the reflected beam with the x-axis (at the center of the fiber) and the y-axis as shown in Fig. 2-27.

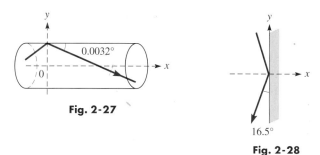

Fig. 2-27

50. A police report stated that a bullet caromed downward off a wall at an angle of 16.5° with the wall, as shown in Fig. 2-28. What is the equation of the path of the bullet after impact?

Ⓦ 51. A survey of the traffic on a particular highway showed that the number of cars passing a particular point each minute varied linearly from 6:30 A.M. to 8:30 A.M. on workday mornings. The study showed that an average of 45 cars passed the point in one minute at 7 A.M. and that 115 cars passed in one minute at 8 A.M. If n is the number of cars passing the point in one minute and t is the number of minutes after 6:30 A.M., find the equation relating n and t, and graph the equation. From the graph, determine n at 6:30 A.M. and at 8:30 A.M. What is the meaning of the slope of the line?

52. In a research project on cancer, a tumor was determined to weigh 30 mg when first discovered. While being treated, it grew smaller by 2 mg each month. Find the equation relating the weight w of the tumor as a function of the time t in months. Graph the equation.

In Exercises 53–56, treat the given nonlinear functions as linear functions in order to sketch their graphs. At times, this can be useful in showing certain values of a function. For example, $y = 2 + 3x^2$ can be shown as a straight line by graphing y as a function of x^2. A table of values for this graph is shown along with the corresponding graph in Fig. 2-29.

x	0	1	2	3	4	5
x^2	0	1	4	9	16	25
y	2	5	14	29	50	77

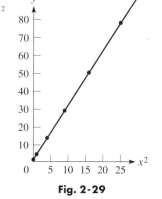

Fig. 2-29

53. The number n of memory cells of a certain computer that can be tested in t seconds is given by $n = 1200\sqrt{t}$. Sketch n as a function of $\sqrt{t}$.

54. The force F (in lb) applied to a lever to balance a certain weight on the opposite side of the fulcrum is given by $F = 40/d$, where d is the distance (in ft) of the force from the fulcrum. Sketch F as a function of $1/d$.

55. A spacecraft is launched such that its altitude h (in km) is given by $h = 300 + 2t^{3/2}$ for $0 \leq t < 100$ s. Sketch this as a linear function.

56. The current i (in A) in a certain electric circuit is given by $i = 6(1 - e^{-t})$. Sketch this as a linear function. (e is a constant, equal to about 2.72.)

In Exercises 57–60, show that the given nonlinear functions are linear when plotted on semilogarithmic or logarithmic paper.

57. A function of the form $y = ax^n$ is straight when plotted on logarithmic paper, since $\log y = \log a + n \log x$ is in the form of a straight line. The variables are $\log y$ and $\log x$; the slope can be found from $(\log y - \log a)/\log x = n$, and the intercept is a. (To get the slope from the graph, it is necessary to measure vertical and horizontal distances between two points. The log y-intercept is found where $\log x = 0$, and this occurs when $x = 1$.) Plot $y = 3x^4$ on logarithmic paper to verify this analysis.

58. A function of the form $y = a(b^x)$ is a straight line on semilogarithmic paper, since $\log y = \log a + x \log b$ is in the form of a straight line. The variables are $\log y$ and x, the slope is $\log b$, and the intercept is a. (To get the slope from the graph, we calculate $(\log y - \log a)/x$ for some set of values x and y. The intercept is read directly off the graph where $x = 0$.) Plot $y = 3(2^x)$ on semilogarithmic paper to verify this analysis.

59. If experimental data are plotted on logarithmic paper and the points lie on a straight line, it is possible to determine the function (see Exercise 57). The following data come from an experiment to determine the functional relationship between the pressure p and the volume V of a gas undergoing an adiabatic (no heat loss) change. From the graph on logarithmic paper, determine p as a function of V.

V (m^3)	0.100	0.500	2.00	5.00	10.0
p (kPa)	20.1	2.11	0.303	0.0840	0.0318

60. If experimental data are plotted on semilogarithmic paper, and the points lie on a straight line, it is possible to determine the function (see Exercise 58). The following data come from an experiment designed to determine the relationship between the voltage across an inductor and the time, after the switch is opened. Determine v as a function of t.

v (V)	40	15	5.6	2.2	0.8
t (ms)	0.0	20	40	60	80

$2\text{-}3$ THE CIRCLE

We have found that we can obtain a general equation that represents a straight line by considering a fixed point on the line and then a general point $P(x, y)$, which can represent any other point on the same line. Mathematically, we can state this as "the line is the **locus** of a point $P(x, y)$ that *moves* from a fixed point with constant slope along the line." That is, the point $P(x, y)$ can be considered as a variable point that moves along the line.

In this way we can define a number of important curves. *A* **circle** *is defined as the locus of a point $P(x, y)$ that moves so that it is always equidistant from a fixed point. We call this fixed distance the* **radius,** *and we call the fixed point the* **center** *of the circle.* Thus, using this definition, calling the fixed point (h, k) and the radius r, we have

$$\sqrt{(x - h)^2 + (y - k)^2} = r$$

or, by squaring both sides, we have

$$(x - h)^2 + (y - k)^2 = r^2 \qquad \text{(2-11)}$$

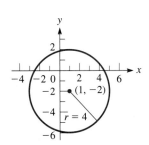

Fig. 2-30

Equation (2-11) is called the **standard equation** *of a circle with center at (h, k) and radius r.* See Fig. 2-30.

■EXAMPLE 1   The equation $(x - 1)^2 + (y + 2)^2 = 16$ represents a circle with center at $(1, -2)$ and a radius of 4. We determine these values by considering the equation of the circle to be in the form of Eq. (2-11) as

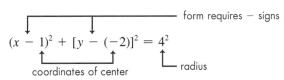

Note carefully the way in which we found the y-coordinate of the center. *We must have a minus sign before each of the coordinates.* Here, to get the y-coordinate, we had to write $+2$ as $-(-2)$. This circle is shown in Fig. 2-31.   ■

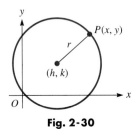

Fig. 2-31

■EXAMPLE 2   Find the equation of the circle with center at $(2, 1)$ and that passes through $(4, 8)$.

In Eq. (2-11) we can determine the equation if we can find h, k, and r for this circle. From the given information, $h = 2$ and $k = 1$. To find r, we use the fact that *all points on the circle must satisfy the equation of the circle.* The point $(4, 8)$ must satisfy Eq. (2-11), with $h = 2$ and $k = 1$. Thus,

$$(4 - 2)^2 + (8 - 1)^2 = r^2 \qquad \text{or} \qquad r^2 = 53$$

Therefore, the equation of the circle is

$$(x - 2)^2 + (y - 1)^2 = 53$$

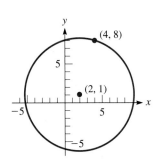

Fig. 2-32

This circle is shown in Fig. 2-32.   ■

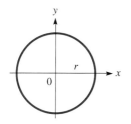

Fig. 2-33

SOLVING A WORD PROBLEM

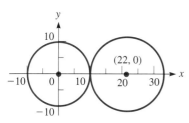

Fig. 2-34

If the center of the circle is at the origin, which means that the coordinates of the center are $(0, 0)$, the equation of the circle (see Fig. 2-33) becomes

$$x^2 + y^2 = r^2 \qquad \text{(2-12)}$$

The following example illustrates an application using this type of circle and one with its center not at the origin.

■EXAMPLE 3   A student is drawing a friction drive in which two circular disks are in contact with each other. They are represented by circles in the drawing. The first has a radius of 10.0 cm, and the second has a radius of 12.0 cm. What is the equation of each circle if the origin is at the center of the first circle and the positive x-axis passes through the center of the second circle? See Fig. 2-34.

Since the center of the smaller circle is at the origin, we can use Eq. (2-12). Given that the radius is 10.0 cm, we have as its equation

$$x^2 + y^2 = 100$$

The fact that the two disks are in contact tells us that they meet at the point $(10.0, 0)$. Knowing that the radius of the larger circle is 12.0 cm tells us that its center is at $(22.0, 0)$. Thus, using Eq. (2-11) with $h = 22.0$, $k = 0$, and $r = 12.0$,

$$(x - 22.0)^2 + (y - 0)^2 = 12.0^2$$

or

$$(x - 22.0)^2 + y^2 = 144$$

as the equation of the larger circle.   ■

SYMMETRY

NOTE▶

NOTE▶

A circle with its center at the origin exhibits an important property of the graphs of many equations. *It is **symmetrical** to the x-axis and also to the y-axis.* Symmetry to the x-axis can be thought of as meaning that the lower half of the curve is a reflection of the upper half, and conversely. It can be shown that *if $-y$ can replace y in an equation without changing the equation, the graph of the equation is **symmetrical to the x-axis**.* Symmetry to the y-axis is similar. *If $-x$ can replace x in the equation without changing the equation, the graph is symmetrical to the y-axis.*

This type of circle is also symmetrical to the origin as well as being symmetrical to both axes. The meaning of symmetry to the origin is that the origin is the midpoint of any two points (x, y) and $(-x, -y)$ that are on the curve. Thus, *if $-x$ can replace x and $-y$ can replace y at the same time, without changing the equation, the graph of the equation is symmetrical to the origin.*

■EXAMPLE 4   The equation of the circle with its center at the origin and with a radius of 6 is $x^2 + y^2 = 36$.

The symmetry of this circle can be shown analytically by the substitutions mentioned above. Replacing x by $-x$, we obtain $(-x)^2 + y^2 = 36$. Since $(-x)^2 = x^2$, this equation can be rewritten as $x^2 + y^2 = 36$. Since this substitution did not change the equation, the graph is symmetrical to the y-axis.

Replacing y by $-y$, we obtain $x^2 + (-y)^2 = 36$, which is the same as $x^2 + y^2 = 36$. This means that the curve is symmetrical to the x-axis.

Replacing x by $-x$ and simultaneously replacing y by $-y$, we obtain $(-x)^2 + (-y)^2 = 36$, which is the same as $x^2 + y^2 = 36$. This means that the curve is symmetrical to the origin. This circle is shown in Fig. 2-35.   ■

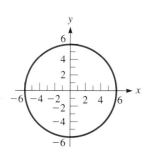

Fig. 2-35

If we multiply out each of the terms in Eq. (2-11), we may combine the resulting terms to obtain

$$x^2 - 2hx + h^2 + y^2 - 2ky + k^2 = r^2$$

$$\mathbf{x^2 + y^2 - 2hx - 2ky + (h^2 + k^2 - r^2) = 0} \qquad \text{(2-13)}$$

Since each of h, k, and r is constant for any given circle, the coefficients of x and y and the term within parentheses in Eq. (2-13) are constants. Equation (2-13) can then be written as

$$\boxed{x^2 + y^2 + Dx + Ey + F = 0} \qquad \text{(2-14)}$$

Equation (2-14) is called the **general equation** *of the circle.* It tells us that any equation which can be written in that form will represent a circle.

■**EXAMPLE 5**   Find the center and radius of the circle

$$x^2 + y^2 - 6x + 8y - 24 = 0$$

We can find this information if we write the given equation in standard form. To do so, *we must complete the square in the x-terms and also in the y-terms.* This is done by first writing the equation in the form

CAUTION ▶

$$(x^2 - 6x \quad) + (y^2 + 8y \quad) = 24$$

To complete the square of the x-terms, we take half of -6, which is -3, square it, and add the result, 9, to each side of the equation. In the same way, we complete the square of the y-terms by adding 16 to each side of the equation, which gives

$$\left(\frac{-6}{2}\right)^2 \qquad \left(\frac{8}{2}\right)^2 \qquad \text{add to both sides}$$

$$(x^2 - 6x + 9) + (y^2 + 8y + 16) = 24 + 9 + 16$$

$$(x - 3)^2 + (y + 4)^2 = 49$$

$$(x - 3)^2 + (y - (-4))^2 = 7^2$$

coordinates of center   radius

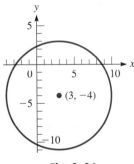

Fig. 2-36

Thus, the center is $(3, -4)$, and the radius is 7 (see Fig. 2-36). ■

SOLVING A WORD PROBLEM

■**EXAMPLE 6**   A certain pendulum is found to swing through an arc of the circle $3x^2 + 3y^2 - 9.60y - 2.80 = 0$. What is the length (in m) of the pendulum, and from what point is it swinging?

We see that this equation represents a circle by dividing through by 3. This gives us $x^2 + y^2 - 3.20y - 2.80/3 = 0$. The length of the pendulum is the radius of the circle, and the point from which it swings is the center. These are found as follows:

$$x^2 + (y^2 - 3.20y + 1.60^2) = 1.60^2 + 2.80/3 \qquad \text{complete squares in both } x\text{- and } y\text{-terms}$$

$$x^2 + (y - 1.60)^2 = 3.493 \qquad \text{standard form}$$

Since $\sqrt{3.493} = 1.87$, the length of the pendulum is 1.87 m. The point from which it is swinging is $(0, 1.60)$. See Fig. 2-37.

Replacing x by $-x$, the equation does not change. Replacing y by $-y$, the equation does change (the $3.20y$ term changes sign). Thus, the circle is symmetric only to the y-axis. ■

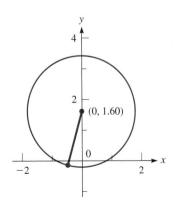

Fig. 2-37

NOTE ▶

The general equation of a circle, Eq. (2-14), does not represent a *function* since there are two values of y for most values of x in the domain. In fact, *it might be necessary to use the quadratic formula to find the two functions to enter into a graphing calculator* in order to view the curve. This is illustrated in the following example.

▌EXAMPLE 7 Display the graph of the circle $3x^2 + 3y^2 + 6y - 20 = 0$ on a graphing calculator.

To fit the form of a quadratic equation in y, we write

$$3y^2 + 6y + (3x^2 - 20) = 0$$

Now, using the quadratic formula to solve for y, we let

$$a = 3 \qquad b = 6 \qquad c = 3x^2 - 20$$

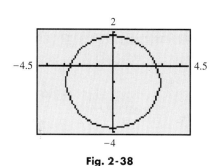

Fig. 2-38

Therefore,

$$y = \frac{-6 \pm \sqrt{6^2 - 4(3)(3x^2 - 20)}}{2(3)}$$

which means we get the *two functions*

$$y_1 = \frac{-6 + \sqrt{276 - 36x^2}}{6} \quad \text{and} \quad y_2 = \frac{-6 - \sqrt{276 - 36x^2}}{6}$$

which are entered into the calculator to get the view shown in Fig. 2-38.

The *window* values were chosen so that the length along the x-axis is about 1.5 times that along the y-axis so as to have less distortion in the circle. There may be gaps at the left and right sides of the circle. ▬

EXERCISES 2-3

In Exercises 1–4, determine the center and the radius of each circle.

1. $(x - 2)^2 + (y - 1)^2 = 25$

2. $(x - 3)^2 + (y + 4)^2 = 49$

3. $(x + 1)^2 + y^2 = 4$ **4.** $x^2 + (y - 6)^2 = 64$

In Exercises 5–20, find the equation of each of the circles from the given information.

5. Center at $(0, 0)$, radius 3   **6.** Center at $(0, 0)$, radius 1

7. Center at $(2, 2)$, radius 4   **8.** Center at $(0, 2)$, radius 2

9. Center at $(-2, 5)$, radius $\sqrt{5}$

10. Center at $(-3, -5)$, radius $2\sqrt{3}$

11. Center at $(12, -15)$, radius 18

12. Center at $(\frac{3}{2}, -2)$, radius $\frac{5}{2}$

13. Center at $(2, 1)$, passes through $(4, -1)$

14. Center at $(-1, 4)$, passes through $(-2, 3)$

15. Center at $(-3, 5)$, tangent to the x-axis

16. Center at $(2, -4)$, tangent to the y-axis

17. Tangent to both axes and the lines $y = 4$ and $x = 4$

18. Tangent to both axes, radius 4, in the second quadrant

19. Center on the line $5x = 2y$, radius 5, tangent to the x-axis

20. The points $(3, 8)$ and $(-3, 0)$ are the ends of a diameter.

In Exercises 21–32, determine the center and radius of each circle. Sketch each circle.

21. $x^2 + (y - 3)^2 = 4$

22. $(x - 2)^2 + (y + 3)^2 = 49$

23. $4(x + 1)^2 + 4(y - 5)^2 = 81$

24. $2(x + 4)^2 + 2(y + 3)^2 = 25$

25. $x^2 + y^2 - 2x - 8 = 0$

26. $x^2 + y^2 - 4x - 6y - 12 = 0$

27. $x^2 + y^2 + 4.20x - 2.60y = 3.51$

28. $x^2 + y^2 + 22x + 14y = 26$

29. $4x^2 + 4y^2 - 16y = 9$

30. $9x^2 + 9y^2 + 18y = 7$

31. $2x^2 + 2y^2 - 4x - 8y - 1 = 0$

32. $3x^2 + 3y^2 - 12x + 4 = 0$

In Exercises 33–36, determine whether the circles with the given equations are symmetric to either axis or to the origin.

33. $x^2 + y^2 = 100$

34. $x^2 + y^2 - 4x - 5 = 0$

35. $3x^2 + 3y^2 + 24y = 8$

36. $5x^2 + 5y^2 - 10x + 20y = 3$

In Exercises 37–48, solve the given problems.

37. Determine whether the circle $x^2 - 6x + y^2 - 7 = 0$ crosses the x-axis.

38. Find the points of intersection of the circle $x^2 + y^2 - x - 3y = 0$ and the line $y = x - 1$.

(W) 39. Find the equation of the locus of a point $P(x, y)$ that moves so that its distance from $(2, 4)$ is twice its distance from $(0, 0)$. Describe the locus.

(W) 40. Find the equation of the locus of a point $P(x, y)$ that moves so that the line joining it and $(2, 0)$ is always perpendicular to the line joining it and $(-2, 0)$. Describe the locus.

41. Use a graphing calculator to view the circle $x^2 + y^2 + 5y - 4 = 0$.

42. Use a graphing calculator to view the circle $2x^2 + 2y^2 + 2y - x - 1 = 0$.

43. In a hoisting device, two of the pulley wheels may be represented by $x^2 + y^2 = 14.5$ and $x^2 + y^2 - 19.6y + 86.0 = 0$. How far apart (in in.) are the wheels?

44. The design of a machine part shows it as a circle represented by the equation $x^2 + y^2 = 42.5$, with a circular hole represented by $x^2 + y^2 + 3.06y - 1.24 = 0$ cut out. What is the least distance (in in.) from the edge of the hole to the edge of the machine part?

45. A wire is rotating in a circular path through a magnetic field to induce an electric current in the wire. The wire is rotating at 60.0 Hz with a constant velocity of 37.7 m/s. Taking the origin at the center of the circle of rotation, find the equation of the path of the wire.

46. A communications satellite remains stationary at an altitude of 22,500 mi over a point on the earth's equator. It therefore rotates once each day about the earth's center. Its velocity is constant, but the horizontal and vertical components, v_H and v_V, of the velocity constantly change. Show that the equation relating v_H and v_V (in mi/h) is that of a circle. The radius of the earth is 3960 mi.

47. In analyzing the stress on a beam, *Mohr's circle* is often used. To form it, normal stress is plotted as the x-coordinate and shear stress is plotted as the y-coordinate. The center of the circle is midway between the minimum and maximum values of normal stress on the x-axis. Find the equation of Mohr's circle if the minimum normal stress is 100×10^{-6} and the maximum normal stress is 900×10^{-6} (stress is unitless). Sketch the graph.

48. A Norman window has the form of a rectangle surmounted by a semicircle. An architect designs a Norman window on a coordinate system as shown in Fig. 2-39. If the circumference of the circular part of the window is on the circle $x^2 + y^2 - 3.00y + 1.25 = 0$, find the area of the window. Measurements are in meters.

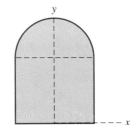

Fig. 2-39

$2\text{-}4$ THE PARABOLA

Another curve important in technical and scientific applications is the parabola. We came across this curve several times when we plotted or displayed curves in the previous chapter. In this section we give the general definition of a parabola and thereby find the general form of its equation.

A **parabola** *is defined as the locus of a point $P(x, y)$ that moves so that it is always equidistant from a given line and a given point. The given line is called the* **directrix,** *and the given point is called the* **focus.** The line through the focus that is perpendicular to the directrix is called the **axis** of the parabola. The point midway between the directrix and focus is the **vertex** of the parabola. Using this definition, we shall find the equation of the parabola for which the focus is the point $(p, 0)$ and the directrix is the line $x = -p$. By choosing the focus and directrix in this manner, we can find a general representation of the equation of a parabola with its vertex at the origin.

According to the definition of the parabola, the distance from a point $P(x, y)$ on the parabola to the focus $(p, 0)$ must equal the distance from $P(x, y)$ to the directrix $x = -p$. The distance from P to the focus can be found by using the distance formula. The distance from P to the directrix is the perpendicular distance, and this can be found as the distance between two points on a line parallel to the x-axis. These distances are indicated in Fig. 2-40.

Thus, we have

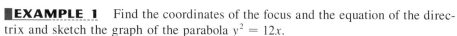

$$\sqrt{(x - p)^2 + (y - 0)^2} = x + p$$

Squaring both sides of this equation, we have

$$(x - p)^2 + y^2 = (x + p)^2$$

or

$$x^2 - 2px + p^2 + y^2 = x^2 + 2px + p^2$$

Simplifying, we obtain

$$\boxed{y^2 = 4px} \qquad (2\text{-}15)$$

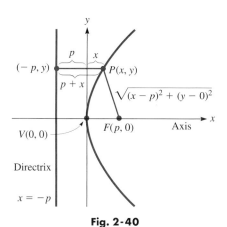

Fig. 2-40

Equation (2-15) is called the **standard form** *of the equation of a parabola with its axis along the x-axis and the vertex at the origin*. Its symmetry to the x-axis can be proven since $(-y)^2 = 4px$ is the same as $y^2 = 4px$.

■**EXAMPLE 1**   Find the coordinates of the focus and the equation of the directrix and sketch the graph of the parabola $y^2 = 12x$.

Since the equation of this parabola fits the form of Eq. (2-15), we know that the vertex is at the origin. The coefficient of 12 tells us that

$$4p = 12, \qquad p = 3$$

Since $p = 3$, the focus is the point $(3, 0)$, and the directrix is the line $x = -3$, as shown in Fig. 2-41.

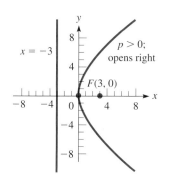

Fig. 2-41

■**EXAMPLE 2**   If the focus is to the left of the origin, with the directrix an equal distance to the right, the coefficient of the x-term is negative. This tells us that the parabola opens to the left, rather than to the right, as is the case when the focus is to the right of the origin. For example, the parabola $y^2 = -8x$ has its vertex at the origin, its focus at $(-2, 0)$, and the line $x = 2$ as its directrix. We determine this from the equation as follows:

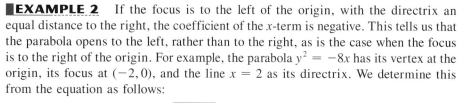

$$y^2 = -8x \qquad 4p = -8, \qquad p = -2$$

Since $p = -2$, we find

the focus is $(-2, 0)$
the directrix is the line $x = -(-2)$, or $x = 2$

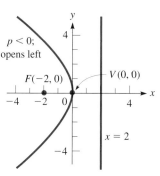

Fig. 2-42

The parabola opens to the left, as shown in Fig. 2-42.

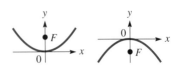

Fig. 2-43

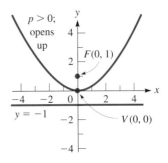

Fig. 2-44

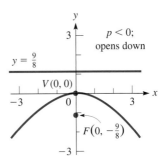

Fig. 2-45

SOLVING A WORD PROBLEM

See the chapter introduction. Since about 1990, many automobile headlights have not been designed with a parabolic reflector as illustrated in this example.

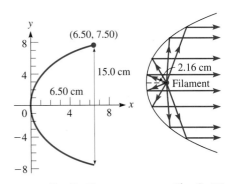

Fig. 2-46   **Fig. 2-47**

If we chose the focus as the point $(0, p)$ and the directrix as the line $y = -p$ (see Fig. 2-43), we would find that the resulting equation is

$$x^2 = 4py \qquad \text{(2-16)}$$

This is the standard form of the equation of a parabola with the y-axis as its axis and the vertex at the origin. Its symmetry to the y-axis can be proved, since $(-x)^2 = 4py$ is the same as $x^2 = 4py$. We note that **the difference between this equation and Eq. (2-15) is that x is squared and y appears to the first power in Eq. (2-16), rather than the reverse, as in Eq. (2-15).**

EXAMPLE 3   The parabola $x^2 = 4y$ fits the form of Eq. (2-16). Therefore, its axis is along the y-axis and its vertex is at the origin. From the equation, we find the value of p, which in turn tells us the location of the vertex and the directrix.

$$x^2 = 4y \qquad 4p = 4, \qquad p = 1$$

Focus $(0, p)$ is $(0, 1)$; directrix $y = -p$ is $y = -1$. The parabola is shown in Fig. 2-44, and we see in this case that it opens upward.

EXAMPLE 4   The parabola $2x^2 = -9y$ fits the form of Eq. (2-16) if we write it in the form

$$x^2 = -\tfrac{9}{2}y$$

Here, we see that $4p = -9/2$. Therefore, its axis is along the y-axis, and its vertex is at the origin. Since $4p = -9/2$, we have

$$p = -\frac{9}{8} \qquad \text{focus}\left(0, -\frac{9}{8}\right) \qquad \text{directrix } y = \frac{9}{8}$$

The parabola opens downward, as shown in Fig. 2-45.

EXAMPLE 5   In calculus it can be shown that a light ray coming from the focus of a parabola will be reflected off the parabolic surface parallel to the axis of the parabola. This property of a parabola involving light reflection has many very useful applications. One of these is in the design of automobile headlights.

An automobile headlight reflector is designed with cross sections of the reflecting surface being equal parabolas. It has an opening of 15.0 cm and is 6.50 cm deep, as shown in Fig. 2-46. Determine where the filament of the bulb should be located so that reflected light rays are parallel in order to create a light beam.

By finding the equation of the parabola we can find the location of the focus, which is the desired location of the filament. Since the parabolic opening is 15.0 cm wide and 6.50 cm deep, the point $(6.50, 7.50)$ is on the parabola.

Since we have placed the parabola with its vertex at the origin and its axis along the x-axis, its general form is given by Eq. (2-15), or $y^2 = 4px$. We can find the value of p by using the fact that $(6.50, 7.50)$ is a point on the parabola and the fact that the coordinates must satisfy the equation. This means that

$$7.50^2 = 4p(6.50) \quad \text{or} \quad p = 2.16$$

The equation of the parabola is $y^2 = 8.64x$. The filament should be at the focus, which means it should be on the axis of the parabola, 2.16 cm from the vertex. In this way the reflected rays are parallel to the axis, as shown in Fig. 2-47.

Equations (2-15) and (2-16) give us the general form of the equation of a parabola with its vertex at the origin and its focus on one of the coordinate axes. The next example shows the use of the definition to find the equation of a parabola that has its vertex at a point other than at the origin.

EXAMPLE 6 Using the definition of the parabola, find the equation of the parabola with its focus at (2, 3) and its directrix the line $y = -1$. See Fig. 2-48.

Choosing a general point $P(x, y)$ on the parabola and equating the distances from this point to (2, 3) and to the line $y = -1$, we have

$$\sqrt{(x - 2)^2 + (y - 3)^2} = y + 1$$

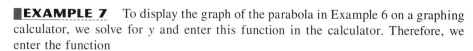

Squaring both sides of this equation and simplifying, we have

$$(x - 2)^2 + (y - 3)^2 = (y + 1)^2$$
$$x^2 - 4x + 4 + y^2 - 6y + 9 = y^2 + 2y + 1$$

or

$$8y = 12 - 4x + x^2$$

We note that this type of equation has appeared frequently in earlier chapters. The x-term and the constant (12 in this case) are characteristic of a parabola that does not have its vertex at the origin if the directrix is parallel to the x-axis.

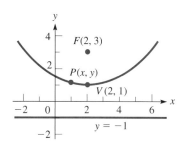

Fig. 2-48

We can readily view a parabola on a graphing calculator. If the axis is along the y-axis, or parallel to it, such as in Examples 3, 4, and 6, we simply solve for y and use this function. However, if the axis is along the x-axis, or parallel to it, as in Examples 1, 2, and 5, we get *two* functions to graph. This is similar to Example 7 on page 41. These cases are shown in the following example.

EXAMPLE 7 To display the graph of the parabola in Example 6 on a graphing calculator, we solve for y and enter this function in the calculator. Therefore, we enter the function

$$y_1 = (12 - 4x + x^2)/8$$

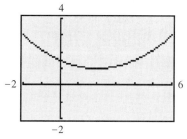

Fig. 2-49

in the calculator, and we get the display shown in Fig. 2-49.

To display the graph of the parabola in Example 1, where $y^2 = 12x$, when we solve for y, we get $y = \pm\sqrt{12x}$. Therefore, we enter the two functions

$$y_1 = \sqrt{12x} \quad \text{and} \quad y_2 = -\sqrt{12x}$$

in the calculator, and we get the display shown in Fig. 2-50.

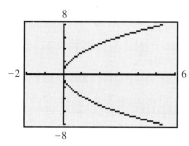

Fig. 2-50

We can conclude that *the equation of a parabola is characterized by the presence of the square of either (but not both) x or y, and a first-power term in the other.* We will consider further the equation of the parabola in Sections 2-7 and 2-8.

The parabola has numerous technical applications. The reflection property illustrated in Example 5 has other important applications, such as the design of a radar antenna. The path of a projectile is parabolic. The cables of a suspension bridge are parabolic. These and other applications are illustrated in the exercises.

— EXERCISES *2-4*

In Exercises 1–12, determine the coordinates of the focus and the equation of the directrix of the given parabolas. Sketch each curve.

1. $y^2 = 4x$

2. $y^2 = 16x$

3. $y^2 = -4x$

4. $y^2 = -16x$

5. $x^2 = 8y$

6. $x^2 = 10y$

7. $x^2 = -4y$

8. $x^2 = -12y$

9. $2y^2 = 5x$

10. $3x^2 = 8y$

11. $y = 0.48x^2$

12. $x = 7.6y^2$

In Exercises 13–20, find the equations of the parabolas satisfying the given conditions.

13. Focus $(3, 0)$, directrix $x = -3$

14. Focus $(-2, 0)$, directrix $x = 2$

15. Focus $(0, 4)$, vertex $(0, 0)$

16. Focus $(-3, 0)$, vertex $(0, 0)$

17. Vertex $(0, 0)$, directrix $y = -0.16$

18. Vertex $(0, 0)$, directrix $y = 2.3$

19. Vertex $(0, 0)$, axis along the y-axis, passes through $(-1, 8)$

20. Vertex $(0, 0)$, symmetric to the x-axis, passes through $(2, -1)$

In Exercises 21–40, solve the given problems.

21. Find the equation of the parabola with focus $(6, 1)$ and directrix $x = 0$ by use of the definition. Sketch the curve.

22. Find the equation of the parabola with focus $(1, 1)$ and directrix $y = 5$ by use of the definition. Sketch the curve.

23. Use a graphing calculator to view the parabola $y^2 + 2x + 8y + 13 = 0$.

24. Use a graphing calculator to view the parabola $y^2 - 2x - 6y + 19 = 0$.

25. The equation of a parabola with vertex (h, k) and axis parallel to the x-axis is $(y - k)^2 = 4p(x - h)$. (This is shown in Section 2-7.) Sketch the parabola for which (h, k) is $(2, -3)$ and $p = 2$.

26. The equation of a parabola with vertex (h, k) and axis parallel to the y-axis is $(x - h)^2 = 4p(y - k)$. (This is shown in Section 2-7.) Sketch the parabola for which (h, k) is $(-1, 2)$ and $p = -3$.

27. The chord of a parabola that passes through the focus and is parallel to the directrix is called the *latus rectum* of the parabola. Find the length of the latus rectum of the parabola $y^2 = 4px$.

28. Find the equation of the circle that has the focus and the vertex of the parabola $x^2 = 8y$ as the ends of a diameter.

29. The Golden Gate Bridge at San Francisco Bay is a suspension bridge, and its supporting cables are parabolic. See Fig. 2-51. With the origin at the low point of the cable, what equation represents the cable if the towers are 4200 ft apart and the maximum sag is 300 ft?

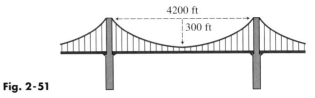

Fig. 2-51

30. The entrance to a building is a parabolic arch 5.6 m high at the center and 7.4 m wide at the base. What equation represents the arch if the vertex is at the top of the arch?

31. The rate of development of heat H (in W) in a resistor of resistance R (in Ω) of an electric circuit is given by $H = Ri^2$, where i is the current (in A) in the resistor. Sketch the graph of H vs. i, if $R = 6.0\ \Omega$.

32. What is the length of the horizontal bar across the parabolically shaped window shown in Fig. 2-52?

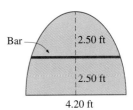

Bar — 2.50 ft

2.50 ft

Fig. 2-52   4.20 ft

33. The primary mirror in the Hubble space telescope has a parabolic cross section, which is shown in Fig. 2-53. What is the focal length (vertex to focus) of the mirror?

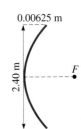

0.00625 m

2.40 m F

Fig. 2-53

34. A rocket is fired horizontally from a plane. Its horizontal distance x and vertical distance y from the point at which it was fired are given by $x = v_0 t$ and $y = \frac{1}{2}gt^2$, where v_0 is the initial velocity of the rocket, t is the time, and g is the acceleration due to gravity. Express y as a function of x and show that it is the equation of a parabola.

35. A wave entering parallel to the axis of a radio wave antenna with a parabolic cross section is reflected through the focus. What is the equation of the parabola representing the antenna with the reflected wave shown in Fig. 2-54?

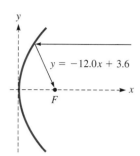

$y = -12.0x + 3.6$

F

Fig. 2-54

36. A wire is fastened 36.0 ft up on each of two telephone poles that are 200 ft apart. Halfway between the poles the wire is 30.0 ft above the ground. Assuming the wire is parabolic, find the height of the wire 50.0 ft from either pole.

37. The total annual fraction f of energy supplied by solar energy to a home is given by $f = 0.065 \sqrt{A}$, where A is the area of the solar collector. Sketch the graph of f as a function of A ($0 < A \le 200$ m^2).

38. The velocity v (in ft/s) of a jet of water flowing from an opening in the side of a certain container is given by $v = 8 \sqrt{h}$, where h is the depth (in ft) of the opening. Sketch a graph of v vs. h.

39. A small island is 4 km from a straight shoreline. A ship channel is equidistant between the island and the shoreline. Write an equation for the channel.

40. Under certain circumstances, the maximum power P (in W) in an electric circuit varies as the square of the voltage of the source E_0 and inversely as the internal resistance R_i (in Ω) of the source. If 10 W is the maximum power for a source of 2.0 V and internal resistance of 0.10 Ω, sketch the graph of P vs. E_0 if R_i remains constant.

$2\text{-}5$ THE ELLIPSE

The next important curve is the ellipse. *An **ellipse** is defined as the locus of a point $P(x, y)$ that moves so that the sum of its distances from two fixed points is constant. These fixed points are the **foci** of the ellipse.* Letting this sum of distances be $2a$ and the foci be the points $(-c, 0)$ and $(c, 0)$, we have

$$\sqrt{(x - c)^2 + y^2} + \sqrt{(x + c)^2 + y^2} = 2a$$

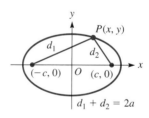

Fig. 2-55

See Fig. 2-55. The ellipse has its center at the origin such that c is the length of the line segment from the center to a focus. We shall also see that a has a special meaning. To simplify this equation we should move one radical to the right and then square each side. This leads to the following steps.

$$\sqrt{(x + c)^2 + y^2} = 2a - \sqrt{(x - c)^2 + y^2}$$
$$(x + c)^2 + y^2 = 4a^2 - 4a\sqrt{(x - c)^2 + y^2} + (\sqrt{(x - c)^2 + y^2})^2$$
$$x^2 + 2cx + c^2 + y^2 = 4a^2 - 4a\sqrt{(x - c)^2 + y^2} + x^2 - 2cx + c^2 + y^2$$
$$4a\sqrt{(x - c)^2 + y^2} = 4a^2 - 4cx$$
$$a\sqrt{(x - c)^2 + y^2} = a^2 - cx$$
$$a^2(x^2 - 2cx + c^2 + y^2) = a^4 - 2a^2cx + c^2x^2$$
$$(a^2 - c^2)x^2 + a^2y^2 = a^2(a^2 - c^2)$$

We now define $a^2 - c^2 = b^2$ (the reason will be shown presently). Therefore,

$$b^2x^2 + a^2y^2 = a^2b^2$$

Dividing through by a^2b^2, we have

$$\frac{x^2}{a^2} + \frac{y^2}{b^2} = 1 \tag{2-17}$$

A graphical analysis of this equation is found on the next page.

For reference, Eq. (2-17) is
$$\frac{x^2}{a^2} + \frac{y^2}{b^2} = 1.$$

The x-intercepts are $(-a, 0)$ and $(a, 0)$. This means that $2a$ (the sum of distances used in the derivation) is also the distance between the x-intercepts. *The points $(a, 0)$ and $(-a, 0)$ are the* **vertices** *of the ellipse, and the line between them is the* **major axis** [see Fig. 2-56(a)]. *Thus, a is the length of the* **semimajor axis.**

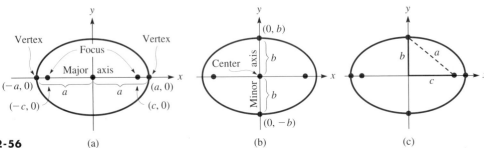

Fig. 2-56   (a)   (b)   (c)

We can now state that *Eq. (2-17) is called the* **standard equation** *of the ellipse with its major axis along the x-axis and its center at the origin.*

The y-intercepts of this ellipse are $(0, -b)$ and $(0, b)$. *The line joining these intercepts is called the* **minor axis** *of the ellipse* [Fig. 2-56(b)], *which means b is the length of the* **semiminor axis.** The intercept $(0, b)$ is equidistant from $(-c, 0)$ and $(c, 0)$. Since the sum of the distances from these points to $(0, b)$ is $2a$, the distance from $(c, 0)$ to $(0, b)$ must be a. Thus, we have a right triangle with line segments of lengths a, b, and c, with a as hypotenuse [Fig. 2-56(c)]. Therefore,

$$a^2 = b^2 + c^2 \qquad (2\text{-}18)$$

is the relation between distances a, b, and c. This also shows why b was defined as it was in the derivation of Eq. (2-17).

If we choose points on the y-axis as the foci, *the standard equation of the ellipse, with its center at the origin and its major axis along the y-axis, is*

$$\frac{y^2}{a^2} + \frac{x^2}{b^2} = 1 \qquad (2\text{-}19)$$

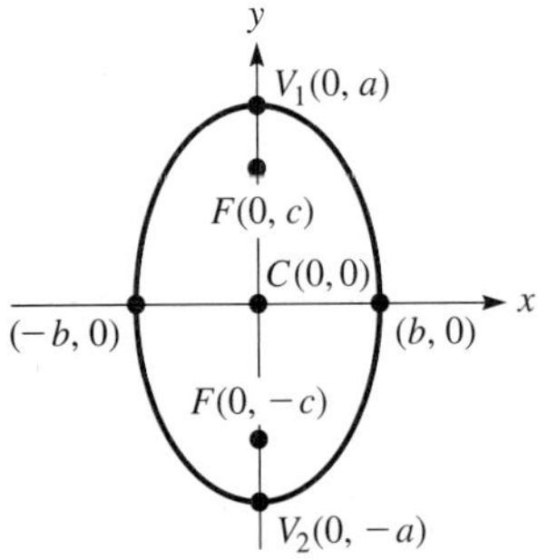

Fig. 2-57

In this case the vertices are $(0, a)$ and $(0, -a)$, the foci are $(0, c)$ and $(0, -c)$, and the ends of the minor axis are $(b, 0)$ and $(-b, 0)$. See Fig. 2-57.

The ellipses represented by Eqs. (2-17) and (2-19) are both symmetric to both axes and to the origin.

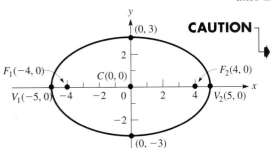

Fig. 2-58

CAUTION

■EXAMPLE 1 The ellipse $\dfrac{x^2}{25} + \dfrac{y^2}{9} = 1$ seems to fit the form of either Eq. (2-17) or Eq. (2-19). Since $a^2 = b^2 + c^2$, we know that *a is always larger than b.* Since the square of the larger number appears under x^2, we know the equation is in the form of Eq. (2-17). Therefore, $a^2 = 25$ and $b^2 = 9$, or $a = 5$ and $b = 3$. This means that the vertices are $(5, 0)$ and $(-5, 0)$ and the minor axis extends from $(0, -3)$ to $(0, 3)$. See Fig. 2-58.

We find c from the relation $c^2 = a^2 - b^2$. This means that $c^2 = 16$ and the foci are $(4, 0)$ and $(-4, 0)$.   -------------■

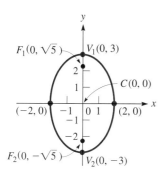

Fig. 2-59

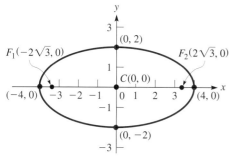

Fig. 2-60

SOLVING A WORD PROBLEM

The power for most space satellites is supplied by solar collectors.

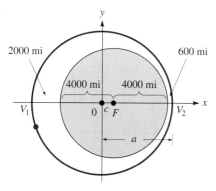

Fig. 2-61

EXAMPLE 2   The ellipse

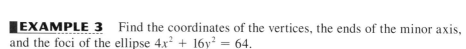

has vertices $(0, 3)$ and $(0, -3)$. The minor axis extends from $(-2, 0)$ to $(2, 0)$. The equation fits the form of Eq. (2-19) since the larger number appears under y^2. Therefore, $a^2 = 9$, $b^2 = 4$, and $c^2 = 5$. The foci are $(0, \sqrt{5})$ and $(0, -\sqrt{5})$. This ellipse is shown in Fig. 2-59.

EXAMPLE 3   Find the coordinates of the vertices, the ends of the minor axis, and the foci of the ellipse $4x^2 + 16y^2 = 64$.

This equation must be put in standard form first, which we do by dividing through by 64. When this is done, we obtain

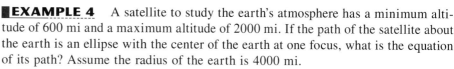

We see that $a^2 = 16$ and $b^2 = 4$, which tells us that $a = 4$ and $b = 2$. Then, $c = \sqrt{16 - 4} = \sqrt{12} = 2\sqrt{3}$. Since a^2 appears under x^2, the vertices are $(4, 0)$ and $(-4, 0)$. The ends of the minor axis are $(0, 2)$ and $(0, -2)$, and the foci are $(2\sqrt{3}, 0)$ and $(-2\sqrt{3}, 0)$. See Fig. 2-60.

EXAMPLE 4   A satellite to study the earth's atmosphere has a minimum altitude of 600 mi and a maximum altitude of 2000 mi. If the path of the satellite about the earth is an ellipse with the center of the earth at one focus, what is the equation of its path? Assume the radius of the earth is 4000 mi.

We set up the coordinate system such that the center of the ellipse is at the origin and the center of the earth is at the right focus, as shown in Fig. 2-61. We know that the distance between vertices is

$$2a = 2000 + 4000 + 4000 + 600 = 10{,}600 \text{ mi}$$
$$a = 5300 \text{ mi}$$

From the right focus to the right vertex is 4600 mi. This tells us

$$c = a - 4600 = 5300 - 4600 = 700 \text{ mi}$$

We can now calculate b^2 as

$$b^2 = a^2 - c^2 = 5300^2 - 700^2 = 2.76 \times 10^7 \text{ mi}^2$$

Since $a^2 = 5300^2 = 2.81 \times 10^7 \text{ mi}^2$, the equation is

$$\frac{x^2}{2.81 \times 10^7} + \frac{y^2}{2.76 \times 10^7} = 1$$

or

$$2.76x^2 + 2.81y^2 = 7.76 \times 10^7$$

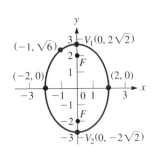

Fig. 2-62

EXAMPLE 5   Find the equation of the ellipse with its center at the origin and an end of its minor axis at $(2, 0)$ and which passes through $(-1, \sqrt{6})$.

Since the center is at the origin and an end of the minor axis is at $(2, 0)$, we know that the ellipse is of the form of Eq. (2-19) and that $b = 2$. Thus, we have

$$\frac{y^2}{a^2} + \frac{x^2}{2^2} = 1$$

In order to find a^2, we use the fact that the ellipse passes through $(-1, \sqrt{6})$. This means that these coordinates satisfy the equation of the ellipse. This gives

$$\frac{(\sqrt{6})^2}{a^2} + \frac{(-1)^2}{4} = 1, \qquad \frac{6}{a^2} = \frac{3}{4}, \qquad a^2 = 8$$

Therefore, the equation of the ellipse, shown in Fig. 2-62, is

$$\frac{y^2}{8} + \frac{x^2}{4} = 1$$

The following example illustrates the use of the definition of the ellipse to find the equation of an ellipse with its center at a point other than the origin.

EXAMPLE 6   Using the definition, find the equation of the ellipse with foci at $(1, 3)$ and $(9, 3)$, with major axis of 10.

Recalling that the sum of distances in the definition equals the length of the major axis, we now use the same method as in the derivation of Eq. (2-17).

$$\sqrt{(x-1)^2 + (y-3)^2} + \sqrt{(x-9)^2 + (y-3)^2} = 10 \qquad \text{use definition of ellipse}$$

$$\sqrt{(x-1)^2 + (y-3)^2} = 10 - \sqrt{(x-9)^2 + (y-3)^2} \qquad \text{isolate a radical}$$

$$x^2 - 2x + 1 + y^2 - 6y + 9 = 100 - 20\sqrt{(x-9)^2 + (y-3)^2} \qquad \text{square both sides and}$$
$$+ x^2 - 18x + 81 + y^2 - 6y + 9 \qquad \text{simplify}$$

$$20\sqrt{(x-9)^2 + (y-3)^2} = 180 - 16x \qquad \text{isolate radical}$$

$$5\sqrt{(x-9)^2 + (y-3)^2} = 45 - 4x \qquad \text{divide by 4}$$

$$25(x^2 - 18x + 81 + y^2 - 6y + 9) = 2025 - 360x + 16x^2 \qquad \text{square both sides}$$

$$9x^2 - 90x + 25y^2 - 150y + 225 = 0 \qquad \text{simplify}$$

The additional x- and y-terms are characteristic of the equation of an ellipse whose center is not at the origin (see Fig. 2-63).

To view this ellipse on a graphing calculator as shown in Fig. 2-64, we solve for y to get the two functions needed. The solutions are $y = \dfrac{15 \pm 3\sqrt{10x - x^2}}{5}$.

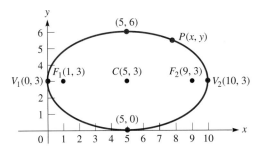

Fig. 2-63

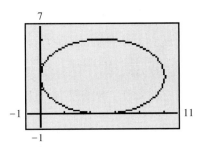

Fig. 2-64

We can conclude that *the equation of an ellipse is characterized by the presence of both an x^2-term and a y^2-term, having different coefficients (in value but not in sign)*. The difference between the equation of an ellipse and that of a circle is that the coefficients of the squared terms in the equation of the circle are the same, whereas those of the ellipse differ. We will consider the equation of the ellipse further in Sections 2-7 and 2-8.

The ellipse has many applications. The orbits of the planets about the sun are elliptical. Gears, cams, and springs are often elliptical in shape. Arches are often constructed in the form of a semiellipse. These and other applications are illustrated in the exercises.

The Polish astronomer Nicolaus Copernicus (1473–1543) is credited as being the first to suggest that the earth revolved about the sun, rather than the previously held belief that the earth was the center of the universe.

EXERCISES *2-5*

In Exercises 1–12, find the coordinates of the vertices and foci of the given ellipses. Sketch each curve.

1. $\dfrac{x^2}{4} + \dfrac{y^2}{1} = 1$

2. $\dfrac{x^2}{100} + \dfrac{y^2}{64} = 1$

3. $\dfrac{x^2}{25} + \dfrac{y^2}{36} = 1$

4. $\dfrac{x^2}{49} + \dfrac{y^2}{81} = 1$

5. $4x^2 + 9y^2 = 36$

6. $x^2 + 36y^2 = 144$

7. $49x^2 + 4y^2 = 196$

8. $25x^2 + y^2 = 25$

9. $8x^2 + y^2 = 16$

10. $2x^2 + 3y^2 = 600$

11. $4x^2 + 25y^2 = 0.25$

12. $9x^2 + 4y^2 = 0.09$

In Exercises 13–20, find the equations of the ellipses satisfying the given conditions. The center of each is at the origin.

13. Vertex $(15, 0)$, focus $(9, 0)$

14. Minor axis 8, vertex $(0, -5)$

15. Focus $(0, 2)$, major axis 6

16. Sum of lengths of major and minor axes 18, focus $(3, 0)$

17. Vertex $(8, 0)$, passes through $(2, 3)$

18. Focus $(0, 2)$, passes through $(-1, \sqrt{3})$

19. Passes through $(2, 2)$ and $(1, 4)$

20. Passes through $(-2, 2)$ and $(1, \sqrt{6})$

In Exercises 21–40, solve the given problems.

21. Find the equation of the ellipse with foci $(-2, 1)$ and $(4, 1)$ and a major axis of 10 by use of the definition. Sketch the curve.

22. Find the equation of the ellipse with foci $(1, 4)$ and $(1, 0)$ that passes through $(4, 4)$ by use of the definition. Sketch the curve.

23. Use a graphing calculator to view the ellipse $4x^2 + 3y^2 + 16x - 18y + 31 = 0$.

24. Use a graphing calculator to view the ellipse $4x^2 + 8y^2 + 4x - 24y + 1 = 0$.

25. The equation of an ellipse with center (h, k) and major axis parallel to the x-axis is $\dfrac{(x - h)^2}{a^2} + \dfrac{(y - k)^2}{b^2} = 1$. (This is shown in Section 2-7.) Sketch the ellipse that has a major axis of 6, a minor axis of 4, and for which (h, k) is $(2, -1)$.

26. The equation of an ellipse with center (h, k) and major axis parallel to the y-axis is $\dfrac{(y - k)^2}{a^2} + \dfrac{(x - h)^2}{b^2} = 1$. (This is shown in Section 2-7.) Sketch the ellipse that has a major axis of 8, a minor axis of 6, and for which (h, k) is $(1, 3)$.

(W) 27. For what values of k does the ellipse $x^2 + ky^2 = 1$ have its vertices on the y-axis? Explain how these values are found.

(W) 28. For what value of k does the ellipse $x^2 + k^2y^2 = 25$ have a focus at $(3, 0)$? Explain how this value is found.

29. Show that the ellipse $2x^2 + 3y^2 - 8x - 4 = 0$ is symmetrical to the x-axis.

30. Show that the ellipse $5x^2 + y^2 - 3y - 7 = 0$ is symmetrical to the y-axis.

31. The *eccentricity e* of an ellipse is defined as $e = c/a$. A cam in the shape of an ellipse can be described by the equation $x^2 + 9y^2 = 81$. Find the eccentricity of this elliptical cam.

32. The planet Pluto moves about the sun in an elliptical orbit, with the sun at one focus. The closest that Pluto approaches the sun is 2.8 billion miles, and the farthest it gets from the sun is 4.6 billion miles. Find the eccentricity of Pluto's orbit. (See Exercise 31.)

33. A draftsman draws a series of triangles with a base from $(-3, 0)$ to $(3, 0)$ and a perimeter of 14 cm (all measurements in centimeters). Find the equation of the curve on which all of the third vertices of the triangles are located.

34. The electric power P (in W) dissipated in a resistance R (in Ω) is given by $P = Ri^2$, where i is the current (in A) in the resistor. Find the equation for the total power of 64 W dissipated in two resistors, with resistances 2.0 Ω and 8.0 Ω, respectively, and with currents i_1 and i_2, respectively. Sketch the graph, assuming that negative values of current are meaningful.

35. An ellipse has a focal property such that a light ray or sound wave emanating from one focus will be reflected through the other focus. Many buildings, such as Statuary Hall in the U.S. Capitol and the Taj Mahal, are built with elliptical ceilings with the property that a sound from one focus is easily heard at the other focus. If a building has a ceiling whose cross sections are part of an ellipse that can be described by the equation $36x^2 + 225y^2 = 8100$ (measurements in meters), how far apart must two persons stand in order to whisper to each other using this focal property?

36. An airplane wing is designed such that a certain cross section is an ellipse 8.40 ft wide and 1.20 ft thick. Find an equation that can be used to describe the perimeter of this cross section.

37. A road passes through a tunnel with a semielliptical cross section 64 ft wide and 18 ft high at the center. What is the height of the tallest vehicle that can pass through the tunnel at a point 22 ft from the center? See Fig. 2-65.

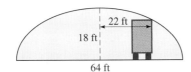

Fig. 2-65

38. An architect designs a window in the shape of an ellipse 4.50 ft wide and 3.20 ft high. Find the perimeter of the window from the formula $p = \pi(a + b)$. This formula gives a good *approximation* for the perimeter when a and b are nearly equal.

39. The vertical cross sections of a horizontal tank 20.0 ft long are ellipses, which can be described by the equation $9x^2 + 20y^2 = 180$, where x and y are measured in feet. The area of an ellipse is $A = \pi ab$. Find the volume of the tank.

40. A laser beam 6.80 mm in diameter is incident on a plane surface at an angle of $62.0°$, as shown in Fig. 2-66. What is the elliptical area that the laser covers on the surface? (See Exercise 39.)

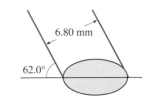

Fig. 2-66

$2\text{-}6$ THE HYPERBOLA

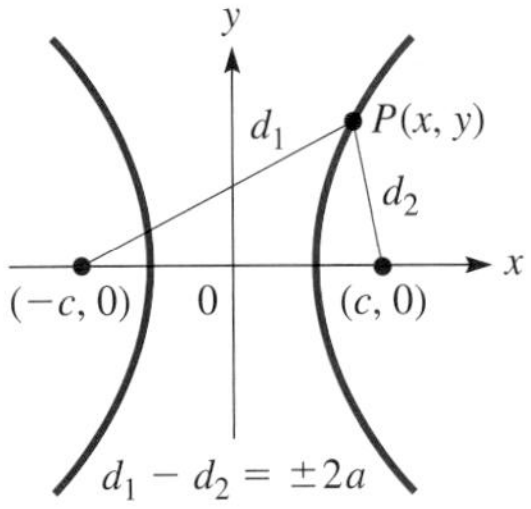

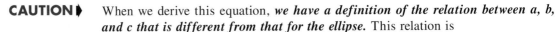

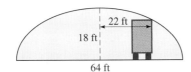

Fig. 2-67

The final curve we shall discuss in detail is the hyperbola. *A* **hyperbola** *is defined as the locus of a point $P(x, y)$ that moves so that the difference of the distances from two fixed points is a constant. These fixed points are the* **foci** *of the hyperbola.* We choose the foci of the hyperbola as the points $(-c, 0)$ and $(c, 0)$ (see Fig. 2-67) and the constant difference to be $2a$. As with the ellipse, these choices make c the length of the line segment from the center to a focus and a (as we will see) the length of the line segment from the center to a vertex. Therefore,

$$\sqrt{(x + c)^2 + y^2} - \sqrt{(x - c)^2 + y^2} = 2a$$

Following the same procedure as in the preceding section, we find the equation of the hyperbola to be

$$\frac{x^2}{a^2} - \frac{y^2}{b^2} = 1 \tag{2-20}$$

CAUTION ▶ When we derive this equation, *we have a definition of the relation between a, b, and c that is different from that for the ellipse.* This relation is

$$c^2 = a^2 + b^2 \tag{2-21}$$

In Eq. (2-20), if we let $y = 0$, we find that the x-intercepts are $(-a, 0)$ and $(a, 0)$, just as they are for the ellipse. *These are the* **vertices** *of the hyperbola.* For $x = 0$, we find that we have imaginary solutions for y, which means there are no points on the curve that correspond to a value of $x = 0$.

To find the meaning of b, we solve Eq. (2-20) for y in a special form:

$$\frac{y^2}{b^2} = \frac{x^2}{a^2} - 1$$

$$= \frac{x^2}{a^2} - \frac{a^2 x^2}{a^2 x^2}$$

$$= \frac{x^2}{a^2}\left(1 - \frac{a^2}{x^2}\right)$$

Multiplying through by b^2 and then taking the square root of each side, we have

$$y^2 = \frac{b^2 x^2}{a^2}\left(1 - \frac{a^2}{x^2}\right)$$

$$y = \pm \frac{bx}{a}\sqrt{1 - \frac{a^2}{x^2}} \qquad (2\text{-}22)$$

We note that, if large values of x are assumed in Eq. (2-22), the quantity under the radical becomes approximately 1. In fact, the larger x becomes, the nearer 1 this expression becomes, since the x^2 in the denominator of a^2/x^2 makes this term nearly zero. Thus, for large values of x, Eq. (2-22) is approximately

$$\boxed{y = \pm \frac{bx}{a}} \qquad (2\text{-}23)$$

NOTE ▶

The symbol ∞ for infinity was first used by the English mathematician John Wallis (1616–1703).

Equation (2-23) is seen to represent two straight lines, each of which passes through the origin. One has a slope of b/a, and the other has a slope of $-b/a$. **These lines are called the asymptotes of the hyperbola.** An **asymptote** *is a line that the curve approaches as one of the variables approaches some particular value.* The concept of an asymptote was introduced in Section 1-4. The fact that the hyperbola *approaches* the asymptote as x becomes large without bound (*approaches* infinity) is designated by

$$y \rightarrow \frac{bx}{a} \quad \text{as} \quad x \rightarrow \pm\infty$$

The symbol ∞ is read as **infinity**, *but it must not be thought of as a number.* It is simply a symbol that stands for a *process* of considering numbers that become large without bound.

Since straight lines are easily sketched, the easiest way to sketch a hyperbola is to draw its asymptotes and then to draw the hyperbola out from each vertex so that it comes closer and closer to each of these asymptotes as x becomes numerically larger. To draw in the asymptotes, the usual procedure is to first draw a small rectangle $2a$ by $2b$, with the origin at the center, as shown in Fig. 2-68. Then straight lines are drawn through opposite vertices of the rectangle. These straight lines are the asymptotes of the hyperbola. Therefore, we see that the significance of the value of b lies in the slope of the asymptotes of the hyperbola.

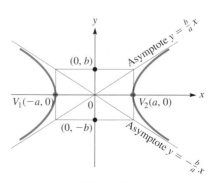

Fig. 2-68

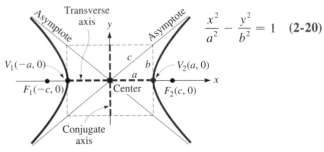

$$\frac{x^2}{a^2} - \frac{y^2}{b^2} = 1 \quad \textbf{(2-20)}$$

Fig. 2-69

Equation (2-20) is called the **standard equation** *of the hyperbola with its center at the origin. It has a* **transverse axis** *of length 2a along the x-axis and a* **conjugate axis** *of length 2b along the y-axis.* This means that a represents the length of the semitransverse axis and b represents the length of the semiconjugate axis. See Fig. 2-69. From the definition of c, it is the length of the line segment from the center to a focus. Also, c is the length of the semidiagonal, as shown in Fig. 2-69. This shows us the geometric meaning of the relation among a, b, and c given in Eq. (2-21).

If the tranverse axis is along the y-axis and the conjugate axis is along the x-axis, the equation of a hyperbola with its center at the origin (see Fig. 2-70) is

$$\frac{y^2}{a^2} - \frac{x^2}{b^2} = 1 \qquad \textbf{(2-24)}$$

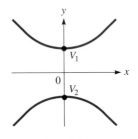

Fig. 2-70

The hyperbolas represented by Eqs. (2-20) and (2-24) are both symmetrical to both axes and to the origin.

EXAMPLE 1 The hyperbola $\dfrac{x^2}{16} - \dfrac{y^2}{9} = 1$

$\qquad\qquad a^2 \uparrow \qquad \uparrow b^2$

fits the form of Eq. (2-20). We know that it fits Eq. (2-20) and not Eq. (2-24) since the x^2-term is the positive term with 1 on the right. From the equation we see that $a^2 = 16$ and $b^2 = 9$, or $a = 4$ and $b = 3$. In turn this means the vertices are $(4, 0)$ and $(\ 4, 0)$ and the conjugate axis extends from $(0, -3)$ to $(0, 3)$.

Since $c^2 = a^2 + b^2$, we find that $c^2 = 25$, or $c = 5$. The foci are $(-5, 0)$ and $(5, 0)$.

Drawing the rectangle and the asymptotes in Fig. 2-71, we then sketch in the hyperbola from each vertex toward each asymptote. ---------■

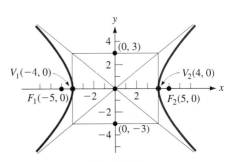

Fig. 2-71

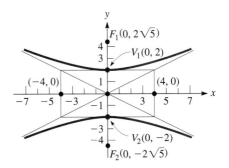

Fig. 2-72

EXAMPLE 2 The hyperbola $\dfrac{y^2}{4} - \dfrac{x^2}{16} = 1$

$\qquad\qquad a^2 \uparrow \qquad \uparrow b^2$

has vertices at $(0, -2)$ and $(0, 2)$. Its conjugate axis extends from $(-4, 0)$ to $(4, 0)$. The foci are $(0, -2\sqrt{5})$ and $(0, 2\sqrt{5})$. We find this directly from the equation since the y^2-term is the positive term with 1 on the right. This means the equation fits the form of Eq. (2-24) with $a^2 = 4$ and $b^2 = 16$. Also, $c^2 = 20$, which means that $c = \sqrt{20} = 2\sqrt{5}$.

Since $2a$ extends along the y-axis, we see that the equations of the asymptotes are $y = \pm(a/b)x$. This is not a contradiction of Eq. (2-23) but an extension of it for a hyperbola with its transverse axis along the y-axis. The ratio a/b gives the slope of the asymptote. The hyperbola is shown in Fig. 2-72. ---------■

EXAMPLE 3   Determine the coordinates of the vertices of the hyperbola

$$4x^2 - 9y^2 = 36$$

First, by dividing through by 36, we have

$$\frac{x^2}{9} - \frac{y^2}{4} = 1 \longleftarrow$$

form requires
— and 1

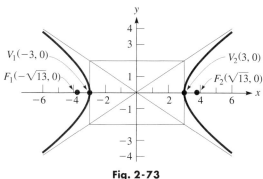

Fig. 2-73

From this form we see that $a^2 = 9$ and $b^2 = 4$. In turn this tells us that $a = 3$, $b = 2$, and $c = \sqrt{9 + 4} = \sqrt{13}$. Since a^2 appears under x^2, the equation fits the form of Eq. (2-20). Therefore, the vertices are $(-3, 0)$ and $(3, 0)$ and the foci are $(-\sqrt{13}, 0)$ and $(\sqrt{13}, 0)$. The hyperbola is shown in Fig. 2-73.   ■

SOLVING A WORD PROBLEM

EXAMPLE 4   In physics it is shown that where the velocity of a fluid is greatest, the pressure is the least. In designing an experiment to study this effect in the flow of water, a pipe is constructed such that its lengthwise cross section is hyperbolic. The pipe is 1.0 m long, 0.2 m in diameter at the narrowest point in the middle, and 0.4 m in diameter at each end. What is the equation that represents the cross section of the pipe as shown in Fig. 2-74?

As shown, the hyperbola has its transverse axis along the y-axis and its center at the origin. This means the general equation is given by Eq. (2-24). Since the radius at the middle of the pipe is 0.1 m, we know that $a = 0.1$ m. Also, since it is 1.0 m long and the radius at the end is 0.2 m, we know the point $(0.5, 0.2)$ is on the hyperbola. This point must satisfy the equation.

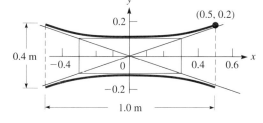

Fig. 2-74

$$\frac{y^2}{a^2} - \frac{x^2}{b^2} = 1 \qquad \text{Eq. (2-24)}$$

point $(0.5, 0.2)$ satisfies equation

$$a = 0.1 \rightarrow \frac{0.2^2}{0.1^2} - \frac{0.5^2}{b^2} = 1$$

$$4 - \frac{0.25}{b^2} = 1, \qquad 3b^2 = 0.25, \qquad b^2 = 0.083$$

$$\frac{y^2}{0.1^2} - \frac{x^2}{0.083} = 1 \qquad \text{substituting } a = 0.1, b^2 = 0.083 \text{ in Eq. (2-24)}$$

$$100y^2 - 12x^2 = 1 \qquad \text{equation of cross section}$$ ■

If we use a graphing calculator to display a hyperbola represented by either Eq. (2-20) or (2-24) we have two functions when we solve for y. One represents the upper half of the hyperbola, and the other represents the lower half. For the hyperbola in Example 3, the functions are $y_1 = \sqrt{(4x^2 - 36)/9}$ and $y_2 = -\sqrt{(4x^2 - 36)/9}$. For the hyperbola in Example 4, they are $y_1 = \sqrt{(12x^2 + 1)/100}$ and $y_2 = -\sqrt{(12x^2 + 1)/100}$.

Equations (2-20) and (2-24) give us the standard forms of the equation of the hyperbola with its center at the origin and its foci on one of the coordinate axes. There is another important equation form that represents a hyperbola, and it is

$$xy = c \qquad\qquad (2\text{-}25)$$

The asymptotes of this hyperbola are the coordinate axes, and the foci are on the line $y = x$ if c is positive or on the line $y = -x$ if c is negative.

The hyperbola represented by Eq. (2-25) is symmetrical to the origin, for if $-x$ replaces x and $-y$ replaces y at the same time, we obtain $(-x)(-y) = c$, or $xy = c$. The equation is unchanged. However, if $-x$ replaces x or if $-y$ replaces y, but not both, the sign on the left is changed. This means it is not symmetrical to either axis. Here, c represents a constant and is not related to the focus.

■**EXAMPLE 5**   Plot the graph of the equation $xy = 4$.

We find the values in the table below and then plot the appropriate points. Here it is permissible to use a limited number of points, since we know the equation represents a hyperbola. Therefore, using $y = 4/x$, we obtain the values

x	-8	-4	-1	$-\frac{1}{2}$	$\frac{1}{2}$	1	4	8
y	$-\frac{1}{2}$	-1	-4	-8	8	4	1	$\frac{1}{2}$

Note that neither x nor y may equal zero. The hyperbola is shown in Fig. 2-75.

If the constant on the right is negative (for example, if $xy = -4$), then the two branches of the hyperbola are in the second and fourth quadrants.   ■

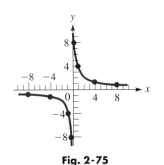

Fig. 2-75

The first reasonable measurement of the speed of light was made by the Danish astronomer Olaf Roemer (1644–1710). He measured the time required for light to come from the moons of Jupiter across the earth's orbit.

The unit of frequency, the Hertz, was named for the German physicist Heinrich Hertz (1857–1894).

■**EXAMPLE 6**   For a light wave, the product of its frequency f of vibration and its wavelength λ is a constant, and this constant is the speed of light c. For green light, for which $f = 600$ THz, $\lambda = 500$ nm. Graph λ as a function of f for any light wave.

From the statement above, we know that $f\lambda = c$, and from the given values we have

$$(600 \text{ THz})(500 \text{ nm}) = (6.0 \times 10^{14} \text{ Hz})(5.0 \times 10^{-7} \text{ m}) = 3.0 \times 10^8 \text{ m/s}$$

which means $c = 3.0 \times 10^8$ m/s. We are to sketch $f\lambda = 3.0 \times 10^8$. Solving for λ as $\lambda = 3.0 \times 10^8/f$, we have the following table (only positive values have meaning).

f (THz)	750	600	500	430
λ (nm)	400	500	600	700

See Fig. 2-76. (Violet light has wavelengths of about 400 nm, orange light has wavelengths of about 600 nm, and red light has wavelengths of about 700 nm.)   ■

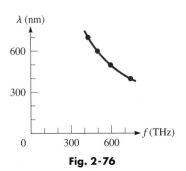

Fig. 2-76

We can conclude that *the equation of a hyperbola is characterized by the presence of both an x^2-term and a y^2-term, having different signs, or by the presence of an xy-term with no squared terms.* We will consider the equation of the hyperbola further in Sections 2-7 and 2-8.

The hyperbola has some very useful applications. The LORAN radio navigation system is based on the use of hyperbolic paths. Some reflecting telescopes use hyperbolic mirrors. The paths of comets that never return to pass by the sun are hyperbolic. Some applications are illustrated in the exercises.

EXERCISES 2-6

In Exercises 1–12, find the coordinates of the vertices and the foci of the given hyperbolas. Sketch each curve.

1. $\dfrac{x^2}{25} - \dfrac{y^2}{144} = 1$

2. $\dfrac{x^2}{16} - \dfrac{y^2}{4} = 1$

3. $\dfrac{y^2}{9} - \dfrac{x^2}{1} = 1$

4. $\dfrac{y^2}{2} - \dfrac{x^2}{2} = 1$

5. $4x^2 - y^2 = 4$

6. $x^2 - 9y^2 = 81$

7. $2y^2 - 5x^2 = 10$

8. $3y^2 - 2x^2 = 300$

9. $4x^2 - y^2 + 4 = 0$

10. $9x^2 - y^2 - 9 = 0$

11. $4x^2 - y^2 = 0.64$

12. $9y^2 - x^2 = 0.36$

In Exercises 13–20, find the equations of the hyperbolas satisfying the given conditions. The center of each is at the origin.

13. Vertex $(3, 0)$, focus $(5, 0)$

14. Vertex $(0, 1)$, focus $(0, \sqrt{3})$

15. Conjugate axis $= 12$, vertex $(0, 10)$

16. Sum of lengths of transverse and conjugate axes 28, focus $(10, 0)$

17. Passes through $(2, 3)$, focus $(2, 0)$

18. Passes through $(8, \sqrt{3})$, vertex $(4, 0)$

19. Passes through $(5, 4)$ and $(3, \frac{4}{5}\sqrt{5})$

20. Passes through $(1, 2)$ and $(2, 2\sqrt{2})$

In Exercises 21–36, solve the given problems.

21. Sketch the graph of the hyperbola $xy = 2$.

22. Sketch the graph of the hyperbola $xy = -4$.

23. Find the equation of the hyperbola with foci $(1, 2)$ and $(11, 2)$, and a transverse axis of 8, by use of the definition. Sketch the curve.

24. Find the equation of the hyperbola with vertices $(-2, 4)$ and $(-2, -2)$, and a conjugate axis of 4, by use of the definition. Sketch the curve.

25. Use a graphing calculator to view the hyperbola $x^2 - 4y^2 + 4x + 32y - 64 = 0$.

26. Use a graphing calculator to view the hyperbola $5y^2 - 4x^2 + 8x + 40y + 56 = 0$.

27. The equation of a hyperbola with center (h, k) and transverse axis parallel to the x-axis is $\dfrac{(x - h)^2}{a^2} - \dfrac{(y - k)^2}{b^2} = 1$. (This is shown in Section 2-7.) Sketch the hyperbola that has a transverse axis of 4, a conjugate axis of 6, and for which (h, k) is $(-3, 2)$.

28. The equation of a hyperbola with center (h, k) and transverse axis parallel to the y-axis is $\dfrac{(y - k)^2}{a^2} - \dfrac{(x - h)^2}{b^2} = 1$. (This is shown in Section 2-7.) Sketch the hyperbola that has a transverse axis of 2, a conjugate axis of 8, and for which (h, k) is $(5, 0)$.

29. Two concentric (same center) hyperbolas are called conjugate hyperbolas if the transverse and conjugate axes of one are, respectively, the conjugate and transverse axes of the other. What is the equation of the hyperbola conjugate to the hyperbola in Exercise 14?

30. As with an ellipse, the *eccentricity e* of a hyperbola is defined as $e = c/a$. Find the eccentricity of the hyperbola $2x^2 - 3y^2 = 24$.

31. A plane flying at a constant altitude of 2000 m is observed from the control tower of an airport. Show that the equation relating the horizontal distance x and direct line distance l from the tower to the plane is that of a hyperbola. Sketch the graph of l as a function of x. See Fig. 2-77.

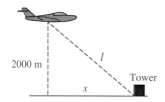

Fig. 2-77

32. Two holes of radius r are drilled from a circular area of radius R such that 24 in.^{2} of material remains. Show that the equation relating R and r is that of a hyperbola.

33. Ohm's law in electricity states that the product of the current i and the resistance R equals the voltage V across the resistance. If a battery of 6.00 V is placed across a variable resistor R, find the equation relating i and R and sketch the graph of i as a function of R.

34. A ray of light directed at one focus of a hyperbolic mirror is reflected toward the other focus. Find the equation that represents the hyperbolic mirror shown in Fig. 2-78.

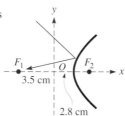

Fig. 2-78

35. A radio signal is sent simultaneously from stations A and B 600 km apart on the Carolina coast. A ship receives the signal from A 1.20 ms before it receives the signal from B. Given that radio signals travel at 300 km/ms, draw a graph showing the possible locations of the ship. This problem illustrates the basis of LORAN.

36. For monochromatic (single-color) light coming from two point sources, curves of maximum intensity occur where the difference in the distances from the sources is an integral number of wavelengths. If a thin translucent film is placed in the plane of the sources, find the equation of the curves of maximum intensity in the film where the difference in paths is two wavelengths and the sources are separated by four wavelengths. Assume the sources are on the x-axis with the origin midway between and use units of one wavelength for both x and y.

2-7 TRANSLATION OF AXES

The equations we have considered for the parabola, the ellipse, and the hyperbola are those for which the center of the ellipse or hyperbola, or vertex of the parabola, is at the origin. In this section we consider, without specific use of the definition, the equations of these curves for the cases in which the axes of the curve are parallel to the coordinate axes. This is done by **translation of axes.**

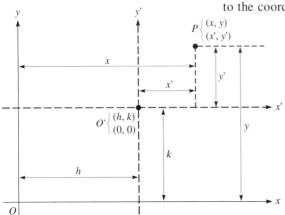

In Fig. 2-79, we choose a point (h, k) in the xy-coordinate plane as the origin of another coordinate system, the $x'y'$-coordinate system. The x'-axis is parallel to the x-axis, and the y'-axis is parallel to the y-axis. Every point now has two sets of coordinates, (x, y) and (x', y'). We see that

$$x = x' + h \quad \text{and} \quad y = y' + k \qquad \textbf{(2-26)}$$

Equations (2-26) can also be written in the form

$$x' = x - h \quad \text{and} \quad y' = y - k \qquad \textbf{(2-27)}$$

Fig. 2-79

■EXAMPLE 1 Find the equation of the parabola with vertex $(2, 4)$ and focus $(4, 4)$.

If we let the origin of the $x'y'$-coordinate system be the point $(2, 4)$, then the point $(4, 4)$ is the point $(2, 0)$ in the $x'y'$-system. This means $p = 2$ and $4p = 8$. See Fig. 2-80. In the $x'y'$-system, the equation is

$$(y')^2 = 8(x')$$

Using Eqs. (2-27), we have

$$(y - 4)^2 = 8(x - 2)$$

└─── coordinates of vertex $(2, 4)$

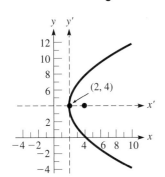

Fig. 2-80

as the equation of the parabola in the xy-coordinate system. ----------■

Following the method of Example 1, by writing the equation of the curve in the $x'y'$-system and then using Eqs. (2-27), we have the following more general forms of the equations of the parabola, ellipse, and hyperbola.

Parabola, vertex (h, k):	$(y - k)^2 = 4p(x - h)$	(axis parallel to x-axis)	**(2-28)**
	$(x - h)^2 = 4p(y - k)$	(axis parallel to y-axis)	**(2-29)**
Ellipse, center (h, k):	$\dfrac{(x - h)^2}{a^2} + \dfrac{(y - k)^2}{b^2} = 1$	(major axis parallel to x-axis)	**(2-30)**
	$\dfrac{(y - k)^2}{a^2} + \dfrac{(x - h)^2}{b^2} = 1$	(major axis parallel to y-axis)	**(2-31)**
Hyperbola, center (h, k):	$\dfrac{(x - h)^2}{a^2} - \dfrac{(y - k)^2}{b^2} = 1$	(transverse axis parallel to x-axis)	**(2-32)**
	$\dfrac{(y - k)^2}{a^2} - \dfrac{(x - h)^2}{b^2} = 1$	(transverse axis parallel to y-axis)	**(2-33)**

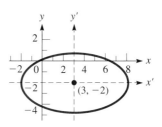

Fig. 2-81

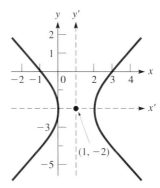

Fig. 2-82

CAUTION ▶

SOLVING A WORD PROBLEM

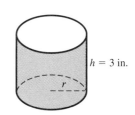

Fig. 2-83

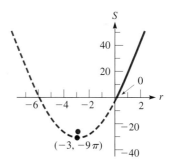

Fig. 2-84

■EXAMPLE 2 Describe the curve of the equation

$$\frac{(x - 3)^2}{25} + \frac{(y + 2)^2}{9} = 1$$

We see that this equation fits the form of Eq. (2-30) with $h = 3$ and $k = -2$. It is the equation of an ellipse with its center at $(3, -2)$ and its major axis parallel to the x-axis. The semimajor axis is $a = 5$, and the semiminor axis is $b = 3$. The ellipse is shown in Fig. 2-81.   ----------■

■EXAMPLE 3 Find the center of the hyperbola
$2x^2 - y^2 - 4x - 4y - 4 = 0$.

To analyze this curve, we first complete the square in the x-terms and in the y-terms. This will allow us to recognize properly the choice of h and k.

$$2x^2 - 4x - y^2 - 4y = 4$$
$$2(x^2 - 2x \quad\quad) - (y^2 + 4y \quad\quad) = 4$$
$$2(x^2 - 2x + 1) - (y^2 + 4y + 4) = 4 + 2 - 4$$

We note here that when we added 1 to complete the square of the x-terms within the parentheses, ***we were actually adding 2 to the left side.*** Thus, we added 2 to the right side. Similarly, when we added 4 to the y-terms within the parentheses, ***we were actually subtracting 4 from the left side.*** Continuing, we have

$$2(x - 1)^2 - (y + 2)^2 = 2$$

coordinates of center $(1, -2)$

$$\frac{(x - 1)^2}{1} - \frac{(y + 2)^2}{2} = 1$$

Therefore, the center of the hyperbola is $(1, -2)$. See Fig. 2-82.   ----------■

■EXAMPLE 4 Cylindrical glass beakers are to be made with a height of 3 in. Express the surface area in terms of the radius of the base, and sketch the curve.

The total surface area S of a beaker is the sum of the area of the base and the lateral surface area of the side. In general, S in terms of the radius r of the base and height h of the side is $S = \pi r^2 + 2\pi rh$. Since $h = 3$ in., we have

$$S = \pi r^2 + 6\pi r$$

which is the desired relationship. See Fig. 2-83.

To get the equation relating S and r, we complete the square of the r terms.

$$S = \pi(r^2 + 6r)$$
$$S + 9\pi = \pi(r^2 + 6r + 9) \quad\quad \text{complete the square}$$
$$S + 9\pi = \pi(r + 3)^2$$

vertex $(-3, -9\pi)$

$$(r + 3)^2 = \frac{1}{\pi}(S + 9\pi)$$

This represents a parabola with vertex $(-3, -9\pi)$. Since $4p = 1/\pi$, $p = 1/(4\pi)$, the focus is $(-3, \frac{1}{4\pi} - 9\pi)$ as shown in Fig. 2-84. The part of the graph for negative r is dashed since only positive values have meaning.   ----------■

EXERCISES 2-7

In Exercises 1–8, describe the curve represented by each equation. Identify the type of curve and its center (or vertex if it is a parabola). Sketch each curve.

1. $(y - 2)^2 = 4(x + 1)$

2. $\dfrac{(x + 4)^2}{4} + \dfrac{(y - 1)^2}{1} = 1$

3. $\dfrac{(x - 1)^2}{4} - \dfrac{(y - 2)^2}{9} = 1$

4. $(y + 5)^2 = -8(x - 2)$

5. $\dfrac{(x + 1)^2}{1} + \dfrac{y^2}{9} = 1$

6. $\dfrac{(y - 4)^2}{16} - \dfrac{(x + 2)^2}{4} = 1$

7. $(x + 3)^2 = -12(y - 1)$

8. $\dfrac{x^2}{0.16} + \dfrac{(y + 1)^2}{0.25} = 1$

In Exercises 9–20, find the equation of each of the curves described by the given information.

9. Parabola: vertex $(-1, 3)$, focus $(3, 3)$

10. Parabola: vertex $(2, -1)$, directrix $y = 3$

11. Parabola: axis and directrix are the coordinate axes; focus $(12, 0)$

12. Parabola: focus $(2, 4)$, directrix $x = 6$

13. Ellipse: center $(-2, 2)$, focus $(-5, 2)$, vertex $(-7, 2)$

14. Ellipse: center $(0, 3)$, focus $(12, 3)$, major axis 26 units

15. Ellipse: vertices $(-2, -3)$ and $(-2, 5)$, end of minor axis $(0, 1)$

16. Ellipse: foci $(1, -2)$ and $(1, 10)$, minor axis 5 units

17. Hyperbola: vertex $(-1, 1)$, focus $(-1, 4)$, center $(-1, 2)$

18. Hyperbola: foci $(2, 1)$ and $(8, 1)$, conjugate axis 6 units

19. Hyperbola: vertices $(2, 1)$ and $(-4, 1)$, focus $(-6, 1)$

20. Hyperbola: center $(1, -4)$, focus $(1, 1)$, transverse axis 8 units

In Exercises 21–36, determine the center (or vertex if the curve is a parabola) of the given curve. Sketch each curve.

21. $x^2 + 2x - 4y - 3 = 0$

22. $y^2 - 2x - 2y - 9 = 0$

23. $4x^2 + 9y^2 + 24x = 0$

24. $2x^2 + 9y^2 + 8x - 72y + 134 = 0$

25. $9x^2 - y^2 + 8y - 7 = 0$

26. $5x^2 - 4y^2 + 20x + 8y = 4$

27. $2x^2 - 4x = 9y - 2$

28. $0.04x^2 + 0.16y^2 = 0.01y$

29. $4x^2 - y^2 + 32x + 10y + 35 = 0$

30. $2x^2 + 2y^2 - 24x + 16y + 95 = 0$

31. $9x^2 + 4y^2 - 12x + 16y + 16 = 0$

32. $5x^2 - 3y^2 - 40x + 95 = 0$

33. $7x^2 - y^2 - 14x - 16y - 64 = 0$

34. $5x^2 - 2y^2 + 12y + 18 = 0$

35. $9x^2 + 9y^2 - 6x - 24y + 14 = 0$

36. $4y^2 - 15x - 12y + 29 = 0$

In Exercises 37–44, solve the given problems.

37. Find the equation of the hyperbola with asymptotes $x - y = -1$ and $x + y = -3$ and vertex $(3, -1)$.

38. The circle $x^2 + y^2 + 4x - 5 = 0$ passes through the foci and the ends of the minor axis of an ellipse that has its major axis along the x-axis. Find the equation of the ellipse.

39. A first parabola has its vertex at the focus of a second parabola and its focus at the vertex of the second parabola. If the equation of the second parabola is $y^2 = 4x$, find the equation of the first parabola.

40. Identify the curve represented by the equation $4y^2 - x^2 - 6x - 2y - 14 = 0$ and then view its graph on a graphing calculator.

41. The stream of water from a fire hose follows a parabolic curve. If the stream from a hose nozzle fastened at the ground reaches a maximum height of 60 ft at a horizontal distance of 95 ft from the nozzle, find the equation that describes the stream. Take the origin at the location of the nozzle. Sketch the graph of the stream.

42. The impedance Z, resistance R, capacitive reactance X_C, and inductive reactance X_L, in an electric circuit are related by the equation $Z = \sqrt{(X_L - X_C)^2 + R^2}$. Sketch the graph of Z as a function of X_L for constant X_C and constant R.

43. Two wheels in a friction drive assembly are congruent ellipses, as shown in Fig. 2-85. The wheels are always in contact, with the center of the left wheel fixed in position and with the right wheel able to move horizontally. Find the equation that can be used to describe the circumference of each wheel in the position shown.

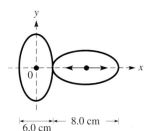

Fig. 2-85

44. An agricultural test station plans to divide a large tract of land into rectangular sections such that the perimeter of each section is 480 m. Express the area A of each section in terms of its width w. Identify the type of curve the equation represents and sketch the graph of A as a function of w. For what value of w is A the greatest?

2-8 THE SECOND-DEGREE EQUATION

The equations of the circle, parabola, ellipse, and hyperbola are all special cases of the same general equation. In this section we discuss this equation and how to identify the particular form it takes when it represents a specific type of curve.

Each of these curves can be represented by a **second-degree equation** *of the form*

$$Ax^2 + Bxy + Cy^2 + Dx + Ey + F = 0 \qquad (2\text{-}34)$$

The coefficients of the second-degree equation terms determine the type of curve that results. Recalling the discussions of the general forms of the equations of the circle, parabola, ellipse, and hyperbola from the previous sections of this chapter, Eq. (2-34) represents the indicated curve for given conditions of *A, B,* and *C,* as follows:

1. If $A = C$, $B = 0$, a circle.
2. If $A \neq C$ (but they have the same sign), $B = 0$, an ellipse.
3. If A and C have different signs, $B = 0$, a hyperbola.
4. If $A = 0$, $C = 0$, $B \neq 0$, a hyperbola.
5. If either $A = 0$ or $C = 0$ (but not both), $B = 0$, a parabola.
   (Special cases, such as a single point or no real locus, can also result.)

Another conclusion about Eq. (2-34) is that, if either $D \neq 0$ or $E \neq 0$ (or both), the center of the curve (or the vertex of a parabola) is not at the origin. If $B \neq 0$, the axis of the curve has been rotated. We have considered only one such case (the hyperbola $xy = c$) in this chapter. Rotation of axes is covered as one of the supplementary topics following the final chapter.

EXAMPLE 1 The equation $2x^2 = 3 - 2y^2$ represents a circle. This can be seen by putting the equation in the form of Eq. (2-34). This form is

$$2x^2 + 2y^2 - 3 = 0$$
$$A = 2 \qquad C = 2$$

We see that $A = C$. Also, since there is no *xy*-term, we know that $B = 0$. This means that the equation represents a circle. If we write it as $x^2 + y^2 = \frac{3}{2}$, we see that it fits the form of Eq. (2-12). The circle is shown in Fig. 2-86.

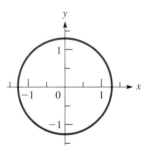

Fig. 2-86

EXAMPLE 2 The equation $3x^2 = 6x - y^2 + 3$ represents an ellipse. Before we analyze the equation, we should put it in the form of Eq. (2-34). For this equation, this form is

$$3x^2 + y^2 - 6x - 3 = 0$$
$$A = 3 \qquad C = 1$$

Here we see that $B = 0$, A and C have the same sign, and $A \neq C$. Therefore, it is an ellipse. The $-6x$ term indicates that the center of the ellipse is not at the origin. The ellipse is shown in Fig. 2-87.

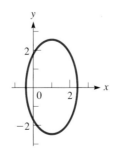

Fig. 2-87

EXAMPLE 3 Identify the curve represented by $2x^2 + 12x = y^2 - 14$. Determine the appropriate quantities for the curve, and sketch the graph.

Writing this equation in the form of Eq. (2-34), we have

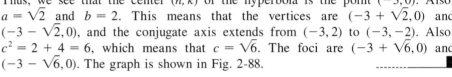

$$2x^2 - y^2 + 12x + 14 = 0$$

$$A = 2 \quad\quad C = -1$$

We identify this equation as representing a hyperbola, since A and C have different signs and $B = 0$. We now write it in the standard form of a hyperbola.

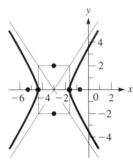

Fig. 2-88

$$2x^2 + 12x - y^2 = -14$$
$$2(x^2 + 6x \quad\quad) - y^2 = -14 \quad\quad \text{complete the square}$$
$$2(x^2 + 6x + 9) - y^2 = -14 + 18$$
$$2(x + 3)^2 - y^2 = 4$$

center is $(-3, 0)$

$$\frac{(x + 3)^2}{2} - \frac{y^2}{4} = 1 \quad\quad \frac{x'^2}{2} - \frac{y'^2}{4} = 1$$

Thus, we see that the center (h, k) of the hyperbola is the point $(-3, 0)$. Also, $a = \sqrt{2}$ and $b = 2$. This means that the vertices are $(-3 + \sqrt{2}, 0)$ and $(-3 - \sqrt{2}, 0)$, and the conjugate axis extends from $(-3, 2)$ to $(-3, -2)$. Also, $c^2 = 2 + 4 = 6$, which means that $c = \sqrt{6}$. The foci are $(-3 + \sqrt{6}, 0)$ and $(-3 - \sqrt{6}, 0)$. The graph is shown in Fig. 2-88.

EXAMPLE 4 Identify the curve represented by $4y^2 - 23 = 4(4x + 3y)$ and find the appropriate important quantities. Then view it on a graphing calculator.

Writing the equation in the form of Eq. (2-34), we have

$$4y^2 - 16x - 12y - 23 = 0$$

Therefore, we recognize the equation as representing a parabola, since $A = 0$ and $B = 0$. Now, writing the equation in the standard form of a parabola, we have

$$4y^2 - 12y = 16x + 23$$
$$4(y^2 - 3y \quad\quad) = 16x + 23 \quad\quad \text{complete the square}$$
$$4\left(y^2 - 3y + \frac{9}{4}\right) = 16x + 23 + 9$$
$$4\left(y - \frac{3}{2}\right)^2 = 16(x + 2)$$

vertex $\left(-2, \frac{3}{2}\right)$

$$\left(y - \frac{3}{2}\right)^2 = 4(x + 2) \quad\quad \text{or} \quad y'^2 = 4x'$$

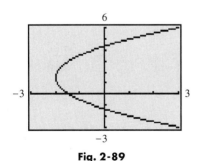

Fig. 2-89

We now note that the vertex is $(-2, 3/2)$ and that $p = 1$. This means that the focus is $(-1, 3/2)$ and the directrix is $x = -3$.

To view the graph of this equation on a graphing calculator, we first solve the equation for y and get $y = (3 \pm 4\sqrt{x + 2})/2$. Entering these two functions in the calculator, we get the view shown in Fig. 2-89.

See Appendix C for a graphing calculator program GRAPHCON. It displays the graph of the conic
$Ax^2 + Cy^2 + Dx + Ey + F = 0$.

In the chapter introduction these curves were referred to as **conic sections.** If a plane is passed through a cone, the intersection of the plane and the cone results in one of these curves; the curve formed depends on the angle of the plane with respect to the axis of the cone. This is shown in Fig. 2-90.

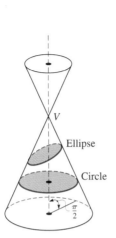
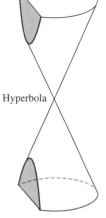
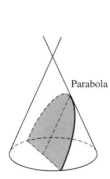

Fig. 2-90

EXERCISES 2-8

In Exercises 1–20, identify each of the equations as representing either a circle, a parabola, an ellipse, a hyperbola, or none of these.

1. $x^2 + 2y^2 - 2 = 0$ **2.** $x^2 - y = 0$

3. $2x^2 - y^2 - 1 = 0$ **4.** $y(y + x^2) = 4$

5. $2x^2 + 2y^2 - 3y - 1 = 0$

6. $x(x - 3) = y(1 - 2y^2)$

7. $2.2x^2 - x - y = 1.6$

8. $2x^2 + 4y^2 - y - 2x = 4$

9. $x^2 = y^2 - 1$ **10.** $32x^2 = 21y - 47y^2$

11. $3.6x^2 = 1.1y - 3.6y^2$ **12.** $y = 3 - 6x^2$

13. $y(3 - 2x) = x(5 - 2y)$ **14.** $x(13 - 5x) = 5y^2$

15. $2xy + x - 3y = 6$ **16.** $(y + 1)^2 = x^2 + y^2 - 1$

17. $2x(x - y) = y(3 - y - 2x)$

18. $2x^2 = x(x - 1) + 4y^2$

19. $x(y + 3x) = x^2 + xy - y^2 + 1$

20. $4x(x - 1) = 2x^2 - 2y^2 + 3$

In Exercises 21–28, identify the curve represented by each of the given equations. Determine the appropriate important quantities for the curve and sketch the graph.

21. $x^2 = 8(y - x - 2)$ **22.** $x^2 = 6x - 4y^2 - 1$

23. $y^2 = 2(x^2 - 2x - 2y)$ **24.** $4x^2 + 4 = 9 - 8x - 4y^2$

25. $y^2 + 42 = 2x(10 - x)$

26. $x^2 - 4y = y^2 + 4(1 - x)$

27. $4(y^2 - 4x - 2) = 5(4y - 5)$

28. $2(2x^2 - y) = 8 - y^2$

In Exercises 29–32, view the curve for each equation on a graphing calculator. In Exercises 29 and 30, identify the type of curve before viewing it. In Exercises 31 and 32, the axis of each curve has been rotated.

29. $x^2 + 2y^2 - 4x + 12y + 14 = 0$

30. $4y^2 - x^2 + 40y - 4x + 60 = 0$

31. $x^2 + 6xy + 9y^2 - 2x + 14y - 10 = 0$

32. $x^2 - xy + y^2 - 6 = 0$

In Exercises 33–36, use the given values to determine the type of curve represented.

33. For the equation $x^2 + ky^2 = a^2$, what type of curve is represented if (a) $k = 1$, (b) $k < 0$, and (c) if $k > 0$ ($k \neq 1$)?

34. For the equation $\dfrac{x^2}{4 - C} - \dfrac{y^2}{C} = 1$, what type of curve is represented if (a) $C < 0$, (b) $0 < C < 4$? (For $C > 4$, see Exercise 36.)

(W) 35. In Eq. (2-34), if $A = C \neq 0$ and $B = D = E = F = 0$, describe the locus.

(W) 36. For the equation in Exercise 34, describe the locus if $C > 4$.

In Exercises 37–40, determine the type of curve from the given information.

37. The diagonal brace in a rectangular metal frame is 3.0 cm longer than the length of one of the sides. Determine the type of curve represented by the equation relating the lengths of the sides of the frame.

38. One circular solar cell has a radius that is 2.0 in. less than the radius r of a second circular solar cell. Determine the type of curve represented by the equation relating the total area A of both cells and r.

39. A flashlight emits a cone of light onto the floor. What type of curve is the perimeter of the lighted area on the floor, if the floor cuts completely through the cone of light?

40. The supersonic jet airliner Concorde creates a conical shock wave behind it. What type of curve is outlined on the surface of a lake by the shock wave if the Concorde is flying horizontally?

CHAPTER EQUATIONS

Distance formula	Fig. 2-2	$d = \sqrt{(x_2 - x_1)^2 + (y_2 - y_1)^2}$	(2-1)
Slope	Fig. 2-4	$m = \dfrac{y_2 - y_1}{x_2 - x_1}$	(2-2)
	Fig. 2-7	$m = \tan \alpha \quad (0° \leq \alpha < 180°)$	(2-3)
	Fig. 2-9	$m_1 = m_2 \quad (\text{for } \parallel \text{ lines})$	(2-4)
	Fig. 2-10	$m_2 = -\dfrac{1}{m_1} \quad \text{or} \quad m_1 m_2 = -1 \quad (\text{for } \perp \text{ lines})$	(2-5)
Straight line	Fig. 2-14	$y - y_1 = m(x - x_1)$	(2-6)
	Fig. 2-17	$x = a$	(2-7)
	Fig. 2-18	$y = b$	(2-8)
	Fig. 2-21	$y = mx + b$	(2-9)
		$Ax + By + C = 0$	(2-10)
Circle	Fig. 2-30	$(x - h)^2 + (y - k)^2 = r^2$	(2-11)
	Fig. 2-33	$x^2 + y^2 = r^2$	(2-12)
		$x^2 + y^2 + Dx + Ey + F = 0$	(2-14)
Parabola	Fig. 2-40	$y^2 = 4px$	(2-15)
	Fig. 2-43	$x^2 = 4py$	(2-16)
Ellipse	Fig. 2-56	$\dfrac{x^2}{a^2} + \dfrac{y^2}{b^2} = 1$	(2-17)
	Fig. 2-56	$a^2 = b^2 + c^2$	(2-18)
	Fig. 2-57	$\dfrac{y^2}{a^2} + \dfrac{x^2}{b^2} = 1$	(2-19)

Hyperbola	Fig. 2-69	$\dfrac{x^2}{a^2} - \dfrac{y^2}{b^2} = 1$	(2-20)
	Fig. 2-69	$c^2 = a^2 + b^2$	(2-21)
	Fig. 2-68	$y = \pm\dfrac{bx}{a}$ (asymptotes)	(2-23)
	Fig. 2-70	$\dfrac{y^2}{a^2} - \dfrac{x^2}{b^2} = 1$	(2-24)
	Fig. 2-75	$xy = c$	(2-25)

Translation of axes	Fig. 2-79	$x = x' + h$ and $y = y' + k$	(2-26)
		$x' = x - h$ and $y' = y - k$	(2-27)

Parabola, vertex (h, k):
$(y - k)^2 = 4p(x - h)$ (axis parallel to x-axis) $\qquad$ (2-28)
$(x - h)^2 = 4p(y - k)$ (axis parallel to y-axis) $\qquad$ (2-29)

Ellipse, center (h, k):
$\dfrac{(x - h)^2}{a^2} + \dfrac{(y - k)^2}{b^2} = 1$ (major axis parallel to x-axis) $\qquad$ (2-30)

$\dfrac{(y - k)^2}{a^2} + \dfrac{(x - h)^2}{b^2} = 1$ (major axis parallel to y-axis) $\qquad$ (2-31)

Hyperbola, center (h, k):
$\dfrac{(x - h)^2}{a^2} - \dfrac{(y - k)^2}{b^2} = 1$ (transverse axis parallel to x-axis) $\qquad$ (2-32)

$\dfrac{(y - k)^2}{a^2} - \dfrac{(x - h)^2}{b^2} = 1$ (transverse axis parallel to y-axis) $\qquad$ (2-33)

Second-degree equation $Ax^2 + Bxy + Cy^2 + Dx + Ey + F = 0$ $\qquad$ (2-34)

REVIEW EXERCISES

In Exercises 1–12, find the equation of the indicated curve subject to the given conditions. Sketch each curve.

1. Straight line: passes through $(1, -7)$ with a slope of 4
2. Straight line: passes through $(-1, 5)$ and $(-2, -3)$
3. Straight line: perpendicular to $3x - 2y + 8 = 0$ and has a y-intercept of $(0, -1)$
4. Straight line: parallel to $2x - 5y + 1 = 0$ and has an x-intercept of $(2, 0)$
5. Circle: center at $(1, -2)$, passes through $(4, -3)$
6. Circle: tangent to the line $x = 3$, center at $(5, 1)$
7. Parabola: focus $(3, 0)$, vertex $(0, 0)$
8. Parabola: directrix $y = -5$, vertex $(0, 0)$
9. Ellipse: vertex $(10, 0)$, focus $(8, 0)$, center $(0, 0)$
10. Ellipse: center $(0, 0)$, passes through $(0, 3)$ and $(2, 1)$
11. Hyperbola: $V(0, 13)$, $C(0, 0)$, conj. axis of 24
12. Hyperbola: $F_1(0, 10)$, $F_2(0, -10)$, $V(0, 8)$

In Exercises 13–24, find the indicated quantities for each of the given equations. Sketch each curve.

13. $x^2 + y^2 + 6x - 7 = 0$, center and radius
14. $x^2 + y^2 - 4x + 2y - 20 = 0$, center and radius
15. $x^2 = -20y$, focus and directrix
16. $y^2 = 24x$, focus and directrix
17. $16x^2 + y^2 = 16$, vertices and foci
18. $2y^2 - 9x^2 = 18$, vertices and foci
19. $2x^2 - 5y^2 = 0.25$, vertices and foci
20. $2x^2 + 25y^2 = 800$, vertices and foci
21. $x^2 - 8x - 4y - 16 = 0$, vertex and focus

22. $y^2 - 4x + 4y + 24 = 0$, vertex and directrix

23. $4x^2 + y^2 - 16x + 2y + 13 = 0$, center

24. $x^2 - 2y^2 + 4x + 4y + 6 = 0$, center

In Exercises 25–28, determine the number of real solutions of the given systems of equations by finding the number of points of intersection from a sketch of the curves.

25. $x^2 + y^2 = 9$
$4x^2 + y^2 = 16$

26. $y = |x|$
$x^2 - y^2 = 1$

27. $x^2 + y^2 - 4y - 5 = 0$
$y^2 - 4x^2 - 4 = 0$

28. $x^2 - 4y^2 + 2x - 3 = 0$
$y^2 - 4x - 4 = 0$

In Exercises 29–32, view the curves of the given equations on a graphing calculator.

29. $x^2 - 4y^2 + 4x + 24y - 48 = 0$

30. $x^2 + 2xy + y^2 - 3x + 8y = 0$

31. $y^2 - 4y + 6x - 8 = 0$

32. $9x^2 + 4y^2 - 72x - 8y + 144 = 0$

In Exercises 33–60, solve the given problems.

33. In two ways show that the line segments joining $(-3, 11)$, $(2, -1)$, and $(14, 4)$ form a right triangle.

34. Show that the altitudes of the triangle with vertices $(2, -4)$, $(3, -1)$, and $(-2, 5)$ meet at a single point.

35. Find the area of the square that can be inscribed in the ellipse $7x^2 + 2y^2 = 18$.

36. For the ellipse shown in Fig. 2-91, show that the product of the slopes of PA and PB is $-b^2/a^2$.

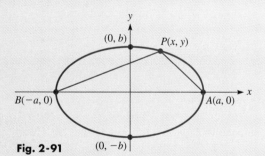

Fig. 2-91

37. By means of the definition of a parabola, find the equation of the parabola with focus at $(3, 1)$ and directrix the line $y = -3$. Find the same equation by the method of translation of axes.

38. For what value of k does $x^2 - ky^2 = 1$ represent an ellipse with vertices on the y-axis?

39. The total resistance R_T of two resistances in series in an electric circuit is the sum of the resistances. If a variable resistor R is in series with a 2.5-Ω resistor, express R_T as a function of R and sketch the graph.

40. The acceleration of an object is defined as the change in velocity v divided by the corresponding change in time t. Find the equation relating the velocity v and time t for an object for which the acceleration is 20 ft/s² and $v = 5.0$ ft/s when $t = 0$ s.

41. One computer printer prints 2500 lines/min for x min, and a second printer prints 1500 lines/min for y min. If they print a total of 37,500 lines together, express y as a function of x and sketch the graph.

42. An airplane touches down when landing at 100 mi/h. Its velocity v while coming to a stop is given by $v = 100 - 20,000t$, where t is the time in hours. Sketch the graph of v vs. t.

43. It takes 2.010 kJ of heat to raise the temperature of 1.000 kg of steam by 1.000°C. In a steam generator, a total of y kJ is used to raise the temperature of 50.00 kg of steam from 100°C to T°C. Express y as a function of T and sketch the graph.

44. The temperature in a certain region is 27°C, and at an altitude of 2500 m above the region it is 12°C. If the equation relating the temperature T and the altitude h is linear, find the equation.

45. The radar gun on a police helicopter 490 ft above a multilane highway is directed vertically down onto the highway. If the radar gun signal is cone-shaped with a vertex angle of 14°, what area of the highway is covered by the signal?

46. One of the most famous Ferris wheels is Vienna's (Austria) *Prata*. Find the equation representing its circumference, given that $c = 191$ m. Place the origin of the coordinate system 2.0 m below the bottom of the wheel, and the center of the wheel on the y-axis.

47. The top horizontal cross section of a dam is parabolic. The open area within this cross section is 80 ft across and 50 ft from front to back. Find the equation of the edge of the open area with the vertex at the origin of the coordinate system and the axis along the x-axis.

48. The *quality factor* Q of a series resonant electric circuit with resistance R, inductance L, and capacitance C is given by $Q = \dfrac{1}{R}\sqrt{\dfrac{L}{C}}$. Sketch the graph relating Q and L for a circuit in which $R = 1000\ \Omega$ and $C = 4.00\ \mu$F.

49. A rectangular parking lot is to have a perimeter of 600 m. Express the area A in terms of the width w and sketch the graph.

50. At very low temperatures, certain metals have an electric resistance of zero. This phenomenon is called *superconductivity*. A magnetic field also affects the superconductivity. A certain level of magnetic field H_T, the threshold field, is related to the thermodynamic temperature T by $H_T/H_0 = 1 - (T/T_0)^2$, where H_0 and T_0 are specifically defined values of magnetic field and temperature. Sketch the graph of H_T/H_0 vs. T/T_0.

51. The electric power P (in W) supplied by a battery is given by $P = 12.0i - 0.500i^2$, where i is the current (in A). Sketch the graph of P vs. i.

52. The Colosseum in Rome is in the shape of an ellipse 188 m long and 156 m wide. Find the area of the Colosseum. ($A = \pi ab$ for an ellipse.)

53. A specialty electronics company makes an ultrasonic device to repel animals. It emits a 20–25 kHz sound (above those heard by people), which is unpleasant to animals. The sound covers an elliptical area starting at the device, with the longest dimension extending 120 ft from the device and the focus of the area 15 ft from the device. Find the area covered by the signal. ($A = \pi ab$)

54. A study indicated that the fraction f of cells destroyed by various dosages d of X-rays is given by the graph in Fig. 2-92. Assuming that the curve is a quarter-ellipse, find the equation relating f and d for $0 \le f \le 1$ and $0 < d \le 10$ units.

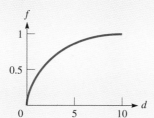

Fig. 2-92

55. A machine-part designer wishes to make a model for an elliptical cam by placing two pins in a design board, putting a loop of string over the pins, and marking off the outline by keeping the string taut. (Note that the definition of the ellipse is being used.) If the cam is to measure 10 cm by 6 cm, how long should the loop of string be and how far apart should the pins be?

56. Soon after reaching the vicinity of the moon, *Apollo 11* (the first spacecraft to land a man on the moon) went into an elliptical lunar orbit. The closest the craft was to the moon in this orbit was 70 mi, and the farthest it was from the moon was 190 mi. What was the equation of the path if the center of the moon was at one of the foci of the ellipse? Assume that the major axis is along the x-axis and that the center of the ellipse is at the origin. The radius of the moon is 1080 mi.

57. The vertical cross section of the cooling tower of a nuclear power plant is hyperbolic, as shown in Fig. 2-93. Find the radius r of the smallest circular horizontal cross section.

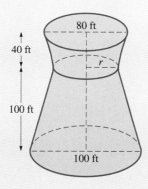

Fig. 2-93

58. Tremors from an earthquake are recorded at the California Institute of Technology (Pasadena, California) 36 s before they are recorded at Stanford University (Palo Alto, California). If the seismographs are 510 km apart and the shock waves from the tremors travel at 5.0 km/s, what is the equation of the curve on which lies the point where the earthquake occurred?

59. A cross section of the roof of a storage building shown in Fig. 2-94 is hyperbolic with the horizontal beam passing through the focus. Determine the equation of the hyperbola such that its center is at the origin of the coordinate system.

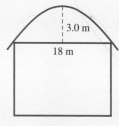

Fig. 2-94

60. A 60-ft rope passes over a pulley 10 ft above the ground, and a crate on the ground is attached at one end. The other end of the rope is held at a level of 4 ft above the ground and is drawn away from the pulley. Express the height of the crate over the ground in terms of the distance the person is from directly below the crate. Sketch the graph of distance and height. See Fig. 2-95. (Neglect the thickness of the crate.)

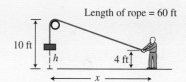

Fig. 2-95

Writing Exercise

61. An electronic recorder located at point P records the sound of a rifle shot and the impact of the bullet striking the target at the same instant. See Fig. 2-96. Write two or three paragraphs explaining why P lies on a branch of a hyperbola.

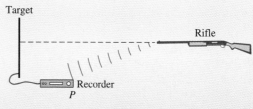

Fig. 2-96

PRACTICE TEST

1. Identify the type of curve represented by the equation $2(x^2 + x) = 1 - y^2$.

2. Sketch the graph of the straight line $4x - 2y + 5 = 0$ by finding its slope and y-intercept.

3. Find the vertex and the focus of the parabola $x^2 = -12y$. Sketch the graph.

4. Find the equation of the circle with center at $(-1, 2)$ and that passes through $(2, 3)$.

5. Find the equation of the straight line that passes through $(-4, 1)$ and $(2, -2)$.

6. Where is the focus of a parabolic reflector that is 12.0 cm across and 4.00 cm deep?

7. A hallway 16 ft wide has a ceiling whose cross section is a semiellipse. The ceiling is 10 ft high at the walls and 14 ft high at the center. Find the height of the ceiling 4 ft from each wall.

8. Find the center and vertices of the conic section $4y^2 - x^2 - 4x - 8y - 4 = 0$. Show the sketch of the curve.

We now start our study of calculus. In this chapter we will develop the methods of **differential calculus.** As we mentioned in our general introduction on page 1, this subject deals with the very important concept of the *rate of change* of one quantity with respect to another. On page 1 we noted some examples of rates of change including *velocity,* which is the rate of change of distance with respect to time.

By the seventeenth century a great deal of information had been discovered about the motion of objects such as the planets and projectiles. However, it had become clear that the traditional mathematics of algebra, geometry, and trigonometry was insufficient to solve many of the problems that were encountered in the study of the motion of these objects. The development of analytic geometry had been of great use, but it was apparent that still more advanced mathematical methods were needed. This led to the development of the branch of mathematics we know as calculus.

Isaac Newton (1642–1727), an English mathematician and physicist, and Gottfried Wilhelm Leibniz (1646–1716), a German mathematician and philosopher, are credited with the creation of calculus. In order to solve problems of motion, as well as certain problems in geometry, each independently developed the basic methods of calculus. A French mathematician, Pierre de Fermat (1601–1665), is also known to have earlier developed some of the methods and concepts of calculus.

Calculus has applications in many different fields of study and today is fundamental for the study of higher mathematics. It has extensive applications in the physical sciences, social sciences, and engineering and is of primary importance in all technical and scientific work. We will illustrate many of these applications throughout the book.

The topic of this chapter, the **derivative**, is the basic concept of differential calculus that is used to measure a rate of change. We will show some basic applications of the derivative in this chapter, and we will develop several very important applications in Chapter 4.

Determining how fast an object is moving is important in physics and in many areas of technology. In Section 3-4 we develop the method of finding the instantaneous velocity of a moving object.

3-1 LIMITS

Before dealing with the rate of change of a function, we first take up the concept of a *limit*. We encountered the idea of a limit when discussing an asymptote to a curve, such as a hyperbola. It is necessary to develop this concept.

CONTINUITY

To help develop the concept of a limit, we first consider briefly the **continuity** of a function. *For a function to be* **continuous at a point,** *the function must exist at the point, and any small change in x produces only a small change in f(x).* In fact, the change in $f(x)$ can be made as small as we wish by restricting the change in x sufficiently, if the function is continuous. Also, *a function is said to be* **continuous over an interval** *if it is continuous at each point in the interval.*

NOTE ▶

If the domain of a function includes the point and values on only one side of the point, it is *continuous at the point* if the definition of continuity holds for that part of the domain.

■**EXAMPLE 1**   The function $f(x) = 3x^2$ is continuous for all values of x. That is, $f(x)$ is defined for all values of x, and a small change in x for any given value of x produces only a small change in $f(x)$. If we choose $x = 2$ and then let x change by 0.1, 0.01, and so on, we obtain the values in the following table.

x	2	2.1	2.01	2.001
$f(x)$	12	13.23	12.1203	12.012003
Change in x		0.1	0.01	0.001
Change in $f(x)$		1.23	0.1203	0.012003

We can see that the change in $f(x)$ is made smaller by the smaller changes in x. This shows that $f(x)$ is continuous at $x = 2$. Since this type of result would be obtained for any other x we may choose, we see that $f(x)$ is continuous for all values, and therefore it is continuous over the interval of all values of x.   ■

■**EXAMPLE 2**   The function $f(x) = \dfrac{1}{x - 2}$ is not continuous at $x = 2$. When we substitute 2 for x, we have division by zero. This means the function is not defined. The condition that the function must exist is not satisfied.   ■

From a graphical point of view, a function that is continuous over an interval has no "breaks" in its graph over that interval. The function is continuous over the interval if we can draw its graph without lifting the marker from the paper. If the function is *discontinuous,* a break occurs because the function is not defined or the definition of the function leads to an instantaneous "jump" in its values.

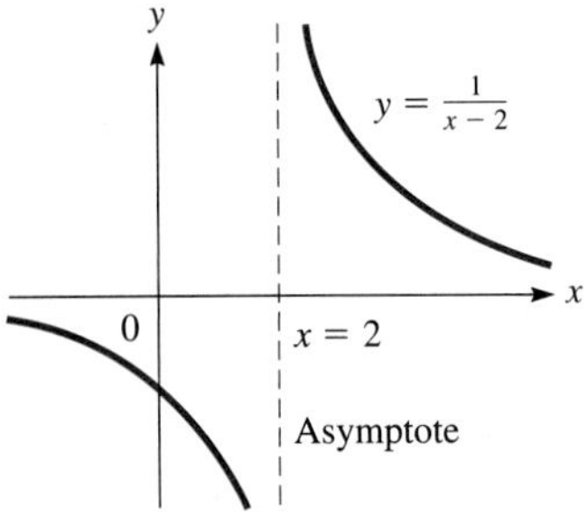

■**EXAMPLE 3**   **(a)** The graph of the function $f(x) = 3x^2$, which we determined to be continuous for all values of x in Example 1, is shown in Fig. 3-1. We see that there are no breaks in the curve.

(b) The graph $f(x) = \dfrac{1}{x - 2}$, which we determined not to be continuous at $x = 2$ in Example 2, is shown in Fig. 3-2. There is a break in the curve for $x = 2$, and this shows that $f(x)$ does not exist at $x = 2$. It is a hyperbola with an asymptote $x = 2$.   ■

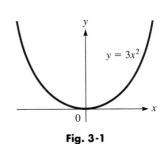

$y = 3x^2$

Fig. 3-1

$y = \dfrac{1}{x - 2}$

$x = 2$

Asymptote

Fig. 3-2

EXAMPLE 4 **(a)** For the function represented in Fig. 3-3, the solid circle at $x = 1$ shows that the point is on the graph. Since it is continuous to the right of the point, it is also continuous at the point. Thus, the function is continuous for $x \geq 1$.

(b) The function represented by the graph in Fig. 3-4 is not continuous at $x = 1$. The function is defined (by the solid circle point) for $x = 1$. However, a small change from $x = 1$ may result in a change of at least 1.5 in $f(x)$, regardless of how small a change in x is made. The small change condition is not satisfied.

(c) The function represented by the graph in Fig. 3-5 is not continuous for $x = -2$. The open circle shows that the point is not part of the graph, and therefore $f(x)$ is not defined for $x = -2$. ▪

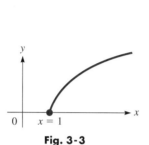

Fig. 3-3

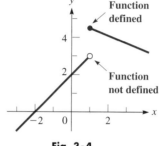

Fig. 3-4

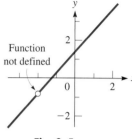

Fig. 3-5

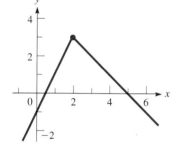

Fig. 3-6

EXAMPLE 5 **(a)** We can define the function in Fig. 3-4 as

$$f(x) = \begin{cases} x + 2 & \text{for } x < 1 \\ -\frac{1}{2}x + 5 & \text{for } x \geq 1 \end{cases}$$

where we note that the equation differs for different parts of the domain.

(b) The graph of the function

$$g(x) = \begin{cases} 2x - 1 & \text{for } x \leq 2 \\ -x + 5 & \text{for } x > 2 \end{cases}$$

is shown in Fig. 3-6. We see that it is a continuous function even though the equation for $x \leq 2$ is different from that for $x > 2$. ▪

In our earlier discussion of the asymptotes of a hyperbola, we used the symbol →, which means "approaches." *When we say that $x \to 2$, we mean that x may take on any value as close to 2 as desired, but it also distinctly means that* **CAUTION ▶** *x cannot be set equal to 2.*

EXAMPLE 6 Consider the behavior of $f(x) = 2x + 1$ as $x \to 2$.

Since we are not to use $x = 2$, we use a calculator to set up tables in order to determine values of $f(x)$, as x gets close to 2.

x	1.000	1.500	1.900	1.990	1.999
$f(x)$	3.000	4.000	4.800	4.980	4.998

x	3.000	2.500	2.100	2.010	2.001
$f(x)$	7.000	6.000	5.200	5.020	5.002

values approach 5

We can see that $f(x)$ approaches 5, as x approaches 2, from above 2 and from below 2. ▪

Limit of a Function

In Example 6, since $f(x) \to 5$ as $x \to 2$, the number 5 is called the limit of $f(x)$ as $x \to 2$. This leads to the meaning of the limit of a function. In general, *the* **limit of a function** $f(x)$ *is that value which the function approaches as x approaches the given value a.* This is written as

$$\lim_{x \to a} f(x) = L \qquad\qquad (3\text{-}1)$$

where L is the value of the limit of the function. Remember, in approaching a, **CAUTION ▶** x may come as arbitrarily close as desired to a, but x **may not equal** a.

An important conclusion can be drawn from the limit in Example 6. The function $f(x)$ is a continuous function, and $f(2)$ equals the value of the limit as $x \to 2$. In general, it is true that

if $f(x)$ is continuous at $x = a$, then the limit as $x \to a$ equals $f(a)$.

In fact, looking back at our definition of continuity, we see that this is what the definition means. That is, a function $f(x)$ is continuous at $x = a$ if *all three* of the following conditions are satisfied:

1. $f(a)$ exists   **2.** $\lim_{x \to a} f(x)$ exists   **3.** $\lim_{x \to a} f(x) = f(a)$

Although we can evaluate the limit for a continuous function as $x \to a$ by evaluating $f(a)$, it is possible that a function is not continuous at $x = a$ and that the limit exists and can be determined. Thus, we must be able to determine the value of a limit without finding $f(a)$. The following example illustrates the evaluation of such a limit.

■EXAMPLE 7   Find $\displaystyle\lim_{x \to 2} \frac{2x^2 - 3x - 2}{x - 2}$.

We note immediately that the function is not continuous at $x = 2$, for division by zero is indicated. Thus, we cannot evaluate the limit by substituting $x = 2$ into the function. Using a calculator to set up tables (see the margin note), we determine the value that $f(x)$ approaches, as x approaches 2.

x	1.000	1.500	1.900	1.990	1.999
$f(x)$	3.000	4.000	4.800	4.980	4.998

x	3.000	2.500	2.100	2.010	2.001
$f(x)$	7.000	6.000	5.200	5.020	5.002

values approach 5

We see that the values obtained are identical to those in Example 6. Since $f(x) \to 5$ as $x \to 2$, we have

$$\lim_{x \to 2} \frac{2x^2 - 3x - 2}{x - 2} = 5$$

Therefore, we see that the limit exists as $x \to 2$, although the function does not exist at $x = 2$.

See the margin note on page 3 for evaluating functions on a calculator. By entering the function as Y_1, it is necessary to enter the function only once for evaluation. See Fig. 3-7.

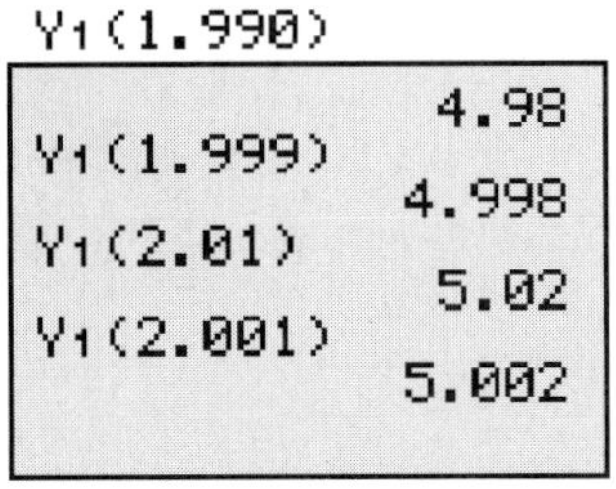

```
Y₁(1.990)
                4.98
Y₁(1.999)
                4.998
Y₁(2.01)
                5.02
Y₁(2.001)
                5.002
```

Fig. 3-7

The reason that the functions in Examples 6 and 7 have the same limit is shown in the following example.

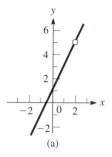

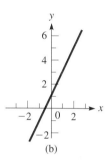

Fig. 3-8

EXAMPLE 8 The function $\dfrac{2x^2 - 3x - 2}{x - 2}$ in Example 7 is the same as the function $2x + 1$ in Example 6, except when $x = 2$. By factoring the numerator of the function of Example 7, we have

$$\frac{2x^2 - 3x - 2}{x - 2} = \frac{(2x + 1)(x - 2)}{x - 2} = 2x + 1$$

The cancellation here is valid, as long as x does not equal 2, for we have division by zero at $x = 2$. Also, in finding the limit as $x \rightarrow 2$, we do not use the value $x = 2$. Therefore,

$$\lim_{x \rightarrow 2} \frac{2x^2 - 3x - 2}{x - 2} = \lim_{x \rightarrow 2}(2x + 1) = 5$$

The limits of the two functions are equal, since, again, in finding the limit, we do not let $x = 2$. The graphs of the two functions are shown in Fig. 3-8(a) and (b). We can see from the graphs that the limits are the same, although one of the functions is not continuous.

If $f(x) = 5$ for $x = 2$ is added to the definition of the function in Example 7, it is then the same as $2x + 1$, and its graph is that in Fig. 3-8(b). ∎

The limit of the function in Example 7 was determined by calculating values near $x = 2$ and by means of an algebraic change in the function. This illustrates that limits may be found through the meaning and definition and through other procedures when the function is not continuous. The following example illustrates a function for which the limit does not exist as x approaches the indicated value.

See Appendix C for the graphing calculator program LIMFUNC. It evaluates $f(x)$ as x approaches a (from values $x > a$).

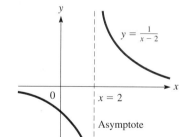

Fig. 3-2

EXAMPLE 9 In trying to find

$$\lim_{x \rightarrow 2} \frac{1}{x - 2}$$

we note that $f(x)$ is not defined for $x = 2$, since we would have division by zero. Therefore, we set up the following table to see how $f(x)$ behaves as $x \rightarrow 2$.

x	3	2.5	2.1	2.01	2.001
$f(x)$	1	2	10	100	1000

$f(x) \rightarrow +\infty$

x	1	1.5	1.9	1.99	1.999
$f(x)$	-1	-2	-10	-100	-1000

$f(x) \rightarrow -\infty$

We see that $f(x)$ gets larger as $x \rightarrow 2$ from above 2 and $f(x)$ gets smaller (large negative values) as $x \rightarrow 2$ from below 2. This may be written as $f(x) \rightarrow +\infty$ as $x \rightarrow 2^+$ and $f(x) \rightarrow -\infty$ as $x \rightarrow 2^-$, but we must remember that ∞ is not a real number. Therefore, the limit as $x \rightarrow 2$ does not exist. The graph of this function is shown in Fig. 3-2, which is shown again for reference. ∎

The following examples further illustrate the evaluation of limits.

EXAMPLE 10   Find $\lim_{x \to 4} (x^2 - 7)$.

Since the function $x^2 - 7$ is continuous at $x = 4$, we may evaluate this limit by substitution. For $f(x) = x^2 - 7$, we have $f(4) = 9$. This means that

$$\lim_{x \to 4} (x^2 - 7) = 9$$

EXAMPLE 11   Find

$$\lim_{x \to 2} \left(\frac{x^2 - 4}{x - 2} \right)$$

Since

$$\frac{x^2 - 4}{x - 2} = \frac{(x - 2)(x + 2)}{x - 2} = x + 2$$

is valid as long as $x \neq 2$, we find that

$$\lim_{x \to 2} \left(\frac{x^2 - 4}{x - 2} \right) = \lim_{x \to 2} (x + 2) = 4$$

Again, we do not have to be concerned with the fact that the cancellation is not valid for $x = 2$. In finding the limit we do not consider the value of $f(x)$ at $x = 2$.

EXAMPLE 12   Find $\lim_{x \to 0} (x \sqrt{x - 3})$.

We see that $x\sqrt{x - 3} = 0$ if $x = 0$, but this function does not have real values for values of x less than 3 other than $x = 0$. *Since x cannot approach 0, $f(x)$ does not approach 0 and the limit does not exist.* The point of this example is that even if $f(a)$ exists, we cannot evaluate the limit by finding $f(a)$, unless $f(x)$ is continuous at $x = a$. Here, $f(a)$ exists but the limit does not exist.

Returning briefly to the discussion of continuity, we again see the need for all three conditions of continuity given on page 72. If $f(a)$ does not exist, the function is discontinuous at $x = a$, and if $f(a)$ does not equal $\lim_{x \to a} f(x)$, a small change in x will not result in a small change in $f(x)$. Also, Example 12 shows another reason to carefully consider the domain of the function in finding the limit.

Limits as x Approaches Infinity

Limits as x approaches infinity are also of importance. However, when dealing with these limits, we must remember that

CAUTION▶   *∞ does not represent a real number and that algebraic operations may not be performed on it.*

Therefore, when we write $x \to \infty$, we know we are to consider values of x that are becoming large without bound. We encountered this concept in Chapter 2 when we discussed the asymptotes of a hyperbola. The following examples illustrate the evaluation of this type of limit.

EXAMPLE 13 The efficiency E of an engine is given by $E = 1 - Q_2/Q_1$, where Q_1 is the heat taken in and Q_2 is the heat ejected by the engine. ($Q_1 - Q_2$ is the work done by the engine.) If, in an engine cycle, $Q_2 = 500$ kJ, determine E as Q_1 becomes large without bound.

We are to find

$$\lim_{Q_1 \to \infty} \left(1 - \frac{500}{Q_1} \right)$$

As Q_1 becomes larger and larger, $500/Q_1$ becomes smaller and smaller and approaches zero. This means $f(Q_1) \to 1$ as $Q_1 \to \infty$. Thus,

$$\lim_{Q_1 \to \infty} \left(1 - \frac{500}{Q_1} \right) = 1$$

We can verify our reasoning and the value of the limit by making a table of values for Q_1 and E as Q_1 becomes large without bound.

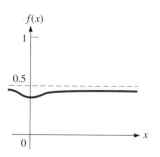

Q_1	500	5000	50,000	500,000	
E	0	0.9	0.99	0.999	values approach 1

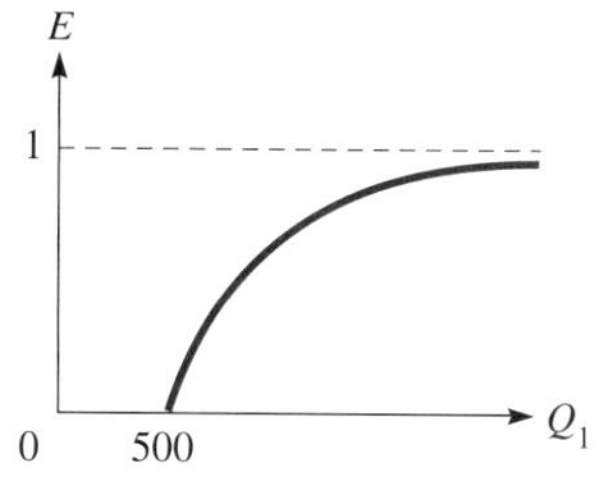

Again we see that $E \to 1$ as $Q_1 \to \infty$. See Fig. 3-9.

This is primarily a theoretical consideration, as there are obvious practical limitations as to how much heat can be supplied to an engine. An engine for which $E = 1$ would operate at 100% efficiency. ∎

Fig. 3-9

EXAMPLE 14 Find $\lim\limits_{x \to \infty} \dfrac{x^2 + 1}{2x^2 + 3}$.

We note that as $x \to \infty$, both the numerator and the denominator become large without bound. Therefore, we use a calculator (see Fig. 3-10) to make a table to see how $f(x)$ behaves as x becomes very large.

x	1	10	100	1000	
$f(x)$	0.4	0.4975369458	0.4999750037	0.49999975	values approach 0.5

From this table we see that $f(x) \to 0.5$ as $x \to \infty$. See Fig. 3-11.

This limit can also be found through algebraic operations and an examination of the resulting algebraic form. If we divide both the numerator and the denominator of **NOTE ▸** the function by x^2, **which is the largest power of x that appears in either the numerator or the denominator,** we have

Fig. 3-10

$$\frac{x^2 + 1}{2x^2 + 3} = \frac{1 + \dfrac{1}{x^2}}{2 + \dfrac{3}{x^2}} \qquad \text{terms} \longrightarrow 0 \\ \text{as } x \longrightarrow \infty$$

Here we see that $1/x^2$ and $3/x^2$ both approach zero as $x \to \infty$. This means that the numerator approaches 1 and the denominator approaches 2. Therefore, we get the same result as above since

$$\lim_{x \to \infty} \frac{x^2 + 1}{2x^2 + 3} = \lim_{x \to \infty} \frac{1 + \dfrac{1}{x^2}}{2 + \dfrac{3}{x^2}} = \frac{1}{2}$$

∎

Fig. 3-11

Calculus was not on a sound mathematical basis until limits were properly developed by the French mathematician Augustin-Louis Cauchy (1789–1857) and others in the mid-1800s.

The definitions and development of continuity and of a limit presented in this section are not mathematically rigorous. However, the development is consistent with a more rigorous development, and the concept of a limit is the principal concern.

EXERCISES 3-1

In Exercises 1–6, determine the values of x for which the function is continuous. If the function is not continuous, determine the reason.

1. $f(x) = 3x - 2$

2. $f(x) = 9 - x^2$

3. $f(x) = \dfrac{2}{x^2 - x}$

4. $f(x) = \dfrac{1}{\sqrt{x}}$

5. $f(x) = \sqrt{\dfrac{x}{x - 2}}$

6. $f(x) = \dfrac{\sqrt{x + 2}}{x}$

In Exercises 7–12, determine the values of x for which the function, as represented by the graphs in Fig. 3-12, is continuous. If the function is not continuous, determine the reason.

7.

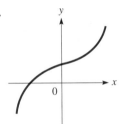

8.

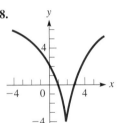

9.

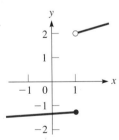

10.

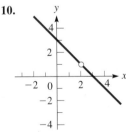

11.

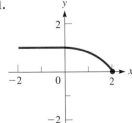

12.

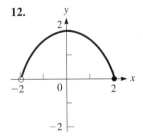

Fig. 3-12

(W) *In Exercises 13–16, graph the function and determine the values of x for which the functions are continuous. Explain.*

13. $f(x) = \begin{cases} x^2 & \text{for } x < 2 \\ 2 & \text{for } x \geq 2 \end{cases}$

14. $f(x) = \begin{cases} \dfrac{x^3 - x^2}{x - 1} & \text{for } x \neq 1 \\ 1 & \text{for } x = 1 \end{cases}$

15. $f(x) = \begin{cases} \dfrac{2x^2 - 18}{x - 3} & \text{for } x < 3 \text{ or } x > 3 \\ 12 & \text{for } x = 3 \end{cases}$

16. $f(x) = \begin{cases} \dfrac{x + 2}{x^2 - 4} & \text{for } x < -2 \\ \dfrac{x}{8} & \text{for } x > -2 \end{cases}$

In Exercises 17–24, evaluate the given function for the values of x shown in the table. Do not change the form of the function. Then, by observing the values obtained, find the indicated limit.

17. Evaluate $f(x) = 3x - 2$.
Find $\lim\limits_{x \to 3}(3x - 2)$.

x	2.900	2.990	2.999	3.001	3.010	3.100
$f(x)$						

18. Evaluate $f(x) = x^2 - 7$.
Find $\lim\limits_{x \to 4}(x^2 - 7)$.

x	3.900	3.990	3.999	4.001	4.010	4.100
$f(x)$						

19. Evaluate $f(x) = \dfrac{x^3 - x}{x - 1}$.
Find $\lim\limits_{x \to 1} \dfrac{x^3 - x}{x - 1}$.

x	0.900	0.990	0.999	1.001	1.010	1.100
$f(x)$						

20. Evaluate $f(x) = \dfrac{x^3 + 2x^2 - 2x + 3}{x + 3}$.

Find $\displaystyle\lim_{x \to -3} \dfrac{x^3 + 2x^2 - 2x + 3}{x + 3}$.

x	-3.100	-3.010	-3.001	-2.999	-2.990	-2.900
$f(x)$						

21. Evaluate $f(x) = \dfrac{2 - \sqrt{x + 2}}{x - 2}$. Find $\displaystyle\lim_{x \to 2} \dfrac{2 - \sqrt{x + 2}}{x - 2}$.

x	1.900	1.990	1.999	2.001	2.010	2.100
$f(x)$						

22. Evaluate $f(x) = \dfrac{2 - \sqrt{x}}{4 - x}$. Find $\displaystyle\lim_{x \to 4} \dfrac{2 - \sqrt{x}}{4 - x}$.

x	3.900	3.990	3.999	4.001	4.010	4.100
$f(x)$						

23. Evaluate $f(x) = \dfrac{2x + 1}{5x - 3}$.

x	10	100	1000
$f(x)$			

Find $\displaystyle\lim_{x \to \infty} \dfrac{2x + 1}{5x - 3}$.

24. Evaluate $f(x) = \dfrac{1 - x^2}{8x^2 + 5}$.

x	10	100	1000
$f(x)$			

Find $\displaystyle\lim_{x \to \infty} \dfrac{1 - x^2}{8x^2 + 5}$.

In Exercises 25–44, evaluate the indicated limits by direct evaluation as in Examples 10–14. Change the form of the function where necessary.

25. $\displaystyle\lim_{x \to 3} (3x - 2)$

26. $\displaystyle\lim_{x \to 4} \sqrt{x^2 - 7}$

27. $\displaystyle\lim_{x \to 2} \dfrac{x^2 - 1}{x + 1}$

28. $\displaystyle\lim_{x \to 5} \left(\dfrac{3}{x^2 + 2} \right)$

29. $\displaystyle\lim_{x \to 0} \dfrac{x^2 + x}{x}$

30. $\displaystyle\lim_{x \to 2} \dfrac{4x^2 - 8x}{x - 2}$

31. $\displaystyle\lim_{x \to -1} \dfrac{x^2 - 1}{3x + 3}$

32. $\displaystyle\lim_{x \to 3} \dfrac{x^2 - 2x - 3}{3 - x}$

33. $\displaystyle\lim_{x \to 1} \dfrac{x^3 - x}{x - 1}$

34. $\displaystyle\lim_{x \to 1/3} \dfrac{3x - 1}{3x^2 + 5x - 2}$

35. $\displaystyle\lim_{x \to 1} \dfrac{(2x - 1)^2 - 1}{2x - 2}$

36. $\displaystyle\lim_{x \to 0} \dfrac{(2 + x)^2 - 4}{x}$

37. $\displaystyle\lim_{x \to -1} \sqrt{x}(x + 1)$

38. $\displaystyle\lim_{x \to 1} (x - 1)\sqrt{x^2 - 4}$

39. $\displaystyle\lim_{x \to \infty} \dfrac{2/x}{1 - 2x}$

40. $\displaystyle\lim_{x \to \infty} \dfrac{6}{1 + \dfrac{2}{x^2}}$

41. $\displaystyle\lim_{x \to \infty} \dfrac{3x^2 + 5}{x^2 - 2}$

42. $\displaystyle\lim_{x \to \infty} \dfrac{x - 1}{7x + 4}$

43. $\displaystyle\lim_{x \to \infty} \dfrac{2x - 6}{x^2 - 9}$

44. $\displaystyle\lim_{x \to \infty} \dfrac{1 - 2x^2}{(4x + 3)^2}$

In Exercises 45 and 46, evaluate the function at 0.1, 0.01, and 0.001 from both sides of the value it approaches. In Exercises 47 and 48, evaluate the function for values of x of 10, 100, and 1000. From these values, determine the limit. Then, by using an appropriate change of algebraic form, evaluate the limit directly and compare values.

45. $\displaystyle\lim_{x \to 0} \dfrac{x^2 - 3x}{x}$

46. $\displaystyle\lim_{x \to 3} \dfrac{2x^2 - 6x}{x - 3}$

47. $\displaystyle\lim_{x \to \infty} \dfrac{2x^2 + x}{x^2 - 3}$

48. $\displaystyle\lim_{x \to \infty} \dfrac{x^2 + 5}{\sqrt{64x^4 + 1}}$

In Exercises 49–56, solve the given problems involving limits.

49. Velocity can be found by dividing the displacement s of an object by the elapsed time t in moving through the displacement. In a certain experiment, the following values were measured for the displacements and elapsed times for the motion of an object. Determine the limiting value of the velocity.

s (cm)	0.480000	0.280000	0.029800	0.0029980	0.00029998
t (s)	0.200000	0.100000	0.010000	0.0010000	0.00010000

50. A rectangular solar panel is to be designed to have an area of 520 cm^2. Express the perimeter p as a function of the width w. Find $\displaystyle\lim_{w \to 20} p$ by evaluating the function for the following values of w (in cm): 19.0, 19.9, 19.99, 20.01, 20.1, and 21.

51. A certain object, after being heated, cools at such a rate that its temperature T (in °C) decreases 10% each minute. If the object is originally heated to 100°C, find $\displaystyle\lim_{t \to 10} T$ and $\displaystyle\lim_{t \to \infty} T$, where t is the time (in min).

52. A 5-Ω resistor and a variable resistor of resistance R are placed in parallel. The expression for the resulting resistance R_T is given by $R_T = \dfrac{5R}{5 + R}$. Determine the limiting value of R_T as $R \to \infty$.

53. Using a calculator, find $\displaystyle\lim_{x \to 0} (1 + x)^{1/x}$. Do you recognize the limiting value?

54. Using a calculator in radian mode, find $\displaystyle\lim_{x \to 0} \dfrac{\sin x}{x}$.

Ⓦ **55.** For $f(x) = x/|x|$, find $\displaystyle\lim_{x \to 0^+} f(x)$ and $\displaystyle\lim_{x \to 0^-} f(x)$, where $\displaystyle\lim_{x \to 0^+}$ means to find the limit as x approaches zero from the right only and $\displaystyle\lim_{x \to 0^-}$ means to find the limit as x approaches zero from the left only. Is $f(x)$ continuous at $x = 0$? Explain.

Ⓦ **56.** Explain why $\displaystyle\lim_{x \to 0^+} 2^{1/x} \neq \lim_{x \to 0^-} 2^{1/x}$. (See Exercise 55.)

3-2 THE SLOPE OF A TANGENT TO A CURVE

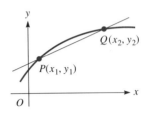

Fig. 3-13

Having developed the basic operations with functions and the concept of a limit, we now turn our attention to a graphical interpretation of the rate of change of a function. This interpretation, basic to an understanding of the calculus, deals with the slope of a line tangent to the graph of a function.

Consider the points $P(x_1, y_1)$ and $Q(x_2, y_2)$ in Fig. 3-13. From Chapter 2 we know that the slope of the line through these points is given by

$$m = \frac{y_2 - y_1}{x_2 - x_1}$$

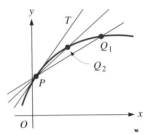

Fig. 3-14

This, however, represents the slope of the line through P and Q and no other line. If we now allow Q to be a point closer to P, the slope of PQ will more closely approximate the slope of a line drawn tangent to the curve at P (see Fig. 3-14). In fact, the closer Q is to P, the better this approximation becomes. It is not possible to allow Q to coincide with P, for then it would not be possible to define the slope of PQ in terms of two distinct points. *The slope of the tangent line, often referred to as the slope of the curve, is the limiting value of the slope of PQ as Q approaches P.*

■**EXAMPLE 1**   Find the slope of a line tangent to the curve $y = x^2 + 3x$ at the point $P(2, 10)$ by finding the limit of slopes of lines PQ as Q approaches P.

We shall let point Q have the x-values of 3.0, 2.5, 2.1, 2.01, and 2.001. Then, using a calculator, we tabulate the necessary values. Since P is the point $(2, 10)$, $x_1 = 2$ and $y_1 = 10$. Thus, using the values of x_2, we tabulate the values of y_2, $y_2 - 10$, $x_2 - 2$ and thereby the values of the slope m.

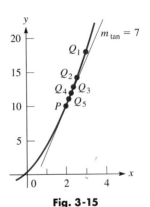

Fig. 3-15

Point	Q_1	Q_2	Q_3	Q_4	Q_5	P
x_2	3.0	2.5	2.1	2.01	2.001	2
y_2	18.0	13.75	10.71	10.0701	10.007001	10
$y_2 - 10$	8.0	3.75	0.71	0.0701	0.007001	
$x_2 - 2$	1.0	0.5	0.1	0.01	0.001	
$m = \dfrac{y_2 - 10}{x_2 - 2}$	8.0	7.5	7.1	7.01	7.001	

See Appendix C for the graphing calculator program SECTOTAN. It displays the graph of a function and four secant lines as the secant line approaches the tangent line.

We see that the slope of PQ approaches the value of 7 as Q approaches P. Therefore, the slope of the tangent line at $(2, 10)$ is 7. See Fig. 3-15.   ∎

With the proper notation, it is possible to express the coordinates of Q in terms of the coordinates of P. If we define the quantities Δx ("delta" x) and Δy ("delta" y) by the equations

$$\Delta x = x_2 - x_1 \tag{3-2}$$
$$\Delta y = y_2 - y_1 \tag{3-3}$$

CAUTION ▶

the coordinates of $Q(x_2, y_2)$ become $(x_1 + \Delta x, y_1 + \Delta y)$. As we see, the quantities Δx and Δy represent the differences in the coordinates of P and Q. **The quantity Δx is not to be thought of as "Δ times x,"** for the symbol Δ used here has no meaning by itself. *The name* **increment** *is given to the difference of the coordinates of two points, and therefore Δx and Δy are the increments in x and y, respectively.*

Using Eqs. (3-2) and (3-3), along with the definition of slope, we can express the slope of PQ as

$$m_{PQ} = \frac{(y_1 + \Delta y) - y_1}{(x_1 + \Delta x) - x_1} = \frac{\Delta y}{\Delta x} \qquad \textbf{(3-4)}$$

By previous discussion, as Q approaches P, the slope of the tangent line is more *nearly* approximated by $\Delta y/\Delta x$.

■EXAMPLE 2   Find the slope of a line tangent to the curve $y = x^2 + 3x$ at the point (2, 10) by the increment method indicated in Eq. (3-4). (This is the same slope as calculated in Example 1.)

As in Example 1, point P has the coordinates (2, 10). Thus, the coordinates of any other point Q can be expressed as $(2 + \Delta x, 10 + \Delta y)$. See Fig. 3-16. The slope of PQ then becomes

$$m_{PQ} = \frac{(10 + \Delta y) - 10}{(2 + \Delta x) - 2} = \frac{\Delta y}{\Delta x}$$

This expression itself does not enable us to find the slope, since values of Δx and Δy are not known. If, however, we can express Δy in terms of Δx, we might derive more information. Both P and Q are on the graph of the function, which means that the coordinates of each must satisfy the function. Using the coordinates of Q, we have

$$(10 + \Delta y) = (2 + \Delta x)^2 + 3(2 + \Delta x)$$
$$= 4 + 4\Delta x + (\Delta x)^2 + 6 + 3\Delta x$$

Subtracting 10 from each side, we have

$$\Delta y = 7\Delta x + (\Delta x)^2$$

As Q approaches P, Δx becomes smaller and smaller. Calculating Δy and the ratio $\Delta y/\Delta x$ for increasingly small Δx, we have the following values.

Δx	0.1	0.01	0.001
Δy	0.71	0.0701	0.007001
$\dfrac{\Delta y}{\Delta x}$	7.1	7.01	7.001

$\Delta y = 7\Delta x + (\Delta x)^2$

We note that the values of $\Delta y/\Delta x$ are the same as those for the slope for points Q_3, Q_4, and Q_5 in Example 1. We also see that as $\Delta x \to 0$, $\Delta y/\Delta x \to 7$.

Now, substituting the expression for Δy into the expression for m_{PQ}, we have

$$m_{PQ} = \frac{7\Delta x + (\Delta x)^2}{\Delta x} = 7 + \Delta x$$

From this expression we can see directly that as $\Delta x \to 0$, $m_{PQ} \to 7$. Using this fact, we can see that the slope of the tangent line is

$$m_{\text{tan}} = \lim_{\Delta x \to 0} m_{PQ} = 7$$

We see that this result agrees with that found in Example 1.

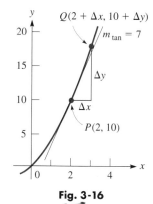

Fig. 3-16

A graphing calculator can also be used to verify the result of Example 2. Graph the function and *zoom* in near the point (2, 10). Then use *trace* to find another point near (2, 10). Calculate the slope between the two points.

■EXAMPLE 3   Find the slope of a line tangent to the curve $y = 4x - x^2$ at the point (x_1, y_1).

The points P and Q are $P(x_1, y_1)$ and $Q(x_1 + \Delta x, y_1 + \Delta y)$. Using the coordinates of Q in the function, we obtain

$$(y_1 + \Delta y) = 4(x_1 + \Delta x) - (x_1 + \Delta x)^2$$
$$= 4x_1 + 4\Delta x - x_1^2 - 2x_1\Delta x - (\Delta x)^2$$

Using the coordinates of P, we obtain

$$y_1 = 4x_1 - x_1^2$$

Subtracting the second equation from the first to solve for Δy, we obtain

$$(y_1 + \Delta y) - y_1 = [4x_1 + 4\Delta x - x_1^2 - 2x_1\Delta x - (\Delta x)^2] - (4x_1 - x_1^2)$$
$$\Delta y = 4\Delta x - 2x_1\Delta x - (\Delta x)^2$$

Dividing through by Δx to obtain an expression for $\Delta y/\Delta x$, we obtain

$$\frac{\Delta y}{\Delta x} = 4 - 2x_1 - \Delta x$$

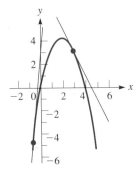

Fig. 3-17

In this last equation, the desired expression is on the left, but all we can determine from $\Delta y/\Delta x$ itself is that the ratio will become one very small number divided by another very small number as $\Delta x \to 0$. The right side, however, approaches $4 - 2x_1$ as $\Delta x \to 0$. This indicates that the slope of a tangent at the point (x_1, y_1) is given by

$$m_{\text{tan}} = 4 - 2x_1$$

This method has an advantage over that used in Example 2. We now have a general expression for the slope of a tangent line for any value x_1. If $x_1 = -1$, $m_{\text{tan}} = 6$, and if $x_1 = 3$, $m_{\text{tan}} = -2$. The tangent lines are shown in Fig. 3-17.   --------■

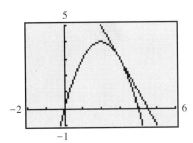

Fig. 3-18

Many calculators have a tangent feature. By entering the function and the value of x for which a tangent line is to be drawn, the calculator displays the curve and the tangent line. See Fig. 3-18 for $y = 4x - x^2$ and the tangent line at $x = 3$.

■EXAMPLE 4   Find the expression for the slope of a line tangent to the curve $y = x^3 + 2$ at the general point (x_1, y_1) and use this expression to find the slope when $x = 1/2$.

Using the points $P(x_1, y_1)$ and $Q(x_1 + \Delta x, y_1 + \Delta y)$, we have the following steps:

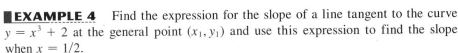

$$y_1 + \Delta y = (x_1 + \Delta x)^3 + 2 \qquad \text{substitute coordinates of } Q$$
$$= x_1^3 + 3x_1^2\Delta x + 3x_1(\Delta x)^2 + (\Delta x)^3 + 2$$
$$y_1 = x_1^3 + 2 \qquad \text{substitute coordinates of } P$$
$$\Delta y = 3x_1^2\Delta x + 3x_1(\Delta x)^2 + (\Delta x)^3 \qquad \text{subtract } y_1 \text{ from } y_1 + \Delta y$$
$$\frac{\Delta y}{\Delta x} = 3x_1^2 + 3x_1\Delta x + (\Delta x)^2 \qquad \text{divide by } \Delta x$$

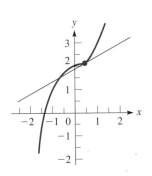

Fig. 3-19

As $\Delta x \to 0$, the right side approaches the value $3x_1^2$. This means that

$$m_{\text{tan}} = 3x_1^2$$

When $x_1 = \frac{1}{2}$, we find that the slope of the tangent is $3(\frac{1}{4}) = \frac{3}{4}$. The curve and this tangent line are indicated in Fig. 3-19.   --------■

In interpreting this analysis as the rate of change of a function, we see that if $\Delta x = 1$ and the corresponding value of Δy is found, it can be said that y changes by the amount Δy as x changes one unit. If x changed by some lesser amount, we could still calculate a ratio for the amount of change in y for the given change in x. Therefore, as long as x changes at all, there will be a corresponding change in y. In this way, *the ratio $\Delta y/\Delta x$ is the* **average rate of change** *of y, which is a function of x, with respect to x. As $\Delta x \to 0$ the limit of the ratio $\Delta y/\Delta x$ gives us the* **instantaneous rate of change** *of y with respect to x.*

EXAMPLE 5   In Example 1, consider points $P(2, 10)$ and $Q_2(2.5, 13.75)$. From P to Q_2, x changes by 0.5 unit and y changes by 3.75 units. This means *the average change in y for a 1-unit change in x is $3.75/0.5 = 7.5$ units.* However, this is not the rate at which y is changing with respect to x *at* most points within this interval.

NOTE ▶   *At* point P, the slope of the tangent line of 7 tells us that y is changing 7 units for a one-unit change in x. However, *this is an instantaneous rate of change at point P* and tells us the rate at which y is changing with respect to x *at P and only at P.*

EXERCISES *3-2*

In Exercises 1–4, use the method of Example 1 to calculate the slope of a line tangent to the graph of each of the given functions. Let Q_1, Q_2, Q_3, and Q_4 have the indicated x-values. Sketch the curve and tangent lines.

1. $y = x^2$; P is $(2, 4)$; let Q have x-values of 1.5, 1.9, 1.99, 1.999.

2. $y = 1 - \frac{1}{2}x^2$; P is $(2, -1)$; let Q have x-values of 1.5, 1.9, 1.99, 1.999.

3. $y = 2x^2 + 5x$; P is $(-2, -2)$; let Q have x-values of -1.5, -1.9, -1.99, -1.999.

4. $y = x^3 + 1$; P is $(-1, 0)$; let Q have x-values of -0.5, -0.9, -0.99, -0.999.

In Exercises 5–8, use the method of Example 2 (divide the expression for Δy by Δx and use the simplified expression for m_{PQ}) to calculate the slope of a line tangent to the graph of each of the given functions for the given points P. (These are the same functions and points as for Exercises 1–4.)

5. $y = x^2$; P is $(2, 4)$.

6. $y = 1 - \frac{1}{2}x^2$; P is $(2, -1)$.

7. $y = 2x^2 + 5x$; P is $(-2, -2)$.

8. $y = x^3 + 1$; P is $(-1, 0)$.

In Exercises 9–20, use the method of Example 3 to find a general expression for the slope of a tangent line to each of the indicated curves. Then find the slopes for the given values of x. Sketch the curves and tangent lines.

9. $y = x^2$; $x = 2$, $x = -1$

10. $y = 1 - \frac{1}{2}x^2$; $x = 2$, $x = -2$

11. $y = 2x^2 + 5x$; $x = -2$, $x = 0.5$

12. $y = 4 - 3x^2$; $x = 0$, $x = 2$

13. $y = x^2 + 4x + 5$; $x = -3$, $x = 2$

14. $y = 2x^2 - 4x$; $x = 1$, $x = 1.5$

15. $y = 6x - x^2$; $x = -2$, $x = 3$

16. $y = x^3 - 2x$; $x = -1$, $x = 0$, $x = 1$

17. $y = x^4$; $x = 0$, $x = 0.5$, $x = 1$

18. $y = 1 - x^4$; $x = 0$, $x = 1$, $x = 2$

19. $y = x^5$; $x = 0$, $x = 0.5$, $x = 1$

20. $y = \dfrac{1}{x}$; $x = 0.5$, $x = 1$, $x = 2$

In Exercises 21–24, use the tangent feature of a graphing calculator to display the curve and the tangent line for the given values of x.

21. $y = 2x - 3x^2$; $x = 0$, $x = 0.5$

22. $y = 3x - x^3$; $x = -2$, $x = 0$, $x = 2$

23. $y = x^6$; $x = 0$, $x = 0.5$, $x = 1$

24. $y = \dfrac{1}{x + 1}$; $x = -0.5$, $x = 0$, $x = 1$

In Exercises 25–28, find the average rate of change of y with respect to x from P to Q. Then compare this with the instantaneous rate of change of y with respect to x at P by finding m_{tan} at P.

25. $y = x^2 + 2$; $P(2, 6)$, $Q(2.1, 6.41)$

26. $y = 1 - 2x^2$; $P(1, -1)$, $Q(1.1, -1.42)$

27. $y = 9 - x^3$; $P(2, 1)$, $Q(2.1, -0.261)$

28. $y = x^3 - 6x$; $P(3, 9)$, $Q(3.1, 11.191)$

3-3 THE DERIVATIVE

We are now ready to establish one of the fundamental definitions of calculus. Generalizing on the method of Example 2 of the preceding section, if the point $P(x, y)$ is held constant while the point $Q(x + \Delta x, y + \Delta y)$ approaches it, then both Δx and Δy approach zero. Both points P and Q lie on the curve of the function $f(x)$, which means the coordinates of each point must satisfy the equation. For P this means $y = f(x)$, and for Q this means $y + \Delta y = f(x + \Delta x)$. Subtracting the first expression from the second, we have

$$(y + \Delta y) - y = f(x + \Delta x) - f(x)$$

$$\Delta y = f(x + \Delta x) - f(x) \qquad (3\text{-}5)$$

In the previous section, we saw that the slope of a line tangent to $f(x)$ at $P(x, y)$ was found as the limiting value of the ratio $\Delta y / \Delta x$ as Δx approaches zero. Formally, *this limiting value of the ratio $\Delta y / \Delta x$ is known as the* **derivative** *of the function.* Therefore, the derivative of a function $f(x)$ is defined as

$$\lim_{\Delta x \to 0} \frac{\Delta y}{\Delta x} = \lim_{\Delta x \to 0} \frac{f(x + \Delta x) - f(x)}{\Delta x} \qquad (3\text{-}6)$$

The process of finding the derivative of a function from its definition is called the *delta-process. The general process of finding a derivative is called* **differentiation.**

> Later in this chapter we will develop formulas that will enable us to differentiate without using the delta-process.

■ EXAMPLE 1   Find the derivative of $y = 2x^2 + 3x$ by the delta-process.

To find Δy, we must first derive the quantity $f(x + \Delta x) - f(x)$. Thus, we first find $y + \Delta y = f(x + \Delta x)$ by replacing y by $y + \Delta y$ and x by $x + \Delta x$.

$$y + \Delta y = 2(x + \Delta x)^2 + 3(x + \Delta x)$$

Then we subtract the original function:

$$y + \Delta y - y = 2(x + \Delta x)^2 + 3(x + \Delta x) - (2x^2 + 3x)$$

Simplifying, we find that the result is

$$\Delta y = 4x\,\Delta x + 2(\Delta x)^2 + 3\,\Delta x$$

Next, dividing through by Δx, we obtain

$$\frac{\Delta y}{\Delta x} = 4x + 2\Delta x + 3$$

As Δx approaches zero, the $2\Delta x$-term on the right approaches zero. Therefore,

$$\lim_{\Delta x \to 0} \frac{\Delta y}{\Delta x} = 4x + 3$$

We see that the derivative of the function $2x^2 + 3x$ is the function $4x + 3$. From the definition of the derivative and from the previous section, this means we can find the slope of a tangent line for any point on the curve of $y = 2x^2 + 3x$ by substituting the x-coordinate into the expression $4x + 3$. For example, the slope of a tangent line is 5 if $x = 1/2$ (at the point $(1/2, 2)$). See Fig. 3-20.   ■

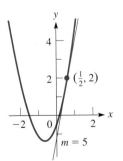

Fig. 3-20

EXAMPLE 2   Find the derivative of the function $y = 6x - 2x^3$ by using the delta-process.

By first replacing y by $y + \Delta y$ and x by $x + \Delta x$, we then have the following steps.

$$y + \Delta y = 6(x + \Delta x) - 2(x + \Delta x)^3 \qquad\qquad\qquad y + \Delta y = f(x + \Delta x)$$

$$= 6x + 6(\Delta x) - 2[x^3 + 3x^2(\Delta x) + 3x(\Delta x)^2 + (\Delta x)^3]$$

$$= 6x + 6(\Delta x) - 2x^3 - 6x^2(\Delta x) - 6x(\Delta x)^2 - 2(\Delta x)^3$$

$$y + \Delta y - y = [6x + 6(\Delta x) - 2x^3 - 6x^2(\Delta x) - 6x(\Delta x)^2 - 2(\Delta x)^3] - (6x - 2x^3) \qquad \text{subtract } y = 6x - 2x^3$$

$$\Delta y = 6(\Delta x) - 6x^2(\Delta x) - 6x(\Delta x)^2 - 2(\Delta x)^3$$

$$\frac{\Delta y}{\Delta x} = \frac{6(\Delta x) - 6x^2(\Delta x) - 6x(\Delta x)^2 - 2(\Delta x)^3}{\Delta x} \qquad\qquad \text{divide by } \Delta x$$

$$= 6 - 6x^2 - 6x(\Delta x) - 2(\Delta x)^2$$

$$\lim_{\Delta x \to 0} \frac{\Delta y}{\Delta x} = \lim_{\Delta x \to 0} [6 - 6x^2 - 6x(\Delta x) - 2(\Delta x)^2] \qquad\qquad \text{find limit as } \Delta x \to 0$$

$$= 6 - 6x^2$$

Therefore, the derivative of $y = 6x - 2x^3$ is $6 - 6x^2$.   ----------- ∎

The derivative of a function is itself a function, and it is possible that it may not be defined for all values of x. *If the value x_0 is in the domain of the derivative, then the function is said to be* **differentiable** *at x_0.* The examples that follow illustrate functions which are not differentiable for all values of x.

EXAMPLE 3   Find the derivative of $y = \dfrac{1}{x}$ by the delta-process.

$$y + \Delta y = \frac{1}{x + \Delta x} \qquad\qquad\qquad y + \Delta y = f(x + \Delta x)$$

$$y + \Delta y - y = \frac{1}{x + \Delta x} - \frac{1}{x} \qquad\qquad\qquad \text{subtract } y = \frac{1}{x}$$

CAUTION ▶

$$\Delta y = \frac{x - (x + \Delta x)}{x(x + \Delta x)} = \frac{-\Delta x}{x(x + \Delta x)} \qquad\qquad \text{combine fractions}$$

$$\frac{\Delta y}{\Delta x} = \frac{-1}{x(x + \Delta x)} \qquad\qquad\qquad \text{divide by } \Delta x$$

$$\lim_{\Delta x \to 0} \frac{\Delta y}{\Delta x} = \lim_{\Delta x \to 0} \frac{-1}{x(x + \Delta x)} \qquad\qquad \text{find limit as } \Delta x \to 0$$

$$= \frac{-1}{x^2}$$

We note that neither the function nor the derivative is defined for $x = 0$. This means the function is not differentiable at $x = 0$.   ----------- ∎

NOTE ▶   In Example 3 it was necessary to combine fractions in the process of finding the derivative. Such algebraic operations must be done with care. *One of the more common sources of errors is the improper handling of fractions.*

■**EXAMPLE 4**   Find the derivative of $y = x^2 + \dfrac{1}{x+1}$ by the delta-process.

$$y + \Delta y = (x + \Delta x)^2 + \frac{1}{x + \Delta x + 1} \qquad\qquad\qquad y + \Delta y = f(x + \Delta x)$$

$$y + \Delta y - y = x^2 + 2x\Delta x + (\Delta x)^2 + \frac{1}{x + \Delta x + 1} - x^2 - \frac{1}{x + 1} \qquad \text{subtract } y = x^2 + \frac{1}{x+1}$$

$$\Delta y = 2x\Delta x + (\Delta x)^2 + \frac{1}{x + \Delta x + 1} - \frac{1}{x + 1} \qquad \begin{array}{l}\text{algebra handled most easily}\\\text{if fractions combined separately}\\\text{from other terms}\end{array}$$

$$= 2x\Delta x + (\Delta x)^2 + \frac{(x + 1) - (x + \Delta x + 1)}{(x + \Delta x + 1)(x + 1)}$$

$$= 2x\Delta x + (\Delta x)^2 - \frac{\Delta x}{(x + \Delta x + 1)(x + 1)}$$

$$\frac{\Delta y}{\Delta x} = 2x + \Delta x - \frac{1}{(x + \Delta x + 1)(x + 1)} \qquad\qquad \text{divide by } \Delta x$$

$$\lim_{\Delta x \to 0} \frac{\Delta y}{\Delta x} = 2x - \frac{1}{(x + 1)^2} \qquad\qquad\qquad\qquad \text{find limit as } \Delta x \to 0$$

We note that this function is not differentiable at $x = -1$.   ----------■

The notation of the derivative in terms of the limit is somewhat awkward for general use. Therefore, other notations are commonly used. They include

$$y' \qquad D_x y \qquad f'(x) \qquad \frac{dy}{dx}$$

The next example illustrates the use of some of these notations for the derivative.

■**EXAMPLE 5**   In Example 2, $y = 6x - 2x^3$, and we found that the derivative is $6 - 6x^2$. Therefore, we may write

$$y' = 6 - 6x^2 \qquad \text{or} \qquad \frac{dy}{dx} = 6 - 6x^2$$

If we had written $f(x) = 6x - 2x^3$, we would then write $f'(x) = 6 - 6x^2$.

If we wish to find the value of the derivative at some point, such as $(-2, 4)$, we write

$$\frac{dy}{dx} = 6 - 6x^2$$

$$\left.\frac{dy}{dx}\right|_{x=-2} = 6 - 6(-2)^2 = 6 - 24$$

$$= -18$$

Note that only the x-coordinate of the point was needed to evaluate the derivative.
    ----------■

```
nDeriv(6X-2X^3,X
,-2)
        -18.000002
```

Fig. 3-21

Many calculators have a feature for evaluating a derivative. Since the evaluation is done by a numerical approximation, it may be denoted as a *numerical derivative*. The display for the evaluation of Example 5 is shown in Fig. 3-21. The result shown is to the default accuracy, but the accuracy can be set using an additional entry of this derivative feature.

See Exercise 35 for a different method of finding this derivative.

EXAMPLE 6   Find dy/dx for the function $y = \sqrt{x}$ by the delta-process.

We first square both sides of the equation, thus obtaining $y^2 = x$. (This is valid only for $y \geq 0$, since the original function $y = \sqrt{x}$ is not defined for $y < 0$.) Replacing y by $y + \Delta y$ and x by $x + \Delta x$, we have

$$(y + \Delta y)^2 = x + \Delta x$$

$$y^2 + 2y\,\Delta y + (\Delta y)^2 - y^2 = x + \Delta x - x \qquad \text{expand left side and subtract } y^2 = x$$

$$2y\,\Delta y + (\Delta y)^2 = \Delta x$$

$$2y\frac{\Delta y}{\Delta x} + \Delta y\frac{\Delta y}{\Delta x} = 1 \qquad \text{divide by } \Delta x$$

$$(2y + \Delta y)\frac{\Delta y}{\Delta x} = 1 \qquad \text{factor}$$

$$\frac{\Delta y}{\Delta x} = \frac{1}{2y + \Delta y} \qquad \text{solve for } \frac{\Delta y}{\Delta x}$$

$$\lim_{\Delta x \to 0}\frac{\Delta y}{\Delta x} = \lim_{\Delta x \to 0}\frac{1}{2y + \Delta y} = \frac{1}{2y} \qquad \text{find limit as } \Delta x \to 0$$

The Δy is omitted, since $\Delta y \to 0$ as $\Delta x \to 0$. Now, substituting in the original function $y = \sqrt{x}$, we obtain the final result:

$$\frac{dy}{dx} = \frac{1}{2\sqrt{x}}$$

The domain of the function is $x \geq 0$. However, since x appears in the denominator of the derivative, the domain of the derivative is $x > 0$. This means that the function is differentiable for $x > 0$.

One might ask why, when we are finding a derivative, we take a limit as Δx approaches zero and do not simply let Δx equal zero. If we did this, the ratio $\Delta y/\Delta x$ would be exactly 0/0, which would then require division by zero. As we know, this

CAUTION▶ is an undefined operation in mathematics, and therefore **Δx *cannot* equal *zero*.** However, it can equal any value as near zero as necessary. This idea is basic in the meaning of the word *limit*.

EXERCISES $3\text{-}3$

In Exercises 1–24, find the derivative of each of the functions by using the delta-process.

1. $y = 3x - 1$

2. $y = 6x + 3$

3. $y = 1 - 2x$

4. $y = 2 - 5x$

5. $y = x^2 - 1$

6. $y = 4 - x^2$

7. $y = 5x^2$

8. $y = -6x^2$

9. $y = x^2 - 7x$

10. $y = x^2 + 4x$

11. $y = 8x - 2x^2$

12. $y = 3x - \frac{1}{2}x^2$

13. $y = x^3 + 4x - 6$

14. $y = 2x - 4x^3$

15. $y = \dfrac{1}{x + 2}$

16. $y = \dfrac{3}{2x + 1}$

17. $y = x + \dfrac{4}{3x}$

18. $y = \dfrac{x}{x - 1}$

19. $y = \dfrac{2}{x^2}$

20. $y = \dfrac{2}{x^2 + 4}$

21. $y = x^4 + x^3 + x^2 + x$

22. $y = \frac{1}{3}x^3 + \frac{1}{2}x^2 + x$

23. $y = x^4 - \dfrac{2}{x}$

24. $y = \dfrac{1}{x} + \dfrac{1}{x^2}$

In Exercises 25–28, find the derivative of each function by using the delta-process. Then evaluate the derivative at the given point. In Exercises 27 and 28, check your result using the derivative evaluation feature of a graphing calculator.

25. $y = 3x^2 - 2x;\ (-1, 5)$ 　　**26.** $y = 9x - x^3;\ (2, 10)$

27. $y = \dfrac{11}{3x + 2};\ (3, 1)$ 　　**28.** $y = x^2 - \dfrac{2}{x};\ (-2, 5)$

In Exercises 29–32, find the derivative of each function by using the delta-process. Then determine the values for which the function is differentiable.

29. $y = 1 + \dfrac{2}{x}$ 　　　　　**30.** $y = \dfrac{5x}{x - 4}$

31. $y = \dfrac{3}{x^2 - 1}$ 　　　　　**32.** $y = \dfrac{2}{x^2 + 1}$

In Exercises 33–36, solve the given problems.

33. Find dy/dx for $y = \sqrt{x + 1}$ by the method of Example 6.

34. Find dy/dx for $y = \sqrt{x^2 + 3}$ by the method of Example 6.

35. Find dy/dx for $y = \sqrt{x}$ by using the delta-process and the function directly. Do not square both sides. (*Hint:* In the expression for Δy, multiply and divide by $\sqrt{x} + \sqrt{x + \Delta x}$. This is rationalizing the numerator.)

(W) 36. Find dy/dx for $y = \sqrt{x - 2}$ by the method outlined in Exercise 35. For what values of x is the function differentiable? Explain.

$3\text{-}4$ THE DERIVATIVE AS AN INSTANTANEOUS RATE OF CHANGE

In Section 3-2 we saw that the slope of a line tangent to a curve at point P was the limiting value of the slope of the line through points P and Q as Q approaches P. In Section 3-3 we defined the limit of the ratio $\Delta y/\Delta x$ as $\Delta x \to 0$ as the derivative. Therefore, *the first meaning we have given to the derivative is the slope of a line tangent to a curve,* as we noted in Example 1 of Section 3-3. The following example further illustrates this meaning of the derivative.

■**EXAMPLE 1**　Find the slope of a line tangent to the curve of $y = 4x - x^2$ at the point $(1, 3)$.

We first find the derivative and then evaluate it at the given point.

$$y + \Delta y = 4(x + \Delta x) - (x + \Delta x)^2$$
$$y + \Delta y - y = 4x + 4\,\Delta x - x^2 - 2x\,\Delta x - (\Delta x)^2 - (4x - x^2)$$
$$\Delta y = 4\,\Delta x - 2x\,\Delta x - (\Delta x)^2$$
$$\frac{\Delta y}{\Delta x} = 4 - 2x - \Delta x$$
$$\lim_{\Delta x \to 0} \frac{\Delta y}{\Delta x} = 4 - 2x \qquad \text{derivative}$$
$$\frac{dy}{dx}\bigg|_{(1, 3)} = 4 - 2(1) = 2 \qquad \text{evaluate derivative}$$

The slope of the tangent line at $(1, 3)$ is 2. Note that only $x = 1$ was needed for the evaluation. The curve and tangent line are shown in Fig. 3-22. ------------■

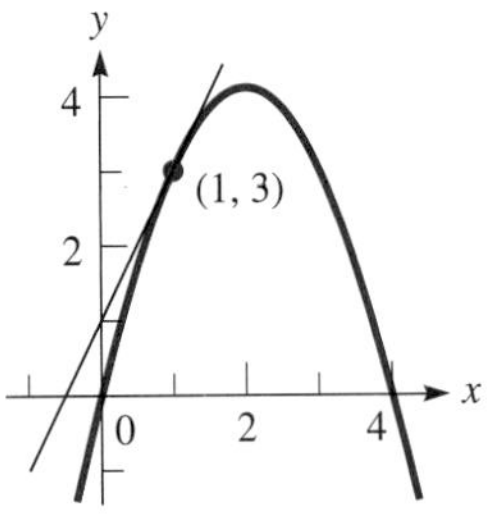

Fig. 3-22

At the end of Section 3-2, we discussed the idea that $\Delta y/\Delta x$ indicates the rate of change of y with respect to x. In defining the derivative as the limit of the ratio of $\Delta y/\Delta x$ as $\Delta x \to 0$, it is a measure of the rate of change of y with respect to x at point P. However, P may represent any point, which means that the value of the derivative changes from one point on a curve to another point.

NOTE We therefore interpret the derivative as the **instantaneous rate of change** of y with respect to x.

EXAMPLE 2   In Examples 1 and 2 of Section 3-2, y is changing at the rate of 7 units for every 1-unit change in x, *when x is 2, and only when x is 2.* Then, in Example 3 of Section 3-2, y is increasing 6 units for every increase of 1 unit of x *when x = −1.* When $x = 3$, y is decreasing 2 units for a 1-unit increase in x. ∎

This gives us a more general meaning of the derivative. If a functional relationship exists between any two variables, then one can be taken to be varying with respect to the other and the derivative gives us the instantaneous rate of change. There are many applications of this principle, one of which is the velocity of an object. We consider here the case of *rectilinear motion*—that is, motion along a straight line.

> The *displacement* of an object is its change in position.

As we have seen, the velocity of an object is found by dividing the change in displacement by the time required for this change. This, however, gives a value only for the **average velocity** for the specified time interval. If the time interval considered becomes smaller and smaller, then the average velocity that is calculated more nearly approximates the **instantaneous velocity** at some particular time. In the limit, the value of the average velocity gives the value of the instantaneous velocity. Using the symbols as defined for the derivative, *the instantaneous velocity of an object moving in rectilinear motion at a particular time t is given by*

INSTANTANEOUS VELOCITY

$$v = \lim_{\Delta t \to 0} \frac{\Delta s}{\Delta t} \qquad (3\text{-}7)$$

where s is the displacement.

In Eq. (3-7) the derivative gives us the instantaneous rate of change of s with respect to t. Therefore, the units of this derivative would be in units of displacement divided by units of time. In general, *units of the derivative of y = f(x) are in units of y divided by units of x.*

> See the chapter introduction.

EXAMPLE 3   Find the instantaneous velocity, when $t = 4$ s (exactly), of a falling object for which the displacement (in ft), which is the distance fallen, is given by $s = 16t^2$ by calculating values of $\Delta s / \Delta t$ and finding the limit as Δt approaches zero.

Here, we shall let t take on values of 3.5, 3.9, 3.99, and 3.999 s. When $t = 4$ s, $s = 256$ ft. Therefore, we calculate Δt by subtracting values of t from 4, and Δs is calculated by subtracting values from 256. The values of velocity are then calculated by using $v = \Delta s / \Delta t$.

		t (s)	3.5	3.9	3.99	3.999
		s (ft)	196.0	243.36	254.7216	255.872016
256 − s	Δs (ft)		60.0	12.64	1.2784	0.127984
4 − t	Δt (s)		0.5	0.1	0.01	0.001
$\dfrac{\Delta s}{\Delta t}$	v (ft/s)		120.0	126.4	127.84	127.984

We can see that the value of v is approaching 128 ft/s, which is therefore the instantaneous velocity when $t = 4$ s. ∎

See the chapter introduction.

EXAMPLE 4 Find the expression for the instantaneous velocity of the object of Example 3, for which $s = 16t^2$, where s is the displacement (in ft) and t is the time (in s). Determine the instantaneous velocity for $t = 2$ s and $t = 4$ s.

The required expression is the derivative of s with respect to t.

$$s + \Delta s = 16(t + \Delta t)^2 = 16t^2 + 32t\,\Delta t + 16(\Delta t)^2$$

$$\Delta s = 32t\,\Delta t + 16(\Delta t)^2$$

$$\frac{\Delta s}{\Delta t} = 32t + 16\,\Delta t$$

$$v = \lim_{\Delta t \to 0} \frac{\Delta s}{\Delta t} = \frac{ds}{dt} = 32t \qquad \text{expression for instantaneous velocity}$$

$$\frac{ds}{dt}\bigg|_{t=2} = 32(2) = 64 \text{ ft/s} \qquad \text{and} \qquad \frac{ds}{dt}\bigg|_{t=4} = 32(4) = 128 \text{ ft/s}$$

We see that the second result agrees with that found in Example 3. ▪

By finding $\lim_{\Delta x \to 0} \Delta y / \Delta x$, we can find the instantaneous rate of change of y with respect to x. The expression $\lim_{\Delta t \to 0} \Delta s / \Delta t$ gives the velocity, or instantaneous rate of change of displacement with respect to time. Generalizing, we can say that

NOTE▶ *the derivative can be interpreted as the instantaneous rate of change of the dependent variable with respect to the independent variable.*

This is true for a differentiable function, no matter what the variables represent.

SOLVING A WORD PROBLEM

EXAMPLE 5 A spherical balloon is being inflated. Find the expression for the instantaneous rate of change of the volume with respect to the radius. Evaluate this rate of change for a radius of 2.00 m.

$$V = \frac{4}{3}\pi r^3 \qquad \text{volume of sphere}$$

$$V + \Delta V = \frac{4\pi}{3}(r + \Delta r)^3 \qquad \text{find derivative}$$

$$= \frac{4\pi}{3}[r^3 + 3r^2\,\Delta r + 3r(\Delta r)^2 + (\Delta r)^3]$$

$$\Delta V = \frac{4\pi}{3}[3r^2\,\Delta r + 3r(\Delta r)^2 + (\Delta r)^3]$$

$$\frac{\Delta V}{\Delta r} = \frac{4\pi}{3}[3r^2 + 3r\,\Delta r + (\Delta r)^2]$$

$$\lim_{\Delta r \to 0}\frac{\Delta V}{\Delta r} = 4\pi r^2 \qquad \text{expression for instantaneous rate of change}$$

$$\frac{dV}{dr}\bigg|_{r=2.00\,\text{m}} = 4\pi(2.00)^2 = 16.0\pi = 50.3 \text{ m}^2 \qquad \text{instantaneous rate of change when } r = 2.00 \text{ m}$$

The instantaneous rate of change of the volume with respect to the radius (dV/dr) for $r = 2.00$ m is 50.3 m³/m (this way of showing the units is more meaningful).

As r increases, dV/dr also increases. This should be expected as the volume of a sphere varies directly as the cube of the radius. ▪

SOLVING A WORD PROBLEM   **EXAMPLE 6**   The power P produced by an electric current i in a resistor varies directly as the square of the current. Given that 1.2 W of power are produced by a current of 0.50 A in a certain resistor, find an expression for the instantaneous rate of change of power with respect to current. Evaluate this rate of change for $i = 2.5$ A.

We must first find the functional relationship between power and current, by solving the indicated problem in variation:

$$P = ki^2 \qquad 1.2 = k(0.50)^2 \qquad k = 4.8 \text{ W/A}^2 \qquad P = 4.8i^2$$

Now, knowing the function, we may determine the expression for the instantaneous rate of change of P with respect to i by using the delta-process:

$$P + \Delta P = 4.8(i + \Delta i)^2 = 4.8[i^2 + 2i(\Delta i) + (\Delta i)^2]$$

$$\Delta P = 4.8[2i(\Delta i) + (\Delta i)^2]$$

$$\frac{\Delta P}{\Delta i} = 4.8(2i + \Delta i)$$

$$\lim_{\Delta i \to 0} \frac{\Delta P}{\Delta i} = 9.6i \qquad \text{expression for instantaneous rate of change}$$

$$\left. \frac{dP}{di} \right|_{i=2.5\,\text{A}} = 9.6(2.5) = 24 \text{ W/A} \qquad \text{instantaneous rate of change when } i = 2.5 \text{ A}$$

This tells us that when $i = 2.5$ A, the rate of change of power with respect to current is 24 W/A. Also, we see that the larger the current is, the greater is the increase in power. This should be expected, since the power varies directly as the square of the current.

EXERCISES *3-4*

In Exercises 1–4, find the slope of a line tangent to the graph of the given equation at the given point. Sketch the graph and the tangent line.

1. $y = x^2 - 1$; (2, 3)   **2.** $y = 2x - x^2$; (−1, −3)

3. $y = \dfrac{16}{3x + 1}$; (−3, −2)   **4.** $y = 3 - \dfrac{16}{x^2}$; (2, −1)

In Exercises 5–8, calculate the instantaneous velocity for the indicated value of the time (in s) of an object for which the displacement (in ft) is given by the indicated function. Use the method of Example 3 and calculate values of $\Delta s/\Delta t$ for the given values of t and find the limit as Δt approaches zero.

5. $s = 4t + 10$; when $t = 3$; use values of t of 2.0, 2.5, 2.9, 2.99, 2.999

6. $s = 6 - 3t$; when $t = 4$; use values of t of 3.0, 3.5, 3.9, 3.99, 3.999

7. $s = 3t^2 - 4t$; when $t = 2$; use values of t of 1.0, 1.5, 1.9, 1.99, 1.999

8. $s = 120t - 16t^2$; when $t = 0.5$; use values of t of 0.4, 0.45, 0.49, 0.499, 0.4999

In Exercises 9–12, use the delta-process to find an expression for the instantaneous velocity of an object moving with rectilinear motion according to the given functions (the same as those for Exercises 5–8) relating s (in ft) and t (in s). Then calculate the instantaneous velocity for the given value of t.

9. $s = 4t + 10$; $t = 3$ **10.** $s = 6 - 3t$; $t = 4$

11. $s = 3t^2 - 4t$; $t = 2$ **12.** $s = 120t - 16t^2$; $t = 0.5$

In Exercises 13–16, use the delta-process to find an expression for the instantaneous velocity of an object moving with rectilinear motion according to the given functions relating s and t.

13. $s = 3t - \dfrac{2}{5t}$ **14.** $s = \dfrac{2t}{t + 2}$

15. $s = 3t^2 - 2t^3$

16. $s = s_0 + v_0 t - \dfrac{1}{2}at^2$ (s_0, v_0, and a are constants.)

In Exercises 17–20, use the delta-process to find an expression for the instantaneous acceleration a of an object moving with rectilinear motion according to the given functions. The instantaneous acceleration of an object is defined as the instantaneous rate of change of velocity with respect to time.

17. $v = 6t^2 - 4t + 2$ **18.** $v = \sqrt{2t + 1}$

19. $s = t^3 + 2t$ (Find v, then find a.)

20. $s = s_0 + v_0 t - \frac{1}{2} at^2$ (s_0, v_0, and a are constants.)
(Find v, then find a.)

In Exercises 21–32, find the indicated instantaneous rates of change.

21. The electric current i at a point in an electric circuit is the instantaneous rate of change of the electric charge q that passes the point, with respect to the time t. Find i in a circuit for which $q = 30 - 2t$.

22. A load L (in N) is distributed along a beam 10 m long such that $L = 5x - 0.5x^2$, where x is the distance from one end of the beam. Find the expression for the instantaneous rate of change of L with respect to x.

23. A rectangular metal plate contracts while cooling. Find the expression for the instantaneous rate of change of the area A of the plate with respect to its width w, if the length of the plate is constantly three times as long as the width.

24. A circular oil spill is increasing in size. Find the instantaneous rate of change of the area A of the spill with respect to its radius r for $r = 240$ m.

25. The total power P (in W) transmitted by an AM radio station is given by $P = 500 + 250m^2$, where m is the modulation index. Find the instantaneous rate of change of P with respect to m for $m = 0.92$.

26. The bottom of a soft-drink can is being designed as an inverted spherical segment, the volume of which is $V = \frac{1}{6}\pi h^3 + 2.00\pi h$, where h is the depth (in cm) of the segment. Find the instantaneous rate of change of V with respect to h for $h = 0.60$ cm.

27. The total solar radiation H (in W/m²) on a particular surface during an average clear day is given by $H = \dfrac{5000}{t^2 + 10}$, where t ($-6 \le t \le 6$) is the number of hours from noon (6 A.M. is equivalent to $t = -6$ h). Find the instantaneous rate of change of H with respect to t at 3 P.M.

28. The value (in thousands of dollars) of a certain car is given by the function $V = \dfrac{48}{t + 3}$, where t is measured in years. Find a general expression for the instantaneous rate of change of V with respect to t and evaluate this expression when $t = 3$ years.

29. Oil in a certain machine is stored in a conical reservoir, for which the radius and height are both 4 cm. Find the instantaneous rate of change of the volume V of oil in the reservoir with respect to the depth d of the oil. See Fig. 3-23.

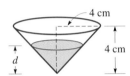

Fig. 3-23

30. The time t required to test a computer memory unit is directly proportional to the square of the number n of memory cells in the unit. For a particular type of unit, $n = 6400$ for $t = 25.0$ s. Find the instantaneous rate of change of t with respect to n for this type of unit for $n = 8000$.

31. A *holograph* (an image formed without using a lens) of concentric circles is formed. The radius r of each circle varies directly as the square root of the wavelength λ of the light used. If $r = 3.72$ cm for $\lambda = 592$ nm, find the expression for the instantaneous rate of change of r with respect to λ.

32. The force F between two electric charges varies inversely as the square of the distance r between them. For two charged particles, $F = 0.12$ N for $r = 0.060$ m. Find the instantaneous rate of change of F with respect to r for $r = 0.120$ m.

$3\text{-}5$ DERIVATIVES OF POLYNOMIALS

The task of finding the derivative of a function can be considerably shortened from that involved in the direct use of the delta-process. We can use the delta-process to derive certain basic formulas for finding derivatives of particular types of functions. These formulas will then be used to find the derivatives. In this section we derive the formulas for finding the derivatives of polynomial functions of the form $f(x) = a_0 x^n + a_1 x^{n-1} + \cdots + a_n$.

First, we find the derivative of a constant. By letting $y = c$ and applying the delta-process to this function, we obtain the desired result:

$$y = c, \qquad y + \Delta y = c, \qquad \Delta y = 0, \qquad \frac{\Delta y}{\Delta x} = 0, \qquad \lim_{\Delta x \to 0} \frac{\Delta y}{\Delta x} = 0$$

From this we conclude that *the derivative of a constant is zero*. This result holds for all constants. Therefore, if $y = c$, $dy/dx = 0$, or

DERIVATIVE OF A CONSTANT

$$\boxed{\frac{dc}{dx} = 0} \tag{3-8}$$

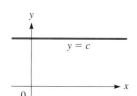

Fig. 3-24

Graphically this means that for any function of the type $y = c$ the slope is always zero. We know that $y = c$ represents a straight line parallel to the x-axis. From the definition of slope, we know that any line parallel to the x-axis has a slope of zero. See Fig. 3-24. We see that the two results are consistent.

Next, we find the derivative of any integral power of x. If $y = x^n$, where n is a positive integer, by using the binomial theorem we have

$$(y + \Delta y) = (x + \Delta x)^n$$

$$= x^n + nx^{n-1}\Delta x + \frac{n(n-1)}{2}x^{n-2}(\Delta x)^2 + \cdots + (\Delta x)^n$$

$$\Delta y = nx^{n-1}\Delta x + \frac{n(n-1)}{2}x^{n-2}(\Delta x)^2 + \cdots + (\Delta x)^n$$

$$\frac{\Delta y}{\Delta x} = nx^{n-1} + \underbrace{\frac{n(n-1)}{2}x^{n-2}\Delta x + \cdots + (\Delta x)^{n-1}}_{\text{each term} \to 0 \text{ as } \Delta x \to 0}$$

$$\lim_{\Delta x \to 0} \frac{\Delta y}{\Delta x} = nx^{n-1}$$

Thus, *the derivative of the nth power of x is*

DERIVATIVE OF POWER OF x

$$\boxed{\frac{dx^n}{dx} = nx^{n-1}} \tag{3-9}$$

▌EXAMPLE 1   Find the derivative of the function $y = -5$.
  Since -5 is a constant, applying Eq. (3-8), we have

$$\frac{dy}{dx} = \frac{d(-5)}{dx} = 0$$

▌EXAMPLE 2   Find the derivative of $y = x^3$.
  Using Eq. (3-9), we have

$$\frac{dy}{dx} = \frac{d(x^3)}{dx} = 3x^{3-1} = 3x^2$$

This result is consistent with those found previously in this chapter.

EXAMPLE 3   Find the derivative of the function $y = x$.

In using Eq. (3-9), we have $n = 1$ since $x = x^1$. This means

$$\frac{dy}{dx} = \frac{d(x)}{dx} = (1)x^{1-1} = (1)(x^0)$$

Since $x^0 = 1$, we have

$$\frac{dy}{dx} = 1$$

Thus, the derivative of $y = x$ is 1, which means that the slope of the line $y = x$ is always 1. This is consistent with our previous discussion of the slope of a straight line. ------------------- ∎

EXAMPLE 4   Find the derivative of $v = r^{10}$.

Here, the dependent variable is v, and the independent variable is r. Therefore,

$$\frac{dv}{dr} = \frac{d(r^{10})}{dr} = 10r^{10-1}$$

$$= 10r^9$$ ------------------- ∎

Next, we find the derivative of a constant times a function of x. If $y = cu$, where $u = f(x)$, we have the following result:

$$y + \Delta y = c(u + \Delta u)$$

(As x increases by Δx, u increases by Δu, since u is a function of x.) Then,

$$\Delta y = c\,\Delta u, \qquad \frac{\Delta y}{\Delta x} = c\frac{\Delta u}{\Delta x}, \qquad \lim_{\Delta x \to 0} \frac{\Delta y}{\Delta x} = c \lim_{\Delta x \to 0} \frac{\Delta u}{\Delta x}$$

Therefore, *the derivative of the product of a constant and a differentiable function of x is the product of the constant and the derivative of the function of x.* This is written as

DERIVATIVE OF A CONSTANT TIMES A FUNCTION

$$\frac{d(cu)}{dx} = c\frac{du}{dx}$$ (3-10)

EXAMPLE 5   Find the derivative of $y = 3x^2$.

In this case, $c = 3$ and $u = x^2$. Thus, $du/dx = 2x$. Therefore,

$$\frac{dy}{dx} = \frac{d(3x^2)}{dx} = 3\frac{d(x^2)}{dx} = 3(2x)$$

$$= 6x$$ ------------------- ∎

Occasionally, the derivative of a constant times a function of x is confused with the derivative of a constant that stands alone. It is necessary to clearly distinguish between a constant that multiplies a function and an isolated constant.

Finally, if the types of functions for which we have found derivatives are added, the result is a polynomial function with more than one term. The derivative of such a function is found by letting $y = u + v$, where u and v are functions of x. Applying the delta-process, since u and v are functions of x, each has an increment corresponding to an increment in x. Thus, we have the following result:

$$y + \Delta y = (u + \Delta u) + (v + \Delta v)$$

$$\Delta y = \Delta u + \Delta v$$

$$\frac{\Delta y}{\Delta x} = \frac{\Delta u}{\Delta x} + \frac{\Delta v}{\Delta x}$$

$$\lim_{\Delta x \to 0} \frac{\Delta y}{\Delta x} = \lim_{\Delta x \to 0} \frac{\Delta u}{\Delta x} + \lim_{\Delta x \to 0} \frac{\Delta v}{\Delta x}$$

This tells us that *the derivative of the sum of differentiable functions of x is the sum of the derivatives of the functions.* This is written as

DERIVATIVE OF A SUM

$$\frac{d(u + v)}{dx} = \frac{du}{dx} + \frac{dv}{dx}$$ (3-11)

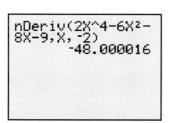

nDeriv(2X^4-6X²-
8X-9,X,-2)
          -48.000016

Fig. 3-25

■EXAMPLE 6 Evaluate the derivative of $f(x) = 2x^4 - 6x^2 - 8x - 9$ at $(-2, 15)$.

First, finding the derivative, we have

$$f'(x) = \frac{d(2x^4)}{dx} - \frac{d(6x^2)}{dx} - \frac{d(8x)}{dx} - \frac{d(9)}{dx}$$

$$= 8x^3 - 12x - 8$$

Since the derivative is a function only of x, we now evaluate it for $x = -2$.

$$f'(x)\Big|_{x=-2} = 8(-2)^3 - 12(-2) - 8 = -48$$

Using the *derivative evaluation* feature of a graphing calculator, we have the display in Fig. 3-25 for this evaluation. We see that the values agree. ■

■EXAMPLE 7 Find the slope of a line tangent to the curve of $y = 4x^7 - x^4$ at the point $(1, 3)$.

We must find and then evaluate the derivative for the value $x = 1$.

$$\frac{dy}{dx} = 28x^6 - 4x^3 \qquad \text{find derivative}$$

$$\frac{dy}{dx}\Big|_{x=1} = 28(1) - 4(1) \qquad \text{evaluate derivative}$$

$$= 24$$

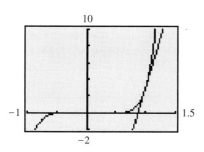

Fig. 3-26

Thus, the slope of the tangent line is 24. Again, we note that **the substitution $x = 1$ must be made after the differentiation has been performed.**

Using the *tangent* feature of a graphing calculator, we have the display for the curve and tangent line in Fig. 3-26. Although it is difficult to approximate a slope as large as 24, we can see that the line is very steep, which tends to verify the result.

  ■

EXAMPLE 8   For each 4.0-s cycle, the displacement s (in cm) of a piston is given by the equation $s = t^3 - 6t^2 + 8t$, where t is the time (in s). Find the instantaneous velocity of the piston for $t = 2.6$ s.

$$s = t^3 - 6t^2 + 8t$$

$$\frac{ds}{dt} = 3t^2 - 12t + 8 \qquad \text{find derivative}$$

$$\left.\frac{ds}{dt}\right|_{t=2.6} = 3(2.6)^2 - 12(2.6) + 8 \qquad \text{evaluate derivative}$$

$$= -2.9 \text{ cm/s}$$

This tells us that the piston is moving at -2.9 cm/s (it is moving in a negative direction) when $t = 2.6$ s.

EXERCISES 3-5

In Exercises 1–16, find the derivative of each of the given functions.

1. $y = x^5$

2. $y = x^{12}$

3. $f(x) = -4x^9$

4. $y = -7x^6$

5. $y = x^4 - 6$

6. $s = 3t^5 + 4$

7. $y = x^2 + 2x$

8. $y = x^3 - 2x^2$

9. $p = 5r^3 - 2r + 1$

10. $y = 6x^2 - 6x + 5$

11. $y = x^8 - 4x^7 - x$

12. $y = 4x^4 - 2x + 9$

13. $f(x) = -6x^7 + 5x^3 + \pi^2$

14. $y = 13x^4 - 6x^3 - x - 1$

15. $y = \frac{1}{3}x^3 + \frac{1}{2}x^2$

16. $f(z) = -\frac{1}{4}z^8 + \frac{1}{2}z^4 - 2^3$

In Exercises 17–20, evaluate the derivative of each of the given functions at the given point. In Exercises 19 and 20, check your result using the derivative evaluation feature of a graphing calculator.

17. $y = 6x^2 - 8x + 1$ $(2, 9)$

18. $s = 2t^3 - 5t^2 + 4$ $(-1, -3)$

19. $y = 2x^3 + 9x - 7$ $(-2, -41)$

20. $y = x^4 - 9x^2 - 5x$ $(3, -15)$

In Exercises 21–24, find the slope of a line tangent to the graph of each of the given functions for the given values of x. Use the tangent feature of a graphing calculator to display the curve and the tangent line to see if the slope is reasonable for the given value of x.

21. $y = 2x^6 - 4x^2$ $(x = -1)$ **22.** $y = 3x^3 - 9x$ $(x = 1)$

23. $y = 35x - 2x^4$ $(x = 2)$

24. $y = x^4 - \frac{1}{2}x^2 + 2$ $(x = -2)$

In Exercises 25–28, determine an expression for the instantaneous velocity of objects moving with rectilinear motion according to the functions given, if s represents displacement in terms of time t.

25. $s = 6t^5 - 5t + 2$

26. $s = 20 + 60t - 4.9t^2$

27. $s = 2 - 6t - 2t^3$

28. $s = s_0 + v_0 t + \frac{1}{2}at^2$

In Exercises 29–32, s represents the displacement and t represents the time for objects moving with rectilinear motion according to the given functions. Find the instantaneous velocity for the given times.

29. $s = 2t^3 - 4t^2$; $t = 4$

30. $s = 120 + 80t - 16t^2$; $t = 2.5$

31. $s = 0.5t^4 - 1.5t^2 + 2.5$; $t = 3$

32. $s = 8t^2 - 10t + 6$; $t = 5$

In Exercises 33–48, solve the given problems by finding the appropriate derivative.

33. For what value(s) of x is the tangent to the graph of $y = 3x^2 - 6x$ parallel to the x-axis? (That is, where is the slope zero?)

34. Find the value of a if the tangent to the graph of $y = ax^2 + 2x$ has a slope of -4 for $x = 2$.

35. For what point(s) on the graph of $y = 3x^2 - 4x$ is the slope of a tangent line equal to 8?

(W) 36. Explain why the curve $y = 5x^3 + 4x - 3$ does not have a tangent line with a slope less than 4.

37. For what value(s) of x is the slope of a line tangent to the graph of $y = 4x^2 + 3x$ equal to the slope of a line tangent to the curve of $y = 5 - 2x^2$?

38. For what value(s) of t is the instantaneous velocity of an object moving according to $s = 5t - 2t^2$ equal to the instantaneous velocity of an object moving according to $s = 3t^2 + 4$?

39. A cylindrical metal container is heated and then allowed to cool. If the radius always equals the height, find an expression for the instantaneous rate of change of the volume V with respect to the radius r.

40. As an ice cube melts uniformly, find the expression for the instantaneous rate of change of the surface area A of the cube with respect to the edge e.

41. The electric power P (in W) as a function of the current i (in A) in a certain circuit is given by $P = 16i^2 + 60i$. Find the instantaneous rate of change of P with respect to i for $i = 0.75$ A.

42. The torque T on the arm of a robotic control mechanism varies directly as the cube of the diameter d of the arm. If $T = 850$ lb·in. for $d = 0.925$ in., find the expression for the instantaneous rate of change of T with respect to d.

43. The electric polarization P of a light wave for high values of the electric field E is given by $P = a(c_1 E + c_2 E^2 + c_3 E^3)$, where a, c_1, c_2, and c_3 are constants. Find the expression for the instantaneous rate of change of P with respect to E.

44. The ends of a 10-ft beam are supported at different levels. The deflection y of the beam is given by $y = kx^2(x^3 + 450x - 3500)$, where x is the horizontal distance from one end and k is a constant. Determine the expression for the instantaneous rate of change of y with respect to x.

45. The altitude h (in m) of a jet as a function of the horizontal distance x (in km) it has traveled is given by $h = 0.000104x^4 - 0.0417x^3 + 4.21x^2 - 8.33x$. Find the instantaneous rate of change of h with respect to x for $x = 120$ km.

46. The force F (in N) exerted by a cam on a lever is given by $F = x^4 - 12x^3 + 46x^2 - 60x + 25$, where x ($1 \leq x \leq 5$) is the distance (in cm) from the center of rotation of the cam to the edge of the cam in contact with the lever (see Fig. 3-27). Find the instantaneous rate of change of F with respect to x when $x = 4.0$ cm.

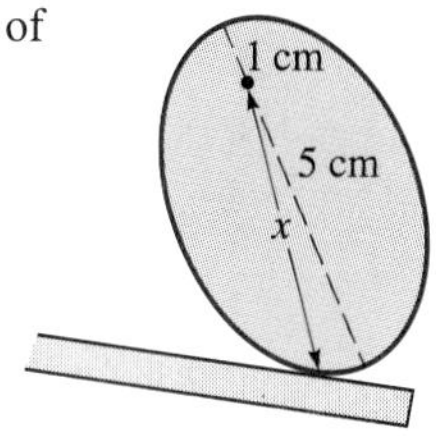

Fig. 3-27

47. Two ball bearings wear down such that the radius r of one is constantly 1.20 mm less than the radius of the other. Find the instantaneous rate of change of the total volume V_T of the two ball bearings with respect to r for $r = 3.30$ mm.

48. An open-top container is to be made from a rectangular piece of cardboard 6.00 in. by 8.00 in. Equal squares of side x are to be cut from each corner, then the sides are to be bent up and taped together. Find the instantaneous rate of change of the volume V of the container with respect to x for $x = 1.75$ in.

$$\underset{3\text{-}6}{} \quad \textbf{DERIVATIVES OF PRODUCTS AND QUOTIENTS OF FUNCTIONS}$$

The formulas developed in the previous section are valid for polynomial functions. However, many functions are not polynomial in form. Some functions can best be expressed as the product of two or more simpler functions, others are the quotient of two simpler functions, and some are expressed as powers of a function. In this section we develop the formula for the derivative of a product of functions and the formula for the derivative of the quotient of two functions.

EXAMPLE 1 The functions $f(x) = x^2 + 2$ and $g(x) = 3 - 2x$ can be combined to form new functions of the types mentioned above. For example, the function

$$p(x) = f(x)g(x) = (x^2 + 2)(3 - 2x)$$

is an example of a function expressed as the product of two simpler functions.
The function

$$q(x) = \frac{g(x)}{f(x)} = \frac{3 - 2x}{x^2 + 2}$$

is an example of a rational function that is the quotient of two other functions.
The function

$$F(x) = [g(x)]^3 = (3 - 2x)^3$$

is an example of a power of a function. ────── ∎

If u and v both represent differentiable functions of x, the derivative of the product of u and v is found by applying the delta-process. This leads to the following result:

$$y = uv$$

$$y + \Delta y = (u + \Delta u)(v + \Delta v) = uv + u\,\Delta v + v\,\Delta u + \Delta u\,\Delta v$$

(Since u and v are functions of x, each has an increment corresponding to an increment in x.) Then we have

$$\Delta y = u\,\Delta v + v\,\Delta u + \Delta u\,\Delta v$$

$$\frac{\Delta y}{\Delta x} = u\frac{\Delta v}{\Delta x} + v\frac{\Delta u}{\Delta x} + \Delta u\frac{\Delta v}{\Delta x}$$

$$\lim_{\Delta x \to 0}\frac{\Delta y}{\Delta x} = u\lim_{\Delta x \to 0}\frac{\Delta v}{\Delta x} + v\lim_{\Delta x \to 0}\frac{\Delta u}{\Delta x} + \lim_{\Delta x \to 0}\left(\Delta u\frac{\Delta v}{\Delta x}\right)$$

(The functions u and v are not affected by Δx approaching zero, but Δu and Δv both approach zero as Δx approaches zero.) Thus,

$$\frac{dy}{dx} = u\frac{dv}{dx} + v\frac{du}{dx} + 0\frac{dv}{dx}$$

We conclude that *the derivative of the product of two differentiable functions equals the first function times the derivative of the second function plus the second function times the derivative of the first function.* This is written as

DERIVATIVE OF A PRODUCT

$$\boxed{\frac{d(uv)}{dx} = u\frac{dv}{dx} + v\frac{du}{dx}}$$

(3-12)

■**EXAMPLE 2**   Find the derivative of the product function in Example 1.

$$p(x) = (x^2 + 2)(3 - 2x) \qquad u = x^2 + 2 \qquad v = 3 - 2x$$

$$\frac{d(uv)}{dx} = \quad u \quad \frac{dv}{dx} \quad + \quad v \quad \frac{du}{dx}$$

$$p'(x) = (x^2 + 2)(-2) + (3 - 2x)(2x) = -2x^2 - 4 + 6x - 4x^2$$

$$= -6x^2 + 6x - 4 \qquad\qquad\qquad ■$$

■**EXAMPLE 3**   Find the derivative of the function $y = (3 - x - 2x^2)(x^4 - x)$.
In this problem, $u = 3 - x - 2x^2$ and $v = x^4 - x$. Hence,

$$\frac{dy}{dx} = (3 - x - 2x^2)(4x^3 - 1) + (x^4 - x)(-1 - 4x)$$

$$= 12x^3 - 3 - 4x^4 + x - 8x^5 + 2x^2 - x^4 - 4x^5 + x + 4x^2$$

$$= -12x^5 - 5x^4 + 12x^3 + 6x^2 + 2x - 3 \qquad\qquad ■$$

In both of these examples, we could have multiplied the functions first and then taken the derivative as a polynomial. However, we shall soon meet functions for which this latter method would not be applicable.

We shall now find the derivative of the quotient of two differentiable functions by applying the delta-process to the function $y = u/v$ as follows:

$$y + \Delta y = \frac{u + \Delta u}{v + \Delta v}$$

$$\Delta y = \frac{u + \Delta u}{v + \Delta v} - \frac{u}{v} = \frac{vu + v\Delta u - uv - u\,\Delta v}{v(v + \Delta v)}$$

$$\frac{\Delta y}{\Delta x} = \frac{v(\Delta u/\Delta x) - u(\Delta v/\Delta x)}{v(v + \Delta v)}$$

$$\lim_{\Delta x \to 0} \frac{\Delta y}{\Delta x} = \frac{v \lim_{\Delta x \to 0} (\Delta u/\Delta x) - u \lim_{\Delta x \to 0} (\Delta v/\Delta x)}{\lim_{\Delta x \to 0} v(v + \Delta v)}$$

$$\frac{dy}{dx} = \frac{v(du/dx) - u(dv/dx)}{v^2}$$

Therefore, *the derivative of the quotient of two differentiable functions equals the denominator times the derivative of the numerator minus the numerator times the derivative of the denominator, all divided by the square of the denominator.*

DERIVATIVE OF A QUOTIENT

$$\frac{d\left(\dfrac{u}{v}\right)}{dx} = \frac{v\dfrac{du}{dx} - u\dfrac{dv}{dx}}{v^2}$$

(3-13)

EXAMPLE 4   Find the derivative of the quotient indicated in Example 1.

$$q(x) = \frac{3 - 2x}{x^2 + 2} \qquad u = 3 - 2x \qquad v = x^2 + 2$$

$$q'(x) = \frac{\overset{v}{(x^2 + 2)}\overset{du/dx}{(-2)} - \overset{u}{(3 - 2x)}\overset{dv/dx}{(2x)}}{(x^2 + 2)^2} = \frac{-2x^2 - 4 - 6x + 4x^2}{(x^2 + 2)^2}$$

$$= \frac{2(x^2 - 3x - 2)}{(x^2 + 2)^2}$$

In the first expression for $q'(x)$, be careful not to cancel the factor of $(x^2 + 2)$, as it is not a factor of both terms of the numerator.

EXAMPLE 5   The stress S on a hollow tube is given by

$$S = \frac{16DT}{\pi(D^4 - d^4)}$$

where T is the tension, D is the outer diameter, and d is the inner diameter of the tube. Find the expression for the instantaneous rate of change of S with respect to D, with the other values being constant.

We are to find the derivative of S with respect to D, and it is found as follows:

$$\frac{dS}{dD} = \frac{\pi(D^4 - d^4)(16T) - 16DT(\pi)(4D^3)}{\pi^2(D^4 - d^4)^2} = \frac{16\pi T(D^4 - d^4 - 4D^4)}{\pi^2(D^4 - d^4)^2}$$

$$= \frac{-16T(3D^4 + d^4)}{\pi(D^4 - d^4)^2}$$

EXAMPLE 6   Evaluate the derivative of $y = \dfrac{3x^2 + x}{1 - 4x}$ at $(2, -2)$.

$$\frac{dy}{dx} = \frac{(1 - 4x)(6x + 1) - (3x^2 + x)(-4)}{(1 - 4x)^2}$$

$$= \frac{6x + 1 - 24x^2 - 4x + 12x^2 + 4x}{(1 - 4x)^2}$$

$$= \frac{-12x^2 + 6x + 1}{(1 - 4x)^2}$$

$$\frac{dy}{dx}\bigg|_{x=2} = \frac{-12(2^2) + 6(2) + 1}{[1 - 4(2)]^2} = \frac{-48 + 12 + 1}{49} = \frac{-35}{49}$$

$$= -\frac{5}{7}$$

```
nDeriv((3X²+X)/(
1-4X),X,2)
          -.7142857027
-5/7
          -.7142857143
```

Fig. 3-28

Checking this result by using the *numerical derivative* feature of a graphing calculator, we have the display shown in Fig. 3-28. We see that the calculator evaluation and the decimal value of $-5/7$ agree. (The difference in the last three decimal places is due to the fact that the numerical derivative evaluation is an approximation.) ∎

EXERCISES 3-6

In Exercises 1–8, find the derivative of each function by using Eq. (3-12). Do not find the product before finding the derivative.

1. $y = x^2(3x + 2)$ **2.** $y = 3x(x^3 + 1)$

3. $y = 6x(3x^2 - 5x)$ **4.** $y = 2x^3(3x^4 + x)$

5. $s = (3t + 2)(2t - 5)$ **6.** $f(x) = (3x - 2)(4x^2 + 3)$

7. $y = (x^4 - 3x^2 + 3)(1 - 2x^3)$

8. $y = (x^3 - 6x)(2 - 4x^3)$

In Exercises 9–12, find the derivative of each function by using Eq. (3-12). Then multiply out each function and find the derivative by treating it as a polynomial. Compare the results.

9. $y = (2x - 7)(5 - 2x)$

10. $f(s) = (5s^2 + 2)(2s^2 - 1)$

11. $y = (x^3 - 1)(2x^2 - x - 1)$

12. $y = (3x^2 - 4x + 1)(5 - 6x^2)$

In Exercises 13–24, find the derivative of each function by using Eq. (3-13).

13. $y = \dfrac{x}{2x + 3}$ **14.** $y = \dfrac{2x}{x + 1}$

15. $y = \dfrac{1}{x^2 + 1}$ **16.** $R = \dfrac{5i + 2}{2i + 3}$

17. $y = \dfrac{x^2}{3 - 2x}$ **18.** $y = \dfrac{2}{3x^2 - 5x}$

19. $y = \dfrac{2x - 1}{3x^2 + 2}$ **20.** $y = \dfrac{2x^3}{4 - x}$

21. $f(x) = \dfrac{3x + 8}{x^2 + 4x + 2}$ **22.** $y = \dfrac{3x}{4x^5 - 3x - 4}$

23. $y = \dfrac{2x^2 - x - 1}{x^3 + 2x^2}$ **24.** $y = \dfrac{3x^3 - x}{2x^2 - 5x + 4}$

In Exercises 25–32, evaluate the derivatives of the given functions for the given values of x. In Exercises 25–28, use Eq. (3-12). In Exercises 27, 28, 31, and 32, check your results using the derivative evaluation feature of a graphing calculator.

25. $y = (3x - 1)(4 - 7x)$, $x = 3$

26. $y = (3x^2 - 5)(2x^2 - 1)$, $x = -1$

27. $y = (2x^2 - x + 1)(4 - 2x - x^2)$, $x = -3$

28. $y = (4x^4 + 0.5x^2 + 1)(3x - 2x^2)$, $x = 0.5$

29. $y = \dfrac{3x - 5}{2x + 3}$, $x = -2$ **30.** $y = \dfrac{2x^2 - 5x}{3x + 2}$, $x = 2$

31. $S = \dfrac{2n^3 - 3n + 8}{2n - 3n^4}$, $n = -1$

32. $y = \dfrac{2x^3 - x^2 - 2}{4x + 3}$, $x = 0.5$

In Exercises 33–48, solve the given problems by finding the appropriate derivatives.

33. Find the derivative of $y = \dfrac{x^2(1 - 2x)}{3x - 7}$ in each of the following two ways. (1) Do *not* multiply out the numerator before finding the derivative. (2) Multiply out the numerator before finding the derivative. Compare the results.

34. Find the derivative of $y = 4x^2 - \dfrac{1}{x - 1}$ in each of the following two ways. (1) Do *not* combine the terms over a common denominator before finding the derivative. (2) Combine the terms over a common denominator before finding the derivative. Compare the results.

35. Find the slope of a line tangent to the graph of the function $y = (4x + 1)(x^4 - 1)$ at the point $(-1, 0)$. Do not multiply the factors together before taking the derivative. Use the derivative evaluation feature of a graphing calculator to check your result.

36. Find the slope of a line tangent to the graph of the function $y = (3x + 4)(1 - 4x)$ at the point $(2, -70)$. Do not multiply the factors together before taking the derivative. Use the derivative evaluation feature of a graphing calculator to check your result.

37. For what value(s) of x is the slope of a tangent to the graph of $y = \dfrac{x}{x^2 + 1}$ equal to zero? View the graph on a graphing calculator to verify the values found.

38. Determine the sign of the derivative of the function $y = \dfrac{2x - 1}{1 - x^2}$ for the following values of x: $-2, -1, 0, 1, 2$. Is the slope of a tangent line to this curve ever negative? View the graph on a graphing calculator to verify your conclusion.

39. During each cycle, the vertical displacement s of the end of a robot arm is given by $s = (t^2 - 8t)(2t^2 + t + 1)$, where t is the time. Find the expression for the instantaneous velocity of the end of the robot arm. See Fig. 3-29.

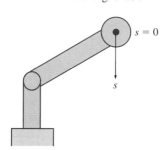

Fig. 3-29

40. The number n of dollars saved by increasing the fuel efficiency of e mi/gal to $e + 6$ mi/gal for a car driven 10,000 mi/year is $n = \dfrac{75{,}000}{e(e + 6)}$, if the cost of gas is \$1.25/gal. Find dn/de.

41. The voltage V at the junction of a 3-Ω resistance and a variable resistance R in a circuit is given by $V = \dfrac{6R + 25}{R + 3}$. Find the instantaneous rate of change of V with respect to R for $R = 7\ \Omega$.

42. During a chemical change, the number n of grams of a compound being formed is given by $n = \dfrac{6t^2}{2t^2 + 3}$, where t is the time (in s). How many grams per second are being formed after 3.0 s?

43. A computer, using data from a refrigeration plant, estimated that in the event of a power failure the temperature T (in °C) in the freezers would be given by $T = \dfrac{2t}{0.05t + 1} - 20$, where t is the number of hours after the power failure. Find the time rate of change of temperature after 6.0 h.

44. The voltage V across a resistor in an electric circuit is the product of the resistance and the current. If the current I (in A) varies with time t (in s) according to the relation $I = 5.00 + 0.01t^2$ and the resistance varies with time according to the relation $R = 15.00 - 0.10t$, find the time rate of change of the voltage when $t = 5.00$ s.

45. The frictional radius r_f of a disc clutch is given by the equation $r_f = \dfrac{2(R^2 + Rr + r^2)}{3(R + r)}$, where R and r are the outer radius and the inner radius of the clutch, respectively. Find the derivative of r_f with respect to R with r constant.

46. In thermodynamics, an equation relating the thermodynamic temperature T, the pressure p, and the volume V of a gas is $T = \left(p + \dfrac{a}{V^2}\right)\left(\dfrac{V - b}{R}\right)$, where a, b, and R are constants. Find the derivative of T with respect to V, assuming p is constant.

47. The electric power P produced by a certain source is given by $P = \dfrac{E^2 r}{R^2 + 2Rr + r^2}$, where E is the voltage of the source, R is the resistance of the source, and r is the resistance in the circuit. Find the derivative of P with respect to r, assuming that the other quantities remain constant.

48. In the theory of lasers, the power P radiated is given by the equation $P = \dfrac{kf^2}{\omega^2 - 2\omega f + f^2 + a^2}$, where f is the field frequency and a, k, and ω are constants. Find the derivative of P with respect to f.

$3\text{-}7$ THE DERIVATIVE OF A POWER OF A FUNCTION

In Example 1 of Section 3-6, we illustrated $y = (3 - 2x)^3$ as the third power of a function of x, where $3 - 2x$ is the function. If we let $u = 3 - 2x$, we can write

$$y = u^3 \quad \text{where } u = 3 - 2x$$

Writing it this way, y is a function of u, and u is a function of x. This means that y *is a function of a function of x, referred to as a* **composite function.** However, y is still a function of x, since u is a function of x.

Since we will often need to find the derivative of a power of a function, we now develop the necessary formula. For $y = f(u)$, where $u = g(x)$, we can express the derivative dy/dx in terms of dy/du and du/dx. If Δx is the increment in x, then Δy and Δu are the corresponding increments in y and u, respectively. We may write

$$\frac{\Delta y}{\Delta x} = \frac{\Delta y}{\Delta u} \frac{\Delta u}{\Delta x}$$

When Δx approaches zero, Δu and Δy both approach zero, for u and y are assumed to be continuous functions of x. Thus,

$$\lim_{\Delta x \to 0} \frac{\Delta y}{\Delta x} = \left(\lim_{\Delta u \to 0} \frac{\Delta y}{\Delta u} \right) \left(\lim_{\Delta x \to 0} \frac{\Delta u}{\Delta x} \right)$$

CHAIN RULE

$$\boxed{\frac{dy}{dx} = \frac{dy}{du} \frac{du}{dx}} \tag{3-14}$$

(Here we have assumed $\Delta u \neq 0$, although it can be shown that this condition is not necessary.) *Equation (3-14) is known as the* **chain rule** *for derivatives.*

Using Eq. (3-14) for $y = u^n$, where u is a differentiable function of x, we have

$$\frac{dy}{dx} = \frac{d(u^n)}{du} \frac{du}{dx} \quad \text{or}$$

DERIVATIVE OF A POWER OF A FUNCTION OF x

$$\boxed{\frac{du^n}{dx} = nu^{n-1} \left(\frac{du}{dx} \right)} \tag{3-15}$$

We use Eq. (3-15) to find the derivative of a power of a differentiable function of x.

■EXAMPLE 1 Find the derivative of $y = (3 - 2x)^3$.

For this function $n = 3$ and $u = 3 - 2x$. Therefore, $du/dx = -2$. This means

$$\frac{du^n}{dx} = n \quad u \quad {}^{n-1} \left(\frac{du}{dx} \right)$$

$$\frac{dy}{dx} = 3(3 - 2x)^2(-2)$$

$$= -6(3 - 2x)^2$$

CAUTION ▶ A common type of error in finding this type of derivative is to omit the du/dx factor; in this case it is the -2. *The derivative is incomplete and therefore incorrect without this factor.* ■

EXAMPLE 2 Find the derivative of $p(x) = (1 - 3x^2)^4$.

In this example, $n = 4$ and $u = 1 - 3x^2$. Hence,

$$p'(x) = 4(1 - 3x^2)^3(-6x) = -24x(1 - 3x^2)^3$$

CAUTION ▶ *(We must not forget the $-6x$.)*

EXAMPLE 3 Find the derivative of $y = \dfrac{(3x - 1)^3}{1 - x}$.

NOTE ▶ Here, we **use the quotient rule in combination with the power rule.** We find the derivative of the numerator by using the power rule.

$$\frac{dy}{dx} = \frac{(1 - x)\overbrace{3(3x - 1)^2(3)}^{\text{derivative of numerator}} - (3x - 1)^3(-1)}{(1 - x)^2}$$

$$= \frac{(3x - 1)^2(9 - 9x + 3x - 1)}{(1 - x)^2} = \frac{(3x - 1)^2(8 - 6x)}{(1 - x)^2}$$

$$= \frac{2(3x - 1)^2(4 - 3x)}{(1 - x)^2}$$

It is normally better to have the derivative in a factored, simplified form, since this is the only form from which useful information may readily be obtained. In this form we can determine where the derivative is undefined (denominator equal to zero) or where the slope is zero (numerator equal to zero); we can also make other **NOTE ▶** required analyses of the derivative. Thus, *all derivatives should be in simplest algebraic form.*

To this point, we have derived the formulas for the derivatives of differentiable functions of x raised to positive integral powers. We shall now establish that these formulas are also valid for any rational number used as an exponent. If $y = u^{p/q}$ and if each side is raised to the qth power, we have $y^q = u^p$. Applying the power rule to each side of this equation, we have

$$qy^{q-1}\left(\frac{dy}{dx}\right) = pu^{p-1}\left(\frac{du}{dx}\right)$$

Solving for dy/dx, we have

$$\frac{dy}{dx} = \frac{pu^{p-1}(du/dx)}{qy^{q-1}} = \frac{p}{q}\frac{u^{p-1}}{(u^{p/q})^{q-1}}\frac{du}{dx} = \frac{p}{q}\frac{u^{p-1}}{u^{p-p/q}}\frac{du}{dx}$$

$$= \frac{p}{q}u^{p-1-p+(p/q)}\frac{du}{dx}$$

Thus,

$$\boxed{\frac{du^{p/q}}{dx} = \frac{p}{q}u^{(p/q)-1}\frac{du}{dx}}$$
(3-16)

We see that in finding the derivative we multiply the function by the rational exponent and subtract 1 from it to find the exponent of the function in the derivative. **NOTE ▶** *This is the same rule as derived for positive integral exponents in Eq. (3-15).*

For reference, Eq. (3-9) is $\dfrac{dx^n}{dx} = nx^{n-1}$.

In deriving Eqs. (3-15) and (3-16), we used Eq. (3-9), which is valid for positive integral exponents. Equation (3-9) can be shown to be also valid for negative integral exponents by using the quotient rule on $1/x^n$, which is equal to x^{-n}. Therefore,

the power rule for derivatives, Eq. (3-15), can be extended to include all rational exponents, positive or negative.

This of course includes all integral exponents, positive and negative. Also, we note that Eq. (3-9) is equivalent to Eq. (3-15) with $u = x$ (since $du/dx = 1$).

■**EXAMPLE 4**   We can now find the derivative of $y = \sqrt{x^2 + 1}$.
By using Eq. (3-16) or Eq. (3-15) and using the fact that $\sqrt{u} = u^{1/2}$ we can derive the result.

$$y = (x^2 + 1)^{1/2}$$
$$\frac{dy}{dx} = \frac{1}{2}(x^2 + 1)^{-1/2}(2x)$$
$$= \frac{x}{(x^2 + 1)^{1/2}}$$

To avoid introducing apparently significant factors into the numerator, we do not usually rationalize such fractions. ------------■

Note that we first rewrote the function in a different, more useful form. This is often an important step to perform before taking the derivative.

Having shown that we may use fractional exponents to find derivatives of roots of functions of x, we may also use them to find derivatives of roots of x itself. Consider the following example.

■**EXAMPLE 5**   Find the derivative of $y = 6\sqrt[3]{x^2}$.
We can write this function as $y = 6x^{2/3}$. In finding the derivative, we may use Eq. (3-9) with $n = \frac{2}{3}$. This gives us

$$y = 6x^{2/3}$$

$$\frac{dy}{dx} = 6\left(\frac{2}{3}\right)x^{-1/3} = \frac{4}{x^{1/3}} = \frac{4}{\sqrt[3]{x}}$$

with the bracket marking $\frac{2}{3} - 1$ pointing to the $-1/3$ exponent.

We could also use Eq. (3-15) with $u = x$ and $n = \frac{2}{3}$. This gives us

$$\frac{dy}{dx} = 6\left(\frac{2}{3}\right)x^{-1/3}(1) = \frac{4}{x^{1/3}} = \frac{4}{\sqrt[3]{x}}$$

with the bracket marking $\dfrac{du}{dx} = \dfrac{dx}{dx} = 1$ pointing to the (1) factor.

This shows us why Eq. (3-9) is equivalent to Eq. (3-15) with $u = x$.
Note that the domain of the function is all real numbers, but the function is not differentiable for $x = 0$. ------------■

In the following examples, we illustrate the use of Eqs. (3-9) and (3-15) for the case in which n is a negative exponent. *Special care must be taken in the case of a negative exponent,* so carefully note the caution in each example.

SOLVING A WORD PROBLEM

EXAMPLE 6   The electric resistance R of a wire varies inversely as the square of its radius r. For a given wire, $R = 4.66\ \Omega$ for $r = 0.150$ mm. Find the derivative of R with respect to r for this wire.

Since R varies inversely as the square of r, we have $R = k/r^2$. Then, using the fact that $R = 4.66\ \Omega$ for $r = 0.150$ mm, we have

$$4.66 = \frac{k}{(0.150)^2}, \qquad k = 0.105\ \Omega \cdot \text{mm}^2$$

which means that

$$R = \frac{0.105}{r^2}$$

We could find the derivative by the quotient rule. However, when the numerator is constant, the derivative is easily found by using negative exponents.

$$R = \frac{0.105}{r^2} = 0.105r^{-2}$$

CAUTION

$$\frac{dR}{dr} = 0.105(-2)r^{-3} \longleftarrow -2 - 1 = -3$$

$$= -\frac{0.210}{r^3}$$

Here we used Eq. (3-9) directly. ----------

EXAMPLE 7   Find the derivative of $y = \dfrac{1}{(1 - 4x)^5}$.

The derivative is found as follows:

$$y = \frac{1}{(1 - 4x)^5} = (1 - 4x)^{-5} \quad \text{use negative exponent}$$

$$\frac{dy}{dx} = (-5)(1 - 4x)^{-6}(-4) \qquad \text{use Eq. (3-15)}$$

$$= \frac{20}{(1 - 4x)^6} \qquad\qquad \text{express result with positive exponent}$$

CAUTION   Remember: *Subtracting* **1** *from* -5 *gives* -6. ----------

NOTE   We now see the value of fractional exponents in calculus. They are useful in many algebraic operations, but they are almost essential in calculus. Without fractional exponents, it would be necessary to develop additional formulas to find the derivatives of radical expressions. In order to find the derivative of an algebraic function, we need only those formulas we have already developed. Often it is necessary to combine these formulas, as we saw in Example 3. Actually, most derivatives are combinations. The problem in finding the derivative is **recognizing the form of the function** with which you are dealing. When you have recognized the form, completing the problem is only a matter of mechanics and algebra. You should now see the importance of being able to handle algebraic operations with ease.

EXAMPLE 8 Evaluate the derivative of

$$y = \frac{x}{\sqrt{1 - 4x}}$$

for $x = -2$.

Here, we have a quotient, and in order to find the derivative of this quotient, we must also use the power rule (and a derivative of a polynomial form). With sufficient practice in taking derivatives, we can recognize the rule to use almost automatically. Thus, we find the derivative:

$$\frac{dy}{dx} = \frac{(1 - 4x)^{1/2}(1) - x(\frac{1}{2})(1 - 4x)^{-1/2}(-4)}{1 - 4x}$$

$$= \frac{(1 - 4x)^{1/2} + \dfrac{2x}{(1 - 4x)^{1/2}}}{1 - 4x} = \frac{\dfrac{(1 - 4x)^{1/2}(1 - 4x)^{1/2} + 2x}{(1 - 4x)^{1/2}}}{1 - 4x}$$

$$= \frac{(1 - 4x) + 2x}{(1 - 4x)^{1/2}(1 - 4x)}$$

$$= \frac{1 - 2x}{(1 - 4x)^{3/2}}$$

Now, evaluating the derivative for $x = -2$, we have

$$\frac{dy}{dx}\bigg|_{x=-2} = \frac{1 - 2(-2)}{[1 - 4(-2)]^{3/2}} = \frac{1 + 4}{(1 + 8)^{3/2}} = \frac{5}{9^{3/2}}$$

$$= \frac{5}{27}$$

EXERCISES 3-7

In Exercises 1–24, find the derivative of each of the given functions.

1. $y = \sqrt{x}$

2. $y = \sqrt[4]{x^3}$

3. $v = \dfrac{3}{t^2}$

4. $y = \dfrac{2}{x^4}$

5. $y = \dfrac{3}{\sqrt[3]{x}}$

6. $y = \dfrac{1}{\sqrt[5]{x^2}}$

7. $y = x\sqrt{x} - \dfrac{1}{x}$

8. $f(x) = 2x^{-3} - 3x^{-2}$

9. $y = (x^2 + 1)^5$

10. $y = (1 - 2x)^4$

11. $y = 2(7 - 4x^3)^8$

12. $y = 3(8x^2 - 1)^6$

13. $y = (2x^3 - 3)^{1/3}$

14. $y = (1 - 6x)^{3/2}$

15. $f(y) = \dfrac{3}{(4 - y^2)^4}$

16. $y = \dfrac{4}{\sqrt{1 - 3x}}$

17. $y = 4(2x^4 - 5)^{3/4}$

18. $r = 5(3\theta^6 - 4)^{2/3}$

19. $y = \sqrt[4]{1 - 8x^2}$

20. $y = \sqrt[3]{4x^6 + 2}$

21. $y = x\sqrt{8x + 5}$

22. $y = x^2(1 - 3x)^5$

23. $f(R) = \sqrt{\dfrac{2R + 1}{4R + 1}}$

24. $y = \left(\dfrac{2x + 1}{3x - 2}\right)^2$

In Exercises 25–28, evaluate the derivatives of the given functions for the given values of x. In Exercises 27 and 28, check your results using the derivative evaluation feature of a graphing calculator.

25. $y = \sqrt{3x + 4}$, $x = 7$

26. $y = (4 - x^2)^{-1}$, $x = -1$

27. $y = \dfrac{\sqrt{x}}{1 - x}$, $x = 4$

28. $y = x^2\sqrt[3]{3x + 2}$, $x = 2$

In Exercises 29–44, solve the given problems by finding the appropriate derivatives.

29. Find the derivative of $y = 1/x^3$ as **(a)** a quotient and **(b)** a negative power of x and show that the results are the same.

30. Find the derivative of $y = \dfrac{2}{4x + 3}$ as **(a)** a quotient and **(b)** a negative power of $4x + 3$ and show that the results are the same.

31. Find any values of x for which the derivative of $y = \dfrac{x^2}{\sqrt{x^2 + 1}}$ is zero. View the graph of the function on a graphing calculator to verify the values found.

32. Find any values of x for which the derivative of $y = \dfrac{x}{\sqrt{4x - 1}}$ is zero. View the graph of the function on a graphing calculator to verify the values found.

33. Find the slope of a line tangent to the parabola $y^2 = 4x$ at the point $(1, 2)$. Use the derivative evaluation feature of a graphing calculator to check your result.

34. Find the slope of a line tangent to the circle $x^2 + y^2 = 25$ at the point $(4, 3)$. Use the derivative evaluation feature of a graphing calculator to check your result.

35. The displacement s (in cm) of a linkage joint of a robot is given by $s = (8t - t^2)^{2/3}$, where t is the time (in s). Find the velocity of the joint for $t = 6.25$ s.

36. Water is slowly rising in a horizontal drainage pipe. The width w of the water as a function of the depth h is $w = \sqrt{2rh - h^2}$, where r is the radius of the pipe. Find dw/dh for $h = 2.25$ in. and $r = 6.00$ in.

37. When the volume of a gas changes very rapidly, an approximate relation is that the pressure P varies inversely as the $3/2$ power of the volume. If P is 300 kPa when $V = 100$ cm³, find the derivative of P with respect to V. Evaluate this derivative for $V = 100$ cm³.

38. The power gain G of a certain antenna is inversely proportional to the square of the wavelength λ (in ft) of the carrier wave. If $G = 5.0 \times 10^4$ for $\lambda = 0.35$ ft, find the derivative of G with respect to λ for $\lambda = 0.35$ ft.

39. The total solar radiation H (in W/m²) on a certain surface during an average clear day is given by

$$H = \frac{4000}{\sqrt{t^6 + 100}}, \quad (-6 < t < 6)$$

where t is the number of hours from noon. Find the rate at which H is changing with time at 4 P.M.

40. In determining the time for a laser beam to go from S to P (see Fig. 3-30), which are in different mediums, it is necessary to find the derivative of the time

$$t = \frac{\sqrt{a^2 + x^2}}{v_1} + \frac{\sqrt{b^2 + (c - x)^2}}{v_2}$$

with respect to x, where a, b, c, v_1, and v_2 are constants. Here, v_1 and v_2 are the velocities of the laser in each medium. Find this derivative.

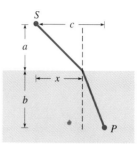

Fig. 3-30

41. The radio waveguide wavelength λ_r is related to its free-space wavelength λ by

$$\lambda_r = \frac{2a\lambda}{\sqrt{4a^2 - \lambda^2}}$$

where a is a constant. Find $d\lambda_r/d\lambda$.

42. The current I in a circuit containing a resistance R and an inductance L is found from the expression

$$I = \frac{V}{\sqrt{R^2 + (\omega L)^2}}$$

Find the expression for the instantaneous rate of change of current with respect to L, assuming that the other quantities remain constant.

43. The length l of a rectangular microprocessor chip is 2 mm longer than its width w. Find the derivative of the length of the diagonal D with respect to w.

44. The trapezoidal structure shown in Fig. 3-31 has an internal support of length l. Find the derivative of l with respect to x.

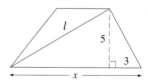

Fig. 3-31

3-8 DIFFERENTIATION OF IMPLICIT FUNCTIONS

To this point the functions we have differentiated have been of the form $y = f(x)$. There are, however, occasions when we need to find the derivative of a function determined by an equation that does not express the dependent variable explicitly in terms of the independent variable.

An equation in which y is not expressed explicitly in terms of x may determine one or more functions. *Any such function, where y is defined implicitly as a function of x, is called an* **implicit function.** Some equations defining implicit functions may be solved to determine the explicit functions, and for others it is not possible to solve for the explicit functions. Also, not all such equations define y as a function of x for real values of x.

EXAMPLE 1 **(a)** The equation $3x + 4y = 5$ is an equation that defines a function, although it is not in explicit form. In solving for y as $y = -\frac{3}{4}x + \frac{5}{4}$, we have the explicit form of the function.

(b) The equation $y^2 + x = 3$ is an equation that defines two functions, although we do not have the explicit forms. When we solve for y, we obtain the explicit functions $y = \sqrt{3 - x}$ and $y = -\sqrt{3 - x}$.

(c) The equation $y^5 + xy^2 + 3x^2 = 5$ defines y as a function of x, although we cannot actually solve for the explicit algebraic form of the function.

(d) The equation $x^2 + y^2 + 4 = 0$ is not satisfied by any pair of real values of x and y.

Even when it is possible to determine the explicit form of a function given in implicit form, it is not always desirable to do so. In some cases the implicit form is more convenient than the explicit form.

The derivative of an implicit function may be found directly without having to solve for the explicit function. Thus, *to find dy/dx when y is defined as an implicit function of x, we differentiate each term of the equation with respect to x, regarding y as a differentiable function of x.* We then solve for dy/dx, which will usually be in terms of x and y.

NOTE ▶

EXAMPLE 2 Find dy/dx if $y^2 + 2x^2 = 5$.

Here, we find the derivative of each term and then solve for dy/dx. Thus,

$$\frac{d(y^2)}{dx} + \frac{d(2x^2)}{dx} = \frac{d(5)}{dx}$$

$$2y^{2-1}\frac{dy}{dx} + 2\left(2x^{2-1}\frac{dx}{dx}\right) = 0$$

$$2y\frac{dy}{dx} + 4x = 0$$

$$\frac{dy}{dx} = -\frac{2x}{y}$$

For reference, Eq. (3-15) is $\frac{du^n}{dx} = nu^{n-1}\frac{du}{dx}$.

CAUTION ▶ ***The factor dy/dx arises from the derivative of the first term as a result of using the derivative of a power of a function of x (Eq. 3-15). The factor dy/dx corresponds to the du/dx of the formula.*** In the second term, no factor of dy/dx appears, since there are no y factors in the term.

▌EXAMPLE 3   Find dy/dx if $3y^4 + xy^2 + 2x^3 - 6 = 0$.

In finding the derivative, we note that the second term is a product, and we must use the product rule for derivatives on it. Thus, we have

$$\frac{d(3y^4)}{dx} + \frac{d(xy^2)}{dx} + \frac{d(2x^3)}{dx} - \frac{d(6)}{dx} = \frac{d(0)}{dx}$$

using product rule

$$12y^3\frac{dy}{dx} + \left[x\left(2y\frac{dy}{dx}\right) + y^2(1)\right] + 6x^2 - 0 = 0$$

$$12y^3\frac{dy}{dx} + 2xy\frac{dy}{dx} + y^2 + 6x^2 = 0 \qquad \text{solve for } \frac{dy}{dx}$$

$$(12y^3 + 2xy)\frac{dy}{dx} = -y^2 - 6x^2$$

$$\frac{dy}{dx} = \frac{-y^2 - 6x^2}{12y^3 + 2xy}$$

▌EXAMPLE 4   Find dy/dx if $2x^3y + (y^2 + x)^3 = x^4$.

In this case we use the product rule on the first term and the power rule on the second term.

$$\frac{d(2x^3y)}{dx} + \frac{d(y^2 + x)^3}{dx} = \frac{d(x^4)}{dx}$$

product ——— power

$$2x^3\left(\frac{dy}{dx}\right) + y(6x^2) + 3(y^2 + x)^2\left(2y\frac{dy}{dx} + 1\right) = 4x^3$$

$$2x^3\frac{dy}{dx} + 6x^2y + 3(y^2 + x)^2\left(2y\frac{dy}{dx}\right) + 3(y^2 + x)^2 = 4x^3$$

$$[2x^3 + 6y(y^2 + x)^2]\frac{dy}{dx} = 4x^3 - 6x^2y - 3(y^2 + x)^2$$

$$\frac{dy}{dx} = \frac{4x^3 - 6x^2y - 3(y^2 + x)^2}{2x^3 + 6y(y^2 + x)^2}$$

▌EXAMPLE 5   Find the slope of a line tangent to the graph of $2y^3 + xy + 1 = 0$ at the point $(-3, 1)$.

Here, we must find dy/dx and evaluate it for $x = -3$ and $y = 1$.

$$\frac{d(2y^3)}{dx} + \frac{d(xy)}{dx} + \frac{d(1)}{dx} = \frac{d(0)}{dx}$$

$$6y^2\frac{dy}{dx} + x\frac{dy}{dx} + y + 0 = 0$$

$$\frac{dy}{dx} = \frac{-y}{6y^2 + x}$$

$$\frac{dy}{dx}\bigg|_{(-3,1)} = \frac{-1}{6(1^2) - 3} = \frac{-1}{6 - 3} = -\frac{1}{3}$$

Thus, the slope is $-\frac{1}{3}$.

—————————————— EXERCISES $3\text{-}8$ ——————————————

In Exercises 1–20, find dy/dx by differentiating implicitly. When applicable, express the result in terms of x and y.

1. $3x + 2y = 5$ **2.** $6x - 3y = 4$

3. $4y - 3x^2 = x$ **4.** $x^5 - 5y = 6 - x$

5. $x^2 - 4y^2 - 9 = 0$ **6.** $x^2 + 2y^2 - 11 = 0$

7. $y^5 = x^2 - 1$ **8.** $y^4 = 3x^3 - x$

9. $y^2 + y = x^2 - 4$ **10.** $2y^3 - y = 7 - x^4$

11. $y + 3xy - 4 = 0$ **12.** $8y - xy - 7 = 0$

13. $xy^3 + 3y + x^2 = 9$ **14.** $y^2x - \dfrac{5y}{x + 1} + 3x = 4$

15. $\dfrac{3x^2}{y^2 + 1} + y = 3x + 1$ **16.** $2x - x^3y^2 = y - x^2 - 1$

17. $(2y - x)^4 + x^2 = y + 3$

18. $(y^2 + 2)^3 = x^4y + 11$

19. $2(x^2 + 1)^3 + (y^2 + 1)^2 = 17$

20. $(2x + 1)(1 - 3y) + y^2 = 13$

In Exercises 21–24, evaluate the derivatives of the given functions at the given points.

21. $3x^3y^2 - 2y^3 = -4;$ $(1, 2)$

22. $2y + 5 - x^2 - y^3 = 0;$ $(2, -1)$

23. $5y^4 + 7 = x^4 - 3y;$ $(3, -2)$

24. $(xy - y^2)^3 = 5y^2 + 22;$ $(4, 1)$

In Exercises 25–32, solve the given problems by using implicit differentiation.

25. Find the slope of a line tangent to the curve that represents the implicit function $xy + y^2 + 2 = 0$ at the point $(-3, 1)$. Use the derivative evaluation feature of a graphing calculator to check your result.

26. Oil moves through a pipeline such that the distance s it moves and the time t are related by $s^3 - t^2 = 7t$. Find the velocity of the oil for $s = 4.01$ m and $t = 5.25$ s.

27. The shelf support shown in Fig. 3-32 is 2.38 ft long. Find the expression for dy/dx in terms of x and y.

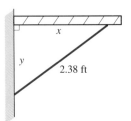

Fig. 3-32

28. An open (no top) right circular cylindrical container of radius r and height h has a total surface area of 940 cm^2. Find dr/dh in terms of r and h.

29. Two resistors, with resistances r and $r + 2$, are connected in parallel. Their combined resistance R is related to r by the equation $r^2 = 2rR + 2R - 2r$. Find dR/dr.

30. The polar moment of inertia I of a rectangular slab of concrete is given by $I = \frac{1}{12}(b^3h + bh^3)$, where b and h are the base and the height, respectively, of the slab. If I is constant, find the expression for db/dh.

31. A formula relating the length L and radius of gyration r of a steel column is $24C^3Sr^3 = 40C^3r^3 + 9LC^2r^2 - 3L^3$, where C and S are constants. Find dL/dr.

32. A computer is programmed to draw the graph of the implicit function $(x^2 + y^2)^3 = 64x^2y^2$. The graph is shown in Fig. 3-33. Find the slope of a line tangent to this curve at $(2.00, 0.56)$ and at $(2.00, 3.07)$.

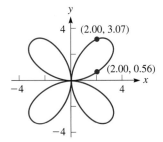

Fig. 3-33

$3\text{-}9$ HIGHER DERIVATIVES

In the previous sections of this chapter we developed certain important formulas for finding the derivative of a function. The derivative itself is a function, and we may therefore take its derivative. In technology and in mathematics, many applications require the use of the derivative of a derivative. Therefore, in this section we develop the concept and necessary notation, as well as show some of the applications.

THE SECOND DERIVATIVE

The derivative of a function is called the **first derivative** *of the function. The derivative of the first derivative is called the* **second derivative.** Since the second derivative is a function, we may find its derivative, which is called the third derivative. We may continue to find the fourth derivative, fifth derivative, and so on (provided each derivative is defined). *The second derivative, third derivative, and so on, are known as* **higher derivatives.**

The notations used for higher derivatives follow closely those used for the first derivative. As shown in Section 3-3, the notations for the first derivative are y', $D_x y$, $f'(x)$, and dy/dx. The notations for the second derivative are y'', $D_x^2 y$, $f''(x)$, and d^2y/dx^2. Similar notations are used for other higher derivatives.

■**EXAMPLE 1**   Find the higher derivatives of $y = 5x^3 - 2x$.
We find the first derivative as

$$\frac{dy}{dx} = 15x^2 - 2 \quad \text{or} \quad y' = 15x^2 - 2$$

Next, we obtain the second derivative by finding the derivative of the first derivative:

$$\frac{d^2y}{dx^2} = 30x \quad \text{or} \quad y'' = 30x$$

Continuing to find the successive derivatives, we have

$$\frac{d^3y}{dx^3} = 30 \quad \text{or} \quad y''' = 30$$

$$\frac{d^4y}{dx^4} = 0 \quad \text{or} \quad y^{(4)} = 0$$

Since the third derivative is a constant, the fourth derivative and all successive derivatives will be zero. This can be shown as $d^n y/dx^n = 0$ for $n \geq 4$. ■

■**EXAMPLE 2**   Find the higher derivatives of $f(x) = x(x^2 - 1)^2$.
Using the product rule, Eq. (3-12), to find the first derivative, we have

$$f'(x) = x(2)(x^2 - 1)(2x) + (x^2 - 1)^2(1)$$
$$= (x^2 - 1)(4x^2 + x^2 - 1) = (x^2 - 1)(5x^2 - 1)$$
$$= 5x^4 - 6x^2 + 1$$

For reference, Eq. (3-12) is
$$\frac{d(uv)}{dx} = u\frac{dv}{dx} + v\frac{du}{dx}.$$

Continuing to find the higher derivatives, we have

$$f''(x) = 20x^3 - 12x$$
$$f'''(x) = 60x^2 - 12$$
$$f^{(4)}(x) = 120x$$
$$f^{(5)}(x) = 120$$
$$f^{(n)}(x) = 0 \quad \text{for } n \geq 6$$

Note that when using the prime ($f'(x)$) notation the nth derivative may be shown as $f^{(n)}(x)$.

All derivatives after the fifth derivative are equal to zero. ■

EXAMPLE 3   Evaluate the second derivative of $y = \dfrac{2}{1-x}$ for $x = -2$.

We write the function as $y = 2(1 - x)^{-1}$ and then find the derivatives.

$$y = 2(1 - x)^{-1}$$

$$\frac{dy}{dx} = 2(-1)(1 - x)^{-2}(-1) = 2(1 - x)^{-2}$$

$$\frac{d^2y}{dx^2} = 2(-2)(1 - x)^{-3}(-1) = 4(1 - x)^{-3} = \frac{4}{(1 - x)^3}$$

Evaluating the second derivative for $x = -2$, we have

$$\frac{d^2y}{dx^2}\bigg|_{x=-2} = \frac{4}{(1 + 2)^3} = \frac{4}{27}$$

The function is not differentiable for $x = 1$. Also, if we continue to find higher derivatives, the expressions will not become zero, as in Examples 1 and 2. ∎

EXAMPLE 4   Find y'' for the implicit function defined by $2x^2 + 3y^2 = 6$.
Differentiating with respect to x, we have

$$2(2x) + 3(2yy') = 0$$
$$4x + 6yy' = 0 \quad \text{or} \quad 2x + 3yy' = 0 \tag{1}$$

CAUTION ▸   Before differentiating again we see that **$3yy'$ is a product,** and we note that the derivative of y' is y''. Thus, differentiating again, we have

$$\overbrace{2 + 3yy'' + 3y'(y')}^{\text{differentiation of } 3yy'} = 0$$
$$2 + 3yy'' + 3(y')^2 = 0 \tag{2}$$

Now, solving Eq. (1) for y' and substituting this into Eq. (2), we have

$$y' = -\frac{2x}{3y}$$

$$2 + 3yy'' + 3\left(-\frac{2x}{3y}\right)^2 = 0$$

$$2 + 3yy'' + \frac{4x^2}{3y^2} = 0$$

$$6y^2 + 9y^3y'' + 4x^2 = 0$$

$$y'' = \frac{-4x^2 - 6y^2}{9y^3} = \frac{-2(2x^2 + 3y^2)}{9y^3}$$

Since $2x^2 + 3y^2 = 6$, we have

$$y'' = \frac{-2(6)}{9y^3} = -\frac{4}{3y^3} \qquad ∎$$

As mentioned earlier, higher derivatives are useful in certain applications. This is particularly true of the second derivative. The first and second derivatives are used in the next chapter for several types of applications, and higher derivatives are used when we discuss infinite series in Chapter 13. An important technical application of the second derivative is shown in the example that follows.

In Section 3-4 we briefly discussed the instantaneous velocity of an object, and in the exercises we mentioned acceleration. From that discussion we recall that the instantaneous velocity is the time rate of change of the displacement, and *the instantaneous acceleration is the time rate of change of the instantaneous velocity.* Therefore, we see that the *acceleration is found from the second derivative of the displacement with respect to time.* Consider the following example.

SOLVING A WORD PROBLEM

■**EXAMPLE 5** For the first 12 s after launch, the height s (in m) of a certain rocket is given by $s = 10\sqrt{t^4 + 25} - 50$. Find the vertical acceleration of the rocket when $t = 10.0$ s.

Since the velocity is found from the first derivative and the acceleration is found from the second derivative, we must find the second derivative and then evaluate it for $t = 10.0$ s.

$$s = 10\sqrt{t^4 + 25} - 50$$

$$v = \frac{ds}{dt} = 10\left(\frac{1}{2}\right)(t^4 + 25)^{-1/2}(4t^3) = \frac{20t^3}{(t^4 + 25)^{1/2}}$$

$$a = \frac{dv}{dt} = \frac{d^2s}{dt^2} = \frac{(t^4 + 25)^{1/2}(60t^2) - 20t^3(\frac{1}{2})(t^4 + 25)^{-1/2}(4t^3)}{t^4 + 25}$$

multiply numerator and denominator by $(t^4 + 25)^{1/2}$

$$= \frac{(t^4 + 25)(60t^2) - 40t^6}{(t^4 + 25)^{3/2}} = \frac{20t^6 + 1500t^2}{(t^4 + 25)^{3/2}}$$

$$= \frac{20t^2(t^4 + 75)}{(t^4 + 25)^{3/2}}$$

Finding the value of the acceleration when $t = 10.0$ s, we have

$$a\Big|_{t=10.0} = \frac{20(10.0^2)(10.0^4 + 75)}{(10.0^4 + 25)^{3/2}} = 20.1 \text{ m/s}^2$$

───────── EXERCISES 3-9 ─────────

In Exercises 1–8, find all the higher derivatives of the given functions.

1. $y = x^3 + x^2$
2. $f(x) = 3x - x^4$
3. $f(x) = x^3 - 6x^4$
4. $s = 2t^5 + 5t^4$
5. $y = (1 - 2x)^4$
6. $f(x) = (3x + 2)^3$
7. $f(r) = r(4r + 1)^3$
8. $y = x(x - 1)^3$

In Exercises 9–28, find the second derivative of each of the given functions.

9. $y = 2x^7 - x^6 - 3x$
10. $y = 6x - 2x^5$
11. $y = 2x + \sqrt{x}$
12. $r = 3\theta^2 - \dfrac{1}{2\sqrt{\theta}}$
13. $f(x) = \sqrt[4]{8x - 3}$
14. $f(x) = \sqrt[3]{6x + 5}$
15. $f(p) = \dfrac{4}{\sqrt{1 + 2p}}$
16. $f(x) = \dfrac{5}{\sqrt{3 - 4x}}$
17. $y = 2(2 - 5x)^4$
18. $y = (4x + 1)^6$
19. $y = (3x^2 - 1)^5$
20. $y = 3(2x^3 + 3)^4$

21. $f(x) = \dfrac{2x}{1 - x}$
22. $f(R) = \dfrac{1 - 3R}{1 + 3R}$
23. $y = \dfrac{x^2}{x + 1}$
24. $y = \dfrac{x}{\sqrt{1 - x^2}}$
25. $x^2 - y^2 = 9$
26. $xy + y^2 = 4$
27. $x^2 - xy = 1 - y^2$
28. $xy = y^2 + 1$

In Exercises 29–34, evaluate the second derivative of the given function for the given value of x.

29. $f(x) = \sqrt{x^2 + 9}$, $x = 4$
30. $f(x) = x - \dfrac{2}{x^3}$, $x = -1$
31. $y = 3x^{2/3} - \dfrac{2}{x}$, $x = -8$
32. $y = 3(1 + 2x)^4$, $x = \dfrac{1}{2}$
33. $y = x(1 - x)^5$, $x = 2$
34. $y = \dfrac{x}{2 - 3x}$, $x = -\dfrac{1}{3}$

In Exercises 35–40, solve the given problems by finding the appropriate derivatives.

35. What is the instantaneous rate of change of the first derivative of y with respect to x for $y = (1 - 2x)^4$ for $x = 1$?

36. The deflection y (in m) of a 5.00-m beam as a function of the distance x (in m) from one end is $y = 0.0001(x^5 - 25x^2)$. Find the value of d^2y/dx^2 (the rate of change at which the slope of the beam changes) where $x = 3.00$ m.

37. A bullet is fired vertically upward. Its distance s (in ft) above the ground is given by $s = 2250t - 16.1t^2$, where t is the time (in s). Find the acceleration of the bullet.

38. In testing the brakes on a new model automobile, it was found that the distance s (in ft) that it traveled under specified conditions after the brakes were applied was given by $s = 57.6t - 1.20t^3$. What were the velocity and the acceleration of the automobile for $t = 4.00$ s?

39. The voltage V induced in an inductor in an electric circuit is given by $V = L(d^2q/dt^2)$, where L is the inductance (in H). Find the expression for the voltage induced in a 1.60-H inductor if $q = \sqrt{2t + 1} - 1$.

40. How fast is the rate of change of solar radiation changing on the surface in Exercise 27 of Section 3-4 at 3 P.M.?

CHAPTER EQUATIONS

Limit of function	$\lim_{x \to a} f(x) = L$	(3-1)
Increments	$\Delta x = x_2 - x_1$	(3-2)
	$\Delta y = y_2 - y_1$	(3-3)
Slope	$m_{PQ} = \dfrac{(y_1 + \Delta y) - y_1}{(x_1 + \Delta x) - x_1} = \dfrac{\Delta y}{\Delta x}$	(3-4)
	$\Delta y = f(x + \Delta x) - f(x)$	(3-5)
Definition of derivative	$\lim_{\Delta x \to 0} \dfrac{\Delta y}{\Delta x} = \lim_{\Delta x \to 0} \dfrac{f(x + \Delta x) - f(x)}{\Delta x}$	(3-6)
Instantaneous velocity	$v = \lim_{\Delta t \to 0} \dfrac{\Delta s}{\Delta t}$	(3-7)
Derivatives of polynomials	$\dfrac{dc}{dx} = 0$	(3-8)
	$\dfrac{dx^n}{dx} = nx^{n-1}$	(3-9)
	$\dfrac{d(cu)}{dx} = c\dfrac{du}{dx}$	(3-10)
	$\dfrac{d(u + v)}{dx} = \dfrac{du}{dx} + \dfrac{dv}{dx}$	(3-11)
Derivative of product	$\dfrac{d(uv)}{dx} = u\dfrac{dv}{dx} + v\dfrac{du}{dx}$	(3-12)
Derivative of quotient	$\dfrac{d\left(\dfrac{u}{v}\right)}{dx} = \dfrac{v\dfrac{du}{dx} - u\dfrac{dv}{dx}}{v^2}$	(3-13)
Chain rule	$\dfrac{dy}{dx} = \dfrac{dy}{du}\dfrac{du}{dx}$	(3-14)
Derivative of power	$\dfrac{du^n}{dx} = nu^{n-1}\left(\dfrac{du}{dx}\right)$	(3-15)
	$\dfrac{du^{p/q}}{dx} = \dfrac{p}{q}u^{(p/q)-1}\dfrac{du}{dx}$	(3-16)

REVIEW EXERCISES

In Exercises 1–12, evaluate the given limits.

1. $\lim\limits_{x \to 4} (8 - 3x)$

2. $\lim\limits_{x \to 3} (2x^2 - 10)$

3. $\lim\limits_{x \to -3} \dfrac{2x + 5}{x - 1}$

4. $\lim\limits_{x \to 1} (x - 1)\sqrt{x^2 + 9}$

5. $\lim\limits_{x \to 2} \dfrac{4x - 8}{x^2 - 4}$

6. $\lim\limits_{x \to 5} \dfrac{x^2 - 25}{3x - 15}$

7. $\lim\limits_{x \to 2} \dfrac{x^2 + 3x - 10}{x^2 - x - 2}$

8. $\lim\limits_{x \to 0} \dfrac{(x - 3)^2 - 9}{x}$

9. $\lim\limits_{x \to \infty} \dfrac{2 + \dfrac{1}{x + 4}}{3 - \dfrac{1}{x^2}}$

10. $\lim\limits_{x \to \infty} \left(7 - \dfrac{1}{x + 1}\right)$

11. $\lim\limits_{x \to \infty} \dfrac{x - 2x^3}{(1 + x)^3}$

12. $\lim\limits_{x \to \infty} \dfrac{2x + 5}{3x^3 - 2x}$

In Exercises 13–20, use the delta-process to find the derivative of each of the given functions.

13. $y = 7 + 5x$

14. $y = 6x - 2$

15. $y = 6 - 2x^2$

16. $y = 2x^2 - x^3$

17. $y = \dfrac{2}{x^2}$

18. $y = \dfrac{1}{1 - 4x}$

19. $y = \sqrt{x + 5}$

20. $y = \dfrac{1}{\sqrt{x}}$

In Exercises 21–36, find the derivative of each of the given functions.

21. $y = 2x^7 - 3x^2 + 5$

22. $y = 8x^7 - 2^5 - x$

23. $y = 4\sqrt{x} - \dfrac{3}{x} + \sqrt{3}$

24. $y = \dfrac{3}{x^2} - 8\sqrt[4]{x}$

25. $f(y) = \dfrac{3y}{1 - 5y}$

26. $y = \dfrac{2x - 1}{x^2 + 1}$

27. $y = (2 - 3x)^4$

28. $y = (2x^2 - 3)^6$

29. $y = \dfrac{3}{(5 - 2x^2)^{3/4}}$

30. $f(Q) = \dfrac{70}{(3Q + 1)^3}$

31. $v = \sqrt{1 + \sqrt{1 + \sqrt{1 + 8s}}}$

32. $y = (x - 1)^3(x^2 - 2)^2$

33. $y = \dfrac{\sqrt{4x + 3}}{2x}$

34. $R = \dfrac{\sqrt{t + 1}}{\sqrt{t - 1}}$

35. $(2x - 3y)^3 = x^2 - y$

36. $x^2 y^2 = x^2 + y^2$

In Exercises 37–40, evaluate the derivatives of the given functions for the given values of x. Check your results using the derivative evaluation feature of a graphing calculator.

37. $y = \dfrac{4}{x} + 2\sqrt[3]{x}$, $x = 8$

38. $y = (3x - 5)^4$, $x = -2$

39. $y = 2x\sqrt{4x + 1}$, $x = 6$

40. $y = \dfrac{\sqrt{2x^2 + 1}}{3x}$, $x = 2$

In Exercises 41–44, find the second derivative of each of the given functions.

41. $y = 3x^4 - \dfrac{1}{x}$

42. $y = \sqrt{1 - 8x}$

43. $y = \dfrac{1 - 3x}{1 + 4x}$

44. $y = 2x(6x + 5)^4$

In Exercises 45–72, solve the given problems.

(W) 45. View the graph of $y = \dfrac{2(x^2 - 4)}{x - 2}$ on a graphing calculator with window values such that y can be evaluated exactly for $x = 2$. (Xmin $= -1$ (or 0), Xmax $= 4$, Ymin $= 0$, Ymax $= 10$ will probably work.) Using the *trace* feature, determine the value of y for $x = 2$. Comment on the accuracy of the view and the value found.

(W) 46. A continuous function $f(x)$ is positive at $x = 0$ and negative for $x = 1$. How many solutions does $f(x) = 0$ have between $x = 0$ and $x = 1$? Explain.

47. The velocity v (in ft/s) of a weight falling in water is given by $v = \dfrac{6(t + 5)}{t + 1}$, where t is the time (in s). What are (a) the initial velocity and (b) the terminal velocity (as $t \to \infty$)?

48. Two lenses of focal lengths f_1 and f_2, separated by a distance d, are used in the study of lasers. The combined focal length f of this lens combination is $f = \dfrac{f_1 f_2}{f_1 + f_2 - d}$. If f_2 and d remain constant, find the limiting value of f as f_1 continues to increase in value.

49. Find the slope of a line tangent to the graph of $y = 7x^4 - x^3$ at $(-1, 8)$. Use the derivative evaluation feature of a graphing calculator to check your result.

50. Find the slope of a line tangent to the graph of $y = \sqrt[3]{3 - 8x}$ at $(-3, 3)$. Use the derivative evaluation feature of a graphing calculator to check your result.

51. The cable of a 200-m suspension bridge can be represented by $y = 0.0015x^2 + C$. At one point the tension is directed along the line $y = 0.3x - 10$. Find the value of C.

52. The displacement s (in cm) of a piston during each 8-second cycle is given by $s = 8t - t^2$, where t is the time (in s). For what value(s) of t is the velocity of the piston 4 cm/s?

53. The reliability R of a computer system measures the probability that the system will be operating properly after t hours. For one system, $R = 1 - kt + \dfrac{k^2t^2}{2} - \dfrac{k^3t^3}{6}$, where k is a constant. Find the expression for the instantaneous rate of change of R with respect to t.

54. The distance s (in ft) traveled by a subway train after the brakes are applied is given by $s = 40t - 5t^2$. How far does it travel, after the brakes are applied, in coming to a stop?

55. The electric field E at a distance r from a point charge is $E = k/r^2$, where k is a constant. Find an expression for the instantaneous rate of change of the electric field with respect to r.

56. The velocity of an object moving with constant acceleration can be found from the equation $v = \sqrt{v_0^2 + 2as}$, where v_0 is the initial velocity, a is the acceleration, and s is the distance traveled. Find dv/ds.

57. The voltage induced in an inductor L is given by $E = L(dI/dt)$, where I is the current in the circuit and t is the time. Find the voltage induced in a 0.4-H inductor if the current I (in A) is related to the time (in s) by $I = t(0.01t + 1)^3$.

58. In studying the energy used by a mechanical robotic device, the equation $v = \dfrac{z}{\alpha(1 - z^2) - \beta}$ is used. If α and β are constants, find dv/dz.

59. The frictional radius r_f of a collar used in a braking system is given by $r_f = \dfrac{2(R^3 - r^3)}{3(R^2 - r^2)}$, where R is the outer radius and r is the inner radius. Find dr_f/dR if r is constant.

60. Water is being drained from a pond such that the volume V (in m^3) of water in the pond after t hours is given by $V = 5000(60 - t)^2$. Find the rate at which the pond is being drained after 4.00 h.

61. The frequency f of a certain electronic oscillator is given by $f = \dfrac{1}{2\pi\sqrt{C(L + 2)}}$, where C is a capacitance and L is an inductance. If C is constant, find df/dL.

62. The volume V of fluid produced in the retina of the eye in reaction to exposure to light of intensity I is given by $V = \dfrac{aI^2}{b - I}$, where a and b are constants. Find dV/dI.

63. The temperature T (in °C) in a freezer as a function of the time t (in h) is given by $T = \dfrac{10(1 - t)}{0.5t + 1}$. Find dT/dt.

64. Under certain conditions, the efficiency e (in %) of an internal combustion engine is given by
$$e = 100\left(1 - \dfrac{1}{(V_1/V_2)^{0.4}}\right)$$
where V_1 and V_2 are the maximum and minimum volumes of air in a cylinder, respectively. Assuming that V_2 is kept constant, find the expression for the instantaneous rate of change of efficiency with respect to V_1.

65. The deflection y of a cantilever beam (clamped at one end and free at the other end) is $y = \dfrac{w}{24EI}(6L^2x^2 - 4Lx^3 + x^4)$. Here, L is the length of the beam, and w, E, and I are constants. Find the first four derivatives of y with respect to x. (Each of these derivatives is useful in analyzing the properties of the beam.)

66. The number n of grams of a compound formed during a certain chemical reaction is given by $n = \dfrac{2t}{t + 1}$, where t is the time (in min). Evaluate d^2n/dt^2 (the rate of increase of the rate at which the compound is being formed) when $t = 4.00$ min.

67. The area of a rectangular patio is to be 75 m^2. Express the perimeter p of the patio as a function of its width w and find dp/dw.

68. A water tank is being designed in the shape of a right circular cylinder with a volume of 100 ft^3. Find the expression for the instantaneous rate of change of the total surface area A of the tank with respect to the radius r of the base.

69. An arch over a walkway can be described by the first-quadrant part of the parabola $y = 4 - x^2$. In order to determine the size and shape of rectangular objects that can pass under the arch, express the area A of a rectangle inscribed under the parabola in terms of x. Find dA/dx.

70. An airplane flies over an observer with a velocity of 400 mi/h and at an altitude of 2640 ft. If the plane flies horizontally in a straight line, find the rate at which the distance x from the observer to the plane is changing 0.600 min after the plane passes over the observer. See Fig. 3-34.

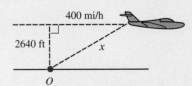

Fig. 3-34

71. A computer analysis showed that a specialized piece of machinery has a value (in dollars) given by $V = 1,500,000/(2t + 10)$, where t is the number of years after the purchase. Calculate the value of dV/dt and d^2V/dt^2 for $t = 5$ years. What is the meaning of these values?

72. The *radius of curvature* at the point (x, y) on the graph of $y = f(x)$ is given by

$$R = \frac{[1 + (y')^2]^{3/2}}{|y''|}$$

A certain roadway follows the parabola $y = 1.2x - x^2$ for $0 < x < 1.2$, where x is measured in miles. Find R for $x = 0.2$ mi and $x = 0.6$ mi. See Fig. 3-35.

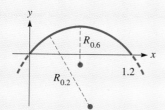

Fig. 3-35

Writing Exercise

73. An engineer designing military rockets uses computer simulation to find the path of a rocket as $y = f(x)$ and the path of an aircraft to be $y = g(x)$. Write two or three paragraphs explaining how the engineer can determine the angle at which the path of the rocket crosses the path of the aircraft.

PRACTICE TEST

1. Find $\lim\limits_{x \to 1} \dfrac{x^2 - x}{x^2 - 1}$.

2. Find $\lim\limits_{x \to \infty} \dfrac{1 - 4x^2}{x + 2x^2}$.

3. Find the slope of a line tangent to the graph of $y = 3x^2 - \dfrac{4}{x^2}$ at $(2, 11)$. Check your result using the derivative evaluation feature of a graphing calculator. Write down the complete value shown on the calculator.

4. The displacement s (in cm) of a pumping machine piston in each cycle is given by $s = t\sqrt{10 - 2t}$, where t is the time (in s). Find the velocity of the piston for $t = 4.00$ s.

5. Find dy/dx: $(1 + y^2)^3 - x^2y = 7x$.

6. Under certain conditions, due to the presence of a charge q, the electric potential V along a line is given by

$$V = \frac{kq}{\sqrt{x^2 + b^2}}$$

where k is a constant and b is the minimum distance from the charge to the line. Find the expression for the instantaneous rate of change of V with respect to x.

7. Find the second derivative of $y = \dfrac{2x}{3x + 2}$.

8. By using the delta-process, find the derivative of $y = 5x - 2x^2$ with respect to x.

APPLICATIONS OF
THE DERIVATIVE

In Section 4-7 we see how to use the derivative in the design of cylindrical containers such as oil storage tanks.

In Chapter 3 we developed the meaning of the derivative of a function and then went on to find several formulas by which we can differentiate functions. We also established the important concept of the derivative as an instantaneous rate of change.

Numerous applications of the derivative were indicated in the examples and exercises of Chapter 3. It was not necessary to develop any of these applications in detail, since finding the derivative was the primary concern. There are, however, certain types of problems in many areas of technology in which the derivative plays a key role in the solution.

These applications include the analysis of the motion of objects and finding the maximum values or minimum values of functions. Such values are useful, for example, in finding the maximum possible income from production or the least amount of material needed in making a product. In this chapter we consider many of these types of applications of the derivative.

4-1 TANGENTS AND NORMALS

The first application of the derivative we consider involves finding the equation of a line that is *tangent* to a given curve and the equation of a line that is *normal* (perpendicular) to a given curve.

Tangent Line

To find the equation of a line tangent to a curve at a given point, we first find the derivative of the function. The derivative is then evaluated at the point, and this gives us the slope of a line tangent to the curve at the point. Then, by using the point-slope form of the equation of a straight line, we find the equation of the tangent line. The following examples illustrate the method.

EXAMPLE 1 Find the equation of the line tangent to the parabola $y = x^2 - 1$ at the point $(-2, 3)$.

The derivative of this function is

$$\frac{dy}{dx} = 2x$$

The value of this derivative for the value $x = -2$ (the y-value of 3 is not used since the derivative does not directly contain y) is

$$\left.\frac{dy}{dx}\right|_{x=-2} = -4$$

which means that the slope of the tangent line at $(-2, 3)$ is -4. Thus, by using the point-slope form of the equation of the straight line we obtain the desired equation.

$$y - 3 = -4(x + 2)$$
$$y - 3 = -4x - 8$$
$$y = -4x - 5$$

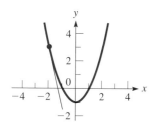

Fig. 4-1

The parabola and the tangent line $y = -4x - 5$ are shown in Fig. 4-1.

EXAMPLE 2 Find the equation of the line tangent to the ellipse $4x^2 + 9y^2 = 40$ at the point $(1, 2)$.

The easiest method of finding the derivative of the function defined by this equation is to treat the equation as an implicit function. In this way we have the following solution:

$$8x + 18yy' = 0 \qquad \text{differentiate implicitly}$$

$$y' = -\frac{4x}{9y}$$

$$y'|_{(1,2)} = -\frac{4}{18} = -\frac{2}{9} \qquad \begin{array}{l}\text{evaluate derivative to find slope}\\\text{of tangent line}\end{array}$$

$$y - 2 = -\frac{2}{9}(x - 1) \qquad \text{point-slope form of tangent line}$$

$$9y - 18 = -2x + 2$$

$$2x + 9y - 20 = 0 \qquad \text{standard form of tangent line}$$

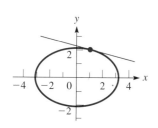

Fig. 4-2

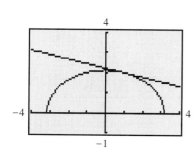

Fig. 4-3

The ellipse and the tangent line $2x + 9y - 20 = 0$ are shown in Fig. 4-2.

To display the function and the tangent line by using the *tangent* feature of a graphing calculator, we would solve for y. Since the point $(1, 2)$ is on the upper half of the ellipse, we would use $y = \frac{2}{3}\sqrt{10 - x^2}$. See Fig. 4-3 for the display. (We could also use the derivative of this function to find the equation of the tangent line.)

Normal Line

About 1700 the word *normal* was adapted from the Latin word *normalis*, which was being used for *perpendicular*.

NOTE ▶

If we wish to obtain the equation of a line normal (perpendicular to a tangent) to a curve, we recall that the slopes of perpendicular lines are negative reciprocals. Thus, the derivative is found and evaluated at the specified point. Since this gives the slope of a tangent line, *we take the negative reciprocal of this number to find the slope of the normal line.* Then, by using the point-slope form of the equation of a straight line, we find the equation of the normal. The following examples illustrate the method.

█EXAMPLE 3   Find the equation of the line normal to the hyperbola $y = 2/x$ at the point $(2, 1)$.

Taking the derivative of this function and evaluating it for $x = 2$, we have

$$\frac{dy}{dx} = -\frac{2}{x^2}; \qquad \left.\frac{dy}{dx}\right|_{x=2} = -\frac{1}{2}$$

Therefore, the slope of a line normal to the curve at $(2, 1)$ is 2. The equation of the normal line is then

$$y - 1 = 2(x - 2)$$

or

$$y = 2x - 3$$

The hyperbola and the normal line are shown in the graphing calculator display shown in Fig. 4-4.   ■

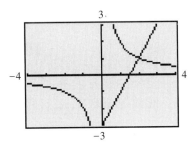

Fig. 4-4

█EXAMPLE 4   Find the y-intercept of the line that is normal to the graph of $y = 2x - \frac{1}{3}x^3$, where $x = 3$.

First, we find the equation of the normal line. We find the y-intercept by writing the equation in slope-intercept form. The solution proceeds as follows:

$$\frac{dy}{dx} = 2 - x^2 \qquad \text{find derivative}$$

$$\left.\frac{dy}{dx}\right|_{x=3} = 2 - 3^2 = -7 \qquad \text{evaluate derivative}$$

$$m_{\text{norm}} = \tfrac{1}{7} \qquad \text{negative reciprocal}$$

$$y|_{x=3} = 2(3) - \tfrac{1}{3}(3^3) = -3 \qquad \text{find } y\text{-coordinate of point}$$

$$y - (-3) = \tfrac{1}{7}(x - 3) \qquad \text{point-slope form of normal line}$$

$$7y + 21 = x - 3$$

$$y = \tfrac{1}{7}x - \tfrac{24}{7} \qquad \text{slope-intercept form}$$

This tells us that the y-intercept is $(0, -\frac{24}{7})$. The curve, normal line, and intercept are shown in Fig. 4-5.   ■

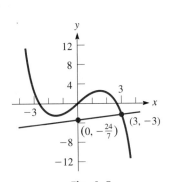

Fig. 4-5

Many of the applications of tangents and normals are geometric. However, there are certain applications in technology, and one of these is shown in the following example. Others are shown in the exercises.

SOLVING A WORD PROBLEM

EXAMPLE 5 In Fig. 4-6 the cross section of a parabolic solar reflector is shown, along with an incident ray of light and the reflected ray. The angle of incidence i is equal to the angle of reflection r, where both angles are measured with respect to the normal to the surface. If the incident ray strikes at the point where the slope of the normal is -1 and the equation of the parabola is $4y = x^2$, what is the equation of the normal line?

If the slope of the normal line is -1, then the slope of a tangent line is $-(\frac{1}{-1}) = 1$. Therefore, we know that the value of the derivative at the point of reflection is 1. This allows us to find the coordinates of the point.

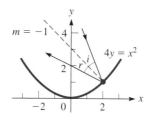

Fig. 4-6

$$4y = x^2$$

$$4\frac{dy}{dx} = 2x, \qquad \frac{dy}{dx} = \frac{1}{2}x \qquad \text{find derivative}$$

$$1 = \frac{1}{2}x \qquad \qquad \text{substitute } \frac{dy}{dx} = 1$$

$$x = 2$$

This means that the x-coordinate of the point of reflection is 2. We can find the y-coordinate by substituting $x = 2$ into the equation of the parabola. Thus, the point is $(2, 1)$. Since the slope is -1, the equation is

$$y - 1 = (-1)(x - 2)$$
$$y = -x + 3$$

We might note that if the incident ray is vertical, for which $i = 45°$, the reflected ray passes through $(0, 1)$, which is the focus of the parabola. This shows the important reflection property of a parabola that *any incident ray parallel to the axis of a parabola passes through the focus.* We first noted this property in our discussion of the parabola in Example 5 of Section 2-4.

EXERCISES *4-1*

In Exercises 1–4, find the equations of the lines tangent to the indicated curves at the given points. In Exercises 1 and 4, sketch the curve and tangent line. In Exercises 2 and 3, use the tangent feature of a graphing calculator to view the curve and tangent line.

1. $y = x^2 + 2$ at $(2, 6)$

2. $y = \frac{1}{3}x^3 - 5x$ at $(3, -6)$

3. $y = \dfrac{1}{x^2 + 1}$ at $(1, \frac{1}{2})$

4. $x^2 + y^2 = 25$ at $(3, 4)$

In Exercises 5–8, find the equations of the lines normal to the indicated curves at the given points. In Exercises 5 and 8, sketch the curve and normal line. In Exercises 6 and 7, use a graphing calculator to view the curve and normal line.

5. $y = 6x - 2x^2$ at $(2, 4)$

6. $y = 8 - x^3$ at $(-1, 9)$

7. $y = \dfrac{6}{(x^2 + 1)^2}$ at $(1, \frac{3}{2})$

8. $x^2 - y^2 = 8$ at $(3, 1)$

In Exercises 9–12, find the equations of the tangent lines and the normal lines to the indicated curves. In Exercises 9 and 10, use a graphing calculator to view the curve and the lines. In Exercises 11 and 12, sketch the curve and lines.

9. $y = \dfrac{1}{\sqrt{x^2 + 1}}$, where $x = \sqrt{3}$

10. $y \doteq \dfrac{4}{(5 - 2x)^2}$, where $x = 2$

11. The parabola with vertex at $(0, 3)$ and focus at $(0, 0)$, where $x = -1$

12. The ellipse with focus at $(4, 0)$, vertex at $(5, 0)$, and center at $(0, 0)$, where $x = 2$

In Exercises 13–16, find the equations of the lines tangent or normal to the given curves and with the given slopes. View the curves and lines on a graphing calculator.

13. $y = x^2 - 2x$, tangent line with slope 2

14. $y = \sqrt{2x - 9}$, tangent line with slope 1

15. $y = (2x - 1)^3$, normal line with slope $-\frac{1}{24}$, $x > 0$

16. $y = \frac{1}{2}x^4 + 1$, normal line with slope 4

In Exercises 17–24, solve the given problems involving tangent and normal lines.

(W) 17. Without actually finding the points of intersection, explain why the parabola $y^2 = 4x$ and the ellipse $2x^2 + y^2 = 6$ intersect at right angles. (*Hint:* Call a point of intersection (a, b).)

18. Find the y-intercept of the line normal to the curve $y = x^{3/4}$, where $x = 16$.

19. A certain suspension cable with supports on the same level is closely approximated as being parabolic in shape. If the supports are 200 ft apart and the sag at the center is 30 ft, what is the equation of the line along which the tension acts (tangentially) at the right support? (Choose the origin of the coordinate system at the lowest point of the cable.)

20. A laser source is 2.00 cm from a spherical surface of radius 3.00 cm, and the laser beam is tangent to the spherical surface. By placing the center of the sphere at the origin and the source on the positive x-axis, find the equation of the line along which the beam shown in Fig. 4-7 is directed.

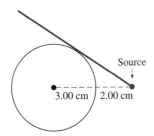

Source

3.00 cm   2.00 cm

Fig. 4-7

21. In an electric field, the lines of force are perpendicular to the curves of equal electric potential. In a certain electric field, a curve of equal potential is $y = \sqrt{2x^2 + 8}$. If the line along which the force acts on an electron has an inclination of 135°, find its equation.

22. A radio wave reflects from a reflecting surface in the same way as a light wave (see Example 5). A certain horizontal radio wave reflects off a parabolic reflector such that the reflected wave is 43.60° below the horizontal, as shown in Fig. 4-8. If the equation of the parabola is $y^2 = 8x$, what is the equation of the normal line through the point of reflection?

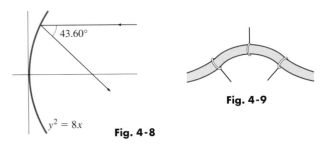

43.60°

$y^2 = 8x$ **Fig. 4-8**

Fig. 4-9

23. In designing a flexible tubing system, the supports for the tubing must be perpendicular to the tubing. If a section of the tubing follows the curve $y = \dfrac{4}{x^2 + 1}$ (-2 dm $< x <$ 2 dm), along which lines must the supports be directed if they are located at $x = -1$, $x = 0$, and $x = 1$? See Fig. 4-9.

24. On a particular drawing, a pulley wheel can be described by the equation $x^2 + y^2 = 100$ (units in cm). The pulley belt is directed along the lines $y = -10$ and $4y - 3x - 50 = 0$ when first and last making contact with the wheel. What are the first and last points on the wheel where the belt makes contact?

Another mathematical development by the English mathematician and physicist Isaac Newton (1642–1727).

4-2 NEWTON'S METHOD FOR SOLVING EQUATIONS

As we know, finding the roots of an equation $f(x) = 0$ is very important in mathematics and in many types of applications. However, for a great many algebraic and nonalgebraic equations, there is no method for finding the roots exactly.

We have shown how equations can be solved graphically, and by using a graphing calculator the roots can be found with great accuracy. In this section we show a method, known as **Newton's method,** which uses the derivative to locate approximately, but very accurately, the real roots of many kinds of equations. It can be used with polynomial equations of any degree and with other algebraic and nonalgebraic equations.

Newton's method is an example of an **iterative method.** In using this type of method, we start with a reasonable guess for a root of the equation. By using the method, we obtain a new value, which is usually a better approximation. This in turn gives a still better approximation. Continuing in this way, using a calculator we can obtain an approximate answer with the required accuracy. Iterative methods in general are easily programmable for use on a computer.

Let us consider a section of the graph of $y = f(x)$ that (a) crosses the x-axis, (b) always has either a positive slope or a negative slope, and (c) has a slope that either becomes greater or becomes smaller as x increases. See Fig. 4-10. The curve in the figure crosses the x-axis at $x = r$, which means that $x = r$ is a root of the equation $f(x) = 0$. If x_1 is sufficiently close to r, a line tangent to the curve at $[x_1, f(x_1)]$ will cross the x-axis at a point $(x_2, 0)$, which is closer to r than is x_1.

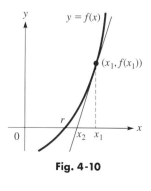

Fig. 4-10

We know that the slope of the tangent line is the value of the derivative at x_1, or $m_{\text{tan}} = f'(x_1)$. Therefore, the equation of the tangent line is

$$y - f(x_1) = f'(x_1)(x - x_1)$$

For the point $(x_2, 0)$ on this line, we have

$$-f(x_1) = f'(x_1)(x_2 - x_1)$$

Solving for x_2, we have the formula

$$\boxed{x_2 = x_1 - \frac{f(x_1)}{f'(x_1)}} \qquad \text{(4-1)}$$

NOTE▶ *Here, x_2 is a second approximation to the root. We can then replace x_1 in Eq. (4-1) by x_2 and find a closer approximation, x_3. This process can be repeated as many times as needed to find the root to the required accuracy. This method lends itself well to the use of a calculator or a computer for finding the root.*

■**EXAMPLE 1** Find the root of $x^2 - 3x + 1 = 0$ between $x = 0$ and $x = 1$.

Here, $f(x) = x^2 - 3x + 1$. Therefore, $f(0) = 1$ and $f(1) = -1$, which indicates that the root may be near the middle of the interval. Since x_1 must be within the interval, we choose $x_1 = 0.5$.

The derivative is

$$f'(x) = 2x - 3$$

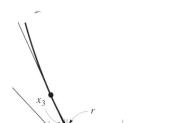

Fig. 4-11

Therefore, $f(0.5) = -0.25$ and $f'(0.5) = -2$, which gives us

$$x_2 = 0.5 - \frac{-0.25}{-2} = 0.375$$

This is a second approximation, which is closer to the actual value of the root. See Fig. 4-11. We can get an even better approximation, x_3, by using the method again with $x_2 = 0.375$, $f(0.375) = 0.015625$, and $f'(0.375) = -2.25$. This gives us

$$x_3 = 0.375 - \frac{0.015625}{-2.25} = 0.3819444$$

Since this is a quadratic equation, we can check this result by using the quadratic formula. Using this formula we find the root is $x = 0.3819660$. Our result using Newton's method is good to three decimal places. Additional accuracy may be obtained by using the method again as many times as needed. ━━━━━ ∎

SOLVING A WORD PROBLEM

■**EXAMPLE 2** A spherical water storage tank holds 500.0 m³. If the outside diameter is 10.0000 m, what is the thickness of the metal of which the tank is made?

Let x = the thickness of the metal. We know that the outside radius of the tank is 5.0000 m. Therefore, using the formula for the volume of a sphere, we have

$$\frac{4\pi}{3}(5.0000 - x)^3 = 500.0$$

$$125.0 - 75.00x + 15.00x^2 - x^3 = 119.366$$

$$x^3 - 15.00x^2 + 75.00x - 5.634 = 0$$

$$f(x) = x^3 - 15.00x^2 + 75.00x - 5.634$$

$$f'(x) = 3x^2 - 30.00x + 75.00$$

Since $f(0) = -5.634$ and $f(0.1) = 1.717$, the root may be closer to 0.1 than to 0.0. Therefore, we let $x_1 = 0.07$. Setting up a table, we have these values:

n	x_n	$f(x_n)$	$f'(x_n)$	$x_n - \dfrac{f(x_n)}{f'(x_n)}$
1	0.07	−0.457157	72.9147	0.0762697508
2	0.0762697508	−0.000581145	72.7293587	0.0762777413

See Appendix C for the graphing calculator program NEWTON. It finds the *n*th approximation of a root of $f(x) = 0$, using Newton's method.

Since $x_2 = x_3 = 0.0763$ to four decimal places, the thickness is 0.0763 m. This means the inside radius of the tank is 4.9237 m, and this value gives an inside volume of 500.0 m³. In using a calculator, the values in the table are more easily found if the values of x_n, $f(x_n)$, and $f'(x_n)$ are stored in memory for each step. ■

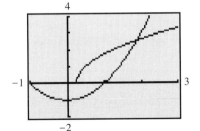

Fig. 4-12

■**EXAMPLE 3** Solve the equation $x^2 - 1 = \sqrt{4x - 1}$.

We can see approximately where the root is by sketching the graphs of $y_1 = x^2 - 1$ and $y_2 = \sqrt{4x - 1}$ or by viewing the graphs on a graphing calculator, as shown in Fig. 4-12. From this view we see that they intersect between $x = 1$ and $x = 2$. Therefore, we choose $x_1 = 1.5$. With

$$f(x) = x^2 - 1 - \sqrt{4x - 1}$$

$$f'(x) = 2x - \frac{2}{\sqrt{4x - 1}}$$

we now find the values in the following table:

This table shows the values used in calculating each approximation. Since the necessary values of x_n, $f(x_n)$, and $f'(x_n)$ can be stored in the memory of the calculator, it is not necessary to write out the table.

n	x_n	$f(x_n)$	$f'(x_n)$	$x_n - \dfrac{f(x_n)}{f'(x_n)}$
1	1.5	−0.98606798	2.1055728	1.9683134
2	1.9683134	0.25256859	3.1737598	1.8887332
3	1.8887332	0.00705269	2.9962957	1.8863794
4	1.8863794	0.00000620	2.9910265	1.8863773

Since $x_5 = x_4 = 1.88638$ to five decimal places, this is the required solution. (Here, rounded-off values of x_n are shown, although additional digits were carried and used.) This value can be verified on the graphing calculator by using the *intersect* (or *zero*) feature. ■

EXERCISES *4-2*

In Exercises 1–4, find the indicated roots of the given quadratic equations by finding x_3 from Newton's method. Compare this root with that obtained by using the quadratic formula.

1. $x^2 - 2x - 5 = 0$ (between 3 and 4)

2. $2x^2 - x - 2 = 0$ (between 1 and 2)

3. $3x^2 - 5x - 1 = 0$ (between −1 and 0)

4. $x^2 + 4x + 2 = 0$ (between −4 and −3)

In Exercises 5–16, find the indicated roots of the given equations to at least four decimal places by using Newton's method. Compare with the value of the root found using a graphing calculator.

5. $x^3 - 6x^2 + 10x - 4 = 0$ (between 0 and 1)

6. $x^3 - 3x^2 - 2x + 3 = 0$ (between 0 and 1)

7. $x^3 + 5x^2 + x - 1 = 0$ (the positive root)

8. $2x^3 + 2x^2 - 11x + 3 = 0$ (the larger positive root)

9. $x^4 - x^3 - 3x^2 - x - 4 = 0$ (between 2 and 3)

10. $2x^4 - 2x^3 - 5x^2 - x - 3 = 0$ (between 2 and 3)

11. $x^4 - 2x^3 - 8x - 16 = 0$ (the negative root)

12. $3x^4 - 3x^3 - 11x^2 - x - 4 = 0$ (the negative root)

13. $2x^2 = \sqrt{2x + 1}$ (the positive real solution)

14. $x^3 = \sqrt{x + 1}$ (the real solution)

15. $x = \dfrac{1}{\sqrt{x + 2}}$ (the real solution)

16. $x^{3/2} = \dfrac{1}{2x + 1}$ (the real solution)

In Exercises 17–24, determine the required values by using Newton's method.

17. Find all the real roots of $x^3 - 2x^2 - 5x + 4 = 0$.

18. Find all the real roots of $x^3 - 2x^2 - 2x - 7 = 0$.

(W) 19. Explain how to find $\sqrt[3]{4}$ by using Newton's method.

(W) 20. In Appendix D, page 496, there is an explanation and example of Newton's method, which was copied directly from *Essays on Several Curious and Useful Subjects in Speculative and Mix'd Mathematicks* by Thomas Simpson. It was published in London in 1740. Explain where the numerical error is in the example and what you think caused the error.

21. A dome in the shape of a spherical segment is to be placed over the top of a sports stadium. If the radius r of the dome is to be 60.0 m and the volume V within the dome is 180,000 m³, find the height h of the dome. See Fig. 4-13. ($V = \frac{1}{6}\pi h(h^2 + 3r^2)$.)

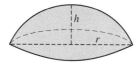

Fig. 4-13

22. The capacitances (in μF) of three capacitors in series are C, $C + 1.00$, and $C + 2.00$. If their combined capacitance is 1.00 μF, their individual values can be found by solving the equation

$$\frac{1}{C} + \frac{1}{C + 1.00} + \frac{1}{C + 2.00} = 1.00$$

Find these capacitances.

23. An oil storage tank has the shape of a right circular cylinder with a hemisphere at each end. See Fig. 4-14. If the volume of the tank is 1500 ft³ and the length l is 12.0 ft, find the radius r.

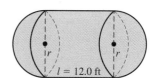

Fig. 4-14

24. A rectangular block of plastic with edges 2.00 cm, 2.00 cm, and 4.00 cm is heated until its volume doubles. By how much does each edge increase if each increases by the same amount?

4-3 CURVILINEAR MOTION

A great many phenomena in technology involve the time rate of change of certain quantities. We encountered one of the most fundamental and most important of these when we discussed velocity, the time rate of change of displacement, in Section 3-4. In this section we further develop the concept of velocity and certain other concepts necessary for the discussion. Other time-rate-of-change problems are discussed in the next section.

When velocity was introduced in Section 3-4, the discussion was limited to rectilinear motion, or motion along a straight line. A more general discussion of velocity is necessary when we discuss the motion of an object in a plane. There are many important applications of motion in a plane, a principal one being the motion of a projectile.

A concept necessary in developing this topic is that of a *vector*. Basically, a **vector** *is a quantity that has direction and magnitude.* For example, velocity, acceleration, and force are vector quantities. Each is specified by giving both its magnitude and its direction. For instance, we cannot determine the effect of a force unless we know the magnitude of the force as well as the direction in which it is acting. *To represent a vector, we draw a line segment with its length proportional to the magnitude of the vector, and in a direction that shows the direction of the vector.*

The addition of two vectors **OA** *and* **AB***, directed from O to A and from A to B, respectively, gives the* **resultant** *vector* **OB***, as shown in Fig. 4-15. The* **initial point** *of the resultant is O and the* **terminal point** *is B.* This is equivalent to making the two vectors being added the sides of a parallelogram, with the diagonal being the resultant (see Fig. 4-16). (Vectors may be subtracted by reversing the direction of the vector being subtracted and proceeding as in addition.) Note that a vector is represented by a letter printed in boldface type, and the same letter in italic (lightface) type represents the magnitude only. In general, *a resultant is a single vector that can replace any number of other vectors and still produce the same physical effect.*

A given vector may be considered to be the sum of other vectors. *The* **components** *of a given vector are vectors for which the resultant is the given vector.* With a vector's initial point at the origin of rectangular coordinates, and direction given by its angle with the positive *x*-axis (*standard position*), we find its **x- and y-components.** *These components are directed along the coordinate axes.* Their terminal points are located where perpendiculars from the terminal point of the given vector cross the axes. *Finding these components is called* **resolving** *the vector into its components.*

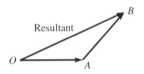

Fig. 4-15

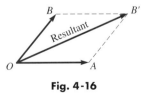

Fig. 4-16

It is assumed here that you have a basic knowledge of angles and the trigonometric functions. If a review is needed, see Section 7-1.

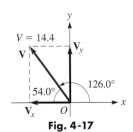

Fig. 4-17

EXAMPLE 1   Resolve a vector 14.4 units long and directed at an angle of 126.0° into its *x*- and *y*-components. See Fig. 4-17.

Placing the initial point of the vector at the origin and putting the angle in standard position, we see that the vector directed along the *x*-axis, V_x, is related to the vector **V** of magnitude V by

$$V_x = V \cos 126.0°$$

magnitude of vector — standard position angle

or in terms of the reference angle by

$$V_x = -V \cos 54.0° \longleftarrow \text{reference angle}$$

directed along negative *x*-axis

since $\cos 126.0° = -\cos 54.0°$. We see that the minus sign shows that the *x*-component is directed in the negative direction—that is, to the left.

Since the vector directed along the *y*-axis, V_y, could also be placed along the vertical dashed line, it is related to the vector **V** by

$$V_y = V \sin 126.0° = V \sin 54.0°$$

Thus, the vectors V_x and V_y have the magnitudes

$$V_x = 14.4 \cos 126.0° = -8.46 \qquad V_y = 14.4 \sin 126.0° = 11.6$$

Therefore, we have resolved the given vector into two components, one directed along the negative *x*-axis and the other along the positive *y*-axis.   ■

Accurate numerical results for the sum of vectors can be found by the use of components. This is illustrated in the following example.

See Appendix C for a graphing calculator program ADDVCTR. It can be used to add vectors.

EXAMPLE 2 Find the resultant of two vectors **A** and **B** such that $A = 1200$, $\theta_A = 270.0°$, $B = 1750$, and $\theta_B = 115.0°$.

We first place the vectors on a coordinate system with the tail of each at the origin as shown in Fig. 4-18(a). We then resolve each vector into its x- and y-components, as shown in Fig. 4-18(b) and as calculated below. (Note that **A** is vertical and has no horizontal component.) Next, the components are combined, as in Fig. 4-18(c) and as calculated. Finally, the magnitude of the resultant and the angle θ (to determine the direction), as shown in Fig. 4-18(d), are calculated.

Some calculators have a specific feature for adding vectors.

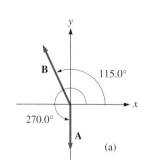

Fig. 4-18

(a)   (b)   (c)   (d)

$$A_x = A \cos 270.0° = 1200 \cos 270.0° = 0$$
$$B_x = B \cos 115.0° = 1750 \cos 115.0° = -739.6$$
$$A_y = A \sin 270.0° = 1200 \sin 270.0° = -1200$$
$$B_y = B \sin 115.0° = 1750 \sin 115.0° = 1586$$

— Fig. 4-18(b)

$$R_x = A_x + B_x = 0 - 739.6 = -739.6$$
$$R_y = A_y + B_y = -1200 + 1586 = 386$$

— Fig. 4-18(c)

$$R = \sqrt{R_x^2 + R_y^2} = \sqrt{(-739.6)^2 + 386^2} = 834$$
$$\tan \theta = \frac{R_y}{R_x} = \frac{386}{-739.6} \qquad \theta = 152.4° \leftarrow 180° - 27.6°$$

— Fig. 4-18(d)

Thus, the resultant has a magnitude of 834 and is directed at a standard-position angle of 152.4°. In finding θ from a calculator (see Fig. 4-19),

the calculator display shows an angle of $-27.6°$. *However, we know* θ *is a second-quadrant angle, since* **R**$_x$ *is negative and* **R**$_y$ *is positive.*

Therefore, we must use 27.6° as a reference angle. For this reason, it is usually advisable to *find the reference angle first* by disregarding the signs of R_x and R_y when finding θ. Thus,

$$\tan \theta_{\text{ref}} = \left| \frac{R_y}{R_x} \right| = \frac{386}{739.6} \qquad \theta_{\text{ref}} = 27.6°$$

In using the calculator, R_x and R_y are each calculated in one step and stored for the calculation of R and θ, as shown in the calculator display of Fig. 4-19. The above values are rounded off, and individual values are given to more clearly show the steps of the method.

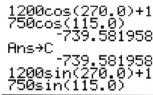

Fig. 4-19

Although vectors can be used to represent many physical quantities, we shall restrict our attention in this section to their use in describing the velocity and the acceleration of an object moving in a plane along a specified path. Such motion is called **curvilinear motion.**

In describing an object undergoing curvilinear motion, it is common to express the x- and y-coordinates of its position separately as functions of time. Equations given in this form—that is, *x and y both given in terms of a third variable (in this case, t)*—are said to be in **parametric form.** *The third variable, t, is called the* **parameter.**

To find the velocity of an object whose coordinates are given in parametric form, we find its x-component of velocity v_x by determining dx/dt and its y-component of velocity v_y by determining dy/dt. These are then evaluated, and the resultant velocity is found from $v = \sqrt{v_x^2 + v_y^2}$. The direction in which the object is moving is found from $\tan \theta = v_y/v_x$.

■**EXAMPLE 3**   If the horizontal distance x that an object has moved is given by $x = 3t^2$ and the vertical distance y is given by $y = 1 - t^2$, find the resultant velocity when $t = 2$.

To find the resultant velocity, we must find v and θ, by first finding v_x and v_y. After the derivatives are found, they are evaluated for $t = 2$. Therefore,

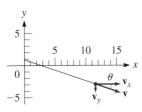
Fig. 4-20

$$v_x = \frac{dx}{dt} = 6t \qquad v_x|_{t=2} = 12 \qquad \text{find velocity components}$$

$$v_y = \frac{dy}{dt} = -2t \qquad v_y|_{t=2} = -4$$

$$v = \sqrt{12^2 + (-4)^2} = 12.6 \qquad \text{magnitude of velocity}$$

$$\tan \theta = \frac{-4}{12} \qquad \theta = -18.4° \qquad \text{direction of motion}$$

The path and the velocity vectors are shown in Fig. 4-20.   ■

■**EXAMPLE 4**   Find the velocity and direction of motion when $t = 2$ of an object moving such that its x- and y-coordinates of position are given by $x = 1 + 2t$ and $y = t^2 - 3t$.

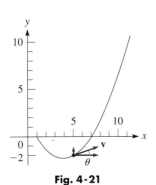
Fig. 4-21

$$v_x = \frac{dx}{dt} = 2 \qquad v_x|_{t=2} = 2 \qquad \text{find velocity components}$$

$$v_y = \frac{dy}{dt} = 2t - 3 \qquad v_y|_{t=2} = 1$$

$$v|_{t=2} = \sqrt{2^2 + 1^2} = 2.24 \qquad \text{magnitude of velocity}$$

$$\tan \theta = \frac{1}{2} \qquad \theta = 26.6° \qquad \text{direction of motion}$$

These quantities are shown in Fig. 4-21.   ■

CAUTION▶   In these examples we note that *we first find the necessary derivatives, and then we evaluate them.* This procedure should always be followed. When a derivative is to be found, it is incorrect to take the derivative of the expression that is the evaluated function.

Acceleration *is the time rate of change of velocity.* Therefore, if the velocity, or its components, is known as a function of time, the acceleration of an object can be found by taking the derivative of the velocity with respect to time. If the displacement is known, the acceleration is found by finding the second derivative with respect to time. Finding the acceleration of an object is illustrated in the following example.

EXAMPLE 5 Find the magnitude and direction of the acceleration when $t = 2$ for an object that is moving such that its x- and y-coordinates of position are given by $x = t^3$ and $y = 1 - t^2$.

$$v_x = \frac{dx}{dt} = 3t^2 \qquad a_x = \frac{dv_x}{dt} = \frac{d^2x}{dt^2} = 6t \qquad a_x|_{t=2} = 12$$

take second derivatives to find acceleration components

$$v_y = \frac{dy}{dt} = -2t \qquad a_y = \frac{dv_y}{dt} = \frac{d^2y}{dt^2} = -2 \qquad a_y|_{t=2} = -2$$

$$a|_{t=2} = \sqrt{12^2 + (-2)^2} = 12.2 \qquad \text{magnitude of acceleration}$$

$$\tan \theta = \frac{a_y}{a_x} = -\frac{2}{12} \qquad \theta = -9.5° \qquad \text{direction of acceleration}$$

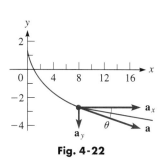

Fig. 4-22

CAUTION ▶

The quadrant in which θ lies is determined from the fact that $\boldsymbol{a_y}$ *is negative and* $\boldsymbol{a_x}$ *is positive.* Thus, θ must be a fourth-quadrant angle (see Fig. 4-22). We see from this example that the magnitude and direction of acceleration are found from its components just as with velocity. ■

We now summarize the equations used to find the velocity and acceleration of an object for which the displacement is a function of time. They indicate how to find the components, as well as the magnitude and direction, of each.

$v_x = \dfrac{dx}{dt}$	$v_y = \dfrac{dy}{dt}$	velocity components	**(4-2)**
$a_x = \dfrac{dv_x}{dt} = \dfrac{d^2x}{dt^2}$	$a_y = \dfrac{dv_y}{dt} = \dfrac{d^2y}{dt^2}$	acceleration components	**(4-3)**
$v = \sqrt{v_x^2 + v_y^2}$	$a = \sqrt{a_x^2 + a_y^2}$	magnitude	**(4-4)**
$\tan \theta_v = \dfrac{v_y}{v_x}$	$\tan \theta_a = \dfrac{a_y}{a_x}$	direction	**(4-5)**

CAUTION ▶

If the curvilinear path an object follows is given with y as a function of x, *the velocity (or acceleration) is found by taking derivatives of each term of the equation with respect to time.* It is assumed that both x and y are functions of time, although these functions are not stated. When finding derivatives we must be careful in using the power rule, Eq. (3-15), so that the factor du/dx is not neglected. In the following examples, we illustrate the use of Eqs. (4-2) to (4-5) in applied situations for which we know the equation of the path of the motion. Again, we must be careful to find the direction of the vector as well as its magnitude in order to have a complete solution.

For reference, Eq. (3-15) is
$$\frac{du^n}{dx} = nu^{n-1}\frac{du}{dx}.$$

SOLVING A WORD PROBLEM

CAUTION ▶

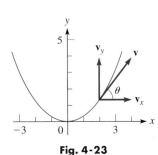

Fig. 4-23

SOLVING A WORD PROBLEM

The first successful helicopter was made in the United States by Igor Sikorsky in 1939.

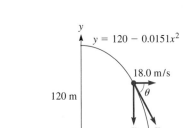

Fig. 4-24

■**EXAMPLE 6**   In a physics experiment, a small sphere is constrained to move along a parabolic path described by $y = \frac{1}{3}x^2$. If the horizontal velocity v_x is constant at 6.00 cm/s, find the velocity at the point (2.00, 1.33). See Fig. 4-23.

Since both y and x change with time, both can be considered to be functions of time. Therefore, we can take derivatives of $y = \frac{1}{3}x^2$ with respect to time.

$$\frac{dy}{dt} = \frac{1}{3}\left(2x\frac{dx}{dt}\right) \longleftarrow \frac{dx^2}{dt} = 2x\frac{dx}{dt}$$

$$v_y = \frac{2}{3}xv_x \qquad\qquad \text{using Eqs. (4-2)}$$

$$v_y = \frac{2}{3}(2.00)(6.00) = 8.00 \text{ cm/s} \qquad \text{substituting}$$

$$v = \sqrt{6.00^2 + 8.00^2} = 10.0 \text{ cm/s} \qquad \text{magnitude [Eqs. (4-4)]}$$

$$\tan\theta = \frac{8.00}{6.00}, \qquad \theta = 53.1° \qquad \text{direction [Eqs. (4-5)]}$$

■**EXAMPLE 7**   A helicopter is flying at 18.0 m/s and at an altitude of 120 m when a rescue marker is released from it. The marker maintains a horizontal velocity and follows a path given by $y = 120 - 0.0151x^2$, as shown in Fig. 4-24. Find the magnitude and direction of the velocity and of the acceleration of the marker 3.00 s after release. This is a typical problem in projectile motion.

From the given information we know that $v_x = dx/dt = 18.0$ m/s. Taking derivatives with respect to time leads to this solution:

$$y = 120 - 0.0151x^2$$

$$\frac{dy}{dt} = -0.0302x\frac{dx}{dt} \qquad\qquad \text{taking derivatives}$$

$$v_y = -0.0302xv_x \qquad\qquad \text{using Eqs. (4-2)}$$

$$x = (3.00)(18.0) = 54.0 \text{ m} \qquad\qquad \text{evaluating at } t = 3.00 \text{ s}$$

$$v_y = -0.0302(54.0)(18.0) = -29.35 \text{ m/s}$$

$$v = \sqrt{18.0^2 + (-29.35)^2} = 34.4 \text{ m/s} \qquad \text{magnitude}$$

$$\tan\theta = \frac{-29.35}{18.0}, \qquad \theta = -58.5° \qquad \text{direction}$$

The velocity is 34.4 m/s and is directed at an angle of 58.5° below the horizontal.

To find the acceleration, we return to the equation $v_y = -0.0302xv_x$. Since v_x is constant, we can substitute 18.0 for v_x to get

$$v_y = -0.5436x$$

Again taking derivatives with respect to time, we have

$$\frac{dv_y}{dt} = -0.5436\frac{dx}{dt}$$

$$a_y = -0.5436v_x \qquad\qquad \text{using Eqs. (4-3) and (4-2)}$$

$$a_y = -0.5436(18.0) = -9.78 \text{ m/s}^2 \qquad \text{evaluating}$$

We know that v_x is constant, which means that $a_x = 0$. Therefore, the acceleration is 9.78 m/s^2 and is directed vertically downward.

In Exercises 1–4, given that the x- and y-coordinates of a moving particle are given by the indicated parametric equations, find the magnitude and direction of the velocity for the specific value of t. Sketch the curves and show the appropriate components of the velocity.

1. $x = 3t$, $y = 1 - t$, $t = 4$

2. $x = \dfrac{5t}{2t + 1}$, $y = 0.1(t^2 + t)$, $t = 2$

3. $x = t(2t + 1)^2$, $y = \dfrac{6}{\sqrt{4t + 3}}$, $t = 0.5$

4. $x = \sqrt{1 + 2t}$, $y = t - t^2$, $t = 4$

In Exercises 5–8, use the parametric equations and values of t of Exercises 1–4 to find the magnitude and direction of the acceleration in each case.

In Exercises 9–24, find the indicated velocities and accelerations.

9. The water from a valve at the bottom of a water tank follows a path described by $y = 4.0 - 0.20x^2$, where units are in meters. If the velocity v_x is constant at 5.0 m/s, find the resultant velocity at the point $(4.0, 0.80)$.

10. A roller mechanism follows a path described by $y = \sqrt{4x + 1}$, where units are in feet. If $v_x = 2x$, find the resultant velocity (in ft/s) at the point $(2.0, 3.0)$.

11. A float is used to test the flow pattern of a stream. It follows a path described by $x = 0.2t^2$, $y = -0.1t^3$, where units of x and y are in feet and t is in minutes. Find the acceleration of the float after 2.0 min.

12. A car on a test track goes into a turn described by $x = 0.2t^3$, $y = 20t - 2t^2$, where x and y are measured in meters and t is in seconds. Find the acceleration of the car at $t = 3.0$ s.

13. A golf ball moves according to the equations $x = 96t$ and $y = 120t - 16t^2$, where distances are in feet and time is in seconds. Find the resultant velocity and acceleration of the golf ball for $t = 6.0$ s.

14. A package of relief supplies is dropped and moves according to the parametric equations $x = 45t$ and $y = -4.9t^2$, where distances are in meters and time is in seconds. Find the magnitude and direction of the velocity and of the acceleration when $t = 3.0$ s.

15. A spacecraft moves along a path described by the parametric equations $x = 10(\sqrt{1 + t^4} - 1)$, $y = 40t^{3/2}$ for the first 100 s after launch. Here, x and y are measured in meters and t is measured in seconds. Find the magnitude and direction of the velocity of the spacecraft 10.0 s and 100 s after launch.

16. A ski jump is designed to follow the path given by the parametric equations $x = 3.50t^2$, $y = 20.0 + 0.120t^4 - 3.00\sqrt{t^4 + 1}$ $(0 \le t \le 4.00$ s), where distances are in meters. Find the velocity of a skier when $t = 4.00$ s. See Fig. 4-25.

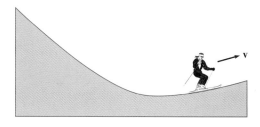

Fig. 4-25

17. Find the resultant acceleration of the spacecraft in Exercise 15 for the specified times.

18. Find the resultant acceleration of the skier in Exercise 16 for $t = 4.00$ s.

19. A rocket follows a path given by $y = x - \frac{1}{90}x^3$ (distances in miles). If the horizontal velocity is given by $v_x = x$, find the magnitude and direction of the velocity when the rocket hits the ground (assume level terrain) if time is in minutes.

20. A shipping route around an island is described by the equation $y = 3x^2 - 0.2x^3$. A ship on this route is moving such that $v_x = 1.2$ km/h, where $x = 3.5$ km. Find the velocity of the ship at this point.

21. A computer's hard disk is 3.50 in. in diameter and rotates at 3600 r/min. Set up the equation for the circumference of the disk, with its center at the origin. Using this equation, find the components v_x and v_y of the velocity at a point on the circumference for $x = 1.20$ in., $y > 0$, and $v_x > 0$.

22. A robot arm joint moves in an elliptical path. The horizontal major axis of the ellipse is 8.0 cm long, and the minor axis is 4.0 cm long. For $-2 < x < 2$ and $y > 0$, the joint moves such that $v_x = 2.5$ cm/s. On this part of its path, find its velocity at $x = -1.5$ cm. Assume the center of the path is at the origin.

23. An airplane ascends such that its gain h in altitude is proportional to the square root of the change x in horizontal distance traveled. If $h = 280$ m for $x = 400$ m and v_x is constant at 350 m/s, find the velocity at this point.

Ⓦ **24.** A meteor traveling toward the earth has a velocity inversely proportional to the square root of the distance from the earth's center. State how its acceleration is related to its distance from the center of the earth.

4-4 RELATED RATES

Any two variables that vary with respect to time and between which a relation is known to exist can have the time rate of change of one expressed in terms of the time rate of change of the other. We do this by taking the derivative with respect to time of the equation that relates the variables, as we did in Examples 6 and 7 of Section 4-3. Since the rates of change are related, this type of problem is referred to as a **related-rate** problem. The following examples illustrate the basic method of solution.

EXAMPLE 1   The voltage of a certain thermocouple as a function of the temperature is given by $E = 2.800T + 0.006T^2$. If the temperature is increasing at the rate of $1.00°C/min$, how fast is the voltage increasing when $T = 100°C$?

Since we are asked to find the time rate of change of voltage, we first take derivatives with respect to time. This gives us

$$\frac{dE}{dt} = 2.800\frac{dT}{dt} + 0.012T\frac{dT}{dt} \longleftarrow \frac{d}{dt}(0.006T^2) = 0.006\left(2T\frac{dT}{dt}\right)$$

The Celsius degree named for the Swedish astronomer Andres Celsius (1701–1744). He designated 100° as the freezing point of water and 0° as the boiling point. These were later reversed.

CAUTION ▶   *again being careful to include the factor dT/dt.* From the given information we know that $dT/dt = 1.00°C/min$ and that we wish to know dE/dt when $T = 100°C$. Thus,

$$\left.\frac{dE}{dt}\right|_{T=100} = 2.800(1.00) + 0.012(100)(1.00) = 4.00 \text{ V/min}$$

NOTE ▶   *The derivative must be taken before values are substituted.* In this problem we are finding the time rate of change of the voltage for a specified value of T. For other values of T, dE/dt would have different values.   ■

EXAMPLE 2   The distance q that an image is from a certain telescope lens in terms of p, the distance of the object from the lens, is given by

$$q = \frac{10p}{p - 10}$$

If the object distance is increasing at the rate of 0.200 cm/s, how fast is the image distance changing when $p = 15.0 \text{ cm}$? See Fig. 4-26.

Taking derivatives with respect to time, we have

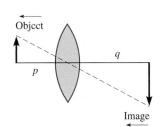

Object

q

p

Image

Fig. 4-26

$$\frac{dq}{dt} = \frac{(p - 10)\left(10\dfrac{dp}{dt}\right) - 10p\left(\dfrac{dp}{dt}\right)}{(p - 10)^2} = \frac{-100\dfrac{dp}{dt}}{(p - 10)^2} \quad \text{—— don't forget the } \frac{dp}{dt}$$

The telescope was invented by a Dutch lens maker Lippershay in about 1608. In 1609 the Italian scientist Galileo (1564–1642) learned of the invention and developed the telescope for astronomical observations. Among his first discoveries were the four largest moons of Jupiter. Since the 1990s the Hubble space telescope has been used to learn a great deal more about our universe.

Now, substituting $p = 15.0$ and $dp/dt = 0.200$, we have

$$\left.\frac{dq}{dt}\right|_{p=15} = \frac{-100(0.200)}{(15.0 - 10)^2}$$

$$= -0.800 \text{ cm/s}$$

Thus, the image distance is decreasing (the significance of the minus sign) at the rate of 0.800 cm/s when $p = 15.0 \text{ cm}$.   ■

In many related-rate problems the function is not given but must be set up according to the statement of the problem. The following examples illustrate this type of problem.

SOLVING A WORD PROBLEM

■EXAMPLE 3 A spherical balloon is being blown up such that its volume increases at the constant rate of 2.00 ft^3/min. Find the rate at which the radius is increasing when it is 3.00 ft. See Fig. 4-27.

We are asked to find the relation between the rate of change of the volume of a sphere with respect to time and the corresponding rate of change of the radius with respect to time. Therefore, we are to ***take derivatives of the expression for the volume of a sphere with respect to time.***

CAUTION ▶

$$V = \frac{4}{3}\pi r^3 \qquad \text{volume of sphere}$$

$$\frac{dV}{dt} = 4\pi r^2\left(\frac{dr}{dt}\right) \qquad \text{take derivatives with respect to time}$$

$$2.00 = 4\pi(3.00)^2\left(\frac{dr}{dt}\right) \qquad \text{substitute } \frac{dV}{dt} = 2.00 \text{ ft}^3/\text{min and } r = 3.00 \text{ ft}$$

$$\frac{dr}{dt}\bigg|_{r=3} = \frac{1}{18.0\pi} \qquad \text{solve for } \frac{dr}{dt}$$

$$= 0.0177 \text{ ft/min}$$

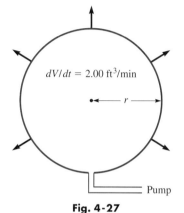

$dV/dt = 2.00 \text{ ft}^3/\text{min}$

r

Pump

Fig. 4-27

SOLVING A WORD PROBLEM

■EXAMPLE 4 The force F of gravity of the earth on a spacecraft varies inversely as the square of the distance r of the spacecraft from the center of the earth. A particular spacecraft weighs 4500 N on the launchpad ($F = 4500$ N for $r = 6370$ km). Find the rate at which F changes later as the spacecraft moves away from the earth at the rate of 12 km/s, where $r = 8500$ km.

First setting up the equation, we have the following solution.

$$F = \frac{k}{r^2} \qquad \text{inverse variation}$$

$$4500 = \frac{k}{6370^2} \qquad \text{substitute } F = 4500 \text{ N}, r = 6370 \text{ km}$$

$$k = 1.83 \times 10^{11} \text{ N} \cdot \text{km}^2 \qquad \text{solve for } k$$

$$F = \frac{1.83 \times 10^{11}}{r^2} \qquad \text{substitute for } k \text{ in equation}$$

$$\frac{dF}{dt} = (1.83 \times 10^{11})(-2)(r^{-3})\frac{dr}{dt} \qquad \text{take derivatives with respect to time}$$

$$= \frac{-3.66 \times 10^{11}}{r^3}\frac{dr}{dt}$$

$$\frac{dF}{dt}\bigg|_{t=8500 \text{ km}} = \frac{-3.66 \times 10^{11}}{8500^3}(12) \qquad \begin{array}{l}\text{evaluate derivative for}\\ r = 8500 \text{ km}, dr/dt = 12 \text{ km/s}\end{array}$$

$$= -7.2 \text{ N/s}$$

Therefore, the gravitational force is decreasing at the rate of 7.2 N/s.

The formula in Example 4 is based on the universal law of gravitation formulated by the great English mathematician and physicist Isaac Newton (1642–1727). Also, the metric unit of force, the newton (N), is named for him.

SOLVING A WORD PROBLEM

EXAMPLE 5   Two cruise ships leave Vancouver, British Columbia, at noon. Ship *A* travels west at 12.0 km/h (before turning toward Alaska), and ship *B* travels south at 16.0 km/h (toward Seattle). How fast are they separating at 2 P.M.?

In Fig. 4-28 we let x = the distance traveled by *A* and y = the distance traveled by *B*. We can find the distance between them, z, from the Pythagorean theorem. Therefore, we are to find dz/dt for $t = 2.00$ h. Even though there are three variables, each is a function of time. This means we can find dz/dt by taking derivatives of each term with respect to time. This gives us

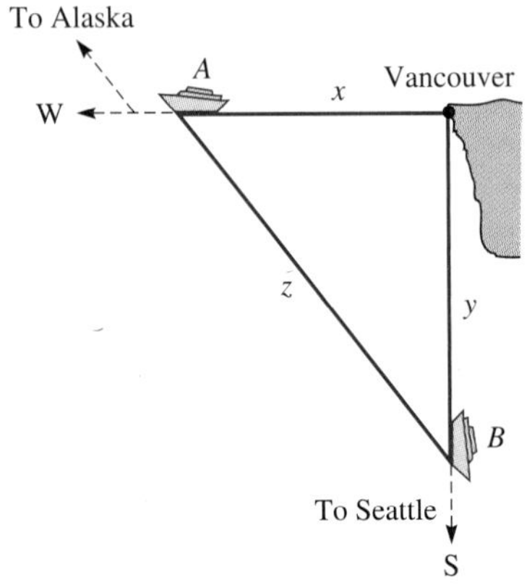

To Alaska

W

A x Vancouver

z y

B

To Seattle

S

Fig. 4-28

$$z^2 = x^2 + y^2 \qquad \text{using Pythagorean theorem}$$

$$2z\frac{dz}{dt} = 2x\frac{dx}{dt} + 2y\frac{dy}{dt} \qquad \text{taking derivatives with respect to time}$$

$$\frac{dz}{dt} = \frac{x(dx/dt) + y(dy/dt)}{z} \qquad \text{solve for } \frac{dz}{dt}$$

At 2 P.M. we have the values

$$x = 24.0 \text{ km} \quad y = 32.0 \text{ km} \quad z = 40.0 \text{ km} \qquad d = rt \text{ and Pythagorean theorem}$$

$$dx/dt = 12.0 \text{ km/h}, \quad dy/dt = 16.0 \text{ km/h} \qquad \text{from statement of problem}$$

$$\left.\frac{dz}{dt}\right|_{z=40} = \frac{(24.0)(12.0) + (32.0)(16.0)}{40.0} = 20.0 \text{ km/h} \qquad \text{substitute values} \blacksquare$$

From these examples, we see that we have the following method of solving a related-rates problem.

Steps for Solving Related-Rates Problems

1. *Identify the variables and rates* in the problem.

2. If possible, *make a sketch* showing the variables.

3. *Determine the equation* relating the variables.

4. *Differentiate with respect to time.*

5. *Solve for the required rate.*

EXERCISES *4-4*

Solve the following problems in related rates.

1. The electric resistance *R* (in Ω) of a certain resistor as a function of the temperature *T* (in °C) is $R = 4.000 + 0.003T^2$. If the temperature is increasing at the rate of 0.100°C/s, find how fast the resistance changes when $T = 150$°C.

2. The kinetic energy *K* (in J) of an object is given by $K = \frac{1}{2}mv^2$, where *m* is the mass (in kg) of the object and *v* is its velocity. If a 250-kg wrecking ball accelerates at 5.00 m/s², how fast is the kinetic energy changing when $v = 30.0$ m/s?

3. A plane flying at an altitude of 2.0 mi is at a direct distance $D = \sqrt{4.0 + x^2}$ from an airport control tower, where *x* is the horizontal distance to the tower. If the plane's speed is 350 mi/h, how fast is *D* changing when $x = 6.2$ mi?

4. A variable resistor *R* and an 8-Ω resistor in parallel have a combined resistance R_T given by $R_T = \dfrac{8R}{8 + R}$. If *R* is changing at 0.30 Ω/min, find the rate at which R_T is changing when $R = 6.0$ Ω.

5. The radius r of a ring of a certain holograph (an image produced without using a lens) is given by $r = \sqrt{0.4\lambda}$, where λ is the wavelength of the light being used. If λ is changing at the rate of 0.10×10^{-7} m/s when $\lambda = 6.0 \times 10^{-7}$ m, find the rate at which r is changing.

6. An earth satellite moves in a path that can be described by $\dfrac{x^2}{28.0} + \dfrac{y^2}{27.6} = 1$, where x and y are in thousands of miles. If $dx/dt = 7750$ mi/h for $x = 2020$ mi and $y > 0$, find dy/dt.

7. The magnetic field B due to a magnet of length l at a distance r is given by $B = \dfrac{k}{[r^2 + (l/2)^2]^{3/2}}$, where k is a constant for a given magnet. Find the expression for the time rate of change of B in terms of the time rate of change of r.

8. An approximate relationship between the pressure p and volume v of the vapor in a diesel engine cylinder is $pv^{1.4} = k$, where k is a constant. At a certain instant, $p = 4200$ kPa, $v = 75$ cm³, and the volume is increasing at the rate of 850 cm³/s. What is the time rate of change of the pressure at this instant?

9. Fatty deposits have decreased the circular cross-sectional opening of a person's artery. A test drug reduces these deposits such that the radius of the opening increases at the rate of 0.020 mm/month. Find the rate at which the area of the opening increases when $r = 1.2$ mm.

10. A computer program increases the side of a square image on the screen at the rate of 0.25 in./s. Find the rate at which the area of the image increases when the edge is 6.50 in.

11. A metal cube dissolves in acid such that an edge of the cube decreases by 0.50 mm/min. How fast is the volume of the cube changing when the edge is 8.20 mm?

12. A light in a garage is 9.50 ft above the floor and 12.0 ft behind the door. If the garage door descends vertically at 1.50 ft/s, how fast is the door's shadow moving toward the garage when the door is 2.00 ft above the floor?

13. One statement of Boyle's law is that the pressure of a gas varies inversely as the volume for constant temperature. If a certain gas occupies 650 cm³ when the pressure is 230 kPa and the volume is increasing at the rate of 20.0 cm³/min, how fast is the pressure changing when the volume is 810 cm³?

14. The tuning frequency f of an electronic tuner is inversely proportional to the square root of the capacitance C in the circuit. If $f = 920$ kHz for $C = 3.5$ pF, find how fast f is changing at this frequency if $dC/dt = 0.3$ pF/s.

15. A spherical metal object is ejected from an earth satellite and reenters the atmosphere. It heats up (until it burns) so that the radius increases at the rate of 5.00 mm/s. What is the time rate of change of volume when the radius is 225 mm?

16. The acceleration due to the gravity g on a spacecraft is inversely proportional to its distance from the center of the earth. At the surface of the earth, $g = 32.2$ ft/s². Given that the radius of the earth is 3960 mi, how fast is g changing on a spacecraft approaching the earth at 4500 ft/s at a distance of 25,500 mi from the surface?

17. A tank in the shape of an inverted cone has a height of 3.60 m and a radius at the top of 1.15 m. Water is flowing into the tank at the rate of 0.50 m³/min. How fast is the level rising when it is 1.80 m deep?

18. A ladder is slipping down along a vertical wall. If the ladder is 10.0 ft long and the top of it is slipping at the constant rate of 10.0 ft/s, how fast is the bottom of the ladder moving along the ground when the bottom is 6.00 ft from the wall?

19. A rope attached to a boat is being pulled in at a rate of 10.0 ft/s. If the water is 20.0 ft below the level at which the rope is being drawn in, how fast is the boat approaching the wharf when 36.0 ft of rope are yet to be pulled in? See Fig. 4-29.

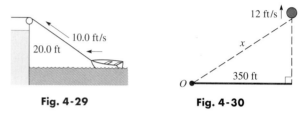

Fig. 4-29   **Fig. 4-30**

20. A weather balloon leaves the ground 350 ft from an observer and rises vertically at 12 ft/s. How fast is the line of sight from the observer to the balloon increasing when the balloon is 250 ft high? See Fig. 4-30.

21. A supersonic jet leaves an airfield traveling due east at 1600 mi/h. A second jet leaves the same airfield at the same time and travels at 1800 mi/h along a line north of east such that it remains due north of the first jet. After a half-hour, how fast are the jets separating?

22. A car passes over a bridge at 15.0 m/s at the same time a boat passes under the bridge at a point 10.5 m directly below the car. If the boat is moving perpendicularly to the bridge at 4.0 m/s, how fast are the car and the boat separating 5.0 s later?

23. A man 6.00 ft tall approaches a street light 15.0 ft above the ground at the rate of 5.00 ft/s. How fast is the end of the man's shadow moving when he is 10.0 ft from the base of the light? See Fig. 4-31.

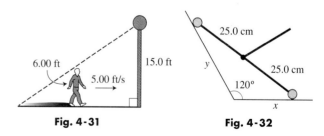

Fig. 4-31   **Fig. 4-32**

24. A roller mechanism, as shown in Fig. 4-32, moves such that the right roller is always in contact with the bottom surface and the left roller is always in contact with the left surface. If the right roller is moving to the right at 1.50 cm/s when $x = 10.0$ cm, how fast is the left roller moving?

4-5 USING DERIVATIVES IN CURVE SKETCHING

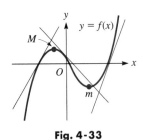

Fig. 4-33

Derivatives can be used effectively in sketching curves. An analysis of the first two derivatives can provide useful information as to the graph of a function. This information, possibly along with two or three key points on the curve, is often sufficient to obtain a good, although approximate, graph of the function. Graphs where this information must be supplemented with other analyses are the subject of the next section.

Considering the function $f(x)$ as shown in Fig. 4-33, we see that as x increases (from left to right) the y-values also increase until the point M is reached. From M to m, the values of y decrease. To the right of m, the values of y again increase. We also note that any tangent line to the left of M or to the right of m will have a positive slope. Any tangent line between M and m will have a negative slope. Since the derivative of a function determines the slope of a tangent line, we can conclude that, *as x increases, y increases if the derivative is positive and decreases if the derivative is negative.* This can be stated as

FUNCTION INCREASING $f(x)$ increases if $f'(x) > 0$

and

FUNCTION DECREASING $f(x)$ decreases if $f'(x) < 0$

CAUTION ▶   *It is always assumed that x is increasing. Also, we assume in our present analysis that f(x) and its derivatives are continuous over the indicated interval.*

■**EXAMPLE 1**   Find those values of x for which the function $f(x) = x^3 - 3x^2$ is increasing and those values for which it is decreasing.

We solve this problem by finding those values of x for which the derivative is positive and those values for which it is negative. The derivative is

$$f'(x) = 3x^2 - 6x = 3x(x - 2)$$

This now becomes a problem of solving an inequality. To find the values of x for which $f(x)$ is increasing, we must solve the inequality

$$3x(x - 2) > 0$$

We now recall that the solution of an inequality consists of *all* values of x which may satisfy it. Normally, this consists of certain intervals of values of x. *These intervals are found by first setting the left side of the inequality equal to zero, thus obtaining the* **critical values** *of x.* The function on the left will have the same *sign* for all values of x less than the leftmost critical value. The sign of the function will also be the same within any given interval between critical values and to the right of the rightmost critical value. Those intervals that give the proper sign will satisfy the inequality. In this case the critical values are $x = 0$ and $x = 2$. Therefore, we have the following analysis:

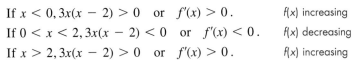

Therefore, the solution of the above inequality is $x < 0$ or $x > 2$, which means that for these values $f(x)$ is increasing. We can also see that for $0 < x < 2$, $f'(x) < 0$, which means that $f(x)$ is decreasing for these values of x. The solution is now complete. The graph of $f(x) = x^3 - 3x^2$ is shown in Fig. 4-34.   ----------■

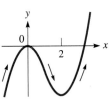

Fig. 4-34

MAXIMUM POINTS
MINIMUM POINTS

The points *M* and *m* in Fig. 4-33 are called a **relative maximum point** and a **relative minimum point,** respectively. *This means that M has a greater y-value than any other point near it and that m has a smaller y-value than any point near it.* This does not necessarily mean that *M* has the greatest *y*-value of any point on the curve or that *m* has the least *y*-value of any point on the curve. However, the points *M* and *m* are the greatest or least values of *y* for that part of the curve (that is why we use the word *relative*). Examination of Fig. 4-33 verifies this point. *The characteristic of both M and m is that the derivative is zero at each point.* (We see that this is so since a tangent line would have a slope of zero at each.) *This is how relative maximum and relative minimum points are located. The derivative is found and then set equal to zero. The solutions of the resulting equation give the x-coordinates of the maximum and minimum points.*

It remains now to determine whether a given value of *x*, for which the derivative is zero, is the coordinate of a maximum or a minimum point (or neither, which is also possible). From the discussion of increasing and decreasing values for *y*, we see that *the derivative changes sign from plus to minus when passing through a relative maximum point and from minus to plus when passing through a relative minimum point.* Thus, we find maximum and minimum points by determining those values of *x* for which the derivative is zero and by properly analyzing the sign change of the derivative. If the sign of the derivative does not change, it is neither a maximum nor a minimum point. This is known as the **first-derivative test for maxima and minima.**

FIRST DERIVATIVE TEST
FOR MAXIMA AND MINIMA

In Fig. 4-35 a diagram for the first-derivative test is shown. The test for a relative maximum is shown in Fig. 4-35(a) and that for a relative minimum is shown in Fig. 4-35(b). For the curves shown in Fig. 4-35, *f*(*x*) and *f*′(*x*) are continuous throughout the interval shown. (Although *f*(*x*) must be continuous, *f*′(*x*) may be discontinuous at the maximum point or the minimum point, and the sign changes of the first-derivative test remain valid.)

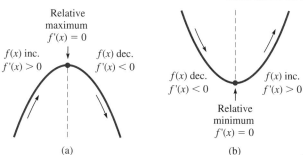

Fig. 4-35

EXAMPLE 2   Find any maximum points and minimum points on the graph of the function

$$y = 3x^5 - 5x^3$$

Finding the derivative and setting it equal to zero, we have

$$y' = 15x^4 - 15x^2 = 15x^2(x^2 - 1) = 15x^2(x - 1)(x + 1)$$

Therefore,

$$15x^2(x - 1)(x + 1) = 0 \quad \text{for } x = 0, \quad x = 1, \quad x = -1$$

Thus, the sign of the derivative is the same for all points to the left of *x* = −1. For these values, *y*′ > 0 (thus, *y* is increasing). For values of *x* between −1 and 0, *y*′ < 0. For values of *x* between 0 and 1, *y*′ < 0. For values of *x* greater than 1, *y*′ > 0. Thus, the curve has a maximum at (−1, 2) and a minimum at (1, −2). The point (0, 0) is neither a maximum nor a minimum, since the sign of the derivative did not change at this value of *x*. The graph of *y* = 3*x*⁵ − 5*x*³ is shown in Fig. 4-36.

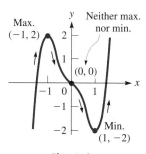

Fig. 4-36

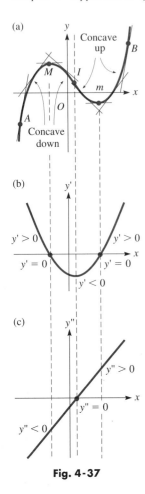

(a)

(b)

(c)

Fig. 4-37

We now look again at the slope of a tangent drawn to a curve. In Fig. 4-37(a), consider the *change* in the values of the slope of a tangent at a point as the point moves from *A* to *B*. At *A* the slope is positive, and as the point moves toward *M*, the slope remains positive but becomes smaller until it becomes zero at *M*. To the right of *M*, the slope is negative and becomes more negative until it reaches *I*. Therefore, *from A to I, the slope continually decreases.* To the right of *I*, the slope remains negative but increases until it becomes zero again at *m*. To the right of *m*, the slope becomes positive and increases to point *B*. Therefore, *from I to B, the slope continually increases.* We say that *the curve is* **concave down** *from A to I and* **concave up** *from I to B.*

The curve in Fig. 4-37(b) is that of the derivative, and it therefore indicates the values of the slope of $f(x)$. If the slope changes, we are dealing with the rate of change of slope or the rate of change of the derivative. This function is the second derivative. The curve in Fig. 4-37(c) is that of the second derivative. We see that *where the second derivative of a function is* **negative,** *the slope is decreasing, or the curve is* **concave down** *(opens down). Where the second derivative is* **positive,** *the slope is increasing, or the curve is* **concave up** *(opens up).* This may be summarized as follows:

If $f''(x) > 0$, the curve is concave up.

If $f''(x) < 0$, the curve is concave down.

We can also now use this information in the determination of maximum and minimum points. By the nature of the definition of maximum and minimum points and of concavity, if $f'(x)$ and $f''(x)$ are continuous, *a curve is concave down at a maximum point and concave up at a minimum point.* We can see these properties when we make a close analysis of the curve in Fig. 4-37. Therefore, at $x = a$,

if $f'(a) = 0$ and $f''(a) < 0$,

then $f(x)$ has a relative maximum at $x = a$, or

if $f'(a) = 0$ and $f''(a) > 0$,

then $f(x)$ has a relative minimum at $x = a$.

SECOND DERIVATIVE TEST FOR MAXIMA AND MINIMA

These statements comprise what is known as the **second-derivative test for maxima and minima.** This test is often easier to use than the first-derivative test. However, it can happen that $y'' = 0$ at a maximum or minimum point, and in such cases it is necessary that we use the first-derivative test.

In using the second-derivative test, we should note that $f''(x)$ is **negative** at a **maximum** point and **positive** at a **minimum** point. This is contrary to a natural inclination to think of "maximum" and "positive" together or "minimum" and "negative" together.

POINTS OF INFLECTION

The points at which the curve changes from concave up to concave down, or from concave down to concave up, are known as **points of inflection.** Thus, point *I* in Fig. 4-37 is a point of inflection. Inflection points are found by determining those values of *x* for which the second derivative changes sign. This is analogous to finding maximum and minimum points by the first-derivative test. In Fig. 4-38, various types of points of inflection are illustrated.

To show that it is necessary for $f(x)$ and its derivatives to be continuous, see Exercise 40 of this section.

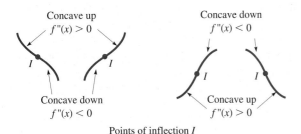

Points of inflection *I*

Fig. 4-38

EXAMPLE 3   Determine the concavity and find any points of inflection of the function $y = x^3 - 3x$.

This requires an inspection and analysis of the second derivative. Therefore, we find the first two derivatives.

$$y' = 3x^2 - 3$$
$$y'' = 6x$$

The second derivative is positive where the function is concave up, and this occurs if $x > 0$. The curve is concave down for $x < 0$, since y'' is negative. Thus, $(0, 0)$ is a point of inflection, since the concavity changes there. The graph of $y = x^3 - 3x$ is shown in Fig. 4-39.

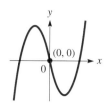

Fig. 4-39

At this point we summarize the information found from the derivatives of a function $f(x)$. See Fig. 4-40.

$f'(x) > 0$ **where** $f(x)$ **increases;** $f'(x) < 0$ **where** $f(x)$ **decreases.**

$f''(x) > 0$ **where the graph of** $f(x)$ **is concave up;** $f''(x) < 0$ **where the graph of** $f(x)$ **is concave down.**

If $f'(x) = 0$ **at** $x = a$**, there is a maximum point if** $f'(x)$ **changes from + to − or if** $f''(a) < 0$**.**

If $f'(x) = 0$ **at** $x = a$**, there is a minimum point if** $f'(x)$ **changes from − to + or if** $f''(a) > 0$**.**

If $f''(x) = 0$ **at** $x = a$**, there is a point of inflection if** $f''(x)$ **changes from + to − or from − to +.**

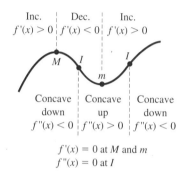

Fig. 4-40

The following examples illustrate how the above information is put together to obtain the graph of a function.

EXAMPLE 4   Sketch the graph of $y = 6x - x^2$.

Finding the first two derivatives, we have

$$y' = 6 - 2x = 2(3 - x)$$
$$y'' = -2$$

We now note that $y' = 0$ for $x = 3$. For $x < 3$, we see that $y' > 0$, which means that y is increasing over this interval. Also, for $x > 3$, we note that $y' < 0$, which means that y is decreasing over this interval.

Since y' changes from positive on the left of $x = 3$ to negative on the right of $x = 3$, the curve has a maximum point where $x = 3$. Since $y = 9$ for $x = 3$, this maximum point is $(3, 9)$.

Since $y'' = -2$, this means that its value remains constant for all values of x. Therefore, there are no points of inflection, and the curve is concave down for all values of x. This also shows that the point $(3, 9)$ is a maximum point.

Summarizing, we know that y is increasing for $x < 3$, y is decreasing for $x > 3$, there is a maximum point at $(3, 9)$, and the curve is always concave down. Using this information, we sketch the curve shown in Fig. 4-41.

From the equation, we know this curve is a parabola. We could also find the maximum point from the material of Section 2-7. However, using derivatives we can find this kind of important information about the graphs of a great many types of functions.

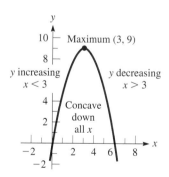

Fig. 4-41

EXAMPLE 5 Sketch the graph of $y = 2x^3 + 3x^2 - 12x$.

Finding the first two derivatives, we have

$$y' = 6x^2 + 6x - 12 = 6(x + 2)(x - 1)$$

$$y'' = 12x + 6 = 6(2x + 1)$$

We note that $y' = 0$ when $x = -2$ and $x = 1$. Using these values in the second derivative, we find that y'' is negative (-18) for $x = -2$ and y'' is positive $(+18)$ when $x = 1$. When $x = -2$, $y = 20$; and when $x = 1$, $y = -7$. Therefore, $(-2, 20)$ is a maximum point, and $(1, -7)$ is a minimum point.

Next we see that $y' > 0$ if $x < -2$ or $x > 1$. Also, $y' < 0$ for the interval $-2 < x < 1$. Therefore, y is increasing if $x < -2$ or $x > 1$, and y is decreasing if $-2 < x < 1$.

Now we note that $y'' = 0$ when $x = -\frac{1}{2}$, $y'' < 0$ when $x < -\frac{1}{2}$, and $y'' > 0$ when $x > -\frac{1}{2}$. When $x = -\frac{1}{2}$, $y = \frac{13}{2}$. Therefore, there is a point of inflection at $(-\frac{1}{2}, \frac{13}{2})$, the curve is concave down if $x < -\frac{1}{2}$, and the curve is concave up if $x > -\frac{1}{2}$.

Finally, by locating the points $(-2, 20)$, $(-\frac{1}{2}, \frac{13}{2})$, and $(1, -7)$, we draw the curve *up* to $(-2, 20)$ and then *down* to $(-\frac{1}{2}, \frac{13}{2})$, with the curve *concave down.* **CAUTION** *Continuing* **down,** *but* **concave up,** we draw the curve to $(1, -7)$, at which point we start *up* and continue up. We now know the key points and the shape of the curve. See Fig. 4-42. For more precision, additional points may be used. ■

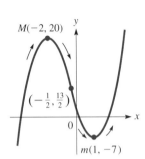

Fig. 4-42

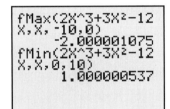

Fig. 4-43

Most graphing calculators have a feature by which we can find the x-value for which a function is maximum (or minimum) over a specified interval. To find a *relative* maximum (or minimum) value, care must be used in choosing the lower and upper endpoints of the interval so as not to include values for which the function may be greater (or less) than the value at the *relative* maximum (or minimum). A typical calculator display (showing the function, the variable, the lower interval value chosen, and the upper interval value chosen) to find the values of x for the maximum point and minimum point of Example 5 is shown in Fig. 4-43.

EXAMPLE 6 Sketch the graph of $y = x^5 - 5x^4$.

The first two derivatives are

$$y' = 5x^4 - 20x^3 = 5x^3(x - 4)$$

$$y'' = 20x^3 - 60x^2 = 20x^2(x - 3)$$

CAUTION We now see that $y' = 0$ when $x = 0$ and $x = 4$. For $x = 0$, $y'' = 0$ also, which means **we cannot use the second-derivative test** for maximum and minimum points for $x = 0$ in this case. For $x = 4$, $y'' > 0$ $(+320)$, which means that $(4, -256)$ is a minimum point.

Next we note that

$$y' > 0 \quad \text{for } x < 0 \quad \text{or} \quad x > 4 \qquad y' < 0 \quad \text{for } 0 < x < 4$$

Thus, by the first-derivative test, there is a maximum point at $(0, 0)$. Also, y is increasing for $x < 0$ or $x > 4$ and decreasing for $0 < x < 4$.

The second derivative indicates that there is a point of inflection at $(3, -162)$. It also indicates that the curve is concave down for $x < 3$ $(x \neq 0)$ and concave up for $x > 3$. There is no point of inflection at $(0, 0)$ since the second derivative does not change sign at $x = 0$.

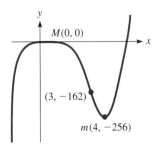

Fig. 4-44

From this information, we sketch the curve in Fig. 4-44. ■

EXERCISES *4-5*

In Exercises 1–4, find those values of x for which the given functions are increasing and those values of x for which they are decreasing.

1. $y = x^2 + 2x$ **2.** $y = 2 + 6x - 3x^2$
3. $y = 12x - x^3$ **4.** $y = x^4 - 6x^2$

In Exercises 5–8, find any maximum or minimum points of the given functions. (These are the same functions as in Exercises 1–4.)

5. $y = x^2 + 2x$ **6.** $y = 2 + 6x - 3x^2$
7. $y = 12x - x^3$ **8.** $y = x^4 - 6x^2$

In Exercises 9–12, find the values of x for which the given function is concave up, the values of x for which it is concave down, and any points of inflection. (These are the same functions as in Exercises 1–4.)

9. $y = x^2 + 2x$ **10.** $y = 2 + 6x - 3x^2$
11. $y = 12x - x^3$ **12.** $y = x^4 - 6x^2$

In Exercises 13–16, sketch the graphs of the given functions by determining the appropriate information and points from the first and second derivatives (see Exercises 1–12). Use a graphing calculator to check the graph.

13. $y = x^2 + 2x$ **14.** $y = 2 + 6x - 3x^2$
15. $y = 12x - x^3$ **16.** $y = x^4 - 6x^2$

In Exercises 17–28, sketch the graphs of the given functions by determining the appropriate information and points from the first and second derivatives. Use a graphing calculator to check the graph. In Exercises 23–28, use the function maximum-minimum feature to check the maximum and minimum points.

17. $y = 12x - 2x^2$ **18.** $y = 3x^2 - 1$
19. $y = 2x^3 + 6x^2$ **20.** $y = x^3 - 9x^2 + 15x + 1$
21. $y = x^3 + 3x^2 + 3x + 2$ **22.** $y = x^3 - 12x + 12$
23. $y = 4x^3 - 24x^2 + 36x$ **24.** $y = x(x - 4)^3$
25. $y = 4x^3 - 3x^4$ **26.** $y = x^5 - 20x^2$
27. $y = x^5 - 5x$ **28.** $y = x^4 + 8x + 2$

(W) In Exercises 29 and 30, view the graphs of y, y′, and y″ together on a graphing calculator. State how the graphs of y′ and y″ are related to the graph of y.

29. $y = x^3 - 12x$ **30.** $y = 24x - 9x^2 - 2x^3$

In Exercises 31–36, sketch the indicated curves by the methods of this section. You may check the graphs by using a graphing calculator.

31. A batter hits a baseball that follows a path given by $y = x - 0.0025x^2$, where distances are in feet. Sketch the graph of the path of the baseball.

32. The angle θ (in degrees) of a robot arm with the horizontal as a function of the time t (in s) is given by $\theta = 10 + 12t^2 - 2t^3$. Sketch the graph for $0 \le t \le 6$ s.

33. An electric circuit is designed such that the resistance R (in Ω) is a function of the current i (in mA) according to $R = 75 - 18i^2 + 8i^3 - i^4$. Sketch the graph if $R \ge 0$ and i can be positive or negative.

34. An analysis of data showed that the mean density d (in mg/cm³) of a calcium compound in the bones of women was given by $d = 0.00181x^3 - 0.289x^2 + 12.2x + 30.4$, where x represents the ages of women ($20 < x < 80$ years). (A woman probably has osteoporosis if $d < 115$ mg/cm³.) Sketch the graph.

35. A rectangular box is made from a piece of cardboard 8 in. by 12 in. by cutting equal squares from each corner and bending up the sides. See Fig. 4-45. Express the volume of the box as a function of the side of the square that is cut out and then sketch the curve of the resulting equation.

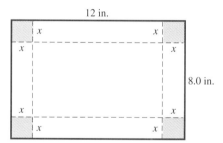

Fig. 4-45

36. A rectangular planter with a square end is to be made from 64 ft² of redwood. Express the volume of soil the planter can hold as a function of the side of the square of the end. Sketch the graph of the resulting function.

In Exercises 37–39, sketch a continuous curve that has the given characteristics. In Exercise 40, perform the required operations.

37. $f(1) = 0$, $f'(x) > 0$ for all x, $f''(x) < 0$ for all x
38. $f(0) = 1$, $f'(x) < 0$ for all x
    $f''(x) < 0$ for $x < 0$, $f''(x) > 0$ for $x > 0$
39. $f(-1) = 0$, $f(2) = 2$; $f'(x) < 0$ for $x < -1$;
    $f'(x) > 0$ for $x > -1$; $f''(x) < 0$ for $0 < x < 2$;
    $f''(x) > 0$ for $x < 0$ or $x > 2$

(W) **40.** Display the graph of $f(x) = x^{2/3}$ for $-2 < x < 2$ on a graphing calculator. Determine the continuity of $f(x)$, $f'(x)$, and $f''(x)$. Discuss the concavity of the curve in relation to the minimum point. (See the last paragraph on page 136.)

$4\text{-}6$ MORE ON CURVE SKETCHING

At this point we combine the information from the derivative with information obtainable from the function itself to sketch the graph. We determine intercepts, symmetry, the behavior of the curve as x becomes large, the vertical asymptotes, and the domain and range of the function. Also, continuity is important in sketching certain functions. We will find that some of these features are of more value than others in graphing any particular curve.

■EXAMPLE 1   Sketch the graph of $y = \dfrac{8}{x^2 + 4}$.

Intercepts: If $x = 0$, $y = 2$, which means $(0, 2)$ is an intercept. If $y = 0$, there is no corresponding value of x, since $2/(x^2 + 1)$ is a fraction greater than zero for all x. This also indicates that all points on the curve are above the x-axis.

Symmetry: For a review of symmetry, see Section 2-3.

The curve is symmetric to the y-axis since $y = \dfrac{8}{(-x)^2 + 4}$ is the same as $y = \dfrac{8}{x^2 + 4}$.

The curve is not symmetric to the x-axis since $-y = \dfrac{8}{x^2 + 4}$ is not the same as $y = \dfrac{8}{x^2 + 4}$.

The curve is not symmetric to the origin since $-y = \dfrac{8}{(-x)^2 + 4}$ is not the same as $y = \dfrac{8}{x^2 + 4}$.

The value in knowing the symmetry is that we should find those portions of the curve on either side of the y-axis reflections of the other. It is possible to use this fact directly or to use it as a check.

Behavior as x becomes large: We note that as $x \to \infty$, $y \to 0$ since $8/(x^2 + 4)$ is always a fraction that is greater than zero but which approaches zero as x becomes larger. Therefore, we see that $y = 0$ is an asymptote. From either the symmetry or the function, we also see that $y \to 0$ as $x \to -\infty$.

Vertical asymptotes: From the discussion of the hyperbola, we recall that an asymptote is a line that a curve approaches. We have already noted that $y = 0$ is an asymptote for this curve. This asymptote, the x-axis, is a horizontal line. *Vertical asymptotes, if any exist, are found by determining those values of x for which the denominator of any term is zero.* Such a value of x makes y undefined. Since $x^2 + 4$ cannot be zero, this curve has no vertical asymptotes. The next example illustrates a curve that has a vertical asymptote.

Domain and range: Since the denominator $x^2 + 4$ cannot be zero, x can take on any value. This means the domain of the function is all values of x. Also, we have noted that $8/(x^2 + 4)$ is a fraction greater than zero. Since $x^2 + 4$ is 4 or greater, y is 2 or less. This tells us that the range of the function is $0 < y \le 2$.

Derivatives: Since $y = \dfrac{8}{x^2 + 4} = 8(x^2 + 4)^{-1}$

$$y' = -8(x^2 + 4)^{-2}(2x) = \frac{-16x}{(x^2 + 4)^2}$$

Since $(x^2 + 4)^2$ is positive for all values of x, the sign of y' is determined by the numerator. Thus, we note that $y' = 0$ for $x = 0$ and that $y' > 0$ for $x < 0$ and $y' < 0$ for $x > 0$. The curve, therefore, is increasing for $x < 0$, is decreasing for $x > 0$, and has a maximum point at $(0, 2)$.

The curve for Example 1 is a special case (with $a = 1$) of the curve known as the *witch of Agnesi.* Its general form is

$$y = \frac{8a^3}{x^2 + 4a^2}$$

It is named for the Italian mathematician Maria Gaetana Agnesi (1718–1799). She wrote the first text that contained analytic geometry, differential and integral calculus, series (see Chapter 13), and differential equations (see Chapter 14). The word *witch* was used due to a mistranslation from Italian to English.

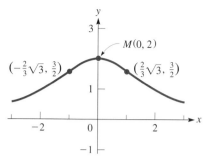

Fig. 4-46

Now finding the second derivative, we have

$$y'' = \frac{(x^2 + 4)^2(-16) + 16x(2)(x^2 + 4)(2x)}{(x^2 + 4)^4} = \frac{-16(x^2 + 4) + 64x^2}{(x^2 + 4)^3}$$

$$= \frac{48x^2 - 64}{(x^2 + 4)^3} = \frac{16(3x^2 - 4)}{(x^2 + 4)^3}$$

We note that y'' is negative for $x = 0$, which confirms that $(0, 2)$ is a maximum point. Also, points of inflection are found for the values of x satisfying $3x^2 - 4 = 0$. Thus, $(-\frac{2}{3}\sqrt{3}, \frac{3}{2})$ and $(\frac{2}{3}\sqrt{3}, \frac{3}{2})$ are points of inflection. The curve is concave up if $x < -\frac{2}{3}\sqrt{3}$ or $x > \frac{2}{3}\sqrt{3}$, and the curve is concave down if $-\frac{2}{3}\sqrt{3} < x < \frac{2}{3}\sqrt{3}$.

Putting this information together, we sketch the curve shown in Fig. 4-46. Note that this curve could have been sketched primarily by use of the fact that $y \to 0$ as $x \to +\infty$ and as $x \to -\infty$ and the fact that a maximum point exists at $(0, 2)$. However, the other parts of the analysis, such as symmetry and concavity, serve as checks and make the curve more accurate.

∎

EXAMPLE 2 Sketch the graph of $y = x + \dfrac{4}{x}$.

Intercepts: If we set $x = 0$, y is undefined. This means that the curve is *not continuous* at $x = 0$ and there are no y-intercepts. If we set $y = 0$, $x + 4/x = (x^2 + 4)/x$ cannot be zero since $x^2 + 4$ cannot be zero. Therefore, there are no intercepts. This may seem to be of little value, but we must realize *this curve does not cross either axis*. This will be of value when we sketch the curve in Fig. 4-47.

Symmetry: In testing for symmetry, we find that the curve is not symmetric to either axis. However, this curve does possess symmetry to the origin. This is determined by the fact that when $-x$ replaces x and at the same time $-y$ replaces y, the equation is the same as the original equation (after simplifying).

Behavior as x becomes large: As $x \to +\infty$ and as $x \to -\infty$, $y \to x$ since $4/x \to 0$. Thus, $y = x$ is an asymptote of the curve.

Vertical asymptotes: As we noted in Example 1, vertical asymptotes exist for values of x for which y is undefined. In this equation, $x = 0$ makes the second term on the right undefined, and therefore y is undefined. In fact, as $x \to 0$ from the positive side, $y \to +\infty$, and as $x \to 0$ from the negative side, $y \to -\infty$. This is derived from the sign of $4/x$ in each case.

Domain and range: Since x cannot be zero, the domain of the function is all x except zero. As for the range, the analysis from the derivatives will show it to be $y \leq -4$, $y \geq 4$.

Derivatives: Finding the first derivative, we have

$$y' = 1 - \frac{4}{x^2} = \frac{x^2 - 4}{x^2}$$

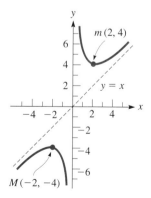

Fig. 4-47

The x^2 in the denominator indicates that the sign of the first derivative is the same as its numerator. The numerator is zero if $x = -2$ or $x = 2$. If $x < -2$ or $x > 2$, then $y' > 0$; and if $-2 < x < 2$, $x \neq 0$, $y' < 0$. Thus, y is increasing if $x < -2$ or $x > 2$, and y is decreasing if $-2 < x < 2$, except at $x = 0$ (y is undefined). Also, $(-2, -4)$ is a maximum point, and $(2, 4)$ is a minimum point. The second derivative is $y'' = 8/x^3$. This cannot be zero, but it is negative if $x < 0$ and positive if $x > 0$. Thus, the curve is concave down if $x < 0$ and concave up if $x > 0$. Using this information, we have the curve shown in Fig. 4-47.

∎

EXAMPLE 3   Sketch the graph of $y = \dfrac{1}{\sqrt{1 - x^2}}$.

Intercepts: If $x = 0$, $y = 1$. If $y = 0$, $1/\sqrt{1 - x^2}$ would have to be zero, but it cannot since it is a fraction with 1 as the numerator for all values of x. Thus, $(0, 1)$ is an intercept.

Symmetry: The curve is symmetric to the y-axis.

Behavior as x becomes large: The values of x cannot be considered beyond 1 or -1, for any value of $x < -1$ or $x > 1$ gives imaginary values for y. Thus, the curve does not exist for values of $x < -1$ or $x > 1$.

Vertical asymptotes: If $x = 1$ or $x = -1$, y is undefined. In each case, as $x \to 1$ (from the left) and as $x \to -1$ (from the right), $y \to +\infty$.

Domain and range: From the analysis of x becoming large and of the vertical asymptotes, we see that the domain is $-1 < x < 1$. Also, since $\sqrt{1 - x^2}$ is 1 or less, $1/\sqrt{1 - x^2}$ is 1 or more, which means the range is $y \geq 1$.

Derivatives:

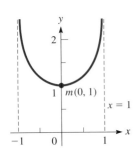

Fig. 4-48

$$y' = -\frac{1}{2}(1 - x^2)^{-3/2}(-2x) = \frac{x}{(1 - x^2)^{3/2}}$$

We see that $y' = 0$ if $x = 0$. If $-1 < x < 0$, $y' < 0$, and also if $0 < x < 1$, $y' > 0$. Thus, the curve is decreasing if $-1 < x < 0$ and increasing if $0 < x < 1$. There is a minimum point at $(0, 1)$.

$$y'' = \frac{(1 - x^2)^{3/2} - x(\frac{3}{2})(1 - x^2)^{1/2}(-2x)}{(1 - x^2)^3} = \frac{(1 - x^2) + 3x^2}{(1 - x^2)^{5/2}}$$

$$= \frac{2x^2 + 1}{(1 - x^2)^{5/2}}$$

The second derivative cannot be zero since $2x^2 + 1$ is positive for all values of x. The second derivative is also positive for all permissible values of x, which means the curve is concave up for these values.

Using this information, we sketch the graph in Fig. 4-48. ■

EXAMPLE 4   Sketch the graph of $y = \dfrac{x}{x^2 - 4}$.

Intercepts: If $x = 0$, $y = 0$, and if $y = 0$, $x = 0$. The only intercept is $(0, 0)$.

Symmetry: The curve is not symmetric to either axis. However, since $-y = -x/[(-x)^2 - 4]$ is the same (after simplifying) as $y = x/(x^2 - 4)$, it is symmetric to the origin.

Behavior as x becomes large: As $x \to +\infty$ and as $x \to -\infty$, $y \to 0$. This means that $y = 0$ is an asymptote.

Vertical asymptotes: If $x = -2$ or $x = 2$, y is undefined. As $x \to -2$, $y \to -\infty$ if $x < -2$ since $x^2 - 4$ is positive, and $y \to +\infty$ if $x > -2$ since $x^2 - 4$ is negative. As $x \to 2$, $y \to -\infty$ if $x < 2$, and $y \to +\infty$ if $x > 2$.

Domain and range: The domain is all real values of x except -2 and 2. As for the range, if $x < -2$, $y < 0$ (the numerator is negative and the denominator is positive). If $x > 2$, $y > 0$ (both numerator and denominator are positive). Since $(0, 0)$ is an intercept, we see that the range is all values of y.

Derivatives:

$$y' = \frac{(x^2 - 4)(1) - x(2x)}{(x^2 - 4)^2} = -\frac{x^2 + 4}{(x^2 - 4)^2}$$

Since $y' < 0$ for all values of x except -2 and 2, the curve is decreasing for all values in the domain.

$$y'' = -\frac{(x^2 - 4)^2(2x) - (x^2 + 4)(2)(x^2 - 4)(2x)}{(x^2 - 4)^4}$$

$$= -\frac{2x(x^2 - 4) - 4x(x^2 + 4)}{(x^2 - 4)^3} = \frac{2x^3 + 24x}{(x^2 - 4)^3} = \frac{2x(x^2 + 12)}{(x^2 - 4)^3}$$

The sign of y'' depends on x and $(x^2 - 4)^3$. If $x < -2$, $y'' < 0$. If $-2 < x < 0$, $y'' > 0$. If $0 < x < 2$, $y'' < 0$. If $x > 2$, $y'' > 0$. This means the curve is concave down for $x < -2$ or $0 < x < 2$ and is concave up for $-2 < x < 0$ or $x > 2$. The curve is sketched in Fig. 4-49.

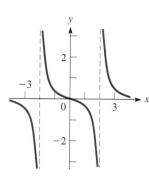

Fig. 4-49

EXERCISES 4-6

In the following exercises, use the method of the examples of this section to sketch the indicated curves. Use a graphing calculator to check the graph.

1. $y = \dfrac{4}{x^2}$

2. $y = \dfrac{2}{x^3}$

3. $y = \dfrac{2}{x + 1}$

4. $y = \dfrac{x}{x - 2}$

5. $y = x^2 + \dfrac{2}{x}$

6. $y = x + \dfrac{4}{x^2}$

7. $y = x - \dfrac{1}{x}$

8. $y = 3x + \dfrac{1}{x^3}$

9. $y = \dfrac{x^2}{x + 1}$

10. $y = \dfrac{9x}{x^2 + 9}$

11. $y = \dfrac{1}{x^2 - 1}$

12. $y = \dfrac{x^2 - 1}{x^3}$

13. $y = \dfrac{4}{x} - \dfrac{4}{x^2}$

14. $y = 4x + \dfrac{1}{\sqrt{x}}$

15. $y = x\sqrt{1 - x^2}$

16. $y = \dfrac{x - 1}{x^2 - 2x}$

17. $y = \dfrac{9x}{9 - x^2}$

18. $y = \dfrac{x^2 - 4}{x^2 + 4}$

19. The combined capacitance C_T (in μF) of a 6-μF capacitance and a variable capacitance C in series is given by $C_T = \dfrac{6C}{6 + C}$. Sketch the graph.

20. The number n of dollars saved by increasing the fuel efficiency of e mi/gal to $e + 6$ mi/gal for a car driven 10,000 mi/year is $n = \dfrac{75{,}000}{e(e + 6)}$, if the cost of gas is \$1.25/gal. Sketch the graph. (See Exercise 40 on page 99).

21. The reliability R of a computer model is found to be $R = \dfrac{200}{\sqrt{t^2 + 40{,}000}}$, where t is the time of operation in hours. ($R = 1$ is perfect reliability, and $R = 0.5$ means there is a 50% chance of a malfunction.) Sketch the graph.

22. The electric power P (in W) produced by a source is given by $P = \dfrac{36R}{R^2 + 2R + 1}$, where R is the resistance in the circuit. Sketch the graph.

23. A cylindrical oil drum is to be made such that it will contain 20 kL. Sketch the area of sheet metal required for construction as a function of the radius of the drum.

24. A fence is to be constructed to enclose a rectangular area of 20,000 m². A previously constructed wall is to be used for one side. Sketch the length of fence to be built as a function of the length of the side of the fence parallel to the wall. See Fig. 4-50.

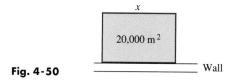

Fig. 4-50

$4\text{-}7$ APPLIED MAXIMUM AND MINIMUM PROBLEMS

Problems from various applied situations frequently occur that require finding a maximum or minimum value of some function. If the function is known, the methods we have already discussed can be used directly. This is discussed in the following example.

> The first gasoline-engine automobile was built by the German engineer Karl Benz (1844–1929) in the 1880s.

■EXAMPLE 1 An automobile manufacturer, in testing a new engine on one of its new models, found that the efficiency e of the engine as a function of the speed s of the car was given by $e = 0.768s - 0.00004s^3$. Here, e is measured in percent and s is measured in km/h. What is the maximum efficiency of the engine?

In order to find a maximum value, we find the derivative of e with respect to s.

$$\frac{de}{ds} = 0.768 - 0.00012s^2$$

We then set the derivative equal to zero in order to find the value of s for which a maximum may occur.

$$0.768 - 0.00012s^2 = 0$$
$$0.00012s^2 = 0.768$$
$$s^2 = 6400$$
$$s = 80.0 \text{ km/h}$$

We know that s must be positive to have meaning in this problem. Therefore, the apparent solution of $s = -80$ is discarded. The second derivative is

$$\frac{d^2e}{ds^2} = -0.00024s$$

which is negative for any positive value of s. Therefore, we have a maximum for $s = 80.0$. Substituting $s = 80.0$ in the function for e, we obtain

$$e = 0.768(80.0) - 0.00004(80.0^3) = 61.44 - 20.48 = 40.96$$

The maximum efficiency is about 41.0%, which occurs for $s = 80.0$ km/h. ■

In many problems for which a maximum or minimum value is to be found, the function is not given. To solve such a problem, we use these steps:

Steps in Solving Applied Maximum and Minimum Problems

1. *Determine the quantity Q to be maximized or minimized.*
2. If possible, *draw a figure illustrating the problem.*
3. *Write an equation for Q in terms of another variable of the problem.*
4. *Take the derivative of the function in step 3.*
5. *Set the derivative equal to zero, and solve the resulting equation.*
6. *Check as to whether the value found in step 5 makes Q a maximum or a minimum.* This might be clear from the statement of the problem, or it might require one of the derivative tests.
7. *Be sure the stated answer is the one the problem required.* Some problems require the maximum or minimum value, and others require values of other variables that give the maximum or minimum value.

CAUTION ▶ ***The principal difficulty that arises in these problems is finding the proper function.*** We must carefully read the problem to find the information needed to set up the function. The following examples illustrate several types of stated problems involving maximum and minimum values.

SOLVING A WORD PROBLEM ▮**EXAMPLE 2**   Find the number that exceeds its square by the greatest amount.

The quantity to be maximized is the difference D between a number x and its square x^2. Therefore, the required function is

$$D = x - x^2$$

Since we want D to be a maximum, we find dD/dx, which is

$$\frac{dD}{dx} = 1 - 2x$$

Setting the derivative equal to zero and solving for x, we have

$$0 = 1 - 2x, \qquad x = \tfrac{1}{2}$$

The second derivative gives $d^2D/dx^2 = -2$, which tells us that the second derivative is always negative. This means that whenever the first derivative is zero it represents a maximum. In many problems it is not necessary to test for maximum or minimum, since the nature of the problem will indicate which must be the case. For example, in this problem we know that numbers greater than 1 do not exceed their squares at all. The same is true for all negative numbers. Thus the answer must be between 0 and 1; in this case it is $x = 1/2$. ▮

SOLVING A WORD PROBLEM ▮**EXAMPLE 3**   A rectangular corral is to be enclosed with 1600 ft of fencing. Find the maximum possible area of the corral.

There are limitless possibilities for rectangles of a perimeter of 1600 ft and differing areas. See Fig. 4-51. For example, if the sides are 700 ft and 100 ft, the area is 70,000 ft^2, or if the sides are 600 ft and 200 ft, the area is 120,000 ft^2. Therefore, we set up a function for the area of a rectangle in terms of its sides x and y.

$$A = xy$$

Another important fact is that the perimeter of the corral is 1600 ft. Therefore, $2x + 2y = 1600$. Solving for y, we have $y = 800 - x$. By using this expression for y, we can express the area in terms of x only. This gives us

$$A = x(800 - x) = 800x - x^2$$

Fig. 4-51

We complete the solution as follows:

$$\frac{dA}{dx} = 800 - 2x \qquad \text{take derivative}$$

$$800 - 2x = 0 \qquad \text{set derivative equal to zero}$$

$$x = 400 \text{ ft}$$

By checking values of the derivative near 400 or by finding the second derivative, we can show that we have a maximum for $x = 400$. This means that $x = 400$ ft and $y = 400$ ft give the maximum area of 160,000 ft^2 for the corral. ▮

EXAMPLE 4 The strength S of a beam with a rectangular cross section is directly proportional to the product of its width w and the square of its depth d. Find the dimensions of the strongest beam that can be cut from a log with a circular cross section that is 16.0 in. in diameter. See Fig. 4-52.

The solution proceeds as follows:

Fig. 4-52

$$S = kwd^2 \qquad \text{direct variation}$$
$$d^2 = 256 - w^2 \qquad \text{Pythagorean theorem}$$
$$S = kw(256 - w^2) \qquad \text{substituting}$$
$$= k(256w - w^3) \qquad S = f(w)$$
$$\frac{dS}{dw} = k(256 - 3w^2) \qquad \text{take derivative}$$
$$0 = k(256 - 3w^2) \qquad \text{set derivative equal to zero}$$
$$3w^2 = 256 \qquad \text{solve for } w$$
$$w = \frac{16}{\sqrt{3}} = 9.24 \text{ in.}$$
$$d = \sqrt{256 - \frac{256}{3}} \qquad \text{solve for } d$$
$$= 13.1 \text{ in.}$$

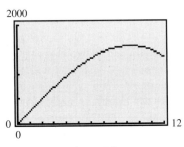

Fig. 4-53

This means that the strongest beam is about 9.24 in. wide and 13.1 in. deep. Since $d^2S/dw^2 = -6kw$ and is negative for $w > 0$ (the only values with meaning in this problem), these dimensions give the maximum strength for the beam.

The solution can also be checked on a graphing calculator. By graphing the equation $S = 256w - w^3$ (using y for S, x for w, and $k = 1$ (*the value of k does not affect the solution*)), we see in Fig. 4-53 that S is a maximum for w between 9 and 10. (The value $w = 9.24$ can be found by using the *trace* and *zoom* (or *maximum*) features.)

EXAMPLE 5 Find the point on the parabola $y = x^2$ that is nearest to the point $(6, 3)$.

In this example we must set up a function for this distance between a general point (x, y) on the parabola and the point $(6, 3)$. This relation is

$$D = \sqrt{(x - 6)^2 + (y - 3)^2}$$

To make it easier to take derivatives, we shall square both sides of this expression. If a nonnegative valued function is a minimum, then so is its square. We shall also use the fact that the point (x, y) is on $y = x^2$ by replacing y by x^2.

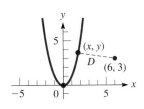

Fig. 4-54

This equation can be solved by use of a calculator (see page 20), or by synthetic division.

$$D^2 = (x - 6)^2 + (x^2 - 3)^2 = x^2 - 12x + 36 + x^4 - 6x^2 + 9$$
$$= x^4 - 5x^2 - 12x + 45$$
$$\frac{dD^2}{dx} = 4x^3 - 10x - 12 \qquad \text{take derivative}$$
$$0 = 2x^3 - 5x - 6 \qquad \text{set derivative equal to zero}$$

We find that the solution to this equation is $x = 2$. Thus, the required point on the parabola is $(2, 4)$. (See Fig. 4-54.) We can show that we have a minimum by analyzing the first derivative, by analyzing the second derivative, or by noting that points at much greater distances exist (therefore, it cannot be a maximum).

SOLVING A WORD PROBLEM

■**EXAMPLE 6** A company determines that it can sell 1000 units of a product per month if the price is $5 for each unit. It also estimates that for each 1¢ reduction in unit price, 10 more units can be sold. Under these conditions, what is the maximum possible income and what price per unit gives this income?

If we let x = the number of units over 1000 sold, the total number of units sold is $1000 + x$. The price for each unit is $5 less 1¢ ($0.01) for each block of 10 units over 1000 that are sold. Thus, the price for each unit is

$$5 - 0.01\left(\frac{x}{10}\right) \quad \text{or} \quad 5 - 0.001x \text{ dollars}$$

The income I is the number of units sold times the price of each unit. Therefore,

$$I = (1000 + x)(5 - 0.001x)$$

Multiplying and finding the first derivative, we have

$$I = 5000 + 4x - 0.001x^2$$

$$\frac{dI}{dx} = 4 - 0.002x \quad \text{take derivative}$$

$$0 = 4 - 0.002x \quad \text{set derivative equal to zero}$$

$$x = 2000$$

We note that if $x < 2000$, the derivative is positive, and if $x > 2000$, the derivative is negative. Therefore, if $x = 2000$, I is at a maximum. This means that the maximum income is derived if 2000 units over 1000 are sold, or 3000 units in all. This in turn means that the maximum income is $9000 and the price per unit is $3. These values are found by substituting $x = 2000$ into the expression for I and for the price. ■

SOLVING A WORD PROBLEM

See the chapter introduction.

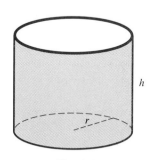

Fig. 4-55

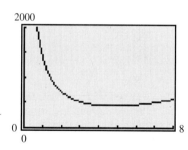

Fig. 4-56

■**EXAMPLE 7** Find the dimensions of a 700-kL cylindrical oil storage tank that can be made with the least cost of sheet metal, assuming there is no wasted sheet metal.

Analyzing the wording of the problem carefully, we see that we are to minimize the surface area of a right circular cylinder with a volume of 700 kL. Therefore, we set up expressions for the surface area and the volume (700 kL = 700 m³).

$$A = 2\pi r^2 + 2\pi rh \qquad V = 700 = \pi r^2 h \qquad \text{(see Fig. 4-55)}$$

We can express the equation for the area in terms of r only by solving the second equation for h and substituting in the first equation. This gives us

$$h = \frac{700}{\pi r^2}, \qquad A = 2\pi r^2 + 2\pi r\left(\frac{700}{\pi r^2}\right) = 2\pi r^2 + \frac{1400}{r}$$

In order to find the minimum value of A, we find dA/dr and set it equal to zero.

$$\frac{dA}{dr} = 4\pi r - \frac{1400}{r^2}, \qquad 4\pi r - \frac{1400}{r^2} = 0, \qquad \frac{4\pi r^3 - 1400}{r^2} = 0$$

$$4\pi r^3 - 1400 = 0, \qquad r^3 = \frac{1400}{4\pi} \qquad \text{numerator must} = 0$$

$$r = \sqrt[3]{\frac{1400}{4\pi}} = 4.81 \text{ m} \qquad h = \frac{700}{\pi r^2} = 9.62 \text{ m}$$

Since dA/dr changes sign from negative to positive at (about) $r = 4.81$ m, A is a minimum. This is also verified by the calculator graph of A vs. r in Fig. 4-56. ■

SOLVING A WORD PROBLEM　▐**EXAMPLE 8**　The illuminance of a light source at any point equals the strength of the source divided by the square of the distance from the source. Two sources, of strengths 8 units and 1 unit, respectively, are 100 m apart. Determine at what point between them the illuminance is the least, assuming that the illuminance at any point is the sum of the illuminances of the two sources.

$$\text{Let } I = \text{the sum of the illuminances and}$$
$$x = \text{the distance from the source of strength 8}$$

Then, we find that

$$I = \frac{8}{x^2} + \frac{1}{(100 - x)^2}$$

is the function relating the illuminance and the distance from the source of strength 8. We must now take a derivative of I with respect to x, set it equal to zero, and solve for x to find the point at which the illuminance is a minimum:

$$\frac{dI}{dx} = -\frac{16}{x^3} + \frac{2}{(100 - x)^3} = \frac{-16(100 - x)^3 + 2x^3}{x^3(100 - x)^3}$$

This function will be zero if the numerator is zero. Therefore, we have

$$2x^3 - 16(100 - x)^3 = 0 \quad \text{or} \quad x^3 = 8(100 - x)^3$$

Taking cube roots of each side, we have

$$x = 2(100 - x) \quad \text{or} \quad x = 66.7 \text{ m}$$

The point where the illuminance is a minimum is 66.7 m from the 8-unit source of illuminance.

EXERCISES 4-7

In the following exercises, solve the given maximum and minimum problems.

1. The height (in ft) of a flare shot upward from the ground is given by $s = 112t - 16.0t^2$, where t is the time (in s). What is the greatest height to which the flare rises?

2. A small oil refinery estimates that its daily profit P (in dollars) from refining x barrels of oil is $P = 8x - 0.02x^2$. How many barrels should be refined for maximum daily profit, and what is the maximum profit?

3. The power output P of a battery of voltage E and internal resistance R is $P = EI - RI^2$, where I is the current. Find the current for which the power is a maximum.

4. In 1998 the projected U.S. Social Security Fund assets S (in dollars × 10^{12}) was given by
   $S = -0.00074t^3 + 0.020t^2 + 0.8$,
   where t is the number of years after 2000. What is the maximum projected fund value, and in what year will it occur? Using a graphing calculator, determine when the fund will run out of money.

5. A company projects that its total savings S (in dollars) by converting to a solar heating system with a solar collector area A (in m²) will be $S = 360A - 0.10A^3$. Find the area that should give the maximum savings and find the amount of the maximum savings.

6. The altitude h (in ft) of a jet that goes into a dive and then again turns upward is given by $h = 16t^3 - 240t^2 + 10,000$, where t is the time (in s) of the dive and turn. What is the altitude of the jet when it turns up out of the dive?

7. The impedance Z (in Ω) in an electric circuit is given by $Z = \sqrt{R^2 + (X_L - X_C)^2}$. If $R = 2500$ Ω and $X_L = 1500$ Ω, what value of X_C makes the impedance a minimum?

8. The electric potential V on the line $3x + 2y = 6$ is given by $V = 3x^2 + 2y^2$. At what point on this line is the potential a minimum?

9. A rectangular hole is to be cut in a wall for a vent. If the perimeter of the hole is 48 in. and the length of the diagonal is a minimum, what are the dimensions of the hole?

10. When two electric resistors R_1 and R_2 are in series, their total resistance (the sum) is 32 Ω. If the same resistors are in parallel, their total resistance (the reciprocal of which equals the sum of the reciprocals of the individual resistances) is the maximum possible for two such resistors. What is the resistance of each?

11. A rectangular microprocessor chip is designed to have an area of 25 mm^2. What must be its dimensions if its perimeter is to be a minimum?

12. A rectangular storage area is to be constructed along the side of a tall building. A security fence is required along the remaining three sides of the area. What is the maximum area that can be enclosed with 800 ft of fencing?

13. Ship *A* is traveling due east at 18.0 km/h as it passes a point 40.0 km due south of ship *B*, which is traveling due south at 16.0 km/h. How much later are the ships nearest each other?

14. An architect is designing a rectangular building in which the front wall costs twice as much per linear meter as the other three walls. The building is to cover 1350 m^2. What dimensions must it have such that the cost of the walls is a minimum?

15. A computer is programmed to display a slowly changing right triangle with its hypotenuse always equal to 12.0 cm. What are the legs of the triangle when it has its maximum area?

16. U.S. Postal Service regulations require that the length plus the girth (distance around) of a package not exceed 108 in. What are the dimensions of the largest rectangular box with square ends that can be mailed?

17. A culvert designed with a semicircular cross section of diameter 6.00 ft is redesigned to have an isosceles trapezoidal cross section by inscribing the trapezoid in the semicircle. See Fig. 4-57. What is the length of the bottom base *b* of the trapezoid if its area is to be maximum?

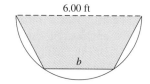

Fig. 4-57

18. A lap pool (a pool for swimming laps) is designed to be seven times as long as it is wide. If the area of the sides and bottom is 980 ft^2, what are the dimensions of the pool if the volume of water it can hold is at maximum?

19. What is the maximum slope of the curve $y = 6x^2 - x^3$?

20. What is the minimum slope of the curve $y = x^5 - 10x^2$?

21. The deflection *y* of a beam of length *L* at a horizontal distance *x* from one end is given by $y = k(2x^4 - 5Lx^3 + 3L^2x^2)$, where *k* is a constant. For what value of *x* does the maximum deflection occur?

22. The potential energy *E* of an electric charge *q* due to another charge q_1 at a distance r_1 is proportional to q_1 and inversely proportional to r_1. If charge *q* is placed directly between two charges of 2.00 nC and 1.00 nC that are separated by 10.0 mm, find the point at which the total potential energy (the sum due to the other two charges) of *q* is a minimum.

23. An open box is to be made from a square piece of cardboard whose sides are 8.00 in. long, by cutting equal squares from the corners and bending up the sides. Determine the side of the square that is to be cut out so that the volume of the box may be a maximum. See Fig. 4-58.

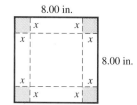

Fig. 4-58

24. A cone-shaped paper cup is to hold 100 cm^3 of water. Find the height and radius of the cup that can be made from the least amount of paper.

25. A race track 400 m long is to be built around an area that is a rectangle with a semicircle at each end. Find the open side of the rectangle if the area of the rectangle is to be a maximum. See Fig. 4-59.

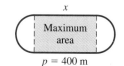

Fig. 4-59 $p = 400$ m

26. A company finds that there is a net profit of $10 for each of the first 1000 units produced each week. For each unit over 1000 produced, there is 2 cents less profit per unit. How many units should be produced each week to net the greatest profit?

27. A beam of rectangular cross section is to be cut from a log 2.00 ft in diameter. The stiffness of the beam varies directly as the width and the cube of the depth. What dimensions will give the beam maximum stiffness? See Fig. 4-60.

Fig. 4-60

28. On a computer simulation, a target plane is projected to be at $(1.20, 7.00)$ (distances in km), and a rocket is fired along the path $y = 8.00 - 2.00x^2$. How far from the target does the rocket pass?

29. An oil pipeline is to be built from a refinery to a tanker loading area. The loading area is 10.0 mi downstream from the refinery and on the opposite side of a river 2.5 mi wide. The pipeline is to run along the river and then cross to the loading area. If the pipeline costs $50,000 per mile alongside the river and $80,000 per mile across the river, find the point *P* (see Fig. 4-61) at which the pipeline should be turned to cross the river if construction costs are to be a minimum.

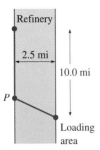

Fig. 4-61

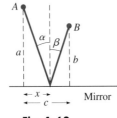

Fig. 4-62

30. A light ray follows a path of least time. If a ray starts at point *A* (see Fig. 4-62) and is reflected off a plane mirror to point *B*, show that the angle of incidence α equals the angle of reflection β. (*Hint:* Set up the expression in terms of *x*, which will lead to sin α = sin β.)

31. A rectangular building covering 7000 m² is to be built on a rectangular lot as shown in Fig. 4-63. If the building is to be 10.0 m from the lot boundary on each side and 20.0 m from the boundary in front and back, find the dimensions of the building if the area of the lot is a minimum.

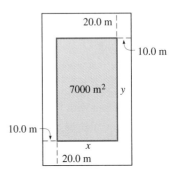

Fig. 4-63

32. A cylindrical cup (no top) is designed to hold 375 cm³ (375 mL). There is no waste in the material used for the sides. However, there is waste in that the bottom is made from a square 2*r* on a side. What are the most economical dimensions for a cup made under these conditions?

$4\text{-}8$ DIFFERENTIALS AND LINEAR APPROXIMATIONS

The symbol *dy/dx* for the derivative was first used by Leibniz (see page 69).

To this point we have used the *dy/dx* notation for the derivative of *y* with respect to *x*, but we have not considered it to be a ratio. In this section we define the quantities *dy* and *dx*, called *differentials*, such that their ratio is equal to the derivative. Then we show that differentials have applications in errors in measurement and in approximating values of functions. Also, in the next chapter we use the differential notation in the development of integration, which is the inverse process of differentiation.

Differentials

We define the **differential** *of a function* $y = f(x)$ *as*

$$dy = f'(x)\, dx \qquad (4\text{-}6)$$

In Eq. (4-6), the quantity *dy is the differential of y, and dx is the differential of x. The differential dx is defined as equal to* Δx, *the increment in x.* In this way we can interpret the derivative as the ratio of the differential of *y* to the differential of *x*.

EXAMPLE 1 Find the differential of $y = 3x^5 - x$.
Since $f(x) = 3x^5 - x$, we find $f'(x) = 15x^4 - 1$. This means that

$$dy = (15x^4 - 1)\,dx$$

with $f'(x)$ labeling $15x^4$ and "the differential of x" labeling dx.

EXAMPLE 2 Find the differential of $s = (2t^3 - 1)^4$.

$$ds = 4(2t^3 - 1)^3(6t^2)\,dt$$
$$= 24t^2(2t^3 - 1)^3\,dt$$

EXAMPLE 3 Find the differential of $y = \dfrac{4x}{x^2 + 4}$.

$$dy = \frac{(x^2 + 4)(4) - (4x)(2x)}{(x^2 + 4)^2}\,dx \qquad \text{using derivative quotient rule}$$

$$= \frac{4x^2 + 16 - 8x^2}{(x^2 + 4)^2}\,dx = \frac{-4x^2 + 16}{(x^2 + 4)^2}\,dx$$

CAUTION

$$= \frac{-4(x^2 - 4)}{(x^2 + 4)^2}\,dx$$

with "don't forget the dx" labeling dx.

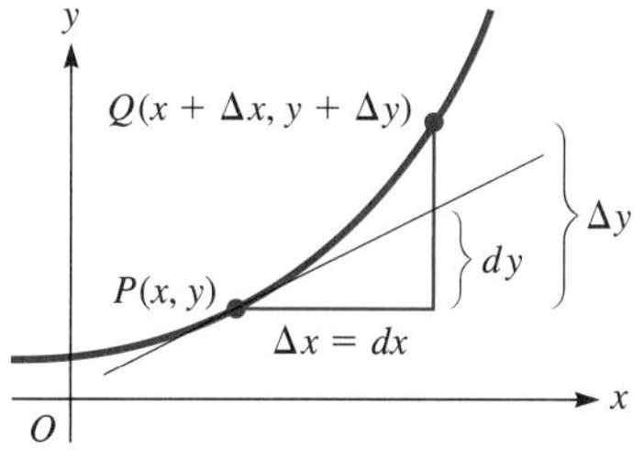

The applications of the differential are based on the fact that the differential of y, dy, closely approximates the increment in y, Δy, if the differential of x, dx, is small. To understand this statement, let us look at Fig. 4-64. Recalling the meaning of Δx and Δy, we see that points $P(x, y)$ and $Q(x + \Delta x, y + \Delta y)$ lie on the graph of $f(x)$. However, $f'(x) = dy/dx$ at P, which means that if we draw a tangent line at P its slope may be indicated by dy/dx. By choosing $\Delta x = dx$, we can see the difference between Δy and dy. We see that as dx becomes smaller, Δy more nearly equals dy.

For given changes in x, it is necessary to use the delta-process to find the exact change, Δy, in y. However, *for small values of Δx, dy can be used to approximate Δy closely.* Generally, dy is much more easily determined than is Δy.

Fig. 4-64

EXAMPLE 4 Calculate Δy and dy for $y = x^3 - 2x$ for $x = 3$ and $\Delta x = 0.1$.
By the delta-process, we find

$$y + \Delta y = (x + \Delta x)^3 - 2(x + \Delta x)$$
$$\Delta y = 3x^2\Delta x + 3x(\Delta x)^2 + (\Delta x)^3 - 2\Delta x$$

Using the given values, we find

$$\Delta y = 3(9)(0.1) + 3(3)(0.01) + (0.001) - 2(0.1) = 2.591$$

The differential of y is

$$dy = (3x^2 - 2)\,dx$$

Since $dx = \Delta x$, we have

$$dy = [3(9) - 2](0.1) = 2.5$$

Thus, $\Delta y = 2.591$ and $dy = 2.5$. In this case, dy is very nearly equal to Δy.

■**EXAMPLE 5**   If finding Δy by the delta-process is complicated or lengthy, we can use a calculator to find the difference between Δy and dy. Therefore, for the function

$$y = f(x) = \sqrt{8x + 3}$$

the differential is

$$dy = \frac{1}{2}(8x + 3)^{-1/2}(8)\,dx = \frac{4\,dx}{\sqrt{8x + 3}}$$

For $x = 2$ and $\Delta x = 0.003$,

$$f(x) = \sqrt{8(2) + 3} = \sqrt{19} = 4.3588989$$

and

$$f(x + \Delta x) = \sqrt{8(2.003) + 3} = \sqrt{19.024} = 4.3616511$$

This means that

$$\Delta y = f(x + \Delta x) - f(x)$$
$$= 4.3616511 - 4.3588989 = 0.0027522$$

Now, calculating the value of dy,

$$dy = \frac{4(0.003)}{\sqrt{8(2) + 3}} = 0.0027530$$

Again, in this example the values of dy and Δy are very nearly equal. ■

Estimating Errors in Measurement

The fact that dy can be used to approximate Δy is useful in finding the error in a result from a measurement, if the data are in error. It is also useful in finding the change in the result if a change is made in the data. Even though such changes can be found by using a calculator, the differential can be used to set up a general expression for the change of a particular function.

SOLVING A WORD PROBLEM   ■**EXAMPLE 6**   The edge of a cube of gold was measured to be 3.850 cm. From this value the volume was found. Later it was discovered that the value of the edge was 0.020 cm too small. By approximately how much was the volume in error?

The volume V of a cube, in terms of an edge e, is $V = e^3$. Since we wish to find the change in V for a given change in e, we want the value of dV for $e = 3.850$ cm and $de = 0.020$ cm.

First, finding the general expression for dV, we have

$$dV = 3e^2\,de$$

Now, evaluating this expression for the given values, we have

$$dV = 3(3.850)^2(0.020) = 0.89 \text{ cm}^3$$

In this case the volume was in error by about 0.89 cm^3. As long as de is small compared with e, we can calculate an error or change in the volume of a cube by calculating the value of $3e^2\,de$. ■

Often when considering the error of a given value or result, *the actual numerical value of the error, the* **absolute error,** is not as important as its size in relation to the size of the quantity itself. *The ratio of the absolute error to the size of the quantity itself is known as the* **relative error,** which is commonly expressed as a percent.

EXAMPLE 7   Referring to Example 6, we see that the absolute error in the edge was 0.020 cm. The relative error in the edge was

$$\frac{de}{e} = \frac{0.020}{3.85} = 0.0052 = 0.52\%$$

The absolute error in the volume was 0.89 cm³, and the original value of the volume was $3.850^3 = 57.07$ cm³. This means the relative error in the volume was

$$\frac{dV}{V} = \frac{0.89}{57.07} = 0.016 = 1.6\%$$

∎

Linear Approximations

Continuing our discussion related to Fig. 4-64, on the graph of the function $f(x)$ at the point $(a, f(a))$, the slope of a tangent line is $f'(a)$. See Fig. 4-65. For a point (x, y) on the tangent line, by using the point-slope form for the equation of a straight line, Eq. (2-6), we have

$$f(x) = f(a) + f'(a)(x - a)$$

For points on $y = f(x)$ near $x = a$, we can use the tangent line to approximate the function, as shown in Fig. 4-65. Therefore, the approximation $f(x) \approx L(x)$ *is the* **linear approximation** *of* $f(x)$ *near* $x = a$, *where the function*

$$\boxed{L(x) = f(a) + f'(a)(x - a)} \tag{4-7}$$

is called the **linearization** *of* $f(x)$ *at* $x = a$.

EXAMPLE 8   Find the linearization of the function $f(x) = \sqrt{2x + 1}$ at $x = 4$. Use it to approximate $\sqrt{9.06}$.
   The solution is as follows.

$$f(x) = \sqrt{2x + 1}$$

$$f'(x) = \tfrac{1}{2}(2x + 1)^{-1/2}(2) = \frac{1}{\sqrt{2x + 1}} \qquad \text{find the derivative}$$

$$f(4) = \sqrt{2(4) + 1} = 3 \qquad f'(4) = \frac{1}{\sqrt{2(4) + 1}} = \frac{1}{3} \qquad \text{evaluate } f(4) \text{ and } f'(4)$$

$$L(x) = 3 + \tfrac{1}{3}(x - 4) = \frac{x + 5}{3} \qquad \text{use Eq. (4-7)}$$

$$\sqrt{2x + 1} \approx \frac{x + 5}{3} \qquad \begin{array}{l} 2x + 1 = 9.06, \\ x = 4.03 \end{array}$$

$$\sqrt{9.06} \approx \frac{4.03 + 5}{3} = \frac{9.03}{3} = 3.01 \qquad \begin{array}{l} \text{by calculator,} \\ \sqrt{9.06} = 3.009983389 \end{array}$$

In Fig. 4-66, a graphing calculator display showing $f(x) = \sqrt{2x + 1}$ and the tangent line (using the *tangent* feature) is shown. We see that the tangent line gives a good approximation of the function when x is near 4.   ∎

For reference, Eq. (2-6) is
$y - y_1 = m(x - x_1)$.

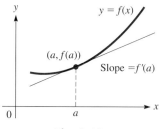

Fig. 4-65

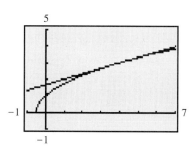

Fig. 4-66

─────────── **EXERCISES** *4-8* ───────────

In Exercises 1–12, find the differential of each of the given functions.

1. $y = x^5 + x$

2. $y = 3x^2 + 6$

3. $V = \dfrac{2}{r^5} + 3\pi^2$

4. $y = 2\sqrt{x} - \dfrac{1}{x}$

5. $s = 2(3t^2 - 5)^4$

6. $y = 5(4 + 3x)^{1/3}$

7. $y = \dfrac{2}{3x^2 + 1}$

8. $R = \sqrt{\dfrac{u}{1 + 2u}}$

9. $y = x^2(1 - x)^3$

10. $y = 6x\sqrt{1 - 4x}$

11. $y = \dfrac{x}{5x + 2}$

12. $y = \dfrac{3x + 1}{\sqrt{2x - 1}}$

In Exercises 13–16, find the values of Δy and dy for the given values of x and dx.

13. $y = 7x^2 + 4x$, $x = 4$, $\Delta x = 0.2$

14. $y = 2x^2 - 3x + 1$, $x = 5$, $\Delta x = 0.15$

15. $y = 2x^3 - 4x$, $x = 2.5$, $\Delta x = 0.05$

16. $y = x - x^4$, $x = 3.2$, $\Delta x = 0.08$

In Exercises 17–20, determine the value of dy for the given values of x and Δx. Compare with values of $f(x + \Delta x) - f(x)$ found by using a calculator.

17. $y = (1 - 3x)^5$, $x = 1$, $\Delta x = 0.01$

18. $y = (x^2 + 2x)^3$, $x = 7$, $\Delta x = 0.02$

19. $y = x\sqrt{1 + 4x}$, $x = 12$, $\Delta x = 0.06$

20. $y = \dfrac{x}{\sqrt{6x - 1}}$, $x = 3.5$, $\Delta x = 0.025$

In Exercises 21–24, find the linearization $L(x)$ of the given functions for the given values of a. Display $f(x)$ and $L(x)$ on the same calculator screen.

21. $f(x) = x^2 + 2x$, $a = 0$

22. $f(x) = 2\sqrt[3]{x}$, $a = 8$

23. $f(x) = \dfrac{1}{2x + 1}$, $a = -1$

24. $g(x) = x\sqrt{2x + 8}$, $a = -2$

In Exercises 25–32, solve the given problems by finding the appropriate differential.

25. The side of a square microprocessor chip is measured as 0.950 cm, and later it is measured as 0.952 cm. What is the difference in the calculations of the area due to the difference in the measurements of the side? See Fig. 4-67.

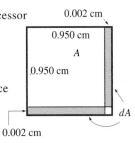

Fig. 4-67

0.002 cm

0.950 cm

0.950 cm

A

0.950 cm

dA

0.002 cm

26. A circular solar cell is made with a radius measured to be 12.00 ± 0.05 cm. (This means that the radius is 12.00 cm with a possible error of no more than 0.05 cm.) What is the maximum possible relative error in the calculation of the area of the cell?

27. The wavelength λ of light is inversely proportional to its frequency f. If $\lambda = 685$ nm for $f = 4.38 \times 10^{14}$ Hz, approximate the change in λ if f increases by 0.20×10^{14} Hz. (These values are for red light.)

28. The velocity of an object rolling down a certain inclined plane is given by $v = \sqrt{100 + 16h}$, where h is the distance (in ft) traveled along the plane by the object. What is the increase in velocity (in ft/s) of an object in moving from 20.0 ft to 20.5 ft along the plane? What is the relative change in the velocity?

29. The radius r of a holograph is directly proportional to the square root of the wavelength λ of the light used. Show that $dr/r = \frac{1}{2}\,d\lambda/\lambda$.

30. The gravitational force F of the earth on an object is inversely proportional to the square of the distance r of the object from the center of the earth. Show that $dF/F = -2dr/r$.

31. Show that an error of 2% in the measurement of the side of a square results in an error of approximately 4% in the calculation of the area.

(**W**) **32.** Explain how to approximate 2.03^4.

In Exercises 33–36, solve the given linearization problems.

33. Linearize $f(x) = \sqrt{2 - x}$ for $a = 1$ and use the result to approximate the value of $\sqrt{1.9}$.

(**W**) **34.** Explain how to approximate $\sqrt[3]{8.03}$, using linearization.

35. The capacitance C (in μF) in an element of an electronic tuner is given by $C = \dfrac{3.6}{\sqrt{1 + 2V}}$, where V is the voltage. Linearize C for $V = 4.0$ V.

36. A 16-Ω resistor is put in parallel with a variable resistor of resistance R. The combined resistance of the two resistors is $R_T = \dfrac{16R}{16 + R}$. See Fig. 4-68. Linearize R_T for $R = 4.0\ \Omega$.

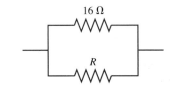

16 Ω

R

Fig. 4-68

CHAPTER EQUATIONS

Newton's method	$x_2 = x_1 - \dfrac{f(x_1)}{f'(x_1)}$	(4-1)

Curvilinear motion

$$v_x = \frac{dx}{dt} \qquad v_y = \frac{dy}{dt} \tag{4-2}$$

$$a_x = \frac{dv_x}{dt} = \frac{d^2x}{dt^2} \qquad a_y = \frac{dv_y}{dt} = \frac{d^2y}{dt^2} \tag{4-3}$$

$$v = \sqrt{v_x^2 + v_y^2} \qquad a = \sqrt{a_x^2 + a_y^2} \tag{4-4}$$

$$\tan \theta_v = \frac{v_y}{v_x} \qquad \tan \theta_a = \frac{a_y}{a_x} \tag{4-5}$$

Curve sketching and maximum and minimum values

$f'(x) > 0$ where $f(x)$ increases; $f'(x) < 0$ where $f(x)$ decreases.

$f''(x) > 0$ where the graph of $f(x)$ is concave up; $f''(x) < 0$ where the graph of $f(x)$ is concave down.

If $f'(x) = 0$ at $x = a$, there is a maximum point if $f'(x)$ changes from $+$ to $-$ or if $f''(a) < 0$.

If $f'(x) = 0$ at $x = a$, there is a minimum point if $f'(x)$ changes from $-$ to $+$ or if $f''(a) > 0$.

If $f''(x) = 0$ at $x = a$, there is a point of inflection if $f''(x)$ changes from $+$ to $-$ or from $-$ to $+$.

Differential	$dy = f'(x)\,dx$	(4-6)
Linearization	$L(x) = f(a) + f'(a)(x - a)$	(4-7)

REVIEW EXERCISES

In Exercises 1–6, find the equations of the tangent and normal lines. Use a graphing calculator to view the curve and the line.

1. Find the equation of the line tangent to the parabola $y = 3x - x^2$ at the point $(-1, -4)$.

2. Find the equation of the line tangent to the curve $y = x^2 - \dfrac{6}{x}$ at the point $(2, 1)$.

3. Find the equation of the line normal to $x^2 - 4y^2 = 9$ at the point $(5, 2)$.

4. Find the equation of the line normal to $y = \dfrac{1}{\sqrt{x - 2}}$ at the point $(6, \frac{1}{2})$.

5. Find the equation of the line tangent to the curve $y = \sqrt{x^2 + 3}$ and that has a slope of $\frac{1}{2}$.

6. Find the equation of the line normal to the curve $y = \dfrac{1}{2x + 1}$ and that has a slope of $\frac{1}{2}$ if $x \ge 0$.

In Exercises 7–12, find the indicated velocities and accelerations.

7. Given that the x- and y-coordinates of a moving particle are given as a function of time t by the parametric equations $x = \sqrt{t} + t$, $y = \frac{1}{12}t^3$, find the magnitude and direction of the velocity when $t = 4$.

8. If the x- and y-coordinates of a moving object as functions of time t are given by $x = 0.1t^2 + 1$, $y = \sqrt{4t + 1}$, find the magnitude and direction of the velocity when $t = 6$.

9. An object moves along the curve $y = 0.5x^2 + x$ such that $v_x = 0.5\sqrt{x}$. Find v_y at $(2, 4)$.

10. A particle moves along the curve of $y = \dfrac{1}{x + 2}$ with a constant velocity in the x-direction of 4 cm/s. Find v_y at $(2, \frac{1}{4})$.

11. Find the magnitude and direction of the acceleration for the particle in Exercise 7.

12. Find the magnitude and direction of the acceleration of the particle in Exercise 10.

In Exercises 13–16, find the indicated roots of the given equations to at least four decimal places by use of Newton's method.

13. $x^3 - 3x^2 - x + 2 = 0$ (between 0 and 1)

14. $2x^3 - 4x^2 - 9 = 0$ (between 2 and 3)

15. $3x^3 - x^2 - 8x - 2 = 0$ (between 1 and 2)

16. $x^4 + 3x^3 + 6x + 4 = 0$ (between -1 and 0)

In Exercises 17–24, sketch the graphs of the given functions by information obtained from the function as well as information obtained from the derivatives. Use a graphing calculator to check the graph.

17. $y = 4x^2 + 16x$

18. $y = x^3 + 2x^2 + x + 1$

19. $y = 27x - x^3$

20. $y = x(6 - x)^3$

21. $y = x^4 - 32x$

22. $y = x^4 + 4x^3 - 16x$

23. $y = \dfrac{x^2}{\sqrt{x^2 - 1}}$

24. $y = x^3 + \dfrac{3}{x}$

In Exercises 25–28, find the differential of each of the given functions.

25. $y = 4x^3 + \dfrac{1}{x}$

26. $y = \dfrac{1}{(2x - 1)^2}$

27. $y = x\sqrt[3]{1 - 3x}$

28. $s = \sqrt{\dfrac{2 + t}{2 - t}}$

In Exercises 29 and 30, evaluate $\Delta y - dy$ for the given functions and values.

29. $y = x^3$, $x = 2$, $\Delta x = 0.1$

30. $y = 6x^2 - x$, $x = 3$, $\Delta x = 0.2$

In Exercises 31 and 32, find the linearization of the given functions for the given values of a.

31. $f(x) = \sqrt{x^4 + 3x^2 + 8}$, $a = 2$

32. $f(x) = x^2(x + 1)^4$, $a = -2$

In Exercises 33–36, solve the given problems by finding the appropriate differentials.

33. A weather balloon 3.500 m in radius becomes covered with a uniform layer of ice 1.2 cm thick. What is the volume of the ice?

34. The total power P (in W) transmitted by an AM radio transmitter is $P = 460 + 230m^2$, where m is the modulation index. What is the change in power if m changes from 0.86 to 0.89?

35. The impedance Z of an electric circuit as a function of the resistance R and the reactance X is given by $Z = \sqrt{R^2 + X^2}$. Derive an expression of the relative error in impedance for an error in R and a given value of X.

36. Show that the relative error in the calculation of the volume of a sphere is approximately three times the relative error in the measurement of the radius.

In Exercises 37–68, solve the given problems.

37. The parabolas $y = x^2 + 2$ and $y = 4x - x^2$ are tangent to each other. Find the equation of the line tangent to them at the point of tangency.

38. Find the equation of the line tangent to the curve of $y = x^4 - 8x$ and perpendicular to the line $4y - x + 5 = 0$.

39. The deflection y (in m) of a beam at a horizontal distance x (in m) from one end is given by $y = k(x^4 - 30x^3 + 1000x)$, where k is a constant. Observing the equation and using Newton's method, find the values of x where the deflection is zero, if the beam is 10.000 m long.

40. The edges of a rectangular water tank are 3.00 ft, 5.00 ft, and 8.00 ft. By Newton's method, determine by how much each edge should be increased equally to double the volume of the tank.

41. A parachutist descends (after the parachute opens) in a path that can be described by $x = 8t$ and $y = -0.15t^2$, where distances are in meters and time is in seconds. Find the parachutist's velocity upon landing if the landing occurs when $t = 12$ s.

42. One of the curves on an automobile test track can be described by $y = 225/x$, where dimensions are in meters. For a car approaching the curve at a constant velocity of 120 km/h, find the x- and y-components of the velocity at $(10.0, 22.5)$.

43. In Fig. 4-69, the tension T supports the 40.0-N weight. The relation between the tension T and the deflection d is

$$d = \frac{1000}{\sqrt{T^2 - 400}}.$$ If the tension is increasing at 2.00 N/s

when $T = 28.0$ N, how fast is the deflection (in cm) changing?

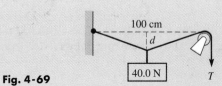

Fig. 4-69

44. The impedance Z (in Ω) in a particular electric circuit is given by $Z = \sqrt{48 + R^2}$, where R is the resistance. If R is increasing at the rate of 0.45 Ω/min for $R = 6.5$ Ω, find the rate at which Z is changing.

45. An analysis of the power output P (in kW/m^3) of a certain turbine showed that it depended on the flow rate r (in m^3/s) of water to the turbine according to the equation $P = 0.030r^3 - 2.6r^2 + 71r - 200$ ($6 \le r \le 30$ m^3/s). Determine the rate for which P is a maximum.

46. The altitude h (in ft) of a certain rocket as a function of the time t (in s) after launching is given by $h = 1600t - 16t^2$. What is the maximum altitude the rocket attains?

47. Sketch a continuous curve having these characteristics:

$$f(0) = 2 \quad f'(x) < 0 \quad \text{for } x < 0 \quad f''(x) > 0 \quad \text{for all } x$$
$$f'(x) > 0 \quad \text{for } x > 0$$

48. Sketch a continuous curve having these characteristics:

$f(0) = 1$ $f'(0) = 0$ $f''(x) < 0$ for $x < 0$

$f'(x) > 0$ for $|x| > 0$ $f''(x) > 0$ for $x > 0$

49. A horizontal cylindrical oil tank (the length is parallel to the ground) of radius 6.00 ft is being emptied. Find how fast the width w of the oil surface is changing when the depth h is 1.50 ft and changing at the rate of 0.250 ft/min.

50. The current I (in A) in a circuit with a resistance R (in Ω) and a battery whose voltage is E and whose internal resistance is r (in Ω) is given by $I = E/(R + r)$. If R changes at the rate of 0.250 Ω/min, how fast is the current changing when $R = 6.25\ \Omega$, if $E = 3.10$ V and $r = 0.230\ \Omega$?

51. The radius of a circular oil spill is increasing at the rate of 15 m/min. How fast is the area of the spill changing when the radius is 400 m?

52. A baseball diamond is a square 90.0 ft on a side. See Fig. 4-70. As a player runs from first base toward second base at 18.0 ft/s, at what rate is the player's distance from home plate increasing when the player is 40.0 ft from first base?

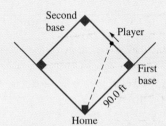

Fig. 4-70

53. A special insulation strip is to be sealed completely around three edges of a rectangular solar panel. If 200 cm of the strip are used, what is the maximum area of the panel?

54. A swimming pool with a rectangular surface of 1200 ft^2 is to have a cement border area that is 12.0 ft wide at each end and 8.00 ft wide at the sides. Find the surface dimensions of the pool if the total area covered is to be a minimum.

55. A study showed that the percent y of persons surviving burns to x percent of the body is given by $y = \dfrac{300}{0.0005x^2 + 2} - 50$. Linearize this function with $a = 50$ and sketch the graphs of y and $L(x)$.

56. A company estimates that the sales S (in dollars) of a new product will be $S = 5000t/(t + 4)^2$, where t is the time (in months) after it is put into production. Sketch the graph of S vs. t.

57. An airplane flying horizontally at 8000 ft is moving toward a radar installation at 680 mi/h. If the plane is directly over a point on the ground 5.00 mi from the radar installation, what is its actual speed? See Fig. 4-71.

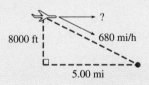

Fig. 4-71

58. The base of a conical machine part is being milled such that the height is decreasing at the rate of 0.050 cm/min. If the part originally had a radius of 1.0 cm and a height of 3.0 cm, how fast is the volume changing when the height is 2.8 cm?

59. The reciprocal of the total capacitance C_T of electric capacitances in series equals the sum of the reciprocals of the individual capacitances. If the sum of two capacitances is 12 μF, find their values if their total capacitance in series is a maximum.

60. A cable is to be from point A to point B on a wall and then to point C. See Fig. 4-72. Where is B located if the total length of cable is a minimum?

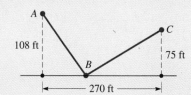

Fig. 4-72

61. A box with a square base and an open top is to be made of 27 ft^2 of cardboard. What is the maximum volume that can be contained within the box?

62. A machine part is to be in the shape of a circular sector of radius r and central angle θ. Find r and θ if the area is one unit and the perimeter is a minimum. See Fig. 4-73.

Fig. 4-73

63. An open drawer for small tools is to be made from a rectangular piece of heavy sheet metal 12.0 in. by 10.0 in., by cutting out equal squares from two corners and bending up the three sides, as shown in Fig. 4-74. Find the side of the square that should be cut out so that the volume of the drawer is a maximum.

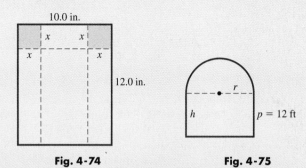

Fig. 4-74   **Fig. 4-75**

64. A Norman window has the form of a rectangle surmounted by a semicircle. Find the dimensions (radius of circular part and height of rectangular part) of the window that will admit the most light if the perimeter of the window is 12 ft. See Fig. 4-75.

65. A pile of sand in the shape of a cone has a radius that always equals the altitude. If 100 ft^3 of sand are poured onto the pile each minute, how fast is the radius increasing when the pile is 10.0 ft high?

66. A person is in a boat 4.0 km from the nearest point P on a straight shoreline. The person wishes to go to point A, which is on the shore, 5.0 km from P. If the person can row at 3.0 km/h and walk at 5.0 km/h, how far along the shore from P toward A should the boat land in order that the person can reach A in the least time?

67. A specially made cylindrical container is made of stainless steel sides and bottom and a silver top. If silver is ten times as expensive as stainless steel, what are the most economical dimensions of the container if it is to hold 314 cm^3?

68. An object is moving in a horizontal circle of a radius 2.00 ft at the rate of 2.00 rad/s. If the object is on the end of a string and the string breaks after 1.05 s, causing the object to travel along a line tangent to the circle, what is the equation of the path of the object after the string breaks? (Choose the origin of the coordinate system at the center of the circle and assume the object started on the positive x-axis moving counterclockwise.)

Writing Exercise

69. A container manufacturer makes various sizes of closed cylindrical plastic containers for shipping liquid products. Write two or three paragraphs explaining how to determine the ratio of the height to radius of the container such that the least amount of plastic is used for each size. Include the reason why it is not necessary to specify the volume of the container in finding this ratio.

PRACTICE TEST

1. Find the equation of the line tangent to the curve $y = x^4 - 3x^2$ at the point $(1, -2)$.

2. For $y = 3x^2 - x$, evaluate (a) Δy, (b) dy, and (c) $\Delta y - dy$ for $x = 3$ and $\Delta x = 0.1$.

3. If the x- and y-coordinates of a moving object as functions of time are given by the parametric equations $x = 3t^2$, $y = 2t^3 - t^2$, find the magnitude and direction of the acceleration when $t = 2$.

4. The electric power (in W) produced by a certain source is given by $P = \dfrac{144r}{(r + 0.6)^2}$, where r is the resistance (in Ω) in the circuit. For what value of r is the power a maximum?

5. Find the root of the equation $x^2 - \sqrt{4x + 1} = 0$ between 1 and 2 to four decimal places by use of Newton's method. Use $x_1 = 1.5$ and find x_3.

6. Linearize the function $y = \sqrt{2x + 4}$ for $a = 6$.

7. Sketch the graph of $y = x^3 + 6x^2$ by finding the values of x for which the function is increasing, decreasing, concave up, and concave down and by finding any maximum points, minimum points, and points of inflection.

8. Sketch the graph of $y = \dfrac{4}{x^2} - x$ by finding the same information as required in Problem 7, as well as intercepts, symmetry, behavior as x becomes large, vertical asymptotes, and the domain and range.

9. Trash is being compacted into a cubical volume. The edge of the cube is decreasing at the rate of 0.50 ft/s. When an edge of the cube is 4.00 ft, how fast is the volume changing?

10. A rectangular field is to be fenced and then divided in half by a fence parallel to two opposite sides. If a total of 6000 m of fencing is used, what is the maximum area that can be fenced?

In the study of physical and technical applications, we frequently find information related to the rate of change of a variable. With such information it is necessary to reverse the process of differentiation in order to determine the functional relationship between the variables. This leads us to the problem of finding the function when we know its derivative. This mathematical procedure is known as *integration*.

As we study integration in this chapter and the next, we shall see that it has many applications in science and technology. It can be applied to finding areas and volumes, as well as to the physical concepts of work, force, and center of mass. Although these applications seem to be quite different, we will show that they have a very similar mathematical formulation. A few of these applications are shown in this chapter, although most of them are developed in the following chapter.

In Section 5-4, integration is used to find the rate of flow of water over an obstacle.

5-1 ANTIDERIVATIVES

As we have noted above, many kinds of problems in many areas, including science and technology, can be solved by reversing the process of finding a derivative or a differential. Therefore, in this section we introduce the basic technique of this procedure. *This reverse process is known as* **antidifferentiation.** In the next section we shall formalize the process, but it is only the basic idea that is the topic of this section. The following examples illustrate the method.

■EXAMPLE 1 Find a function for which the derivative is $8x^3$. That is, find an antiderivative of $8x^3$.

We know that when we find the derivative of a constant times a power of x, we multiply the constant coefficient by the power of x and reduce the power of x by 1. Therefore, in this case the power of x must have been 4 before the differentiation was performed.

If we let the derivative function be $f(x) = 8x^3$ and then let its antiderivative function be $F(x) = ax^4$ (by increasing the power of x in $f(x)$ by 1), we can find the value of a by equating the derivative of $F(x)$ to $f(x)$. This gives us

$$F'(x) = 4ax^3 = 8x^3, \qquad 4a = 8, \qquad a = 2$$

This means that $F(x) = 2x^4$.

■EXAMPLE 2 Find an antiderivative of $v^2 + 2v$.

As for the v^2, we know that the power of v required in an antiderivative is 3. Also, to make the coefficient correct, we must multiply by $\frac{1}{3}$. The $2v$ should be recognized as the derivative of v^2. Therefore, we have as an antiderivative $\frac{1}{3}v^3 + v^2$.

In Examples 1 and 2, we note that we could add any constant to the antiderivative given as the result and still have a correct antiderivative. This is due to the fact that the derivative of a constant is zero. This is considered further in the following section. For the examples and exercises in this section, we will not include any such constants in the results.

NOTE▶ We note that when we find an antiderivative of a given function, we obtain another function. Thus, *we can define an* **antiderivative** *of the function $f(x)$ to be a function $F(x)$ such that $F'(x) = f(x)$.*

■EXAMPLE 3 Find an antiderivative of the function $f(x) = \sqrt{x} - \dfrac{2}{x^3}$.

Since we wish to find an antiderivative of $f(x)$, we know that $f(x)$ is the derivative of the required function.

Considering the term $\sqrt{x}$, we first write it as $x^{1/2}$. To have x to the $\frac{1}{2}$ power in the derivative, we must have x to the $\frac{3}{2}$ power in the antiderivative. Knowing that the derivative of $x^{3/2}$ is $\frac{3}{2}x^{1/2}$, we write $x^{1/2}$ as $\frac{2}{3}(\frac{3}{2}x^{1/2})$. Thus, the first term of the antiderivative is $\frac{2}{3}x^{3/2}$.

As for the term $-2/x^3$, we write it as $-2x^{-3}$. This we recognize as the derivative of x^{-2}, or $1/x^2$.

This means that an antiderivative of the function

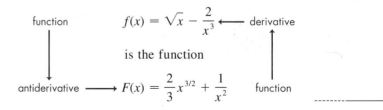

is the function

A great many functions of which we must find an antiderivative are not polynomials or simple powers of x. It is these functions that may cause more difficulty in the general process of antidifferentiation. Pay special attention to the following examples, for they illustrate a type of problem that you will find to be very important.

■**EXAMPLE 4**   Find an antiderivative of the function $f(x) = 3(x^3 - 1)^2(3x^2)$.
Noting that we have a power of $x^3 - 1$ in the derivative, it is reasonable that the antiderivative may include a power of $x^3 - 1$. Since, in the derivative, $x^3 - 1$ is raised to the power 2, the antiderivative would then have $x^3 - 1$ raised to the power 3. Noting that the derivative of $(x^3 - 1)^3$ is $3(x^3 - 1)^2(3x^2)$, the desired antiderivative is

$$F(x) = (x^3 - 1)^3$$

CAUTION ▶   We note that ***the factor of $3x^2$ does not appear in the antiderivative.*** It is included in the process of finding a derivative. Therefore, it must be present for $(x^3 - 1)^3$ to be the proper antiderivative, but it must be excluded in the process of antidifferentiation. ■

■**EXAMPLE 5**   Find an antiderivative of the function $f(x) = (2x + 1)^{1/2}$.
Here, we note a power of $2x + 1$ in the derivative, which infers that the antiderivative has a power of $2x + 1$. Since, in finding a derivative, 1 is subtracted from the power of $2x + 1$, we should add 1 in finding the antiderivative. Thus, we should have $(2x + 1)^{3/2}$ as part of the antiderivative. Finding a derivative of $(2x + 1)^{3/2}$, we obtain $\frac{3}{2}(2x + 1)^{1/2}(2) = 3(2x + 1)^{1/2}$. This differs from the given derivative by the factor of 3. Thus, if we write $(2x + 1)^{1/2} = \frac{1}{3}[3(2x + 1)^{1/2}]$, we have the required antiderivative as

$$F(x) = \tfrac{1}{3}(2x + 1)^{3/2}$$

Checking, the derivative of $\frac{1}{3}(2x + 1)^{3/2}$ is $\frac{1}{3}(\frac{3}{2})(2x + 1)^{1/2}(2) = (2x + 1)^{1/2}$. ■

EXERCISES *5-1*

In Exercises 1–8, determine the value of a that makes $F(x)$ an antiderivative of $f(x)$.

1. $f(x) = 3x^2,\ F(x) = ax^3$ **2.** $f(x) = 5x^4,\ F(a) = ax^5$

3. $f(x) = 18x^5,\ F(x) = ax^6$ **4.** $f(x) = 40x^7,\ F(x) = ax^8$

5. $f(x) = 9\sqrt{x},\ F(x) = ax^{3/2}$ **6.** $f(x) = 10x^{1/4},\ F(x) = ax^{5/4}$

7. $f(x) = \dfrac{1}{x^2},\ F(x) = \dfrac{a}{x}$ **8.** $f(x) = \dfrac{6}{x^4},\ F(x) = \dfrac{a}{x^3}$

In Exercises 9–32, find antiderivatives of the given functions.

9. $f(x) = \frac{5}{2}x^{3/2}$ **10.** $f(x) = \frac{4}{3}x^{1/3}$

11. $f(t) = 6t^3 + 2$ **12.** $f(x) = 12x^5 + 2x$

13. $f(x) = 2x^2 - x$ **14.** $f(x) = x^2 - 5$

15. $f(x) = 2\sqrt{x} + 3$ **16.** $f(s) = 9\sqrt[3]{s} - 3$

17. $f(x) = -\dfrac{7}{x^6}$ **18.** $f(x) = \dfrac{8}{x^5}$

19. $f(v) = 4v + 3\pi^2$ **20.** $f(x) = \dfrac{1}{2\sqrt{x}} + 4$

21. $f(x) = x^2 + 2 + x^{-2}$ **22.** $f(x) = x\sqrt{x} - x^{-3}$

23. $f(x) = 6(2x + 1)^5(2)$ **24.** $f(R) = 3(R^2 + 1)^2(2R)$

25. $f(p) = 4(p^2 - 1)^3(2p)$ **26.** $f(x) = 5(2x^4 + 1)^4(8x^3)$

27. $f(x) = x^3(2x^4 + 1)^4$ **28.** $f(x) = x(1 - x^2)^7$

29. $f(x) = \frac{3}{2}(6x + 1)^{1/2}(6)$ **30.** $f(y) = \frac{5}{4}(1 - y)^{1/4}(-1)$

31. $f(x) = (3x + 1)^{1/3}$ **32.** $f(x) = (4x + 3)^{1/2}$

5-2 THE INDEFINITE INTEGRAL

In the previous section, in developing the basic technique of finding an antiderivative, we noted that the results given are not unique. That is, we could have added any constant to the answers and the results would still have been correct. Again, this is the case since the derivative of a constant is zero.

■EXAMPLE 1 The derivatives of x^3, $x^3 + 4$, $x^3 - 7$, and $x^3 + 4\pi$ are all $3x^2$. This means that any of the functions listed, as well as others, would be a proper answer to the problem of finding an antiderivative of $3x^2$. --------■

From Section 4-8, we know that the differential of a function $F(x)$ can be written as $d[F(x)] = F'(x)\,dx$. Therefore, since finding a differential of a function is closely related to finding the derivative, so is the antiderivative closely related to the process of finding the function for which the differential is known.

The notation used for finding the general form of the antiderivative, the **indefinite integral,** *is written in terms of the differential. Thus, the indefinite integral of a function f(x), for which dF(x)/dx = f(x), or dF(x) = f(x) dx, is defined as*

$$\int f(x)\,dx = F(x) + C \tag{5-1}$$

Here, f(x) is called the **integrand,** *F(x) + C is the indefinite integral, and C is an arbitrary constant, called the* **constant of integration.** It represents any of the constants that may be attached to an antiderivative to have a proper result. We must have additional information beyond a knowledge of the differential to assign a specific value to C. *The symbol ∫ is the* **integral sign,** *and it indicates that the inverse of the differential is to be found. Determining the indefinite integral is called* **integration,** which we can see is essentially the same as finding an antiderivative.

■EXAMPLE 2 In performing the integration

$$\underset{\text{integrand}}{\underbrace{\int 5x^4\,dx}} = \underset{\text{indefinite integral}}{\underbrace{x^5 + \overset{\text{constant of integration}}{C}}}$$

we might think that the inclusion of this constant C would affect the derivative of the function x^5. However, the only effect of the C is to raise or lower the curve. The slope of $x^5 + 2$, $x^5 - 2$, or any function of the form $x^5 + C$ is the same for any given value of x. As Fig. 5-1 shows, tangents drawn to the curves are all parallel for the same value of x. --------■

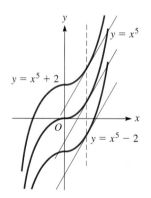

Fig. 5-1

At this point we shall derive some basic formulas for integration. Since we know $d(cu)/dx = c(du/dx)$, where u is a function of x and c is a constant, we can write

$$\int c\,du = c\int du = cu + C \tag{5-2}$$

Also, since the derivative of a sum of functions equals the sum of the derivatives, we write

$$\int (du + dv) = u + v + C \tag{5-3}$$

To find the differential of a power of a function, we multiply by the power, subtract 1 from it, and multiply by the differential of the function. *To find the integral, we reverse this procedure to get the power formula for integration:*

$$\int u^n \, du = \frac{u^{n+1}}{n+1} + C \qquad (n \neq -1) \tag{5-4}$$

We must be able to recognize the proper form and the component parts to use these formulas. Unless you do this and have a good knowledge of differentiation, you will have trouble using Eq. (5-4). Most of the difficulty, if it exists, arises from an improper identification of *du*.

EXAMPLE 3   Integrate: $\int 6x \, dx$.

We must identify *u*, *n*, *du*, and any multiplying constants. Noting that 6 is a multiplying constant, we identify *x* as *u*, which means *dx* must be *du* and $n = 1$.

$$\int 6x \, dx = 6 \int x^1 \, dx = 6\left(\frac{x^2}{2}\right) + C = 3x^2 + C$$

do not forget the constant of integration

We see that our result checks since the differential of $3x^2 + C$ is $6x \, dx$. ∎

EXAMPLE 4   Integrate: $\int (5x^3 - 6x^2 + 1) \, dx$.

Here, we must use a combination of Eqs. (5-2), (5-3), and (5-4). Therefore,

$$\int (5x^3 - 6x^2 + 1) \, dx = \int 5x^3 \, dx + \int (-6x^2) \, dx + \int dx$$

$$= 5 \int x^3 \, dx - 6 \int x^2 \, dx + \int dx$$

In the first integral, $u = x$, $n = 3$, and $du = dx$. In the second, $u = x$, $n = 2$, and $du = dx$. The third uses Eq. (5-2) directly, with $c = 1$ and $du = dx$. This means

$$5 \int x^3 \, dx - 6 \int x^2 \, dx + \int dx = 5\left(\frac{x^4}{4}\right) - 6\left(\frac{x^3}{3}\right) + x + C$$

$$= \frac{5}{4}x^4 - 2x^3 + x + C$$

∎

EXAMPLE 5 Integrate: $\int \left(\sqrt{r} - \dfrac{1}{r^3} \right) dr.$

In order to use Eq. (5-4) we must first write $\sqrt{r} = r^{1/2}$ and $1/r^3 = r^{-3}.$

$$\int \left(\sqrt{r} - \frac{1}{r^3} \right) dr = \int r^{1/2}\, dr - \int r^{-3}\, dr = \frac{1}{\frac{3}{2}} r^{3/2} - \frac{1}{-2} r^{-2} + C$$

$$= \frac{2}{3} r^{3/2} + \frac{1}{2} r^{-2} + C = \frac{2}{3} r^{3/2} + \frac{1}{2r^2} + C \quad\blacksquare$$

EXAMPLE 6 Integrate: $\int (x^2 + 1)^3 (2x\, dx).$

We first note that $n = 3$, for this is the power involved in the function being integrated. If $n = 3$, then $x^2 + 1$ must be u. If $u = x^2 + 1$, then $du = 2x\, dx$. Thus, the integral is in proper form for integration *as it stands*. Using the power formula,

$$\int (x^2 + 1)^3 (2x\, dx) = \frac{(x^2 + 1)^4}{4} + C$$

CAUTION ▶ *It must be emphasized that the entire quantity* **(2x dx)** *must be equated to du.* Normally, u and n are recognized first, and then du is derived from u.

Showing the use of u directly, we can write the integration as

$$\int (x^2 + 1)^3 (2x\, dx) = \int u^3\, du = \frac{1}{4} u^4 + C = \frac{(x^2 + 1)^4}{4} + C \quad\blacksquare$$

Again we note that a good knowledge of differential forms is essential for the proper recognition of *u* and *du*.

EXAMPLE 7 Integrate: $\int x^2 \sqrt{x^3 + 2}\, dx.$

We first note that $n = \frac{1}{2}$ and u is then $x^3 + 2$. Since $u = x^3 + 2$, $du = 3x^2\, dx$. Now we group $3x^2\, dx$ as du. Since there is no 3 under the integral sign, we introduce one. In order not to change the numerical value, we also introduce a factor of $\frac{1}{3}$, normally before the integral sign. In this way we take advantage of the fact that *a constant (and only a constant) factor may be moved across the integral sign.*

$$\int x^2 \sqrt{x^3 + 2}\, dx = \frac{1}{3} \int 3x^2 \sqrt{x^3 + 2}\, dx = \frac{1}{3} \int (x^3 + 2)^{1/2} (3x^2\, dx)$$

Here we indicate the proper grouping to have the proper form of Eq. (5-5).

$$\int x^2 \sqrt{x^3 + 2}\, dx = \frac{1}{3} \int (x^3 + 2)^{1/2} (3x^2\, dx) = \frac{1}{3} \left(\frac{2}{3} \right) (x^3 + 2)^{3/2} + C$$

$$= \frac{2}{9} (x^3 + 2)^{3/2} + C$$

The $1/\frac{3}{2}$ was written as $\frac{2}{3}$, since this form is more convenient with fractions.

With $u = x^3 + 2$ and using u directly in the integration, we can write

$$\int x^2 \sqrt{x^3 + 2}\, dx = \int (x^3 + 2)^{1/2} (x^2\, dx)$$

$$= \int u^{1/2} \left(\frac{1}{3} du \right) = \frac{1}{3} \int u^{1/2}\, du \qquad \text{integrating in terms of } u$$

$$= \frac{1}{3} \left(\frac{2}{3} \right) u^{3/2} + C = \frac{2}{9} u^{3/2} + C$$

$$= \frac{2}{9} (x^3 + 2)^{3/2} + C \qquad \text{substituting } x^3 + 2 = u \quad\blacksquare$$

Evaluating the Constant of Integration

To find the constant of integration, we need information such as a set of values that satisfy the function. A point through which the curve passes would provide the necessary information. This is illustrated in the following examples.

EXAMPLE 8   Find y in terms of x, given that $dy/dx = 3x - 1$ and the curve passes through $(1, 4)$.

We write the equation as

$$dy = (3x - 1)\,dx$$

and then indicate and perform the integration:

$$\int dy = \int (3x - 1)\,dx$$

$$y = \frac{3}{2}x^2 - x + C$$

Since the curve passes through $(1, 4)$, the coordinates must satisfy the equation. Thus,

$$4 = \frac{3}{2} - 1 + C \quad \text{or} \quad C = \frac{7}{2}$$

This means that the solution is

$$y = \frac{3}{2}x^2 - x + \frac{7}{2} \quad \text{or} \quad 2y = 3x^2 - 2x + 7$$

The graph of this parabola is shown in Fig. 5-2.   ▬

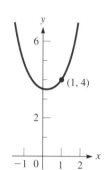

Fig. 5-2

The *displacement* of an object is its change in position.

EXAMPLE 9   The time rate of change of the displacement (velocity) of a robot arm is $ds/dt = t\sqrt{9 - t^2}$. Find the expression for the displacement as a function of time if $s = 0$ cm when $t = 0$ s.

We start the solution by writing

$$ds = t\sqrt{9 - t^2}\,dt$$

$$\int ds = \int t(9 - t^2)^{1/2}\,dt$$

To integrate the expression on the right, we recognize that $n = \frac{1}{2}$ and $u = 9 - t^2$, and therefore $du = -2t\,dt$. This means we need a -2 with the t and dt to form the proper du, which in turn means we place a $-\frac{1}{2}$ before the integral sign.

$$\int ds = \int t(9 - t^2)^{1/2}\,dt = -\frac{1}{2}\int (9 - t^2)^{1/2}(-2t\,dt)$$

$$s = -\frac{1}{2}\left(\frac{2}{3}\right)(9 - t^2)^{3/2} + C = -\frac{1}{3}(9 - t^2)^{3/2} + C$$

Since $s = 0$ cm when $t = 0$ s, we have

$$0 = -\frac{1}{3}(9)^{3/2} + C \qquad \text{or} \qquad C = 9$$

This means that the expression for the displacement is

$$s = -\frac{1}{3}(9 - t^2)^{3/2} + 9$$ ▬

We have discussed the integration of certain basic functions. Many other methods are used to integrate other functions, and some of these are discussed in Chapters 9 and 10. Also, there are many functions for which integral forms do not exist.

— EXERCISES 5-2 —

In Exercises 1–32, integrate each of the given expressions.

1. $\int 2x\,dx$

2. $\int 5x^4\,dx$

3. $\int x^7\,dx$

4. $\int 0.6y^5\,dy$

5. $\int 2x^{3/2}\,dx$

6. $\int 6\sqrt[3]{x}\,dx$

7. $\int x^{-4}\,dx$

8. $\int \dfrac{4}{\sqrt{x}}\,dx$

9. $\int (x^2 - x^5)\,dx$

10. $\int (1 - 3x)\,dx$

11. $\int (9x^2 + x + 3)\,dx$

12. $\int x(x - 2)^2\,dx$

13. $\int \left(\dfrac{t^2}{2} - \dfrac{2}{t^2}\right)dt$

14. $\int \dfrac{3x^2 - 4}{x^2}\,dx$

15. $\int \sqrt{x}(x^2 - x)\,dx$

16. $\int (3R\sqrt{R} - 5R^2)\,dR$

17. $\int (2x^{-2/3} + 3^{-2})\,dx$

18. $\int (x^{1/3} + x^{1/5} + x^{-1/7})\,dx$

19. $\int (1 + 2s^2)^2\,ds$

20. $\int (x^2 + 4x + 4)^{1/3}\,dx$

21. $\int (x^2 - 1)^5(2x\,dx)$

22. $\int (x^3 - 2)^6(3x^2\,dx)$

23. $\int (x^4 + 3)^4(4x^3\,dx)$

24. $\int (1 - 2x)^{1/3}(-2\,dx)$

25. $\int (2\theta^5 + 5)^7\theta^4\,d\theta$

26. $\int 6x^2(1 - x^3)^{4/3}\,dx$

27. $\int \sqrt{8x + 1}\,dx$

28. $\int \dfrac{dV}{(0.3 + 2V)^3}$

29. $\int \dfrac{x\,dx}{\sqrt{6x^2 + 1}}$

30. $\int \dfrac{2x^2\,dx}{\sqrt{2x^3 + 1}}$

31. $\int \dfrac{x - 1}{\sqrt{x^2 - 2x}}\,dx$

32. $\int (x^2 - x)\left(x^3 - \dfrac{3}{2}x^2\right)^8\,dx$

In Exercises 33–36, find y in terms of x.

33. $\dfrac{dy}{dx} = 6x^2$, curve passes through $(0, 2)$

34. $\dfrac{dy}{dx} = 8x + 1$, curve passes through $(-1, 4)$

35. $\dfrac{dy}{dx} = x^2(1 - x^3)^5$, curve passes through $(1, 5)$

36. $\dfrac{dy}{dx} = 2x^3(x^4 - 6)^4$, curve passes through $(2, 10)$

In Exercises 37–44, find the required equations.

37. Find the equation of the curve whose slope is $-x\sqrt{1 - 4x^2}$ and that passes through $(0, 7)$.

(W) 38. Explain why the integral $\int (x^3 - 1)^2\,dx$ cannot be integrated by letting $u = x^3 - 1$ and $du = 3x^2\,dx$.

39. The time rate of change of electric current in a circuit is given by $di/dt = 4t - 0.6t^2$. Find the expression for the current as a function of time if $i = 2$ A when $t = 0$ s.

40. The rate of change of the frequency f of an electronic oscillator with respect to the inductance L is $df/dL = 80(4 + L)^{-3/2}$. Find f as a function of L if $f = 80$ Hz for $L = 0$ H.

41. At a given site, the rate of change of the annual fraction f of energy supplied by solar energy with respect to the solar collector area A (in m²) is $\dfrac{df}{dA} = \dfrac{0.005}{\sqrt{0.01A + 1}}$. Find f as a function of A if $f = 0$ for $A = 0$ m².

42. An analysis of a company's records shows that in a day the rate of change of profit p (in dollars) in producing x generators is $\dfrac{dp}{dx} = \dfrac{600(30 - x)}{\sqrt{60x - x^2}}$. Find the profit in producing x generators if a loss of \$5000 is incurred if none are produced.

43. Find the equation of the curve for which the second derivative is 6. The curve passes through $(1, 2)$ with a slope of 8.

(W) 44. The second derivative of a function is $12x^2$. Explain how to find the function if its curve passes through the points $(1, 6)$ and $(2, 21)$. Find the function.

5-3 THE AREA UNDER A CURVE

In geometry there are formulas and methods for finding the areas of regular figures. By means of integration it is possible to find the area between curves for which we know the equations. The next example illustrates the basic idea behind the method.

EXAMPLE 1 Approximate the area in the first quadrant to the left of the line $x = 4$ and under the parabola $y = x^2 + 1$. First, make this approximation by inscribing two rectangles of equal width under the parabola and finding the sum of the areas of these rectangles. Then, improve the approximation by repeating the process with eight rectangles.

See Appendix C for the graphing calculator program AREAUNCV. It approximates the area under a curve by summing the areas of *n* rectangles.

The area to be approximated is shown in Fig. 5-3(a). The area with two rectangles inscribed under the curve is shown in Fig. 5-3(b). The first approximation, admittedly small, of the area can be found by adding the areas of the two rectangles. Both rectangles have a width of 2. The left rectangle is 1 unit high, and the right rectangle is 5 units high. Thus, the area of the two rectangles is

$$A = 2(1 + 5) = 12$$

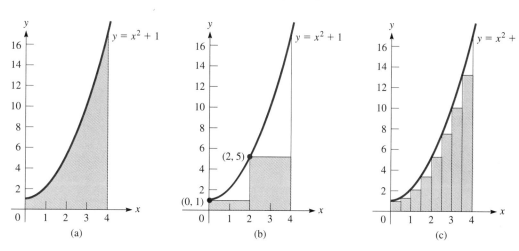

Fig. 5-3   (a)   (b)   (c)

Table 5-1

Number of rectangles n	Total area of rectangles
8	21.5
100	25.0144
1,000	25.301 344
10,000	25.330 134

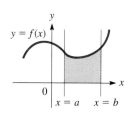

Fig. 5-4

A much better approximation is found by inscribing the eight rectangles as shown in Fig. 5-3(c). Each of these rectangles has a width of $\frac{1}{2}$. The leftmost rectangle has a height of 1. The next has a height of $\frac{5}{4}$, which is determined by finding y for $x = \frac{1}{2}$. The next rectangle has a height of 2, which is found by evaluating y for $x = 1$. Finding the heights of all rectangles and multiplying their sum by $\frac{1}{2}$ gives the area of the eight rectangles as

$$A = \frac{1}{2}\left(1 + \frac{5}{4} + 2 + \frac{13}{4} + 5 + \frac{29}{4} + 10 + \frac{53}{4}\right) = \frac{43}{2} = 21.5$$

An even better approximation could be obtained by inscribing more rectangles under the curve. The greater the number of rectangles, the more nearly the sum of their areas equals the area under the curve. See Table 5-1. By using integration later in this section, we determine the *exact* area to be $\frac{76}{3} = 25\frac{1}{3}$. --------■

We now develop the basic method used to find the area under a curve, which is the area bounded by the curve, the *x*-axis, and the lines $x = a$ and $x = b$. See Fig. 5-4. We assume here that $f(x)$ is never negative in the interval $a < x < b$. In Chapter 6 we will extend the method such that $f(x)$ may be negative.

In finding the area under a curve, we consider the sum of the areas of inscribed rectangles, as the number of rectangles is assumed to increase without bound. The reason for this last condition is that, as we saw in Example 1, as the number of rectangles increases, the approximation of the area is better.

EXAMPLE 2 Find the area under the straight line $y = 2x$, above the x-axis, and to the left of the line $x = 4$.

NOTE ▶ Since this figure is a right triangle, the area can easily be found. However, the *method* we shall use here is the important concept. We first subdivide the interval from $x = 0$ to $x = 4$ into n inscribed rectangles of Δx in width. The extremities of the intervals are labeled $a, x_1, x_2, \ldots, b (= x_n)$, as shown in Fig. 5-5, where

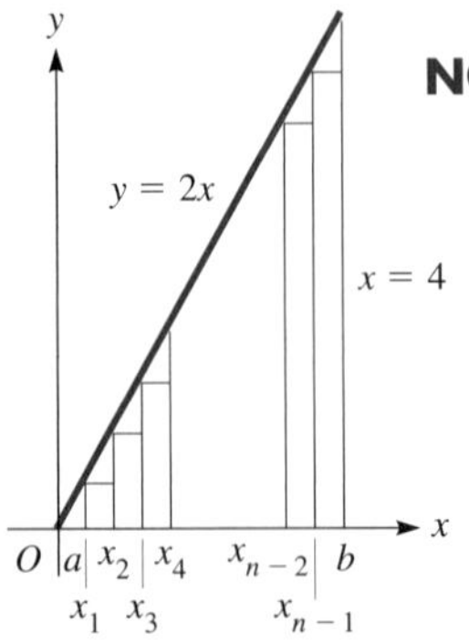

$$x_1 = \Delta x, \qquad x_2 = 2\Delta x, \ldots, \qquad x_{n-1} = (n-1)\Delta x, \qquad b = n\Delta x$$

The area of each of these n rectangles is as follows:

Fig. 5-5

First $f(a)\Delta x$, where $f(a) = f(0) = 2(0) = 0$ is the height.
Second $f(x_1)\Delta x$, where $f(x_1) = 2(\Delta x) = 2\Delta x$ is the height.
Third $f(x_2)\Delta x$, where $f(x_2) = 2(2\Delta x) = 4\Delta x$ is the height.
Fourth $f(x_3)\Delta x$, where $f(x_3) = 2(3\Delta x) = 6\Delta x$ is the height.

$$\vdots$$

Last $f(x_{n-1})\Delta x$, where $f[(n-1)\Delta x] = 2(n-1)\Delta x$ is the height.

These areas are summed up as follows:

NOTE ▶
$$A_n = \boxed{f(a)\Delta x + f(x_1)\Delta x + f(x_2)\Delta x + \cdots + f(x_{n-1})\Delta x} \tag{5-5}$$

$$= 0 + 2\Delta x(\Delta x) + 4\Delta x(\Delta x) + \cdots + 2[n-1]\Delta x]\Delta x$$
$$= 2(\Delta x)^2[1 + 2 + 3 + \cdots + (n-1)]$$

Now, $b = n\Delta x$, or $4 = n\Delta x$, or $\Delta x = 4/n$. Thus,

$$A_n = 2\left(\frac{4}{n}\right)^2[1 + 2 + 3 + \cdots + (n-1)]$$

The sum of the arithmetic sequence $1 + 2 + 3 + \cdots + n - 1$ is

$$s = \frac{n-1}{2}(1 + n - 1) = \frac{n(n-1)}{2} = \frac{n^2 - n}{2}$$

Now the expression for the sum of the areas can be written as

$$A_n = \frac{32}{n^2}\left(\frac{n^2 - n}{2}\right) = 16\left(1 - \frac{1}{n}\right)$$

This expression is an approximation of the actual area under consideration. The larger n becomes, the better the approximation. If we let $n \to \infty$ (which is equivalent to letting $\Delta x \to 0$), the limit of this sum will equal the area in question.

This checks with the geometric result.

$$A = \lim_{n\to\infty} 16\left(1 - \frac{1}{n}\right) = 16 \qquad 1/n \to 0 \text{ as } n \to \infty$$

NOTE ▶ *The area under the curve is the limit of the sum of the areas of the inscribed rectangles, as the number of rectangles approaches infinity.*

The method indicated in Example 2 illustrates the interpretation of finding an area as a summation process, although it should not be considered as a proof. However, we shall find that integration proves to be a much more useful method for finding an area. Let us now see how integration can be used directly.

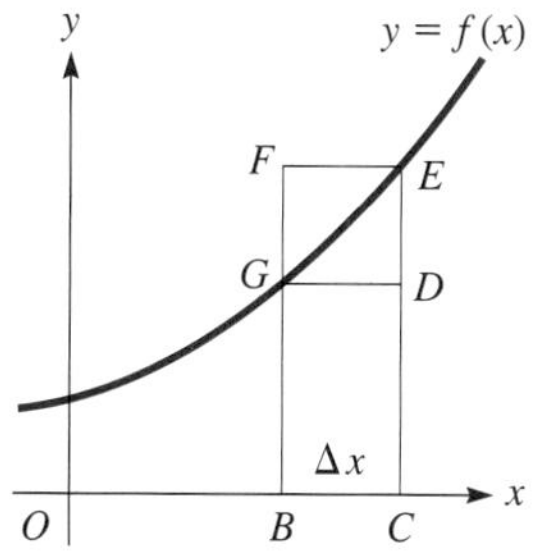

Let ΔA represent the area $BCEG$ under the curve, as indicated in Fig. 5-6. We see that the following inequality is true for the indicated areas:

$$A_{BCDG} < \Delta A < A_{BCEF}$$

If the point G is now designated as (x, y) and E as $(x + \Delta x, y + \Delta y)$, we have $y\,\Delta x < \Delta A < (y + \Delta y)\,\Delta x$. Dividing through by Δx, we have

$$y < \frac{\Delta A}{\Delta x} < y + \Delta y$$

Now we take the limit as $\Delta x \to 0$ (Δy then also approaches zero). This results in

$$\frac{dA}{dx} = y \tag{5-6}$$

This is true since the left member of the inequality is y and the right member approaches y. Also remember that

$$\lim_{\Delta x \to 0} \frac{\Delta A}{\Delta x} = \frac{dA}{dx}$$

We shall now use Eq. (5-6) to show the method of finding the complete area under a curve. We now let $x = a$ be the left boundary of the desired area and $x = b$ be the right boundary (Fig. 5-7). The area under the curve to the right of $x = a$ and bounded on the right by the line GB is now designated as A_{ax}. From Eq. (5-6) we have

$$dA_{ax} = [y\,dx]_a^x \quad \text{or} \quad A_{ax} = \left[\int y\,dx\right]_a^x$$

where $[\ ]_a^x$ is the notation used to indicate the boundaries of the area. Thus,

$$A_{ax} = \left[\int f(x)\,dx\right]_a^x = [F(x) + C]_a^x \tag{5-7}$$

But we know that if $x = a$, then $A_{aa} = 0$. Thus, $0 = F(a) + C$, or $C = -F(a)$. Therefore,

$$A_{ax} = \left[\int f(x)\,dx\right]_a^x = F(x) - F(a) \tag{5-8}$$

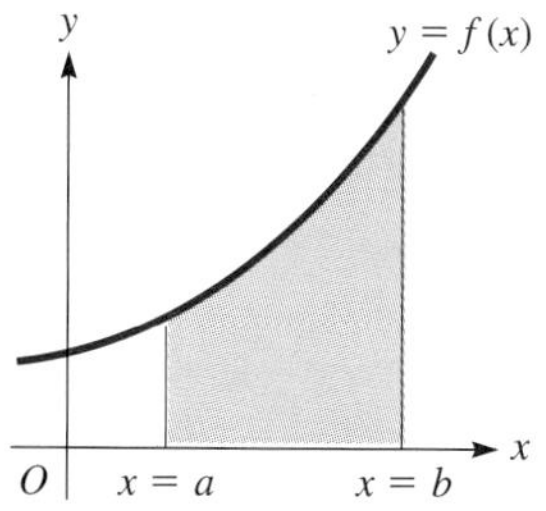

Now, to find the area under the curve that reaches from a to b, we write

$$A_{ab} = F(b) - F(a) \tag{5-9}$$

Thus, the area under the curve that reaches from a to b is given by

$$A_{ab} = \left[\int f(x)\,dx\right]_a^b = F(b) - F(a) \tag{5-10}$$

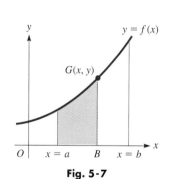

Fig. 5-6

Fig. 5-7

Fig. 5-8

NOTE▶ This shows that *the area under the curve may be found by integrating the function $f(x)$ to find the function $F(x)$, which is then evaluated at each boundary value. The area is the difference between these values of $F(x)$.* See Fig. 5-8.

In Example 2, we found an area under a curve by finding the limit of the sum of the areas of the inscribed rectangles as the number of rectangles approaches infinity. Equation (5-10) expresses the area under a curve in terms of integration. We can now see that we have obtained an area by summation and also expressed it in terms of integration. Therefore, we conclude that

summations can be evaluated by integration.

Also, we have seen the connection between the problem of finding the slope of a tangent to a curve (differentiation) and the problem of finding an area under a curve (integration). We would not normally suspect that these two problems would have solutions that lead to reverse processes. We have also seen that integration has much more application than originally anticipated.

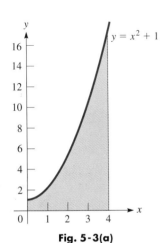

Fig. 5-3(a)

■**EXAMPLE 3**   Find the area under the curve $y = x^2 + 1$ between the y-axis and the line $x = 4$. This is the area of Example 1 shown in Fig. 5-3(a), which is shown again here for reference.

Using Eq. (5-10), we note that $f(x) = x^2 + 1$. This means that

$$\int (x^2 + 1)\, dx = \frac{1}{3}x^3 + x + C$$

Therefore, with $F(x) = \frac{1}{3}x^3 + x$, the area is given by

$$A_{0,4} = F(4) - F(0) \qquad \text{using Eq. (5-10)}$$

$$= \left[\frac{1}{3}(4^3) + 4\right] - \left[\frac{1}{3}(0^3) + 0\right] \qquad \text{evaluating } F(x) \text{ at } x = 4 \text{ and } x = 0$$

$$= \frac{1}{3}(64) + 4 = \frac{76}{3}$$

We note that this is about 4 square units greater than the value obtained by the approximation in Example 1. This result means that the exact area is $25\frac{1}{3}$, as stated at the end of Example 1.   ■

■**EXAMPLE 4**   Find the area under the curve $y = x^3$ that is between the lines $x = 1$ and $x = 2$.

In Eq. (5-10), $f(x) = x^3$. Therefore,

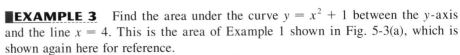

$$\int x^3\, dx = \frac{1}{4}x^4 + C$$

$$A_{1,2} = F(2) - F(1) = \left[\frac{1}{4}(2^4)\right] - \left[\frac{1}{4}(1^4)\right] \qquad \text{using Eq. (5-10) and evaluating}$$

$$= 4 - \frac{1}{4} = \frac{15}{4}$$

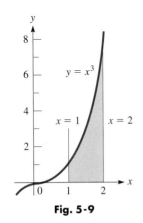

Fig. 5-9

Therefore, we see that the required area is 15/4. Again, as in Example 3, this is an exact area, not an approximation. The curve and the area are shown in Fig. 5-9.   ■

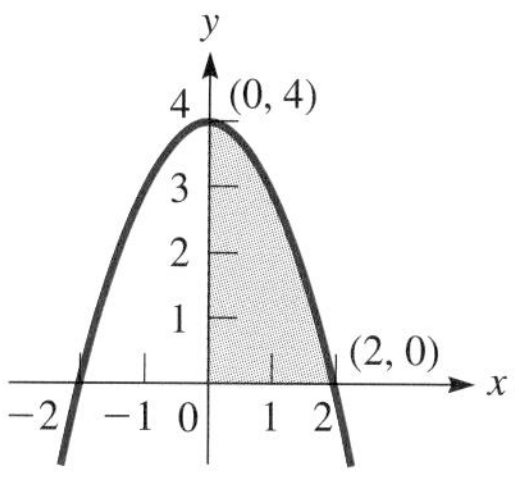

EXAMPLE 5 Find the area under the curve $y = 4 - x^2$ that lies in the first quadrant.

By solving the equation $4 - x^2 = 0$, we determine that the area to be found extends from $x = 0$ to $x = 2$ (see Fig. 5-10). Thus,

$$\int (4 - x^2)\, dx = \overbrace{4x - \frac{x^3}{3}}^{F(x)} + C$$

$$A_{0,2} = \left(8 - \frac{8}{3}\right) - (0 - 0) \qquad \text{using Eq. (5-10) and evaluating}$$

$$= \frac{16}{3}$$

Therefore, the exact area is 16/3.

We can see from these examples that we do not have to include the constant of integration when we are finding areas. Any constant that might be added to $F(x)$ would cancel out when $F(a)$ is subtracted from $F(b)$.

Fig. 5-10

EXERCISES 5-3

In Exercises 1–10, find the approximate area under the given curves by dividing the indicated intervals into n subintervals and then add up the areas of the inscribed rectangles. There are two values of n for each exercise and therefore two approximations for each area. The height of each rectangle may be found by evaluating the function for the proper value of x. See Example 1.

1. $y = 3x$, between $x = 0$ and $x = 3$, for
(a) $n = 3$ ($\Delta x = 1$), (b) $n = 10$ ($\Delta x = 0.3$)

2. $y = 2x + 1$, between $x = 0$ and $x = 2$, for
(a) $n = 4$ ($\Delta x = 0.5$), (b) $n = 10$ ($\Delta x = 0.2$)

3. $y = x^2$, between $x = 0$ and $x = 2$, for
(a) $n = 5$ ($\Delta x = 0.4$), (b) $n = 10$ ($\Delta x = 0.2$)

4. $y = x^2 + 2$, between $x = 0$ and $x = 3$, for
(a) $n = 3$ ($\Delta x = 1$), (b) $n = 10$ ($\Delta x = 0.3$)

5. $y = 4x - x^2$, between $x = 1$ and $x = 4$, for
(a) $n = 6$, (b) $n = 10$

6. $y = 1 - x^2$, between $x = 0.5$ and $x = 1$, for
(a) $n = 5$, (b) $n = 10$

7. $y = \dfrac{1}{x^2}$, between $x = 1$ and $x = 5$, for (a) $n = 4$,
(b) $n = 8$

8. $y = \sqrt{x}$, between $x = 1$ and $x = 4$, for (a) $n = 3$,
(b) $n = 12$

9. $y = \dfrac{1}{\sqrt{x + 1}}$, between $x = 3$ and $x = 8$, for
(a) $n = 5$, (b) $n = 10$

10. $y = 2x\sqrt{x^2 + 1}$, between $x = 0$ and $x = 6$, for
(a) $n = 6$, (b) $n = 12$

In Exercises 11–20, find the exact area under the given curves between the indicated values of x. The functions are the same as those for which approximate areas were found in Exercises 1–10.

11. $y = 3x$, between $x = 0$ and $x = 3$

12. $y = 2x + 1$, between $x = 0$ and $x = 2$

13. $y = x^2$, between $x = 0$ and $x = 2$

14. $y = x^2 + 2$, between $x = 0$ and $x = 3$

15. $y = 4x - x^2$, between $x = 1$ and $x = 4$

16. $y = 1 - x^2$, between $x = 0.5$ and $x = 1$

17. $y = \dfrac{1}{x^2}$, between $x = 1$ and $x = 5$

18. $y = \sqrt{x}$, between $x = 1$ and $x = 4$

19. $y = \dfrac{1}{\sqrt{x + 1}}$, between $x = 3$ and $x = 8$

(W) 20. $y = 2x\sqrt{x^2 + 1}$, between $x = 0$ and $x = 6$
Explain the reason for the difference between this result and the two values found in Exercise 10.

$5\text{-}4$ THE DEFINITE INTEGRAL

Using reasoning similar to that in the preceding section, *we define the* **definite integral** *of a function f(x) as*

$$\int_a^b f(x)\,dx = F(b) - F(a) \qquad \text{(5-11)}$$

where $F'(x) = f(x)$. **We call this a** definite integral *because the final result of integrating and evaluating is a* **number.** (The *indefinite* integral had an arbitrary constant in the result.) *The numbers a and b are called the* **lower limit** *and the* **upper limit,** *respectively. We can see that the value of a definite integral is found by evaluating the antiderivative (found by integration) at the upper limit and subtracting the value of this function at the lower limit.*

LIMITS OF INTEGRATION

We know, from the analysis in the preceding section, that *this definite integral can be interpreted as a summation process*, where the size of the subdivision approaches a limit of zero. *It is this interpretation of integration that we shall apply to many kinds of problems.*

That integration is equivalent to the limit of a sum is the reason that Leibniz (see page 69) used an elongated S for the integral sign. It stands for the Latin word for *sum*.

EXAMPLE 1 Evaluate the integral $\int_0^2 x^4\,dx$.

upper limit

$$\int_0^2 x^4\,dx = \frac{x^5}{5}\bigg|_0^2 = \frac{2^5}{5} - 0 = \frac{32}{5}$$

lower limit $f(x)$ $F(x)$

Note that a vertical line—with the limits written at the top and the bottom—is the way the value is indicated after integration, but before evaluation. ∎

EXAMPLE 2 Evaluate $\int_1^3 (x^{-2} - 1)\,dx$.

upper limit   subtract   lower limit

$$\int_1^3 (x^{-2} - 1)\,dx = -\frac{1}{x} - x\bigg|_1^3 = \left(-\frac{1}{3} - 3\right) - (-1 - 1)$$

$$= -\frac{10}{3} + 2 = -\frac{4}{3}$$ ∎

EXAMPLE 3 Evaluate $\int_0^1 5z(z^2 + 1)^5\,dz$.

For purposes of integration, $n = 5$, $u = z^2 + 1$, and $du = 2z\,dz$. Hence,

$$\int_0^1 5z(z^2 + 1)^5\,dz = \frac{5}{2}\int_0^1 (z^2 + 1)^5(2z\,dz)$$

$$= \frac{5}{2}\left(\frac{1}{6}\right)(z^2 + 1)^6\bigg|_0^1$$

$$= \frac{5}{12}(2^6 - 1^6) = \frac{5(63)}{12} = \frac{105}{4}$$ ∎

EXAMPLE 4 Evaluate $\int_{-1}^{3} x(1 - 3x^2)^{1/3}\,dx$.

For purposes of integration, $n = \frac{1}{3}$, $u = 1 - 3x^2$, and $du = -6x\,dx$. Thus,

$$\int_{-1}^{3} x(1 - 3x^2)^{1/3}\,dx = -\frac{1}{6}\int_{-1}^{3}(1 - 3x^2)^{1/3}(-6x\,dx)$$

$$= -\frac{1}{6}\left(\frac{3}{4}\right)(1 - 3x^2)^{4/3}\Big|_{-1}^{3} \qquad \text{integrate}$$

$$= -\frac{1}{8}(-26)^{4/3} + \frac{1}{8}(-2)^{4/3} \qquad \text{evaluate}$$

$$= \frac{1}{8}(2\sqrt[3]{2} - 26\sqrt[3]{26}) = \frac{1}{4}(\sqrt[3]{2} - 13\sqrt[3]{26})$$

The result shown is the exact value. The approximate value to three decimal places is -9.313. ∎

EXAMPLE 5 Evaluate $\int_{0.1}^{2.7} \dfrac{dx}{\sqrt{4x + 1}}$.

In order to integrate we have $n = -\frac{1}{2}$, $u = 4x + 1$, and $du = 4\,dx$. Therefore,

$$\int_{0.1}^{2.7} \frac{dx}{\sqrt{4x + 1}} = \int_{0.1}^{2.7}(4x + 1)^{-1/2}\,dx = \frac{1}{4}\int_{0.1}^{2.7}(4x + 1)^{-1/2}(4\,dx)$$

$$= \frac{1}{4}\left(\frac{1}{\frac{1}{2}}\right)(4x + 1)^{1/2}\Big|_{0.1}^{2.7} = \frac{1}{2}(4x + 1)^{1/2}\Big|_{0.1}^{2.7} \qquad \text{integrate}$$

$$= \frac{1}{2}(\sqrt{11.8} - \sqrt{1.4}) = 1.126 \qquad \text{evaluate} \quad ∎$$

EXAMPLE 6 Evaluate $\int_{0}^{4} \dfrac{x + 1}{(x^2 + 2x + 2)^3}\,dx$

For purposes of integration,

$$n = -3 \qquad u = x^2 + 2x + 2 \qquad du = (2x + 2)\,dx = 2(x + 1)\,dx$$

Therefore,

$$\int_{0}^{4}(x^2 + 2x + 2)^{-3}(x + 1)\,dx = \frac{1}{2}\int_{0}^{4}(x^2 + 2x + 2)^{-3}[2(x + 1)\,dx]$$

$$= \frac{1}{2}\left(\frac{1}{-2}\right)(x^2 + 2x + 2)^{-2}\Big|_{0}^{4} \qquad \text{integrate}$$

$$= -\frac{1}{4}(16 + 8 + 2)^{-2} + \frac{1}{4}(0 + 0 + 2)^{-2} \qquad \text{evaluate}$$

$$= \frac{1}{4}\left(-\frac{1}{26^2} + \frac{1}{2^2}\right) = \frac{1}{4}\left(\frac{1}{4} - \frac{1}{676}\right)$$

$$= \frac{1}{4}\left(\frac{168}{676}\right) = \frac{21}{338}$$

```
fnInt((X+1)/(X²+
2X+2)^3,X,0,4)
         .0621301775
21/338
         .0621301775
```

Fig. 5-11

The value of a definite integral can be found on a calculator. Figure 5-11 shows a typical display using the *numerical integral* feature to evaluate the integral of this example. Also shown is the check that the fraction above gives the same result. ∎

The definition of the definite integral is valid regardless of the source of $f(x)$. That is, we may apply the definite integral whenever we want to sum a function in a manner similar to that which we use to find an area.

The following example illustrates an application of the definite integral. In Chapter 6 we shall see that the definite integral has applications in many areas of science and technology.

See the chapter introduction.

EXAMPLE 7 The rate of flow Q (in ft³/s) of water over a certain dam is found by evaluating the definite integral in the equation $Q = \int_0^{1.25} 240\sqrt{1.50 - y}\,dy$. See Fig. 5-12. Find Q.

The solution is as follows:

$$Q = \int_0^{1.25} 240\sqrt{1.50 - y}\,dy = -240\int_0^{1.25}(1.50 - y)^{1/2}(-dy)$$

$$= -240\left(\frac{2}{3}\right)(1.50 - y)^{3/2}\Big|_0^{1.25} \qquad \text{integrate}$$

$$= -160[(1.50 - 1.25)^{3/2} - (1.50 - 0)^{3/2}] \qquad \text{evaluate}$$

$$= -160(0.25^{3/2} - 1.50^{3/2}) = 274 \text{ ft}^3/\text{s}$$

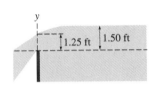

Fig. 5-12

EXERCISES 5-4

In Exercises 1–32, evaluate the given definite integrals.

1. $\int_0^1 2x\,dx$

2. $\int_0^2 3x^2\,dx$

3. $\int_1^4 x^{5/2}\,dx$

4. $\int_4^9 (p^{3/2} - 3)\,dp$

5. $\int_3^6 \left(\frac{1}{\sqrt{x}} + 2\right)dx$

6. $\int_{1.2}^{1.6}\left(5 + \frac{6}{x^4}\right)dx$

7. $\int_{-1.6}^{0.7}(1 - x)^{1/3}\,dx$

8. $\int_1^5 \sqrt{2x - 1}\,dx$

9. $\int_{-2}^2 (T - 2)(T + 2)\,dT$

10. $\int_1^2 (3x^5 - 2x^3)\,dx$

11. $\int_{0.5}^{2.2}(\sqrt[3]{x} - 2)\,dx$

12. $\int_{2.7}^{5.3}\left(\frac{1}{x\sqrt{x}} + 4\right)dx$

13. $\int_0^4 (1 - \sqrt{x})^2\,dx$

14. $\int_1^4 \frac{y + 4}{\sqrt{y}}\,dy$

15. $\int_{-2}^{-1} 2x(4 - x^2)^3\,dx$

16. $\int_0^1 x(3x^2 - 1)^3\,dx$

17. $\int_0^4 \frac{x\,dx}{\sqrt{x^2 + 9}}$

18. $\int_{0.2}^{0.7} x^2(x^3 + 2)^{3/2}\,dx$

19. $\int_{2.75}^{3.25}\frac{dx}{\sqrt[3]{6x + 1}}$

20. $\int_2^6 \frac{2\,dx}{\sqrt{4x + 1}}$

21. $\int_1^3 \frac{2x\,dx}{(2x^2 + 1)^3}$

22. $\int_{12.6}^{17.2}\frac{3\,dx}{(6x - 1)^2}$

23. $\int_3^7 \sqrt{16t^2 + 8t + 1}\,dt$

24. $\int_{-5}^1 \sqrt{6 - 2x}\,dx$

25. $\int_0^2 2x(9 - 2x^2)^2\,dx$

26. $\int_{-1}^2 V(V^3 + 1)\,dV$

27. $\int_0^1 (x^2 + 3)(x^3 + 9x + 6)^2\,dx$

28. $\int_2^3 \frac{x^2 + 1}{(x^3 + 3x)^2}\,dx$

29. $\int_{-1}^2 \frac{8x - 2}{(2x^2 - x + 1)^3}\,dx$

30. $\int_{-3}^{-2} (3x^2 - 2)\sqrt[3]{2x^3 - 4x + 1}\,dx$

31. $\int_{\sqrt{5}}^3 2z\sqrt[4]{z^4 + 8z^2 + 16}\,dz$

32. $\int_{-2}^0 (\sqrt{2x + 4} - \sqrt[3]{3x + 8})\,dx$

In Exercises 33–36, solve the given problems.

33. The work W (in ft·lb) in winding up an 80-ft cable is $W = \int_0^{80}(1000 - 5x)\,dx$. Evaluate W.

W **34.** The total volume V of liquid flowing through a certain pipe of radius R is $V = k(R^2\int_0^R r\,dr - \int_0^R r^3\,dr)$, where k is a constant. Evaluate V and explain why R, but not r, can be to the left of the integral sign.

35. The surface area A (in m²) of a certain parabolic radio-wave reflector is $A = 4\pi\int_0^2 \sqrt{3x + 9}\,dx$. Evaluate A.

36. The total force (in N) on the circular end of a water tank is $F = 19,600\int_0^5 y\sqrt{25 - y^2}\,dy$. Evaluate F.

5-5 NUMERICAL INTEGRATION: THE TRAPEZOIDAL RULE

For data and functions that cannot be directly integrated by available methods, it is possible to develop numerical methods of integration. These numerical methods are of greater importance today since they are readily adaptable for use on a calculator or computer. There are a great many such numerical techniques for approximating the value of a definite integral. In this section we develop one of these, the trapezoidal rule. In the following section, another numerical method is discussed.

We know from Sections 5-3 and 5-4 that we can interpret a definite integral as the area under a curve. We shall therefore show how to approximate the value of the integral by approximating the appropriate area by a set of trapezoids. The basic idea here is very similar to that used when rectangles were inscribed under a curve. However, the use of trapezoids reduces the error and provides a better approximation.

The area to be found is subdivided into n intervals of equal width. Perpendicular lines are then dropped from the curve (or points, if only a given set of numbers is available). If the points on the curve are joined by straight-line segments, the area of successive parts under the curve is approximated by finding the areas of the trapezoids formed. However, if these points are not too far apart, the approximation will be very good (see Fig. 5-13). From geometry we recall that the area of a trapezoid equals one-half the product of the sum of the bases times the altitude. For these trapezoids, the bases are the y-coordinates, and the altitudes are Δx. Therefore, when we indicate the sum of these trapezoidal areas, we have

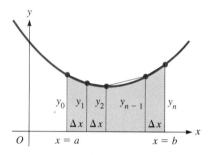

Fig. 5-13

$$A_T = \frac{1}{2}(y_0 + y_1)\Delta x + \frac{1}{2}(y_1 + y_2)\Delta x + \frac{1}{2}(y_2 + y_3)\Delta x + \cdots$$
$$+ \frac{1}{2}(y_{n-2} + y_{n-1})\Delta x + \frac{1}{2}(y_{n-1} + y_n)\Delta x$$

We note, when this addition is performed, that the result is

$$A_T = \left(\frac{1}{2}y_0 + y_1 + y_2 + \cdots + y_{n-1} + \frac{1}{2}y_n\right)\Delta x \qquad \textbf{(5-12)}$$

The y-values to be used either are derived from the function $y = f(x)$ or are the y-coordinates of a set of data.

Since A_T approximates the area under the curve, it also approximates the value of the definite integral, or

$$\int_a^b f(x)\,dx \approx \frac{\Delta x}{2}(y_0 + 2y_1 + 2y_2 + \cdots + 2y_{n-1} + y_n) \qquad \textbf{(5-13)}$$

See Appendix C for the graphing calculator program TRAPRULE. It evaluates an integral using the trapezoidal rule with n intervals.

Equation (5-13) is known as the **trapezoidal rule.** It can be used to find the the approximate value of a definite integral, even one that is defined by a table of values. If the value of n is sufficiently large, the approximation is usually very good.

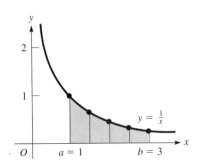

Fig. 5-14

Table 5-2

Number of trapezoids n	Total area of trapezoids
4	1.1166667
100	1.0986419
1,000	1.0986126
10,000	1.0986123

EXAMPLE 1 Approximate the value of $\int_1^3 \frac{1}{x}\,dx$ by the trapezoidal rule. Let $n = 4$.

We are to approximate the area under $y = 1/x$ from $x = 1$ to $x = 3$ by dividing the area into four trapezoids. This area is found by applying Eq. (5-12), which is the approximate value of the integral, as shown in Eq. (5-13). Figure 5-14 shows the graph. In this example, $f(x) = 1/x$, and

$$\Delta x = \frac{3 - 1}{4} = \frac{1}{2} \qquad y_0 = f(a) = f(1) = 1$$

$$y_1 = f\left(\frac{3}{2}\right) = \frac{2}{3} \qquad y_2 = f(2) = \frac{1}{2}$$

$$y_3 = f\left(\frac{5}{2}\right) = \frac{2}{5} \qquad y_n = y_4 = f(b) = f(3) = \frac{1}{3}$$

$$A_T = \frac{1/2}{2}\left[1 + 2\left(\frac{2}{3}\right) + 2\left(\frac{1}{2}\right) + 2\left(\frac{2}{5}\right) + \frac{1}{3}\right]$$

$$= \frac{1}{4}\left(\frac{15 + 20 + 15 + 12 + 5}{15}\right) = \frac{1}{4}\left(\frac{67}{15}\right) = \frac{67}{60}$$

Therefore,

$$\int_1^3 \frac{1}{x}\,dx \approx \frac{67}{60}$$

We cannot perform this integration directly by methods developed up to this point. As we increase the number of trapezoids, the value becomes more accurate. See Table 5-2. The actual value to seven decimal places is 1.0986123. ━━━■

EXAMPLE 2 Approximate the value of $\int_0^1 \sqrt{x^2 + 1}\,dx$ by the trapezoidal rule. Let $n = 5$.

Figure 5-15 shows the graph. In this example, $f(x) = \sqrt{x^2 + 1}$, and

$$\Delta x = \frac{1 - 0}{5} = 0.2$$

$y_0 = f(0) = 1$ $y_1 = f(0.2) = \sqrt{1.04} = 1.0198039$

$y_2 = f(0.4) = \sqrt{1.16} = 1.0770330$ $y_3 = f(0.6) = \sqrt{1.36} = 1.1661904$

$y_4 = f(0.8) = \sqrt{1.64} = 1.2806248$ $y_5 = f(1) = \sqrt{2.00} = 1.4142136$

Hence, we have

$$A_T = \frac{0.2}{2}[1 + 2(1.0198039) + 2(1.0770330) + 2(1.1661904)$$

$$+ 2(1.2806248) + 1.4142136]$$

$$= 1.150 \qquad \text{(rounded off)}$$

This means that

$$\int_0^1 \sqrt{x^2 + 1}\,dx \approx 1.150 \qquad \text{the actual value is 1.148 to three decimal places}$$

The entire calculation can be done on a calculator without tabulating values, as the display shows in Fig. 5-16.

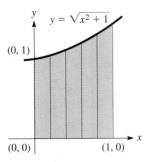

Fig. 5-15

```
(.2/2)(1+2(√(1.0
4)+√(1.16)+√(1.3
6)+√(1.64))+√(2)
)
          1.150151774
```

Fig. 5-16

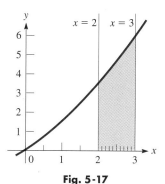

Fig. 5-17

Fig. 5-18

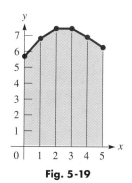

Fig. 5-19

EXAMPLE 3   Approximate the value of the integral $\int_{2}^{3} x\sqrt{x+1}\,dx$ by using the trapezoidal rule. Use $n = 10$.

In Fig. 5-17 the graph of the function and the area used in the trapezoidal rule are shown. From the given values, we have $\Delta x = \frac{3-2}{10} = 0.1$. Therefore,

$$y_0 = f(2) = 2\sqrt{3} = 3.4641016 \qquad y_1 = f(2.1) = 2.1\sqrt{3.1} = 3.6974315$$
$$y_2 = f(2.2) = 2.2\sqrt{3.2} = 3.9354796 \qquad y_3 = f(2.3) = 2.3\sqrt{3.3} = 4.1781575$$
$$y_4 = f(2.4) = 2.4\sqrt{3.4} = 4.4253813 \qquad y_5 = f(2.5) = 2.5\sqrt{3.5} = 4.6770717$$
$$y_6 = f(2.6) = 2.6\sqrt{3.6} = 4.9331532 \qquad y_7 = f(2.7) = 2.7\sqrt{3.7} = 5.1935537$$
$$y_8 = f(2.8) = 2.8\sqrt{3.8} = 5.4582048 \qquad y_9 = f(2.9) = 2.9\sqrt{3.9} = 5.7270411$$
$$y_{10} = f(3) = 3\sqrt{4} = 6.0000000$$

$$A_T = \frac{0.1}{2}[3.4641016 + 2(3.6974315) + \cdots + 2(5.7270411) + 6.0000000]$$

$$= 4.6958$$

Therefore, $\int_{2}^{3} x\sqrt{x+1}\,dx \approx 4.6958$. We see from the calcualtor display shown in Fig. 5-18 that the value is 4.6954 to four decimal places.   ∎

EXAMPLE 4   The following points were found empirically.

x	0	1	2	3	4	5
y	5.68	6.75	7.32	7.35	6.88	6.24

Approximate the value of the integral of the function defined by these points between $x = 0$ and $x = 5$ by the trapezoidal rule.

In order to find A_T, we use the values of y_0, y_1, and so on, directly from the table. We also note that $\Delta x = 1$. The graph is shown in Fig. 5-19. Therefore, we have

$$A_T = \frac{1}{2}[5.68 + 2(6.75) + 2(7.32) + 2(7.35) + 2(6.88) + 6.24] = 34.26$$

Although we do not know the algebraic form of the function, we can state that

$$\int_{0}^{5} f(x)\,dx \approx 34.26$$

∎

EXERCISES 5-5

In Exercises 1–4, (a) approximate the value of each of the given integrals by use of the trapezoidal rule, using the given value of n and (b) check by direct integration.

1. $\int_{0}^{2} 2x^2\,dx$, $n = 4$ **2.** $\int_{0}^{1} (1-x^2)\,dx$, $n = 3$

3. $\int_{1}^{4} (1+\sqrt{x})\,dx$, $n = 6$ **4.** $\int_{3}^{8} \sqrt{1+x}\,dx$, $n = 5$

In Exercises 5–12, approximate the value of each of the given integrals by use of the trapezoidal rule, using the given value of n.

5. $\int_{2}^{3} \frac{1}{2x}\,dx$, $n = 2$ **6.** $\int_{2}^{6} \frac{dx}{x+3}$, $n = 4$

7. $\int_{0}^{5} \sqrt{25-x^2}\,dx$, $n = 5$ **8.** $\int_{0}^{2} \sqrt{x^3+1}\,dx$, $n = 4$

9. $\int_{1}^{5} \frac{1}{x^2+x}\,dx$, $n = 10$ **10.** $\int_{2}^{4} \frac{1}{x^2+1}\,dx$, $n = 10$

11. $\int_{0}^{4} 2^x\,dx$, $n = 12$ **12.** $\int_{0}^{1.5} 10^x\,dx$, $n = 15$

In Exercises 13 and 14, approximate the values of the integrals defined by the given sets of points.

13. $\int_{2}^{14} y\,dx$

x	2	4	6	8	10	12	14
y	0.67	2.34	4.56	3.67	3.56	4.78	6.87

14. $\int_{1.4}^{3.2} y\,dx$

x	1.4	1.7	2.0	2.3	2.6	2.9	3.2
y	0.18	7.87	18.23	23.53	24.62	20.93	20.76

In Exercises 15 and 16, solve the given problems by using the trapezoidal rule.

15. A force F that a distributed electric charge has on a point charge is $F = k \int_{0}^{2} \dfrac{dx}{(4 + x^2)^{3/2}}$, where x is the distance along the distributed charge and k is a constant. With $n = 8$, approximate F in terms of k.

16. The length L (in ft) of telephone wire needed (considering the sag) between two poles exactly 100 ft apart is $L = 2\int_{0}^{100} \sqrt{1.6 \times 10^{-7} x^2 + 1}\,dx$. With $n = 10$, approximate L (to six significant digits).

5-6 SIMPSON'S RULE

The numerical method of integration developed in this section is also readily programmable for use on a computer or easily usable with the necessary calculations done on a calculator. It is obtained by interpreting the definite integral as the area under a curve, as we did in developing the trapezoidal rule, and by approximating the curve by a set of parabolic arcs. The use of parabolic arcs, rather than chords as with the trapezoidal rule, usually gives a better approximation.

Since we will be using parabolic arcs, we first derive a formula for the area that is under a parabolic arc. The curve shown in Fig. 5-20 represents the parabola $y = ax^2 + bx + c$. The points shown on this curve are $(-h, y_0)$, $(0, y_1)$, and (h, y_2). The area under the parabola is given by

$$A = \int_{-h}^{h} y\,dx = \int_{-h}^{h} (ax^2 + bx + c)\,dx = \frac{ax^3}{3} + \frac{bx^2}{2} + cx \Big|_{-h}^{h}$$

$$= \frac{2}{3}ah^3 + 2ch$$

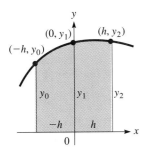

Fig. 5-20

or

$$A = \frac{h}{3}(2ah^2 + 6c) \tag{5-14}$$

The coordinates of the three points also satisfy the equation $y = ax^2 + bx + c$. This means that

$$y_0 = ah^2 - bh + c$$
$$y_1 = c$$
$$y_2 = ah^2 + bh + c$$

By finding the sum of $y_0 + 4y_1 + y_2$, we have

$$y_0 + 4y_1 + y_2 = 2ah^2 + 6c \tag{5-15}$$

Substituting Eq. (5-15) into Eq. (5-14), we have

$$A = \frac{h}{3}(y_0 + 4y_1 + y_2) \tag{5-16}$$

We note that the area depends only on the distance h and the three y-coordinates.

Now let us consider the area under the curve in Fig. 5-21. If a parabolic arc is passed through the points (x_0, y_0), (x_1, y_1), and (x_2, y_2), we may use Eq. (5-16) to approximate the area under the curve between x_0 and x_2. We also note that the distance h used in finding Eq. (5-16) is the difference in the x-coordinates, or $h = \Delta x$. Therefore, the area under the curve between x_0 and x_2 is

$$A_1 = \frac{\Delta x}{3}(y_0 + 4y_1 + y_2)$$

Similarly, if a parabolic arc is passed through the three points starting with (x_2, y_2), the area between x_2 and x_4 is

$$A_2 = \frac{\Delta x}{3}(y_2 + 4y_3 + y_4)$$

The sum of these areas is

$$A_1 + A_2 = \frac{\Delta x}{3}(y_0 + 4y_1 + 2y_2 + 4y_3 + y_4) \tag{5-17}$$

CAUTION ▶

We can continue this procedure until the approximate value of the entire area has been found. We must note, however, that *the number of intervals n of width Δx must be even.* Therefore, generalizing on Eq. (5-17) and recalling again that the value of the definite integral is the area under the curve, we have

$$\int_a^b f(x)\, dx \approx \frac{\Delta x}{3}(y_0 + 4y_1 + 2y_2 + 4y_3 + 2y_4 + \cdots + 4y_{n-1} + y_n) \tag{5-18}$$

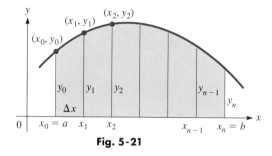

Fig. 5-21

Although named for the English mathematician Thomas Simpson (1710–1761), he did not discover the rule. It was well known when he included it in some of his many books on mathematics (see page 496).

Equation (5-18) is known as **Simpson's rule.** As previously noted, it usually provides a somewhat better approximation for the value of a definite integral than the trapezoidal rule. Simpson's rule is also used in some calculator models for the evaluation of definite integrals.

■**EXAMPLE 1** Approximate the value of the integral $\int_0^1 \dfrac{dx}{x+1}$ by Simpson's rule. Let $n = 2$.

In Fig. 5-22 the graph of the function and the area used are shown. We are to approximate the integral by using Eq. (5-18). We therefore note that $f(x) = 1/(x+1)$. Also, $x_0 = a = 0$, $x_1 = 0.5$, and $x_2 = b = 1$. This is due to the fact that $n = 2$ and $\Delta x = 0.5$ since the total interval is 1 unit (from $x = 0$ to $x = 1$). Therefore,

$$y_0 = \frac{1}{0+1} = 1.0000 \qquad y_1 = \frac{1}{0.5+1} = 0.6667 \qquad y_2 = \frac{1}{1+1} = 0.5000$$

Substituting, we have

$$\int_0^1 \frac{dx}{x+1} \approx \frac{0.5}{3}[1.0000 + 4(0.6667) + 0.5000]$$

$$\approx 0.694$$

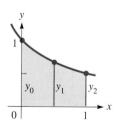

Fig. 5-22

To three decimal places, the actual value of the integral is 0.693. We will consider the method of integrating this function in a later chapter. ------------ ■

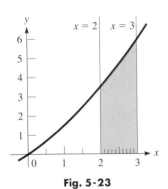

Fig. 5-23

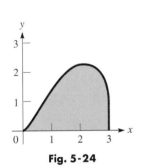

Fig. 5-24

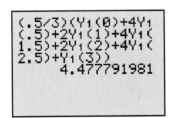

Fig. 5-25

■**EXAMPLE 2**   Approximate the value of $\int_{2}^{3} x\sqrt{x+1}\,dx$ by Simpson's rule. Use $n = 10$.

Since the necessary values for this function are shown in Example 3 of Section 5-5, we shall simply tabulate them here. ($\Delta x = 0.1$.) See Fig. 5-23.

$$y_0 = 3.4641016 \qquad y_1 = 3.6974315 \qquad y_2 = 3.9354796 \qquad y_3 = 4.1781575$$
$$y_4 = 4.4253813 \qquad y_5 = 4.6770717 \qquad y_6 = 4.9331532 \qquad y_7 = 5.1935537$$
$$y_8 = 5.4582048 \qquad y_9 = 5.7270411 \qquad y_{10} = 6.0000000$$

Therefore, we evaluate the integral as follows:

$$\int_{2}^{3} x\sqrt{x+1}\,dx \approx \frac{0.1}{3}[3.4641016 + 4(3.6974315) + 2(3.9354796)$$
$$+ 4(4.1781575) + 2(4.4253813) + 4(4.6770717)$$
$$+ 2(4.9331532) + 4(5.1935537) + 2(5.4582048)$$
$$+ 4(5.7270411) + 6.0000000]$$
$$\approx \frac{0.1}{3}(140.86156) = 4.6953854$$

This result agrees with the actual value to the eight significant digits shown. The value we obtained with the trapezoidal rule was 4.6958. ∎

■**EXAMPLE 3**   The rear stabilizer of a certain aircraft is shown in Fig. 5-24. The area A (in m^2) of one side of the stabilizer is $A = \int_{0}^{3} (3x^2 - x^3)^{0.6}\,dx$. Find this area using Simpson's rule with $n = 6$.

Here, we note that $f(x) = (3x^2 - x^3)^{0.6}$, $a = 0$, $b = 3$, and $\Delta x = \dfrac{3-0}{6} = 0.5$.

Therefore,

$$y_0 = f(0) = [3(0)^2 - 0^3]^{0.6} = 0$$
$$y_1 = f(0.5) = [3(0.5)^2 - 0.5^3]^{0.6} = 0.7542720$$
$$y_2 = f(1) = [3(1)^2 - 1^3]^{0.6} = 1.5157166$$
$$y_3 = f(1.5) = [3(1.5)^2 - 1.5^3]^{0.6} = 2.0747428$$
$$y_4 = f(2) = [3(2)^2 - 2^3]^{0.6} = 2.2973967$$
$$y_5 = f(2.5) = [3(2.5)^2 - 2.5^3]^{0.6} = 1.9811165$$
$$y_6 = f(3) = [3(3)^2 - 3^3]^{0.6} = 0$$
$$A = \int_{0}^{3} (3x^2 - x^3)^{0.6}\,dx \approx \frac{0.5}{3}[0 + 4(0.7542720) + 2(1.5157166)$$
$$+ 4(2.0747428) + 2(2.2973967) + 4(1.9811165) + 0]$$
$$\approx 4.4777920 \text{ m}^2$$

Thus, the area of one side of the stabilizer is 4.478 m^2 (rounded off).

As with the trapezoidal rule, since the calculation can be done completely on a calculator, it is not necessary to record the above values. Using $y = (3x^2 - x^3)^{0.6}$, the display in Fig. 5-25 shows the calculator evaluation of the area. ∎

EXERCISES 5-6

In Exercises 1–4, (a) approximate the value of each of the given integrals by use of Simpson's rule, using the given value of n and (b) check by direct integration.

1. $\int_0^2 (1 + x^3)\, dx$, $n = 2$ **2.** $\int_0^8 x^{1/3}\, dx$, $n = 2$

3. $\int_1^4 (2x + \sqrt{x})\, dx$, $n = 6$ **4.** $\int_0^2 x\sqrt{x^2 + 1}\, dx$, $n = 4$

In Exercises 5–12, approximate the value of each of the given integrals by use of Simpson's rule, using the given values of n. Exercises 5–10 are the same as Exercises 5–10 (except 7) of Section 5-5.

5. $\int_2^3 \dfrac{1}{2x}\, dx$, $n = 2$ **6.** $\int_2^6 \dfrac{dx}{x + 3}$, $n = 4$

7. $\int_0^5 \sqrt{25 - x^2}\, dx$, $n = 4$ **8.** $\int_0^2 \sqrt{x^3 + 1}\, dx$, $n = 4$

9. $\int_1^5 \dfrac{1}{x^2 + x}\, dx$, $n = 10$ **10.** $\int_2^4 \dfrac{1}{x^2 + 1}\, dx$, $n = 10$

11. $\int_{-4}^5 (2x^4 + 1)^{0.1}\, dx$, $n = 6$ **12.** $\int_0^{2.4} \dfrac{dx}{(4 + \sqrt{x})^{3/2}}$, $n = 8$

In Exercises 13 and 14, approximate the values of the integrals of the functions defined by the given sets of points by using Simpson's rule. These are the same as Exercises 13 and 14 of Section 5-5.

13. $\int_2^{14} y\, dx$

x	2	4	6	8	10	12	14
y	0.67	2.34	4.56	3.67	3.56	4.78	6.87

14. $\int_{1.4}^{3.2} y\, dx$

x	1.4	1.7	2.0	2.3	2.6	2.9	3.2
y	0.18	7.87	18.23	23.53	24.62	20.93	20.76

In Exercises 15 and 16, solve the given problems, using Simpson's rule.

15. The distance $\overline{x}$ (in in.) from one end of a barrel plug (with vertical cross section) to its center of mass, as shown in Fig. 5-26, is $\overline{x} = 0.9129 \int_0^3 x\sqrt{0.3 - 0.1x}\, dx$. Find $\overline{x}$ with $n = 12$.

Fig. 5-26   Center of mass

16. The average value of the electric current i_{av} (in A) in a circuit for the first 4 s is $i_{av} = \frac{1}{4}\int_0^4 (4t - t^2)^{0.2}\, dt$. Find i_{av} with $n = 10$.

CHAPTER EQUATIONS

Indefinite integral	$\displaystyle\int f(x)\, dx = F(x) + C$	**(5-1)**
Integrals	$\displaystyle\int c\, du = c\int du = cu + C$	**(5-2)**
	$\displaystyle\int (du + dv) = u + v + C$	**(5-3)**
Power formula	$\displaystyle\int u^n\, du = \dfrac{u^{n+1}}{n + 1} + C \qquad (n \neq -1)$	**(5-4)**
Area under a curve	$A_{ab} = \left[\displaystyle\int f(x)\, dx\right]_a^b = F(b) - F(a)$	**(5-10)**
Definite integral	$\displaystyle\int_a^b f(x)\, dx = F(b) - F(a)$	**(5-11)**
Trapezoidal rule	$\displaystyle\int_a^b f(x)\, dx \approx \dfrac{\Delta x}{2}(y_0 + 2y_1 + 2y_2 + \cdots + 2y_{n-1} + y_n)$	**(5-13)**
Simpson's rule	$\displaystyle\int_a^b f(x)\, dx \approx \dfrac{\Delta x}{3}(y_0 + 4y_1 + 2y_2 + 4y_3 + 2y_4 + \cdots + 4y_{n-1} + y_n)$	**(5-18)**

REVIEW EXERCISES

In Exercises 1–24, evaluate the given integrals.

1. $\int (4x^3 - x)\, dx$

2. $\int (5 + 3x^2)\, dx$

3. $\int \sqrt{u}\,(u^2 + 2)\, du$

4. $\int x(x - 3x^4)\, dx$

5. $\int_1^4 \left(\dfrac{\sqrt{x}}{2} + \dfrac{2}{\sqrt{x}} \right) dx$

6. $\int_1^2 \left(x + \dfrac{1}{x^2} \right) dx$

7. $\int_0^2 x(4 - x)\, dx$

8. $\int_0^1 2t(2t + 1)^2\, dt$

9. $\int \left(3 + \dfrac{2}{x^3} \right) dx$

10. $\int \left(3\sqrt{x} + \dfrac{1}{2\sqrt{x}} - \dfrac{1}{4} \right) dx$

11. $\int_{-2}^5 \dfrac{dx}{\sqrt[3]{x^2 + 6x + 9}}$

12. $\int_{0.35}^{0.85} x(\sqrt{1 - x^2} + 1)\, dx$

13. $\int \dfrac{dn}{(2 - 5n)^3}$

14. $\int \dfrac{1}{x^2} \sqrt{1 + \dfrac{1}{x}}\, dx$

15. $\int 3(7 - 2x)^{3/4}\, dx$

16. $\int (y^3 + 3y^2 + 3y + 1)^{2/3}\, dy$

17. $\int_0^2 \dfrac{3x\, dx}{\sqrt[3]{1 + 2x^2}}$

18. $\int_1^6 \dfrac{2\, dx}{(3x - 2)^{3/4}}$

19. $\int x^2(1 - 2x^3)^4\, dx$

20. $\int 3x^3(1 - 5x^4)^{1/3}\, dx$

21. $\int \dfrac{(2 - 3x^2)\, dx}{(2x - x^3)^2}$

22. $\int \dfrac{x^2 - 3}{\sqrt{6 + 9x - x^3}}\, dx$

23. $\int_1^3 (x^2 + x + 2)(2x^3 + 3x^2 + 12x)\, dx$

24. $\int_0^2 (4x + 18x^2)(x^2 + 3x^3)^2\, dx$

In Exercises 25 and 26, find the required equations.

25. Find the equation of the curve that passes through $(-1, 3)$ for which the slope is given by $3 - x^2$.

26. Find the equation of the curve that passes through $(1, -2)$ for which the slope is $x(x^2 + 1)^2$.

In Exercises 27 and 28, perform the integrations as directed.

(W) **27.** Perform the integration $\int (1 - 2x)\, dx$ (a) term by term, labeling the constant of integration as C_1, and then (b) by letting $u = 1 - 2x$, using the general power rule and labeling the constant of integration as C_2. Is $C_1 = C_2$? Explain.

(W) **28.** Following the methods (a) and (b) in Exercise 27, perform the integration $\int (3x + 2)\, dx$. In (b) let $u = 3x + 2$. Is $C_1 = C_2$? Explain.

In Exercises 29 and 30, use Eq. (5-10) to find the indicated areas.

29. The area under $y = 6x - 1$ between $x = 1$ and $x = 3$

30. The first-quadrant area under $y = 8x - x^4$

In Exercises 31 and 32, solve the given problems by using the trapezoidal rule.

31. Approximate $\int_1^3 \dfrac{dx}{2x - 1}$ with $n = 4$.

32. Approximate the value of the integral of the function defined by the following set of points.

x	6.0	9.0	12	15	18	21
y	2.0	1.2	0.2	1.0	6.0	12

In Exercises 33 and 34, solve the given problems by using Simpson's rule.

33. Approximate $\int_1^3 \dfrac{dx}{2x - 1}$ with $n = 4$ (see Exercise 31).

34. Approximate the value of the integral of the function defined by the following set of points.

x	1.0	1.4	1.8	2.2	2.6	3.0	3.4
y	1.45	1.89	2.66	3.50	3.22	3.04	2.44

In Exercises 35–40, use the function $y = x\sqrt[3]{2x^2 + 1}$ and approximate the area under the curve between $x = 1$ and $x = 4$ by the indicated method. See Fig. 5-27.

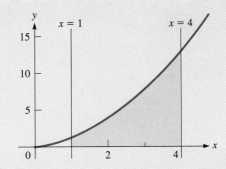

Fig. 5-27

35. Find the sum of the areas of three inscribed rectangles.

36. Find the sum of the areas of six inscribed rectangles.

37. Use the trapezoidal rule with $n = 3$.

38. Use the trapezoidal rule with $n = 6$.

39. Use Simpson's rule with $n = 6$.

40. Use integration (for the exact area).

In Exercises 41–44, find the area of the archway, as shown in Fig. 5-28, by the indicated method. The archway can be described as the area bounded by the elliptical arc $y = 4 + \sqrt{1 + 8x - 2x^2}$, $x = 0$, $x = 4$, *and* $y = 0$, *where dimensions are in meters.*

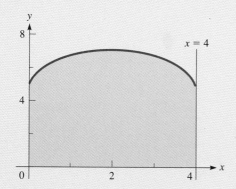

Fig. 5-28

41. Find the sum of the areas of eight inscribed rectangles.

42. Use the trapezoidal rule with $n = 8$.

43. Use Simpson's rule with $n = 8$.

44. Use the integration evaluation feature on a calculator.

In Exercises 45–48, solve the given problems by integration.

45. The deflection y of a certain beam at a distance x from one end is given by $dy/dx = k(2L^3 - 12Lx + 2x^4)$, where k is a constant and L is the length of the beam. Find y as a function of x if $y = 0$ for $x = 0$.

46. The total electric charge Q on a charged sphere is given by $Q = k \int \left(r^2 - \dfrac{r^3}{R} \right) dr$, where k is a constant, r is the distance from the center of the sphere, and R is the radius of the sphere. Find Q as a function of r if $Q = Q_0$ for $r = R$.

47. Part of the deck of a boat is the parabolic area shown in Fig. 5-29. The area A (in m^2) is $A = 2 \int_0^5 \sqrt{5 - y}\, dy$. Evaluate A.

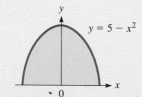

Fig. 5-29

48. The distance s (in in.) through which a cam follower moves in 4 s is $s = \int_0^4 t\sqrt{4 + 9t^2}\, dt$. Evaluate s.

Writing Exercise

49. A computer science student is writing a program to find a good approximation for the value of π by using the formula $A = \pi r^2$ for a circle. The value of π is to be found by approximating the area of a circle with a given radius. Write two or three paragraphs explaining how the value of π can be approximated in this way. Include any equations and values that may be used, but do not actually make the calculations.

PRACTICE TEST

1. Find an antiderivative of $f(x) = 2x - (1 - x)^4$.

2. Integrate: $\int x\sqrt{1 - 2x^2}\, dx$.

3. Find y in terms of x if $dy/dx = (6 - x)^4$ and the curve passes through $(5, 2)$.

4. Approximate the area under $y = \dfrac{1}{x + 2}$ between $x = 1$ and $x = 4$ (above the x-axis) by inscribing six rectangles and finding the sum of their areas.

5. Evaluate $\displaystyle\int_1^4 \frac{dx}{x + 2}$ by using the trapezoidal rule with $n = 6$.

6. Evaluate the definite integral of Problem 5 by using Simpson's rule with $n = 6$.

7. The total electric current i (in A) to pass a point in the circuit between $t = 1$ s and $t = 3$ s is $i = \displaystyle\int_1^3 \left(t^2 + \frac{1}{t^2} \right) dt$. Evaluate i.

 **APPLICATIONS OF INTEGRATION**

Integration is useful in many areas of science, engineering, and technology. It has numerous important applications in electricity, mechanics, hydrostatics, architecture, machine design, and business, as well as in geometry and other areas of physics.

Some of these applications were briefly indicated in the previous chapter. In this chapter we develop the necessary concepts for setting up integrals for several of these types of applications.

In the first section of this chapter we present some important applications of the indefinite integral, with emphasis on the motion of an object and the voltage across a capacitor. In the remaining sections we present uses of the definite integral related to applications in geometry, mechanics, work by a variable force, and force due to liquid pressure.

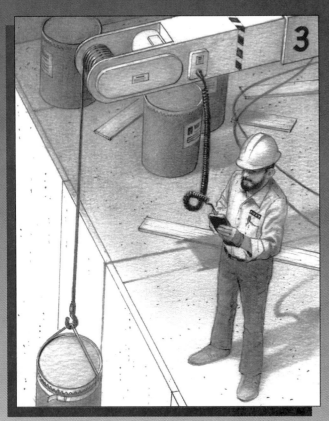

In Section 6-6 integration is used to find the work done in winding up a cable.

6-1 APPLICATIONS OF THE INDEFINITE INTEGRAL

In the examples of this section we show two basic applications of the indefinite integral. Some other applications are shown in the exercises.

Velocity and Displacement

The first application deals with the motion of an object. The concepts of velocity as a first derivative and acceleration as a second derivative were introduced in Chapters 3 and 4. Here, we apply integration to the problem of finding the displacement and velocity as a function of time, when we know the relationship between acceleration and time, and certain values of displacement and velocity. As we saw in Section 5-2, these values are needed for finding the values of the constants of integration that are introduced.

Recalling now that the acceleration a of an object is given by $a = dv/dt$, we can find the expression for the velocity v in terms of a, the time t, and the constant of integration. We write

$$dv = a\,dt$$

or

$$\boxed{v = \int a\,dt} \qquad \textbf{(6-1)}$$

If the acceleration is constant, we have

$$\boxed{v = at + C_1} \qquad \textbf{(6-2)}$$

Of course, Eq. (6-1) can be used in general to find the velocity as a function of time so long as we know the acceleration as a function of time. However, since the case of constant acceleration is often encountered, Eq. (6-2) is often encountered. If the velocity is known for a specified time, the constant C_1 may be evaluated.

■EXAMPLE 1   Find the expression for the velocity if $a = 12t$, given that $v = 8$ when $t = 1$.

Using Eq. (6-1), we have

$$v = \int (12t)\,dt = 6t^2 + C_1$$

Substituting the known values, we obtain

$$8 = 6 + C_1 \quad \text{or} \quad C_1 = 2$$

Thus, $v = 6t^2 + 2$.   ------------■

■EXAMPLE 2   For an object falling under the influence of gravity, the acceleration due to gravity is essentially constant. Its value is -32 ft/s^2. (The negative sign is chosen so that *all quantities directed up are positive and **all quantities directed down are negative**.*) Find the expression for the velocity of an object under the influence of gravity if $v = v_0$ when $t = 0$.

NOTE ▶

We write

$$v = \int (-32)\,dt \qquad \text{substitute } a = -32 \text{ into Eq. (6-1)}$$
$$= -32t + C_1 \qquad \text{integrate}$$
$$v_0 = -32(0) + C_1 \qquad \text{substitute given values}$$
$$C_1 = v_0 \qquad \text{solve for } C_1$$
$$v = v_0 - 32t \qquad \text{substitute}$$

The velocity v_0 is called the *initial velocity*. If the object is given an initial upward velocity of 100 ft/s, $v_0 = 100$ ft/s. If the object is dropped, $v_0 = 0$. If the object is given an initial downward velocity of 100 ft/s, $v_0 = -100$ ft/s.   ------------■

Once we have the expression for velocity, we can then integrate to find the expression for displacement s in terms of the time. Since $v = ds/dt$, we can write $ds = v\,dt$, or

$$s = \int v\,dt \qquad \text{(6-3)}$$

Consider the following examples.

SOLVING A WORD PROBLEM

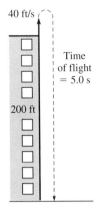

40 ft/s

Time of flight = 5.0 s

200 ft

Fig. 6-1

■EXAMPLE 3   A ball is thrown vertically from the top of a building 200 ft high and hits the ground 5.0 s later. What initial velocity was the ball given?

Measuring vertical distances from the ground, we know that $s = 200$ ft when $t = 0$ and that $v = v_0 - 32t$. Thus,

$$s = \int (v_0 - 32t)\,dt = v_0 t - 16t^2 + C \qquad \text{integrate}$$

$$200 = v_0(0) - 16(0) + C, \qquad C = 200 \qquad \text{evaluate } C$$

$$s = v_0 t - 16t^2 + 200$$

We also know that $s = 0$ when $t = 5.0$ s. Thus,

$$0 = v_0(5.0) - 16(5.0)^2 + 200 \qquad \text{substitute given values}$$

$$5.0v_0 = 200$$

$$v_0 = 40 \text{ ft/s}$$

This means that the initial velocity was 40 ft/s upward. See Fig. 6-1. ∎

■EXAMPLE 4   During the initial stage of launching a spacecraft vertically, the acceleration a (in m/s^2) of the spacecraft is $a = 6t^2$. Find the height s of the spacecraft after 6.0 s if $s = 12$ m for $t = 0.0$ s and $v = 16$ m/s for $t = 2.0$ s.

First, we use Eq. (6-1) to get an expression for the velocity:

$$v = \int 6t^2\,dt = 2t^3 + C_1 \qquad \text{integrate}$$

$$16 = 2(2.0)^3 + C_1, \qquad C_1 = 0 \qquad \text{evaluate } C_1$$

$$v = 2t^3$$

We now use Eq. (6-3) to get an expression for the displacement:

$$s = \int 2t^3\,dt = \tfrac{1}{2}t^4 + C_2 \qquad \text{integrate}$$

$$12 = \tfrac{1}{2}(0.0)^4 + C_2, \qquad C_2 = 12 \qquad \text{evaluate } C_2$$

$$s = \tfrac{1}{2}t^4 + 12$$

Now, finding s for $t = 6.0$ s, we have

$$s = \tfrac{1}{2}(6.0)^4 + 12 = 660 \text{ m} \qquad ∎$$

Voltage Across a Capacitor

The second basic application of the indefinite integral we shall discuss comes from the field of electricity. By definition, *the current i in an electric circuit equals the time rate of change of the charge q (in coulombs) that passes a given point in the circuit*, or

$$i = \frac{dq}{dt} \tag{6-4}$$

Rewriting this expression in differential notation as $dq = i\,dt$ and integrating both sides of the equation, we have

$$q = \int i\,dt \tag{6-5}$$

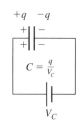

Fig. 6-2

Now, the voltage V_C across a capacitor with capacitance C (see Fig. 6-2) is given by $V_C = q/C$. By combining equations, the voltage V_C is given by

$$V_C = \frac{1}{C}\int i\,dt \tag{6-6}$$

Here, V_C is measured in volts, C in farads, i in amperes, and t in seconds.

The ampere (A) is named for the French physicist Andre Ampere (1775–1836).

The volt (V) is named for the Italian physicist Alessandro Volta (1745–1827).

The farad (F) is named for the British physicist Michael Faraday (1791–1867).

The coulomb (C) is named for the French physicist Charles Coulomb (1736–1806).

■**EXAMPLE 5**   The current in a certain electric circuit as a function of time is given by $i = 6t^2 + 4$. Find an expression for the amount of charge q that passes a point in the circuit as a function of time. Assuming that $q = 0$ when $t = 0$, determine the total charge that passes the point in 2 s.

Since $q = \int i\,dt$, we have

$$q = \int (6t^2 + 4)\,dt \qquad \text{substitute into Eq. (6-5)}$$
$$= 2t^3 + 4t + C \qquad \text{integrate}$$

This last expression is the desired expression giving charge as a function of time. We note that when $t = 0$, then $q = C$, which means that the constant of integration represents the initial charge, or the charge that passed a given point before we started timing. Using q_0 to represent this charge, we have

$$q = 2t^3 + 4t + q_0$$

Now, returning to the second part of the problem, we see that $q_0 = 0$. Therefore, evaluating q for $t = 2$ s, we have

$$q = 2(8) + 4(2) = 24 \text{ C}$$

[Here, the symbol C represents coulombs and is not the C for capacitance of Eq. (6-6) or the constant of integration.] This is the charge that passes any specified point in the circuit in 2 s.

SOLVING A WORD PROBLEM

EXAMPLE 6 The voltage across a 5.0-μF capacitor is zero. What is the voltage after 20 ms if a current of 75 mA charges the capacitor?

Since the current is 75 mA, we know that $i = 0.075$ A $= 7.5 \times 10^{-2}$ A. We see

CAUTION ▶ that *we must use the proper power of 10 that corresponds to each prefix.* Since $5.0 \ \mu\text{F} = 5.0 \times 10^{-6}$ F, we have

$$V_C = \frac{1}{5.0 \times 10^{-6}} \int 7.5 \times 10^{-2} \, dt \qquad \text{substituting into Eq. (6-6)}$$

$$= (1.5 \times 10^4) \int dt$$

$$= (1.5 \times 10^4)t + C_1 \qquad \text{integrate}$$

From the given information we know that $V_C = 0$ when $t = 0$. Thus,

$$0 = (1.5 \times 10^4)(0) + C_1 \quad \text{or} \quad C_1 = 0 \qquad \text{evaluate } C_1$$

This means that

$$V_C = (1.5 \times 10^4)t$$

Evaluating this expression for $t = 20 \times 10^{-3}$ s, we have

$$V_C = (1.5 \times 10^4)(20 \times 10^{-3})$$

$$= 30 \times 10 = 300 \text{ V}$$

SOLVING A WORD PROBLEM

EXAMPLE 7 A certain capacitor is measured to have a voltage of 100 V across it. At this instant a current as a function of time given by $i = 0.06\sqrt{t}$ is sent through the circuit. After 0.25 s, the voltage is measured to be 140 V. What is the capacitance of the capacitor?

Substituting $i = 0.06\sqrt{t}$, we find that

$$V_C = \frac{1}{C} \int (0.06\sqrt{t} \, dt) = \frac{0.06}{C} \int t^{1/2} \, dt \qquad \text{using Eq. (6-6)}$$

$$= \frac{0.04}{C} t^{3/2} + C_1 \qquad \text{integrate}$$

From the given information we know that $V_C = 100$ V when $t = 0$. Thus,

$$100 = \frac{0.04}{C}(0) + C_1 \quad \text{or} \quad C_1 = 100 \text{ V} \qquad \text{evaluate } C_1$$

This means that

$$V_C = \frac{0.04}{C} t^{3/2} + 100$$

We also know that $V_C = 140$ V when $t = 0.25$ s. Therefore,

$$140 = \frac{0.04}{C}(0.25)^{3/2} + 100$$

$$40 = \frac{0.04}{C}(0.125)$$

or

$$C = 1.25 \times 10^{-4} \text{ F} = 125 \ \mu\text{F}$$

EXERCISES $6\text{-}1$

1. What is the velocity (in ft/s) of a wrench 2.5 s after it is dropped from a building platform?

2. A hoop is started upward along an inclined plane at 16 ft/s. If the acceleration of the hoop is 5.0 ft/s^2 downward along the plane, find the velocity of the hoop after 6.0 s.

3. A conveyor belt 8.00 m long moves at 0.25 m/s. If a package is placed at one end, find its displacement from the other end as a function of time.

4. During each cycle, the velocity v (in mm/s) of a piston is $v = 6t - 6t^2$, where t is the time (in s). Find the displacement s of the piston after 0.75 s if the initial displacement is zero.

5. While in the barrel of a tennis ball machine, the acceleration a (in ft/s^2) of a ball is $a = 90\sqrt{1 - 4t}$, where t is the time (in s). If $v = 0$ for $t = 0$, find the velocity of the ball as it leaves the barrel at $t = 0.25$ s.

6. A person skis down a slope with an acceleration (in m/s^2) given by $a = \dfrac{600t}{(60 + 0.5t^2)^2}$, where t is the time (in s). Find the skier's velocity as a function of time if $v = 0$ when $t = 0$.

7. A rocket is fired vertically upward. When it reaches an altitude of 16,500 m, the engines cut off and the rocket is moving upward at 450 m/s. What will be its altitude 3.00 s later $(a = -9.80$ m/s$^2)$?

8. A flare is ejected vertically upward from the ground at 50 ft/s. Find the height of the flare after 2.5 s.

9. What must be the nozzle velocity of the water from a fire hose if it is to reach a point 90 ft directly above the nozzle?

10. An arrow is shot upward with a vertical velocity of 120 ft/s from the edge of a cliff. If it hits the ground below after 9.0 s, how high is the cliff?

11. In coming to a stop, the acceleration of a car is $-12t$. If it is traveling at 96.0 ft/s when the brakes are applied, how far does it travel while stopping?

12. A hoist mechanism raises a crate with an acceleration (in m/s^2) $a = \sqrt{1 + 0.2t}$, where t is the time in seconds. Find the displacement of the crate as a function of time if $v = 0$ m/s and $s = 2$ m for $t = 0$ s.

13. The electric current in a microprocessor circuit is 0.230 μA. How many coulombs pass a given point in the circuit in 1.50 ms?

14. The electric current (in mA) in a computer circuit as a function of time (in s) is $i = 0.3 - 0.2t$. What total charge passes a point in the circuit in 0.050 s?

15. In an amplifier circuit, the current i (in A) changes with time t (in s) according to $i = 0.06t\sqrt{1 + t^2}$. If 0.015 C of charge have passed a point in the circuit at $t = 0$, find the total charge to have passed the point at $t = 0.25$ s.

W 16. The current i (in μA) in a certain microprocessor circuit is given by $i = 8 - t$, where t is the time (in μs) and $0 \le t \le 20$ μs. If $q_0 = 0$, for what value of t, greater than zero, is $q = 0$? What interpretation can be given to this result?

17. The voltage across a 2.5-μF capacitor in a copying machine is zero. What is the voltage after 12 ms if a current of 25 mA charges the capacitor?

18. The voltage across an 8.50-nF capacitor in an FM receiver circuit is zero. Find the voltage after 2.00 μs if a current (in mA) $i = 0.042t$ charges the capacitor.

19. The voltage across a 3.75-μF capacitor in a television circuit is 4.50 mV. Find the voltage after 0.565 ms if a current (in μA) $i = \sqrt[3]{1 + 6t}$ further charges the capacitor.

20. A current $i = t/\sqrt{t^2 + 1}$ (in A) is sent through an electric dryer circuit containing a previously uncharged 2.0-μF capacitor. How long does it take for the capacitor voltage to reach 120 V?

21. The angular velocity ω is the time rate of change of the angular displacement θ of a rotating object. See Fig. 6-3. In testing the shaft of an engine, its angular velocity is $\omega = 16t + 0.50t^2$, where t is the time (in s) of rotation. Find the angular displacement through which the shaft turns in 10.0 s.

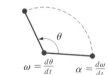

Fig. 6-3 $\omega = \dfrac{d\theta}{dt}$ $\alpha = \dfrac{d\omega}{dt}$

22. The angular acceleration α is the time rate of change of angular velocity ω of a rotating object. See Fig. 6-3. When starting up, the angular acceleration of a helicopter blade is $\alpha = \sqrt{8t + 1}$. Find the expression for θ if $\omega = 0$ and $\theta = 0$ for $t = 0$.

23. An inductor in an electric circuit is essentially a coil of wire in which the voltage is affected by a changing current. By definition, the voltage caused by the changing current is given by $V_L = L(di/dt)$, where L is the inductance (in H). If $V_L = 12.0 - 0.2t$ for a 3.0-H inductor, find the current in the circuit after 20 s if the initial current was zero.

24. If the inner and outer walls of a container are at different temperatures, the rate of change of temperature with respect to the distance from one wall is a function of the distance from the wall. Symbolically, this is stated as $dT/dx = f(x)$, where T is the temperature. If x is measured from the outer wall, at 20°C, and $f(x) = 72x^2$, find the temperature at the inner wall if the container walls are 0.5 cm thick.

25. Surrounding an electrically charged particle is an electric field. The rate of change of electric potential with respect to the distance from the particle creating the field equals the negative of the value of the electric field. That is, $dV/dx = -E$, where E is the electric field. If $E = k/x^2$, where k is a constant, find the electric potential at a distance x_1 from the particle, if $V \to 0$ as $x \to \infty$.

26. The rate of change of the vertical deflection y with respect to the horizontal distance x from one end of a beam is a function of x. For a particular beam, this function is $k(x^5 + 1350x^3 - 7000x^2)$, where k is a constant. Find y as a function of x.

27. Freshwater is flowing into a brine solution, with an equal volume of mixed solution flowing out. The amount of salt in the solution decreases, but more slowly as time increases. Under certain conditions, the time rate of change of mass of salt (in g/min) is given by $-1/\sqrt{t + 1}$. Find the mass m of salt as a function of time if 1000 g were originally present. Under these conditions, how long would it take for all the salt to be removed?

28. A holograph of a circle is formed. The rate of change of the radius r of the circle with respect to the wavelength λ of the light used is inversely proportional to the square root of λ. If $dr/d\lambda = 3.55 \times 10^4$ and $r = 4.08$ cm for $\lambda = 574$ nm, find r as a function of λ.

$6\text{-}2$ AREAS BY INTEGRATION

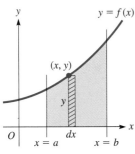

Fig. 6-4

In Section 5-3 we introduced the method of finding the area under a curve by integration. We also showed that the area can be determined by finding the sum of the areas of inscribed rectangles, whose widths approach zero as the number of rectangles approaches infinity (infinitesimally thin rectangles). This means that integration can be interpreted as an infinite summation process. *The applications of the definite integral use this summation interpretation of the integral.* We now develop a general procedure for finding the area for which the bounding curves are known by using integration and summing the areas of inscribed rectangles.

The first step is to make a sketch of the area. Next, a representative **element of area** dA (a typical rectangle) is drawn. In Fig. 6-4 the width of the element is dx. The length of the element is determined by the y-coordinate of the point on the curve. Thus, the length is y. The area of this element is $y\,dx$, which in turn means that $dA = y\,dx$, or

$$A = \int_a^b y\,dx = \int_a^b f(x)\,dx \qquad (6\text{-}7)$$

This formula states that the elements of area are to be summed (this is the meaning of the integral sign) from a (the left boundary) to b (the right boundary).

■**EXAMPLE 1**   Find the area bounded by $y = 2x^2$, $y = 0$, $x = 1$, and $x = 2$.

This area is shown in Fig. 6-5. The rectangle shown is the representative element. Its area is $y\,dx$. The elements are to be summed from $x = 1$ to $x = 2$.

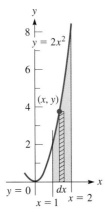

Fig. 6-5

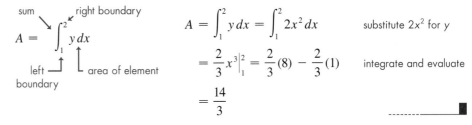

$$A = \int_1^2 y\,dx \qquad \qquad A = \int_1^2 y\,dx = \int_1^2 2x^2\,dx \qquad \text{substitute } 2x^2 \text{ for } y$$

$$= \frac{2}{3}x^3\Big|_1^2 = \frac{2}{3}(8) - \frac{2}{3}(1) \qquad \text{integrate and evaluate}$$

$$= \frac{14}{3}$$

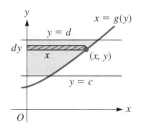

Fig. 6-6

In Figs. 6-4 and 6-5, the elements are vertical. It is also possible to use horizontal elements, and many problems are simplified by using them. In using horizontal elements, the length (longest dimension) is measured in terms of the x-coordinate of the point on the curve, and the width becomes dy. In Fig. 6-6 the area of the element is $x\,dy$, which means $dA = x\,dy$, or

$$A = \int_c^d x\,dy = \int_c^d g(y)\,dy \tag{6-8}$$

In using Eq. (6-8), the elements are summed from c (the lower boundary) to d (the upper boundary). In the following example, the area is found by use of both vertical and horizontal elements of area.

■**EXAMPLE 2** Find the area in the first quadrant bounded by $y = 9 - x^2$.

The area to be found is shown in Fig. 6-7. First, using the vertical element of length y and width dx, we have

$$A = \int_0^3 y\,dx \qquad \text{sum of areas of elements}$$

$$= \int_0^3 (9 - x^2)\,dx \qquad \text{substitute } 9 - x^2 \text{ for } y$$

$$= \left(9x - \frac{x^3}{3}\right)\Bigg|_0^3 \qquad \text{integrate}$$

$$= (27 - 9) - 0 = 18 \qquad \text{evaluate}$$

Now, using the horizontal element of length x and width dy, we have

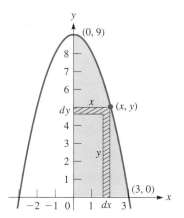

Fig. 6-7

$$A = \int_0^9 x\,dy \qquad \text{sum of areas of elements}$$

$$= \int_0^9 \sqrt{9 - y}\,dy = -\int_0^9 (9 - y)^{1/2}(-dy) \qquad \text{substitute } \sqrt{9 - y} \text{ for } x$$

$$= -\frac{2}{3}(9 - y)^{3/2}\Bigg|_0^9 \qquad \text{integrate}$$

$$= -\frac{2}{3}(9 - 9)^{3/2} + \frac{2}{3}(9 - 0)^{3/2} \qquad \text{evaluate}$$

$$= \frac{2}{3}(27) = 18$$

CAUTION ▶

Note that the limits for the vertical elements were 0 and 3, while those for the horizontal elements were 0 and 9. These limits are determined by the direction in which the elements are summed. As we have noted, *vertical elements are summed from left to right, and horizontal elements are summed from bottom to top.* Doing it in this way means that the summation will be done in a positive direction. ■

The choice of vertical or horizontal elements is determined by (1) which one leads to the simplest solution or (2) the form of the resulting integral. In some problems it makes little difference which is chosen. However, our present methods of integration do not include many types of integrals.

Area Between Two Curves

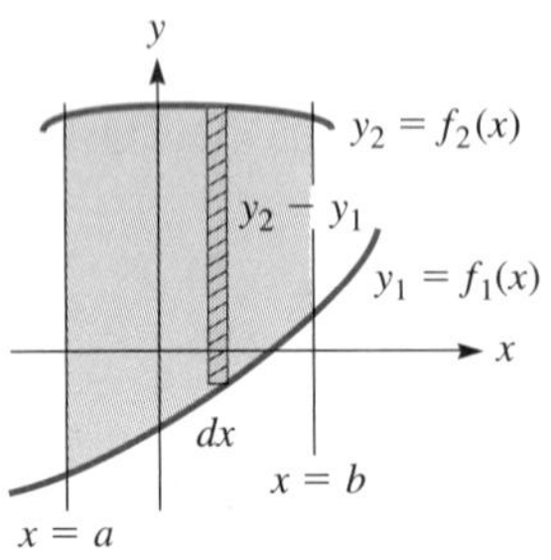

It is also possible to find the area between two curves when one of the curves is not an axis. In such a case, the length of the element of area becomes the difference in the y- or x-coordinates, depending on whether a vertical element or a horizontal element is used.

In Fig. 6-8, by using vertical elements, the element of area is bounded on the bottom by $y_1 = f_1(x)$ and on the top by $y_2 = f_2(x)$. The length of the element is $y_2 - y_1$, and its width is dx. Thus, the area is

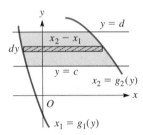

Fig. 6-8

$$A = \int_a^b (y_2 - y_1)\, dx \tag{6-9}$$

In Fig. 6-9, by using horizontal elements, the element of area is bounded on the left by $x_1 = g_1(y)$ and on the right by $x_2 = g_2(y)$. The length of the element is $x_2 - x_1$, and its width is dy. Thus, the area is

Fig. 6-9

$$A = \int_c^d (x_2 - x_1)\, dy \tag{6-10}$$

The following examples show the use of Eqs. (6-9) and (6-10) to find the indicated areas.

EXAMPLE 3   Find the area bounded by $y = x^2$ and $y = x + 2$.

First, by sketching each curve, we see that the area to be found is that shown in Fig. 6-10. The points of intersection of these curves are found by solving the equations simultaneously. The solution for the x-values is shown at the left. We then find the y-coordinates by substituting into either equation. The substitution shows the points of intersection to be $(-1, 1)$ and $(2, 4)$.

$$x^2 = x + 2$$
$$x^2 - x - 2 = 0$$
$$(x + 1)(x - 2) = 0$$
$$x = -1, 2$$

Here, we choose vertical elements, since they are all bounded at the top by the line $y = x + 2$ and at the bottom by the parabola $y = x^2$. (If we were to choose horizontal elements, the bounding curves are different above $(-1, 1)$ from below this point. Choosing horizontal elements would then require two separate integrals for solution.) Therefore, using vertical elements, we have

$$A = \int_{-1}^{2} (y_{\text{line}} - y_{\text{parabola}})\, dx \qquad \text{using Eq. (6-9)}$$

$$= \int_{-1}^{2} (x + 2 - x^2)\, dx = \left(\frac{x^2}{2} + 2x - \frac{x^3}{3} \right)\Bigg|_{-1}^{2}$$

$$= \left(2 + 4 - \frac{8}{3} \right) - \left(\frac{1}{2} - 2 + \frac{1}{3} \right)$$

$$= \frac{10}{3} + \frac{7}{6} = \frac{27}{6}$$

$$= \frac{9}{2}$$

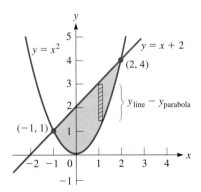

Fig. 6-10

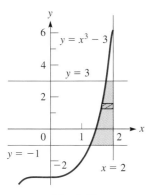

Fig. 6-11

EXAMPLE 4 Find the area bounded by the curve $y = x^3 - 3$ and the lines $x = 2$, $y = -1$, and $y = 3$.

Sketching the curve and lines, we show the area in Fig. 6-11. Horizontal elements are better, since they avoid having to evaluate the area in two parts. Therefore, we have

$$A = \int_{-1}^{3} (x_{\text{line}} - x_{\text{cubic}}) \, dy = \int_{-1}^{3} (2 - \sqrt[3]{y + 3}) \, dy \qquad \text{using Eq. (6-10)}$$

$$= 2y - \frac{3}{4}(y + 3)^{4/3} \Big|_{-1}^{3} = \left[6 - \frac{3}{4}(6^{4/3}) \right] - \left[-2 - \frac{3}{4}(2^{4/3}) \right]$$

$$= 8 - \frac{9}{2}\sqrt[3]{6} + \frac{3}{2}\sqrt[3]{2} = 1.713$$

As we see, the choice of horizontal elements leads to limits of -1 and 3. If we had chosen vertical elements, the limits would have been $\sqrt[3]{2}$ and $\sqrt[3]{6}$ for the area to the left of $(\sqrt[3]{6}, 3)$, and $\sqrt[3]{6}$ and 2 to the right of this point. ▬

CAUTION ▶

It is important to set up the element of area so that its length is positive. If the difference is taken incorrectly, the result will show a negative area. ***Getting positive lengths can be ensured for vertical elements if we subtract y of the lower curve from y of the upper curve. For horizontal elements we should subtract x of the left curve from x of the right curve.*** This important point is illustrated in the following example.

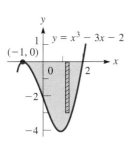

Fig. 6-12

EXAMPLE 5 Find the area bounded by $y = x^3 - 3x - 2$ and the x-axis.

Sketching the graph, we find that $y = x^3 - 3x - 2$ has a maximum point at $(-1, 0)$, a minimum point at $(1, -4)$, and an intercept at $(2, 0)$. The graph is shown in Fig. 6-12, and we see that the area is *below* the x-axis. In using vertical elements, we see that the top is the x-axis $(y = 0)$ and the bottom is the curve $y = x^3 - 3x - 2$. Therefore, we have

$$A = \int_{-1}^{2} [0 - (x^3 - 3x - 2)] \, dx = \int_{-1}^{2} (-x^3 + 3x + 2) \, dx$$

$$= -\frac{1}{4}x^4 + \frac{3}{2}x^2 + 2x \Big|_{-1}^{2}$$

$$= \left[-\frac{1}{4}(2^4) + \frac{3}{2}(2^2) + 2(2) \right] - \left[-\frac{1}{4}(-1)^4 + \frac{3}{2}(-1)^2 + 2(-1) \right]$$

$$= \frac{27}{4} = 6.75$$

If we had simply set up the area as $A = \int_{-1}^{2} (x^3 - 3x - 2) \, dx$, we would have found $A = -6.75$. The negative sign shows that the area is below the x-axis. Again, we avoid any complications with negative areas by making the length of the element positive.

Also note that since

$$0 - (x^3 - 3x - 2) = -(x^3 - 3x - 2)$$

an area bounded on top by the x-axis can be found by setting up the area as being "under" the curve and using the negative of the function. ▬

NOTE ▶

We must *be very careful if the bounding curves of an area cross.* In such a case, for part of the area one curve is above the area, and for a different part of the area this same curve is below the area. When this happens, *two integrals must be used* to find the area. The following example illustrates the necessity of using this procedure.

■EXAMPLE 6   Find the area between $y = x^3 - x$ and the x-axis.

We note from Fig. 6-13 that the area to the left of the origin is above the axis and the area to the right is below. If we find the area from

$$A = \int_{-1}^{1} (x^3 - x)\, dx = \left. \frac{x^4}{4} - \frac{x^2}{2} \right|_{-1}^{1}$$

$$= \left(\frac{1}{4} - \frac{1}{2} \right) - \left(\frac{1}{4} - \frac{1}{2} \right) = 0$$

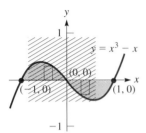

Fig. 6-13

we see that the apparent area is zero. From the figure we know this is not correct. Noting that the y-values (of the area) are negative to the right of the origin, we set up the integrals

$$A = \int_{-1}^{0} (x^3 - x)\, dx + \int_{0}^{1} [0 - (x^3 - x)]\, dx$$

$$= \left. \left(\frac{x^4}{4} - \frac{x^2}{2} \right) \right|_{-1}^{0} - \left. \left(\frac{x^4}{4} - \frac{x^2}{2} \right) \right|_{0}^{1}$$

$$= 0 - \left(\frac{1}{4} - \frac{1}{2} \right) - \left(\frac{1}{4} - \frac{1}{2} \right) + 0 = \frac{1}{2} \quad \text{──────■}$$

The area under a curve can be given many different interpretations. This is illustrated in the following example and in some of the exercises that follow.

The metric unit of energy, the joule, was named for the English physicist James Prescott Joule (1818–1899).

■EXAMPLE 7   Measurements of solar radiation on a particular surface indicated that the rate r, in joules per hour, at which solar energy is received during the day is given by the equation $r = 3600(12t^2 - t^3)$, where t is the time in hours. Since r is a rate, we may write $r = dE/dt$, where E is the energy, in joules, received at the surface. This means that $dE = 3600(12t^2 - t^3)\, dt$, and we can find the total energy by evaluating the definite integral

$$E = 3600 \int_{0}^{12} (12t^2 - t^3)\, dt$$

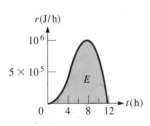

Fig. 6-14

This integral can be interpreted as being the area under $f(t) = 3600(12t^2 - t^3)$ from $t = 0$ to $t = 12$, as shown in Fig. 6-14 (the limits of integration are found from the t-intercepts of the curve). Evaluating this integral, we have

$$E = 3600 \int_{0}^{12} (12t^2 - t^3)\, dt = 3600 \left. \left(4t^3 - \frac{1}{4}t^4 \right) \right|_{0}^{12}$$

$$= 3600 \left[4(12^3) - \frac{1}{4}(12^4) - 0 \right]$$

$$= 6.22 \times 10^6 \text{ J}$$

Fig. 6-15

Therefore, 6.22 MJ of energy were received in 12 h.

The display in Fig. 6-15 shows the calculator evaluation for this integral, where x is used for t in using the calculator. It can be seen that the values check. ──────■

EXERCISES 6-2

In Exercises 1–24, find the areas bounded by the indicated curves.

1. $y = 4x$, $y = 0$, $x = 1$
2. $y = 4 - 2x$, $y = 0$, $x = 0$
3. $y = x^2$, $y = 0$, $x = 2$
4. $y = 3x^2$, $y = 0$, $x = 3$
5. $y = 6 - 4x$, $x = 0$, $y = 0$, $y = 3$
6. $y = x^2 + 2$, $x = 0$, $y = 4$ $(x > 0)$
7. $y = x^2 - 4$, $y = 0$, $x = 4$
8. $y = x^2 - 2x$, $y = 0$
9. $y = x^{-2}$, $y = 0$, $x = 2$, $x = 3$
10. $y = 16 - x^2$, $y = 0$, $x = 1$, $x = 2$
11. $y = \sqrt{x}$, $x = 0$, $y = 1$, $y = 3$
12. $y = 2\sqrt{x + 1}$, $x = 0$, $y = 4$
13. $y = 2/\sqrt{x}$, $x = 0$, $y = 1$, $y = 4$
14. $x = y^2 - y$, $x = 0$
15. $y = 4 - 2x$, $x = 0$, $y = 0$, $y = 3$
16. $y = x$, $y = 2 - x$, $x = 0$
17. $y = x^2$, $y = 2 - x$, $x = 0$ $(x \geq 0)$
18. $y = x^2$, $y = 2 - x$, $y = 1$
19. $y = x^4 - 8x^2 + 16$, $y = 16 - x^4$
20. $y = \sqrt{x - 1}$, $y = 3 - x$, $y = 0$
21. $y = x^2 + 5x$, $y = 3 - x^2$
22. $y = x^3$, $y = x^2 + 4$, $x = -1$
23. $y = x^5$, $x = -1$, $x = 2$, $y = 0$
24. $y = x^2 + 2x - 8$, $y = x + 4$

In Exercises 25–28, find the areas bounded by the indicated curves, using (a) vertical elements, and (b) horizontal elements.

25. $y = 8x$, $x = 0$, $y = 4$ 26. $y = x^3$, $x = 0$, $y = 3$
27. $y = x^4$, $y = 8x$ 28. $y = 4x$, $y = x^3$

In Exercises 29–36, various interpretations of areas are illustrated.

29. Certain physical quantities are often represented as an area under a curve. By definition, power is the time rate of change of performing work. Thus, $p = dw/dt$, or $dw = p\,dt$. Therefore, if $p = 12t - 4t^2$, find the work (in J) performed in 3 s by finding the area under the curve of p vs. t. See Fig. 6-16.

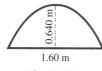

Fig. 6-16

30. The total electric charge Q (in C) to pass a point in the circuit from time t_1 to t_2 is $Q = \int_{t_1}^{t_2} i\,dt$, where i is the current (in A). Find Q if $t_1 = 1$ s, $t_2 = 4$ s, and $i = 0.0032t\sqrt{t^2 + 1}$.

31. Since the displacement s, velocity v, and time t of a moving object are related by $s = \int v\,dt$, it is possible to represent the change in displacement as an area. A rocket is launched such that its vertical velocity v (in km/s) as a function of time t (in s) is $v = 1 - 0.01\sqrt{2t + 1}$. Find the change in vertical displacement from $t = 10$ s to $t = 100$ s.

32. The total cost C (in dollars) of production can be interpreted as an area. If the cost per unit C' (in dollars per unit) of producing x units is given by $100/(0.01x + 1)^2$, find the total cost of producing 100 units by finding the area under the curve of C' vs. x.

33. A cam is designed such that one face of it is described as being the area between the curves $y = x^3 - 2x^2 - x + 2$ and $y = x^2 - 1$ (units in cm). Show that this description does *not* uniquely describe the face of the cam. Find the area of the face of the cam, if a complete description requires that $x \leq 1$.

34. Using CAD (computer-assisted design), an architect programs a computer to sketch the shape of a swimming pool designed between the curves

$$y = \frac{800x}{(x^2 + 10)^2} \qquad y = 0.5x^2 - 4x \qquad x = 8$$

(dimensions in m). Find the area of the surface of the pool.

35. A window is designed to be the area between a parabolic section and a straight base, as shown in Fig. 6-17. What is the area of the window?

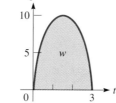

Fig. 6-17

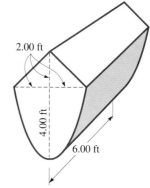

Fig. 6-18

36. The vertical ends of a fuel storage tank have a parabolic bottom section and a triangular top section, as shown in Fig. 6-18. What volume does the tank hold?

6-3 VOLUMES BY INTEGRATION

Consider a region in the xy-plane and its representative element of area, as shown in Fig. 6-19(a). When the region is revolved about the x-axis, it is said to generate a **solid of revolution,** which is also shown in the figure. We shall now show methods of finding volumes of solids that are generated in this way.

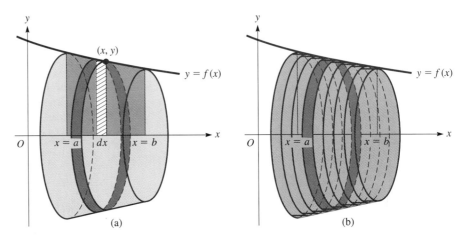

Fig. 6-19

As the region revolves about the x-axis, so does its representative element, which generates a solid for which the volume is known—an infinitesimally thin cylindrical **disk.** The volume of a right circular cylinder is π times its radius squared times its height (in this case, the thickness) of the cylinder. Since the element is revolved about the x-axis, the y-coordinate of the point on the curve that touches the element is the radius. Also, the thickness is dx. This disk, the representative **element of volume,** has a volume of $dV = \pi y^2 \, dx$. Summing these elements of volume from left to right, as shown in Fig. 6-19(b), we have for the total volume

> This is the sum of the volumes of the disks whose thicknesses approach zero as the number of disks approaches infinity.

$$V = \pi \int_a^b y^2 \, dx = \pi \int_a^b [f(x)]^2 \, dx \qquad (6\text{-}11)$$

The element of volume is a ***disk,*** *and by use of Eq. (6-11) we can find the volume of the solid generated by a region bounded by the x-axis, which is revolved about the x-axis.*

EXAMPLE 1 Find the volume of the solid generated by revolving the region bounded by $y = x^2$, $x = 2$, and $y = 0$ about the x-axis. See Fig. 6-20.

From the figure we see that the radius of the disk is y and its thickness is dx. The elements are summed from left ($x = 0$) to right ($x = 2$).

$$V = \pi \int_0^2 y^2 \, dx \qquad \text{using Eq. (6-11)}$$

$$= \pi \int_0^2 (x^2)^2 \, dx = \pi \int_0^2 x^4 \, dx \qquad \text{substitute } x^2 \text{ for } y$$

$$= \frac{\pi}{5} x^5 \Big|_0^2 = \frac{32\pi}{5} \qquad \text{integrate and evaluate}$$

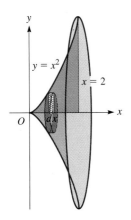

Fig. 6-20

Since π is used in Eq. (6-11), it is common to leave results in terms of π. In applied problems, a decimal result would normally be given. ━━━━━━■

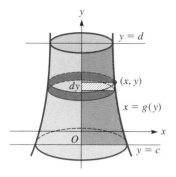

Fig. 6-21

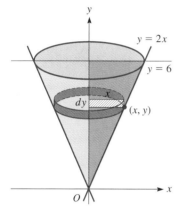

Fig. 6-22

If a region bounded by the y-axis is revolved about the y-axis, the volume of the solid generated is given by

$$V = \pi \int_c^d x^2 \, dy \qquad (6\text{-}12)$$

In this case the radius of the element of volume, a **disk,** is the x-coordinate of the point on the curve, and the thickness of the disk is dy, as shown in Fig. 6-21. One should always be careful to identify the radius and the thickness properly.

EXAMPLE 2 Find the volume of the solid generated by revolving the region bounded by $y = 2x$, $y = 6$, and $x = 0$ about the y-axis.

Figure 6-22 shows the volume to be found. Note that the radius of the disk is x and its thickness is dy.

$$V = \pi \int_0^6 x^2 \, dy \qquad \text{using Eq. (6-12)}$$

$$= \pi \int_0^6 \left(\frac{y}{2}\right)^2 dy = \frac{\pi}{4} \int_0^6 y^2 \, dy \qquad \text{substitute } \frac{y}{2} \text{ for } x$$

$$= \frac{\pi}{12} y^3 \Big|_0^6 = 18\pi \qquad \text{integrate and evaluate}$$

Since this volume is a right circular cone, it is possible to check the result:

$$V = \frac{1}{3} \pi r^2 h = \frac{1}{3} \pi (3^2)(6) = 18\pi$$

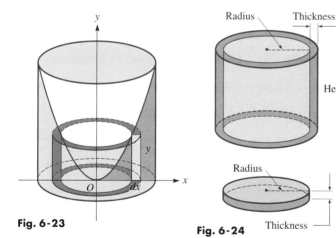

Fig. 6-23

Fig. 6-24

If the region in Fig. 6-23 is revolved about the y-axis, the element of area $y \, dx$ generates a different element of volume from that when it is revolved about the x-axis. In Fig. 6-23, this element of volume is a **cylindrical shell.** *The total volume is made up of an infinite number of concentric shells.* When the volumes of these shells are summed, we have the total volume generated. Thus we must now find the approximate volume dV of the representative shell. By finding the circumference of the base and multiplying this by the height, we obtain an expression for the surface area of the shell. Then, by multiplying this by the thickness of the shell, we find its volume. The volume of the representative **shell** shown in Fig. 6-24(a) is

SHELL

$$dV = 2\pi(\text{radius}) \times (\text{height}) \times (\text{thickness}) \qquad (6\text{-}13)$$

Similarly, the volume of a **disk** is (see Fig. 6-24(b))

DISK

$$dV = \pi(\text{radius})^2 \times (\text{thickness}) \qquad (6\text{-}14)$$

It is generally better to remember the formulas for the elements of volume in the general forms given in Eqs. (6-13) and (6-14), and not in the specific forms such as Eqs. (6-11) and (6-12) (both of these use *disks*). If we remember the formulas in this way, we can readily apply these methods to finding any such volume of a solid of revolution.

EXAMPLE 3 Use the method of cylindrical shells to find the volume of the solid generated by revolving the first-quadrant region bounded by $y = 4 - x^2$, $x = 0$, and $y = 0$ about the y-axis.

From Fig. 6-25, we identify the radius, the height, and the thickness of the shell.

$$\text{radius} = x \qquad \text{height} = y \qquad \text{thickness} = dx$$

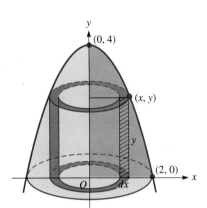

Fig. 6-25

CAUTION ▶ *The fact that the elements of area that generate the shells go from $x = 0$ to $x = 2$ determines the limits of integration as 0 and 2.* Therefore,

$$V = 2\pi \int_0^2 xy\,dx \longleftarrow \text{thickness} \qquad \text{using Eq. (6-13)}$$
$$\text{radius} \quad \text{height}$$

$$= 2\pi \int_0^2 x(4 - x^2)\,dx = 2\pi \int_0^2 (4x - x^3)\,dx \qquad \text{substitute } 4 - x^2 \text{ for } y$$

$$= 2\pi\left(2x^2 - \frac{1}{4}x^4\right)\Big|_0^2 \qquad \text{integrate}$$

$$= 8\pi \qquad \text{evaluate}$$

We can find the volume shown in Example 3 by using disks, as we show in the following example.

EXAMPLE 4 Use the method of disks to find the volume of the solid indicated in Example 3.

From Fig. 6-26 we identify the radius and the thickness of the disk.

$$\text{radius} = x \qquad \text{thickness} = dy$$

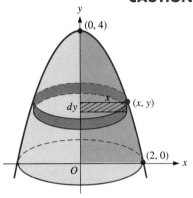

Fig. 6-26

CAUTION ▶ *Since the elements of area that generate the disks go from $y = 0$ to $y = 4$, the limits of integration are 0 and 4.* Thus,

$$V = \pi \int_0^4 x^2\,dy \qquad \text{using Eq. (6-14)}$$
$$\text{radius} \quad \text{thickness}$$

$$= \pi \int_0^4 (4 - y)\,dy \qquad \text{substitute } \sqrt{4 - y} \text{ for } x$$

$$= \pi\left(4y - \frac{1}{2}y^2\right)\Big|_0^4 \qquad \text{integrate}$$

$$= 8\pi \qquad \text{evaluate}$$

We see that the volume of 8π using disks agrees with the result we obtained using shells in Example 3.

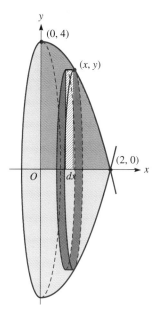

Fig. 6-27

EXAMPLE 5 By using disks, find the volume of the solid generated if the first-quadrant region bounded by $y = 4 - x^2$, $x = 0$, and $y = 0$ is revolved about the x-axis. (This is the same region as used in Examples 3 and 4.)

For the disk in Fig. 6-27, we have

$$\text{radius} = y \qquad \text{thickness} = dx$$

and the limits of integration are $x = 0$ and $x = 2$. This gives us

$$V = \pi \int_0^2 y^2 \, dx \qquad\qquad \text{using Eq. (6-14)}$$

$$= \pi \int_0^2 (4 - x^2)^2 \, dx \qquad \text{substitute } 4 - x^2 \text{ for } y$$

$$= \pi \int_0^2 (16 - 8x^2 + x^4) \, dx$$

$$= \pi \left(16x - \frac{8}{3}x^3 + \frac{1}{5}x^5 \right) \Bigg|_0^2 \qquad \text{integrate}$$

$$= \frac{256\pi}{15} \qquad\qquad\qquad \text{evaluate}$$

We will now show how to set up the integral to find the volume of the solid shown in Example 5 by using cylindrical shells. As it turns out, we are not able at this point to integrate the expression that arises, but we are still able to set up the proper integral.

EXAMPLE 6 Use the method of cylindrical shells to find the volume of the solid indicated in Example 5.

From Fig. 6-28, we see for the shell we have

$$\text{radius} = y \qquad \text{height} = x \qquad \text{thickness} = dy$$

Since the elements go from $y = 0$ to $y = 4$, the limits of integration are 0 and 4. Hence,

$$V = 2\pi \int_0^4 xy \, dy \qquad\qquad \text{using Eq. (6-13)}$$

$$= 2\pi \int_0^4 \sqrt{4 - y} \,(y \, dy) \qquad \text{substitute } \sqrt{4 - y} \text{ for } x$$

$$= \frac{256\pi}{15}$$

The method of performing the integration $\int \sqrt{4 - y} \,(y \, dy)$ has not yet been discussed. We present the answer here for the reader's information to show that the volume found in this example is the same as that found in Example 5.

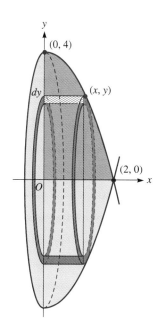

Fig. 6-28

In the next example we show how to find the volume of the solid generated if a region is revolved about a line other than one of the axes. We will see that a proper choice of the radius, height, and thickness for Eq. (6-13) leads to the result.

EXAMPLE 7   Find the volume of the solid generated if the region in Example 3 is revolved about the line $x = 2$.

Shells are convenient, since the volume of a shell can be expressed as a single integral. We can find the radius, height, and thickness of the shell from Fig. 6-29. We carefully note that *the radius is not x but $2 - x$,* since the region is revolved about $x = 2$. We see that

$$\text{radius} = 2 - x \qquad \text{height} = y \qquad \text{thickness} = dx$$

CAUTION ▶

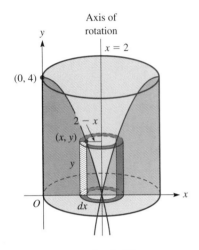

Since the elements that generate the shells go from $x = 0$ to $x = 2$, the limits of integration are 0 and 2. This means we have

$$V = 2\pi \int_0^2 (2 - x)y\,dx \longleftarrow \text{thickness} \qquad \text{using Eq. (6-13)}$$

$$= 2\pi \int_0^2 (2 - x)(4 - x^2)\,dx \qquad \text{substitute } 4 - x^2 \text{ for } y$$

$$= 2\pi \int_0^2 (8 - 2x^2 - 4x + x^3)\,dx$$

$$= 2\pi\left(8x - \frac{2}{3}x^3 - 2x^2 + \frac{1}{4}x^4\right)\Bigg|_0^2 \qquad \text{integrate}$$

$$= \frac{40\pi}{3} \qquad \text{evaluate}$$

Fig. 6-29

Fig. 6-30

(If the region had been revolved about the line $x = 3$, the only difference in the integral would have been that $r = 3 - x$. Everything else, including the limits, would have remained the same.)

The display in Fig. 6-30 shows the calculator evaluation for this integral. The first form of the integral was used to avoid any possible errors that might be made in the simplification. We see that the values agree.   ■

EXERCISES 6-3

In Exercises 1–4, find the volume of the solid generated by revolving the region bounded by $y = 2 - x$, $x = 0$, and $y = 0$ about the indicated axis, using the indicated element of volume.

1. x-axis   (disks)
2. y-axis   (disks)
3. y-axis   (shells)
4. x-axis   (shells)

In Exercises 5–14, find the volume of the solid generated by revolving the region bounded by the given curves about the x-axis. Use the indicated method in each case.

5. $y = x$, $y = 0$, $x = 2$ (disks)
6. $y = \sqrt{x}$, $x = 0$, $y = 2$ (shells)
7. $y = 3\sqrt{x}$, $y = 0$, $x = 4$ (disks)
8. $y = 2x - x^2$, $y = 0$ (disks)
9. $y = x^3$, $y = 8$, $x = 0$ (shells)
10. $y = x^2$, $y = x$ (shells)
11. $y = x^2 + 1$, $x = 0$, $x = 3$, $y = 0$ (disks)

12. $y = 6 - x - x^2$, $x = 0$, $y = 0$ (quadrant I)   (disks)
13. $x = 4y - y^2 - 3$, $x = 0$ (shells)
14. $y = x^4$, $x = 0$, $y = 1$, $y = 2$ (shells)

In Exercises 15–24, find the volume of the solid generated by revolving the region bounded by the given curves about the y-axis. Use the indicated method in each case.

15. $y = x^{1/3}$, $x = 0$, $y = 2$ (disks)
16. $y = \sqrt{x^2 - 1}$, $y = 0$, $x = 3$ (shells)
17. $y = 2\sqrt{x}$, $x = 0$, $y = 2$ (disks)
18. $y^2 = x$, $y = 4$, $x = 0$ (disks)
19. $x^2 - 4y^2 = 4$, $x = 3$ (shells)
20. $y = 3x^2 - x^3$, $y = 0$ (shells)
21. $x = 6y - y^2$, $x = 0$ (disks)
22. $x^2 + 4y^2 = 4$ (quadrant I)   (disks)
23. $y = \sqrt{4 - x^2}$ (quadrant I)   (shells)
24. $y = 8 - x^3$, $x = 0$, $y = 0$ (shells)

In Exercises 25–32, find the indicated volumes by integration.

25. Find the volume of the solid generated if the region of Exercise 8 is revolved about the line $x = 2$.

26. Find the volume of the solid generated if the region bounded by $y = \sqrt{x}$ and $y = x/2$ is revolved about the line $y = 4$.

27. Derive the formula for the volume of a right circular cone of radius r and height h by revolving the region bounded by $y = (r/h)x$, $y = 0$, and $x = h$ about the x-axis.

(W) 28. Explain how to derive the formula for the volume of a sphere by using the disk method.

29. The capillary tube shown in Fig. 6-31 has circular horizontal cross sections of inner radius 1.1 mm. What is the volume of the liquid in the tube above the level of liquid outside the tube if the top of the liquid in the center vertical cross section is described by the equation $y = x^4 + 1.5$, as shown?

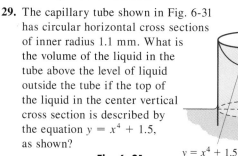

Fig. 6-31

$y = x^4 + 1.5$ $r = 1.1$ mm

30. A commercial dirigible used for outdoor advertising has a helium-filled balloon in the shape of an ellipse revolved about its major axis. If the balloon is 124 ft long and 36.0 ft in diameter, what volume of helium is required to fill it? See Fig. 6-32.

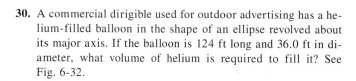

Fig. 6-32

124 ft   36.0 ft

31. A hole 2.00 cm in diameter is drilled through the center of a spherical lead weight 6.00 cm in diameter. How much lead is removed?

32. All horizontal cross sections of a keg 4.00 ft tall are circular, and the sides of the keg are parabolic. The diameter at the top and the bottom is 2.00 ft, and the diameter in the middle is 3.00 ft. Find the volume that the keg holds.

6-4 CENTROIDS

In the study of mechanics, a very important property of an object is its center of mass. In this section we explain the meaning of center of mass and then show how integration is used to determine the center of mass for regions and solids of revolution.

If a mass m is at a distance d from a specified point O, the **moment** *of the mass about O is defined as md.* If several masses $m_1, m_2, \ldots, m_n$ are at distances $d_1, d_2, \ldots, d_n$, respectively, from point O, the total moment (as a group) about O is defined as $m_1 d_1 + m_2 d_2 + \cdots + m_n d_n$. *The* **center of mass** *is that point $\overline{d}$ units from O at which all the masses could be concentrated to get the same total moment.* Therefore $\overline{d}$ is defined by the equation

$$m_1 d_1 + m_2 d_2 + \cdots + m_n d_n = (m_1 + m_2 + \cdots + m_n)\overline{d} \qquad (6\text{-}15)$$

The moment of a mass is a measure of its tendency to rotate about a point. A weight far from the point of balance of a long rod is more likely to make the rod turn than if the same weight were placed near the point of balance. It is easier to open a door if you push near the doorknob than if you push near the hinges. This is the type of physical property that the moment of mass measures.

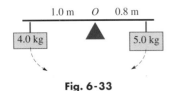

Fig. 6-33

■EXAMPLE 1 One of the simplest and most basic illustrations of moments and center of mass is seen in balancing a long rod with masses of different sizes, one on either side of the balance point.

In Fig. 6-33 a mass of 5.0 kg is hung from the rod 0.8 m to the right of point O. We see that this 5.0-kg mass tends to turn the rod clockwise. A mass placed on the opposite side of O will tend to turn the rod counterclockwise. Neglecting the mass of the rod, in order to balance the rod at O, the moments must be equal in magnitude but opposite in sign. Therefore, a 4.0-kg mass would have to be placed 1.0 m to the left.

Thus, with $d_1 = 0.8$ m, $d_2 = -1.0$ m, we see that

$$(5.0 + 4.0)\overline{d} = 5.0(0.8) + 4.0(-1.0) = 4.0 - 4.0$$
$$\overline{d} = 0.0 \text{ m}$$

CAUTION ▶

The center of mass of the combination of the 5.0-kg mass and the 4.0-kg mass is at O. Also, note that *we must use **directed distances** in finding moments.* ■

■EXAMPLE 2 A mass of 3.0 g is placed at $(2.0, 0)$ on the x-axis (distances in cm). Another mass of 6.0 g is placed at $(5.0, 0)$, and a third mass of 7.0 g is placed at $(6.0, 0)$. See Fig. 6-34. Find the center of mass of these three masses.

Taking the reference point as the origin, we find $d_1 = 2.0$ cm, $d_2 = 5.0$ cm, and $d_3 = 6.0$ cm. Thus, $m_1d_1 + m_2d_2 + m_3d_3 = (m_1 + m_2 + m_3)\overline{d}$ becomes

$$3.0(2.0) + 6.0(5.0) + 7.0(6.0) = (3.0 + 6.0 + 7.0)\overline{d} \quad \text{or} \quad \overline{d} = 4.9 \text{ cm}$$

Center of mass

Fig. 6-34

This means that the center of mass of the three masses is at $(4.9, 0)$. Therefore, a mass of 16.0 g placed at this point has the same moment as the three masses as a unit. ■

SOLVING A WORD PROBLEM

■EXAMPLE 3 Find the center of mass of the flat metal plate that is shown in Fig. 6-35.

We first note that the center of mass is not *on* either axis. This can be seen from the fact that the major portion of the area is in the first quadrant. *We shall therefore measure the moments with respect to each axis to find the point that is the center of mass. This point is also called the **centroid** of the plate.*

The easiest method of finding this centroid is to divide the plate into rectangles, as indicated by the dashed line in Fig. 6-35, and assume that we may consider the mass of each rectangle to be concentrated at its center. In this way the center of the left rectangle is at $(-1.0, 1.0)$ (distances in in.), and the center of the right rectangle is at $(2.5, 2.0)$. The mass of each rectangle area, assumed uniform, is proportional to its area. The area of the left rectangle is 8.0 in.^{2}, and that of the right rectangle is 12.0 in.^{2}. Thus, taking moments with respect to the y-axis, we have

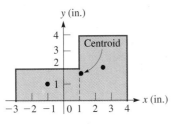

Fig. 6-35

Since the center of mass does not depend on the desity of the metal, we have assumed the constant of proportunality to be 1.

$$8.0(-1.0) + 12.0(2.5) = (8.0 + 12.0)\overline{x}$$

where $\overline{x}$ is the x-coordinate of the centroid. Solving for $\overline{x}$, we have $\overline{x} = 1.1$ in.

Now taking moments with respect to the x-axis, we have

$$8.0(1.0) + 12.0(2.0) = (8.0 + 12.0)\overline{y}$$

where $\overline{y}$ is the y-coordinate of the centroid. Thus, $\overline{y} = 1.6$ in. This means that the coordinates of the centroid, the center of mass, are $(1.1, 1.6)$. This may be interpreted as meaning that a plate of this shape would balance on a single support under this point. As an approximate check, we note from the figure that this point appears to be a reasonable balance point for the plate. ■

Centroid of a Thin, Flat Plate by Integration

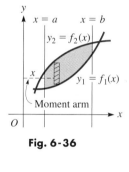

Fig. 6-36

If a thin, flat plate covers the region bounded by $y_1 = f_1(x)$, $y_2 = f_2(x)$, $x = a$, and $x = b$, as shown in Fig. 6-36, the moment of the mass of the element of area about the y-axis is given by $(k\,dA)x$, where k is the mass per unit area. In this expression, $k\,dA$ is the mass of the element, and x is its distance (moment arm) from the y-axis. The element dA may be written as $(y_2 - y_1)\,dx$, which means that the moment may be written as $kx(y_2 - y_1)\,dx$. If we then sum up the moments of all the elements and express this as an integral (which, of course, means infinite sum), we have $k\int_a^b x(y_2 - y_1)\,dx$. If we consider all the mass of the plate to be concentrated at one point $\bar{x}$ units from the y-axis, the moment would be $(kA)\bar{x}$, where kA is the mass of the entire plate and $\bar{x}$ is the distance the center of mass is from the y-axis. By the previous discussion these two expressions should be equal. This means $k\int_a^b x(y_2 - y_1)\,dx = kA\bar{x}$. Since k appears on each side of the equation, we divide it out (we are assuming that the mass per unit area is constant). The area A is found by the integral $\int_a^b (y_2 - y_1)\,dx$. Therefore, the x-coordinate of the centroid of the plate is given by

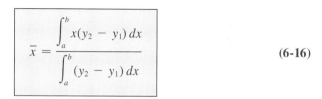

$$\bar{x} = \frac{\displaystyle\int_a^b x(y_2 - y_1)\,dx}{\displaystyle\int_a^b (y_2 - y_1)\,dx} \tag{6-16}$$

Equation (6-16) gives us the x-coordinate of the centroid of the plate if vertical elements are used. Note that

CAUTION▶   *the two integrals in Eq. (6-16) must be evaluated separately.*

We cannot cancel out the apparent common factor $y_2 - y_1$, and we cannot combine quantities and perform only one integration. The two integrals must be evaluated separately first. Then any possible cancellations of factors common to the numerator and the denominator may be made.

Following the same reasoning that we used in developing Eq. (6-16), if a thin plate covering the region bounded by the functions $x_1 = g_1(y)$, $x_2 = g_2(y)$, $y = c$, and $y = d$, as shown in Fig. 6-37, the y-coordinate of the centroid of the plate is given by the equation

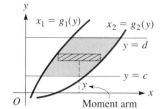

Fig. 6-37

$$\bar{y} = \frac{\displaystyle\int_c^d y(x_2 - x_1)\,dy}{\displaystyle\int_c^d (x_2 - x_1)\,dy} \tag{6-17}$$

In this equation, horizontal elements are used.

In applying Eqs. (6-16) and (6-17), we should keep in mind that each denominator of the right-hand sides gives the area of the plate and that, once we have found this area, we may use it for both $\bar{x}$ and $\bar{y}$. In this way, we can avoid having to set up and perform one of the indicated integrations. Also, in finding the coordinates of the centroid, we should look for and utilize any symmetry the region may have.

EXAMPLE 4   Find the coordinates of the centroid of a thin plate covering the region bounded by the parabola $y = x^2$ and the line $y = 4$.

We sketch a graph indicating the region and an element of area (see Fig. 6-38). The curve is a parabola whose axis is the y-axis. Since the region is symmetric to the y-axis, the centroid must be on this axis. This means that the x-coordinate of the centroid is zero, or $\bar{x} = 0$. To find the y-coordinate of the centroid, we have

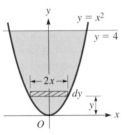

Fig. 6-38

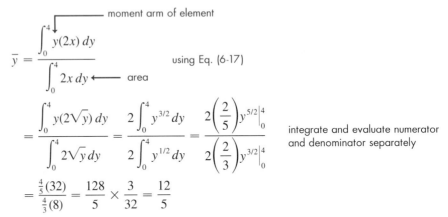

$$\bar{y} = \frac{\displaystyle\int_0^4 y(2x)\,dy}{\displaystyle\int_0^4 2x\,dy} \quad\longleftarrow \text{area}$$

using Eq. (6-17)

$$= \frac{\displaystyle\int_0^4 y(2\sqrt{y})\,dy}{\displaystyle\int_0^4 2\sqrt{y}\,dy} = \frac{2\displaystyle\int_0^4 y^{3/2}\,dy}{2\displaystyle\int_0^4 y^{1/2}\,dy} = \frac{2\left(\dfrac{2}{5}\right)y^{5/2}\Big|_0^4}{2\left(\dfrac{2}{3}\right)y^{3/2}\Big|_0^4}$$

integrate and evaluate numerator and denominator separately

$$= \frac{\tfrac{4}{5}(32)}{\tfrac{4}{3}(8)} = \frac{128}{5} \times \frac{3}{32} = \frac{12}{5}$$

The coordinates of the centroid are $(0, \tfrac{12}{5})$. This plate would balance if a single pointed support were to be put under this point. ■

SOLVING A WORD PROBLEM

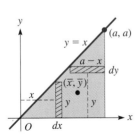

Fig. 6-39

EXAMPLE 5   Find the coordinates of the centroid of an isosceles right triangular plate with side a.

We must first set up the region in the xy-plane. One choice is to place the triangle with one vertex at the origin and the right angle on the x-axis (see Fig. 6-39). Since each side is a, the hypotenuse passes through the point (a, a). The equation of the hypotenuse is $y = x$. The x-coordinate of the centroid is found by using Eq. (6-16):

$$\bar{x} = \frac{\displaystyle\int_0^a xy\,dx}{\displaystyle\int_0^a y\,dx} = \frac{\displaystyle\int_0^a x(x)\,dx}{\displaystyle\int_0^a x\,dx} = \frac{\displaystyle\int_0^a x^2\,dx}{\tfrac{1}{2}x^2\big|_0^a} = \frac{\tfrac{1}{3}x^3\big|_0^a}{\dfrac{a^2}{2}} = \frac{\dfrac{a^3}{3}}{\dfrac{a^2}{2}} = \frac{2a}{3}$$

The y-coordinate of the centroid is found by using Eq. (6-17):

$$\bar{y} = \frac{\displaystyle\int_0^a y(a-x)\,dy}{\dfrac{a^2}{2}} = \frac{\displaystyle\int_0^a y(a-y)\,dy}{\dfrac{a^2}{2}} = \frac{\displaystyle\int_0^a (ay - y^2)\,dy}{\dfrac{a^2}{2}}$$

$$= \frac{\dfrac{ay^2}{2} - \dfrac{y^3}{3}\Big|_0^a}{\dfrac{a^2}{2}} = \frac{\dfrac{a^3}{6}}{\dfrac{a^2}{2}} = \frac{a}{3}$$

Thus, the coordinates of the centroid are $(\tfrac{2}{3}a, \tfrac{1}{3}a)$. The results indicate that the center of mass is $\tfrac{1}{3}a$ units from each of the equal sides. ■

Centroid of a Solid of Revolution

Another figure for which we wish to find the centroid is a solid of revolution. If the density of the solid is constant, the centroid is on the axis of revolution. The problem that remains is to find just where on the axis the centroid is located.

If a region bounded by the x-axis, as shown in Fig. 6-40, is revolved about the x-axis, a vertical element of area generates a disk element of volume. The center of mass of the disk is at its center, and we may consider its mass concentrated there. The moment about the y-axis of a typical element is $x(k)(\pi y^2\,dx)$, where x is the moment arm, k is the density, and $\pi y^2\,dx$ is the volume. The sum of the moments of the elements can be expressed as an integral; it equals the volume times the density times the x-coordinate of the centroid of the volume. Since π and the density k would appear on each side of the equation, they cancel and need not be written. Therefore,

$$\bar{x} = \frac{\int_a^b xy^2\,dx}{\int_a^b y^2\,dx} \tag{6-18}$$

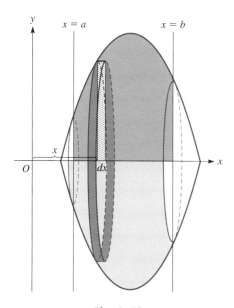

Fig. 6-40

is the equation for the x-coordinate of the centroid of a solid of revolution about the x-axis.

In the same manner, we may find that the y-coordinate of the centroid of a solid of revolution about the y-axis is

$$\bar{y} = \frac{\int_c^d yx^2\,dy}{\int_c^d x^2\,dy} \tag{6-19}$$

EXAMPLE 6 Find the coordinates of the centroid of the solid generated by revolving the first-quadrant region under the curve $y = 4 - x^2$ about the x-axis.

Since the curve (see Fig. 6-41) is rotated about the x-axis, the centroid is on the x-axis, which means that $\bar{y} = 0$. We find the x-coordinate as follows:

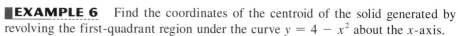

$$\bar{x} = \frac{\int_0^2 xy^2\,dx}{\int_0^2 y^2\,dx} \quad \text{using Eq. (6-18)}$$

$$= \frac{\int_0^2 x(4 - x^2)^2\,dx}{\int_0^2 (4 - x^2)^2\,dx} = \frac{\int_0^2 (16x - 8x^3 + x^5)\,dx}{\int_0^2 (16 - 8x^2 + x^4)\,dx}$$

$$= \frac{8x^2 - 2x^4 + \frac{1}{6}x^6\big|_0^2}{16x - \frac{8}{3}x^3 + \frac{1}{5}x^5\big|_0^2} = \frac{32 - 32 + \frac{64}{6}}{32 - \frac{64}{3} + \frac{32}{5}} = \frac{5}{8}$$

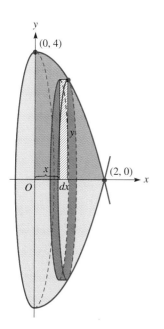

Fig. 6-41

The coordinates of the centroid are $(\frac{5}{8}, 0)$.

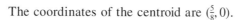

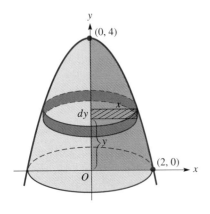

Fig. 6-42

Fig. 6-43

SOLVING A WORD PROBLEM

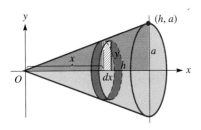

Fig. 6-44

■EXAMPLE 7   Find the coordinates of the centroid of the solid generated by revolving the first-quadrant region under the curve $y = 4 - x^2$ about the y-axis. (This is the same region as in Example 6.)

Since the curve is rotated about the y-axis, $\bar{x} = 0$ (see Fig. 6-42). The y-coordinate is

$$\bar{y} = \frac{\displaystyle\int_0^4 y x^2 \, dy}{\displaystyle\int_0^4 x^2 \, dy} \qquad \text{using Eq. (6-19)}$$

with "moment arm" labeling the y in the numerator.

$$= \frac{\displaystyle\int_0^4 y(4 - y)\, dy}{\displaystyle\int_0^4 (4 - y)\, dy} = \frac{\displaystyle\int_0^4 (4y - y^2)\, dy}{\displaystyle\int_0^4 (4 - y)\, dy} = \frac{2y^2 - \dfrac{1}{3}y^3 \Big|_0^4}{4y - \dfrac{1}{2}y^2 \Big|_0^4}$$

$$= \frac{32 - \dfrac{64}{3}}{16 - 8} = \frac{4}{3}$$

The coordinates of the centroid are $(0, \tfrac{4}{3})$.   -----------■

Figure 6-43 shows the calculator evaluation for this integral. Note that it is not necessary to evaluate each integral separately, as the division can be performed in one calculation.

■EXAMPLE 8   Find the centroid of a solid right circular cone of radius a and altitude h.

To generate a right circular cone, we may revolve a right triangle about one of its legs (Fig. 6-44). Placing the leg of length h along the x-axis, we rotate the right triangle whose hypotenuse is given by $y = (a/h)x$ about the x-axis. The x-coordinate of the centroid is

$$\bar{x} = \frac{\displaystyle\int_0^h x y^2 \, dx}{\displaystyle\int_0^h y^2 \, dx} \qquad \text{using Eq. (6-18)}$$

with "moment arm" labeling the x in the numerator.

$$= \frac{\displaystyle\int_0^h x\left[\left(\dfrac{a}{h}\right)x\right]^2 dx}{\displaystyle\int_0^h \left[\left(\dfrac{a}{h}\right)x\right]^2 dx} = \frac{\left(\dfrac{a^2}{h^2}\right)\left(\dfrac{1}{4}x^4\right)\Big|_0^h}{\left(\dfrac{a^2}{h^2}\right)\left(\dfrac{1}{3}x^3\right)\Big|_0^h} = \frac{3}{4}h$$

Therefore, the centroid is located along the altitude $\tfrac{3}{4}$ of the way from the vertex, or $\tfrac{1}{4}$ of the way from the base.   -----------■

EXERCISES 6-4

In Exercises 1–4, find the center of mass (in cm) of the particles with the given masses located at the given points on the x-axis.

1. 5.0 g at $(1.0, 0)$, 8.5 g at $(4.2, 0)$, 3.6 g at $(2.5, 0)$

2. 2.3 g at $(1.3, 0)$, 6.5 g at $(5.8, 0)$, 1.2 g at $(9.5, 0)$

3. 42 g at $(-3.5, 0)$, 24 g at $(0, 0)$, 15 g at $(2.6, 0)$, 84 g at $(3.7, 0)$

4. 550 g at $(-42, 0)$, 230 g at $(-27, 0)$, 470 g at $(16, 0)$, 120 g at $(22, 0)$

In Exercises 5–8, find the coordinates (to 0.01 in.) of the centroids of the uniform flat-plate machine parts shown.

5. Fig. 6-45(a)   6. Fig. 6-45(b)

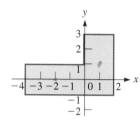

7. Fig. 6-45(c)   8. Fig. 6-45(d)

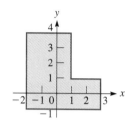

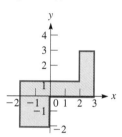
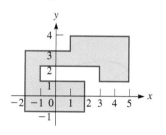

Fig. 6-45

In Exercises 9–28, find the coordinates of the centroids of the given figures. Each region is covered by a thin plate.

9. The region bounded by $y = x^2$ and $y = 2$

10. The semicircular region in Fig. 6-46

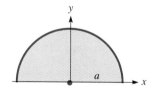

Fig. 6-46

11. The region bounded by $y = 4 - x$ and the axes

12. The region bounded by $y = x^3$, $x = 2$, and the x-axis

13. The region bounded by $y = x^2$ and $y = x^3$

14. The region bounded by $y^2 = x$, $y = 2$, and $x = 0$

15. The region bounded by $y = 2x$, $y = 3x$, and $y = 6$

16. The region bounded by $y = x^{2/3}$, $x = 8$, and $y = 0$

17. The solid generated by revolving the region bounded by $y = x^3$, $y = 0$, and $x = 1$ about the x-axis

18. The solid generated by revolving the region bounded by $y = 2 - 2x$, $x = 0$, and $y = 0$ about the y-axis

19. The solid generated by revolving the region in the first quadrant bounded by $y^2 = 4x$, $y = 0$, and $x = 1$ about the y-axis

20. The solid generated by revolving the region bounded by $y = x^2$, $x = 2$, and the x-axis about the x-axis

21. The solid generated by revolving the region bounded by $y^2 = 4x$ and $x = 1$ about the x-axis

22. The solid generated by revolving the region bounded by $x^2 - y^2 = 9$, $y = 4$, and the x-axis about the y-axis

(W) 23. Explain how to find the centroid of a right trianglular plate with legs a and b. Find the location of the centroid.

24. Find the location of the centroid of a hemisphere of radius a.

25. A lens with semielliptical vertical cross sections and circular horizontal cross sections is shown in Fig. 6-47. For proper installation in an optical device, its centroid must be known. Locate its centroid.

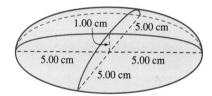

Fig. 6-47

26. A sanding machine disc can be described as the solid generated by rotating the region bounded by $y^2 = 4/x$, $y = 1$, $y = 2$, and the y-axis about the y-axis (measurements in in.). Locate the centroid of the disc.

27. A highway marking pylon has the shape of a frustum of a cone. Find its centroid if the radii of its bases are 5.00 cm and 20.0 cm and the height between bases is 60.0 cm.

28. A floodgate is in the shape of an isosceles trapezoid. Find the location of the centroid of the floodgate if the upper base is 20 m, the lower base is 12 m, and the height between bases is 6.0 m. See Fig. 6-48.

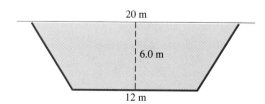

Fig. 6-48

$6\text{-}5$ MOMENTS OF INERTIA

In the discussion of rotational motion in physics, an important quantity is the **moment of inertia** of an object. The moment of inertia of an object rotating about an axis is analogous to the mass of a moving object. In each case, *the moment of inertia or mass is the measure of the tendency of the object to resist a change in motion.*

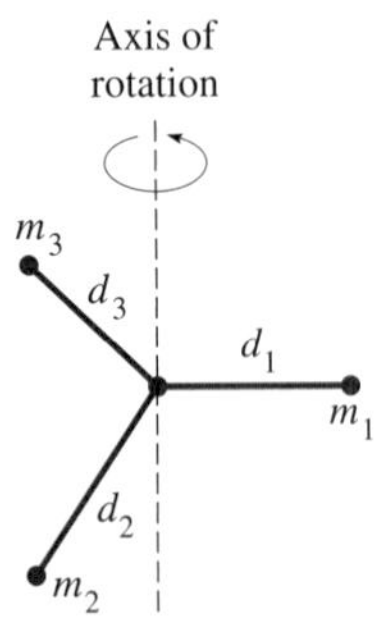

Suppose that a particle of mass m is rotating about some point: We define its moment of inertia as md^2, where d is the distance from the particle to the point. If a group of particles of masses $m_1, m_2, \ldots, m_n$ are rotating about an axis, as shown in Fig. 6-49, the moment of inertia I with respect to the axis of the group is

$$I = m_1 d_1^2 + m_2 d_2^2 + \cdots + m_n d_n^2$$

where the d's are the respective distances of the particles from the axis. If all the masses were at the same distance R from the axis of rotation, so that the total moment of inertia were the same, we would have

$$m_1 d_1^2 + m_2 d_2^2 + \cdots + m_n d_n^2 = (m_1 + m_2 + \cdots + m_n)R^2 \qquad (6\text{-}20)$$

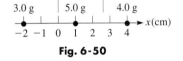

Fig. 6-49

RADIUS OF GYRATION

where R is called the **radius of gyration.**

Fig. 6-50

EXAMPLE 1 Find the moment of inertia and the radius of gyration of the array of three masses, one of 3.0 g at $(-2.0, 0)$, another of 5.0 g at $(1.0, 0)$, and the third of 4.0 g at $(4.0, 0)$, with respect to the origin (distances in cm). See Fig. 6-50.

The moment of inertia of the array is

$$I = 3.0(-2.0)^2 + 5.0(1.0)^2 + 4.0(4.0)^2 = 81 \text{ g} \cdot \text{cm}^2$$

The radius of gyration is found from $I = (m_1 + m_2 + m_3)R^2$. Thus,

$$81 = (3.0 + 5.0 + 4.0)R^2, \qquad R^2 = \frac{81}{12}, \qquad R = 2.6 \text{ cm}$$

Therefore, a mass of 12.0 g placed at $(2.6, 0)$ (or $(-2.6, 0)$) has the same rotational inertia about the origin as the array of masses as a unit. ▬

Moment of Inertia of a Thin, Flat Plate

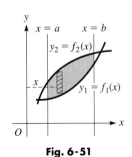

Fig. 6-51

If a thin, flat plate covering the region is bounded by the curves of the functions $y_1 = f_1(x)$, $y_2 = f_2(x)$ and the lines $x = a$ and $x = b$, as shown in Fig. 6-51, the moment of inertia of this plate with respect to the y-axis, I_y, is given by the sum of the moments of inertia of the individual elements. The mass of each element is $k(y_2 - y_1)\,dx$, where k is the mass per unit area and $(y_2 - y_1)\,dx$ is the area of the element. The distance of the element from the y-axis is x. Representing this sum as an integral, we have

$$I_y = k \int_a^b x^2(y_2 - y_1)\,dx \qquad (6\text{-}21)$$

To find the radius of gyration of the plate with respect to the y-axis, R_y, we would first find the moment of inertia, divide this by the mass of the area, and take the square root of this result.

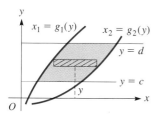

Fig. 6-52

In the same manner, the moment of inertia of a thin plate, with respect to the x-axis, bounded by $x_1 = g_1(y)$ and $x_2 = g_2(y)$ is given by

$$I_x = k \int_c^d y^2(x_2 - x_1)\, dy \qquad (6\text{-}22)$$

We find the radius of gyration of the plate with respect to the x-axis, R_x, in the same manner as we find it with respect to the y-axis (see Fig. 6-52).

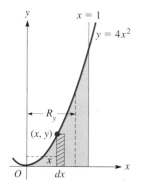

Fig. 6-53

■**EXAMPLE 2** Find the moment of inertia and the radius of gyration of the plate covering the region bounded by $y = 4x^2$, $x = 1$, and the x-axis with respect to the y-axis.

We find the moment of inertia of this plate (see Fig. 6-53) as follows:

$$I_y = k \int_0^1 x^2 y\, dx \qquad \text{using Eq. (6-21)}$$
$$= k \int_0^1 x^2(4x^2)\, dx = 4k \int_0^1 x^4\, dx$$
$$= 4k\left(\frac{1}{5}x^5\right)\Big|_0^1 = \frac{4k}{5}$$

To find the radius of gyration, we first determine the mass of the plate:

$$m = k \int_0^1 y\, dx = k \int_0^1 (4x^2)\, dx \qquad m = kA$$
$$= 4k\left(\frac{1}{3}x^3\right)\Big|_0^1 = \frac{4k}{3}$$
$$R_y^2 = \frac{I_y}{m} = \frac{4k}{5} \times \frac{3}{4k} = \frac{3}{5} \qquad R_y^2 = I_y/m$$
$$R_y = \sqrt{\frac{3}{5}} = \frac{\sqrt{15}}{5}$$

SOLVING A WORD PROBLEM

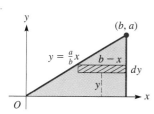

Fig. 6-54

■**EXAMPLE 3** Find the moment of inertia of a right trianglular plate with sides a and b with respect to side b. Assume that $k = 1$.

Placing the triangle as shown in Fig. 6-54, we see that the equation of the hypotenuse is $y = (a/b)x$. The moment of inertia is

$$I_x = \int_0^a y^2(b - x)\, dy \qquad \text{using Eq. (6-22)}$$
$$= \int_0^a y^2\left(b - \frac{b}{a}y\right) dy = b \int_0^a \left(y^2 - \frac{1}{a}y^3\right) dy$$
$$= b\left(\frac{1}{3}y^3 - \frac{1}{4a}y^4\right)\Big|_0^a$$
$$= b\left(\frac{a^3}{3} - \frac{a^3}{4}\right) = \frac{ba^3}{12}$$

Moment of Inertia of a Solid

CAUTION ▶ In applications, among the most important moments of inertia are those of solids of revolution. Since *all parts of an element of mass should be at the same distance from the axis,* the most convenient element of volume to use is the cylindrical shell. In Fig. 6-55, if the region bounded by the curves $y_1 = f_1(x)$, $y_2 = f_2(x)$, $x = a$, and $x = b$ is revolved about the y-axis, the moment of inertia of the element of volume is $k[2\pi x(y_2 - y_1)\,dx](x^2)$, where k is the density, $2\pi x(y_2 - y_1)\,dx$ is the volume of the element, and x^2 is the square of the distance from the x-axis. Expressing the sum of the elements as an integral, *the moment of inertia of the solid with respect to the y-axis, I_y, is*

> Note carefully that Eq. (6-23) gives the moment of inertia with respect to the y-axis and that $(y_2 - y_1)$ is the height of the shell (see Fig. 6-24).

$$I_y = 2\pi k \int_a^b (y_2 - y_1)x^3\,dx \qquad (6\text{-}23)$$

The radius of gyration of the solid with respect to the y-axis, R_y, is found by determining (1) the moment of inertia, (2) the mass of the solid, and (3) the square root of the quotient of the moment of inertia divided by the mass.

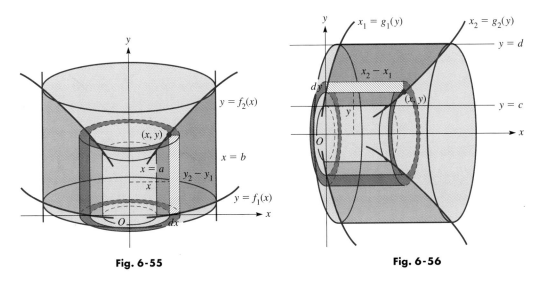

Fig. 6-55   **Fig. 6-56**

The moment of inertia of the solid (see Fig. 6-56) generated by revolving the region bounded by $x_1 = g_1(y)$, $x_2 = g_2(y)$, $y = c$, and $y = d$ about the x-axis, I_x, is given by

> Note carefully that Eq. (6-24) gives the moment of inertia with respect to the x-axis and that $(x_2 - x_1)$ is the height of the shell (see Fig. 6-24).

$$I_x = 2\pi k \int_c^d (x_2 - x_1)y^3\,dy \qquad (6\text{-}24)$$

The radius of gyration of the solid with respect to the x-axis, R_x, is found in the same manner as R_y.

EXAMPLE 4 Find the moment of inertia and the radius of gyration with respect to the x-axis of the solid generated by revolving the region bounded by the curves of $y^3 = x$, $y = 2$, and the y-axis about the x-axis. See Fig. 6-57.

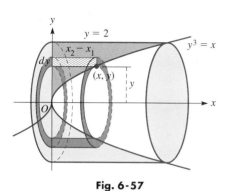

Fig. 6-57

distance from element to axis

$$I_x = 2\pi k \int_0^2 (x_2 - x_1)y^3\,dy \qquad \text{using Eq. (6-24)}$$

$$= 2\pi k \int_0^2 (y^3)y^3\,dy \qquad x_2 - x_1 = y^3 - 0 = y^3$$

$$= 2\pi k\left(\frac{1}{7}y^7\right)\Big|_0^2 = \frac{256\pi k}{7}$$

$$m = 2\pi k \int_0^2 xy\,dy \qquad \text{mass} = k \times \text{volume}$$

$$= 2\pi k \int_0^2 y^3 y\,dy = 2\pi k\left(\frac{1}{5}y^5\right)\Big|_0^2 = \frac{64\pi k}{5}$$

$$R_x^2 = \frac{256\pi k}{7} \times \frac{5}{64\pi k} = \frac{20}{7} \qquad R_x^2 = I_x/m$$

$$R_x = \sqrt{\frac{20}{7}} = \frac{2}{7}\sqrt{35}$$

SOLVING A WORD PROBLEM

EXAMPLE 5 As noted at the beginning of this section, the moment of inertia is important when studying the rotational motion of an object. For this reason, the moments of inertia of various objects are calculated, and the formulas tabulated. Such formulas are usually expressed in terms of the mass of the object.

Among the objects for which the moment of inertia is important is a solid disk. Find the moment of inertia of a disk with respect to its axis and in terms of its mass.

To generate a disk (see Fig. 6-58), we rotate the region bounded by the axes, $x = r$ and $y = b$, about the y-axis. We then have

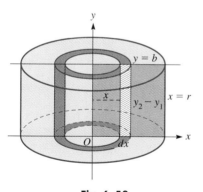

Fig. 6-58

distance from element to axis

$$I_y = 2\pi k \int_0^r (y_2 - y_1)x^3\,dx \qquad \text{using Eq. (6-23)}$$

$$= 2\pi k \int_0^r (b)x^3\,dx = 2\pi kb \int_0^r x^3\,dx \qquad y_2 - y_1 = b - 0 = b$$

$$= 2\pi kb\left(\frac{1}{4}x^4\right)\Big|_0^r = \frac{\pi kbr^4}{2}$$

The mass of the disk is $k(\pi r^2)b$. Rewriting the expression for I_y, we have

$$I_y = \frac{(\pi kbr^2)r^2}{2} = \frac{mr^2}{2}$$

Due to the limited methods of integration available at this point, we cannot integrate the expressions for the moments of inertia of circular plates or of a sphere. These will be introduced in Section 10-3 in the exercises, by which point the proper method of integration will have been developed.

EXERCISES $6\text{-}5$

In Exercises 1–4, find the moment of inertia (in g·cm²) and the radius of gyration (in cm) with respect to the origin of each of the given arrays of masses located at the given points on the x-axis.

1. 5.0 g at $(2.4, 0)$, 3.2 g at $(3.5, 0)$

2. 3.4 g at $(-1.5, 0)$, 6.0 g at $(2.1, 0)$, 2.6 g at $(3.8, 0)$

3. 45.0 g at $(-3.80, 0)$, 90.0 g at $(0.00, 0)$, 62.0 g at $(5.50, 0)$

4. 564 g at $(-45.0, 0)$, 326 g at $(-22.5, 0)$, 720 g at $(15.4, 0)$, 205 g at $(64.0, 0)$

In Exercises 5–24, find the indicated moment of inertia or radius of gyration.

5. Find the moment of inertia of a plate covering the first-quadrant region bounded by $y^2 = x$, $x = 4$, and the x-axis with respect to the x-axis.

6. Find the moment of inertia of a plate covering the region bounded by $y = 2x$, $x = 1$, $x = 2$, and the x-axis with respect to the y-axis.

7. Find the radius of gyration of a plate covering the region bounded by $y = x^3$, $x = 2$, and the x-axis with respect to the y-axis.

8. Find the radius of gyration of a plate covering the first-quadrant region bounded by $y^2 = 1 - x$ with respect to the x-axis.

9. Find the moment of inertia of a right trianglular plate with sides a and b with respect to side a in terms of the mass of the plate.

10. Find the moment of inertia of a rectangular plate of sides a and b with respect to side a. Express the result in terms of the mass of the plate.

11. Find the radius of gyration of a plate covering the region bounded by $y = x^2$, $x = 2$, and the x-axis with respect to the x-axis.

12. Find the radius of gyration of a plate covering the region bounded by $y^2 = x^3$, $y = 8$, and the y-axis with respect to the y-axis.

13. Find the radius of gyration of the plate of Exercise 12 with respect to the x-axis.

14. Find the radius of gyration of a plate covering the first-quadrant region bounded by $x = 1$, $y = 2 - x$, and the y-axis with respect to the y-axis.

15. Find the moment of inertia with respect to its axis of the solid generated by revolving the region bounded by $y^2 = x$, $y = 2$, and the y-axis about the x-axis.

16. Find the radius of gyration with respect to its axis of the solid generated by revolving the first-quadrant region under the curve $y = 4 - x^2$ about the y-axis.

17. Find the radius of gyration with respect to its axis of the solid generated by revolving the region bounded by $y = 2x - x^2$ and the x-axis about the y-axis.

18. Find the radius of gyration with respect to its axis of the solid generated by revolving the region bounded by $y = 2x$ and $y = x^2$ about the y-axis.

19. Find the moment of inertia in terms of its mass of a right circular cone of radius r and height h with respect to its axis. See Fig. 6-59.

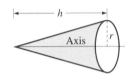

Fig. 6-59

20. Find the moment of inertia in terms of its mass of a circular hoop of radius r and of negligible thickness with respect to its center.

21. A rotating drill head is in the shape of a right circular cone. Find the moment of inertia of the drill head with respect to its axis if its radius is 0.600 cm, its height is 0.800 cm, and its mass is 3.00 g. (See Exercise 19.)

22. Find the moment of inertia (in kg·m²) of a rectangular door 2 m high and 1 m wide with respect to its hinges if $k = 3$ kg/m². (See Exercise 10.)

23. Find the moment of inertia of a flywheel with respect to its axis if its inner radius is 4.0 cm, its outer radius is 6.0 cm, and its mass is 1.2 kg. See Fig. 6-60.

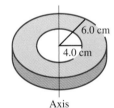

Fig. 6-60   Axis

(W) 24. A cantilever beam is supported only at its left end, as shown in Fig. 6-61. Explain how to find the formula for the moment of inertia of this beam with respect to a vertical axis through its left end if its length is L and its mass is m. (Consider the mass to be distributed evenly along the beam. This is not an area or volume type of problem.) Find the formula for the moment of inertia.

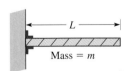

Fig. 6-61

$6\text{-}6$ WORK BY A VARIABLE FORCE

The next application of integration we shall demonstrate involves the physical problem of work done by a variable force. The concept of work is useful for comparing the amounts of mechanical energy required to perform various physical tasks. Applications arise in many fields of technology, and some of these are illustrated in the examples and exercises of this section.

In physics, **work** *is defined as the product of a constant force times the distance through which it acts.* When we consider the work done in stretching a spring, the first thing we recognize is that the more the spring is stretched, the greater is the force necessary to stretch it. Thus the force varies. However, if we are stretching the spring a distance Δx, where we are considering the limit as $\Delta x \to 0$, the force can be considered as approaching a constant over Δx. Adding the products of force$_1$ times Δx_1, force$_2$ times Δx_2, and so forth, we see that the total work is the sum of these products. Therefore the work can be expressed as a definite integral in the form given by

$$W = \int_a^b f(x)\, dx$$ (6-25)

CAUTION ▶

where $f(x)$ is the force as a function of the distance the spring is stretched. The ***limits a and b refer to the initial and final distances the spring is stretched* from its normal length.**

One problem remains: We must find the function $f(x)$. From physics we learn that the force required to stretch a spring is proportional to the amount it is stretched (Hooke's law). If a spring is stretched x units from its natural length, then $f(x) = kx$. From conditions stated for a particular spring, the value of k may be determined. Thus, $W = \int_a^b kx\, dx$ is the formula for finding the total work done in stretching a spring. Here a and b are the initial and final amounts the spring is stretched from its natural length.

Named for the English physicist Robert Hooke (1635–1703). The law assumes the spring is stretched only within its elastic limit.

SOLVING A WORD PROBLEM

■**EXAMPLE 1** A spring of natural length 12 in. requires a force of 6.0 lb to stretch it 2.0 in. See Fig. 6-62. Find the work done in stretching it 6.0 in.

From Hooke's law, we find the constant k for the spring as

$$f(x) = kx, \qquad 6.0 = k(2.0), \qquad k = 3.0 \text{ lb/in.}$$

Since the spring is to be stretched 6.0 in., $a = 0$ (it starts unstretched) and $b = 6.0$ (it is 6.0 in. longer than its natural length). Therefore, the work done in stretching it is

$$W = \int_0^{6.0} 3.0x\, dx = 1.5x^2 \Big|_0^{6.0} \qquad \text{using Eq. (6-25)}$$

$$= 54 \text{ lb} \cdot \text{in.}$$ ■

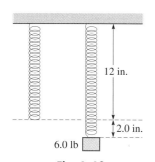

12 in.

2.0 in.

6.0 lb

Fig. 6-62

Another technical application of work by a variable force is found in the field of electricity. On the following page is an illustration that deals with the motion of an electric charge through an electric field that is created by another nearby electric charge.

Electric charges are of two types, designated as positive and negative. A basic law is that charges of the same sign repel each other and charges of opposite signs attract each other. *The force between charges is proportional to the product of their charges and inversely proportional to the square of the distance between them.*

The force $f(x)$ between electric charges is therefore given by

$$f(x) = \frac{kq_1q_2}{x^2} \qquad\qquad \textbf{(6-26)}$$

when q_1 and q_2 are the charges (in coulombs), x is the distance (in meters), the force is in newtons, and $k = 9.0 \times 10^9\ \mathrm{N \cdot m^2/C^2}$. For other systems of units, the numerical value of k is different. We can find the work done when electric charges move toward each other or when they separate by use of Eq. (6-26) in Eq. (6-25).

SOLVING A WORD PROBLEM   ▌**EXAMPLE 2**   Find the work done when two α-particles, $q = 0.32$ aC each, move until they are 10 nm apart, if they were originally separated by 1.0 m.

From the given information, we have for each α-particle

$$q = 0.32\ \mathrm{aC} = 0.32 \times 10^{-18}\ \mathrm{C} = 3.2 \times 10^{-19}\ \mathrm{C}$$

Since the particles start 1.0 m apart and are moved to 10 nm apart, $a = 1.0$ m and $b = 10 \times 10^{-9}$ m $= 10^{-8}$ m. The work done is

$$f(x) \text{ from Eq. (6-26)}$$

$$W = \int_{1.0}^{10^{-8}} \frac{9.0 \times 10^9 (3.2 \times 10^{-19})^2}{x^2}\, dx \qquad \text{using Eq. (6-25)}$$

$$= 9.2 \times 10^{-28} \int_{1.0}^{10^{-8}} \frac{dx}{x^2} = 9.2 \times 10^{-28}\left(-\frac{1}{x}\right)\Bigg|_{1.0}^{10^{-8}}$$

$$= -9.2 \times 10^{-28}(10^8 - 1) = -9.2 \times 10^{-20}\ \mathrm{J}$$

Since $10^8 \gg 1$, where $\gg$ means "much greater than," the 1 may be neglected in the calculation. The minus sign in the result means that work must be done *on* the system to move the particles toward each other. If free to move, they tend to separate. ▐

Following are other types of problems involving work by a variable force.

SOLVING A WORD PROBLEM   ▌**EXAMPLE 3**   Find the work done in winding up 60.0 ft of a 100-ft cable that weighs 4.00 lb/ft. See Fig. 6-63.

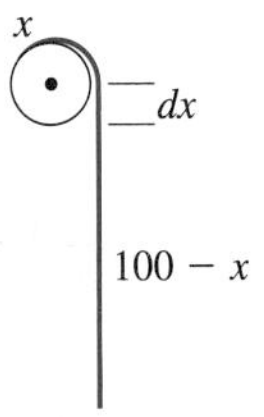

First, we let x denote the length of cable that has been wound up at any time. Then, the force required to raise the remaining cable equals the weight of the cable that has not yet been wound up. This weight is the product of the unwound cable length, $100 - x$, and its weight per unit length, 4.00 lb/ft, or

$$f(x) = 4.00(100 - x)$$

Since 60.0 ft of cable are to be wound up, $a = 0$ (none is initially wound up) and $b = 60.0$ ft. The work done is

$$W = \int_0^{60.0} 4.00(100 - x)\, dx \qquad \text{using Eq. (6-25)}$$

$$= \int_0^{60.0} (400 - 4.00x)\, dx = 400x - 2.00x^2\Big|_0^{60.0} = 16{,}800\ \mathrm{ft \cdot lb}$$ ▐

See the chapter introduction.

Fig. 6-63

SOLVING A WORD PROBLEM

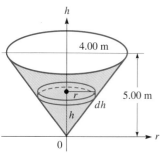

Fig. 6-64

EXAMPLE 4   A conical tank with a circular cross section and vertex at the bottom has a vertical axis. The radius of the top is 4.00 m, and the depth is 5.00 m. The tank is filled with water. Find the work required to pump out all the water over the top. The force of gravity on each cubic meter of water is 9800 N. See Fig. 6-64.

Each horizontal layer of water is a circular disk dh meters thick (just as in finding volumes by use of disk elements). This means that the weight of each disk of water is $9800[\pi(\text{radius}^2)(\text{thickness})]$. From similar triangles in the figure, we see that

$$r/h = 4.00/5.00 \qquad \text{or} \qquad r = 0.800h$$

The distance each disk must be raised in order to empty the tank is $5.00 - h$ meters. Therefore, summing the products of weight and distance, we find the work done.

$$W = \int_0^{5.00} (9800\pi r^2\, dh)(5.00 - h) = 9800\pi \int_0^{5.00} (0.800h)^2(5.00 - h)\, dh$$

$$= 6270\pi \int_0^{5.00} (5.00h^2 - h^3)\, dh$$

$$= 6270\pi \left(\frac{5.00}{3}h^3 - \frac{1}{4}h^4 \right) \Bigg|_0^{5.00} = 6270\pi(52.1)$$

$$= 1.03 \times 10^6 \text{ N} \cdot \text{m} = 1.03 \text{ MJ}$$

EXERCISES 6-6

1. The spring of a spring balance is 8.0 in. long with no weight on the balance and 9.5 in. long with 6.0 lb on the balance. How much work is done in stretching it from 8.0 in. to 10.0 in.?

2. How much work is done in stretching the spring of Exercise 1 from a length of 10.0 in. to 12.0 in.?

3. A force of 25 N stretches the spring of a lever-spring mechanism by 16 mm. How much work is done by the 25-N force?

4. A force of 1200 lb compresses a spring from its natural length of 18 in. to a length of 16 in. How much work is done in compressing it from 16 in. to 14 in.?

5. If another force of 25 N is applied to the spring of Exercise 3, how much work is done by the additional force?

6. An electron has a 1.6×10^{-19} C negative charge. How much work is done in separating two electrons from 1.0 pm to 4.0 pm?

7. How much work is done in separating an electron (see Exercise 6) and an oxygen nucleus (positive 1.3×10^{-18} C charge) from a distance of 2.0 μm to a distance of 1.0 m?

8. Find the work done in separating a 0.8-nC negative charge and a 0.6-nC negative charge to a distance of 200 mm from 100 mm.

9. The gravitational force F between two objects x units apart is $F = k/x^2$. If two objects are 1 cm apart, find the work done (in terms of k) in separating them to 1 m apart.

10. If two objects are 10 ft apart, find the work done (in terms of k) in separating them to 100 ft apart. See Exercise 9.

11. Find the work in winding up a 200-ft cable weighing 100 lb.

12. In Exercise 11, if an object weighing 50 lb is attached at the end of the cable, find the work done in winding it up.

13. Find the work done in winding up 20 m of a 25-m rope on which the force of gravity is 6.0 N/m.

14. A chain is being unwound from a winch. The force of gravity on it is 12.0 N/m. When 20 m have been unwound, how much work is done by gravity in unwinding another 30 m?

15. At lift-off, a rocket weighs 32.5 tons, including fuel. It is fired vertically, and, during the first stage of ascent, the fuel is consumed at 1.25 tons per 1000 ft of ascent. How much work is done in lifting the rocket to an altitude of 12,000 ft?

16. While descending, a 550-N weather balloon enters freezing rain in which ice forms on the balloon at the rate of 7.50 N per 100 m of descent. Find the work done on the balloon during the first 1000 m of descent through the freezing rain.

17. Find the work done in pumping out water from the top of a cylindrical tank 3.00 ft in radius and 10.0 ft high, if the tank is initially full. Water weighs 62.4 lb/ft³.

18. Find the work done in pumping the water from the tank in Exercise 17, if the tank is initially half full.

19. Find the work done in emptying the tank in Example 4, if the water is pumped to a level 1.00 m above the top of the tank and the water is initially 4.00 m deep.

20. A hemispherical tank of radius 10.0 ft is full of oil. Find the work done in pumping out the oil over the top (the flat side) of the tank. The oil weighs 60.0 lb/ft³.

6-7 FORCE DUE TO LIQUID PRESSURE

In physics it is shown that at any point within a liquid the pressure is the same in all directions. Also, it is shown that *the pressure is directly proportional to the depth below the surface of the liquid.* If the weight per unit volume is w, the pressure p at depth h is given by

$$p = wh \tag{6-27}$$

For water, $w = 62.4 \text{ lb/ft}^3$ or $w = 9800 \text{ N/m}^3$. *Since pressure is force per unit area, the force on an area A at a given depth is*

$$F = pA \tag{6-28}$$

Substituting the expression from Eq. (6-27) into Eq. (6-28), we have

$$F = whA \tag{6-29}$$

as the force on an area A at depth h.

Let us now assume that the plate shown in Fig. 6-65 is submerged vertically in water. At points on the plate at different depths the pressure will be different. Therefore, in using Eq. (6-29) to find the force on the plate, we must choose an area A such that all of its points are essentially at the same depth. We approximate this by choosing the horizontal rectangular strip of length l and width dh. The force on this strip is $wh(l\,dh)$. By using integration to sum the forces on all strips from the top of the plate to the bottom, we find *the total force on the plate is given by*

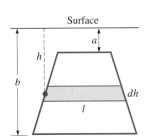

Surface

Fig. 6-65

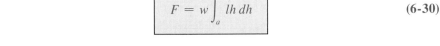

$$F = w \int_a^b lh\,dh \tag{6-30}$$

Here, l is the length of the element of area, h is the depth of the element of area, w is the weight per unit volume of the liquid, a is the depth of the top, and b is the depth of the bottom of the area on which the force is exerted.

SOLVING A WORD PROBLEM

▌EXAMPLE 1 A vertical floodgate of a dam is 5.00 ft wide and 4.00 ft high. Find the force on the floodgate if its upper edge is 3.00 ft below the surface of the water. See Fig. 6-66.

Each element of area of the floodgate has a length of 5.00 ft, which means that $l = 5.00$ ft. Since the top of the gate is 3.00 ft below the surface, $a = 3.00$ ft, and since the gate is 4.00 ft high, $b = 7.00$ ft. Using $w = 62.4 \text{ lb/ft}^3$, we have the force on the gate as

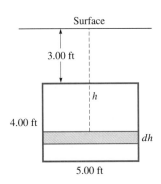

Fig. 6-66

$$F = 62.4 \int_{3.00}^{7.00} 5.00h\,dh \qquad \text{using Eq. (6-30)}$$

$$= 312 \int_{3.00}^{7.00} h\,dh$$

$$= 156h^2 \Big|_{3.00}^{7.00} = 156(49.0 - 9.00)$$

$$= 6240 \text{ lb}$$

SOLVING A WORD PROBLEM

NOTE ▶

▌EXAMPLE 2   The vertical end of a water tank is in the shape of a right triangle, as shown in Fig. 6-67. What is the force on the end of the tank?

In setting up the figure, it is convenient to use coordinate axes. It is also convenient to have the y-axis directed downward, since we integrate from the top to the bottom of the area. The equation of the line OA is $y = \frac{1}{2}x$. Thus, we see that the length of an element of area of the end of the tank is $4.0 - x$, the depth of the element of area is y, the top of the tank is $y = 0$, and the bottom is $y = 2.0$ m. Therefore, the force on the end of the tank is ($w = 9800$ N/m^3).

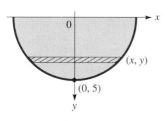

Fig. 6-67

$$F = 9800 \int_0^{2.0} \overset{\text{length}}{\overset{\downarrow}{(4.0}} - x)\overset{\text{depth}}{\overset{\downarrow}{(y)}}(dy) \qquad \text{using Eq. (6-30)}$$

$$= 9800 \int_0^{2.0} (4.0 - 2y)(y\, dy)$$

$$= 19{,}600 \int_0^{2.0} (2.0y - y^2)\, dy = 19{,}600\left(1.0y^2 - \frac{1}{3}y^3\right)\Bigg|_0^{2.0}$$

$$= 26{,}100 \text{ N}$$

SOLVING A WORD PROBLEM

▌EXAMPLE 3   The part of the vertical end of a boat that is under the waterline is a semicircle with a radius of 5.00 ft, as shown in Fig. 6-68. Find the force exerted on the end of the boat by the water.

Again, we select our axes such that the y-axis is directed downward. The equation of the circle is $x^2 + y^2 = 25$. The length of an element is $2x$. The depth of the element is y, the top of the water is at $y = 0$, and the bottom is at $y = 5.00$ ft. Therefore, the force on the end of the boat is

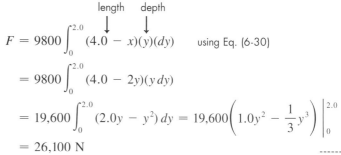

Fig. 6-68

$$F = 62.4 \int_0^{5.00} (2x)(y)\, dy = 62.4 \int_0^{5.00} 2y\sqrt{25 - y^2}\, dy$$

$$= -\frac{2}{3}(62.4)(25 - y^2)^{3/2}\Bigg|_0^{5.00}$$

$$= \frac{2}{3}(62.4)(125) = 5200 \text{ lb}$$

EXERCISES 6-7

1. One end of a spa is a vertical rectangular wall 12.0 ft wide. What is the force exerted on this wall by the water if it is 2.50 ft deep?

2. Find the total force on the sides of a cubic tank 2.0 m on an edge if the tank is full of water.

3. Find the force on a rectangular floodgate 15.0 ft wide and 6.00 ft deep if the upper edge is 2.00 ft below the surface of the water.

4. Find the force on one end of a cubical container 6.00 cm on an edge if the container is filled with mercury. The density of mercury is 133 kN/m^3.

5. A rectangular sea aquarium observation window is 10.0 ft wide and 5.00 ft high. What is the force on this window if the upper edge is 4.00 ft below the surface of the water? The density of seawater is 64.0 lb/ft^3.

6. Another observation window in the aquarium is 20.0 ft below (bottom edge to top edge) the window of Exercise 5. What is the force on the lower window? (The windows have the same dimensions.)

7. The end of a trough in the shape of an isosceles right triangle has a horizontal upper edge of 8.00 ft and is 8.00 ft deep. Find the force on the end if the trough is full of water.

8. Find the force on the end of the trough in Exercise 7 if the depth of the water is 5.00 ft.

9. A right triangular plate of base 2.0 m and height 1.0 m is submerged vertically in water, as shown in Fig. 6-69. Find the force on one side of the plate.

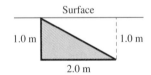

Fig. 6-69

10. Find the force on the plate in Exercise 9 if the top vertex is 3.0 m below the surface.

11. The end of a tank is in the shape of a trapezoid, as shown in Fig. 6-70. Find the force on the end of the tank if it is full of gasoline. The density of gasoline is 42.0 lb/ft³.

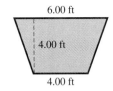

Fig. 6-70

12. A horizontal tank has vertical circular ends, each with a radius of 4.00 ft. It is filled to a depth of 4.00 ft with oil of density 60.0 lb/ft³. Find the force on one end of the tank.

(W) 13. A water-tight cubical box with an edge of 2.00 m is suspended in water such that the top surface is 1.00 m below water level. Find the total force on the top of the box and the total force on the bottom of the box. What meaning can you give to the difference of these two forces?

14. Find the force on the region bounded by $x = 2y - y^2$ and the y-axis if the upper point of the area is at the surface of the water. All distances are in feet.

15. A small dam is in the shape of the region bounded by $y = x^2$ and $y = 20$ (distances in ft). Find the force on the region below $y = 4$ if the surface of the water is at the top of the dam.

16. A horizontal cylindrical tank has vertical elliptical ends with the minor axis horizontal. The major axis is 6.0 ft, and the minor axis is 4.0 ft. Find the force on one end if the tank is half-filled with fuel oil of density 50 lb/ft³.

6-8 OTHER APPLICATIONS

To illustrate the great variety of applications of integration, we shall demonstrate three more types of applied problems in this section. The formulas will be given without proof, and then examples of their use will be given. The first application is that of finding the length of an arc of a curve.

Length of Arc

If $y = f(x)$ is the equation of a given curve, *the **length of arc** s of the curve from $x = a$ to $x = b$ is given by*

$$s = \int_a^b \sqrt{1 + \left(\frac{dy}{dx}\right)^2}\, dx \qquad (6\text{-}31)$$

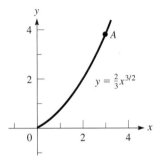

Fig. 6-71

■EXAMPLE 1   Find the length of arc of the curve $y = \frac{2}{3}x^{3/2}$ from $x = 0$ to $x = 3$.

First, we identify $a = 0$ and $b = 3$ (see Fig. 6-71). Next, we see that we are to find the derivative, which is $dy/dx = x^{1/2}$. Using Eq. (6-31), we find the length of arc.

$$s = \int_0^3 \sqrt{1 + (x^{1/2})^2}\, dx = \int_0^3 \sqrt{1 + x}\, dx = \frac{2}{3}(1 + x)^{3/2} \Big|_0^3$$

$$= \frac{2}{3}[4^{3/2} - 1^{3/2}] = \frac{2}{3}(8 - 1) = \frac{14}{3}$$

The distance from O to A, moving along the curve, is 14/3 units.   ■

Area of a Surface of Revolution

The second application of integration in this section is that of find the area of a surface of revolution. *If the arc of a curve is rotated about the x-axis, it generates a surface of revolution from x = a to x = b for which the area is given by*

$$S = 2\pi \int_a^b y \sqrt{1 + \left(\frac{dy}{dx}\right)^2}\, dx \qquad \text{(6-32)}$$

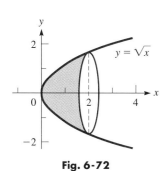

Fig. 6-72

EXAMPLE 2   Find the area of the surface generated by rotating about the x-axis the arc of the parabola $y = \sqrt{x}$ from $x = 0$ to $x = 2$. See Fig. 6-72.

First we note that $a = 0$ and $b = 2$. The derivative of $y = \sqrt{x}$ is $\frac{dy}{dx} = \frac{1}{2\sqrt{x}}$. By substituting in Eq. (6-32), we find the area of the surface.

$$S = 2\pi \int_0^2 x^{1/2} \sqrt{1 + \left(\frac{1}{2x^{1/2}}\right)^2}\, dx = 2\pi \int_0^2 x^{1/2} \sqrt{1 + \frac{1}{4x}}\, dx$$

$$= \pi \int_0^2 \sqrt{4x + 1}\, dx = \frac{\pi}{4}\left(\frac{2}{3}\right)(4x + 1)^{3/2}\Big|_0^2 = \frac{\pi}{6}(9^{3/2} - 1^{3/2})$$

$$= \frac{\pi}{6}(27 - 1) = \frac{13\pi}{3}\ \text{square units}$$

Average Value of a Function

The third application of integration shown in this section is that of the *average value of a function*. In general, *an average is found by summing up the quantities to be averaged and then dividing by the total number of them.* Generalizing on this and using integration for the summation, *the* **average value** *of a function y with respect to x from x = a to x = b is given by*

$$y_{\text{av}} = \frac{\displaystyle\int_a^b y\, dx}{b - a} \qquad \text{(6-33)}$$

SOLVING A WORD PROBLEM

EXAMPLE 3   The velocity v (in ft/s) of an object falling under the influence of gravity as a function of the time t (in s) is given by $v = 32t$. What is the average velocity of the object with respect to time for the first 3.0 s?

In this case, we want the average value of the function v from $t = 0$ to $t = 3.0$ s.

$$v_{\text{av}} = \frac{\displaystyle\int_0^{3.0} v\, dt}{3.0 - 0} \qquad \text{using Eq. (6-33)}$$

$$= \frac{\displaystyle\int_0^{3.0} 32t\, dt}{3.0} = \frac{16t^2}{3.0}\Big|_0^{3.0}$$

$$= 48\ \text{ft/s}$$

This result can be interpreted as meaning that an average velocity of 48 ft/s for 3.0 s would result in the same distance, 144 ft, being traveled by the object as that with the variable velocity. Since $s = \int v\, dt$, the numerator represents the distance traveled.

SOLVING A WORD PROBLEM   ■**EXAMPLE 4**   The power P developed in a certain resistor as a function of the current i is $P = 6i^2$. What is the average power (in W) with respect to the current as the current changes from 2.0 A to 5.0 A?

In this case, we are to find the average value of the function P from $i = 2.0$ A to $i = 5.0$ A. This average value of P is

$$P_{av} = \frac{\int_{2.0}^{5.0} P\, di}{5.0 - 2.0} \qquad \text{using Eq. (6-33)}$$

$$= \frac{6 \int_{2.0}^{5.0} i^2\, di}{3.0} = \frac{2i^3}{3.0}\Big|_{2.0}^{5.0} = \frac{2(125 - 8.0)}{3.0} = 78 \text{ W}$$ ■

In general, it might be noted that the average value of y with respect to x is that value of y which, when multiplied by the length of the interval for x, gives the same area as that under the graph of y as a function of x.

─────────────────── EXERCISES **6-8** ───────────────────

In Exercises 1–4, find the length of arc of each of the given curves.

1. $y = \frac{2}{3}x^{3/2}$ from $x = 3$ to $x = 8$

2. $y = \frac{2}{3}(x - 1)^{3/2}$ from $x = 1$ to $x = 4$

3. $y = \frac{2}{3}(x^2 + 1)^{3/2}$ from $x = 0$ to $x = 3$

4. $y = \frac{1}{6}x^3 + \frac{1}{2x}$ from $x = 1$ to $x = 3$

In Exercises 5–8, find the area generated by rotating the given arc about the x-axis.

5. $y = x$ from $x = 0$ to $x = 2$

6. $y = \frac{4}{3}x$ from $x = 0$ to $x = 3$

7. $y = 2\sqrt{x}$ from $x = 0$ to $x = 3$

8. $y = x^3$ from $x = 1$ to $x = 3$

In Exercises 9–16, solve the given problems.

9. The electric current i (in A) as a function of the time t (in s) for a certain circuit is given by $i = 4t - t^2$. Find the average value of the current with respect to time for the first 4.0 s.

10. The temperature T (in °C) recorded in a city during a given day approximately followed the curve of $T = 0.00100t^4 - 0.280t^2 + 25.0$, where t is the number of hours from noon ($-12 \text{ h} \le t \le 12 \text{ h}$). What was the average temperature during the day?

11. The efficiency e (in %) of an automobile engine is given by $e = 0.768s - 0.00004s^3$, where s is the speed (in km/h) of the car. Find the average efficiency with respect to the speed for $s = 30.0$ km/h to $s = 90.0$ km/h. (See Example 1 of Section 4-7.)

Ⓦ **12.** Find the average value of the volume of a sphere with respect to the radius. Explain the meaning of the result.

13. The cable of a bridge can be described by the equation $y = 0.04x^{3/2}$ from $x = 0$ to $x = 100$ ft. Find the length of the cable. See Fig. 6-73.

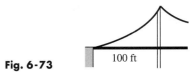

Fig. 6-73   100 ft

14. A rocket takes off in a path described by the equation $y = \frac{2}{3}(x^2 - 1)^{3/2}$. Find the distance traveled by the rocket for $x = 1.0$ km to $x = 3.0$ km.

15. Find the formula for the lateral surface area of a right circular cone of radius r and height h.

16. The grinding surface of a grinding machine can be described as the surface generated by rotating the curve $y = 0.2x^3$ from $x = 0$ to $x = 2.0$ cm about the x axis. Find the grinding surface area.

CHAPTER EQUATIONS

Velocity
$$v = \int a\, dt \tag{6-1}$$
$$v = at + C_1 \tag{6-2}$$

Displacement
$$s = \int v\, dt \tag{6-3}$$

Electric current
$$i = \frac{dq}{dt} \tag{6-4}$$

Electric charge
$$q = \int i\, dt \tag{6-5}$$

Voltage across capacitor
$$V_C = \frac{1}{C} \int i\, dt \tag{6-6}$$

Area
$$A = \int_a^b y\, dx = \int_a^b f(x)\, dx \tag{6-7}$$
$$A = \int_c^d x\, dy = \int_c^d g(y)\, dy \tag{6-8}$$
$$A = \int_a^b (y_2 - y_1)\, dx \tag{6-9}$$
$$A = \int_c^d (x_2 - x_1)\, dy \tag{6-10}$$

Volume
$$V = \pi \int_a^b y^2\, dx = \pi \int_a^b [f(x)]^2\, dx \tag{6-11}$$
$$V = \pi \int_c^d x^2\, dy \tag{6-12}$$

Shell
$$dV = 2\pi(\text{radius}) \times (\text{height}) \times (\text{thickness}) \tag{6-13}$$
Disk
$$dV = \pi(\text{radius})^2 \times (\text{thickness}) \tag{6-14}$$

Center of mass
$$m_1 d_1 + m_2 d_2 + \cdots + m_n d_n = (m_1 + m_2 + \cdots + m_n)\overline{d} \tag{6-15}$$

Centroid of flat plate
$$\overline{x} = \frac{\displaystyle\int_a^b x(y_2 - y_1)\, dx}{\displaystyle\int_a^b (y_2 - y_1)\, dx} \tag{6-16}$$

$$\overline{y} = \frac{\displaystyle\int_c^d y(x_2 - x_1)\, dy}{\displaystyle\int_c^d (x_2 - x_1)\, dy} \tag{6-17}$$

Centroid of solid of revolution

$$\bar{x} = \frac{\int_a^b xy^2 \, dx}{\int_a^b y^2 \, dx} \qquad (6\text{-}18)$$

$$\bar{y} = \frac{\int_c^d yx^2 \, dy}{\int_c^d x^2 \, dy} \qquad (6\text{-}19)$$

Radius of gyration

$$m_1 d_1^2 + m_2 d_2^2 + \cdots + m_n d_n^2 = (m_1 + m_2 + \cdots + m_n)R^2 \qquad (6\text{-}20)$$

Moment of inertia of flat plate

$$I_y = k \int_a^b x^2(y_2 - y_1) \, dx \qquad (6\text{-}21)$$

$$I_x = k \int_c^d y^2(x_2 - x_1) \, dy \qquad (6\text{-}22)$$

Moment of inertia of solid of revolution

$$I_y = 2\pi k \int_a^b (y_2 - y_1)x^3 \, dx \qquad (6\text{-}23)$$

$$I_x = 2\pi k \int_c^d (x_2 - x_1)y^3 \, dy \qquad (6\text{-}24)$$

Work

$$W = \int_a^b f(x) \, dx \qquad (6\text{-}25)$$

Force between electric charges

$$f(x) = \frac{kq_1 q_2}{x^2} \qquad (6\text{-}26)$$

Force due to liquid pressure

$$F = w \int_a^b lh \, dh \qquad (6\text{-}30)$$

Length of arc

$$s = \int_a^b \sqrt{1 + \left(\frac{dy}{dx}\right)^2} \, dx \qquad (6\text{-}31)$$

Area of a surface of revolution

$$S = 2\pi \int_a^b y \sqrt{1 + \left(\frac{dy}{dx}\right)^2} \, dx \qquad (6\text{-}32)$$

Average value

$$y_{av} = \frac{\int_a^b y \, dx}{b - a} \qquad (6\text{-}33)$$

REVIEW EXERCISES

1. How long after it is dropped does a baseball reach a speed of 140 ft/s (the speed of a good fastball)?

2. If the velocity v (in m/s) of a subway train after the brakes are applied can be expressed as $v = \sqrt{400 - 20t}$, where t is the time in seconds, how far does it travel in coming to a stop?

3. A weather balloon is rising at the rate of 20 ft/s when a small metal part drops off. If the balloon is 200 ft high at this instant, when will the part hit the ground?

4. A float is dropped into a river at a point where it is flowing at 5.0 ft/s. How far does the float travel in 30 s if it accelerates downstream at 0.020 ft/s²?

5. The electric current i (in A) in a circuit as a function of the time t (in s) is $i = 0.25(2\sqrt{t} - t)$. Find the total charge to pass a point in the circuit in 2.0 s.

6. The current i (in A) in a certain electric circuit is given by $i = \sqrt{1 + 4t}$, where t is the time (in s). Find the charge that passes a given point from $t = 1.0$ s to $t = 3.0$ s if $q_0 = 0$.

7. The voltage across a 5.5-nF capacitor in an FM radio receiver is zero. What is the voltage after 25 μs if a current of 12 mA charges the capacitor?

8. The initial voltage across a capacitor is zero, and $V_C = 2.50$ V after 8.00 ms. If a current $i = t/\sqrt{t^2 + 1}$, where i is the current (in A) and t is the time (in s), charges the capacitor, find the capacitance C of the capacitor.

9. The distribution of weight on a cable is not uniform. If the slope of the cable at any point is given by $dy/dx = 20 + 0.025x^2$ and if the origin of the coordinate system is at the lowest point, find the equation that gives the curve described by the cable.

10. The time rate of change of the reliability R (in %) of a computer system is $dR/dt = -2.5(0.05t + 1)^{-1.5}$, where t is the time (in h). If $R = 100$ for $t = 0$, find R for $t = 100$ h.

11. Find the area between $y = \sqrt{1 - x}$ and the coordinate axes.

12. Find the area bounded by $y = 3x^2 - x^3$ and the x-axis.

13. Find the area bounded by $y^2 = 2x$ and $y = x - 4$.

14. Find the area bounded by $y = 1/(2x + 1)^2$, $y = 0$, $x = 1$, and $x = 2$.

15. Find the area between $y = x^2$ and $y = x^3 - 2x^2$.

16. Find the area between $y = x^2 + 2$ and $y = 3x^2$.

17. Find the volume of the solid generated by revolving the region bounded by $y = 3 + x^2$ and the line $y = 4$ about the x-axis.

18. Find the volume of the solid generated by revolving the region bounded by $y = 8x - x^4$ and the x-axis about the x-axis.

19. Find the volume of the solid generated by revolving the region bounded by $y = x^3 - 4x^2$ and the x-axis about the y-axis.

20. Find the volume of the solid generated by revolving the region bounded by $y = x$ and $y = 3x - x^2$ about the y-axis.

21. Find the volume generated by revolving an ellipse about its major axis.

22. A hole of radius 1.00 cm is bored along the diameter of a sphere of radius 4.00 cm. Find the volume of the material that is removed from the sphere.

23. Find the centroid of the a plate covering the region bounded by $y^2 = x^3$ and $y = 2x$.

24. Find the centroid of the a plate covering the region bounded by $y = 2x - 4$, $x = 1$, and $y = 0$.

25. Find the centroid of the solid generated by revolving the region bounded by $y = \sqrt{x}$, $x = 1$, $x = 4$, and $y = 0$ about the x-axis.

26. Find the centroid of the solid generated by revolving the region bounded by $yx^4 = 1$, $y = 1$, and $y = 4$ about the y-axis.

27. Find the moment of inertia of a plate covering the region bounded by $y = 3x - x^2$ and $y = x$ with respect to the y-axis.

28. Find the radius of gyration of a plate covering the first-quadrant region bounded by $y = 8 - x^3$ with respect to the y-axis.

29. Find the moment of inertia with respect to its axis of a lead bullet that can be defined by revolving the region bounded by $y = 3.00x^{0.10}$, $x = 0$, $x = 20.0$, and $y = 0$ about the x-axis (all measurements in mm). The density of lead is 0.0114 g/mm^3.

30. Find the radius of gyration with respect to its axis of a rotating machine part that can be defined by revolving the region bounded by $y = 1/x$, $x = 1.00$, $x = 4.00$, and $y = 0.25$ (all measurements in in.) about the x-axis.

31. A pail and its contents weigh 80 lb. The pail is attached to the end of a 100-ft rope that weighs 10 lb and is hanging vertically. How much work is done in winding up the rope with the pail attached?

32. The gravitational force (in lb) of the earth on a satellite (the weight of the satellite) is given by $F = 10^{11}/x^2$, where x is the vertical distance (in mi) from the center of the earth to the satellite. How much work is done in moving the satellite from the earth's surface to an altitude of 2000 mi? The radius of the earth is 3960 mi.

33. The rear stabilizer of a certain aircraft can be described as the region under the curve $y = 3x^2 - x^3$, as shown in Fig. 6-74. Find the x-coordinate (in m) of the centroid of the stabilizer.

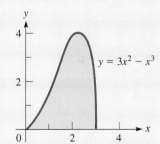

Fig. 6-74

34. The water in a spherical tank 20.0 m in radius is 15.0 m deep at the deepest point. How much water is in the tank?

35. The nose cone of a rocket has the shape of a semiellipse revolved about its major axis, as shown in Fig. 6-75. What is the volume of the nose cone?

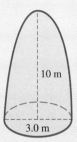

Fig. 6-75

36. The deck area of a boat is a parabolic section as shown in Fig. 6-76. What is the area of the deck?

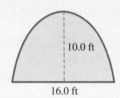

Fig. 6-76   10.0 ft   16.0 ft

37. A cylindrical chemical waste–holding tank 4.50 ft in radius has a depth of 3.25 ft. Find the total force on the circular side of the tank when it is filled with liquid with a density of 68.0 lb/ft^3.

38. A section of a dam is in the shape of a right triangle. The base of the triangle is 6.00 ft and is in the surface of the water. If the triangular section goes to a depth of 4.00 ft, find the force on it. See Fig. 6-77.

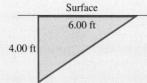

Surface

6.00 ft

4.00 ft

Fig. 6-77

39. The electric resistance of a wire is inversely proportional to the square of its radius. If a certain wire has a resistance of 0.30 Ω when its radius is 2.0 mm, find the average value of the resistance with respect to the radius if the radius changes from 2.0 mm to 2.1 mm.

40. A horizontal straight section of pipe is supported at its center by a vertical wire as shown in Fig. 6-78. Find the formula for the moment of inertia of the pipe with respect to an axis along the wire if the pipe is of length L and mass m.

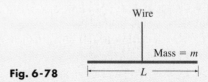

Wire

Mass = m

Fig. 6-78 L

Writing Exercise

41. A tub for holding liquids has a circular top of radius a. All cross sections of the tub that are perpendicular to a fixed diameter of the top are squares. Write one or two paragraphs explaining how to derive the formula that gives the volume of the tub. What is the formula?

PRACTICE TEST

In Problems 1–3, use the region bounded by $y = \frac{1}{4}x^2$, $y = 0$, and $x = 2$.

1. Find the area.

2. Find the coordinates of the centroid of a thin plate covering the given region.

3. Find the volume of the solid if the given region is revolved about the x-axis.

In Problems 4 and 5, use the first-quadrant region bounded by $y = x^2$, $x = 0$, and $y = 9$.

4. Find the volume of the solid if the given region is revolved about the x-axis.

5. Find the moment of inertia of a thin plate covering the region with respect to the y-axis.

6. The velocity v of an object as a function of the time t is $v = 60 - 4t$. Find the expression for the displacement s if $s = 10$ for $t = 0$.

7. The natural length of a spring is 8.0 cm. A force of 12 N stretches it to a length of 10.0 cm. How much work is done in stretching it from a length of 10.0 cm to a length of 14.0 cm?

8. A vertical rectangular floodgate is 6.00 ft wide and 2.00 ft high. Find the force on the gate if its upper edge is 1.00 ft below the surface of the water ($w = 62.4$ lb/ft^3).

DIFFERENTIATION OF THE TRIGONOMETRIC AND INVERSE TRIGONOMETRIC FUNCTIONS

In our development of differentiation and integration in the previous chapters, we have used only algebraic functions. We have not yet used the trigonometric, inverse trigonometric, exponential, or logarithmic functions, for the derivative of each of these is a special form. These functions are the most important of the **transcendental** (nonalgebraic) **functions.** In this chapter we will develop formulas for the derivatives of the trigonometric and the inverse trigonometric functions. In the next chapter we will take up the derivatives of the exponential and logarithmic functions. In later chapters we take up integration that involves these functions.

Many important applications in technology and science are periodic in nature; that is, these phenomena have properties that repeat after definite periods of time. Applications of this type of periodic behavior arise in many fields. Some examples are alternating-current theory (filtering electronic signals in communications), acoustics (mixing musical sounds on a tape in a recording studio), meteorology (studying seasonal temperatures of an area), and optics (studying light waves). The properties of these phenomena may be represented by the trigonometric functions, particularly the sine and cosine functions.

In Section 7-7 we use the derivative of an inverse trigonometric function in analyzing the motion of a rocket.

7-1 THE TRIGONOMETRIC FUNCTIONS

In this section and the next, we present a summary of angle measurement and the trigonometric functions. This summary of the properties of these functions is included for purposes of review and reference. Later, in Section 7-5, we will review the inverse trigonometric functions and their properties. This review material is essential to the development of the derivatives of these functions, which is presented in the other sections of this chapter.

The term *transcendental* was introduced by the Swiss mathematician Leonhard Euler (1707–1783). His works in mathematics and other fields filled about 70 volumes.

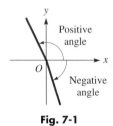

Fig. 7-1

An **angle** *in* **standard position** *is one with its vertex at the origin and its initial side the positive x-axis. If the angle is measured counterclockwise it is* **positive,** *and if it is measured clockwise it is* **negative** *(see Fig. 7-1).*

There are two widely used measurements of angles. These are the **degree** and the **radian**. The degree is used primarily in triangle solution, and the radian is used in other types of applications, such as periodic motion, and when graphing the trigonometric functions.

The **degree** *is defined as* $\frac{1}{360}$ *of a complete revolution.* One method of subdividing a degree is to use decimal parts of a degree. Another traditional way is to divide each degree into 60 equal parts called *minutes,* and each minute into 60 equal parts called *seconds.* The symbols °, ′, ″ are used to designate degrees, minutes, and seconds, respectively. Although most calculators can use angles in degrees, minutes, and seconds, they are programmed to automatically use degrees and decimal parts.

The use of 360 for degree measure comes from the ancient Babylonians and their number system based on 60. However, the specific reason for the choice of 360 is not known. (One theory is that 360 is divisible by many smaller numbers and is close to the number of days in a year.)

A **radian** *is the measure of an angle with its vertex at the center of a circle and with an intercepted arc on the circle equal in length to the radius of the circle.* See Fig. 7-2.

Since the radius may be laid off 2π times along the circumference, it follows that there are 2π radians in one complete rotation. Also, there are 360° in one complete rotation. Therefore, 360° is *equivalent* to 2π radians. It then follows that the relation between degrees and radians is 2π rad $= 360°$, or

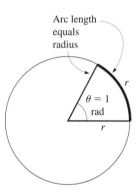

Fig. 7-2

$$\boxed{\pi \text{ rad} = 180°} \tag{7-1}$$

We see from Eq. (7-1) that we convert angle measurements from degrees to radians or from radians to degrees as follows:

> ## Procedure for Converting Angle Measurements
> **1.** *To convert an angle measured in degrees to the same angle measured in* radians, **multiply the number of degrees by $\pi/180°$.**
>
> **2.** *To convert an angle measured in radians to the same angle measured in degrees,* **multiply the number of radians by $180°/\pi$.**

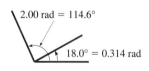

Fig. 7-3

■EXAMPLE 1   converting degrees to radians

(a) $18.0° = \left(\dfrac{\pi}{180°}\right)(18.0°) = \dfrac{\pi}{10.0} = \dfrac{3.14}{10.0} = 0.314$ rad

degrees cancel   (See Fig. 7-3.)

converting radians to degrees

(b) 2.00 rad $= \left(\dfrac{180°}{\pi}\right)(2.00) = \dfrac{360°}{3.14} = 114.6°$ (See Fig. 7-3.)

In multiplying by $\pi/180°$ or $180°/\pi$, we are actually only multiplying by 1 because π rad $= 180°$. Although the unit of measurement is different, *the angle is the same.* See Appendix B on unit conversions. ■

The symbol π (a Greek letter), which we use as a number, was first used in this way as a number in the 1700s.

Due to the nature of the definition of the radian, it is very common to express radians in terms of π. Expressing angles in terms of π is illustrated in the following example.

$\frac{3\pi}{4}$ rad $= 135°$

$45° = \frac{\pi}{4}$ rad

Fig. 7-4

EXAMPLE 2

converting degrees to radians

(a) $45° = \left(\dfrac{\pi}{180°}\right)(45°) = \dfrac{\pi}{4}$ rad   (See Fig. 7-4.)

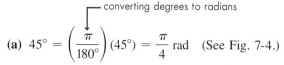

converting radians to degrees

(b) $\dfrac{3\pi}{4}$ rad $= \left(\dfrac{180°}{\pi}\right)\left(\dfrac{3\pi}{4}\right) = 135°$ (See Fig. 7-4.)

We wish now to make a very important point. Since π is a special way of writing the number (slightly greater than 3) that is the ratio of the circumference of a circle to its diameter, it is the ratio of one distance to another. Thus, radians really have no

NOTE ▶ units, and *radian measure amounts to measuring angles in terms of real numbers.* It is this property of radians that makes them useful in many situations. Therefore, when radians are being used, it is customary that no units are indicated for the angle.

CAUTION ▶ *When no units are indicated, the radian is understood to be the unit of angle measurement.*

$3.80 = 218°$

Fig. 7-5

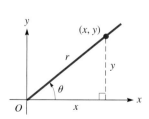

EXAMPLE 3   **(a)** $60° = \left(\dfrac{\pi}{180°}\right)(60.0°) = \dfrac{\pi}{3.00} = 1.05$

no units indicates radian measure

(b) $3.80 = \left(\dfrac{180°}{\pi}\right)(3.80) = 218°$

Since no units are indicated for 1.05 and 3.80 in this example, they are known to be in radian measure. Here, we must know that 3.80 is an angle measure. See Fig. 7-5.

The six **trigonometric functions** of an angle θ in standard position are *sine θ, cosine θ, tangent θ, cotangent θ, secant θ,* and *cosecant θ.* They are defined in terms of the coordinates (x, y) of a point on its terminal side, and r, the distance of the point from the origin. Here, *r is called the* **radius vector,** *and is always* **positive.** The definitions and standard abbreviations of the functions are as follows (see Fig. 7-6):

Fig. 7-6

sine of θ: $\sin \theta = \dfrac{y}{r}$	*cosine of θ:* $\cos \theta = \dfrac{x}{r}$
tangent of θ: $\tan \theta = \dfrac{y}{x}$	*cotangent of θ:* $\cot \theta = \dfrac{x}{y}$
secant of θ: $\sec \theta = \dfrac{r}{x}$	*cosecant of θ:* $\csc \theta = \dfrac{r}{y}$

(7-2)

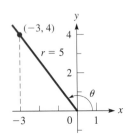

Fig. 7-7

EXAMPLE 4 Find the values of sin θ, cos θ, and tan θ for the angle θ as shown in Fig. 7-7.

We see that the terminal side of θ passes through $(-3, 4)$. Therefore, we find r from the Pythagorean theorem as

$$r = \sqrt{(-3)^2 + 4^2} = \sqrt{25} = 5$$

Now using the values $x = -3$, $y = 4$, and $r = 5$, the values of the functions of θ are

$$\sin \theta = \frac{4}{5} \qquad \cos \theta = -\frac{3}{5} \qquad \tan \theta = -\frac{4}{3}$$

The values of the trigonometric functions sin θ, cos θ, and tan θ are programmed into graphing calculators. To find the values of these functions, we use the keys so marked. Also, from the definitions, we see that csc θ *is the reciprocal of* sin θ, sec θ *is the reciprocal of* cos θ, and cot θ *is the reciprocal of* tan θ. Therefore, the reciprocal key, x^{-1}, is also used in finding values of these reciprocal functions.

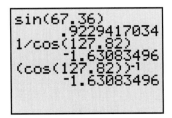

Fig. 7-8

EXAMPLE 5 (a) With a graphing calculator *in degree mode,* to find the value of sin 67.36°, we enter the function and then the angle, just as it is written. The first two lines of the display Fig. 7-8 show that sin 67.36° = 0.9229 (rounded off).

(b) To evaluate sec 127.82°, we enter sec 127.82° as 1/cos 127.82° (see the third and fourth lines of Fig. 7-8), or as (cos 127.82°)$^{-1}$ (see the fifth and sixth lines of Fig. 7-8). Each of these displays shows that sec 127.82° = −1.631 (rounded off).

To find the angle for a given value of the function, we make use of the **inverse trigonometric functions,** which are discussed in detail in Section 7-5. For calculator purposes of the moment, it is sufficient to note the notation that is used. The notation for "the angle whose sine is x" is sin^{-1} x. Similar meanings are given to cos^{-1} x and tan^{-1} x.

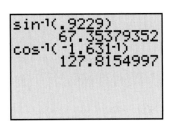

Fig. 7-9

EXAMPLE 6 (a) With a graphing calculator *in degree mode,* if sin θ = 0.9229, we find θ by finding sin^{-1} 0.9229. See the first two lines of the calculator display in Fig. 7-9, which shows that θ = 67.35°. The difference in the angle from that in Example 5 is due to rounded-off values.

(b) If sec θ = −1.631, we find θ by finding cos^{-1}(−1.631^{-1}). In the third and fourth lines of Fig. 7-9, we see that θ = 127.82° (rounded off).

In Examples 5 and 6 we used the calculator in degree mode. The calculator can also be used when the angle is expressed in radians by using radian mode. When the calculator is in radian mode, it uses values in radians directly. Therefore, when evaluating trigonometric functions on a calculator, *always be careful to have the calculator in the proper mode.*

From the definitions, we can see that the signs of the trigonometric functions differ depending on the signs of x and y. This means that the signs depend on the quadrant in which the terminal side lies. It is important that we properly analyze an angle and the sign of its function, for even

CAUTION ▶ *the calculator will not necessarily give us directly the required angle*

for a given value of a function. However, since the angle in standard position is coterminal with some positive angle less than 360°, we need consider only positive angles less than 360° in order to determine the appropriate sign.

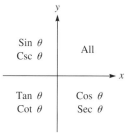

Sin θ | All
Csc θ |

Tan θ | Cos θ
Cot θ | Sec θ

Positive functions

Fig. 7-10

From the definitions, we see that the signs of the trigonometric functions in the four quadrants are as follows:

All functions of first-quadrant angles are positive. Sin θ and csc θ are positive for second-quadrant angles. Tan θ and cot θ are positive for third-quadrant angles. Cos θ and sec θ are positive for fourth-quadrant angles. All others are negative.

This is shown in Fig. 7-10.

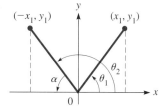

Fig. 7-11

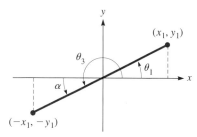

Fig. 7-12

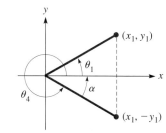

Fig. 7-13

By use of congruent triangles, it can be seen from Figs. 7-11, 7-12, and 7-13 that $\alpha = \theta_1$, and therefore values of functions of α are the same as those of θ_1. Also, the *absolute values* of functions of θ_2, θ_3, and θ_4 are the same as those of α for the respective quadrants. Therefore,

$$
\begin{aligned}
F(\theta_2) &= \pm F(\alpha), & \alpha &= 180° - \theta_2 & \theta_2 \text{ in second} \\
& & &= \pi - \theta_2, & \text{quadrant} \\
F(\theta_3) &= \pm F(\alpha), & \alpha &= \theta_3 - 180° & \theta_3 \text{ in third} \\
& & &= \theta_3 - \pi, & \text{quadrant} \\
F(\theta_4) &= \pm F(\alpha), & \alpha &= 360° - \theta_4 & \theta_4 \text{ in fourth} \\
& & &= 2\pi - \theta_4, & \text{quadrant}
\end{aligned}
$$

(7-3)

Here, F represents any of the trigonometric functions, and α *is the* **reference angle**, *which is the acute angle formed by the terminal side of the angle and the x-axis.* The sign to be used depends on the sign of the function in the given quadrant. By using the reference angle, we can express the function of any angle in terms of the same function of an acute angle.

EXAMPLE 7 Illustrations using Eqs. (7-3) follow:

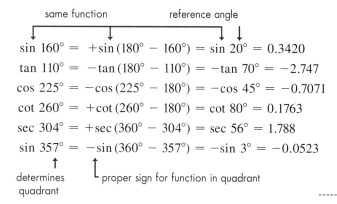

same function   reference angle

$$\sin 160° = +\sin (180° - 160°) = \sin 20° = 0.3420$$
$$\tan 110° = -\tan (180° - 110°) = -\tan 70° = -2.747$$
$$\cos 225° = -\cos (225° - 180°) = -\cos 45° = -0.7071$$
$$\cot 260° = +\cot (260° - 180°) = \cot 80° = 0.1763$$
$$\sec 304° = +\sec (360° - 304°) = \sec 56° = 1.788$$
$$\sin 357° = -\sin (360° - 357°) = -\sin 3° = -0.0523$$

determines   proper sign for function in quadrant
quadrant

QUADRANTAL ANGLES

Using Eqs. (7-3) we may find the value of any function when the terminal side of the angle lies *in* one of the quadrants. In the following table we show the values of the functions of *an angle for which the terminal side is along one of the axes, a* **quadrantal angle**. The values are found using the definitions, recalling that $r > 0$.

θ	$\sin \theta$	$\cos \theta$	$\tan \theta$	$\cot \theta$	$\sec \theta$	$\csc \theta$
0°	0.000	1.000	0.000	undef.	1.000	undef.
90°	1.000	0.000	undef.	0.000	undef.	1.000
180°	0.000	−1.000	0.000	undef.	−1.000	undef.
270°	−1.000	0.000	undef.	0.000	undef.	−1.000
360°	Same as the functions of 0° (same terminal side)					

Graphs of the Sine and Cosine Functions

One of the clearest ways to demonstrate the variation in the trigonometric functions is by means of their graphs. Since the graphs of the sine and cosine functions are of primary importance, we shall restrict our attention here to them.

In graphing the trigonometric functions, it is normal to **express the angle in radians.** In this way both x and y are expressed as numbers. Following are the important quantities related to their graphs.

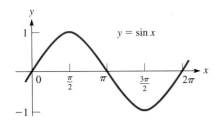

Fig. 7-14

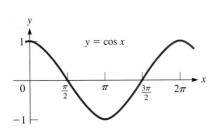

Fig. 7-15

> **Important Quantities to Determine for Sketching Graphs of**
> $y = a \sin(bx + c)$ **and** $y = a \cos(bx + c)$
>
> Amplitude $= |a|$
>
> Period $= \dfrac{2\pi}{b}$
>
> Displacement $= -\dfrac{c}{b}$

The **amplitude** *is the maximum number of units the curve may be from the x-axis. A function has a* **period** P ($P > 0$) *if, for all x,* $F(x) = F(x + P)$. *The* **displacement** *(or* **phase shift***) shifts the graph of* $y = a \sin bx$ *or* $y = a \cos bx$ *to the left if* $c > 0$, *or shifts it to the right if* $c < 0$. *This means that the displacement is negative if* $c > 0$, *and the displacement is positive if* $c < 0$.

By knowing the basic shape of the sine curve and the cosine curve, which are shown in Figs. 7-14 and 7-15, and using the information obtained from the equation, we may sketch the graphs of the sine and cosine functions. One other useful value is the *one-fourth period distance.* The following examples illustrate the method.

EXAMPLE 8   Sketch the graph of the function $y = -\cos(2x + \frac{\pi}{6})$.

First we determine that

(1) the amplitude is 1

(2) the period is $\frac{2\pi}{2} = \pi$

(3) the displacement is $-\frac{\pi}{6} \div 2 = -\frac{\pi}{12}$

We now make a table of important values, noting that the curve starts repeating π units to the right of $-\frac{\pi}{12}$. See Fig. 7-16.

x	$-\frac{\pi}{12}$	$\frac{\pi}{6}$	$\frac{5\pi}{12}$	$\frac{2\pi}{3}$	$\frac{11\pi}{12}$
y	−1	0	1	0	−1

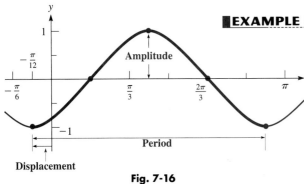

Fig. 7-16

See Appendix C for a graphing calculator program SINECURV. It displays the graphs of $y = \sin x$, $y = 2 \sin x$, $y = \sin 2x$, and $y = 2 \sin(2x - \pi/3)$.

In Example 8, we note that the effect of the negative sign is to invert the curve shown in Fig. 7-16 from that shown in Fig. 7-15. Also, the heavy portion of the graph in Fig. 7-16 is called a *cycle* of the curve. A **cycle** is *any section of the graph that includes exactly one period.* Following now is an example that illustrates the graph of a trigonometric function in an applied problem and that displays the graph on a graphing calculator.

EXAMPLE 9   The cross section of a certain water wave is $y = 0.7 \sin(\frac{\pi}{2}x + \frac{\pi}{4})$, where x and y are measured in feet. Display two cycles of y vs. x on a graphing calculator.

From the values $a = 0.7$ ft, $b = \pi/2$ ft^{-1} (this means 1/ft, or per foot), and $c = \pi/4$, we can find the amplitude, period, and displacement:

(1) amplitude $= 0.7$ ft

(2) period $= \dfrac{2\pi}{\frac{\pi}{2}} = 4$ ft

(3) displacement $= -\dfrac{\frac{\pi}{4}}{\frac{\pi}{2}} = -0.5$ ft

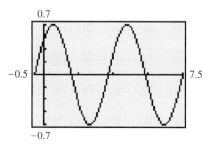

Fig. 7-17

Using these values, we choose the following values for the *window* settings.

(1) Xmin $= -0.5$ (the displacement is -0.5 ft)

(2) Xmax $= 7.5$ (the period is 4 ft, and we want two periods, starting at $x = -0.5$); $(-0.5 + 8 = 7.5)$

(3) Ymin $= -0.7$, Ymax $= 0.7$ (the amplitude is 0.7 ft)

The graphing calculator view is shown in Fig. 7-17. The negative values of x have the significance of giving points to the wave to the left of the origin. (When *time* is used, no actual physical meaning is generally given to negative values of t.)

EXERCISES *7-1*

In Exercises 1 and 2, change the given angles to equal angles expressed in decimal form. In Exercises 3 and 4, change the given angles to equal angles expressed to the nearest minute.

1. 15°12′

2. 301°16′

3. 315.8°

4. 24.92°

In Exercises 5–8, express the given angle measurements in terms of π.

5. 15°, 150°

6. 12°, 225°

7. 75°, 330°

8. 36°, 315°

In Exercises 9–12, the given numbers express angle measurement. Express the measure of each angle in terms of degrees.

9. $\dfrac{2\pi}{5}, \dfrac{3\pi}{2}$

10. $\dfrac{3\pi}{10}, \dfrac{5\pi}{6}$

11. $\dfrac{\pi}{18}, \dfrac{7\pi}{4}$

12. $\dfrac{7\pi}{15}, \dfrac{4\pi}{3}$

In Exercises 13 and 14, express the given angles in radian measure. In Exercises 15 and 16, express the given numbers in terms of angle measure to the nearest 0.1°.

13. 23.0°

14. 178.5°

15. 0.750

16. 16.4

In Exercises 17 through 28, find the values of the indicated trigonometric functions.

17. cos 214°

18. sin 286°

19. csc 137°

20. sec 476°

21. $\sin \dfrac{\pi}{4}$

22. $\cos \dfrac{\pi}{6}$

23. $\tan \dfrac{5\pi}{12}$

24. $\sin \dfrac{7\pi}{18}$

25. $\cot \dfrac{5\pi}{6}$

26. $\tan \dfrac{4\pi}{3}$

27. cos 4.596

28. cot 3.27

In Exercises 29–36, find θ for $0 \le \theta < 2\pi$.

29. $\sin \theta = 0.3090$

30. $\cos \theta = -0.9135$

31. $\tan \theta = -0.2126$

32. $\sin \theta = -0.0436$

33. $\cos \theta = 0.6742$

34. $\cot \theta = 1.860$

35. $\sec \theta = -1.307$

36. $\csc \theta = 3.940$

In Exercises 37–44, sketch the graph of each of the given functions. Check each on a graphing calculator.

37. $y = 3 \sin x$

38. $y = 2 \cos 2x$

39. $y = -4 \sin 3x$

40. $y = 6 \cos \pi x$

41. $y = 2 \sin\left(3x - \dfrac{\pi}{2}\right)$

42. $y = 4 \cos\left(\dfrac{x}{2} + \dfrac{\pi}{8}\right)$

43. $y = \cos\left(\pi x - \dfrac{\pi}{2}\right)$

44. $y = \sin(\pi x + 1)$

In Exercises 45 and 46, sketch the indicated curves. In Exercises 47 and 48, use a graphing calculator to view the indicated curves.

45. A wave traveling in a string may be represented by the equation $y = A \sin 2\pi\left(\dfrac{t}{T} - \dfrac{x}{\lambda}\right)$. Here, A is the amplitude, t is the time the wave has traveled, x is the distance from the origin, T is the time required for the wave to travel one *wavelength* λ (the Greek letter lambda). Sketch three cycles of the wave for which $A = 2.00$ cm, $T = 0.100$ s, $\lambda = 20.0$ cm, and $x = 5.00$ cm.

46. The electric current i (in μA) in a certain circuit is given by $i = 3.8 \cos 2\pi (t + 0.20)$, where t is the time in seconds. Sketch three cycles of this function.

47. A certain satellite circles the earth such that its distance y, in miles north or south (altitude is not considered) from the equator, is $y = 4500 \cos (0.025t - 0.25)$, where t is the time (in min) after launch. View two cycles of the graph.

48. In performing a test on a patient, a medical technician used an ultrasonic signal given by the equation $I = A \sin (\omega t + \theta)$. View two cycles of the graph of I vs. t if $A = 5$ nW/m^2, $\omega = 2 \times 10^5$ rad/s, and $\theta = 0.4$.

(W) *In Exercises 49–52, give the specific form of the equation by evaluating a, b, and c through an inspection of the given curve. Explain how a, b, and c are found.*

49. $y = a \sin (bx + c)$
Fig. 7-18

50. $y = a \cos (bx + c)$
Fig. 7-18

51. $y = a \cos (bx + c)$
Fig. 7-19

52. $y = a \sin (bx + c)$
Fig. 7-19

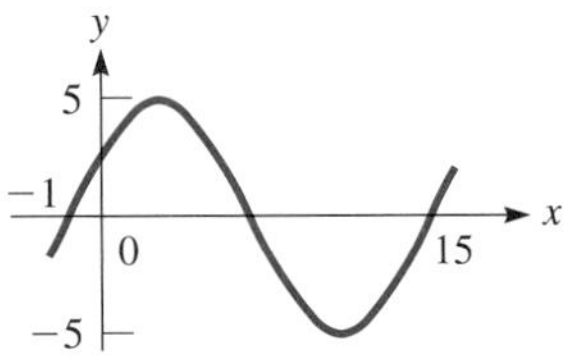

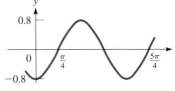

Fig. 7-18   **Fig. 7-19**

7-2   BASIC TRIGONOMETRIC RELATIONS

Referring to the definitions of the trigonometric functions, Eqs. (7-2), we see that a number of relations exist among the functions. Certain other important relationships may also be established. We shall find that these have very definite application, since certain problems rely on a change of form for solution. The ability to effectively use these basic relations depends to a large extent on being familiar with them and recognizing them in different forms. Therefore, in this section we shall review the important trigonometric relations.

Directly from the definitions we observe that

For reference, Eqs. (7-2) are

$$\sin \theta = \frac{y}{r} \quad \cos \theta = \frac{x}{r} \quad \tan \theta = \frac{y}{x}$$

$$\cot \theta = \frac{x}{y} \quad \sec \theta = \frac{r}{x} \quad \csc \theta = \frac{r}{y}$$

$$\boxed{\sin \theta = \frac{1}{\csc \theta} \qquad \cos \theta = \frac{1}{\sec \theta} \qquad \tan \theta = \frac{1}{\cot \theta}} \qquad (7\text{-}4)$$

Also, since $\tan \theta = \dfrac{y}{x} = \dfrac{y/r}{x/r}$, we have

$$\boxed{\tan \theta = \frac{\sin \theta}{\cos \theta} \qquad \cot \theta = \frac{\cos \theta}{\sin \theta}} \qquad (7\text{-}5)$$

Certain important relations can be found by use of the definitions and the Pythagorean relation $x^2 + y^2 = r^2$. By dividing each term by r^2, we obtain Eq. (7-6). By dividing each term of the Pythagorean relation by x^2, we obtain Eq. (7-7). By dividing each term of the Pythagorean relation by y^2, we obtain Eq. (7-8).

$$\sin^2 \theta + \cos^2 \theta = 1 \tag{7-6}$$
$$1 + \tan^2 \theta = \sec^2 \theta \tag{7-7}$$
$$1 + \cot^2 \theta = \csc^2 \theta \tag{7-8}$$

Many other relationships that exist among the trigonometric functions may be proven valid by the basic identities (7-4) to (7-8). The following example illustrates the method of proving such trigonometric identities.

■EXAMPLE 1   Prove the identity $\sec^2 x + \csc^2 x = \sec^2 x \csc^2 x$.

Here we note the presence of $\sec^2 x$ and $\csc^2 x$ on each side. This suggests the possible use of the square relationships. By replacing the $\sec^2 x$ on the right-hand side by $1 + \tan^2 x$, we can create $\csc^2 x$ plus another term. The left-hand side is the $\csc^2 x$ plus another term, so this procedure should help. Thus, changing only the right side,

$$\sec^2 x + \csc^2 x = \sec^2 x \csc^2 x$$
$$= (1 + \tan^2 x)(\csc^2 x) \qquad \text{using Eq. (7-7)}$$
$$= \csc^2 x + \tan^2 x \csc^2 x \qquad \text{multiplying}$$
$$= \csc^2 x + \left(\frac{\sin^2 x}{\cos^2 x}\right)\left(\frac{1}{\sin^2 x}\right) \qquad \text{using Eqs. (7-5) and (7-4)}$$
$$= \csc^2 x + \frac{1}{\cos^2 x} \qquad \text{cancel } \sin^2 x$$
$$= \csc^2 x + \sec^2 x \qquad \text{using Eq. (7-4)}$$

We could have used many other variations of this procedure, and they would have been perfectly valid.   ━━━━━━■

Referring to Fig. 7-20 and the definitions of the functions we see, for example, that $\sin \theta = y/r$ and $\sin(-\theta) = -y/r$. In this way we arrive at the important relations concerning negative angles.

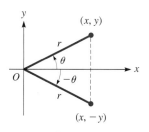

Fig. 7-20

$$\sin(-\theta) = -\sin \theta \qquad \cos(-\theta) = \cos \theta \tag{7-9}$$
$$\tan(-\theta) = -\tan \theta$$

Also, from the definitions we can see that $\cot(-\theta) = -\cot \theta$, $\sec(-\theta) = \sec \theta$, and $\csc(-\theta) = -\csc \theta$.

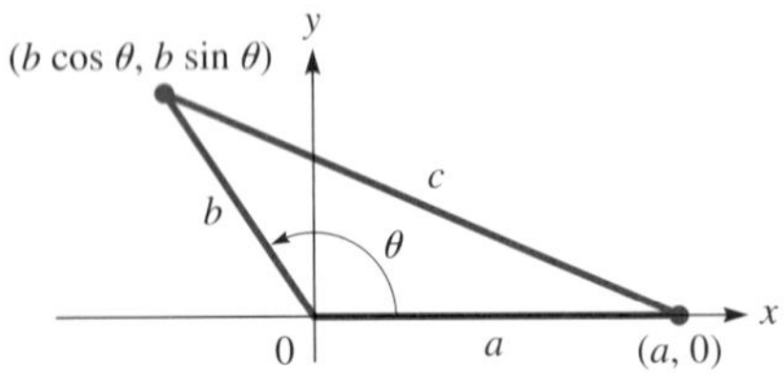

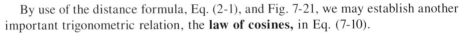

By use of the distance formula, Eq. (2-1), and Fig. 7-21, we may establish another important trigonometric relation, the **law of cosines,** in Eq. (7-10).

$$c^2 = (b \cos \theta - a)^2 + (b \sin \theta)^2$$
$$= b^2 \cos^2\theta - 2ab \cos \theta + a^2 + b^2 \sin^2\theta$$
$$= b^2(\cos^2\theta + \sin^2\theta) + a^2 - 2ab \cos \theta$$

$$\boxed{c^2 = a^2 + b^2 - 2ab \cos \theta} \qquad \textbf{(7-10)}$$

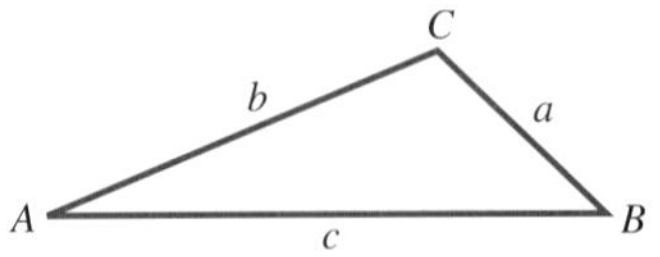

Another important relation between the angles and sides of a triangle, as shown in Fig. 7-22, is the **law of sines.** It is given, without proof, in Eq. (7-11).

$$\boxed{\frac{a}{\sin A} = \frac{b}{\sin B} = \frac{c}{\sin C}} \qquad \textbf{(7-11)}$$

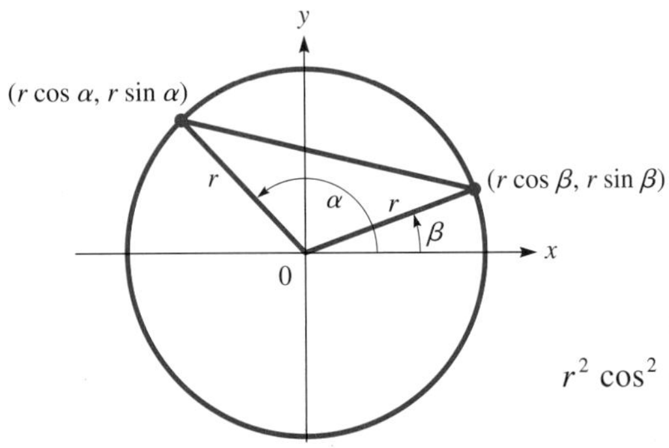

Fig. 7-22

We shall now develop a number of important equations, using the distance formula, the law of cosines, and some of the other relations already established. Also, some of those we derive will be in turn used to derive others. Since there are numerous uses of the various equations of the section in the following material, you will gain facility in the use of the relationships by following the development.

From Fig. 7-23, we have

$$(r \cos \alpha - r \cos \beta)^2 + (r \sin \alpha - r \sin \beta)^2 = r^2 + r^2 - 2(r)(r) \cos(\alpha - \beta)$$
$$r^2 \cos^2 \alpha - 2r^2 \cos \alpha \cos \beta + r^2 \cos^2 \beta + r^2 \sin^2 \alpha - 2r^2 \sin \alpha \sin \beta + r^2 \sin^2 \beta$$
$$= 2r^2 - 2r^2 \cos(\alpha - \beta)$$
$$\cos^2 \alpha + \sin^2 \alpha - 2(\cos \alpha \cos \beta + \sin \alpha \sin \beta) + \cos^2 \beta + \sin^2 \beta = 2 - 2 \cos(\alpha - \beta)$$

Fig. 7-23

Rearranging terms after using Eq. (7-6) we obtain

$$\boxed{\cos(\alpha - \beta) = \cos \alpha \cos \beta + \sin \alpha \sin \beta} \qquad \textbf{(7-12)}$$

Equation (7-12) is one of the important **addition formulas,** the remainder of which we shall now develop.

Writing $\cos(\alpha + \beta) = \cos[\alpha - (-\beta)]$, and using Eqs. (7-9) and (7-12), we have

$$\boxed{\cos (\alpha + \beta) = \cos \alpha \cos \beta - \sin \alpha \sin \beta} \qquad \textbf{(7-13)}$$

In Eq. (7-12), if we let $\alpha = 90°$, the result is

$$\cos(90° - \beta) = \cos 90° \cos \beta + \sin 90° \sin \beta$$

Since $\sin 90° = 1$, $\cos 90° = 0$, we obtain

$$\boxed{\cos(90° - \beta) = \sin \beta} \qquad \textbf{(7-14)}$$

Since β is a general representation of an angle, and $\cos \beta$ is a function of β, we may substitute other expressions for angles for β. Therefore, replacing β by $90° - \beta$ in Eq. (7-14), we have

$$\cos[90° - (90° - \beta)] = \sin(90° - \beta)$$

or

$$\sin(90° - \beta) = \cos \beta \qquad\qquad (7\text{-}15)$$

Now, replacing β by $\alpha + \beta$ in Eq. (7-14), we obtain

$$\sin(\alpha + \beta) = \cos[(90° - \alpha) - \beta]$$
$$= \cos(90° - \alpha) \cos \beta + \sin(90° - \alpha) \sin \beta$$

and my using Eqs. (7-14) and (7-15), we arrive at the equation

$$\sin(\alpha + \beta) = \sin \alpha \cos \beta + \cos \alpha \sin \beta \qquad\qquad (7\text{-}16)$$

If we replace β by $-\beta$ in Eq. (7-16) we have

$$\sin(\alpha - \beta) = \sin \alpha \cos \beta - \cos \alpha \sin \beta \qquad\qquad (7\text{-}17)$$

By dividing the right side of Eq. (7-16) by that of Eq. (7-13), we can determine an expression for $\tan(\alpha + \beta)$; by dividing the right side of Eq. (7-17) by that of Eq. (7-12), we can determine an expression for $\tan(\alpha - \beta)$. These formulas can be written together, as

$$\tan(\alpha \pm \beta) = \frac{\tan \alpha \pm \tan \beta}{1 \mp \tan \alpha \tan \beta} \qquad\qquad (7\text{-}18)$$

The formula for $\tan(\alpha + \beta)$ uses the upper signs, and the formula for $\tan(\alpha - \beta)$ uses the lower signs.

■EXAMPLE 2   Show that $\sin\left(\dfrac{\pi}{4} + x\right) \cos\left(\dfrac{\pi}{4} + x\right) = \dfrac{1}{2}(\cos^2 x - \sin^2 x)$.

The solution is as follows:

$$\sin\left(\frac{\pi}{4} + x\right)\cos\left(\frac{\pi}{4} + x\right) = \left(\sin \frac{\pi}{4} \cos x + \cos \frac{\pi}{4} \sin x\right)\left(\cos \frac{\pi}{4} \cos x - \sin \frac{\pi}{4} \sin x\right) \quad \text{using Eqs. (7-16) and (7-13)}$$

$$= \sin \frac{\pi}{4} \cos \frac{\pi}{4} \cos^2 x - \sin^2 \frac{\pi}{4} \sin x \cos x + \cos^2 \frac{\pi}{4} \sin x \cos x - \sin^2 x \sin \frac{\pi}{4} \cos \frac{\pi}{4} \quad \text{expanding}$$

$$= \frac{\sqrt{2}}{2}\frac{\sqrt{2}}{2} \cos^2 x - \left(\frac{\sqrt{2}}{2}\right)^2 \sin x \cos x + \left(\frac{\sqrt{2}}{2}\right)^2 \sin x \cos x - \frac{\sqrt{2}}{2}\frac{\sqrt{2}}{2} \sin^2 x \quad \text{evaluating}$$

$$= \frac{1}{2} \cos^2 x - \frac{1}{2} \sin^2 x = \frac{1}{2}(\cos^2 x - \sin^2 x)$$

Letting $A = \alpha + \beta$ and $B = \alpha - \beta$ leads to $\alpha = \frac{1}{2}(A + B)$ and $\beta = \frac{1}{2}(A - B)$. Using these in Eqs. (7-16) and (7-17) and subtracting, we obtain

$$\sin A - \sin B = 2 \cos \frac{A + B}{2} \sin \frac{A - B}{2} \qquad \text{(7-19)}$$

We will find this formula of use in the following section.

By letting $\alpha = \beta$ in Eqs. (7-16), (7-13), and (7-18) and simplifying the results, we have the following **double-angle formulas.**

$$\sin 2\alpha = 2 \sin \alpha \cos \alpha \qquad \text{(7-20)}$$
$$\cos 2\alpha = \cos^2 \alpha - \sin^2 \alpha \qquad \text{(7-21)}$$
$$= 2 \cos^2 \alpha - 1 \qquad \text{(7-22)}$$
$$= 1 - 2 \sin^2 \alpha \qquad \text{(7-23)}$$
$$\tan 2\alpha = \frac{2 \tan \alpha}{1 - \tan^2 \alpha} \qquad \text{(7-24)}$$

Eq. (7-6) is used to determine the form of Eqs. (7-22) and (7-23).

By replacing 2α by α in Eqs. (7-22) and (7-23) leads to the **half-angle formulas.**

$$\sin \frac{\alpha}{2} = \pm \sqrt{\frac{1 - \cos \alpha}{2}} \qquad \text{(7-25)}$$
$$\cos \frac{\alpha}{2} = \pm \sqrt{\frac{1 + \cos \alpha}{2}} \qquad \text{(7-26)}$$

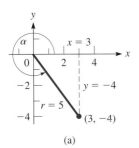

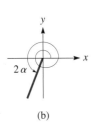

Fig. 7-24

In each of Eqs. (7-25) and (7-26), the sign chosen from the $\pm$ sign depends on the quadrant in which $\alpha/2$ lies.

EXAMPLE 3   Knowing that $\cos \alpha = 3/5$ for an angle in the fourth quadrant, we then determine from Fig. 7-24(a) that $\sin \alpha = -4/5$. Therefore, we have

$$\sin 2\alpha = 2 \sin \alpha \cos \alpha \qquad \text{Eq. (7-20)}$$
$$= 2\left(-\frac{4}{5}\right)\left(\frac{3}{5}\right) = -\frac{24}{25}$$

In Fig. 7-24(b) the angle 2α is shown. It is a third-quadrant angle, which verifies the sign of the result. (Since $\cos \alpha = 3/5$, $\alpha \approx 307°$ and $2\alpha \approx 614°$, which is a third-quadrant angle.)

Two useful relations are found from the definition of a radian. From geometry we know that *the length of an arc of a circle is proportional to the central angle* and that the length of arc for a complete circle is the circumference. Letting s represent the length of arc, we have $s = 2\pi r$ for a complete circle. Thus, *the length of a circular arc with cental angle θ, measured in radians is* (see Fig. 7-25)

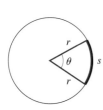

Fig. 7-25

$$s = \theta r \quad (\theta \text{ in radians}) \qquad \text{(7-27)}$$

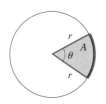

Fig. 7-26

Another geometric fact is that *the area of a circular sector is proportional to the central angle.* The area of a complete circle is $A = \pi r^2$, which can be written in the form $A = \frac{1}{2}(2\pi)r^2$. Since the central angle for a complete circle is 2π, *the area of a sector of a circle in terms of the radius r and the central angle θ, in radians, is*

$$A = \frac{1}{2}\theta r^2 \quad (\theta \text{ in radians}) \qquad (7\text{-}28)$$

See Fig. 7-26.

■EXAMPLE 4 **(a)** The area of a sector of a circle with central angle 218° and a radius of 5.25 in. (see Fig. 7-27(a)) is

$$\overbrace{\qquad\qquad}^{\theta \text{ in radians}}$$
$$A = \frac{1}{2}(218)\left(\frac{\pi}{180}\right)(5.25)^2 = 52.4 \text{ in.}^2$$

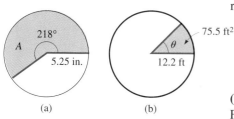

(a)   (b)

Fig. 7-27

(b) Given that the area of a sector is 75.5 ft² and the radius is 12.2 ft (see Fig. 7-27(b)), we find the central angle by solving for θ and then substituting.

$$\theta = \frac{2A}{r^2} = \frac{2(75.5)}{(12.2)^2} = 1.01 \overset{\text{no units indicates radian measure}}{}$$

This means that the central angle is 1.01 rad, or 57.9°.

For clarity of denoting angle measurement, the symbol rad is used at times to indicate radians, rather than show no units.

EXERCISES $7\text{-}2$

In Exercises 1–24, prove the given trigonometric identities.

1. $\dfrac{\cot x}{\cos x} = \csc x$

2. $\dfrac{\sin x}{\tan x} = \cos x$

3. $\tan y \csc y = \sec y$

4. $\sin x \cot x = \cos x$

5. $\cos^2 x - \sin^2 x = 1 - 2\sin^2 x$

6. $\cos^2 \theta(1 + \tan^2 \theta) = 1$

7. $\sin x \tan x + \cos x = \sec x$

8. $\tan x + \cot x = \sec x \csc x$

9. $\tan x(\tan x + \cot x) = \sec^2 x$

10. $\tan^2 y \sec^2 y - \tan^4 y = \tan^2 y$

11. $\dfrac{\sec \theta}{\cos \theta} - \dfrac{\tan \theta}{\cot \theta} = 1$

12. $\dfrac{\csc \theta}{\sin \theta} - \dfrac{\cot \theta}{\tan \theta} = 1$

13. $\sin (x + y) \sin (x - y) = \sin^2 x - \sin^2 y$

14. $\cos (x + y) \cos (x - y) = \cos^2 x - \sin^2 y$

15. $\cos (x - y) + \sin (x + y) = (\cos x + \sin x)(\cos y + \sin y)$

16. $2 \sin x + \sin 2x = \dfrac{2 \sin^3 x}{1 - \cos x}$

17. $\cos^4 x - \sin^4 x = \cos 2x$

18. $2 \csc 2x \tan x = \sec^2 x$

19. $\dfrac{\sin 3x}{\sin x} + \dfrac{\cos 3x}{\cos x} = 4 \cos 2x$

20. $\dfrac{\sin 3x}{\sin x} - \dfrac{\cos 3x}{\cos x} = 2$

21. $\dfrac{1 - \cos \alpha}{2 \sin(\alpha/2)} = \sin \dfrac{\alpha}{2}$

22. $\tan \dfrac{\alpha}{2} = \pm \sqrt{\dfrac{1 - \cos \alpha}{1 + \cos \alpha}}$

23. $2 \sin^2 \dfrac{x}{2} + \cos x = 1$

24. $2 \cos^2 \dfrac{\theta}{2} \sec \theta = \sec \theta + 1$

In Exercises 25–36, solve the given problems.

25. When designing a solar energy collector, it is necessary to account for the latitude and longitude of the location, the angle of the sun, and the angle of the collector. In doing this, the equation

$$\cos \theta = \cos A \cos B \cos C + \sin A \sin B$$

is used. If $\theta = 90°$, show that $\cos C = -\tan A \tan B$.

26. In studying the gravitational force between two objects, the expression $(r - R \cos \theta)^2 + (R \sin \theta)^2$ occurs. Show that this expression can be written as $r^2 - 2rR \cos \theta + R^2$.

27. An alternating electric current i is given by the equation $i = i_0 \sin(\omega t + \alpha)$. Show that this can be written as $i = i_1 \sin \omega t + i_2 \cos \omega t$, where $i_1 = i_0 \cos \alpha$ and $i_2 = i_0 \sin \alpha$.

28. A weight w is held in equilibrium by forces F, directed below the positive x-axis at angle θ, and T, directed to the left of the positive y-axis at angle α. The equations of equilibrium are $F \cos \theta = T \sin \alpha$ and $w + F \sin \theta = T \cos \alpha$. Show that
$$w = \frac{T \cos(\theta + \alpha)}{\cos \theta}.$$

29. The instantaneous electric power p in an inductor is given by the equation $p = vi \sin \omega t \sin(\omega t - \pi/2)$. Show that this equation can be written as $p = -\frac{1}{2} vi \sin 2\omega t$.

30. In the study of the stress at a point in a bar, the equation $s = a \cos^2 \theta + b \sin^2 \theta - 2t \sin \theta \cos \theta$ arises. Show that this equation can be written as
$s = \frac{1}{2}(a + b) + \frac{1}{2}(a - b)\cos 2\theta - t \sin 2\theta$.

31. In electronics, in order to find the *root-mean-square current* in a circuit, it is necessary to express $\sin^2 \omega t$ in terms of $\cos 2\omega t$. Show how this is done.

32. In studying interference patterns of radio signals, the expression $2E^2 - 2E^2 \cos(\pi - \theta)$ arises. Show that this can be written as $4E^2 \cos^2(\theta/2)$.

33. Two streets meet at an angle of 82.0°. What is the length of the piece of curved curbing at the intersection if it is constructed along the arc of a circle 15.0 ft in radius? See Fig. 7-28.

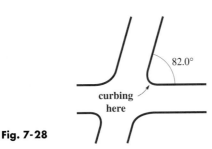

Fig. 7-28

34. Through what angle does the drum in Fig. 7-29 turn in order to lower the crate 18.5 ft?

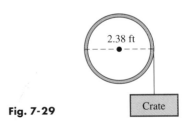

Fig. 7-29

35. What is the floor area of the hallway shown in Fig. 7-30? The outside and inside of the hallway are circular arcs.

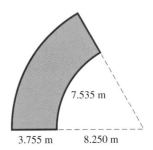

Fig. 7-30   3.755 m   8.250 m

36. Two equal beams of light illuminate the area shown in Fig. 7-31. What area is lit by both beams?

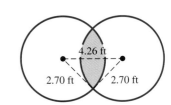

Fig. 7-31

7-3 DERIVATIVES OF THE SINE AND COSINE FUNCTIONS

If we find the derivative of the sine function, we can use it to find the derivatives of the other trigonometric and inverse trigonometric functions. Therefore, we now find the derivative of the sine function by using the delta-process.

Let $y = \sin u$, where u is a function of x and is expressed in radians. If u changes by Δu, y then changes by Δy. Thus,

$$y + \Delta y = \sin (u + \Delta u)$$
$$\Delta y = \sin (u + \Delta u) - \sin u$$
$$\frac{\Delta y}{\Delta u} = \frac{\sin (u + \Delta u) - \sin u}{\Delta u}$$

For reference, Eq. (7-19) is
$\sin A - \sin B = 2 \sin \frac{1}{2}(A - B)\cos \frac{1}{2}(A + B)$.

Referring now to Eq. (7-19), we have

$$\frac{\Delta y}{\Delta u} = \frac{2 \sin \frac{1}{2}(u + \Delta u - u)\cos \frac{1}{2}(u + \Delta u + u)}{\Delta u}$$

$$= \frac{\sin(\Delta u/2)\cos[u + (\Delta u/2)]}{\Delta u/2}$$

Looking ahead to the next step of letting $\Delta u \to 0$, we see that the numerator and denominator both approach zero. This situation is precisely the same as that in which we were finding the derivatives of the algebraic functions. To find the limit, we must find

$$\lim_{\Delta u \to 0} \frac{\sin(\Delta u/2)}{\Delta u/2}$$

since these are the factors that cause the numerator and the denominator to approach zero.

In finding this limit, we let $\theta = \Delta u/2$ for convenience of notation. This means that we are to determine $\lim_{\theta \to 0} \dfrac{\sin \theta}{\theta}$. Of course, it would be convenient to know before proceeding if this limit does actually exist. Therefore, by using a calculator, we can develop a table of values of $\dfrac{\sin \theta}{\theta}$ as θ becomes very small.

θ (radians)	0.5	0.1	0.05	0.01	0.001
$\dfrac{\sin \theta}{\theta}$	0.9588511	0.9983342	0.9995834	0.9999833	0.9999998

We see from this table that the limit of $\dfrac{\sin \theta}{\theta}$, as $\theta \to 0$, appears to be 1.

In order to prove that $\lim_{\theta \to 0} \dfrac{\sin \theta}{\theta} = 1$, we use a geometric approach. Considering Fig. 7-32, we see that the following inequality is true:

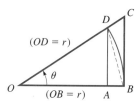

$(OD = r)$

$(OB = r)$

Fig. 7-32

Area triangle *OBD* < area sector *OBD* < area triangle *OBC*

$$\frac{1}{2}r(r \sin \theta) < \frac{1}{2}r^2\theta < \frac{1}{2}r(r \tan \theta) \quad \text{or} \quad \sin \theta < \theta < \tan \theta$$

Remembering that we want to find the limit of $(\sin \theta)/\theta$, we next divide through by $\sin \theta$ and then take reciprocals:

$$1 < \frac{\theta}{\sin \theta} < \frac{1}{\cos \theta} \quad \text{or} \quad 1 > \frac{\sin \theta}{\theta} > \cos \theta$$

When we consider the limit as $\theta \to 0$, we see that the left member remains 1 and the right member approaches 1. Thus, $(\sin \theta)/\theta$ must approach 1. This means

$$\boxed{\lim_{\theta \to 0} \frac{\sin \theta}{\theta} = \lim_{\Delta u \to 0} \frac{\sin(\Delta u/2)}{\Delta u/2} = 1}$$

(7-29)

Using the result in Eq. (7-29) in the expression for $\Delta y/\Delta u$, we have

$$\lim_{\Delta u \to 0} \frac{\Delta y}{\Delta u} = \lim_{\Delta u \to 0} \left[\cos\left(u + \frac{\Delta u}{2} \right) \frac{\sin(\Delta u/2)}{\Delta u/2} \right] = \cos u$$

or

$$\frac{dy}{du} = \cos u \qquad\qquad (7\text{-}30)$$

However, we want the derivative of y with respect to x. This requires the use of the chain rule, Eq. (3-14), which we repeat here for reference.

$$\boxed{\frac{dy}{dx} = \frac{dy}{du}\frac{du}{dx}} \qquad\qquad (7\text{-}31)$$

Combining Eqs. (7-30) and (7-31), we have ($y = \sin u$).

$$\boxed{\frac{d(\sin u)}{dx} = \cos u \frac{du}{dx}} \qquad\qquad (7\text{-}32)$$

EXAMPLE 1   Find the derivative of $y = \sin 2x$.
In this example, $u = 2x$. Thus

$$\frac{dy}{dx} = \frac{d(\sin 2x)}{dx} = \cos 2x \frac{d(2x)}{dx} = (\cos 2x)(2) \qquad \text{using Eq. (7-32)}$$

$$= 2 \cos 2x$$

EXAMPLE 2   Find the derivative of $y = 2 \sin(x^2)$.
In this example, $u = x^2$, which means that $du/dx = 2x$. Hence,

$$\frac{dy}{dx} = 2[\cos(x^2)](2x) \qquad \text{using Eq. (7-32)}$$

$$= 4x \cos(x^2)$$

CAUTION▶   It is important here, just as it is *in finding the derivatives of powers of all functions, to remember to include the factor du/dx.*

EXAMPLE 3   Find the derivative of $r = \sin^2 \theta$.
This example is a combination of the use of the power rule, Eq. (3-15) and the derivative of the sine function Eq. (7-32). Since $\sin^2 \theta$ means $(\sin \theta)^2$, in using the power rule, we have $u = \sin \theta$. Thus,

$$\frac{dr}{d\theta} = 2(\sin \theta)\frac{d \sin \theta}{d\theta} \qquad \text{using Eq. (3-15)}$$

$$= 2 \sin \theta \cos \theta \qquad \text{using Eq. (7-32)}$$

$$= \sin 2\theta \qquad \text{using identity (Eq. (7-20))}$$

For reference, Eq. (3-15) is
$$\frac{du^n}{dx} = nu^{n-1}\left(\frac{du}{dx}\right).$$

EXAMPLE 4 Find the derivative of $y = 2 \sin^3(2x^4)$.

In the general power rule, $u = \sin(2x^4)$. In the derivative of the sine function, $u = 2x^4$. Thus, we have

$$\frac{dy}{dx} = 2(3) \sin^2(2x^4) \frac{d(\sin 2x^4)}{dx} \qquad \text{using Eq. (3-15)}$$

$$= 6 \sin^2(2x^4) \cos(2x^4) \frac{d(2x^4)}{dx} \qquad \text{using Eq. (7-32)}$$

$$= 6 \sin^2(2x^4) \cos(2x^4)(8x^3)$$

$$= 48x^3 \sin^2(2x^4) \cos(2x^4) \qquad \blacksquare$$

In order to find the derivative of the cosine function, we write it in the form $\cos u = \sin\left(\frac{\pi}{2} - u\right)$. Thus, if $y = \sin\left(\frac{\pi}{2} - u\right)$, we have

$$\frac{dy}{dx} = \cos\left(\frac{\pi}{2} - u\right) \frac{d(\frac{\pi}{2} - u)}{dx} = \cos\left(\frac{\pi}{2} - u\right)\left(-\frac{du}{dx}\right)$$

$$= -\cos\left(\frac{\pi}{2} - u\right)\frac{du}{dx}$$

Since $\cos\left(\frac{\pi}{2} - u\right) = \sin u$, we have

$$\frac{d(\cos u)}{dx} = -\sin u \frac{du}{dx} \qquad\qquad (7\text{-}33)$$

EXAMPLE 5 The electric power p developed in a resistor of an amplifier circuit is $p = 25 \cos^2 120\pi t$, where t is the time. Find the expression for the time rate of change of power.

From Chapter 3, we know that we are to find the derivative dp/dt. Therefore,

$$p = 25 \cos^2 120\pi t$$

$$\frac{dp}{dt} = 25(2 \cos 120\pi t) \frac{d \cos 120\pi t}{dt} \qquad \text{using Eq. (3-15)}$$

$$= 50 \cos 120\pi t(-\sin 120\pi t) \frac{d(120\pi t)}{dt} \qquad \text{using Eq. (7-33)}$$

$$= (-50 \cos 120\pi t \sin 120\pi t)(120\pi)$$

$$= -6000\pi \cos 120\pi t \sin 120\pi t$$

$$= -3000\pi \sin 240\pi t \qquad \text{using Eq. (7-20)} \qquad \blacksquare$$

For reference, Eq. (7-20) is $\sin 2\alpha = 2 \sin \alpha \cos \alpha$.

EXAMPLE 6 Find the derivative of $y = \sqrt{1 + \cos 2x}$.

$$y = (1 + \cos 2x)^{1/2}$$

$$\frac{dy}{dx} = \frac{1}{2}(1 + \cos 2x)^{-1/2} \frac{d(1 + \cos 2x)}{dx} \qquad \text{using Eq. (3-15)}$$

$$= \frac{1}{2}(1 + \cos 2x)^{-1/2}(-\sin 2x)(2) \qquad \text{using Eq. (7-33)}$$

$$= -\frac{\sin 2x}{\sqrt{1 + \cos 2x}} \qquad\qquad \blacksquare$$

EXAMPLE 7 Find the differential of $y = \sin 2x \cos x^2$.

From Section 4-8, we recall that the differential of a function $y = f(x)$ is $dy = f'(x)\,dx$. Thus, using the derivative product rule and the derivatives of the sine and cosine functions, we arrive at the following result:

$$y = \sin 2x \cos x^2 \qquad y = (\sin 2x)(\cos x^2)$$
$$dy = [\sin 2x(-\sin x^2)(2x) + \cos x^2(\cos 2x)(2)]\,dx$$
$$= (-2x \sin 2x \sin x^2 + 2 \cos 2x \cos x^2)\,dx \qquad \text{---------}■$$

EXAMPLE 8 Find the slope of a line tangent to the curve of $y = 5 \sin 3x$, where $x = 0.2$.

Here we are to find the derivative of $y = 5 \sin 3x$ and then evaluate the derivative for $x = 0.2$. Therefore, we have the following:

$$y = 5 \sin 3x$$
$$\frac{dy}{dx} = 5(\cos 3x)(3) = 15 \cos 3x \qquad \text{find derivative}$$
$$\left.\frac{dy}{dx}\right|_{x=0.2} = 15 \cos 3(0.2) = 15 \cos 0.6 \qquad \text{evaluate}$$
$$= 12.38$$

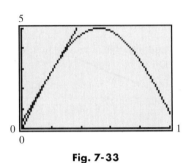

Fig. 7-33

CAUTION ▶ In evaluating the slope, we must remember that $x = 0.2$ means the **values are in radians.** Therefore, the slope is 12.38.

As we showed in Chapter 3, the *tangent* feature of a graphing calculator can be used to display a curve and the line tangent at a specific point. In Fig. 7-33 the calculator display for this function and its tangent at $x = 0.2$ is shown.

Also, reviewing the use of the calculator, the evaluation of this derivative using the *numerical derivative* feature is shown in the calculator display in Fig. 7-34. ■

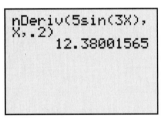

Fig. 7-34

═══════════════ EXERCISES **7-3** ═══════════════

In Exercises 1–32, find the derivatives of the given functions.

1. $y = \sin(x + 2)$

2. $y = 3 \sin 4x$

3. $y = 2 \sin(2x^3 - 1)$

4. $y = 5 \sin(2 - 3t)$

5. $y = 6 \cos \frac{1}{2}x$

6. $y = \cos(1 - x)$

7. $y = 2 \cos(3x - 1)$

8. $y = 4 \cos(6x^2 + 5)$

9. $r = \sin^2 3\pi\theta$

10. $y = 3 \sin^3(2x^4 + 1)$

11. $y = 3 \cos^3(5x + 2)$

12. $y = 4 \cos^2 \sqrt{x}$

13. $y = x \sin 3x$

14. $y = x^2 \sin 2x$

15. $y = 3x^3 \cos 5x$

16. $y = 0.5\theta \cos(2\theta + \pi/4)$

17. $y = \sin x^2 \cos 2x$

18. $y = 6 \sin x \cos 4x$

19. $y = \sqrt{1 + \sin 4x}$

20. $y = (x - \cos^2 x)^4$

21. $r = \dfrac{\sin(3t - \pi/3)}{2t}$

22. $y = \dfrac{2x + 3}{\sin 4x}$

23. $y = \dfrac{2 \cos x^2}{3x - 1}$

24. $y = \dfrac{\cos^2 3x}{1 + 2 \sin^2 2x}$

25. $y = 2 \sin^2 3x \cos 2x$

26. $y = \cos^3 4x \sin^2 2x$

27. $s = \sin(\sin 2t)$

28. $z = 0.2 \cos(\sin 3\phi)$

29. $y = \sin^3 x - \cos 2x$

30. $y = x \sin x + \cos x$

31. $p = \dfrac{1}{\sin s} + \dfrac{1}{\cos s}$

32. $y = 2x \sin x + 2 \cos x - x^2 \cos x$

In Exercises 33–52, solve the given problems.

33. Using a graphing calculator, (a) display the graph of $y = (\sin x)/x$ to verify that $(\sin \theta)/\theta \to 1$ as $\theta \to 0$ and (b) verify the values for $(\sin \theta)/\theta$ in the table on page 239.

34. Evaluate $\lim\limits_{\theta \to 0} (\tan \theta)/\theta$. (Use the fact that $\lim\limits_{\theta \to 0} (\sin \theta)/\theta = 1$.)

(W) 35. On a calculator, find the values of (a) cos 1.0000 and (b) (sin 1.0001 − sin 1.0000)/0.0001. Compare the values and give the meaning of each in relation to the derivative of the sine function where $x = 1$.

(W) 36. On a calculator, find the values of (a) −sin 1.0000 and (b) (cos 1.0001 − cos 1.0000)/0.0001. Compare the values and give the meaning of each in relation to the derivative of the cosine function where $x = 1$.

37. On the graph of $y = \sin x$ in Fig. 7-35, draw tangent lines at the indicated points and determine the slopes of these tangent lines. Then plot the values of these slopes for the same values of x and join the points with a smooth curve. Compare the resulting curve with $y = \cos x$. (Note the meaning of the derivative as the slope of a tangent line.)

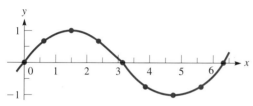

Fig. 7-35

38. Repeat the instructions given in Exercise 37 for the graph of $y = \cos x$ in Fig. 7-36. Compare the resulting curve with $y = \sin x$. (Be careful in this comparison and remember the difference between $y = \sin x$ and the derivative of $y = \cos x$. As in Exercise 37, note the meaning of the derivative as the slope of a tangent line.)

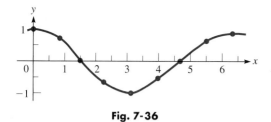

Fig. 7-36

39. Find the derivative of the implicit function $\sin(xy) + \cos 2y = x^2$.

40. Find the derivative of the implicit function $x \cos 2y + \sin x \cos y = 1$.

41. Show that $\dfrac{d^4 \sin x}{dx^4} = \sin x$.

42. If $y = \cos 2x$, show that $\dfrac{d^2y}{dx^2} = -4y$.

43. Evaluate the differential of $y = 3 \sin 2x$ for $x = \frac{\pi}{8}$ and $dx = 0.02$.

44. Find the linearization $L(x)$ of the function $f(x) = \sin(\cos x)$ for $a = \pi/2$.

45. Find the slope of a line tangent to the curve of $y = x \cos 2x$ where $x = 1.20$. Verify the result by using the *derivative-evaluating* feature of a graphing calculator.

46. Find the slope of a line tangent to the graph of $y = \dfrac{2 \sin 3x}{x}$ where $x = 0.15$. Verify the result by using the *derivative-evaluating* feature of a graphing calculator.

47. The voltage V in a certain electric circuit as a function of the time t (in s) is given by $V = 3.0 \sin 188t \cos 188t$. How fast is the voltage changing when $t = 2.0$ ms?

48. A water slide at an amusement park follows the curve (y in m) $y = 2.0 + 2.0 \cos (0.53x + 0.40)$ for $0 \le x \le 5.0$ m. Find the angle with the horizontal of the slide for $x = 2.5$ m.

49. The blade of a saber saw moves vertically up and down, and its displacement y (in cm) is given by $y = 1.85 \sin 36\pi t$, where t is the time (in s). Find the velocity of the blade for $t = 0.0250$ s.

50. The current i (in A) in an amplifier circuit as a function of the time t (in s) is given by $i = 0.10 \cos(120\pi t + \pi/6)$. Find the expression for the voltage across a 2.0-mH inductor in the circuit. (See Exercise 23 of Section 6-1 on page 189.)

51. In testing a heat-seeking rocket, it is always moving directly toward a remote-controlled aircraft. At a certain instant the distance r (in km) from the rocket to the aircraft is $r = \dfrac{100}{1 - \cos \theta}$, where θ is the angle between their directions of flight. Find $dr/d\theta$ for $\theta = 120°$. See Fig. 7-37.

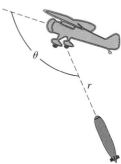

Fig. 7-37

52. The number N of reflections of a light ray passing through an optic fiber of length L and diameter d is $N = \dfrac{L \sin \theta}{d \sqrt{n^2 - \sin^2 \theta}}$. Here, n is the index of refraction of the fiber and θ is the angle between the light ray and the fiber's axis. Find $dN/d\theta$.

7-4 DERIVATIVES OF THE OTHER TRIGONOMETRIC FUNCTIONS

We can find the derivatives of the other trigonometric functions by expressing the functions in terms of the sine and cosine. After we perform the differentiation, we use trigonometric relations to put the derivative in a convenient form.

We obtain the derivative of $\tan u$ by expressing $\tan u$ as $\sin u/\cos u$. Therefore, letting $y = \sin u/\cos u$, by employing the quotient rule, we have

$$\frac{dy}{dx} = \frac{\cos u[\cos u(du/dx)] - \sin u[-\sin u(du/dx)]}{\cos^2 u}$$

$$= \frac{\cos^2 u + \sin^2 u}{\cos^2 u}\frac{du}{dx} = \frac{1}{\cos^2 u}\frac{du}{dx} = \sec^2 u\frac{du}{dx}$$

$$\boxed{\frac{d(\tan u)}{dx} = \sec^2 u\frac{du}{dx}} \tag{7-34}$$

We find the derivative of $\cot u$ by letting $y = \cos u/\sin u$ and again using the quotient rule:

$$\frac{dy}{dx} = \frac{\sin u[-\sin u(du/dx)] - \cos u[\cos u(du/dx)]}{\sin^2 u}$$

$$= \frac{-\sin^2 u - \cos^2 u}{\sin^2 u}\frac{du}{dx}$$

$$\boxed{\frac{d(\cot u)}{dx} = -\csc^2 u\frac{du}{dx}} \tag{7-35}$$

To obtain the derivative of $\sec u$, we let $y = 1/\cos u$. Then,

$$\frac{dy}{dx} = -(\cos u)^{-2}\left[(-\sin u)\left(\frac{du}{dx}\right)\right] = \frac{1}{\cos u}\frac{\sin u}{\cos u}\frac{du}{dx}$$

$$\boxed{\frac{d(\sec u)}{dx} = \sec u \tan u\frac{du}{dx}} \tag{7-36}$$

We obtain the derivative of $\csc u$ by letting $y = 1/\sin u$. And so,

$$\frac{dy}{dx} = -(\sin u)^{-2}\left(\cos u\frac{du}{dx}\right) = -\frac{1}{\sin u}\frac{\cos u}{\sin u}\frac{du}{dx}$$

$$\boxed{\frac{d(\csc u)}{dx} = -\csc u \cot u\frac{du}{dx}} \tag{7-37}$$

Note the convenient forms of these derivatives that are obtained by using basic trigonometric identities.

■EXAMPLE 1 Find the derivative of $y = 2 \tan 8x$.
The derivative is

$$\frac{dy}{dx} = 2(\sec^2 8x)(8) \qquad \text{using Eq. (7-34)}$$

(with $\frac{du}{dx}$ indicated above the term)

$$= 16 \sec^2 8x$$

■EXAMPLE 2 Find the derivative of $y = 3 \sec^2 4x$.
Using the power rule and Eq. (7-36), we have

$$\frac{dy}{dx} = 3(2)(\sec 4x)\frac{d(\sec 4x)}{dx} \qquad \text{using } \frac{du^n}{dx} = nu^{n-1}\frac{du}{dx}$$

$$= 6(\sec 4x)(\sec 4x \tan 4x)(4) \qquad \text{using } \frac{d \sec u}{dx} = \sec u \tan u \frac{du}{dx}$$

$$= 24 \sec^2 4x \tan 4x$$

■EXAMPLE 3 Find the derivative of $y = t \csc^3 2t$.
Using the power rule, the product rule, and Eq. (7-37), we have

$$\frac{dy}{dt} = t(3 \csc^2 2t)(-\csc 2t \cot 2t)(2) + (\csc^3 2t)(1)$$

$$= \csc^3 2t(-6t \cot 2t + 1)$$

■EXAMPLE 4 Find the derivative of $y = (\tan 2x + \sec 2x)^3$.
Using the power rule and Eqs. (7-34) and (7-36), we have

$$\frac{dy}{dx} = 3(\tan 2x + \sec 2x)^2[\sec^2 2x(2) + \sec 2x \tan 2x(2)]$$

$$= 3(\tan 2x + \sec 2x)^2(2 \sec 2x)(\sec 2x + \tan 2x)$$

$$= 6 \sec 2x(\tan 2x + \sec 2x)^3$$

■EXAMPLE 5 Find the differential of $r = \sin 2\theta \tan \theta^2$.
Here, we are to find the derivative of the given function and multiply by $d\theta$. Therefore, using the product rule along with Eqs. (7-32) and (7-34), we have

$$dr = [(\sin 2\theta)(\sec^2 \theta^2)(2\theta) + (\tan \theta^2)(\cos 2\theta)(2)] \, d\theta$$

$$= (2\theta \sin 2\theta \sec^2 \theta^2 + 2 \cos 2\theta \tan \theta^2) \, d\theta \qquad \text{don't forget the } d\theta.$$

■EXAMPLE 6 Find dy/dx if $\cot 2x - 3 \csc xy = y^2$.
In finding the derivative of this implicit function, we must be careful not to forget the factor dy/dx when it occurs. The derivative is found as follows:

$$\cot 2x - 3 \csc xy = y^2$$

$$(-\csc^2 2x)(2) - 3(-\csc xy \cot xy)\left(x\frac{dy}{dx} + y\right) = 2y\frac{dy}{dx}$$

$$3x \csc xy \cot xy\frac{dy}{dx} - 2y\frac{dy}{dx} = 2 \csc^2 2x - 3y \csc xy \cot xy$$

$$\frac{dy}{dx} = \frac{2 \csc^2 2x - 3y \csc xy \cot xy}{3x \csc xy \cot xy - 2y}$$

EXAMPLE 7 Evaluate the derivative of $y = \dfrac{2x}{1 - \cot 3x}$, where $x = 0.25$.

Finding the derivative, we have

$$\frac{dy}{dx} = \frac{(1 - \cot 3x)(2) - 2x(\csc^2 3x)(3)}{(1 - \cot 3x)^2}$$

$$= \frac{2 - 2\cot 3x - 6x\csc^2 3x}{(1 - \cot 3x)^2}$$

Now, substituting $x = 0.25$, we have

$$\left.\frac{dy}{dx}\right|_{x=0.25} = \frac{2 - 2\cot 0.75 - 6(0.25)\csc^2 0.75}{(1 - \cot 0.75)^2}$$

$$= -626.0$$

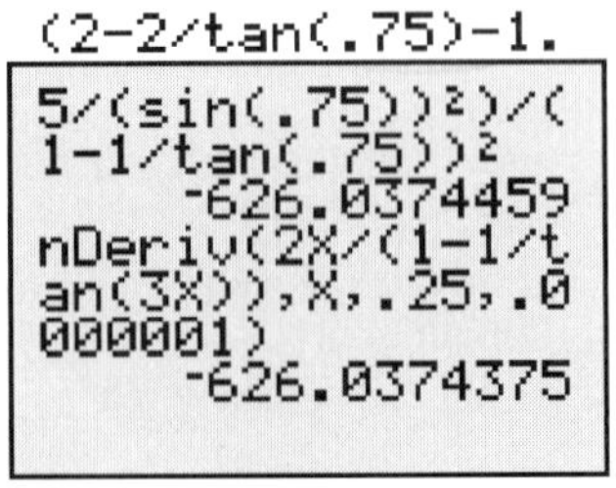

In using the calculator we note again that we must have it in **radian mode.** To evaluate the above expression for dy/dx, we must use the reciprocals of tan 0.75 and sin 0.75 for cot 0.75 and csc 0.75, respectively, as shown in the first four lines of the display in Fig. 7-38. The last four lines of Fig. 7-38 show a check of our result using the *numerical derivative* feature of the calculator. ⎯⎯⎯⎯⎯■

```
(2-2/tan(.75)-1.
5/(sin(.75))²)/(
1-1/tan(.75))²
      -626.0374459
nDeriv(2X/(1-1/t
an(3X)),X,.25,.0
000001)
      -626.0374375
```

Fig. 7-38

EXERCISES 7-4

In Exercises 1–32, find the derivatives of the given functions.

1. $y = \tan 5x$

2. $y = 3\tan(3x + 2)$

3. $y = 5\cot(0.25\pi - \theta)$

4. $y = 3\cot 6x$

5. $y = 3\sec 2x$

6. $y = \sec\sqrt{1 - x}$

7. $y = -3\csc\sqrt{2x + 3}$

8. $h = 0.5\csc(1 - 2\pi t)$

9. $y = 5\tan^2 3x$

10. $y = 2\tan^2(x^2)$

11. $y = 2\cot^4\frac{1}{2}x$

12. $y = \cot^2(1 - x^2)$

13. $y = \sqrt{\sec 4x}$

14. $y = 0.8\sec^3 5u$

15. $y = 3\csc^4 7x$

16. $y = \csc^2(2x^2)$

17. $r = t^2\tan 0.5t$

18. $y = 3x\sec 4x$

19. $y = 4\cos x\csc x^2$

20. $y = \frac{1}{2}\sin 2x\sec x$

21. $y = \dfrac{\csc x}{x}$

22. $u = \dfrac{\cot 0.25z}{2z}$

23. $y = \dfrac{2\cos 4x}{1 + \cot 3x}$

24. $y = \dfrac{\tan^2 3x}{2 + \sin x^2}$

25. $y = \frac{1}{3}\tan^3 x - \tan x$

26. $y = \csc 2x - 2\cot 2x$

27. $r = \tan(\sin 2\theta)$

28. $y = x\tan x + \sec^2 2x$

29. $y = \sqrt{2x + \tan 4x}$

30. $y = (1 - \csc^2 3x)^3$

31. $x\sec y - 2y = \sin 2x$

32. $3\cot(x + y) = \cos y^2$

In Exercises 33–36, find the differentials of the given functions.

33. $y = 4\tan^2 3x$

34. $y = 2.5\sec^3 2t$

35. $y = \tan 4x\sec 4x$

36. $y = 2x\cot 3x$

In Exercises 37–48, solve the given problems.

Ⓦ **37.** On a calculator, find the values of (a) $\sec^2 1.0000$ and (b) $(\tan 1.0001 - \tan 1.0000)/0.0001$. Compare the values and give the meaning of each in relation to the derivative of $\tan x$ where $x = 1$.

Ⓦ **38.** On a calculator, find the values of (a) $\sec 1.0000\tan 1.0000$ and (b) $(\sec 1.0001 - \sec 1.0000)/0.0001$. Compare the values and give the meaning of each in relation to the derivative of $\sec x$ where $x = 1$.

39. Find the derivative of each member of the identity $1 + \tan^2 x = \sec^2 x$ and show that the results are equal.

40. Find the derivative of each member of the identity $1 + \cot^2 x = \csc^2 x$ and show that the results are equal.

41. Find the slope of a line tangent to the graph of $y = 2\cot 3x$ where $x = \pi/12$. Verify the result by using the *numerical derivative* feature of a graphing calculator.

42. Find the slope of a line normal to the graph of $y = \csc\sqrt{2x + 1}$ where $x = 0.45$. Verify the result by using the *numerical derivative* feature of a graphing calculator.

43. Show that $y = 2 \tan x - \sec x$ satisfies
$$\frac{dy}{dx} = \frac{2 - \sin x}{\cos^2 x}.$$

44. Show that $y = \cos^3 x \tan x$ satisfies
$$\cos x \frac{dy}{dx} + 3y \sin x - \cos^2 x = 0.$$

45. The vertical displacement y (in cm) of the end of an industrial robot arm for each cycle is $y = 2t^{1.5} - \tan 0.1t$, where t is the time (in s). Find its vertical velocity for $t = 15$ s.

46. The electric charge q (in C) passing a given point in a circuit is given by $q = t \sec \sqrt{0.2t^2 + 1}$, where t is the time (in s). Find the current i (in A) for $t = 0.80$ s. ($i = dq/dt$.)

47. An observer to a rocket launch was 1000 ft from the take-off position. The observer found the angle of elevation of the rocket as a function of time to be $\theta = 3t/(2t + 10)$. Therefore, the height h (in ft) of the rocket was $h = 1000 \tan \dfrac{3t}{2t + 10}$. Find the time rate of change of height after 5.0 s. See Fig. 7-39.

48. A surveyor measures the distance between two markers to be 378.00 m. Then, moving along a line equidistant from the markers, the distance d from the surveyor to each marker is $d = 189.00 \csc \frac{1}{2}\theta$, where θ is the angle between the lines of sight to the markers. See Fig. 7-40. By using differentials, find the change in d if θ changes from 98.20° to 98.45°.

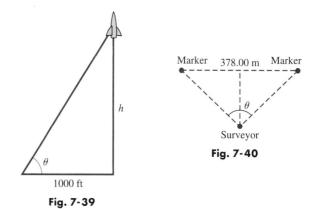

Fig. 7-39

Fig. 7-40

7-5   THE INVERSE TRIGONOMETRIC FUNCTIONS

There are applications in which we wish to determine how an angle changes. In these cases it can be convenient to have a function in which the angle is the dependent variable. To determine such a function, it is necessary that we are able to solve for the angle in a trigonometric function. Functions have been defined for this purpose, and as we shall see, they have significant uses and applications in calculus.

Therefore, we define the **inverse trigonometric functions.** *The* **inverse sine** *of x is defined by*

Other notations for the inverse sine are Arcsin x, arcsin x, and $\operatorname{Sin}^{-1} x$.

$$y = \sin^{-1} x \qquad \left(-\frac{\pi}{2} \le y \le \frac{\pi}{2}\right) \tag{7-38}$$

where *y is the angle whose sine is x.* This means that x is the value of the sine of the angle y, or $x = \sin y$. (It is necessary to show the range as $-\pi/2 \le y \le \pi/2$, as we will see shortly.)

CAUTION ▶   *In Eq. (7-38), the -1 is* **not** *an exponent.*

The -1 in $\sin^{-1} x$ *denotes the inverse function.* We first introduced this notation on page 228 when we were finding the angle with a known value of one of the trigonometric functions.

The inverse sine function is so named because, *two functions are* **inverse functions** *if we can solve for the independent variable in terms of the dependent variable in one function and arrive at the functional relationship expressed by the other.* Definitions for the other inverse trigonometric functions have meanings similar to that for the inverse sine as shown in Eq. (7-38).

EXAMPLE 1   (a) $y = \cos^{-1} x$ is read as "y is the angle whose cosine is x." In this case, $x = \cos y$.

(b) $y = \tan^{-1} 2x$ is read as "y is the angle whose tangent is $2x$." In this case, $2x = \tan y$.

(c) $y = \csc^{-1}(1 - x)$ is read as "y is the angle whose cosecant is $1 - x$." In this case, $1 - x = \csc y$, or $x = 1 - \csc y$.

We have seen that $y = \sin^{-1} x$ means that $x = \sin y$. From our previous work with the trigonometric functions, we know that there is an unlimited number of possible values of y for a given value of x in $x = \sin y$. Consider the following example.

EXAMPLE 2   (a) For $x = \sin y$, we know that

$$\sin \frac{\pi}{6} = \frac{1}{2} \quad \text{and} \quad \sin \frac{5\pi}{6} = \frac{1}{2}$$

In fact, $x = \frac{1}{2}$ also for values of y of $-\frac{7\pi}{6}, \frac{13\pi}{6}, \frac{17\pi}{6}$, and so on.

(b) For $x = \cos y$, we know that

$$\cos 0 = 1 \quad \text{and} \quad \cos 2\pi = 1$$

In fact, the $\cos y = 1$ for y equal to any even multiple of π.

From Chapter 1, we know that *a function may have **only one** value of the dependent variable for a given value of the independent variable.* To have only one value of y for each value of x in the domain of the inverse trigonometric functions, it is not possible to include all values of y in the range. Therefore, *the range of each of the inverse trigonometric functions is defined as follows:*

$$-\frac{\pi}{2} \leq \sin^{-1} x \leq \frac{\pi}{2} \qquad 0 \leq \cos^{-1} x \leq \pi \qquad -\frac{\pi}{2} < \tan^{-1} x < \frac{\pi}{2} \qquad 0 < \cot^{-1} x < \pi$$

$$0 \leq \sec^{-1} x \leq \pi \quad \left(\sec^{-1} x \neq \frac{\pi}{2}\right) \qquad -\frac{\pi}{2} \leq \csc^{-1} x \leq \frac{\pi}{2} \quad (\csc^{-1} x \neq 0)$$

(7-39)

For any of the inverse trigonometric functions $y = f(x)$, we must use a value of y in the range as defined in Eqs. (7-39) that corresponds to a given value of x in the domain. We will discuss the domains and the reasons for these definitions, along with the graphs of the inverse trigonometric functions, following the next two examples.

EXAMPLE 3   (a) $\sin^{-1}\left(\frac{1}{2}\right) = \frac{\pi}{6}$ first-quadrant angle

This is the only value of the function that lies within the defined range. The value $\frac{5\pi}{6}$ is not correct, even though $\sin(\frac{5\pi}{6}) = \frac{1}{2}$, since $\frac{5\pi}{6}$ lies outside the defined range.

(b) $\cos^{-1}\left(-\frac{1}{2}\right) = \frac{2\pi}{3}$ second-quadrant angle

Other values such as $\frac{4\pi}{3}$ and $-\frac{2\pi}{3}$ are not correct, since they are not within the defined range for the function $\cos^{-1} x$.

EXAMPLE 4 $\tan^{-1}(-1) = -\dfrac{\pi}{4}$ fourth-quadrant angle

CAUTION▶ This is the only value within the defined range for the function $\tan^{-1} x$. We must remember that **when x is negative for $\sin^{-1} x$ and $\tan^{-1} x$, the value of y is a fourth-quadrant angle, expressed as a negative angle.** This is a direct result of the definition. (The single exception is $\sin^{-1}(-1) = -\pi/2$, which is a quadrantal angle and is not *in* the fourth quadrant.)   ▬▬▬▬▬▬▬■

The ranges of the inverse trigonometric functions are chosen so that if x is positive, the resulting value is an angle in the first quadrant. However, care must be taken in choosing the range for negative values of x.

For the ranges of $y = \sin^{-1} x$ and $y = \cos^{-1} x$, we note that the *domain* of each is $-1 \le x \le 1$, since the sine and cosine functions take on only these values. For each value in this domain, we use only one value of y in the range. Although the domain of $y = \tan^{-1} x$ is all real numbers, we still use only one value of y in the range.

Considering negative values of x, we cannot use second-quadrant angles for the range of $\sin^{-1} x$ since the sine of a second-quadrant angle is positive. To have a continuous range, we express the fourth-quadrant angles as negative angles. This range is also chosen for $\tan^{-1} x$, for similar reasons. The range for $\cos^{-1} x$ cannot be chosen this way since the cosine of fourth-quadrant angles is positive. For a continuous range of values for $\cos^{-1} x$, the second-quadrant angles are chosen for negative values of x.

As for the other functions, we chose values of the range such that if x is positive, the result is also a first-quadrant angle. As for negative values of x, it rarely makes any difference, since either positive values of x arise, or we can use another function. Our definitions, however, are those that are generally used.

The graphs of the inverse trigonometric functions can be used to show the domains and ranges. We can obtain the graph of the inverse sine function by first sketching the sine curve $x = \sin y$ *along the y-axis.* We then mark the specific part of this curve for which $-\frac{\pi}{2} \le y \le \frac{\pi}{2}$ as the graph of the inverse sine function. The graphs of the other inverse trigonometric functions are found in the same manner. In Figs. 7-41, 7-42, and 7-43, the graphs of $x = \sin y$, $x = \cos y$, and $x = \tan y$, respectively, are shown. The heavier, colored portions indicate the graphs of the respective inverse trigonometric functions.

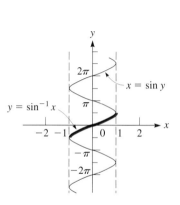

Fig. 7-41

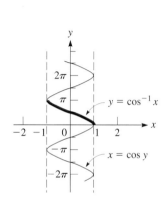

Fig. 7-42

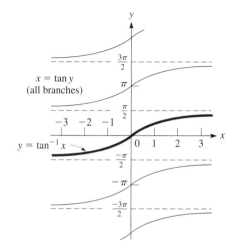

Fig. 7-43

The following examples further illustrate the values and meanings of the inverse trigonometric functions.

EXAMPLE 5   (a) $\sin^{-1}(-\sqrt{3}/2) = -\pi/3$ (b) $\cos^{-1}(-1) = \pi$
(c) $\tan^{-1} 0 = 0$ (d) $\tan^{-1}(\sqrt{3}) = \pi/3$

Using a calculator in radian mode, we find the following values:

(e) $\sin^{-1} 0.6294 = 0.6808$ (f) $\sin^{-1}(-0.1568) = -0.1574$
(g) $\cos^{-1}(-0.8026) = 2.5024$ (h) $\tan^{-1}(-1.9268) = -1.0921$

We note that the calculator gives values that are in the defined range for each function. ----------■

EXAMPLE 6   Given that $y = \pi - \sec^{-1} 2x$, solve for x.
We first find the expression for $\sec^{-1} 2x$ and then use the meaning of the inverse secant. The solution follows.

$$y = \pi - \sec^{-1} 2x$$
$$\sec^{-1} 2x = \pi - y \qquad \text{solve for } \sec^{-1} 2x$$
$$2x = \sec(\pi - y) \qquad \text{use meaning of inverse secant}$$
$$x = -\frac{1}{2} \sec y \qquad \sec(\pi - y) = -\sec y$$

Since the values of $\sec^{-1} 2x$ are restricted, so are the resulting values that are obtained for y. It should be carefully noted that the function $\sec^{-1} 2x$ is not the same as the function $2 \sec^{-1} x$. ----------■

EXAMPLE 7   The instantaneous power p in an electric inductor is given by the equation $p = vi \sin \omega t \cos \omega t$. Solve for t.
Noting the product $\sin \omega t \cos \omega t$ suggests using $\sin 2\alpha = 2 \sin \alpha \cos \alpha$. Then, using the meaning of the inverse sine, we can complete the solution.

$$p = vi \sin \omega t \cos \omega t$$
$$= \frac{1}{2} vi \sin 2\omega t \qquad \text{using double-angle formula}$$
$$\sin 2\omega t = \frac{2p}{vi}$$
$$2\omega t = \sin^{-1}\left(\frac{2p}{vi}\right) \qquad \text{using meaning of inverse sine}$$
$$t = \frac{1}{2\omega} \sin^{-1}\left(\frac{2p}{vi}\right) \qquad \qquad \text{----------■}$$

If we know the value of one of the inverse functions, we can find the trigonometric functions of the angle. If general relations are desired, a representative triangle is very useful. The following examples illustrate these methods.

EXAMPLE 8   Find $\cos(\sin^{-1} 0.5)$.

Knowing that the values of inverse trigonometric functions are *angles,* we see that $\sin^{-1} 0.5$ is a first-quadrant angle. Thus, we find $\sin^{-1} 0.5 = \pi/6$. The problem is now to find $\cos(\pi/6)$. This is, of course, $\sqrt{3}/2$, or 0.8660. Thus,

$$\cos(\sin^{-1} 0.5) = \cos(\pi/6) = 0.8660.$$

EXAMPLE 9   **(a)** $\sin(\cot^{-1} 1) = \sin(\pi/4)$ first-quadrant angle

$$= \frac{\sqrt{2}}{2} = 0.7071$$

(b) $\tan[\cos^{-1}(-1)] = \tan \pi$ quadrantal angle

$$= 0$$

(c) $\cos[\sin^{-1}(-0.2395)] = 0.9709$ using a calculator

EXAMPLE 10   Find $\sin(\tan^{-1} x)$.

We know that $\tan^{-1} x$ is another way of stating "the angle whose tangent is x." Thus, let us draw a right triangle (as in Fig. 7-44) and label one of the acute angles as θ, the side opposite θ as x, and the side adjacent to θ as 1. In this way we see that, by definition, $\tan \theta = \frac{x}{1}$, or $\theta = \tan^{-1} x$, which means θ is the desired angle. By the Pythagorean theorem, the hypotenuse of this triangle is $\sqrt{x^2 + 1}$. Now we find that $\sin \theta$, which is the same as $\sin(\tan^{-1} x)$, is $x/\sqrt{x^2 + 1}$. Thus,

$$\sin(\tan^{-1} x) = \frac{x}{\sqrt{x^2 + 1}}$$

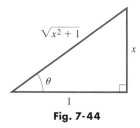

Fig. 7-44

EXAMPLE 11   Find $\cos(2 \sin^{-1} x)$.

From Fig. 7-45, we see that $\theta = \sin^{-1} x$. From the double-angle formulas, we have

$$\cos 2\theta = 1 - 2 \sin^2 \theta$$

Thus, since $\sin \theta = x$, we have

$$\cos(2 \sin^{-1} x) = 1 - 2x^2$$

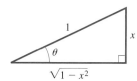

Fig. 7-45

EXAMPLE 12   A triangular brace of sides a, b, and c supports a shelf, as shown in Fig. 7-46. Find the expression for the angle between sides b and c.

The law of cosines leads to the solution, which follows.

$$a^2 = b^2 + c^2 - 2bc \cos A$$ law of cosines

$$2bc \cos A = b^2 + c^2 - a^2$$ solving for $\cos A$

$$\cos A = \frac{b^2 + c^2 - a^2}{2bc}$$

$$A = \cos^{-1}\left(\frac{b^2 + c^2 - a^2}{2bc}\right)$$ using meaning of inverse cosine

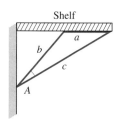

Fig. 7-46

EXERCISES 7-5

In Exercises 1–8, write down the meaning of each of the given equations. See Example 1.

1. $y = \tan^{-1} x$ **2.** $y = \sec^{-1} x$

3. $y = \cot^{-1} 3x$ **4.** $y = \csc^{-1} 4x$

5. $y = 2 \sin^{-1} x$ **6.** $y = 3 \tan^{-1} x$

7. $y = 5 \cos^{-1}(2x - 1)$ **8.** $y = 4 \sin^{-1}(3x + 2)$

In Exercises 9–28, evaluate the given expressions.

9. $\cos^{-1} 0.5$ **10.** $\sin^{-1} 1$

11. $\tan^{-1} 1$ **12.** $\cos^{-1} 0$

13. $\tan^{-1}(-\sqrt{3})$ **14.** $\sin^{-1}(-0.5)$

15. $\sec^{-1} 2$ **16.** $\cot^{-1} \sqrt{3}$

17. $\sin^{-1}(-\sqrt{2}/2)$ **18.** $\cos^{-1}(-\sqrt{3}/2)$

19. $\csc^{-1}\sqrt{2}$ **20.** $\cot^{-1}(-\sqrt{3})$

21. $\tan(\sin^{-1} 0)$ **22.** $\csc(\tan^{-1} 1)$

23. $\sin(\tan^{-1} \sqrt{3})$ **24.** $\tan[\sin^{-1}(\sqrt{2}/2)]$

25. $\cos[\tan^{-1}(-1)]$ **26.** $\sec[\cos^{-1}(-0.5)]$

27. $\cos(2 \sin^{-1} 1)$ **28.** $\sin(2 \tan^{-1} 2)$

In Exercises 29–32, find the exact value of x.

29. $\tan^{-1} x = \sin^{-1} \frac{2}{5}$ **30.** $\cot^{-1} x = \cos^{-1} \frac{1}{3}$

31. $\sec^{-1} x = -\sin^{-1}(-\frac{1}{2})$ **32.** $\sin^{-1} x = -\tan^{-1}(-1)$

In Exercises 33–44, use a calculator to evaluate the given expressions.

33. $\tan^{-1}(-3.7321)$ **34.** $\cos^{-1}(-0.6561)$

35. $\sin^{-1}(-0.8326)$ **36.** $\tan^{-1} 0.2846$

37. $\cos^{-1} 0.1291$ **38.** $\sin^{-1} 0.2119$

39. $\tan^{-1} 8.2614$ **40.** $\sin^{-1}(-0.8881)$

41. $\tan[\cos^{-1}(-0.6281)]$ **42.** $\cos[\tan^{-1}(-1.2256)]$

43. $\sin[\tan^{-1}(-0.2297)]$ **44.** $\tan[\sin^{-1}(-0.3019)]$

In Exercises 45–52, solve the given equations for x.

45. $y = \sin 3x$ **46.** $y = \cos(x - \pi)$

47. $y = \tan^{-1}(x/4)$ **48.** $y = 2 \sin^{-1}(x/6)$

49. $y = 1 + 3 \sec 3x$ **50.** $4y = 5 - 2 \csc 8x$

51. $1 - y = \cos^{-1}(1 - x)$ **52.** $2y = \cot^{-1} 3x - 5$

In Exercises 53–60, find an algebraic expression for each of the given expressions.

53. $\tan(\sin^{-1} x)$ **54.** $\sin(\cos^{-1} x)$

55. $\cos(\sec^{-1} x)$ **56.** $\cot(\cot^{-1} x)$

57. $\sec(\csc^{-1} 3x)$ **58.** $\tan(\sin^{-1} 2x)$

59. $\sin(2 \sin^{-1} x)$ **60.** $\cos(2 \tan^{-1} x)$

In Exercises 61–64, solve the given problems with the use of the inverse trigonometric functions.

61. In the analysis of ocean tides, the equation $y = A \cos 2(\omega t + \phi)$ is used. Solve for t.

62. For an object of weight w on an inclined plane that is at an angle θ to the horizontal, the equation relating w and θ is $\mu w \cos \theta = w \sin \theta$, where μ is the coefficient of friction between the surfaces in contact. Solve for θ.

63. The electric current in a certain circuit is given by $i = I_m[\sin(\omega t + \alpha)\cos \phi + \cos(\omega t + \alpha)\sin \phi]$. Solve for t.

64. The time t as a function of the displacement d of a piston is given by $t = \dfrac{1}{2\pi f} \cos^{-1} \dfrac{d}{A}$. Solve for d.

In Exercises 65 and 66, prove that the given expressions are equal. Use the relation for $\sin(\alpha + \beta)$ and show that the sine of the sum of the angles on the left equals the sine of the angle on the right.

65. $\sin^{-1} \dfrac{3}{5} + \sin^{-1} \dfrac{5}{13} = \sin^{-1} \dfrac{56}{65}$

66. $\tan^{-1} \dfrac{1}{3} + \tan^{-1} \dfrac{1}{2} = \dfrac{\pi}{4}$

In Exercises 67 and 68, verify the given expressions.

67. $\sin^{-1} 0.5 + \cos^{-1} 0.5 = \pi/2$

68. $\tan^{-1} \sqrt{3} + \cot^{-1} \sqrt{3} = \pi/2$

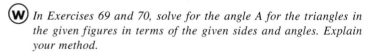 *In Exercises 69 and 70, solve for the angle A for the triangles in the given figures in terms of the given sides and angles. Explain your method.*

69. Fig. 7-47   **70.** Fig. 7-48

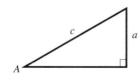 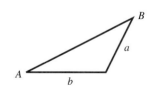

In Exercises 71 and 72, derive the given expressions.

71. The height of the Statue of Liberty is 151 ft. See Fig. 7-49. From the deck of a boat at a horizontal distance d from the statue, the angles of elevation of the top of the statue and the top of its pedestal are α and β, respectively. Show that

$$\alpha = \tan^{-1}\left(\frac{151}{d} + \tan \beta\right).$$

151 ft

Fig. 7-49

72. Show that the length L of the pulley belt shown in Fig. 7-50 is $L = 24 + 11\pi + 10 \sin^{-1}\dfrac{5}{13}$.

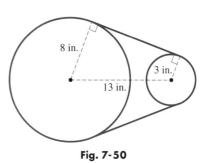

8 in.

3 in.

13 in.

Fig. 7-50

7-6 DERIVATIVES OF THE INVERSE TRIGONOMETRIC FUNCTIONS

To obtain the derivative of $y = \sin^{-1} u$, we first solve for u in the form $u = \sin y$ and then take derivatives with respect to x:

$$\frac{du}{dx} = \cos y \frac{dy}{dx}$$

Solving this equation for dy/dx, we obtain

$$\frac{dy}{dx} = \frac{1}{\cos y}\frac{du}{dx} = \frac{1}{\sqrt{1 - \sin^2 y}}\frac{du}{dx} = \frac{1}{\sqrt{1 - u^2}}\frac{du}{dx}$$

We choose the positive square root since $\cos y > 0$ for $-\frac{\pi}{2} < y < \frac{\pi}{2}$, which is the range of the defined values of $\sin^{-1} u$. Therefore, we obtain the following result:

$$\boxed{\frac{d(\sin^{-1} u)}{dx} = \frac{1}{\sqrt{1 - u^2}}\frac{du}{dx}} \qquad (7\text{-}40)$$

Note that the derivative of the inverse sine function is an algebraic function.

■EXAMPLE 1 Find the derivative of $y = \sin^{-1} 4x$.

$$\frac{dy}{dx} = \frac{1}{\sqrt{1 - (4x)^2}}\overset{\overset{\displaystyle \frac{du}{dx}}{}}{(4)} = \frac{4}{\sqrt{1 - 16x^2}} \qquad \text{using Eq. (7-40)}$$

We find the derivative of the inverse cosine function by letting $y = \cos^{-1} u$ and by following the same procedure as that used in finding the derivative of $\sin^{-1} u$:

$$u = \cos y, \qquad \frac{du}{dx} = -\sin y \frac{dy}{dx}$$

$$\frac{dy}{dx} = -\frac{1}{\sin y}\frac{du}{dx} = -\frac{1}{\sqrt{1 - \cos^2 y}}\frac{du}{dx}$$

Therefore,

$$\boxed{\frac{d(\cos^{-1} u)}{dx} = -\frac{1}{\sqrt{1 - u^2}}\frac{du}{dx}} \tag{7-41}$$

The positive square root is chosen here since $\sin y > 0$ for $0 < y < \pi$, which is the range of the defined values of $\cos^{-1} u$. We note that the derivative of the inverse cosine is the negative of the derivative of the inverse sine.

By letting $y = \tan^{-1} u$, solving for u, and taking derivatives, we find the derivative of the inverse tangent function:

$$u = \tan y, \qquad \frac{du}{dx} = \sec^2 y \frac{dy}{dx}, \qquad \frac{dy}{dx} = \frac{1}{\sec^2 y}\frac{du}{dx} = \frac{1}{1 + \tan^2 y}\frac{du}{dx}$$

Therefore,

$$\boxed{\frac{d(\tan^{-1} u)}{dx} = \frac{1}{1 + u^2}\frac{du}{dx}} \tag{7-42}$$

We can see that the derivative of the inverse tangent is an algebraic function also.

The inverse sine, inverse cosine, and inverse tangent prove to be of the greatest importance in applications and in further development of mathematics. Therefore, the formulas for the derivatives of the other inverse functions are not presented here, although they are included in the exercises.

SOLVING A WORD PROBLEM

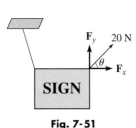

Fig. 7-51

■EXAMPLE 2   A 20-N force acts on a sign as shown in Fig. 7-51. Express the angle θ as a function of the x-component F_x of the force and then find the expression for the instantaneous rate of change of θ with respect to F_x.

From the figure we see that $F_x = 20 \cos \theta$. Solving for θ, we have $\theta = \cos^{-1}(F_x/20)$. To find the instantaneous rate of change of θ with respect to F_x, we are to take the derivative $d\theta/dF_x$:

$$\theta = \cos^{-1}\frac{F_x}{20} = \cos^{-1} 0.05F_x$$

$$\frac{d\theta}{dF_x} = -\frac{1}{\sqrt{1 - (0.05F_x)^2}}(0.05) \qquad \text{using Eq. (7-41)}$$

$$= \frac{-0.05}{\sqrt{1 - 0.0025F_x^2}}$$

EXAMPLE 3   Find the derivative of $y = (x^2 + 1)\tan^{-1} x - x$.

Using the product rule along with Eq. (7-42) on the first term, we have

$$\frac{dy}{dx} = (x^2 + 1)\left(\frac{1}{1 + x^2}\right)(1) + (\tan^{-1} x)(2x) - 1$$

using Eq. (7-42)

$$= 2x \tan^{-1} x$$

EXAMPLE 4   Find the differential of $y = x(\tan^{-1} 2x)^2$.

Using the product rule, the power rule Eq. (3-15), and Eq. (7-42), along with the meaning of the differential, we have

using Eq. (3-15)

$$dy = \left[x(2)(\tan^{-1} 2x)\left(\frac{1}{1 + (2x)^2}\right)(2) + (\tan^{-1} 2x)^2(1) \right] dx$$

using Eq. (7-42)

$$= \tan^{-1} 2x\left(\frac{4x}{1 + 4x^2} + \tan^{-1} 2x\right) dx$$

EXAMPLE 5   Find the derivative of $y = x \sin^{-1} 2x + \frac{1}{2}\sqrt{1 - 4x^2}$.

We write

using Eq. (7-40)   using Eq. (3-15)

$$\frac{dy}{dx} = x\left(\frac{2}{\sqrt{1 - 4x^2}}\right) + \sin^{-1} 2x + \frac{1}{2}\left(\frac{1}{2}\right)(1 - 4x^2)^{-1/2}(-8x)$$

$$= \frac{2x}{\sqrt{1 - 4x^2}} + \sin^{-1} 2x - \frac{2x}{\sqrt{1 - 4x^2}}$$

$$= \sin^{-1} 2x$$

EXAMPLE 6   Find the slope of a tangent to the curve $y = \dfrac{\tan^{-1} x}{x^2 + 1}$, where $x = 3.60$. In Fig. 7-52 the calculator display of the function and the tangent line is shown.

Here we are to find the derivative and then evaluate it for $x = 3.60$.

$$\frac{dy}{dx} = \frac{(x^2 + 1)\left(\dfrac{1}{1 + x^2}\right)(1) - (\tan^{-1} x)(2x)}{(x^2 + 1)^2} \qquad \text{take derivative}$$

$$= \frac{1 - 2x \tan^{-1} x}{(x^2 + 1)^2}$$

$$\left.\frac{dy}{dx}\right|_{x=3.60} = \frac{1 - 2(3.60)(\tan^{-1} 3.60)}{(3.60^2 + 1)^2} \qquad \text{evaluate}$$

$$= -0.0429$$

This result can be checked by using the *numerical derivative* feature of a graphing calculator.

For reference, Eq. (3-15) is
$$\frac{du^n}{dx} = nu^{n-1}\left(\frac{du}{dx}\right).$$

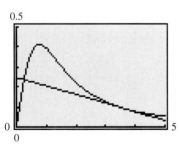

Fig. 7-52

EXERCISES 7-6

In Exercises 1–32, find the derivatives of the given functions.

1. $y = \sin^{-1} x^2$

2. $y = \sin^{-1}(1 - x^2)$

3. $y = 2 \sin^{-1} 3x^3$

4. $y = \sin^{-1} \sqrt{1 - 2x}$

5. $y = 3.6 \cos^{-1} 0.5s$

6. $\theta = 0.2 \cos^{-1} 5t$

7. $y = 2 \cos^{-1} \sqrt{2 - x}$

8. $y = 3 \cos^{-1}(x^2 + 0.5)$

9. $y = \tan^{-1} \sqrt{x}$

10. $y = \tan^{-1}(1 - x)$

11. $y = 6 \tan^{-1}(1/x)$

12. $y = 4 \tan^{-1} 3x^4$

13. $y = 5x \sin^{-1} x$

14. $y = x^2 \cos^{-1} x$

15. $v = 0.4u \tan^{-1} 2u$

16. $y = (x^2 + 1)\sin^{-1} 4x$

17. $y = \dfrac{3x - 1}{\sin^{-1} 2x}$

18. $\theta = \dfrac{\tan^{-1} 2r}{\pi r}$

19. $y = \dfrac{\sin^{-1} 2x}{\cos^{-1} 2x}$

20. $y = \dfrac{x^2 + 1}{\tan^{-1} x}$

21. $y = 2(\cos^{-1} 4x)^3$

22. $r = 0.5(\sin^{-1} 3t)^4$

23. $y = [\sin^{-1}(4x + 1)]^2$

24. $y = \sqrt{\sin^{-1}(x - 1)}$

25. $y = \tan^{-1}\left(\dfrac{1 - t}{1 + t}\right)$

26. $y = \dfrac{1}{\cos^{-1} 2x}$

27. $y = \dfrac{1}{1 + 4x^2} - \tan^{-1} 2x$

28. $y = \sin^{-1} x - \sqrt{1 - x^2}$

29. $y = 3(4 - \cos^{-1} 2x)^3$

30. $\sin^{-1}(x + y) + y = x^2$

31. $2 \tan^{-1} xy + x = 3$

32. $y = \sqrt{1 - \sin^{-1} 4x}$

In Exercises 33–48, solve the given problems.

(W) 33. On a calculator, find the values of (a) $1/\sqrt{1 - 0.5^2}$ and (b) $(\sin^{-1} 0.5001 - \sin^{-1} 0.5000)/0.0001$. Compare the values and give the meaning of each in relation to the derivative of $\sin^{-1} x$ where $x = 0.5$.

(W) 34. On a calculator, find the values of (a) $1/(1 + 0.5^2)$ and (b) $(\tan^{-1} 0.5001 - \tan^{-1} 0.5000)/0.0001$. Compare the values and give the meaning of each in relation to the derivative of $\tan^{-1} x$ where $x = 0.5$.

35. Find the differential of the function $y = (\sin^{-1} x)^3$.

36. Find the linearization $L(x)$ of the function $f(x) = 2x \cos^{-1} x$ for $a = 0$.

37. Find the slope of a line tangent to the curve $y = x/\tan^{-1} x$ at $x = 0.80$. Verify the result by using the *numerical derivative* feature of a graphing calculator.

(W) 38. Explain what is wrong with a problem that requires finding the derivative of $y = \sin^{-1}(x^2 + 1)$.

39. Use a graphing calculator to display the graphs of $y = \sin^{-1} x$ and $y = 1/\sqrt{1 - x^2}$. By roughly estimating slopes of tangent lines of $y = \sin^{-1} x$, note that $y = 1/\sqrt{1 - x^2}$ gives reasonable values for the derivative of $y = \sin^{-1} x$.

40. Use a graphing calculator to display the graphs of $y = \tan^{-1} x$ and $y = 1/(1 + x^2)$. By roughly estimating slopes of tangent lines of $y = \tan^{-1} x$, note that $y = 1/(1 + x^2)$ gives reasonable values for the derivative of $y = \tan^{-1} x$.

41. Find the second derivative of the function $y = \tan^{-1} 2x$.

42. Show that $\dfrac{d(\cot^{-1} u)}{dx} = -\dfrac{1}{1 + u^2}\dfrac{du}{dx}$.

43. Show that $\dfrac{d(\sec^{-1} u)}{dx} = \dfrac{1}{\sqrt{u^2(u^2 - 1)}}\dfrac{du}{dx}$.

44. Show that $\dfrac{d(\csc^{-1} u)}{dx} = -\dfrac{1}{\sqrt{u^2(u^2 - 1)}}\dfrac{du}{dx}$.

45. In the analysis of the waveform of an AM radio wave, the equation $t = \dfrac{1}{\omega} \sin^{-1} \dfrac{A - E}{mE}$ arises. Find dt/dm, assuming that the other quantities are constant.

46. An equation that arises in the theory of solar collectors is $\alpha = \cos^{-1} \dfrac{2f - r}{r}$. Find the expression for $d\alpha/dr$ if f is constant.

47. As a person approaches a building of height h, the angle of elevation of the top of the building is a function of the person's distance from the building. Express the angle of elevation θ in terms of h and the distance x from the building and then find $d\theta/dx$. Assume the person's height is negligible to that of the building. See Fig. 7-53.

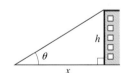

Fig. 7-53

48. A triangular metal frame is being designed as shown in Fig. 7-54. Express the angle A as a function of x and evaluate dA/dx for $x = 6$ cm.

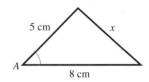

Fig. 7-54

7-7 APPLICATIONS

With our development of the formulas for the derivatives of the trigonometric and inverse trigonometric functions, it is now possible for us to apply these derivatives in the same manner as we applied the derivatives of algebraic functions. We can now use trigonometric and inverse trigonometric functions to solve tangent and normal line, Newton's method, time rate of change, curve tracing, maximum and minimum, and differential application problems. The following examples illustrate the use of these functions in these types of problems.

■**EXAMPLE 1** Sketch the curve $y = \sin^2 x - \dfrac{x}{2}$ $(0 \le x \le 2\pi)$.

First, by setting $x = 0$, we see that the only easily obtainable intercept is $(0, 0)$. Replacing x by $-x$ and y by $-y$, we find that the curve is not symmetric to either axis or to the origin. Also, since x does not appear in a denominator, there are no vertical asymptotes. We are considering only the restricted domain $0 \le x \le 2\pi$. (Without this restriction, the domain is all x and the range is all y.)

To find the information from the derivatives, we write

$$\frac{dy}{dx} = 2 \sin x \cos x - \frac{1}{2} = \sin 2x - \frac{1}{2}$$

$$\frac{d^2y}{dx^2} = 2 \cos 2x$$

Maximum and minimum points will occur for $\sin 2x = \frac{1}{2}$. Thus, we have possible maximum and minimum points for

$$2x = \frac{\pi}{6}, \frac{5\pi}{6}, \frac{13\pi}{6}, \frac{17\pi}{6} \quad \text{or} \quad x = \frac{\pi}{12}, \frac{5\pi}{12}, \frac{13\pi}{12}, \frac{17\pi}{12}$$

Now, using the second derivative, we find that d^2y/dx^2 is positive for $x = \frac{\pi}{12}$ and $x = \frac{13\pi}{12}$ and is negative for $x = \frac{5\pi}{12}$ and $x = \frac{17\pi}{12}$. Thus, the maximum points are $(\frac{5\pi}{12}, 0.279)$ and $(\frac{17\pi}{12}, -1.29)$. Minimum points are $(\frac{\pi}{12}, -0.064)$ and $(\frac{13\pi}{12}, -1.63)$. Inflection points occur for $\cos 2x = 0$, or

$$2x = \frac{\pi}{2}, \frac{3\pi}{2}, \frac{5\pi}{2}, \frac{7\pi}{2} \quad \text{or} \quad x = \frac{\pi}{4}, \frac{3\pi}{4}, \frac{5\pi}{4}, \frac{7\pi}{4}$$

Therefore, the points of inflection are $(\frac{\pi}{4}, 0.11)$, $(\frac{3\pi}{4}, -0.68)$, $(\frac{5\pi}{4}, -1.46)$, and $(\frac{7\pi}{4}, -2.25)$. Using this information, we sketch the curve in Fig. 7-55.

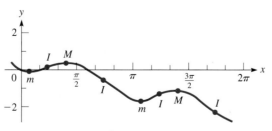

Fig. 7-55

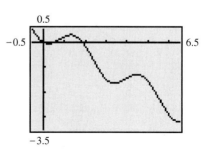

Fig. 7-56

See Fig. 7-56 for a graphing calculator display of this graph.

EXAMPLE 2 By using Newton's method, solve the equation $2x - 1 = 3 \cos x$.

First we locate the required root approximately by sketching $y_1 = 2x - 1$ and $y_2 = 3 \cos x$. Using the graphing calculator view shown in Fig. 7-57, we see that they intersect between $x = 1$ and $x = 2$, near $x = 1.2$. Therefore, using $x_1 = 1.2$, with

$$f(x) = 2x - 1 - 3 \cos x$$
$$f'(x) = 2 + 3 \sin x$$

we use Eq. (4-1), which is

$$x_2 = x_1 - \frac{f(x_1)}{f'(x_1)}$$

To find x_2, we have

$$f(x_1) = 2(1.2) - 1 - 3 \cos 1.2 = 0.3129267$$
$$f'(x_1) = 2 + 3 \sin 1.2 = 4.7961173$$
$$x_2 = 1.2 - \frac{0.3129267}{4.7961173} = 1.1347542$$

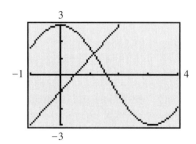

Fig. 7-57

Applying this procedure again for the next approximation, we find $x_3 = 1.1342366$, which is accurate to the value shown. Again, when using the calculator it is not necessary to list the values of $f(x_1)$ and $f'(x_1)$, as the complete calculation can be done directly on the calculator.

If we use the *intersect* feature of a graphing calculator, we find that we obtain the value of x_3 given above. ∎

SOLVING A WORD PROBLEM

EXAMPLE 3 Logs with a circular cross section 4.0 ft in diameter are cut in half lengthwise. Find the largest rectangular cross-sectional area that can then be cut from one of the halves.

First, we draw Fig. 7-58 to help set up the necessary equation. From the figure we see that $x = 2.0 \cos \theta$ and $y = 2.0 \sin \theta$, which means that the area of the rectangle inscribed within the semicircular area is

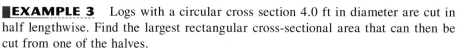

$$A = (2x)y = 2(2.0 \cos \theta)(2.0 \sin \theta) = 8.0 \cos \theta \sin \theta$$
$$= 4.0 \sin 2\theta \qquad \text{using trigonometric identity (Eq. (7-20))}$$

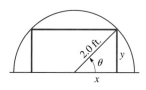

Fig. 7-58

Now, taking the derivative and setting it equal to zero, we have

$$\frac{dA}{d\theta} = (4.0 \cos 2\theta)(2) = 8.0 \cos 2\theta$$

$$8.0 \cos 2\theta = 0, \qquad 2\theta = \frac{\pi}{2}, \qquad \theta = \frac{\pi}{4} \qquad \cos \frac{\pi}{2} = 0$$

(Using $2\theta = \frac{3\pi}{2}$, $\theta = \frac{3\pi}{4}$ leads to the same solution.) Since the minimum area is zero, we have the maximum area when $\theta = \frac{\pi}{4}$, and this maximum area is

$$A = 4.0 \sin 2\left(\frac{\pi}{4}\right) = 4.0 \sin \frac{\pi}{2} = 4.0 \text{ ft}^2 \qquad \sin \frac{\pi}{2} = 1$$

Therefore, the largest rectangular cross-sectional area is 4.0 ft^2. ∎

SOLVING A WORD PROBLEM

See the chapter introduction.

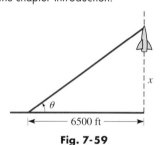

Fig. 7-59

EXAMPLE 4   A rocket is taking off vertically at a distance of 6500 ft from an observer. If, when the angle of elevation is 38.4°, it is changing at the rate of 5.0°/s, how fast is the rocket ascending?

From Fig. 7-59 we see that

$$\tan \theta = \frac{x}{6500}, \quad \text{or} \quad \theta = \tan^{-1} \frac{x}{6500}$$

Taking derivatives with respect to time, we have

$$\frac{d\theta}{dt} = \frac{1}{1 + (x/6500)^2} \frac{dx/dt}{6500}$$

$$= \frac{6500 \, dx/dt}{6500^2 + x^2}$$

When evaluating this expression, we must remember to express angles in radians. Therefore, this means that $d\theta/dt = 5.0°/s = 0.0873$ rad/s. Substituting this value and $x = 6500 \tan 38.4° = 5150$ ft, we have

$$0.0873 = \frac{6500 \, dx/dt}{6500^2 + 5150^2}$$

$$\frac{dx}{dt} = 924 \text{ ft/s}$$

EXAMPLE 5   A particle is moving so that its x- and y-coordinates are given by $x = \cos 2t$ and $y = \sin 2t$. Find the magnitude and direction of its velocity when $t = \pi/8$.

Taking derivatives with respect to time, we have

$$v_x = \frac{dx}{dt} = -2 \sin 2t \qquad v_y = \frac{dy}{dt} = 2 \cos 2t$$

$$v_x \big|_{t=\pi/8} = -2 \sin 2\left(\frac{\pi}{8}\right) = -2\left(\frac{\sqrt{2}}{2}\right) = -\sqrt{2} \qquad \text{evaluating}$$

$$v_y \big|_{t=\pi/8} = 2 \cos 2\left(\frac{\pi}{8}\right) = 2\left(\frac{\sqrt{2}}{2}\right) = \sqrt{2}$$

$$v = \sqrt{v_x^2 + v_y^2} = \sqrt{2 + 2} = 2 \qquad \text{magnitude}$$

$$\tan \theta = \frac{v_y}{v_x} = \frac{\sqrt{2}}{-\sqrt{2}} = -1$$

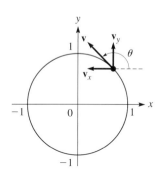

Fig. 7-60

This curve can be plotted by assuming values of *t* to find the necessary values of *x* and *y*.

Since v_x is negative and v_y is positive, $\theta = 135°$. Note that in this example θ is the angle, in standard position, between the horizontal and the resultant velocity.

By plotting the curve we note that it is a circle (see Fig. 7-60). Thus, the object is moving about a circle in a counterclockwise direction. It can be determined in another way that the curve is a circle. If we square each of the expressions defining x and y and then add these, we have the equation $x^2 + y^2 = \cos^2 2t + \sin^2 2t = 1$. This is a circle of radius 1.

SOLVING A WORD PROBLEM

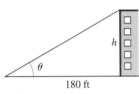

180 ft

Fig. 7-61

■**EXAMPLE 6**   At a point 180 ft from the base of a building on level ground, the angle of elevation of the top of the building is 30.00°. What would be the error in calculating the height of the building due to an error 0.25° in the measurement of the angle?

From Fig. 7-61 we see that $h = 180 \tan \theta$. To find the error in h, we must find the differential dh. Therefore, we have

$$dh = 180 \sec^2 \theta \, d\theta$$

The possible error in θ is 0.25°, which in *radian measure* is $0.25\pi/180$, which is the value we should use in the calculation of dh. We can use the value $\theta = 30.00°$ if we have the calculator set in degree mode. Calculating dh, we have

$$dh = \frac{180(0.25\pi/180)}{\cos^2 30.00°} = 1.05 \text{ ft}$$

In using the calculator we divide by $\cos^2 30.00°$ since $\sec \theta = 1/\cos \theta$.

We see that an error of 0.25° in the angle can result in an error of over 1 ft in the calculated value of the height.

--------------■

━━━━━━━━━━━━━━━━━━━━━ EXERCISES 7-7 ━━━━━━━━━━━━━

1. Show that the slopes of the sine and cosine curves are negatives of each other at the points of intersection.

2. Show that the tangent curve is always increasing (when the tangent is defined.)

3. Show that the curve $y = \tan^{-1} x$ is always increasing.

4. Sketch the graph of $y = \sin x + \cos x \, (0 \leq x \leq 2\pi)$. Check the graph on a graphing calculator.

5. Sketch the graph of $y = x - \tan x \, (-\frac{\pi}{2} < x < \frac{\pi}{2})$. Check the graph on a graphing calculator.

6. Sketch the graph of $y = 2 \sin x + \sin 2x \, (0 \leq x \leq 2\pi)$. Check the graph on a graphing calculator.

7. Find the equation of the line tangent to the curve $y = \sin 2x$ at $x = \frac{5\pi}{8}$.

8. Find the equation of a line normal to the curve $y = \tan^{-1}(x/2)$ at $x = 3$.

9. By Newton's method, find the positive root of the equation $x^2 - 4 \sin x = 0$. Verify the result by using the *intersect* (or *zero*) feature of a graphing calculator.

10. By Newton's method, find the smallest positive root of the equation $\tan x = 2x$. Verify the result by using the *intersect* (or *zero*) feature of a graphing calculator.

11. Find the maximum value of the function $y = 6 \cos x - 8 \sin x$.

12. Find the minimum value of the function $y = \cos 2x + 2 \sin x$.

13. In studying water waves, the vertical displacement y (in ft) of a wave was determined to be $y = 0.50 \sin 2t + 0.30 \cos t$, where t is the time (in s). Find the velocity and the acceleration for $t = 0.40$ s.

14. At 30°N latitude the number of hours h of daylight each day during the year is given approximately by the equation $h = 12.1 + 2.0 \sin[\frac{\pi}{6}(x - 2.7)]$, where x is measured in months ($x = 0.5$ is Jan. 15, etc.). Find the date of the longest day and the date of the shortest day. (Cities near 30°N are Houston, Texas, and Cairo, Egypt.)

15. Find the time rate of change of the horizontal component T_x of the constant 46.6-lb tension shown in Fig. 7-62 if $d\theta/dt = 0.36°/\text{s}$ for $\theta = 14.2°$.

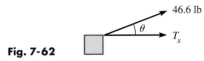

Fig. 7-62

16. The *apparent power* P_a (in W) in an electric circuit whose power is P and whose impedance phase angle is θ is given by $P_a = P \sec \theta$. Given that P is constant at 12 W, find the time rate of change of P_a if θ is changing at the rate of 0.050 rad/min, when $\theta = 40.0°$.

17. A point on the rim of a 5.25-in. computer floppy disk can be described by the equations $x = 2.625 \cos 12\pi t$ and $y = 2.625 \sin 12\pi t$. Find the velocity of the point for $t = 1.250$ s.

18. A machine is programmed to move an etching tool such that the position (in cm) of the tool is given by $x = 2 \cos 3t$ and $y = \cos 2t$, where t is the time (in s). Find the velocity of the tool for $t = 4.1$ s.

19. Find the acceleration of the point on the floppy disk of Exercise 17 for $t = 1.250$ s.

20. Find the acceleration of the tool of Exercise 18 for $t = 4.1$ s.

21. A person observes an object dropped from the top of a building 100 ft away. If the top of the building is 200 ft above the person's eye level, how fast is the angle of elevation of the object changing after 1.0 s? (The distance the object drops is given by $s = 16t^2$.) See Fig. 7-63.

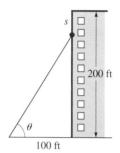

Fig. 7-63

22. A car passes directly under a police helicopter 450 ft above a straight and level highway. After the car has traveled another 50.0 ft, the angle of depression of the car from the helicopter is decreasing at the rate of 0.215 rad/s. What is the speed of the car?

23. A searchlight is 225 ft from a straight wall. As the beam moves along the wall, the angle between the beam and the perpendicular to the wall is increasing at the rate of 1.5°/s. How fast is the length of the beam increasing when it is 315 ft long? See Fig. 7-64.

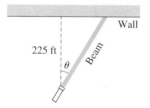

Fig. 7-64

24. In a modern hotel, where the elevators are directly observable from the lobby area (and a person can see from the elevators), a person in the lobby observes one of the elevators rising at the rate of 12.0 ft/s. If the person was 50.0 ft from the elevator when it left the lobby, how fast is the angle of elevation of the line of sight to the elevator increasing 10.0 s later?

25. If a block is placed on a plane inclined with the horizontal at an angle θ such that the block just moves down the plane, the coefficient of friction μ is given by $\mu = \tan \theta$. Use differentials to find the change in μ if θ changes from 20° to 21°.

26. The electric power p (in W) developed in a resistor in an FM receiver circuit is $p = 0.0307 \cos^2 120\pi t$, where t is the time (in s). Linearize p for $t = 0.0010$ s.

27. A surveyor measures two sides and the included angle of a triangular parcel of land to be 82.04 m, 75.37 m, and 38.38°. What error is caused in the calculation of the third side by an error of 0.15° in the angle?

28. To connect the four vertices of a square with the minimum amount of electric wire requires using the wiring pattern shown in Fig. 7-65. Find θ for the total length of wire ($L = 4x + y$) to be a minimum.

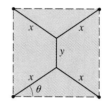

Fig. 7-65

29. The strength S of a rectangular beam is directly proportional to the product of its width w and the square of its depth d. Use trigonometric functions to find the dimensions of the strongest beam that can be cut from a circular log 16.0 in. in diameter. (See Example 4 on page 146).

30. An architect is designing a window in the shape of an isosceles triangle with a perimeter of 60 in. What is the vertex angle of the window of greatest area?

31. A wall is 6.0 ft high and 4.0 ft from a building. What is the length of the shortest pole that can touch the building and the ground beyond the wall? (*Hint:* From Fig. 7-66 it can be shown that $y = 6.0 \csc \theta + 4.0 \sec \theta$.)

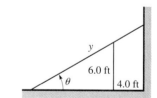

Fig. 7-66

32. The television screen at a sports arena is vertical and 8.0 ft high. The lower edge is 25.0 ft above an observer's eye level. If the best view of the screen is obtained when the angle subtended by the screen at eye level is a maximum, how far from directly below the screen must the observer's eye be? See Fig. 7-67.

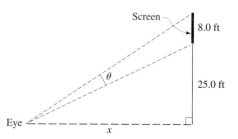

Fig. 7-67

CHAPTER EQUATIONS

Radian-degree conversion $\qquad$ $\pi \text{ rad} = 180°$ $\qquad$ (7-1)

Trigonometric functions $\qquad$ $\sin \theta = \dfrac{y}{r} \qquad \cos \theta = \dfrac{x}{r} \qquad \tan \theta = \dfrac{y}{x}$ $\qquad$ (7-2)

$$\cot \theta = \dfrac{x}{y} \qquad \sec \theta = \dfrac{r}{x} \qquad \csc \theta = \dfrac{r}{y}$$

Basic trigonometric identities $\qquad$ $\sin \theta = \dfrac{1}{\csc \theta} \qquad \cos \theta = \dfrac{1}{\sec \theta} \qquad \tan \theta = \dfrac{1}{\cot \theta}$ $\qquad$ (7-4)

$$\tan \theta = \dfrac{\sin \theta}{\cos \theta} \qquad \cot \theta = \dfrac{\cos \theta}{\sin \theta} \qquad (7\text{-}5)$$

$$\sin^2 \theta + \cos^2 \theta = 1 \qquad (7\text{-}6)$$

$$1 + \tan^2 \theta = \sec^2 \theta \qquad (7\text{-}7)$$

$$1 + \cot^2 \theta = \csc^2 \theta \qquad (7\text{-}8)$$

Trigonometric functions of negative angles $\qquad$ $\sin(-\theta) = -\sin \theta \qquad \cos(-\theta) = \cos \theta$

$$\tan(-\theta) = -\tan \theta \qquad (7\text{-}9)$$

Law of cosines $\qquad$ $c^2 = a^2 + b^2 - 2ab \cos \theta$ $\qquad$ (7-10)

Law of sines $\qquad$ $\dfrac{a}{\sin A} = \dfrac{b}{\sin B} = \dfrac{c}{\sin C}$ $\qquad$ (7-11)

Sum and difference formulas $\qquad$ $\sin(\alpha + \beta) = \sin \alpha \cos \beta + \cos \alpha \sin \beta$ $\qquad$ (7-16)

$$\cos(\alpha + \beta) = \cos \alpha \cos \beta - \sin \alpha \sin \beta \qquad (7\text{-}13)$$

$$\sin(\alpha - \beta) = \sin \alpha \cos \beta - \cos \alpha \sin \beta \qquad (7\text{-}17)$$

$$\cos(\alpha - \beta) = \cos \alpha \cos \beta + \sin \alpha \sin \beta \qquad (7\text{-}12)$$

$$\tan(\alpha \pm \beta) = \dfrac{\tan \alpha \pm \tan \beta}{1 \mp \tan \alpha \tan \beta} \qquad (7\text{-}18)$$

$$\sin A - \sin B = 2 \cos \dfrac{A + B}{2} \sin \dfrac{A - B}{2} \qquad (7\text{-}19)$$

Double-angle formulas $\qquad$ $\sin 2\alpha = 2 \sin \alpha \cos \alpha$ $\qquad$ (7-20)

$$\cos 2\alpha = \cos^2 \alpha - \sin^2 \alpha \qquad (7\text{-}21)$$

$$= 2 \cos^2 \alpha - 1 \qquad (7\text{-}22)$$

$$= 1 - 2 \sin^2 \alpha \qquad (7\text{-}23)$$

$$\tan 2\alpha = \dfrac{2 \tan \alpha}{1 - \tan^2 \alpha} \qquad (7\text{-}24)$$

Half-angle formulas $\qquad$ $\sin \dfrac{\alpha}{2} = \pm \sqrt{\dfrac{1 - \cos \alpha}{2}}$ $\qquad$ (7-25)

$$\cos \dfrac{\alpha}{2} = \pm \sqrt{\dfrac{1 + \cos \alpha}{2}} \qquad (7\text{-}26)$$

Circular arc length	$s = \theta r$ (θ in radians)	
Circular sector area	$A = \dfrac{1}{2}\,\theta r^2$ (θ in radians)	
Limit of $\dfrac{\sin\theta}{\theta}$ as $\theta \to 0$	$\displaystyle\lim_{\theta\to 0}\frac{\sin\theta}{\theta} = \lim_{\Delta u\to 0}\frac{\sin(\Delta u/2)}{\Delta u/2} = 1$	
Chain rule	$\dfrac{dy}{dx} = \dfrac{dy}{du}\dfrac{du}{dx}$	
Derivatives	$\dfrac{d(\sin u)}{dx} = \cos u\,\dfrac{du}{dx}$	(7-32)
	$\dfrac{d(\cos u)}{dx} = -\sin u\,\dfrac{du}{dx}$	(7-33)
	$\dfrac{d(\tan u)}{dx} = \sec^2 u\,\dfrac{du}{dx}$	(7-34)
	$\dfrac{d(\cot u)}{dx} = -\csc^2 u\,\dfrac{du}{dx}$	(7-35)
	$\dfrac{d(\sec u)}{dx} = \sec u\,\tan u\,\dfrac{du}{dx}$	(7-36)
	$\dfrac{d(\csc u)}{dx} = -\csc u\,\cot u\,\dfrac{du}{dx}$	(7-37)
Inverse trigonometric functions	$y = \sin^{-1} x \quad \left(-\dfrac{\pi}{2} \le y \le \dfrac{\pi}{2}\right)$	(7-38)

$$-\frac{\pi}{2} \le \sin^{-1} x \le \frac{\pi}{2} \qquad 0 \le \cos^{-1} x \le \pi \qquad -\frac{\pi}{2} < \tan^{-1} x < \frac{\pi}{2} \qquad 0 < \cot^{-1} x < \pi$$

$$0 \le \sec^{-1} x \le \pi \;\left(\sec^{-1} x \ne \frac{\pi}{2}\right) \qquad -\frac{\pi}{2} \le \csc^{-1} x \le \frac{\pi}{2} \;(\csc^{-1} x \ne 0)$$

(7-39)

Derivatives	$\dfrac{d(\sin^{-1} u)}{dx} = \dfrac{1}{\sqrt{1-u^2}}\dfrac{du}{dx}$	(7-40)
	$\dfrac{d(\cos^{-1} u)}{dx} = -\dfrac{1}{\sqrt{1-u^2}}\dfrac{du}{dx}$	(7-41)
	$\dfrac{d(\tan^{-1} u)}{dx} = \dfrac{1}{1+u^2}\dfrac{du}{dx}$	(7-42)

REVIEW EXERCISES

In Exercises 1–24, find the derivative of the given functions.

1. $y = 3\cos(4x - 1)$

2. $y = 4\sec(1 - x^3)$

3. $u = 0.2\tan\sqrt{3 - 2v}$

4. $y = 5\sin(1 - 6x)$

5. $y = \csc^2(3x + 2)$

6. $r = \cot^2 5\pi\theta$

7. $y = 3\cos^4 x^2$

8. $y = 2\sin^3\sqrt{x}$

9. $y = 3\tan^{-1}\left(\dfrac{x}{3}\right)$

10. $y = 0.4\cos^{-1}(2\pi t + 1)$

11. $y = \sin^{-1}(\cos x)$

12. $y = \sin(\tan^{-1} x)$

13. $y = \sqrt{\csc 4x + \cot 4x}$

14. $y = \cos^2(\tan x)$

15. $y = (1 + \sin 2x)^4$

16. $y = \sqrt{\dfrac{1 + \cos 2x}{2}}$

$\dfrac{x^2}{\tan^{-1} 2x}$

18. $y = \dfrac{\sin^{-1} x}{4x}$

$y^2 \sin 2x + \tan x = 0$

20. $y = x(\sin^{-1} x)^2 + 2\sqrt{1 - x^2}\, \sin^{-1} x - 2x$

21. $y = x \cos^{-1} x - \sqrt{1 - x^2}$ **22.** $x + \sec^2(xy) = 1$

23. $x \sin 2y = y \cos 2x$ **24.** $\tan^{-1}(y/x) = x^2 y$

In Exercises 25–48, solve the given problems.

25. Sketch the graph of $y = x - \cos x$. Check the graph by displaying it on a graphing calculator.

26. Sketch the graph of $y = \sin 4x + \cos 2x$. Check the graph by displaying it on a graphing calculator.

27. Find the equation of the line tangent to the curve $y = 4 \cos^2(x^2)$ at $x = 1$.

28. Find the equation of the line normal to the curve $y = \tan^{-1} x$ at $x = 1$.

29. Find the derivative of each member of the identity $\sin^2 x + \cos^2 x = 1$ and show that the results give another identity.

30. Find the derivative of each member of the identity

$\sin(x + 1) = \sin x \cos 1 + \cos x \sin 1$

and show that the results give another identity.

31. By Newton's method find the positive root of the equation $\sin x = 1 - x$.

32. By Newton's method, solve the equation $x^2 = \tan^{-1} x$. Check the solution by displaying the graph on a calculator and using the *intersect* (or *zero*) feature.

33. An analysis of temperature records for Sydney, Australia, indicates that the average daily temperature (in °C) during the year is given approximately by $T = 17.2 + 5.2 \cos[\frac{\pi}{6}(x - 0.50)]$, where x is measured in months ($x = 0.5$ is Jan. 15, etc.). What is the *daily* time rate of change of temperature on March 1?
(*Hint:* 12 months/365 days = 0.033 month/day = dx/dt.)

34. Power can be defined as the time rate of doing work. If work is being done in an electric circuit according to $W = 10 \cos 2t$, find P as a function of t.

35. In the theory of making images by holography, an expression used for the light-intensity distribution is $I = kE_0^2 \cos^2 \frac{1}{2}\theta$, where k and E_0 are constants and θ is the phase angle between two light waves. Find the expression for $dI/d\theta$.

36. In the design of a cone-type clutch, an equation that relates the cone angle θ and the applied force F is $\theta = \sin^{-1}(Ff/R)$, where R is the frictional resistance and f is the coefficient of friction. For constant R and f, find $d\theta/dF$.

37. If we neglect air resistance, the range R of a bullet fired at an angle θ with the horizontal is $R = \dfrac{v_0^2}{g} \sin 2\theta$, where v_0 is the initial velocity and g is the acceleration due to gravity. Find θ for the maximum range. See Fig. 7-68.

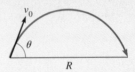

Fig. 7-68 R

38. An object attached to a cord of length l, as shown in Fig. 7-69, moves in a circular path. The angular velocity ω is given by $\omega = \sqrt{g/(l \cos \theta)}$. By use of differentials, find the approximate change in ω if θ changes from 32.50° to 32.75°, given that $g = 9.800$ m/s^2 and $l = 0.6375$ m.

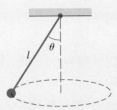

Fig. 7-69

39. Find the magnitude and direction of the velocity of an object that moves so that its x- and y-coordinates are given by $x = \sin 2\pi t$ and $y = 2 \cos^2 \pi t$ when $t = 13/12$.

40. Find the magnitude and direction of the acceleration of the object in Exercise 39.

41. A football is thrown horizontally (assume no arc) at 56 ft/s parallel to the sideline. A TV camera is 92 ft from the path of the football. Find $d\theta/dt$, the rate at which the camera must turn to follow the ball when $\theta = 15°$. See Fig. 7-70.

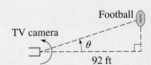

TV camera Football

Fig. 7-70 92 ft

42. A force P (in lb) at an angle θ above the horizontal drags a 50-lb box across a level floor. The coefficient of friction between the floor and the box is constant and equals 0.20. The magnitude of the force P is given by $P = \dfrac{(0.20)(50)}{0.20 \sin \theta + \cos \theta}$. Find θ such that P is a minimum.

43. A jet is flying at 880 ft/s directly away from the control tower of an airport. If the jet is at a constant altitude of 6800 ft, how fast is the angle of elevation of the jet from the control tower changing when it is 13.0°?

44. When a wheel rolls along a straight line, a point P on the circumference traces a curve called a *cycloid*. See Fig. 7-71. The parametric equations of a cycloid are $x = r(\theta - \sin\theta)$ and $y = r(1 - \cos\theta)$. Find the velocity of the point on the rim of a wheel for which $r = 5.500$ cm and $d\theta/dt = 0.12$ rad/s for $\theta = 35.0°$. [An inverted cycloid is the path of least time of descent (the *brachistochrone*) of an object acted on only by gravity.]

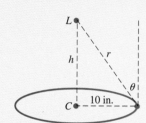

Fig. 7-71

45. The illuminance from a point source of light varies directly as the cosine of the angle of incidence (measured from the perpendicular) and inversely as the square of the distance r from the source. How high above the center of a circle of radius 10.0 in. should a light be placed so that the illuminance at the circumference will be a maximum? See Fig. 7-72.

Fig. 7-72   **Fig. 7-73**

46. A Y-shaped metal bracket is to be made such that its height is 10.0 cm and its width across the top is 6.00 cm. What shape will require the least amount of material? See Fig. 7-73.

47. A gutter is to be made from a sheet of metal 12 in. wide by turning up strips of width 4 in. along each side to make equal angles θ with the vertical. Sketch a graph of the cross-sectional area A as a function of θ. See Fig. 7-74.

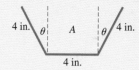

Fig. 7-74

48. A conical filter is made from a circular piece of wire mesh of radius 24.0 cm by cutting out a sector with central angle θ and then taping the cut edges of the remaining piece together (see Fig. 7-75). What is the maximum possible volume the resulting filter can hold?

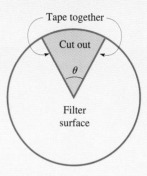

Fig. 7-75

Writing Exercise

49. To find the area of the largest rectangular microprocessor chip with a perimeter of 40 mm, it is possible to use either an algebraic function or a trigonometric function. Write two or three paragraphs to explain how each type of function can be used to find the required area.

PRACTICE TEST

In Problems 1–3, find the derivative of each of the functions.

1. $y = \tan^3 2x + \tan^{-1} 2x$ **2.** $y = 2(3 + \cot 4x)^3$

3. $y\sec 2x = \sin^{-1} 3y$

4. Find the differential of the function $y = \dfrac{\cos^2(3x + 1)}{x}$.

5. Find the slope of a tangent to the curve of $y = x\cos^{-1} 2x$ for $x = 0.1$. Also record the value shown by the *numerical derivative* feature of a graphing calculator.

6. The equation $T = \dfrac{A}{1 + B\sin^2(\theta/2)}$ is used in the study of the transmission of light, where T and θ are the variables. Find the expression for the time rate of change of T.

7. Sketch the graph of $y = x - \cos x$ for $-\pi < x < 2\pi$.

8. A balloon leaves the ground 250 ft from an observer and rises at the rate of 5.0 ft/s. How fast is the angle of elevation of the balloon increasing after 8.0 s?

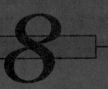
DIFFERENTIATION OF THE EXPONENTIAL AND LOGARITHMIC FUNCTIONS

In this chapter we develop the formulas for the derivatives of two more important types of functions, the *exponential function* and the *logarithmic function*.

Historically, logarithms were first developed (in the early seventeenth century) to help with the lengthy calculations that were needed in areas such as navigation and astronomy. With the extensive use of calculators and computers, they are not used for such purposes today. However, these functions are very important in many scientific and technical applications, as well as in areas of advanced mathematics.

To illustrate the importance of exponential functions, many applications may be cited. These functions are used extensively in electronics, mechanical systems, thermodynamics, and nuclear physics. They are also used in biology in studying population growth and in business to calculate compound interest.

The use of the logarithmic function, which is closely related to the exponential function, is also extensive. The basic units used to measure intensity of sound and those used to measure the intensity of earthquakes are defined in terms of logarithms. In chemistry, the distinction between a base and an acid is defined in terms of logarithms. In electrical transmission lines, power gains and losses are measured in terms of logarithmic units.

In the first section of this chapter, we review the properties of the exponential and logarithmic functions. In the remaining sections, we develop formulas for their derivatives and show applications of these derivatives.

The growth of population can often be measured by use of an exponential function. In Section 8-4 we show how the rate of population growth is measured.

8-1 EXPONENTIAL AND LOGARITHMIC FUNCTIONS

To now we have dealt with exponents in the form x^n, where n is a rational number and a constant. We shall now deal with expressions of the form b^x, where b is a constant and x is any real number. The main difference is that in the second expression *the exponent is a variable*. Therefore, we define the **exponential function** *to be*

$$y = b^x \qquad (8\text{-}1)$$

In Eq. (8-1) x is called the **logarithm** *of the number y to the base b.* In our work with the exponential function, we shall restrict all numbers to the real number system. *This leads us to choose the base as a positive number other than* 1. We know that 1 raised to any power will result in 1, which would make y a constant regardless of the value of x. Negative numbers for b would result in imaginary values for y if x were any fractional exponent with an even integer for its denominator.

EXAMPLE 1 $y = 2^x$ is an exponential function, where x is the logarithm of y to the base 2. This means that 2 raised to a given power gives the corresponding value of y.

If $x = 3$, $y = 2^3 = 8$; this means that 3 is the logarithm of 8 to the base 2. If $x = 4$, $y = 2^4 = 16$; this means that 4 is the logarithm of 16 to the base 2.

If $x = \frac{1}{2}$, $y = 2^{1/2} \approx 1.41$. Here $\frac{1}{2}$ is the logarithm of 1.41 to the base 2.

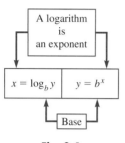

Fig. 8-1

Using the definition of a logarithm, we may express x in terms of y in the form

$$x = \log_b y \qquad\qquad (8\text{-}2)$$

This equation is read in accordance with the definition of x in Eq. (8-1); that is, *x equals the logarithm of y to the base b.* This means that x is the power to which the base b must be raised in order to equal the number y; that is, x is a logarithm, and

CAUTION *a logarithm is an exponent.* Note that Eqs. (8-1) and (8-2) state the same relationship but in a different way. Equation (8-1) is the **exponential form,** and Eq. (8-2) is the **logarithmic form.** See Fig. 8-1.

EXAMPLE 2 The equation $y = 2^x$ would be written as $x = \log_2 y$ if we put it in logarithmic form. When we choose values of y to find the corresponding values of x from this equation, we ask ourselves "2 raised to what power x gives y?"

This means that if $y = 8$, we know that $2^3 = 8$, and therefore $x = 3$.

EXAMPLE 3 **(a)** $3^2 = 9$ in logarithmic form is $2 = \log_3 9$.

(b) $4^{-1} = 1/4$ in logarithmic form is $-1 = \log_4(1/4)$.

CAUTION Remember, *the exponent may be negative.* The *base* must be positive.

When we are working with functions, we must keep in mind that a function is defined by the operation being performed on the independent variable, and not by the letter chosen to represent it. However, for consistency, it is standard practice to let y represent the dependent variable and x represent the independent variable. Therefore, *the* **logarithmic function** *is*

$$y = \log_b x \qquad\qquad (8\text{-}3)$$

As with the exponential function, $b > 0$ and $b \neq 1$.

NOTE Equations (8-2) and (8-3) do not represent different *functions,* due to the difference in location of the variables, since they represent the *same operation* on the independent variable that appears in each. However, Eq. (8-3) expresses the function with the standard dependent and independent variables.

We note that for the exponential function $y = b^x$ and the logarithmic function $y = \log_b x$, if we change the form of one and then interchange the variables x and y, we then obtain the other function. This means that *the exponential function and the logarithmic function are inverse functions.* This follows from the definition of inverse functions given on page 247 in the discussion of the inverse trigonometric functions.

We shall now show the graphs of an exponential function and a logarithmic function. In this way we can show their properties.

■EXAMPLE 4   Plot the graph of $y = 2^x$.

Assuming values for x and then finding the corresponding values for y, we obtain the following table.

x	-3	-2	-1	0	1	2	3
y	$\frac{1}{8}$	$\frac{1}{4}$	$\frac{1}{2}$	1	2	4	8

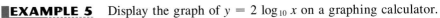

From these values we plot the curve, as shown in Fig. 8-2. We note that the x-axis is an asymptote of the curve. ----------■

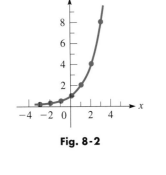

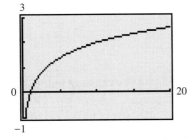

Fig. 8-2

The log keys on a calculator are the *log* key (for $\log_{10} x$) and the *ln* key (for $\log_e x$). The number e is an irrational number equal to approximately 2.718. This number is used widely in certain applications and for theoretical purposes. In the next section we shall see why e is chosen for this purpose. The following example illustrates the graph of the logarithmic function on a graphing calculator.

■EXAMPLE 5   Display the graph of $y = 2 \log_{10} x$ on a graphing calculator.

On a graphing calculator, we set $y_1 = 2 \log x$, and the curve is displayed as shown in Fig. 8-3. The settings for Xmin and Xmax were selected since the domain is $x > 0$ and $\log_{10} x$ increases very slowly. The settings for Ymin and Ymax were selected since $\log_{10} 0.1 = -1$ and y does not reach 4 until $x = 100$. ----------■

Fig. 8-3

Any exponential or logarithmic curve, where $b > 1$, will be similar in shape to those shown in Examples 4 and 5. From these examples we can see that these curves have the following basic properties.

Basic Properties of Exponential and Logarithmic Curves

1. *If $0 < x < 1$, $\log_b x < 0$; if $x = 1$, $\log_b 1 = 0$; if $x > 1$, $\log_b x > 0$.*

2. *If $x > 1$, x increases more rapidly than $\log_b x$.*

3. *For all values of x, $b^x > 0$.*

4. *If $x > 1$, b^x increases more rapidly than x.*

From the graphs in Figs. 8-2 and 8-3 and the above analysis, we note that the domain of the exponential function $y = b^x$ is all real numbers and that its range is $y > 0$. For the logarithmic function $y = \log_b x$, the domain is $x > 0$, and the range is all real numbers.

A logarithm is an exponent. However, a historical curiosity is that logarithms were developed before exponents were used.

Since a logarithm is an exponent, it must follow the laws of exponents. Those laws that are of the greatest importance at this time are listed here for reference.

$$b^u b^v = b^{u+v} \tag{8-4}$$

$$\frac{b^u}{b^v} = b^{u-v} \tag{8-5}$$

$$(b^u)^n = b^{nu} \tag{8-6}$$

We will use these laws of exponents to derive certain useful properties of logarithms.

If we let $u = \log_b x$ and $v = \log_b y$ and write these equations in exponential form, we have $x = b^u$ and $y = b^v$. Forming the product of x and y, we obtain

$$xy = b^u b^v = b^{u+v} \quad \text{or} \quad xy = b^{u+v}$$

Writing this last equation in logarithmic form yields $u + v = \log_b xy$, or

LOGARITHM OF A PRODUCT

$$\log_b xy = \log_b x + \log_b y \tag{8-7}$$

Equation (8-7) states the property that *the logarithm of the product of two numbers is equal to the sum of the logarithms of the numbers.*

Using the same definitions of u and v to form the quotient of x and y, we then have

$$\frac{x}{y} = \frac{b^u}{b^v} = b^{u-v} \quad \text{or} \quad \frac{x}{y} = b^{u-v}$$

Writing this last equation in logarithmic form, we have $u - v = \log_b(x/y)$, or

LOGARITHM OF A QUOTIENT

$$\log_b \left(\frac{x}{y} \right) = \log_b x - \log_b y \tag{8-8}$$

Equation (8-8) states the property that *the logarithm of the quotient of two numbers is equal to the logarithm of the numerator minus the logarithm of the denominator.*

If we again let $u = \log_b x$ and write this in exponential form, we have $x = b^u$. To find the nth power of x, we write

$$x^n = (b^u)^n = b^{nu}$$

Expressing this equation in logarithmic form yields $nu = \log_b(x^n)$, or

LOGARITHM OF A POWER

$$\log_b(x^n) = n \log_b x \tag{8-9}$$

Equation (8-9) states that *the logarithm of the nth power of a number is equal to n times the logarithm of the number.* The exponent n may be integral or fractional.

Previously we showed that the base b of logarithms must be a positive number. Since $x = b^u$ and $y = b^v$, this means that x and y are also positive numbers. Therefore, *the properties of logarithms that have just been derived are valid only for positive values of x and y.*

In advanced mathematics the logarithms of negative and imaginary numbers are defined.

■EXAMPLE 6 **(a)** Using Eq. (8-7) we may express $\log_4 15$ as a sum of logarithms.

$$\log_4 15 = \log_4(3 \times 5) = \log_4 3 + \log_4 5 \qquad \begin{array}{l}\text{logarithm of product}\\ \text{sum of logarithms}\end{array}$$

(b) Using Eq. (8-8) we may express $\log_4\left(\frac{5}{3}\right)$ as the difference of logarithms.

$$\log_4\left(\frac{5}{3}\right) = \log_4 5 - \log_4 3 \qquad \begin{array}{l}\text{logarithm of quotient}\\ \text{difference of logarithms}\end{array}$$

(c) Using Eq. (8-9) we may express $\log_4(t^2)$ as twice $\log_4 t$.

$$\log_4(t^2) = 2\log_4 t \qquad \begin{array}{l}\text{logarithm of power}\\ \text{multiple of logarithm}\end{array}$$

(d) For the equation $\log_b y = 2\log_b x + \log_b a$, we may use Eq. (8-9) and then Eq. (8-7), to write

$$\log_b y = \log_b(x^2) + \log_b a = \log_b(ax^2)$$

Since we have the logarithm to the base b of different expressions on each side of the resulting equation, the expressions must be equal. Therefore,

$$y = ax^2 \qquad \text{-------------■}$$

Two bases of logarithms are widely used. Base 10 logarithms were used widely for calculational purposes before the extensive use of calculators. They are still used in certain specific scientific applications. Base e logarithms are widely used in many technical and scientific applications. We noted previously that a calculator uses two different logarithm keys, the *log* key for base 10 logarithms and the *ln* key for base e logarithms.

The notation **log** x *is used for logarithms to the base 10.* Note that no base is shown in this notation. *The notation* **ln** x *is used for logarithms to the base e.* Base e logarithms are called *natural logarithms* (thus, the n in the notation). They are used extensively in calculus. As we noted before, we will see the reason for using the number e when we discuss the derivative of the logarithmic function in the next section.

Since more than one base is important, at times it is useful to change a logarithm from one base to another. If $u = \log_b x$, then $b^u = x$. Taking logarithms of both sides of this last expression to the base a, we have

$$\log_a b^u = \log_a x, \qquad u\log_a b = \log_a x, \qquad u = \frac{\log_a x}{\log_a b}$$

However, $u = \log_b x$, which means that

$$\boxed{\log_b x = \frac{\log_a x}{\log_a b}} \qquad (8\text{-}10)$$

Logarithms were invented by the Scottish mathematician John Napier (1550–1617), who published a table of base e logarithms. The English mathematician Henry Briggs (1561–1631) realized base 10 logarithms made calculations much easier. For many years he laboriously developed a table of base 10 logarithms, which was completed after his death. Logarithms were received enthusiastically by mathematicians and scientists as a long-needed tool for lengthy calculations. They were used commonly for calculations until the 1970s when the scientific calculator came into use.

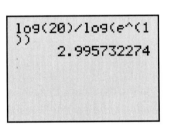

Fig. 8-4

■EXAMPLE 7 Change log 20 to a logarithm with base e; that is, find ln 20.
Using Eq. (8-10) with $a = 10$, $b = e$, and $x = 20$, we have

$$\log_e 20 = \frac{\log_{10} 20}{\log_{10} e} \qquad \text{or} \qquad \ln 20 = \frac{\log 20}{\log e} = 2.996$$

This means that $e^{2.996} = 20$. See Fig. 8-4. $\qquad \text{-------------■}$

EXERCISES $8\text{-}1$

In Exercises 1–12, change those in exponential form to logarithmic form and those in logarithmic form to exponential form.

1. $4^4 = 256$

2. $3^{-2} = \frac{1}{9}$

3. $\log_8 16 = \frac{4}{3}$

4. $\log_7 (\frac{1}{49}) = -2$

5. $8^{1/3} = 2$

6. $81^{3/4} = 27$

7. $\log_{0.5} 16 = -4$

8. $\log_{243} 3 = \frac{1}{5}$

9. $12^0 = 1$

10. $(\frac{1}{2})^{-2} = 4$

11. $\log_8(\frac{1}{512}) = -3$

12. $\log_{1/32} (\frac{1}{8}) = \frac{3}{5}$

In Exercises 13–16, determine the value of the unknown.

13. $\log_7 y = 3$

14. $\log_b 4 = \frac{2}{3}$

15. $\log 10^{0.2} = x$

16. $\log_8 y = -\frac{2}{3}$

In Exercises 17–20, plot the graphs of the given functions.

17. $y = 3^x$

18. $y = (1/3)^x$

19. $y = \log_2 x$

20. $y = 2 \log_4 x$

In Exercises 21–28, express each as a single logarithm.

21. $\log_5 9 - \log_5 3$

22. $\log_2 3 + \log_2 x$

23. $\log_6 x^2 + \log_6 \sqrt{x}$

24. $\log_7 3^4 - \log_7 9$

25. $2 \log_b 2 + \log_b y$

26. $3 \log_b 4 - \log_b x$

27. $2 \log_2 x + \log_2 3 - \log_2 5$

28. $\log_4 7 - 2 \log_4 t + \log_4 a$

In Exercises 29–32, evaluate the given logarithm by use of logarithms to base 10.

29. $\ln 3$

30. $\ln 0.9$

31. $\log_7 42$

32. $\log_{12} 122$

In Exercises 33–48, solve the given problems.

33. The magnitudes (visual brightnesses), m_1 and m_2, of two stars are related to their (actual) brightnesses, b_1 and b_2, by the equation $m_1 - m_2 = 2.5 \log_{10}(b_2/b_1)$. Solve for b_2.

34. The velocity v of a rocket at the point at which its fuel is completely burned is given by $v = u \log_e(w_0/w)$, where u is the exhaust velocity, w_0 is the lift-off weight, and w is the burnout weight. Solve for w.

35. An equation relating the number N of atoms of radium at any time t in terms of the number N_0 of atoms at $t = 0$ is $\ln(N/N_0) = -kt$, where k is a constant. Solve for N.

36. The charge q on a capacitor is given by $q = q_0(1 - e^{-at})$, where q_0 is the initial charge, a is a constant, and t is the time. Solve for t.

37. A projection of the annual growth rate p (in %) of the number of users of the Internet is $p = 10(1.2^{-t} + 1)$, where t is the number of years after 2000. Sketch the graph of this function from 2000 to 2010.

38. The current i (in A) in a certain electric circuit is given by $i = 16(1 - e^{-250t})$, where t is the time in seconds. Using appropriate values of t, sketch the graph of this function.

39. The time t (in ps) required for N calculations by a certain computer design is $t = N + \log_2 N$. Sketch the graph of this function.

40. An original amount of 100 mg of radium radioactively decomposes such that N mg remain after t years. The function relating t and N is $t = 2350(\log_e 100 - \log_e N)$. Sketch the graph.

41. The distance x traveled by a motorboat in t seconds after the engine is cut off is given by $x = k^{-1} \ln (kv_0t + 1)$, where v_0 is the velocity of the boat at the time the engine is cut and k is a constant. Find how long it takes a boat to go 150 m if $v_0 = 12.0$ m/s and $k = 6.80 \times 10^{-3}$/m.

42. The electric current i (in A) in a circuit containing a 1-H inductor, a 10-Ω resistor, and a 6-V battery is a function of the time t (in s) given by $i = 0.6(1 - e^{-10t})$. Solve for t as a function of i.

43. Computations can be made by use of logarithms. Use Eq. (8-7) and the logarithms (to base 10) of 6.700 and 3400 to calculate the value of the product (6.700)(3400). The final product will require raising 10 to the appropriate power.

Ⓦ **44.** Find the product in Exercise 43 by use of natural logarithms. What change in the procedure from that of Exercise 43 is required?

45. Graphs of inverse functions displayed on the same coordinate system are mirror images of each other across the line $y = x$. Show this by displaying the graphs of $y = e^x$, $y = \ln x$, and $y = x$ on a graphing calculator. Choose settings such that the scales on the axes are about equally spaced.

46. Repeat the instructions of Exercise 45 for the functions $y = 10^{x/2}$, $y = 2 \log x$, and $y = x$.

47. The **hyperbolic sine of u** is defined as

$$\sinh u = \frac{1}{2}(e^u - e^{-u})$$

and the **hyperbolic cosine of u** is defined as

$$\cosh u = \frac{1}{2}(e^u + e^{-u})$$

These functions are called hyperbolic functions since, if $x = \cosh u$ and $y = \sinh u$, x and y satisfy the equation of the hyperbola $x^2 - y^2 = 1$. Verify this by substituting the appropriate exponential expressions into the equation of the hyperbola and simplifying.

48. From the definitions of $\sinh u$ and $\cosh u$ (see Exercise 47), find the values of $\sinh 0.25$ and $\cosh 0.25$ and show that $\cosh^2 0.25 - \sinh^2 0.25 = 1$.

$8\text{-}2$ DERIVATIVE OF THE LOGARITHMIC FUNCTION

We shall first find the derivative of the logarithmic function. Then in the next section we can use this derivative to find the derivative of the exponential function. Again, we shall use the delta-process.

If we let $y = \log_b u$, where u is a function of x, we have

$$y + \Delta y = \log_b(u + \Delta u)$$

$$\Delta y = \log_b(u + \Delta u) - \log_b u = \log_b\left(\frac{u + \Delta u}{u}\right) = \log_b\left(1 + \frac{\Delta u}{u}\right)$$

$$\frac{\Delta y}{\Delta u} = \frac{\log_b(1 + \Delta u/u)}{\Delta u} = \frac{1}{u}\frac{u}{\Delta u}\log_b\left(1 + \frac{\Delta u}{u}\right)$$

$$= \frac{1}{u}\log_b\left(1 + \frac{\Delta u}{u}\right)^{u/\Delta u}$$

(We multiply and divide by u for purposes of evaluating the limit, as we shall now show.)

Before we can evaluate $\lim_{\Delta u \to 0} \Delta y/\Delta u$, we must determine

$$\lim_{\Delta u \to 0}\left(1 + \frac{\Delta u}{u}\right)^{u/\Delta u}$$

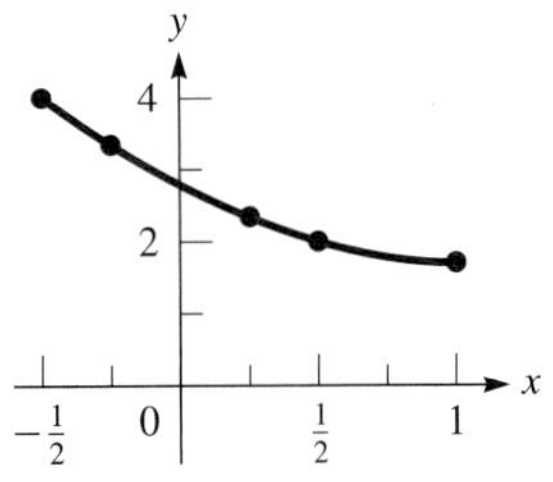

We can see that the exponent becomes unbounded, but the number being raised to this exponent approaches 1. Therefore, we shall investigate this limiting value.

To find an approximate value, let us graph the function $y = (1 + x)^{1/x}$ (for purposes of graphing, we let $\Delta u/u = x$). We construct a table of values and then graph the function (see Fig. 8-5).

Fig. 8-5

x	-0.5	-0.25	$+0.25$	$+0.50$	$+1.00$
y	4.00	3.16	2.44	2.25	2.00

Only these values are shown, since we are interested in the y-value corresponding to $x = 0$. We see from the graph that this value is approximately 2.7. Choosing very small values of x, we may obtain these values:

x	0.1	0.01	0.001	0.0001
y	2.5937	2.7048	2.7169	2.71815

NOTE ▶

By methods we will develop in Chapter 13, it can be shown that this value is about 2.7182818. *The limiting value is the irrational number e,* which is the base of natural logarithms we introduced in the previous section.

Returning to the derivative of the logarithmic function, we have

$$\lim_{\Delta u \to 0}\frac{\Delta y}{\Delta u} = \lim_{\Delta u \to 0}\left[\frac{1}{u}\log_b\left(1 + \frac{\Delta u}{u}\right)^{u/\Delta u}\right] = \frac{1}{u}\log_b e$$

Therefore,

$$\frac{dy}{du} = \frac{1}{u}\log_b e$$

The number e is named for the Swiss mathematician Leonhard Euler (1707–1783).

For reference, Eq. (7-31) is
$$\frac{dy}{dx} = \frac{dy}{du}\frac{du}{dx}.$$

Combining this equation with Eq. (7-31), we have

$$\frac{d(\log_b u)}{dx} = \frac{1}{u}\log_b e\,\frac{du}{dx} \qquad \text{(8-11)}$$

At this point we see that if we choose e as the basis of a system of logarithms, the above formula becomes

$$\frac{d(\ln u)}{dx} = \frac{1}{u}\frac{du}{dx} \qquad \text{(8-12)}$$

The choice of e as the base b makes $\log_b e = \log_e e = 1$; thus, this factor does not appear in Eq. (8-12). We now see why the number e is chosen as the base for a system of logarithms, the natural logarithms. The notation $\ln u$ is the same as that used in Section 8-1 for natural logarithms.

■EXAMPLE 1 Find the derivative of $y = \log 4x$.
Using Eq. (8-11) we have

$$\frac{dy}{dx} = \frac{1}{4x}(\log e)(\overset{\overset{\textstyle \frac{du}{dx}}{\textstyle |}}{4})$$

$$= \frac{1}{x}\log e \qquad (\log e = 0.4343)$$

■EXAMPLE 2 Find the derivative of $s = \ln 3t^4$.
Using Eq. (8-12) we have (with $u = 3t^4$)

$$\frac{ds}{dt} = \frac{1}{3t^4}(12t^3) \quad\underset{\frac{du}{dt}}{\Big\uparrow}$$

$$= \frac{4}{t}$$

■EXAMPLE 3 Find the derivative of $y = \ln \tan 4x$.
Using Eq. (8-12) along with the derivative of the tangent, we have

$$\frac{dy}{dx} = \frac{1}{\tan 4x}(\sec^2 4x)(4) \quad\underset{\frac{d\tan 4x}{dx}}{\Big\uparrow}$$

$$= \frac{\cos 4x}{\sin 4x}\frac{4}{\cos^2 4x} \qquad \text{using trigonometric relations}$$

$$= \frac{1}{\sin 4x}\frac{4}{\cos 4x}$$

$$= 4\csc 4x \sec 4x$$

NOTE ▶   Frequently, we can find the derivative of a logarithmic function more simply if we *use the properties of logarithms to simplify the expression before the derivative is found.* The following examples illustrate this.

For reference, Eqs. (8-7), (8-8), and (8-9) are:

$$\log_b xy = \log_b x + \log_b y$$

$$\log_b \left(\frac{x}{y}\right) = \log_b x - \log_b y$$

$$\log_b x^n = n \log_b x$$

■EXAMPLE 4   Find the derivative of $y = \ln \dfrac{x - 1}{x + 1}$.

In this example it is easier to differentiate if we first write y in the form

$$y = \ln(x - 1) - \ln(x + 1)$$

by using the properties of logarithms (Eq. (8-8)). Hence,

$$\frac{dy}{dx} = \frac{1}{x - 1} - \frac{1}{x + 1} = \frac{x + 1 - x + 1}{(x - 1)(x + 1)}$$

$$= \frac{2}{x^2 - 1}$$

■EXAMPLE 5   **(a)** Find the derivative of $y = \ln(1 - 2x)^3$.

First, using Eq. (8-9), we rewrite the equation as $y = 3 \ln(1 - 2x)$. Then we have

$$\frac{dy}{dx} = 3\left(\frac{1}{1 - 2x}\right)(-2) = \frac{-6}{1 - 2x}$$

(b) Find the derivative of $y = \ln^3(1 - 2x)$.

First, we note that

$$y = \ln^3(1 - 2x) = [\ln(1 - 2x)]^3$$

where $\ln^3(1 - 2x)$ is usually the preferred notation.

NOTE ▶   Next, we must be careful to distinguish this function from that in part (a). For $y = \ln^3(1 - 2x)$, it is the logarithm of $1 - 2x$ that is being cubed, whereas for $y = \ln(1 - 2x)^3$, it is $1 - 2x$ that is being cubed.

Now, finding the derivative of $y = \ln^3(1 - 2x)$, we have

$$\frac{dy}{dx} = 3[\ln^2(1 - 2x)]\left(\frac{1}{1 - 2x}\right)(-2)$$

$$= -\frac{6 \ln^2(1 - 2x)}{1 - 2x} \qquad \underset{dx}{\underline{}} \; \frac{d \ln(1 - 2x)}{dx}$$

■EXAMPLE 6   Evaluate the derivative of $y = \ln[(\sin 2x)(\sqrt{x^2 + 1})]$ for $x = 0.375$.

First, using the properties of logarithms, we rewrite the function as

$$y = \ln \sin 2x + \frac{1}{2} \ln(x^2 + 1)$$

Now we have

$$\frac{dy}{dx} = \frac{1}{\sin 2x}(\cos 2x)(2) + \frac{1}{2}\left(\frac{1}{x^2 + 1}\right)(2x) \qquad \text{take the derivative}$$

$$= 2 \cot 2x + \frac{x}{x^2 + 1}$$

$$\left.\frac{dy}{dx}\right|_{x=0.375} = 2 \cot 0.750 + \frac{0.375}{0.375^2 + 1} = 2.48 \qquad \text{evaluate}$$

EXERCISES 8-2

In Exercises 1–32, find the derivatives of the given functions.

1. $y = \log x^2$

2. $y = \log_2 6x$

3. $y = 2 \log_5(3x + 1)$

4. $y = 3 \log_7(x^2 + 1)$

5. $u = 0.2 \ln(1 - 3x)$

6. $y = 2 \ln(3x^2 - 1)$

7. $y = 2 \ln \tan 2x$

8. $y = \ln \sin^2 x$

9. $y = \ln \sqrt{x}$

10. $y = 5 \ln \sqrt{4x - 3}$

11. $y = \ln(x^2 + 2x)^3$

12. $s = [\ln(2t^3 - t)]^2$

13. $v = 3[t + \ln t^2]^2$

14. $y = x^2 \ln 2x$

15. $y = \dfrac{3x}{\ln(2x + 1)}$

16. $y = \dfrac{8 \ln x}{x}$

17. $y = \ln(\ln x)$

18. $r = 0.5 \ln \cos(\pi\theta^2)$

19. $y = \ln \dfrac{2x}{1 + x}$

20. $y = \ln(x\sqrt{x + 1})$

21. $y = \sin \ln x$

22. $y = \tan^{-1} \ln 2x$

23. $u = 3v \ln^2 2v$

24. $h = 0.1s \ln^4 s$

25. $y = \ln(x \tan x)$

26. $y = \ln(x + \sqrt{x^2 - 1})$

27. $y = \ln \dfrac{x^2}{x + 2}$

28. $y = \sqrt{x + \ln 3x}$

29. $y = \sqrt{x^2 + 1} - \ln \dfrac{1 + \sqrt{x^2 + 1}}{x}$

30. $3 \ln xy + \sin y = x^2$

31. $y = x - \ln^2(x + y)$

32. $y = \ln(x + \ln x)$

In Exercises 33–48, solve the given problems.

(W) 33. On a calculator find the value of $(\ln 2.0001 - \ln 2.0000)/0.0001$ and compare it with 0.5. Give the meanings of the value found and 0.5 in relation to the derivative of $\ln x$, where $x = 2$.

(W) 34. On a calculator find the value of $(\ln 0.5001 - \ln 0.5000)/0.0001$ and compare it with 2. Give the meanings of the value found and 2 in relation to the derivative of $\ln x$, where $x = 0.5$.

35. Using a graphing calculator, (a) display the graph of $y = (1 + x)^{1/x}$ to verify that $(1 + x)^{1/x} \to 2.718$ as $x \to 0$ and (b) verify the values for $(1 + x)^{1/x}$ in the tables on page 272.

36. Find the second derivative of the function $y = x^2 \ln x$.

37. Evaluate the derivative of $y = \sin^{-1} 2x + \sqrt{1 - 4x^2}$, where $x = 0.250$.

38. Evaluate the derivative of $y = \ln \sqrt{\dfrac{2x + 1}{3x + 1}}$, where $x = 2.75$.

39. Find the linearization $L(x)$ for the function $f(x) = 2 \ln \tan x$ for $a = \pi/4$.

40. Find the differential of the function $y = 6 \log_x 2$.

41. Find the slope of a line tangent to the curve $y = \tan^{-1} 2x + \ln(4x^2 + 1)$, where $x = 0.625$. Verify the result by using the *numerical derivative* feature of a graphing calculator.

42. Find the slope of a line tangent to the curve $y = x \ln 2x$ at $x = 2$. Verify the result by using the *numerical derivative* feature of a graphing calculator.

(W) 43. Find the derivative of $y = x^x$ by first taking logarithms of each side of the equation. Explain why Eq. (3-15) cannot be used to find the derivative of this function.

(W) 44. Find the derivative of $y = (\sin x)^x$ by first taking logarithms of each side of the equation. Explain why Eq. (3-15) cannot be used to find the derivative of this function.

45. If the loudness b (in decibels) of a sound of intensity I is given by $b = 10 \log(I/I_0)$, where I_0 is a constant, find the expression for db/dt in terms of dI/dt.

46. The time t for a particular computer system to process N bits of data is directly proportional to $N \ln N$. Find the expression for dt/dN.

47. When air friction is considered, the time t (in s) it takes a certain falling object to attain a velocity v (in ft/s) is given by $t = 5 \ln \dfrac{16}{16 - 0.1v}$. Find dt/dv for $v = 100$ ft/s.

48. The electric potential V at a point P at a distance x from an electric charge distributed along a wire of length $2a$ (see Fig. 8-6) is $V = k \ln \dfrac{\sqrt{a^2 + x^2} + a}{\sqrt{a^2 + x^2} - a}$, where k is a constant. Find the expression for the electric field E, which is defined as $E = -dV/dx$.

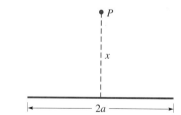

Fig. 8-6

$8\text{-}3$ DERIVATIVE OF THE EXPONENTIAL FUNCTION

To obtain the derivative of the exponential function, we let $y = b^u$ and then take natural logarithms of both sides:

$$\ln y = \ln b^u = u \ln b$$

$$\frac{1}{y}\frac{dy}{dx} = \ln b\frac{du}{dx}$$

$$\frac{dy}{dx} = y \ln b\frac{du}{dx}$$

Thus,

$$\frac{d(b^u)}{dx} = b^u \ln b\left(\frac{du}{dx}\right) \tag{8-13}$$

If we let $b = e$, Eq. (8-13) becomes

$$\frac{d(e^u)}{dx} = e^u\left(\frac{du}{dx}\right) \tag{8-14}$$

The simplicity of Eq. (8-14) compared with Eq. (8-13) again shows the advantage of choosing e as the basis of natural logarithms. It is for this reason that e appears so often in applications of calculus.

■**EXAMPLE 1**   Find the derivative of $y = e^x$.
   Using Eq. (8-14) we have

$$\frac{dy}{dx} = e^x(1) = e^x \qquad \text{with } u = x$$

with the arrow labeled $\dfrac{du}{dx}$ pointing to the (1).

We see that the derivative of the function e^x equals itself. This exponential function is widely used in applications of calculus.   ■

For reference, Eq. (3-15) is
$$\frac{du^n}{dx} = nu^{n-1}\left(\frac{du}{dx}\right).$$

We should note carefully that Eq. (3-15) is used with a variable raised to a constant exponent, whereas with Eqs. (8-13) and (8-14) we are finding the derivative of a constant raised to a variable exponent.

$$\underset{\text{variable constant}}{\frac{du^n}{dx}} = nu^{n-1}\left(\frac{du}{dx}\right) \qquad \underset{\text{constant variable}}{\frac{db^u}{dx}} = b^u \ln b\left(\frac{du}{dx}\right)$$

In the following example we must note carefully this difference in the type of function that leads us to use either Eq. (3-15) or Eq. (8-13) in order to find the derivative.

EXAMPLE 2   Find the derivatives of $y = (4x)^2$ and $y = 2^{4x}$.

Using Eq. (3-15) we have   Using Eq. (8-13) we have

$$y = (4x)^2 \qquad \frac{du}{dx}$$

$$\frac{dy}{dx} = 2(4x)^1(4)$$

$$= 32x$$

$$y = 2^{4x} \qquad \frac{du}{dx}$$

$$\frac{dy}{dx} = 2^{4x}(\ln 2)(4)$$

$$= (4 \ln 2)(2^{4x})$$

We continue now with additional examples of the use of Eq. (8-14).

EXAMPLE 3   Find the derivative of $y = \ln \cos e^{2x}$.

Using Eq. (8-14) along with the derivatives of the logarithmic and cosine functions, we have

$$\frac{dy}{dx} = \frac{1}{\cos e^{2x}} \frac{d \cos e^{2x}}{dx} \qquad \text{using } \frac{d \ln u}{dx} = \frac{1}{u}\frac{du}{dx}$$

$$= \frac{1}{\cos e^{2x}}(-\sin e^{2x})\frac{de^{2x}}{dx} \qquad \text{using } \frac{d \cos u}{dx} = -\sin u \frac{du}{dx}$$

$$= -\frac{\sin e^{2x}}{\cos e^{2x}}(e^{2x})(2) \qquad \text{using } \frac{de^u}{dx} = e^u \frac{du}{dx}$$

$$= -2e^{2x} \tan e^{2x} \qquad \text{using } \frac{\sin \theta}{\cos \theta} = \tan \theta$$

EXAMPLE 4   Find the derivative of $r = \theta e^{\tan \theta}$.

Here we use Eq. (8-14) with the derivatives of a product and the tangent.

$$\frac{dr}{d\theta} = \theta e^{\tan \theta}(\sec^2 \theta) + e^{\tan \theta}(1)$$

$$= e^{\tan \theta}(\theta \sec^2 \theta + 1)$$

EXAMPLE 5   Find the derivative of $y = (e^{1/x})^2$.

In this example we use Eqs. (3-15) and (8-14).

$$\frac{dy}{dx} = 2(e^{1/x})(e^{1/x})\left(-\frac{1}{x^2}\right)$$

using Eqs. (8-14) and (3-15) to find $\frac{du}{dx}$ of Eq. (3-15)

$$= \frac{-2(e^{1/x})^2}{x^2} = \frac{-2e^{2/x}}{x^2}$$

This problem could have also been solved by first writing the function as $y = e^{2/x}$, which is an equivalent form determined by the laws of exponents. When we use this form, the derivative becomes

using Eq. (3-15) to find $\frac{du}{dx}$ of Eq. (8-14)

$$\frac{dy}{dx} = e^{2/x}\left(-\frac{2}{x^2}\right) = \frac{-2e^{2/x}}{x^2}$$

This change in form of the function simplifies the steps necessary for finding the derivative.

■EXAMPLE 6   Find the derivative of $y = (3e^{4x} + 4x^2 \ln x)^3$.

Using the general power rule (Eq. (3-15)) for derivatives, the derivative of the exponential function (Eq. (8-14)), the derivative of a product (Eq. (3-12)), and the derivative of a logarithm (Eq. (8-12)), we have

$$\frac{dy}{dx} = 3(3e^{4x} + 4x^2 \ln x)^2\left[12e^{4x} + 4x^2\left(\frac{1}{x}\right) + (\ln x)(8x)\right]$$

$$= 3(3e^{4x} + 4x^2 \ln x)^2(12e^{4x} + 4x + 8x \ln x) \qquad \text{-------}■$$

■EXAMPLE 7   Find the slope of a line tangent to the curve $y = \dfrac{3e^{2x}}{x^2 + 1}$, where $x = 1.275$.

Here we are to find the derivative and then evaluate it for $x = 1.275$. The solution is as follows:

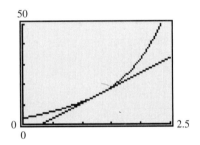

50

0
0   2.5

Fig. 8-7

$$\frac{dy}{dx} = \frac{(x^2 + 1)(3e^{2x})(2) - 3e^{2x}(2x)}{(x^2 + 1)^2} \qquad \text{take the derivative}$$

$$= \frac{6e^{2x}(x^2 - x + 1)}{(x^2 + 1)^2}$$

$$\left.\frac{dy}{dx}\right|_{x=1.275} = \frac{6e^{2(1.275)}(1.275^2 - 1.275 + 1)}{(1.275^2 + 1)^2} \qquad \text{evaluate}$$

$$= 15.05$$

Using the *tangent* feature of a graphing calculator, the function and the tangent line are shown in Fig. 8-7. The value of the derivative can be checked by using the *numerical derivative* feature. --------■

EXERCISES 8-3

In Exercises 1–32, find the derivatives of the given functions.

1. $y = 3^{2x}$

2. $y = 3^{1-x}$

3. $y = 4^{6x}$

4. $y = 10^{x^2}$

5. $y = e^{\sqrt{x}}$

6. $r = 0.3e^{\theta^2}$

7. $y = 4e^t(e^{2t} - e^t)$

8. $y = 0.2 \ln(e^{5x} + 1)$

9. $y = xe^{-x}$

10. $y = 5x^2e^{2x}$

11. $y = xe^{\sin x}$

12. $y = 4e^x \sin \frac{1}{2}x$

13. $r = \dfrac{2(e^{2s} - e^{-2s})}{e^{2s}}$

14. $u = \dfrac{e^{0.5v}}{2v}$

15. $y = e^{-3x} \sin 4x$

16. $y = (\cos 2x)(e^{x^2-1})$

17. $y = \dfrac{2e^{3x}}{4x + 3}$

18. $y = \dfrac{7 \ln 2x}{e^{2x} + 2}$

19. $y = \ln(e^{x^2} + 4)$

20. $p = (3e^{2n} + e^2)^3$

21. $y = (2e^{2x})^3 \sin x^2$

22. $y = (e^{3/x} \cos x)^2$

23. $u = 4\sqrt{\ln 2t + e^{2t}}$

24. $y = (2e^{x^2} + x^2)^3$

25. $y = xe^{xy} + \sin y$

26. $y = 4e^{-2/x} \ln y + 1$

27. $y = 3e^{2x} \ln x$

28. $r = 0.4e^{2\theta} \ln \cos \theta$

29. $y = \ln \sin 2e^{6x}$

30. $y = 6 \tan e^{x+1}$

31. $y = 2 \sin^{-1} e^{2x}$

32. $y = \tan^{-1} e^{3x}$

In Exercises 33–48, solve the given problems.

(W) 33. On a calculator, find the values of (a) e and (b) $(e^{1.0001} - e^{1.0000})/0.0001$. Compare the values and give the meaning of each in relation to the derivative of e^x, where $x = 1$.

(W) 34. On a calculator, find the values of (a) e^2 and (b) $(e^{2.0001} - e^{2.0000})/0.0001$. Compare the values and give the meaning of each in relation to the derivative of e^x, where $x = 2$.

35. Find the slope of a line tangent to the curve $y = e^{-2x} \cos 2x$ for $x = 0.625$. Verify the result by using the *numerical derivative* feature of a graphing calculator.

36. Find the slope of a line tangent to the curve $y = \dfrac{e^{-x}}{1 + \ln 4x}$ for $x = 1.842$. Verify the result by using the *numerical derivative* feature of a graphing calculator.

37. Find the differential of the function $y = \dfrac{2e^{4x}}{x + 2}$.

38. Find the linearization of the function $f(x) = \dfrac{6e^{4x}}{2x + 3}$ for $a = 0$.

39. Use a graphing calculator to display the graph of $y = e^x$. By roughly estimating slopes of tangent lines, note that it is reasonable that these values are equal to the y-coordinates of the points at which these estimates are made. (*Remember:* For $y = e^x$, $dy/dx = e^x$ also.)

40. Use a graphing calculator to display the graphs of $y = e^{-x}$ and $y = -e^{-x}$. By roughly estimating slopes of tangent lines of $y = e^{-x}$, note that $y = -e^{-x}$ gives reasonable values for the derivative of $y = e^{-x}$.

41. Show that $y = xe^{-x}$ satisfies the equation $(dy/dx) + y = e^{-x}$.

42. Show that $y = e^{-x} \sin x$ satisfies the equation
$$\dfrac{d^2y}{dx^2} + 2\dfrac{dy}{dx} + 2y = 0.$$

43. For $y = \dfrac{e^{2x} - 1}{e^{2x} + 1}$, show that $\dfrac{dy}{dx} = 1 - y^2$.

44. If $e^x + e^y = e^{x+y}$, show that $dy/dx = -e^{y-x}$.

45. The reliability R ($0 \le R \le 1$) of a certain computer system is given by $R = e^{-0.002t}$, where t is the time of operation (in h). Find dR/dt for $t = 100$ h.

46. A thermometer is taken from a freezer at $-16°C$ and placed in a room at $24°C$. The temperature T of the thermometer as a function of the time t (in min) after removal is given by $T = 8.0(3.0 - 5.0e^{-0.50t})$. How fast is the temperature changing when $t = 6.0$ min?

47. For the electric circuit shown in Fig. 8-8, the current i (in A) is given by $i = 4.42e^{-66.7t} \sin 226t$, where t is the time (in s). Find the expression for di/dt.

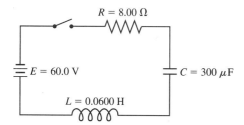

$R = 8.00\ \Omega$

$E = 60.0$ V

$C = 300\ \mu$F

$L = 0.0600$ H

Fig. 8-8

48. Under certain assumptions of limitations to population growth, the population P (in billions) of the world is given by the *logistic equation* $P = \dfrac{10}{1 + 0.65e^{-0.060t}}$, where t is the number of years after the year 2000. Find the expression for dP/dt.

*In Exercise 47 of Section 8-1, we defined the **hyperbolic functions**. We repeat here the definitions and show the graphs of these functions. In Exercises 49–52, find the derivatives of the indicated hyperbolic functions.*

*The **hyperbolic sine** of u is defined as*

$$\sinh u = \dfrac{1}{2}(e^u - e^{-u})$$

Figure 8-9 shows the graph of $y = \sinh x$.

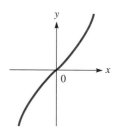

Fig. 8-9

*The **hyperbolic cosine** of u is defined as*

$$\cosh u = \dfrac{1}{2}(e^u + e^{-u})$$

Figure 8-10 shows the graph of $y = \cosh x$.

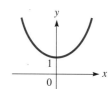

Fig. 8-10

49. Show that
$$\dfrac{d}{dx} \sinh u = \cosh u \dfrac{du}{dx}$$
where u is a function of x.

50. Show that
$$\dfrac{d}{dx} \cosh u = \sinh u \dfrac{du}{dx}$$
where u is a function of x.

51. Referring to Exercise 49, find the derivative of the function $y = x \sinh 2x$.

52. Show that
$$\dfrac{d^2 \sinh x}{dx^2} = \sinh x \quad \text{and} \quad \dfrac{d^2 \cosh x}{dx^2} = \cosh x$$

8-4 APPLICATIONS

The following examples show applications of the logarithmic and exponential functions to curve tracing, Newton's method, and time-rate-of-change problems. Certain other applications are indicated in the exercises.

EXAMPLE 1   Sketch the graph of the function $y = x \ln x$.

First, we note that x cannot be zero since $\ln x$ is not defined at $x = 0$. Since $\ln 1 = 0$, we have an intercept at $(1, 0)$. There is no symmetry to the axes or origin, and there are no vertical asymptotes. Also, because $\ln x$ is defined only for $x > 0$, the domain is $x > 0$.

Finding the first two derivatives, we have

$$\frac{dy}{dx} = x\left(\frac{1}{x}\right) + \ln x = 1 + \ln x \qquad \frac{d^2y}{dx^2} = \frac{1}{x}$$

The first derivative is zero if $\ln x = -1$, or $x = e^{-1}$. The second derivative is positive for this value of x. Thus, there is a minimum point at $(1/e, -1/e)$. Since the domain is $x > 0$, the second derivative indicates that the curve is always concave up. In turn, we now see that the range of the function is $y \geq -1/e$. The graph is shown in Fig. 8-11.

Although the curve approaches the origin as x approaches zero, the origin is not included on the graph of the function.

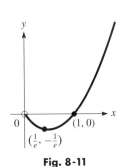

Fig. 8-11

EXAMPLE 2   The electric current i (in mA) in an electric circuit as a function of the time t (in s) is given by $i = e^{-t} \cos t$. Sketch the graph of i vs. t for $0 \leq t \leq 6$ s.

Since $\cos t = 0$ for $t = \pi/2$ and $t = 3\pi/2$, we know that $i = 0$ for $t = 1.571$ s and $t = 4.712$ s. The factor e^{-t} is always positive, and $e^{-t} = 1$ for $t = 0$, which means that $(0, 1)$ is an intercept. There is no symmetry to the axes or the origin, and there are no vertical asymptotes.

Next, finding the first derivative, we have

$$\frac{di}{dt} = -e^{-t} \sin t - e^{-t} \cos t = -e^{-t}(\sin t + \cos t)$$

Setting the derivative equal to zero, since e^{-t} is always positive, we have

$$\sin t + \cos t = 0, \qquad \tan t = -1, \qquad t = 2.356 \text{ s}, 5.498 \text{ s}$$

Now, finding the second derivative, we have

$$\frac{d^2i}{dt^2} = -e^{-t}(\cos t - \sin t) - e^{-t}(-1)(\sin t + \cos t) = 2e^{-t} \sin t$$

The sign of the second derivative depends only on $\sin t$. Thus, $d^2i/dt^2 > 0$ for $t = 2.356$ s and $d^2i/dt^2 < 0$ for $t = 5.498$ s. Hence, $(2.356, -0.067)$ is a minimum, and $(5.498, 0.003)$ is a maximum. Also, from the second derivative, points of inflection occur for $t = 0$ s and $t = 3.142$ s since $\sin t = 0$ for these values. The graph is shown in Fig. 8-12.

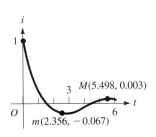

Fig. 8-12

EXAMPLE 3 Find the root of the equation $e^{2x} - 4 \cos x = 0$, which lies between 0 and 1, by using Newton's method.

Here,

$$f(x) = e^{2x} - 4 \cos x$$
$$f'(x) = 2e^{2x} + 4 \sin x$$

We find that $f(0) = -3$ and $f(1) = 5.2$. Therefore, we choose $x_1 = 0.5$. Using Eq. (4-1), which is

$$x_2 = x_1 - \frac{f(x_1)}{f'(x_1)}$$

we have these values:

$$f(x_1) = e^{2(0.5)} - 4 \cos 0.5 = -0.7920484$$
$$f'(x_1) = 2e^{2(0.5)} + 4 \sin 0.5 = 7.3542658$$
$$x_2 = 0.5 - \frac{-0.7920484}{7.3542658} = 0.6076992$$

Using the method again, we find $x_3 = 0.5979751$, which is correct to three decimal places.

SOLVING A WORD PROBLEM

See the chapter introduction.

EXAMPLE 4 One model for population growth is that the population P at time t is given by $P = P_0 e^{kt}$, where P_0 is the *initial* population ($t = 0$, when timing starts for the population being considered) and k is a constant. Show that the instantaneous time rate of change of population is directly proportional to the population present at time t.

To find the time rate of change, we find the derivative dP/dt:

$$\frac{dP}{dt} = (P_0 e^{kt})(k) = kP_0 e^{kt}$$
$$= kP \qquad \text{since } P = P_0 e^{kt}$$

Thus, we see that population growth increases as the population increases.

EXAMPLE 5 A rocket is moving such that the only force on it is due to gravity and its mass is decreasing at a constant rate r. If it moves vertically, its velocity v as a function of the time t is given by

$$v = v_0 - gt - k \ln \left(1 - \frac{rt}{m_0} \right)$$

where v_0 is the initial velocity, g is the acceleration due to gravity, t is the time, m_0 is the initial mass, and k is a constant. Determine the expression for the acceleration.

Since the acceleration is the time rate of change of the velocity, we must find dv/dt. Therefore,

$$\frac{dv}{dt} = -g - k \frac{1}{1 - \dfrac{rt}{m_0}} \left(\frac{-r}{m_0} \right) = -g + \frac{km_0}{m_0 - rt} \left(\frac{r}{m_0} \right)$$

$$= \frac{kr}{m_0 - rt} - g$$

EXERCISES 8-4

In Exercises 1–12, sketch the graphs of the given functions. Check each by displaying the graph on a graphing calculator.

1. $y = \ln \cos x$

2. $y = \dfrac{2 \ln x}{x}$

3. $y = xe^{-x}$

4. $y = \dfrac{e^x}{x}$

5. $y = \ln \dfrac{1}{x^2 + 1}$

6. $y = \ln \dfrac{e}{x}$

7. $y = 4e^{-x^2}$

8. $y = x - e^x$

9. $y = \ln x - x$

10. $y = e^{-x} \sin x$

11. $y = \frac{1}{2}(e^x - e^{-x})$ (See Exercises 49–52 of Section 8-3.)

12. $y = \frac{1}{2}(e^x + e^{-x})$ (See Exercises 49–52 of Section 8-3.)

In Exercises 13–32, solve the given problems by finding the appropriate derivative.

13. Find the equation of the line tangent to the graph of $y = x^2 \ln x$ at the point $(1, 0)$.

14. Find the equation of the line tangent to the graph of $y = \tan^{-1} 2x$, where $x = 1$.

15. Find the equation of the line normal to the graph of $y = 2 \sin \frac{1}{2} x$, where $x = 3\pi/2$.

16. Find the equation of the line normal to the graph of $y = e^{2x}/x$, at $x = 1$.

17. By Newton's method, solve the equation $x^2 - 2 + \ln x = 0$. Check the result by using the *intersect* (or *zero*) feature of a graphing calculator.

18. By Newton's method, solve the equation $e^{-2x} - \tan^{-1} x = 0$. Check the solution by using the *intersect* (or *zero*) feature of a graphing calculator.

19. The power supply P (in W) in a satellite is $P = 100e^{-0.005t}$, where t is measured in days. Find the time rate of change of power after 100 days.

20. The electric current i (in mA) through an inductor of 0.50 H as a function of time t (in s) is $i = e^{-5.0t} \sin 120\pi t$. Find the voltage across the inductor for $t = 1.0$ ms. (See Exercise 23 of Section 6-1 on page 189.)

21. The vapor pressure p and thermodynamic temperature T of a gas are related by the equation $\ln p = \dfrac{a}{T} + b \ln T + c$, where a, b, and c are constants. Find the expression for dp/dT.

22. The charge q on a capacitor in a circuit containing the capacitor of capacitance C, a resistance R, and a source of voltage E is given by $q = CE(1 - e^{-t/RC})$. Show that this equation satisfies the equation $R\dfrac{dq}{dt} + \dfrac{q}{C} = E$.

23. Assuming that force is proportional to acceleration, show that a particle moving along the x-axis, so that its displacement $x = ae^{kt} + be^{-kt}$, has a force acting on it which is proportional to its displacement.

24. The radius of curvature at a point on a curve is given by

$$R = \frac{[1 + (dy/dx)^2]^{3/2}}{d^2y/dx^2}$$

A roller mechanism moves along the path defined by $y = \ln \sec x$ (-1.5 dm $\le x \le 1.5$ dm). Find the radius of curvature of this path for $x = 0.85$ dm.

25. Sketch the graph of $y = \ln \sec x$, marking that part which is the path of the roller mechanism of Exercise 24.

26. In an electronic device, the maximum current density i_m as a function of the temperature T is given by $i_m = AT^2 e^{k/T}$, where A and k are constants. Find the expression for a small change in i_m for a small change in T.

27. A meteorologist sketched the path of the jet stream on a map of the northern United States and southern Canada on which all latitudes were parallel and all longitudes were parallel and equally spaced. A computer analysis showed this path to be $y = 6.0e^{-0.020x} \sin 0.20x$ ($0 \le x \le 60$), where the origin is 125.0°W, 45.0°N and (60, 0) is 65.0°W, 45.0°N. Find the locations of the maximum and minimum latitudes of the jet stream between 65°W and 125°W for that day.

28. The reliability R ($0 \le R \le 1$) of a certain computer system after t hours of operation is found from $R = 3e^{-0.004t} - 2e^{-0.006t}$. Use Newton's method to find how long the system operates to have a reliability of 0.8 (80% probability that there will be no system failure).

29. An object on the end of a spring is moving so that its displacement (in cm) from the equilibrium position is given by $y = e^{-0.5t}(0.4 \cos 6t - 0.2 \sin 6t)$. Find the expression for the velocity of the object. What is the velocity when $t = 0.26$ s? The motion described by this equation is called *damped harmonic motion*.

30. A package of weather instruments is propelled into the air to an altitude of about 7 km. A parachute then opens, and the package returns to the surface. The altitude y of the package as a function of the time t (in min) is given by $y = \dfrac{10t}{e^{0.4t} + 1}$. Find the vertical velocity of the package for $t = 8.0$ min.

31. The speed s of signaling by use of a certain communications cable is directly proportional to $x^2 \ln x^{-1}$, where x is the ratio of the radius of the core of the cable to the thickness of the surrounding insulation. For what value of x is s a maximum?

32. A computer is programmed to inscribe a series of rectangles in the first quadrant under the curve of $y = e^{-x}$. What is the area of the largest rectangle that can be inscribed?

CHAPTER EQUATIONS

Exponential function	$y = b^x$	(8-1)
Logarithmic function	$y = \log_b x$	(8-3)
Properties of logarithms	$\log_b xy = \log_b x + \log_b y$	(8-7)
	$\log_b \left(\dfrac{x}{y} \right) = \log_b x - \log_b y$	(8-8)
	$\log_b (x^n) = n \log_b x$	(8-9)
Changing base of logarithms	$\log_b x = \dfrac{\log_a x}{\log_a b}$	(8-10)
Derivatives	$\dfrac{d(\log_b u)}{dx} = \dfrac{1}{u} \log_b e \dfrac{du}{dx}$	(8-11)
	$\dfrac{d(\ln u)}{dx} = \dfrac{1}{u} \dfrac{du}{dx}$	(8-12)
	$\dfrac{d(b^u)}{dx} = b^u \ln b \dfrac{du}{dx}$	(8-13)
	$\dfrac{d(e^u)}{dx} = e^u \dfrac{du}{dx}$	(8-14)

REVIEW EXERCISES

In Exercises 1–24, find the derivative of the given functions.

1. $y = 3 \ln(x^2 + 1)$

2. $y = \ln(1 - 5x)^3$

3. $y = (e^{x-3})^2$

4. $y = e^{\sqrt{1-x}}$

5. $r = 2 \ln(\csc t^2)$

6. $y = \ln(3 + \sin x^2)$

7. $y = [\ln(3 + \sin x)]^2$

8. $y = \ln(3 + \sin x)^2$

9. $R = 6 e^{\sin 2\theta}$

10. $y = 3 e^{\sec 3x}$

11. $y = x^3 e^x$

12. $y = 2x e^{4x}$

13. $y = \ln(x - e^{-x})^2$

14. $p = \ln\sqrt{\sin 2\phi}$

15. $y = \dfrac{\cos^2 x}{e^{3x} + 1}$

16. $y = \dfrac{\ln \sqrt{3x + 1}}{3x + 1}$

17. $y = x^2 \ln x$

18. $s = (2t + 1)\ln(t^3 + 3)$

19. $y = \dfrac{e^{2x}}{x^2 + 1}$

20. $y = \dfrac{\ln x}{2x + 3}$

21. $v = 3e^{-\theta} \sec 2\theta$

22. $y = e^{3x} \ln x$

23. $x^2 \ln y = y + x$

24. $y = x^2 (e^{\cos^2 x})^2$

In Exercises 25–44, solve the given problems by finding the appropriate derivatives.

25. Find the equation of the line tangent to the graph of $y = \ln \cos x$ at $x = \frac{\pi}{6}$.

26. Find the equation of the line normal to the graph of $y = e^{x^2}$ at $x = \frac{1}{2}$.

27. Sketch the graph of $y = \ln(1 + x)$.

28. Sketch the graph of $y = x(\ln x)^2$.

29. By Newton's method, solve the equation $e^x - x^2 = 0$. Check the solution by displaying the graph on a calculator and using the *intersect* (or *zero*) feature.

30. In the study of atomic spectra, it is necessary to solve the equation $x = 5(1 - e^{-x})$ for x. Use Newton's method to find the solution.

31. Find the area of the largest rectangle that can be inscribed under the curve of $y = e^{-x^2}$ in the first quadrant.

32. Find the acceleration of an object, whose x- and y-components of displacement are $x = e^{-t} \sin 2t$ and $y = e^{-t} \cos 2t$, when $t = \pi/4$.

33. An Earth-orbiting satellite is launched such that its altitude (in mi) is given by $y = 150(1 - e^{-0.05t})$, where t is the time (in min). Find the vertical velocity of the satellite for $t = 10.0$ min.

34. The value V of a bank account in which \$1000 is deposited and then earns 6% annual interest, compounded continuously (daily compounding approximates this, and some banks actually use continuous compounding), is $V = 1000e^{0.06t}$ after t years. How fast is the account growing after exactly 2 years?

35. In determining how to divide files on the hard disk of a computer, we can use the equation $n = xN \log_x N$. Sketch the graph of n vs. x for $1 < x \le 10$ if $N = 8$.

36. Under certain conditions, the potential V (in V) due to a magnet is given by $V = -k \ln\left(1 + \dfrac{L}{x}\right)$, where L is the length of the magnet and x is the distance from the point where the potential is measured. Find the expression for dV/dx.

37. An analysis of samples of air for a city showed that the number of parts per million p of sulfur dioxide on a certain day was $p = 0.05 \ln(2 + 24t - t^2)$, where t is the hour of the day. Using differentials, find the approximate change in the amount of sulfur dioxide between 10 A.M. and 10:30 A.M.

38. According to Newton's law of cooling (Isaac Newton, again), the rate at which a body cools is proportional to the difference in temperature between it and the surrounding medium. By use of this law, the temperature T (in °F) of an engine coolant as a function of the time t (in min) is $T = 80 + 120(0.5)^{0.200t}$. The coolant was initially at 200°F, and the air temperature was 80°F. Linearize this function for $t = 5.00$ min and display the graphs of $T = f(t)$ and $L(t)$ on a graphing calculator.

39. The charge q on a certain capacitor in an amplifier circuit as a function of time t is given by
$q = e^{-0.1t}(0.2 \sin 120\pi t + 0.8 \cos 120\pi t)$. The current i in the circuit is the instantaneous time rate of change of the charge. Find the expression for i as a function of t.

40. A projection of the number n (in millions) of users of the Internet is $n = 160 - 140e^{-0.30t}$, where t is the number of years after 2000. What is the projected annual rate of increase in 2010?

41. An architect designs an arch of height y (in m) over a walkway by the curve of the equation $y = 3.00e^{-0.500x^2}$. What are the dimensions of the largest rectangular passage area under the arch?

42. The current i in an electric circuit with a resistance R and an inductance L is $i = i_0 e^{-Rt/L}$, where i_0 is the initial current. Show that the time rate of change of the current is directly proportional to the current.

43. Show that the equation of the hyperbolic cosine function

$$y = \frac{H}{w} \cosh \frac{wx}{H} \qquad (w \text{ and } H \text{ are constants})$$

satisfies the equation
$$\frac{d^2y}{dx^2} = \frac{w}{H} \sqrt{1 + \left(\frac{dy}{dx}\right)^2}$$

(see Exercise 49 on page 279). A *catenary* is the curve of a uniform cable hanging under its own weight and is in the shape of a hyperbolic cosine curve. Also, this shape (inverted) was chosen for the St. Louis Gateway Arch (shown in Fig. 8-13) and makes the arch self-supporting.

Fig. 8-13

44. The displacement y (in in.) of a weight on a spring in water is given by $y = 3.0te^{-0.20t}$, where t is the time (in s). What is the maximum displacement? (For this type of displacement, the motion is called *critically damped,* as the weight returns to its equilibrium position as quickly as possible without oscillating.)

Writing Exercise

45. Assume that inflation makes the dollar worth 5% less in purchasing value each year than the previous year. Write one or two paragraphs explaining how to use derivatives to determine the approximate change in purchasing value of \$1000 during the nth year after it is set aside in a bank safe deposit box.

PRACTICE TEST

In Problems 1–3, find the derivative of each of the functions.

1. $y = 2x^2 e^{x/2}$

2. $y = \ln \cos(1 - x^2)$

3. $x \ln y = 2 \sin^{-1} x$

4. Find the differential of the function $y = \dfrac{3x}{2 - e^{-x}}$.

5. Find the slope of a tangent to the graph of
$y = \ln \dfrac{2x - 1}{1 + x^2}$ for $x = 2$.

6. Find the expression for the time rate of change of electric current that is given by the equation $i = 8e^{-t} \sin 10t$, where t is the time.

7. Sketch the graph of the function $y = xe^x$.

8. A business determines that the gross income received by selling x items per week is $I = 100xe^{-x/10}$. How many items should be sold per week to make the gross income a maximum?

9 INTEGRATION BY STANDARD FORMS

Having developed the derivatives of the basic transcendental functions, we can now expand considerably the functions that we are able to integrate. In addition to the transcendental functions, we will find that it is now possible to develop methods for integrating many algebraic functions that we were unable to integrate previously.

In this chapter we expand the use of the general power formula for integration for use with integrands that include transcendental functions. We then develop additional standard forms for algebraic and transcendental integrands. In the next chapter we will develop some basic methods of integration. In using all the standard forms and basic methods, *recognition of the integral form* is of the greatest importance.

As we have seen in Chapters 5 and 6, there are numerous applications of integration in geometry, science, and various fields of technology. Additional examples of these applications are found in the examples and exercises of this chapter.

In Section 9-5 we show an application of integration that is important in the design of electric appliances, such as an electric heater.

9-1 THE GENERAL POWER FORMULA

The first formula for integration that we will discuss is the general power formula, and we will expand its use to include transcendental integrands. It was first introduced with the integration of basic algebraic forms in Chapter 5 and is repeated for reference on the next page.

For reference, we repeat the general power formula for integration, Eq. (5-4):

$$\int u^n \, du = \frac{u^{n+1}}{n+1} + C \qquad (n \neq -1) \tag{9-1}$$

CAUTION ▶   In applying Eq. (9-1) to transcendental integrands, as well as with algebraic integrands, *we must properly recognize the quantities u, n, and du.* This requires familiarity with the differential forms of Chapters 3, 7, and 8.

▌EXAMPLE 1   Integrate: $\int \sin^3 x \cos x \, dx$.
Since $d(\sin x) = \cos x \, dx$, we note that this integral fits the form of Eq. (9-1) for $u = \sin x$. Thus, with $u = \sin x$, we have $du = \cos x \, dx$, which means that this integral is of the form $\int u^3 \, du$. Therefore, the integration can now be completed.

$$\int \sin^3 x \cos x \, dx = \int \sin^3 x \overset{\displaystyle du}{(\cos x \, dx)}$$

$$= \frac{1}{4} \sin^4 x + C \longleftarrow \begin{array}{l}\text{do not forget the}\\\text{constant of integration}\end{array}$$

CAUTION ▶   We note here that *the factor* cos *x is a necessary part of the du* in order to have the proper form of integration *and therefore does not appear in the final result.*
We check our result by finding the derivative of $\frac{1}{4} \sin^4 x + C$, which is

$$\frac{d}{dx}\left(\frac{1}{4} \sin^4 x + C\right) = \frac{1}{4} (4) \sin^3 x \cos x = \sin^3 x \cos x \qquad ▄$$

▌EXAMPLE 2   Integrate: $\int 2\sqrt{1 + \tan \theta} \, \sec^2 \theta \, d\theta$.
Here we note that $d(1 + \tan \theta) = \sec^2 \theta \, d\theta$, which means that the integral fits the form of Eq. (9-1) with

$$u = 1 + \tan \theta \qquad du = \sec^2 \theta \, d\theta \qquad n = \tfrac{1}{2}$$

The integral is of the form $\int u^{1/2} \, du$. Thus,

$$\int 2\sqrt{1 + \tan \theta}\,(\sec^2 \theta \, d\theta) = 2\int \underset{\displaystyle u}{(1 + \tan \theta)^{1/2}} \overset{\displaystyle du}{(\sec^2 \theta \, d\theta)}$$

$$= 2\left(\frac{2}{3}\right)(1 + \tan \theta)^{3/2} + C = \frac{4}{3} (1 + \tan \theta)^{3/2} + C \qquad ▄$$

▌EXAMPLE 3   Integrate: $\int \ln x \left(\dfrac{dx}{x}\right)$.
By noting that $d(\ln x) = \dfrac{dx}{x}$, we have

$$u = \ln x \qquad du = \frac{dx}{x} \qquad n = 1$$

This means that the integral is of the form $\int u \, du$. Thus,

$$\int \ln x \left(\frac{dx}{x}\right) = \frac{1}{2} (\ln x)^2 + C = \frac{1}{2} \ln^2 x + C \qquad ▄$$

EXAMPLE 4 Find the value of $\int_0^{0.5} \dfrac{\sin^{-1} x}{\sqrt{1 - x^2}}\, dx$.

For purposes of integrating, we see that

$$u = \sin^{-1} x \qquad du = \frac{dx}{\sqrt{1 - x^2}} \qquad n = 1$$

$$\int_0^{0.5} \frac{\sin^{-1} x}{\sqrt{1 - x^2}}\, dx = \int_0^{0.5} \sin^{-1} x \left(\frac{dx}{\sqrt{1 - x^2}}\right) \qquad \int u\, du$$

$$= \frac{(\sin^{-1} x)^2}{2} \Big|_0^{0.5} \qquad\qquad \text{integrate}$$

$$= \frac{(\frac{\pi}{6})^2}{2} - 0 = \frac{\pi^2}{72} \qquad\qquad \text{evaluate}$$

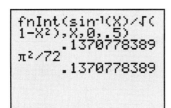

Fig. 9-1

The display in Fig. 9-1 shows the calculator check for this result.

EXAMPLE 5 Find the first-quadrant area bounded by $y = \dfrac{e^{2x}}{\sqrt{e^{2x} + 1}}$ and $x = 1.5$.

The area to be found is shown in Fig. 9-2. The area of the representative element is $y\, dx$. Therefore, the area is found by evaluating the integral

$$\int_0^{1.5} \frac{e^{2x}\, dx}{\sqrt{e^{2x} + 1}}$$

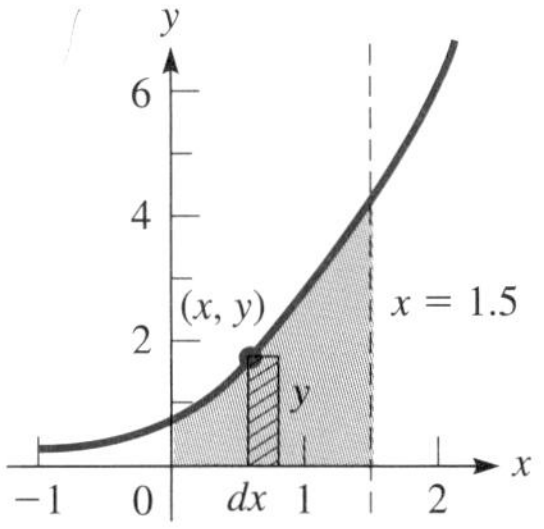

For the purpose of integration, $n = -\frac{1}{2}$, $u = e^{2x} + 1$, and $du = 2e^{2x}\, dx$. Therefore,

$$\int_0^{1.5} (e^{2x} + 1)^{-1/2} e^{2x}\, dx = \frac{1}{2} \int_0^{1.5} (e^{2x} + 1)^{-1/2} (2e^{2x}\, dx)$$

$$= \frac{1}{2}(2)(e^{2x} + 1)^{1/2} \Big|_0^{1.5} = (e^{2x} + 1)^{1/2} \Big|_0^{1.5}$$

$$= \sqrt{e^3 + 1} - \sqrt{2} = 3.178$$

This means that the area is 3.178 square units.

Fig. 9-2

EXERCISES 9-1

In Exercises 1–24, integrate each of the given functions.

1. $\displaystyle\int \sin^4 x \cos x\, dx$

2. $\displaystyle\int \cos^5 x(-\sin x\, dx)$

3. $\displaystyle\int 0.4\sqrt{\cos \theta}\, \sin \theta\, d\theta$

4. $\displaystyle\int 8 \sin^{1/3} x \cos x\, dx$

5. $\displaystyle\int 4 \tan^2 x \sec^2 x\, dx$

6. $\displaystyle\int \sec^3 x(\sec x \tan x)\, dx$

7. $\displaystyle\int_0^{\pi/8} \cos 2x \sin 2x\, dx$

8. $\displaystyle\int_{\pi/6}^{\pi/4} 3\sqrt{\cot x}\, \csc^2 x\, dx$

9. $\displaystyle\int (\sin^{-1} x)^3 \left(\frac{dx}{\sqrt{1 - x^2}}\right)$

10. $\displaystyle\int \frac{20(\cos^{-1} 2t)^4\, dt}{\sqrt{1 - 4t^2}}$

11. $\displaystyle\int \frac{5 \tan^{-1} 5x}{1 + 25x^2}\, dx$

12. $\displaystyle\int \frac{\sin^{-1} 4x\, dx}{\sqrt{1 - 16x^2}}$

13. $\displaystyle\int [\ln (x + 1)]^2 \frac{dx}{x + 1}$

14. $\displaystyle\int 0.8(3 + 2 \ln u)^3 \frac{du}{u}$

15. $\displaystyle\int_0^{1/2} \frac{\ln (2x + 3)}{2x + 3}\, dx$

16. $\displaystyle\int_1^e \frac{(1 - 2 \ln x)\, dx}{x}$

17. $\displaystyle\int (4 + e^x)^3 e^x\, dx$

18. $\displaystyle\int 2\sqrt{1 - e^{-x}}\, (e^{-x}\, dx)$

19. $\displaystyle\int \frac{\sqrt{(1 + e^{-r})(1 - e^{-r})}}{e^{2r}}\, dr$

20. $\displaystyle\int \frac{(1 + 3e^{-2x})^4\, dx}{e^{2x}}$

21. $\displaystyle\int (1 + \sec^2 x)^4 (\sec^2 x \tan x\, dx)$

22. $\int (e^x + e^{-x})^{1/4}(e^x - e^{-x})\,dx$

23. $\int_0^{\pi/6} \dfrac{\tan x}{\cos^2 x}\,dx$ **24.** $\int_{\pi/3}^{\pi/2} \dfrac{\sin \theta\,d\theta}{\sqrt{1 + \cos \theta}}$

In Exercises 25–32, solve the given problems by integration.

25. Find the area under the curve $y = \dfrac{1 + \tan^{-1} 2x}{1 + 4x^2}$ from $x = 0$ to $x = 2$.

26. Find the first-quadrant area bounded by $y = \dfrac{\ln(4x + 1)}{4x + 1}$ and $x = 2$.

27. The general expression for the slope of a given curve is $(\ln x)^2/x$. If the curve passes through $(1, 2)$, find its equation.

28. Find the equation of the curve for which $dy/dx = (1 + \tan 2x)^2 \sec^2 2x$ if the curve passes through $(2, 1)$.

29. In the development of the expression for the total pressure P on a wall due to molecules with mass m and velocity v strik-ing the wall, the equation $P = mnv^2 \int_0^{\pi/2} \sin \theta \cos^2 \theta\,d\theta$ is found. The symbol n represents the number of molecules per unit volume, and θ represents the angle between a perpendicular to the wall and the direction of the molecule. Find the expression for P.

30. The solar energy E passing through a hemispherical surface per unit time, per unit area, is $E = 2\pi I \int_0^{\pi/2} \cos \theta \sin \theta\,d\theta$, where I is the solar intensity and θ is the angle at which it is directed (from the perpendicular). Evaluate this integral.

31. After an electric power interruption, the current i in a circuit is given by $i = 3(1 - e^{-t})^2(e^{-t})$, where t is the time. Find the expression for the total electric charge q to pass a point in the circuit if $q = 0$ for $t = 0$.

32. A space vehicle is launched vertically from the ground such that its velocity v (in km/s) is given by $v = [\ln^2(t^3 + 1)]\dfrac{t^2}{t^3 + 1}$, where t is the time (in s). Find the altitude of the vehicle after 10.0 s.

$9\text{-}2$ THE BASIC LOGARITHMIC FORM

The general power formula for integration, Eq. (9-1), is valid for all values of n except $n = -1$. If n were set equal to -1, this would cause the result to be undefined. When we obtained the derivative of the logarithmic function, we found

$$\frac{d(\ln u)}{dx} = \frac{1}{u}\frac{du}{dx}$$

which means the differential of the logarithmic form is $d(\ln u) = du/u$. Reversing the process, we then determine that $\int du/u = \ln u + C$. In other words, when the exponent of the expression being integrated is -1, the expression is a logarithmic form.

Logarithms are defined only for positive numbers. Thus, $\int du/u = \ln u + C$ is valid if $u > 0$. If $u < 0$, then $-u > 0$. In this case, $d(-u) = -du$, or $\int (-du)/(-u) = \ln(-u) + C$. However, $\int du/u = \int (-du)/(-u)$. These results can be combined into a single form using the absolute value of u. Therefore,

$$\int \frac{du}{u} = \ln |u| + C \tag{9-2}$$

EXAMPLE 1   Integrate: $\int \dfrac{dx}{x + 1}$.

Since $d(x + 1) = dx$, this integral fits the form of Eq. (9-2) with $u = x + 1$ and $du = dx$. Therefore, we have

$$\int \frac{dx}{x + 1} = \ln |x + 1| + C$$

EXAMPLE 2 Newton's law of cooling states that the rate at which an object cools is directly proportional to the difference in its temperature T and the temperature of the surrounding medium. By use of this law, the time t (in min) a certain object takes to cool from 80°C to 50°C in air at 20°C is found to be

$$t = -9.8 \int_{80}^{50} \frac{dT}{T - 20}$$

Find the value of t.

We see that the integral fits Eq. (9-2) with $u = T - 20$ and $du = dT$. Thus,

$$t = -9.8 \int_{80}^{50} \frac{dT}{T - 20} \quad \longleftarrow du$$
$$\longleftarrow u$$
$$= -9.8 \ln |T - 20| \Big|_{80}^{50} \qquad \text{integrate}$$
$$= -9.8(\ln 30 - \ln 60) \qquad \text{evaluate}$$
$$= -9.8 \ln \frac{30}{60} = -9.8 \ln (0.50) \qquad \ln x - \ln y = \ln \frac{x}{y}$$
$$= 6.8 \text{ min}$$ ∎

EXAMPLE 3 Integrate: $\int \frac{\cos x}{\sin x}\, dx$.

We note that $d(\sin x) = \cos x\, dx$. This means that this integral fits the form of Eq. (9-2) with $u = \sin x$ and $du = \cos x\, dx$. Thus,

$$\int \frac{\cos x}{\sin x}\, dx = \int \frac{\cos x\, dx}{\sin x} \quad \longleftarrow du$$
$$\longleftarrow u$$
$$= \ln |\sin x| + C$$ ∎

EXAMPLE 4 Integrate: $\int \frac{x\, dx}{4 - x^2}$.

This integral fits the form of Eq. (9-2) with $u = 4 - x^2$ and $du = -2x\, dx$. This means that we must introduce a factor of -2 into the numerator and a factor of $-\frac{1}{2}$ before the integral. Therefore,

$$\int \frac{x\, dx}{4 - x^2} = -\frac{1}{2} \int \frac{-2x\, dx}{4 - x^2} \quad \longleftarrow du$$
$$\longleftarrow u$$
$$= -\frac{1}{2} \ln |4 - x^2| + C$$

CAUTION ▶ We should note that if the quantity $4 - x^2$ were raised to any power other than that in the example, we would have to employ the general power formula for integration. For example,

$$\int \frac{x\, dx}{(4 - x^2)^2} = -\frac{1}{2} \int \frac{-2x\, dx}{(4 - x^2)^2} \quad \longleftarrow du$$
$$u^2$$
$$= -\frac{1}{2} \frac{(4 - x^2)^{-1}}{-1} + C = \frac{1}{2(4 - x^2)} + C$$ ∎

EXAMPLE 5   Integrate: $\int \dfrac{e^{4x}\,dx}{1 + 3e^{4x}}$.

Since $d(1 + 3e^{4x})/dx = 12e^{4x}$, we see that we can use Eq. (9-2) with $u = 1 + 3e^{4x}$ and $du = 12e^{4x}\,dx$. Therefore, we write

$$\int \frac{e^{4x}\,dx}{1 + 3e^{4x}} = \frac{1}{12} \int \frac{12e^{4x}\,dx}{1 + 3e^{4x}} \qquad \text{introduce factors of } 12$$

$$= \frac{1}{12} \ln|1 + 3e^{4x}| + C \qquad \text{integrate}$$

$$= \frac{1}{12} \ln(1 + 3e^{4x}) + C \qquad 1 + 3e^{4x} > 0 \text{ for all } x \qquad \blacksquare$$

EXAMPLE 6   Evaluate: $\displaystyle\int_0^{\pi/8} \dfrac{\sec^2 2\theta}{1 + \tan 2\theta}\,d\theta$.

Since $d(1 + \tan 2\theta) = 2\sec^2 2\theta\,d\theta$, we see that we can use Eq. (9-2) with $u = 1 + \tan 2\theta$ and $du = 2\sec^2 2\theta\,d\theta$. Therefore, we have

$$\int_0^{\pi/8} \frac{\sec^2 2\theta}{1 + \tan 2\theta}\,d\theta = \frac{1}{2} \int_0^{\pi/8} \frac{2\sec^2 2\theta\,d\theta}{1 + \tan 2\theta} \qquad \text{introduce factors of } 2$$

$$= \frac{1}{2} \ln|1 + \tan 2\theta|\Big|_0^{\pi/8} \qquad \text{integrate}$$

$$= \frac{1}{2}(\ln|1 + 1| - \ln|1 + 0|) \qquad \text{evaluate}$$

$$= \frac{1}{2}(\ln 2 - \ln 1) = \frac{1}{2}(\ln 2 - 0)$$

$$= \frac{1}{2} \ln 2 \qquad \blacksquare$$

EXAMPLE 7   Find the volume within the piece of tapered tubing shown in Fig. 9-3, which can be described as the volume generated by revolving the region bounded by the curve $y = \dfrac{3}{\sqrt{4x + 3}}$, $x = 2.50$ in., and the axes about the x-axis.

The volume can be found by setting up only one integral by using a disk element of volume, as shown. The volume is found as follows:

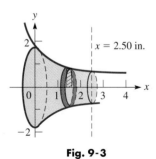

Fig. 9-3

$$V = \pi \int_0^{2.50} y^2\,dx = \pi \int_0^{2.50} \left(\frac{3}{\sqrt{4x + 3}}\right)^2 dx$$

$$= \pi \int_0^{2.50} \frac{9\,dx}{4x + 3} = \frac{9\pi}{4} \int_0^{2.50} \frac{4\,dx}{4x + 3}$$

$$= \frac{9\pi}{4} \ln(4x + 3)\Big|_0^{2.50} = \frac{9\pi}{4}(\ln 13.0 - \ln 3)$$

$$= \frac{9\pi}{4} \ln \frac{13.0}{3} = 10.4 \text{ in.}^3$$

Therefore, the volume within this piece of tubing is about 10.4 in.^{3}. $\blacksquare$

EXERCISES 9-2

In Exercises 1–28, integrate each of the given functions.

1. $\int \dfrac{dx}{1 + 4x}$

2. $\int \dfrac{dx}{1 - 4x}$

3. $\int \dfrac{2x\,dx}{4 - 3x^2}$

4. $\int \dfrac{4\sqrt{u}\,du}{1 + u\sqrt{u}}$

5. $\int_0^2 \dfrac{dx}{8 - 3x}$

6. $\int_{-1}^3 \dfrac{2x^3\,dx}{x^4 + 1}$

7. $\int \dfrac{0.4\csc^2 2\theta\,d\theta}{\cot 2\theta}$

8. $\int \dfrac{7\sin x}{\cos x}\,dx$

9. $\int_0^{\pi/2} \dfrac{\cos x\,dx}{1 + \sin x}$

10. $\int_0^{\pi/4} \dfrac{\sec^2 x\,dx}{4 + \tan x}$

11. $\int \dfrac{e^{-x}}{1 - e^{-x}}\,dx$

12. $\int \dfrac{5e^{3x}}{1 - e^{3x}}\,dx$

13. $\int \dfrac{1 + e^x}{x + e^x}\,dx$

14. $\int \dfrac{3e^t\,dt}{\sqrt{e^{2t} + 4e^t + 4}}$

15. $\int \dfrac{\sec x\tan x\,dx}{1 + 4\sec x}$

16. $\int \dfrac{\sin 2x}{1 - \cos^2 x}\,dx$

17. $\int_1^3 \dfrac{1 + x}{4x + 2x^2}\,dx$

18. $\int_1^2 \dfrac{4x + 6x^2}{x^2 + x^3}\,dx$

19. $\int \dfrac{0.5\,dr}{r\ln r}$

20. $\int \dfrac{dx}{x(1 + 2\ln x)}$

21. $\int \dfrac{2 + \sec^2 x}{2x + \tan x}\,dx$

22. $\int \dfrac{x + \cos 2x}{x^2 + \sin 2x}\,dx$

23. $\int \dfrac{2\,dx}{\sqrt{1 - 2x}}$

24. $\int \dfrac{4x\,dx}{(1 + x^2)^2}$

25. $\int \dfrac{x + 2}{x^2}\,dx$

26. $\int \dfrac{3v^2 - 2v}{v^2}\,dv$

27. $\int_0^{\pi/12} \dfrac{\sec^2 3x}{4 + \tan 3x}\,dx$

28. $\int_1^2 \dfrac{x^2 + 1}{x^3 + 3x}\,dx$

In Exercises 29–40, solve the given problems by integration.

29. Find the area bounded by $y(x + 1) = 1$, $x = 0$, $y = 0$, and $x = 2$. See Fig. 9-4.

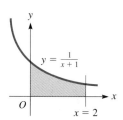

Fig. 9-4

30. Find the area bounded by $xy = 9$, $x = 1$, $x = 2$, and $y = 0$.

31. Find the volume generated by revolving the region bounded by $y = 1/(x^2 + 1)$, $x = 0$, $x = 1$, and $y = 0$ about the y-axis. Use shells.

32. Find the volume of the solid generated by revolving the region bounded by $y = \dfrac{2}{\sqrt{3x + 1}}$, $x = 0$, $x = 3.5$, and $y = 0$ about the x-axis.

33. The general expression for the slope of a curve is $\dfrac{\sin x}{3 + \cos x}$. If the curve passes through the point $(\pi/3, 2)$, find its equation.

34. Find the average value of the function $xy = 4$ from $x = 1$ to $x = 2$.

35. Under ideal conditions, the natural law of population growth is that population increases at a rate proportional to the population P present at any time t. This leads to the equation $t = \dfrac{1}{k}\int \dfrac{dP}{P}$. Assuming ideal conditions for the United States, if $P = 249$ million in 1990 ($t = 0$) and $P = 275$ million in 2000 ($t = 10$ years), find the population that is projected in 2020 ($t = 30$ years).

36. In determining the temperature that is absolute zero ($0\,\text{K}$, or about $-273°\text{C}$), the equation $\ln T = -\int \dfrac{dr}{r - 1}$ is used. Here, T is the thermodynamic temperature, and r is the ratio between certain specific vapor pressures. If $T = 273.16$ K for $r = 1.3361$, find T as a function of r (if $r > 1$ for all T).

37. The time t and electric current i for a certain circuit with a voltage E, a resistance R, and an inductance L is given by $t = L\int \dfrac{di}{E - iR}$. If $t = 0$ for $i = 0$, integrate and express i as a function of t.

38. Conditions are often such that a force proportional to the velocity tends to retard the motion of an object moving through a resisting medium. Under such conditions, the acceleration of a certain object moving down an inclined plane is given by $20 - v$. This leads to the equation $t = \int \dfrac{dv}{20 - v}$. If the object starts from rest, find the expression for the velocity as a function of time.

39. An architect designs a wall panel that can be described as the first-quadrant area bounded by $y = \dfrac{50}{x^2 + 20}$ and $x = 3.00$. If the area of the panel is 6.61 m², find the x-coordinate (in m) of the centroid of the panel.

40. The electric power p developed in a certain resistor is given by $p = 3\int \dfrac{\sin \pi t}{2 + \cos \pi t}\,dt$, where t is the time. Express p as a function of t.

$9\text{-}3$ THE EXPONENTIAL FORM

In deriving the derivative for the exponential function we obtained the result $de^u/dx = e^u(du/dx)$. This means that the differential of the exponential form is $d(e^u) = e^u\,du$. Reversing this form to find the proper form of the integral for the exponential function, we have

$$\int e^u\,du = e^u + C \tag{9-3}$$

EXAMPLE 1   Integrate: $\int xe^{x^2}\,dx$.

Since $d(x^2) = 2x\,dx$, we can write this integral in the form of Eq. (9-3) with $u = x^2$ and $du = 2x\,dx$. Thus,

$$\int xe^{x^2}\,dx = \frac{1}{2}\int e^{x^2}(2x\,dx)$$

$$= \frac{1}{2}\,e^{x^2} + C \quad\blacksquare$$

EXAMPLE 2   For an electric circuit containing a direct voltage source E, a resistance R, and an inductance L, the current i and time t are related by $ie^{Rt/L} = \dfrac{E}{L}\int e^{Rt/L}\,dt$. See Fig. 9-5. If $i = 0$ for $t = 0$, perform the integration and then solve for i as a function of t.

For this integral, we see that $u = \dfrac{Rt}{L}$, which means that $du = \dfrac{R\,dt}{L}$. The solution is then as follows:

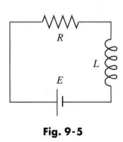

Fig. 9-5

$$ie^{Rt/L} = \frac{E}{L}\int e^{Rt/L}\,dt = \frac{E}{L}\left(\frac{L}{R}\right)\int e^{Rt/L}\left(\frac{R\,dt}{L}\right) \qquad \text{introduce factor } \frac{R}{L}$$

$$= \frac{E}{R}\,e^{Rt/L} + C \qquad \text{integrate}$$

$$0(e^0) = \frac{E}{R}\,e^0 + C, \qquad C = -\frac{E}{R} \qquad i = 0 \text{ for } t = 0; \text{ evaluate } C$$

$$ie^{Rt/L} = \frac{E}{R}\,e^{Rt/L} - \frac{E}{R} \qquad \text{substitute for } C$$

$$i = \frac{E}{R} - \frac{E}{R}\,e^{-Rt/L} = \frac{E}{R}\left(1 - e^{-Rt/L}\right) \qquad \text{solve for } i \quad\blacksquare$$

EXAMPLE 3   Integrate: $\displaystyle\int \frac{dx}{e^{3x}}$.

This integral can be put in proper form by writing it as $\int e^{-3x}\,dx$. In this form, $u = -3x$, $du = -3\,dx$. Thus,

$$\int \frac{dx}{e^{3x}} = \int e^{-3x}\,dx = -\frac{1}{3}\int e^{-3x}(-3\,dx) = -\frac{1}{3}\,e^{-3x} + C \quad\blacksquare$$

EXAMPLE 4 Integrate: $\int \frac{4e^{3x} - 3e^x}{e^{x+1}}\, dx$.

This can be put in the proper form for integration and then integrated, as follows:

$$\begin{aligned}\int \frac{4e^{3x} - 3e^x}{e^{x+1}}\, dx &= \int \frac{4e^{3x}}{e^{x+1}}\, dx - \int \frac{3e^x}{e^{x+1}}\, dx \\ &= 4\int e^{3x-(x+1)}\, dx - 3\int e^{x-(x+1)}\, dx \qquad \text{using Eq. (1-5)} \\ &= 4\int e^{2x-1}\, dx - 3\int e^{-1}\, dx \\ &= \frac{4}{2}\int e^{2x-1}(2\, dx) - \frac{3}{e}\int dx \\ &= 2e^{2x-1} - \frac{3}{e}x + C\end{aligned}$$

For reference, Eq. (1-5) is $\frac{a^m}{a^n} = a^{m-n}$.

EXAMPLE 5 Evaluate: $\int_0^{\pi/2} (\sin 2\theta)(e^{\cos 2\theta})\, d\theta$.

With $u = \cos 2\theta$ and $du = -2 \sin 2\theta\, d\theta$, we have

$$\begin{aligned}\int_0^{\pi/2} (\sin 2\theta)(e^{\cos 2\theta})\, d\theta &= -\frac{1}{2}\int_0^{\pi/2} (e^{\cos 2\theta})(-2 \sin 2\theta\, d\theta) \\ &= -\frac{1}{2} e^{\cos 2\theta}\Big|_0^{\pi/2} \qquad \text{integrate} \\ &= -\frac{1}{2}\left(\frac{1}{e} - e\right) = 1.175 \qquad \text{evaluate}\end{aligned}$$

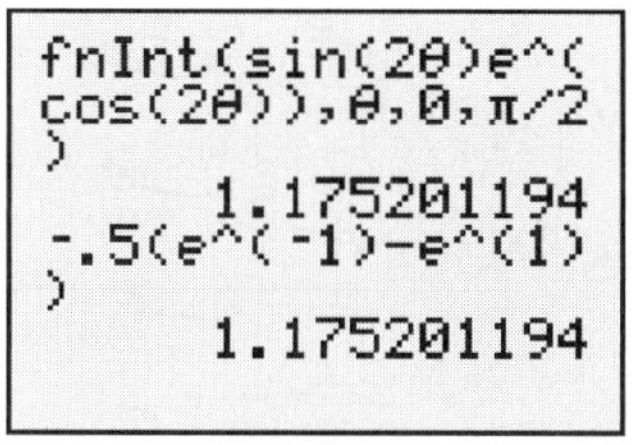

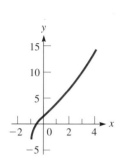

fnInt(sin(2θ)e^(
cos(2θ)),θ,0,π/2
)
1.175201194
-.5(e^(-1)-e^(1)
)
1.175201194
Fig. 9-6

The calculator check for this evaluation is shown in Fig. 9-6.

EXAMPLE 6 Find the equation of the curve for which $\frac{dy}{dx} = \frac{e^{\sqrt{x+1}}}{\sqrt{x+1}}$ if the curve passes through (0, 1).

The solution of this problem requires that we integrate the given function and then evaluate the constant of integration. Hence,

$$dy = \frac{e^{\sqrt{x+1}}}{\sqrt{x+1}}\, dx \qquad \int dy = \int \frac{e^{\sqrt{x+1}}}{\sqrt{x+1}}\, dx$$

For purposes of integrating the right-hand side,

$$u = \sqrt{x+1} \quad \text{and} \quad du = \frac{1}{2\sqrt{x+1}}\, dx$$

$$\begin{aligned}y &= 2\int e^{\sqrt{x+1}}\left(\frac{1}{2\sqrt{x+1}}\, dx\right) \\ &= 2e^{\sqrt{x+1}} + C\end{aligned}$$

Letting $x = 0$ and $y = 1$, we have $1 = 2e + C$, or $C = 1 - 2e$. This means that the equation is

$$y = 2e^{\sqrt{x+1}} + 1 - 2e$$

y
15
10
5
-2
0
2
4
-5
x
Fig. 9-7

The graph of this function is shown in Fig. 9-7.

EXERCISES $9-3$

In Exercises 1–24, integrate each of the given functions.

1. $\int e^{7x}(7\,dx)$

2. $\int e^{x^4}(4x^3\,dx)$

3. $\int e^{2x+5}\,dx$

4. $\int 2e^{-4x}\,dx$

5. $\int_{-2}^{2} 6e^{s/2}\,ds$

6. $\int_{1}^{2} 3e^{4x}\,dx$

7. $\int 6x^2 e^{x^3}\,dx$

8. $\int xe^{-x^2}\,dx$

9. $\int_{1}^{4} \frac{e^{\sqrt{x}}}{\sqrt{x}}\,dx$

10. $\int_{0}^{1} 4(\ln e^u)e^{-2u^2}\,du$

11. $\int 4(\sec\theta\tan\theta)e^{2\sec\theta}\,d\theta$

12. $\int (\sec^2 x)e^{\tan x}\,dx$

13. $\int \sqrt{e^{2y}+e^{3y}}\,dy$

14. $\int (e^x - e^{-x})^2\,dx$

15. $\int_{1}^{3} 3e^{2x}(e^{-2x}-1)\,dx$

16. $\int_{0}^{0.5} \frac{3e^{3x+1}}{e^x}\,dx$

17. $\int \frac{2\,dx}{\sqrt{x}\,e^{\sqrt{x}}}$

18. $\int \frac{4\,dx}{\sec x\,e^{\sin x}}$

19. $\int \frac{e^{\tan^{-1}x}}{x^2+1}\,dx$

20. $\int \frac{e^{\sin^{-1}2x}\,dx}{\sqrt{1-4x^2}}$

21. $\int \frac{e^{\cos 3x}\,dx}{\csc 3x}$

22. $\int 2e^{r^2+\ln r}\,dr$

23. $\int_{0}^{\pi} (\sin 2x)e^{\cos^2 x}\,dx$

24. $\int_{0}^{2} \frac{dt}{1+e^t}$

In Exercises 25–36, solve the given problems by integration.

25. Find the area bounded by $y=3e^x$, $x=0$, $y=0$, and $x=2$.

W 26. Find the area bounded by $x=a$, $x=b$, $y=0$, and $y=e^x$. Explain the meaning of the result.

27. Find the volume generated by revolving the region bounded by $y=e^{x^2}$, $x=1$, $y=0$, and $x=2$ about the y-axis. See Fig. 9-8.

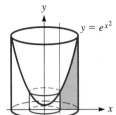

$y = e^{x^2}$

Fig. 9-8

28. Find the equation of the curve for which $dy/dx = \sqrt{e^{x+3}}$ if the curve passes through $(1,0)$.

29. Find the average value of the function $y=e^{2x}$ from $x=0$ to $x=4$.

30. Find the moment of inertia with respect to the y-axis of the first-quadrant region bounded by $y=e^{x^3}$, $x=1$, and the axes.

31. Using Eq. (8-13), show that $\int b^u\,du = \dfrac{b^u}{\ln b} + C$
$(b>0, b\neq 1)$.

32. Find the first-quadrant area bounded by $y=2^x$ and $x=3$. See Exercise 31.

33. For an electric circuit containing a voltage source E, a resistance R, and a capacitance C, an equation relating the charge q on the capacitor and the time t is $qe^{t/RC} = \dfrac{E}{R}\int e^{t/RC}\,dt$. See Fig. 9-9. If $q=0$ for $t=0$, perform the integration and then solve for q as a function of t.

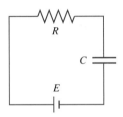

R

C

E

Fig. 9-9

34. In the theory dealing with energy propagation of lasers, the equation $E = a\int_{0}^{I_0} e^{-Tx}\,dx$ is used. Here a, I_0, and T are constants. Evaluate this integral.

35. An object at the end of a spring is immersed in liquid. Its velocity (in cm/s) is then described by the equation $v = 2e^{-2t} + 3e^{-5t}$, where t is the time (in s). Such motion is called *overdamped*. Find the displacement s as a function of t if $s=-1.6$ cm for $t=0$.

36. The force F (in lb) exerted by a robot programmed to staple carton sections together is given by $F = 6\int e^{\sin\pi t}\cos\pi t\,dt$, where t is the time (in s). Find F as a function of t if $F=0$ for $t=1.5$ s.

$9-4$ BASIC TRIGONOMETRIC FORMS

In this section we discuss the integrals of the six trigonometric functions and the trigonometric integrals that arise directly from reversing the formulas for differentiation. Other trigonometric forms will be discussed later.

By reversing differentiation formulas, these integral formulas are obtained:

$$\int \sin u \, du = -\cos u + C \qquad (9\text{-}4)$$

$$\int \cos u \, du = \sin u + C \qquad (9\text{-}5)$$

$$\int \sec^2 u \, du = \tan u + C \qquad (9\text{-}6)$$

$$\int \csc^2 u \, du = -\cot u + C \qquad (9\text{-}7)$$

$$\int \sec u \tan u \, du = \sec u + C \qquad (9\text{-}8)$$

$$\int \csc u \cot u \, du = -\csc u + C \qquad (9\text{-}9)$$

EXAMPLE 1   Integrate: $\int x \sec^2 x^2 \, dx$.
  With $u = x^2$ and $du = 2x \, dx$, we have

$$\int x \sec^2 x^2 \, dx = \frac{1}{2} \int (\sec^2 x^2)(2x \, dx)$$

$$= \frac{1}{2} \tan x^2 + C \qquad \text{using Eq. (9-6)} \qquad ∎$$

EXAMPLE 2   Integrate: $\int \dfrac{\tan 2x}{\cos 2x} \, dx$.

By using the basic identity $\sec \theta = 1/\cos \theta$, we can transform this integral to form $\int \sec 2x \tan 2x \, dx$. In this form, $u = 2x$ and $du = 2 \, dx$. Therefore,

$$\int \frac{\tan 2x}{\cos 2x} \, dx = \int \sec 2x \tan 2x \, dx = \frac{1}{2} \int \sec 2x \tan 2x (2 \, dx)$$

$$= \frac{1}{2} \sec 2x + C \qquad \text{using Eq. (9-8)} \qquad ∎$$

EXAMPLE 3   The vertical velocity v (in cm/s) of the end of a vibrating rod is given by $v = 80 \cos 20\pi t$, where t is the time in seconds. Find the vertical displacement y (in cm) as a function of t if $y = 0$ for $t = 0$.
  Since $v = dy/dt$, we have the following solution.

$$\frac{dy}{dt} = 80 \cos 20\pi t$$

$$\int dy = \int 80 \cos 20\pi t \, dt = \frac{80}{20\pi} \int (\cos 20\pi t)(20\pi \, dt) \quad \text{set up integration}$$

$$y = \frac{4}{\pi} \sin 20\pi t + C \qquad \qquad \text{using Eq. (9-5)}$$

$$0 = 1.27 \sin 0 + C, \qquad C = 0 \quad \text{evaluate } C$$

$$y = 1.27 \sin 20\pi t \qquad \qquad \text{solution} \qquad ∎$$

To find the integrals for the other trigonometric functions, we must change them to a form for which the integral can be determined by methods previously discussed. We can accomplish this by using the basic trigonometric relations.

The formula for $\int \tan u\,du$ is found by expressing the integral in the form $\int (\sin u/\cos u)\,du$. We recognize this as being a logarithmic form, where the u of the logarithmic form is $\cos u$ in this integral. The differential of $\cos u$ is $-\sin u\,du$. Therefore, we have

$$\int \tan u\,du = \int \frac{\sin u}{\cos u}\,du = -\int \frac{-\sin u\,du}{\cos u} = -\ln|\cos u| + C$$

The formula for $\int \cot u\,du$ is found by writing it in the form $\int (\cos u/\sin u)\,du$. In this manner we obtain the result

$$\int \cot u\,du = \int \frac{\cos u}{\sin u}\,du = \int \frac{\cos u\,du}{\sin u} = \ln|\sin u| + C$$

The formula for $\int \sec u\,du$ is found by writing it in the form

$$\int \frac{\sec u(\sec u + \tan u)}{\sec u + \tan u}\,du$$

We see that this form is also a logarithmic form, since

$$d(\sec u + \tan u) = (\sec u \tan u + \sec^2 u)\,du$$

The right side of this equation is the expression appearing in the numerator of the integral. Thus,

$$\int \sec u\,du = \int \frac{\sec u(\sec u + \tan u)\,du}{\sec u + \tan u} = \int \frac{\sec u \tan u + \sec^2 u}{\sec u + \tan u}\,du$$
$$= \ln|\sec u + \tan u| + C$$

To obtain the formula for $\int \csc u\,du$, we write it in the form

$$\int \frac{\csc u(\csc u - \cot u)\,du}{\csc u - \cot u}$$

Thus, we have

$$\int \csc u\,du = \int \frac{\csc u(\csc u - \cot u)}{\csc u - \cot u}\,du = \int \frac{(-\csc u \cot u + \csc^2 u)\,du}{\csc u - \cot u}$$
$$= \ln|\csc u - \cot u| + C$$

Summarizing these results, we have the following integrals:

$$\int \tan u\,du = -\ln|\cos u| + C \tag{9-10}$$

$$\int \cot u\,du = \ln|\sin u| + C \tag{9-11}$$

$$\int \sec u\,du = \ln|\sec u + \tan u| + C \tag{9-12}$$

$$\int \csc u\,du = \ln|\csc u - \cot u| + C \tag{9-13}$$

EXAMPLE 4 Integrate: $\int \tan 4\theta \, d\theta$.

Noting that $u = 4\theta$, $du = 4 \, d\theta$, we have

$$\int \tan 4\theta \, d\theta = \frac{1}{4} \int \tan 4\theta (4 \, d\theta) \qquad \text{introducing factors of 4}$$

$$= -\frac{1}{4} \ln |\cos 4\theta| + C \qquad \text{using Eq. (9-10)}$$

EXAMPLE 5 Integrate: $\int \dfrac{\sec e^{-x} \, dx}{e^x}$.

In this integral, $u = e^{-x}$, $du = -e^{-x} \, dx$. Therefore,

$$\int \frac{\sec e^{-x} \, dx}{e^x} = -\int (\sec e^{-x})(-e^{-x} \, dx) \qquad \text{introducing} - \text{sign}$$

$$= -\ln |\sec e^{-x} + \tan e^{-x}| + C \qquad \text{using Eq. (9-12)}$$

EXAMPLE 6 Evaluate: $\displaystyle\int_{\pi/6}^{\pi/4} \dfrac{1 + \cos x}{\sin x} \, dx$.

The solution is as follows:

For reference, Eqs. (7-4) are

$$\sin \theta = \frac{1}{\csc \theta}, \quad \cos \theta = \frac{1}{\sec \theta},$$

$$\tan \theta = \frac{1}{\cot \theta}$$

Eqs. (7-5) are

$$\tan \theta = \frac{\sin \theta}{\cos \theta}, \quad \cot \theta = \frac{\cos \theta}{\sin \theta}$$

$$\int_{\pi/6}^{\pi/4} \frac{1 + \cos x}{\sin x} \, dx = \int_{\pi/6}^{\pi/4} \csc x \, dx + \int_{\pi/6}^{\pi/4} \cot x \, dx \qquad \text{using Eqs. (7-4) and (7-5)}$$

$$= \ln |\csc x - \cot x| \Big|_{\pi/6}^{\pi/4} + \ln |\sin x| \Big|_{\pi/6}^{\pi/4} \qquad \text{integrating}$$

$$= \ln |\sqrt{2} - 1| - \ln |2 - \sqrt{3}| + \ln \left| \frac{1}{2} \sqrt{2} \right| - \ln \left| \frac{1}{2} \right| \qquad \text{evaluating}$$

$$= \ln \frac{(\frac{1}{2}\sqrt{2})(\sqrt{2} - 1)}{(\frac{1}{2})(2 - \sqrt{3})} = \ln \frac{2 - \sqrt{2}}{2 - \sqrt{3}}$$

$$= 0.782$$

EXERCISES 9-4

In Exercises 1–24, integrate each of the given functions.

1. $\displaystyle\int \cos 2x \, dx$

2. $\displaystyle\int 4 \sin (2 - x) \, dx$

3. $\displaystyle\int 0.3 \sec^2 3\theta \, d\theta$

4. $\displaystyle\int \csc 2x \cot 2x \, dx$

5. $\displaystyle\int \sec \tfrac{1}{2}x \tan \tfrac{1}{2}x \, dx$

6. $\displaystyle\int e^x \csc^2 (e^x) \, dx$

7. $\displaystyle\int_{0.5}^{1} x^2 \cot x^3 \, dx$

8. $\displaystyle\int_{0}^{1} 6 \sin \tfrac{1}{2}t \sec \tfrac{1}{2}t \, dt$

9. $\displaystyle\int 3\phi \sec^2 \phi^2 \cos \phi^2 \, d\phi$

10. $\displaystyle\int 2 \csc 3x \, dx$

11. $\displaystyle\int \frac{\sin(1/x)}{x^2} \, dx$

12. $\displaystyle\int \frac{3 \, dx}{\sin 4x}$

13. $\displaystyle\int_{0}^{\pi/6} \frac{dx}{\cos^2 2x}$

14. $\displaystyle\int_{0}^{1} \frac{2e^s \, ds}{\sec e^s}$

15. $\displaystyle\int \frac{\sec 5x}{\cot 5x} \, dx$

16. $\displaystyle\int \frac{\sin 2x}{\cos^2 x} \, dx$

17. $\displaystyle\int \sqrt{\tan^2 2x + 1} \, dx$

18. $\displaystyle\int 5(\tan u)(\ln \cos u) \, du$

19. $\displaystyle\int \frac{1 + \sin 2x}{\tan 2x} \, dx$

20. $\displaystyle\int \frac{1 - \cot^2 x}{\cos^2 x} \, dx$

21. $\displaystyle\int \frac{1 - \sin x}{1 + \cos x} \, dx$

22. $\displaystyle\int \frac{1 + \sec^2 x}{x + \tan x} \, dx$

23. $\displaystyle\int_{0}^{\pi/9} \sin 3x(\csc 3x + \sec 3x) \, dx$

24. $\displaystyle\int_{\pi/4}^{\pi/3} (1 + \sec x)^2 \, dx$

In Exercises 25–32, solve the given problems by integration.

25. Find the area bounded by $y = 2 \tan x$, $x = \frac{\pi}{4}$, and $y = 0$.

26. Find the area under the curve $y = \sin x$ from $x = 0$ to $x = \pi$.

◎27. Find the volume generated by revolving the region bounded by $y = \sec x$, $x = 0$, $x = \frac{\pi}{3}$, and $y = 0$ about the x-axis.

28. Find the volume generated by revolving the region bounded by $y = \cos x^2$, $x = 0$, $y = 0$, and $x = 1$ about the y-axis.

29. The angular velocity ω (in rad/s) of a pendulum is $\omega = -0.25 \sin 2.5t$. Find the angular displacement θ as a function of t if $\theta = 0.10$ for $t = 0$.

◎ 30. If the current i (in A) in a certain electric circuit is given by $i = 110 \cos 377t$, find the expression for the voltage across a 500-μF capacitor as a function of time. The initial voltage is zero. Show that the voltage across the capacitor is 90° out of phase with the current.

31. A fin on a wind-direction indicator has a shape that can be described as the region bounded by $y = \tan x^2$, $y = 0$, and $x = 1$. Find the x-coordinate (in m) of the centroid of the fin if its area is 0.3984 m².

32. A force is given as a function of the distance from the origin as $F = \dfrac{2 + \tan x}{\cos x}$. Express the work done by this force as a function of x if $W = 0$ for $x = 0$.

9-5 OTHER TRIGONOMETRIC FORMS

The basic trigonometric relations developed in Chapter 7 provide the means by which many other integrals involving trigonometric functions may be integrated. By use of the square relations—Eqs. (7-6), (7-7), and (7-8)—and the equations for the cosine of the double angle—Eqs. (7-22) and (7-23)—it is possible to transform integrals involving powers of the trigonometric functions into integrable form. We repeat these equations here for reference:

$$\cos^2 x + \sin^2 x = 1 \qquad\qquad\qquad\text{(9-14)}$$
$$1 + \tan^2 x = \sec^2 x \qquad\qquad\qquad\text{(9-15)}$$
$$1 + \cot^2 x = \csc^2 x \qquad\qquad\qquad\text{(9-16)}$$
$$2\cos^2 x = 1 + \cos 2x \qquad\qquad\qquad\text{(9-17)}$$
$$2\sin^2 x = 1 - \cos 2x \qquad\qquad\qquad\text{(9-18)}$$

CAUTION▶   *To integrate a product of powers of the sine and cosine, we use Eq. (9-14)* ***if at least one of the powers is odd.*** The method is based on transforming the integral so that it is made up of powers of either the sine or cosine and the first power of the other. In this way this first power becomes a factor of du.

▮EXAMPLE 1   Integrate: $\int \sin^3 x \cos^2 x\, dx$.

Since $\sin^3 x = \sin^2 x \sin x = (1 - \cos^2 x)\sin x$, it is possible to write this integral with powers of $\cos x$ along with $\sin x\, dx$. In this way we can have $-\sin x\, dx$ as the necessary du of the integral. Therefore,

$$\int \sin^3 x \cos^2 x\, dx = \int (1 - \cos^2 x)(\sin x)(\cos^2 x)\, dx \qquad \text{using Eq. (9-14)}$$

$$= \int (\cos^2 x - \cos^4 x)(\sin x\, dx)$$

$$= \int \cos^2 x(\sin x\, dx) - \int \cos^4 x(\sin x\, dx)$$

$$= -\int \cos^2 x(-\sin x\, dx) + \int \cos^4 x(-\sin x\, dx)$$

$$= -\frac{1}{3}\cos^3 x + \frac{1}{5}\cos^5 x + C \qquad\qquad\qquad ▮$$

EXAMPLE 2 Integrate: $\int \cos^5 2x\, dx$.

Since $\cos^5 2x = \cos^4 2x \cos 2x = (1 - \sin^2 2x)^2 \cos 2x$, it is possible to write this integral with powers of sin $2x$ along with cos $2x\, dx$. Thus, with the introduction of a factor of 2, $(\cos 2x)(2\, dx)$ is the necessary du of the integral. Thus,

$$\int \cos^5 2x\, dx = \int (1 - \sin^2 2x)^2 \cos 2x\, dx \qquad \text{using Eq. (9-14)}$$

$$= \int (1 - 2\sin^2 2x + \sin^4 2x)\cos 2x\, dx$$

$$= \int \cos 2x\, dx - \int 2\sin^2 2x \cos 2x\, dx + \int \sin^4 2x \cos 2x\, dx$$

$$= \frac{1}{2}\int \cos 2x(2\, dx) - \int \sin^2 2x(2\cos 2x\, dx) + \frac{1}{2}\int \sin^4 2x(2\cos 2x\, dx)$$

$$= \frac{1}{2}\sin 2x - \frac{1}{3}\sin^3 2x + \frac{1}{10}\sin^5 2x + C$$

CAUTION ▶

*In products of powers of the sine and cosine, **if the powers to be integrated are even, we use Eqs. (9-17) and (9-18) to transform the integral.*** Those most commonly met are $\int \cos^2 u\, du$ and $\int \sin^2 u\, du$. Consider the following examples.

EXAMPLE 3 Integrate: $\int \sin^2 2x\, dx$.

CAUTION ▶

Using Eq. (9-18) in the form $\sin^2 2x = \frac{1}{2}(1 - \cos 4x)$, this integral can be transformed into a form that can be integrated. (Here we note *the x of Eq. (9-18) is treated as* **2x** for this integral.) Therefore, we write

$$\int \sin^2 2x\, dx = \int \left[\frac{1}{2}(1 - \cos 4x)\right] dx \qquad \text{using Eq. (9-18)}$$

$$= \frac{1}{2}\int dx - \frac{1}{8}\int \cos 4x(4\, dx)$$

$$= \frac{x}{2} - \frac{1}{8}\sin 4x + C$$

NOTE ▶

To integrate even powers of the secant, powers of the tangent, or products of the secant and tangent, we use Eq. (9-15) to transform the integral. In transforming, the forms we look for are ***powers of the tangent with sec² x,*** which becomes part of du, or ***powers of the secant along with sec x tan x,*** which becomes part of du in this case. Similar transformations are made when we integrate powers of the cotangent and cosecant, with the use of Eq. (9-16).

EXAMPLE 4 Integrate: $\int \sec^3 t \tan t\, dt$.

By writing $\sec^3 t \tan t$ as $\sec^2 t(\sec t \tan t)$, we can use the sec t tan $t\, dt$ as the du of the integral. Thus,

$$\int \sec^3 t \tan t\, dt = \int (\sec^2 t)(\sec t \tan t\, dt)$$

$$= \frac{1}{3}\sec^3 t + C$$

■EXAMPLE 5 Integrate: $\int \tan^5 x \, dx$.

Since $\tan^5 x = \tan^3 x \tan^2 x = \tan^3 x (\sec^2 x - 1)$, we can write this integral with powers of $\tan x$ along with $\sec^2 x \, dx$. Thus, $\sec^2 x \, dx$ becomes the necessary du of the integral. It is necessary to replace $\tan^2 x$ with $\sec^2 x - 1$ twice during the integration. Therefore,

$$\int \tan^5 x \, dx = \int \tan^3 x (\sec^2 x - 1) \, dx \qquad \text{using Eq. (9-15)}$$

$$= \int \tan^3 x (\sec^2 x \, dx) - \int \tan^3 x \, dx$$

$$= \frac{1}{4} \tan^4 x - \int \tan x (\sec^2 x - 1) \, dx \qquad \text{using Eq. (9-15) again}$$

$$= \frac{1}{4} \tan^4 x - \int \tan x (\sec^2 x \, dx) + \int \tan x \, dx$$

$$= \frac{1}{4} \tan^4 x - \frac{1}{2} \tan^2 x - \ln |\cos x| + C \qquad \blacksquare$$

■EXAMPLE 6 Integrate: $\int \csc^4 2x \, dx$.

By writing $\csc^4 2x = \csc^2 2x \csc^2 2x = \csc^2 2x (1 + \cot^2 2x)$, we can write this integral with powers of $\cot 2x$ along with $\csc^2 2x \, dx$, which becomes part of the necessary du of the integral. Thus,

$$\int \csc^4 2x \, dx = \int \csc^2 2x (1 + \cot^2 2x) \, dx \qquad \text{using Eq. (9-16)}$$

$$= \frac{1}{2} \int \csc^2 2x (2 \, dx) - \frac{1}{2} \int \cot^2 2x (-2 \csc^2 2x \, dx)$$

$$\qquad\qquad\qquad\qquad\qquad\qquad\qquad\qquad \underline{\qquad\quad} du$$

$$= -\frac{1}{2} \cot 2x - \frac{1}{6} \cot^3 2x + C \qquad \blacksquare$$

■EXAMPLE 7 Integrate: $\int_0^{\pi/4} \frac{\tan^3 x}{\sec^3 x} \, dx$.

This integral requires the use of several trigonometric relationships to obtain integrable forms.

$$\int_0^{\pi/4} \frac{\tan^3 x}{\sec^3 x} \, dx = \int_0^{\pi/4} \frac{(\sec^2 x - 1) \tan x}{\sec^3 x} \, dx \qquad \text{using Eq. (9-15)}$$

$$= \int_0^{\pi/4} \frac{\tan x}{\sec x} \, dx - \int_0^{\pi/4} \frac{\tan x \, dx}{\sec^3 x} \qquad \frac{\tan x}{\sec x} = \frac{\sin x}{\cos x \sec x} = \sin x$$

$$= \int_0^{\pi/4} \sin x \, dx - \int_0^{\pi/4} \cos^2 x \sin x \, dx \qquad \frac{\tan x}{\sec^3 x} = \frac{\tan x}{\sec x \sec^2 x} = \sin x \cos^2 x$$

$$= -\cos x + \frac{1}{3} \cos^3 x \Big|_0^{\pi/4} \qquad \text{integrate}$$

$$= -\frac{\sqrt{2}}{2} + \frac{1}{3} \left(\frac{\sqrt{2}}{2}\right)^3 - \left(-1 + \frac{1}{3}\right) \qquad \text{evaluate}$$

$$= \frac{8 - 5\sqrt{2}}{12} = 0.0774$$

```
fnInt((tan(X))^3
(cos(X)^3),X,0,π
/4)
         .0774110157
(8-5√(2))/12
         .0774110157
```

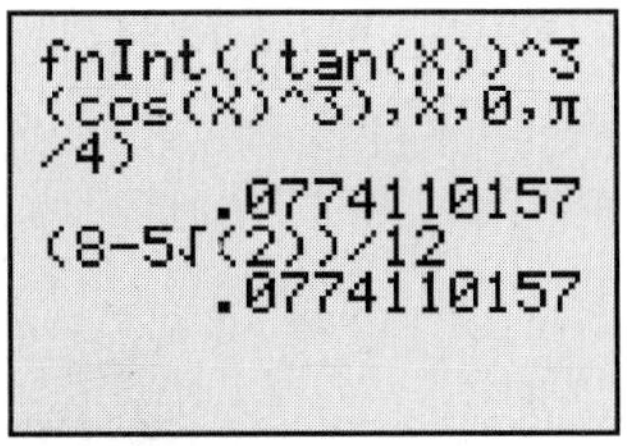

Fig. 9-10

See Fig. 9-10 for a calculator check of this result. $\qquad \blacksquare$

■**EXAMPLE 8**   The *root-mean-square value of a function* with respect to x is defined by

$$y_{\text{rms}} = \sqrt{\frac{1}{T} \int_0^T y^2 \, dx} \qquad (9\text{-}19)$$

Usually the value of T that is of importance is the period of the function. Find the root-mean-square value of the electric current i (in A) used in a home electric heater, for which $i = 17.7 \cos 120\pi t$, for one period.

The period is $\frac{2\pi}{120\pi} = \frac{1}{60.0}$ s. Thus, we must find the square root of the integral

$$\frac{1}{1/60.0} \int_0^{1/60.0} (17.7 \cos 120\pi t)^2 \, dt = 18{,}800 \int_0^{1/60.0} \cos^2 120\pi t \, dt$$

Evaluating this integral, we have

$$18{,}800 \int_0^{1/60.0} \cos^2 120\pi t \, dt = 9400 \int_0^{1/60.0} (1 + \cos 240\pi t) \, dt$$

$$= 9400 t \Big|_0^{1/60.0} + \frac{9400}{240\pi} \int_0^{1/60.0} \cos 240\pi t (240\pi \, dt)$$

$$= 156.7 + \frac{235}{6\pi} \sin 240\pi t \Big|_0^{1/60.0} = 156.7$$

This means the root-mean-square current is

$$i_{\text{rms}} = \sqrt{156.7} = 12.5 \text{ A}$$

This value of the current, often referred to as the *effective current,* is the value of direct current that would produce the same quantity of heat energy in the same time. It is important in the design of electric heaters and other electric appliances.   ∎

— **EXERCISES** *9-5* —

In Exercises 1–28, integrate each of the given functions.

1. $\int \sin^2 x \cos x \, dx$

2. $\int \sin x \cos^5 x \, dx$

3. $\int \sin^3 2x \, dx$

4. $\int 3 \cos^3 x \, dx$

5. $\int 4(\cos^4 \theta - \sin^4 \theta) \, d\theta$

6. $\int \sin^3 x \cos^6 x \, dx$

7. $\int_0^{\pi/4} 5 \sin^5 x \, dx$

8. $\int_{\pi/3}^{\pi/2} 10 \sin t (1 - \cos 2t)^2 \, dt$

9. $\int \sin^2 x \, dx$

10. $\int \cos^2 2x \, dx$

11. $\int 2(1 + \cos 3\phi)^2 \, d\phi$

12. $\int_0^1 \sin^2 4x \, dx$

13. $\int \tan^3 x \, dx$

14. $\int \frac{6 \cot^2 y}{\tan y} \, dy$

15. $\int_0^{\pi/4} \tan x \sec^4 x \, dx$

16. $\int \cot 4x \csc^4 4x \, dx$

17. $\int \tan^4 2x \, dx$

18. $\int 4 \cot^4 x \, dx$

19. $\int 0.5 \sin s \sin 2s \, ds$

20. $\int \sqrt{\tan x} \sec^4 x \, dx$

21. $\int (\sin x + \cos x)^2 \, dx$

22. $\int (\tan 2x + \cot 2x)^2 \, dx$

23. $\int \frac{1 - \cot x}{\sin^4 x} \, dx$

24. $\int \frac{(\sin u + \sin^2 u)^2}{\sec u} \, du$

25. $\int_{\pi/6}^{\pi/4} \cot^5 x \, dx$

26. $\int_{\pi/6}^{\pi/3} \frac{2 \, dx}{1 + \sin x}$

27. $\int \sec^6 x \, dx$

28. $\int \tan^7 x \, dx$

In Exercises 29–40, solve the given problems by integration.

29. Find the volume generated by revolving the region bounded by $y = \sin x$ and $y = 0$, from $x = 0$ to $x = \pi$, about the x-axis.

30. Find the volume generated by revolving the region bounded by $y = \tan^3(x^2)$, $y = 0$, and $x = \frac{\pi}{4}$ about the y-axis.

31. Find the area bounded by $y = \sin x$, $y = \cos x$, and $x = 0$ in the first quadrant.

32. Find the length of the curve $y = \ln \cos x$ from $x = 0$ to $x = \frac{\pi}{3}$.

W 33. Show that $\int \sin x \cos x \, dx$ can be integrated in two ways. Explain the difference in the answers.

W 34. Show that $\int \sec^2 x \tan x \, dx$ can be integrated in two ways. Explain the difference in the answers.

35. In the study of the rate of radiation by an accelerated charge, the following integral must be evaluated: $\int_0^\pi \sin^3 \theta \, d\theta$. Find the value of the integral.

36. In finding the volume of a special O-ring for a space vehicle, the integral $\int \dfrac{\sin^2 \theta}{\cos^2 \theta} \, d\theta$ must be evaluated. Perform this integration.

37. For a voltage $V = 340 \sin 120\pi t$, show that the root-mean-square voltage for one period is 240 V.

38. For a current $i = i_0 \sin \omega t$, show that the root-mean-square current for one period is $i_0/\sqrt{2}$.

39. In the analysis of the intensity of light from a certain source, the equation $I = A \int_{-a/2}^{a/2} \cos^2 [b\pi(c - x)] \, dx$ is used. Here, A, a, b, and c are constants. Evaluate this integral. (The simplification is quite lengthy.)

40. In the study of the lifting force L due to a stream of fluid passing around a cylinder, the equation $L = k \int_0^{2\pi} (a \sin \theta + b \sin^2 \theta - b \sin^3 \theta) \, d\theta$ is used. Here, k, a, and b are constants, and θ is the angle from the direction of flow. Evaluate the integral.

$9\text{-}6$ INVERSE TRIGONOMETRIC FORMS

For reference, Eq. (7-40) is
$$\frac{d(\sin^{-1} u)}{dx} = \frac{1}{\sqrt{1 - u^2}} \frac{du}{dx}.$$

Using Eq. (7-40) we can find the differential of $\sin^{-1}(u/a)$, where a is constant:

$$d\left(\sin^{-1} \frac{u}{a}\right) = \frac{1}{\sqrt{1 - (u/a)^2}} \frac{du}{a} = \frac{a}{\sqrt{a^2 - u^2}} \frac{du}{a} = \frac{du}{\sqrt{a^2 - u^2}}$$

Reversing this differentiation formula, we have the important integration formula

$$\int \frac{du}{\sqrt{a^2 - u^2}} = \sin^{-1} \frac{u}{a} + C \qquad (9\text{-}20)$$

By finding the differential of $\tan^{-1}(u/a)$, we have

$$d\left(\tan^{-1} \frac{u}{a}\right) = \frac{1}{1 + (u/a)^2} \frac{du}{a} = \frac{a^2}{a^2 + u^2} \frac{du}{a} = \frac{a \, du}{a^2 + u^2}$$

Now, reversing this differential, we have

$$\int \frac{du}{a^2 + u^2} = \frac{1}{a} \tan^{-1} \frac{u}{a} + C \qquad (9\text{-}21)$$

This shows one of the principal uses of the inverse trigonometric functions: They provide a solution to the integration of important algebraic functions.

EXAMPLE 1 Integrate: $\int \dfrac{dx}{\sqrt{9 - x^2}}$.

This integral fits the form of Eq. (9-20) with $u = x$, $du = dx$, and $a = 3$. Thus,

$$\int \frac{dx}{\sqrt{9 - x^2}} = \int \frac{dx}{\sqrt{3^2 - x^2}}$$

$$= \sin^{-1} \frac{x}{3} + C$$

EXAMPLE 2 The volume flow rate Q (in m^3/s) of a constantly flowing liquid is given by $Q = 24 \int_0^2 \dfrac{dx}{6 + x^2}$, where x is the distance from the center of flow. Find the value of Q.

For the integral, we see that it fits Eq. (9-21) with $u = x$, $du = dx$, and $a = \sqrt{6}$.

$$Q = 24 \int_0^2 \frac{dx}{6 + x^2} = 24 \int_0^2 \frac{dx}{(\sqrt{6})^2 + x^2}$$

$$= \frac{24}{\sqrt{6}} \tan^{-1} \frac{x}{\sqrt{6}} \bigg|_0^2$$

$$= \frac{24}{\sqrt{6}} \left(\tan^{-1} \frac{2}{\sqrt{6}} - \tan^{-1} 0 \right)$$

$$= 6.71 \text{ m}^3/\text{s}$$

EXAMPLE 3 Integrate: $\displaystyle\int \frac{dx}{\sqrt{25 - 4x^2}}$.

This integral fits the form of Eq. (9-20) with $u = 2x$, $du = 2\,dx$, and $a = 5$. Thus, in order to have the proper du, we must include a factor of 2 in the numerator, and therefore we also place a $\frac{1}{2}$ before the integral. This leads to

$$\int \frac{dx}{\sqrt{25 - 4x^2}} = \frac{1}{2} \int \frac{2\,dx}{\sqrt{5^2 - (2x)^2}} \longleftarrow du$$
$$\phantom{\int \frac{dx}{\sqrt{25 - 4x^2}} = \frac{1}{2} \int \frac{2\,dx}{\sqrt{5^2 - (2x)^2}}} \uparrow\!\!\rule{0pt}{0pt}\, u$$

$$= \frac{1}{2} \sin^{-1} \frac{2x}{5} + C$$

EXAMPLE 4 Integrate: $\displaystyle\int_{-1}^3 \frac{dx}{x^2 + 6x + 13}$.

At first glance, it does not appear that this integral fits any of the forms presented up to this point. However, by writing the denominator in the form **CAUTION ▶** $(x^2 + 6x + 9) + 4 = (x + 3)^2 + 2^2$, *we recognize that $u = x + 3$, $du = dx$,* and $a = 2$. Thus,

$$\int_{-1}^3 \frac{dx}{x^2 + 6x + 13} = \int_{-1}^3 \frac{dx}{(x + 3)^2 + 2^2} \longleftarrow du$$
$$\phantom{\int_{-1}^3 \frac{dx}{x^2 + 6x + 13} = \int_{-1}^3 \frac{dx}{(x + 3)^2 + 2^2}} \uparrow\!\!\rule{0pt}{0pt}\, u$$

$$= \frac{1}{2} \tan^{-1} \frac{x + 3}{2} \bigg|_{-1}^3 \qquad \text{integrate}$$

$$= \frac{1}{2} (\tan^{-1} 3 - \tan^{-1} 1) \qquad \text{evaluate}$$

$$= 0.2318$$

Now we can see the use of completing the square when we are transforming integrals into proper form.

■EXAMPLE 5   Integrate: $\int \dfrac{2r + 5}{r^2 + 9} \, dr$.

By writing this integral as the sum of two integrals, we may integrate each of these separately:

$$\int \frac{2r + 5}{r^2 + 9} \, dr = \int \frac{2r \, dr}{r^2 + 9} + \int \frac{5 \, dr}{r^2 + 9}$$

The first integral is a logarithmic form, and the second is an inverse tangent form. For the first, $u = r^2 + 9$, $du = 2r \, dr$. For the second, $u = r$, $du = dr$, $a = 3$.

$$\int \frac{2r \, dr}{r^2 + 9} + 5 \int \frac{dr}{r^2 + 9} = \ln |r^2 + 9| + \frac{5}{3} \tan^{-1} \frac{r}{3} + C \quad \blacksquare$$

CAUTION ▶   The inverse trigonometric integral forms show very well the importance of *proper recognition of the form of the integral.* It is important that these forms are not confused with those of the general power rule or the logarithmic form.

■EXAMPLE 6   The integral $\int \dfrac{dx}{\sqrt{1 - x^2}}$ is of the inverse sine form with $u = x$, $du = dx$, and $a = 1$. Thus,

$$\int \frac{dx}{\sqrt{1 - x^2}} = \sin^{-1} x + C$$

The integral $\int \dfrac{x \, dx}{\sqrt{1 - x^2}}$ is not of the inverse sine form due to the factor of x in the numerator. It is integrated by use of the general power rule, with $u = 1 - x^2$, $du = -2x \, dx$, and $n = -\frac{1}{2}$. Thus,

$$\int \frac{x \, dx}{\sqrt{1 - x^2}} = -\sqrt{1 - x^2} + C$$

The integral $\int \dfrac{x \, dx}{1 - x^2}$ is of the basic logarithmic form with $u = 1 - x^2$ and $du = -2x \, dx$. If $1 - x^2$ is raised to any power other than 1 in the denominator, we would use the general power rule. To be of the inverse sine form, we would have the square root of $1 - x^2$ and no factor of x, as in the first illustration. Thus,

$$\int \frac{x \, dx}{1 - x^2} = -\frac{1}{2} \ln |1 - x^2| + C \quad \blacksquare$$

■EXAMPLE 7   The following integrals are of the form indicated.

$$\int \frac{dx}{1 + x^2} \qquad \text{Inverse tangent form} \qquad u = x, \, du = dx$$

$$\int \frac{x \, dx}{1 + x^2} \qquad \text{Logarithmic form} \qquad u = 1 + x^2, \, du = 2x \, dx$$

$$\int \frac{x \, dx}{\sqrt{1 + x^2}} \qquad \text{General power form} \qquad u = 1 + x^2, \, du = 2x \, dx$$

$$\int \frac{dx}{1 + x} \qquad \text{Logarithmic form} \qquad u = 1 + x, \, du = dx \quad \blacksquare$$

There are a number of integrals whose forms appear to be similar to those in Examples 6 and 7, but which do not fit the forms we have discussed. They include

$$\int \frac{dx}{\sqrt{x^2 - 1}} \qquad \int \frac{dx}{\sqrt{1 + x^2}} \qquad \int \frac{dx}{1 - x^2} \qquad \int \frac{dx}{x\sqrt{1 + x^2}}$$

We will develop methods to integrate some of these forms, and all of them can be integrated by the tables discussed in Section 10-6.

EXERCISES 9-6

In Exercises 1–24, integrate each of the given functions.

1. $\int \frac{dx}{\sqrt{4 - x^2}}$

2. $\int \frac{dx}{\sqrt{49 - x^2}}$

3. $\int \frac{dx}{64 + x^2}$

4. $\int \frac{6p^2\,dp}{4 + p^6}$

5. $\int \frac{dx}{\sqrt{1 - 16x^2}}$

6. $\int_0^1 \frac{2\,dx}{\sqrt{9 - 4x^2}}$

7. $\int_0^2 \frac{3e^{-t}\,dt}{1 + 9e^{-2t}}$

8. $\int_1^3 \frac{dx}{49 + 4x^2}$

9. $\int_0^{0.4} \frac{2\,dx}{\sqrt{4 - 5x^2}}$

10. $\int \frac{dx}{2\sqrt{x}\,\sqrt{1 - x}}$

11. $\int \frac{8x\,dx}{9x^2 + 16}$

12. $\int \frac{4y\,dy}{\sqrt{25 - 16y^2}}$

13. $\int_1^e \frac{3\,du}{u[1 + (\ln u)^2]}$

14. $\int_0^1 \frac{4x\,dx}{1 + x^4}$

15. $\int \frac{e^x\,dx}{\sqrt{1 - e^{2x}}}$

16. $\int \frac{\sec^2 x\,dx}{\sqrt{1 - \tan^2 x}}$

17. $\int \frac{dx}{x^2 + 2x + 2}$

18. $\int \frac{2\,dx}{x^2 + 8x + 17}$

19. $\int \frac{4\,dx}{\sqrt{-4x - x^2}}$

20. $\int \frac{0.3\,ds}{\sqrt{2s - s^2}}$

21. $\int_{\pi/6}^{\pi/2} \frac{2\cos 2\theta\,d\theta}{1 + \sin^2 2\theta}$

22. $\int_{-4}^0 \frac{dx}{x^2 + 4x + 5}$

23. $\int \frac{2 - x}{\sqrt{4 - x^2}}\,dx$

24. $\int \frac{3 - 2x}{1 + 4x^2}\,dx$

In Exercises 25–28, identify the form of each integral as being inverse sine, inverse tangent, logarithmic, or general power, as in Examples 6 and 7. Do not integrate. In each part (a), explain how the choice was made.

25. (a) $\int \frac{2\,dx}{4 + 9x^2}$ (b) $\int \frac{2\,dx}{4 + 9x}$ (c) $\int \frac{2x\,dx}{\sqrt{4 + 9x^2}}$

26. (a) $\int \frac{2x\,dx}{4 - 9x^2}$ (b) $\int \frac{2\,dx}{\sqrt{4 - 9x}}$ (c) $\int \frac{2x\,dx}{4 + 9x^2}$

27. (a) $\int \frac{2x\,dx}{\sqrt{4 - 9x^2}}$ (b) $\int \frac{2\,dx}{\sqrt{4 - 9x^2}}$ (c) $\int \frac{2\,dx}{4 - 9x}$

28. (a) $\int \frac{2\,dx}{9x^2 + 4}$ (b) $\int \frac{2x\,dx}{\sqrt{9x^2 - 4}}$ (c) $\int \frac{2x\,dx}{9x^2 - 4}$

In Exercises 29–36, solve the given problems by integration.

29. Find the area bounded by $y(1 + x^2) = 1$, $x = 0$, $y = 0$, and $x = 2$.

30. Find the area bounded by $y\sqrt{4 - x^2} = 1$, $x = 0$, $y = 0$, and $x = 1$. See Fig. 9-11.

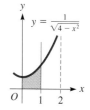

Fig. 9-11

31. To find the electric field E from an electric charge distributed uniformly over the entire xy-plane at a distance d from the plane, it is necessary to evaluate the integral $kd \int \frac{dx}{d^2 + x^2}$. Here, x is the distance from the origin to the element of charge. Perform the indicated integration.

32. An oil storage tank can be described as the volume generated by revolving the region bounded by $y = 24/\sqrt{16 + x^2}$, $x = 0$, $y = 0$, and $x = 3$ about the x-axis. Find the volume (in m^3) of the tank.

33. In dealing with the theory for simple harmonic motion, it is necessary to solve the equation $\frac{dx}{\sqrt{A^2 - x^2}} = \sqrt{\frac{k}{m}}\,dt$ (k, m, and A are constants). Determine the solution if $x = x_0$ when $t = 0$.

34. During each cycle, the velocity v (in ft/s) of a robotic welding device is given by $v = 2t - \frac{12}{2 + t^2}$, where t is the time (in s). Find the expression for the displacement s (in ft) as a function of t if $s = 0$ for $t = 0$.

35. Find the moment of inertia with respect to the y-axis for the region bounded by $y = 1/(1 + x^6)$, the x-axis, $x = 1$, and $x = 2$.

36. Find the length of arc along the curve $y = \sqrt{1 - x^2}$ between $x = 0$ and $x = 1$.

CHAPTER EQUATIONS

Integrals

$$\int u^n\, du = \frac{u^{n+1}}{n+1} + C \qquad (n \neq -1) \tag{9-1}$$

$$\int \frac{du}{u} = \ln|u| + C \tag{9-2}$$

$$\int e^u\, du = e^u + C \tag{9-3}$$

$$\int \sin u\, du = -\cos u + C \tag{9-4}$$

$$\int \cos u\, du = \sin u + C \tag{9-5}$$

$$\int \sec^2 u\, du = \tan u + C \tag{9-6}$$

$$\int \csc^2 u\, du = -\cot u + C \tag{9-7}$$

$$\int \sec u \tan u\, du = \sec u + C \tag{9-8}$$

$$\int \csc u \cot u\, du = -\csc u + C \tag{9-9}$$

$$\int \tan u\, du = -\ln|\cos u| + C \tag{9-10}$$

$$\int \cot u\, du = \ln|\sin u| + C \tag{9-11}$$

$$\int \sec u\, du = \ln|\sec u + \tan u| + C \tag{9-12}$$

$$\int \csc u\, du = \ln|\csc u - \cot u| + C \tag{9-13}$$

Trigonometric relations

$$\cos^2 x + \sin^2 x = 1 \tag{9-14}$$
$$1 + \tan^2 x = \sec^2 x \tag{9-15}$$
$$1 + \cot^2 x = \csc^2 x \tag{9-16}$$
$$2\cos^2 x = 1 + \cos 2x \tag{9-17}$$
$$2\sin^2 x = 1 - \cos 2x \tag{9-18}$$

Root-mean-square value

$$y_{\text{rms}} = \sqrt{\frac{1}{T}\int_0^T y^2\, dx} \tag{9-19}$$

Integrals

$$\int \frac{du}{\sqrt{a^2 - u^2}} = \sin^{-1}\frac{u}{a} + C \tag{9-20}$$

$$\int \frac{du}{a^2 + u^2} = \frac{1}{a}\tan^{-1}\frac{u}{a} + C \tag{9-21}$$

REVIEW EXERCISES

In Exercises 1–36, integrate the given functions.

1. $\int e^{-2x}\, dx$

2. $\int e^{\cos 2x} \sin x \cos x\, dx$

3. $\int \dfrac{dx}{x(\ln 2x)^2}$

4. $\int_1^8 y^{1/3}\sqrt{y^{4/3}+9}\, dy$

5. $\int_0^{\pi/2} \dfrac{4 \cos \theta\, d\theta}{1 + \sin \theta}$

6. $\int \dfrac{\sec^2 x\, dx}{2 + \tan x}$

7. $\int \dfrac{2\, dx}{25 + 49x^2}$

8. $\int \dfrac{dx}{\sqrt{1 - 4x^2}}$

9. $\int_0^{\pi/2} \cos^3 2x\, dx$

10. $\int_0^{\pi/8} \sec^3 2x \tan 2x\, dx$

11. $\int_0^2 \dfrac{x\, dx}{4 + x^2}$

12. $\int_1^e \dfrac{\ln v^2\, dv}{\ln e^v}$

13. $\int (\sin t + \cos t)^2 \sin t\, dt$

14. $\int \dfrac{\sin^3 x\, dx}{\sqrt{\cos x}}$

15. $\int \dfrac{e^x\, dx}{1 + e^{2x}}$

16. $\int \sec^4 3x \tan 3x\, dx$

17. $\int \sec^4 3x\, dx$

18. $\int \dfrac{(1 - \cos^2 \theta)\, d\theta}{1 + \cos 2\theta}$

19. $\int_0^{0.5} \dfrac{2e^{2x} - 3e^x}{e^{2x}}\, dx$

20. $\int \dfrac{4 - e^{\sqrt{x}}}{\sqrt{x}\, e^{\sqrt{x}}}\, dx$

21. $\int \dfrac{3x\, dx}{4 + x^4}$

22. $\int_1^3 \dfrac{2\, dx}{\sqrt{x}(1 + x)}$

23. $\int_1^3 \dfrac{x\, dx}{2(3 + x^2)}$

24. $\int \dfrac{e^x\, dx}{\sqrt{9 - e^{2x}}}$

25. $\int \dfrac{2 + 4e^{2x}}{x + e^{2x}}\, dx$

26. $\int \dfrac{\tan^2 x\, dx}{\tan x - x}$

27. $\int \dfrac{dx}{\sqrt{5 - (4x + x^2)}}$

28. $\int_{1/2}^{e/2} \dfrac{(4 + \ln 2u)^3\, du}{u}$

29. $\int_0^{\pi/6} 3 \sin^2 3\phi\, d\phi$

30. $\int \sin^4 x\, dx$

31. $\int e^{2x} \cos e^{2x}\, dx$

32. $\int \dfrac{3\, dx}{x^2 + 6x + 10}$

33. $\int_1^e \dfrac{3 \cos(\ln x)\, dx}{x}$

34. $\int_1^3 \dfrac{2\, dx}{x^2 - 2x + 5}$

35. $\int \dfrac{x^2 - 1}{x + 2}\, dx$

36. $\int \dfrac{\log_x 2\, dx}{x \ln x}$

In Exercises 37–52, solve the given problems by integration.

W **37.** Show that $\int e^x(e^x + 1)^2\, dx$ can be integrated in two ways. Explain the difference in the answers.

W **38.** Show that $\int \frac{1}{x}(1 + \ln x)\, dx$ can be integrated in two ways. Explain the difference in the answers.

39. Show that $\int \dfrac{1}{1 + \sin x}\, dx$ can be integrated by multiplying the numerator and denominator of the integrand by $1 - \sin x$.

40. Find the equation of the curve for which $dy/dx = e^x(2 - e^x)^2$, if the curve passes through $(0, 4)$.

41. Find the equation of the curve for which $dy/dx = \sec^4 x$, if the curve passes through the origin.

42. Find the area bounded by $y = 4e^{2x}$, $x = 1.5$, and the axes.

43. Find the length of arc along the curve of $y = \ln \sin x$ from $x = \pi/3$ to $x = 2\pi/3$.

44. The change in the thermodynamic entity of entropy ΔS may be expressed as $\Delta S = \int (c_v/T)\, dT$, where c_v is the heat capacity at constant volume and T is the temperature. For increased accuracy, c_v is often given by the equation $c_v = a + bT + cT^2$, where a, b, and c are constants. Express ΔS as a function of temperature.

45. When we consider the resisting force of the air, the velocity v (in ft/s) of a falling brick in terms of the time t (in s) is given by $dv/(32 - 0.5v) = dt$. If $v = 0$ when $t = 0$, find v as a function of t.

46. The power delivered to an electric circuit is given by $P = ei$, where e and i are the instantaneous voltage and the instantaneous current in the circuit, respectively. The mean power, averaged over a period $2\pi/\omega$, is given by $P_{\text{av}} = \dfrac{\omega}{2\pi}\int_0^{2\pi/\omega} ei\, dt$. If $e = 20 \cos 2t$ and $i = 3 \sin 2t$, find the average power over a period of $\pi/4$.

47. Find the root-mean-square value of the electric current i for one period if $i = 2 \sin t$.

48. In atomic theory, when finding the number n of atoms per unit volume of a substance, we use the equation $n = A \int_0^\pi e^{a \cos \theta} \sin \theta\, d\theta$. Perform the indicated integration.

49. In the study of the effects of an electric field on molecular orientation, the integral $\int_0^\pi (1 + k \cos \theta)\cos \theta \sin \theta\, d\theta$ is used. Evaluate this integral.

50. Find the distance from the origin after 0.5 s of the particle for which the velocity v (in cm/s) as a function of the time is given by $v = 4 \cos 3t$. Find the distance by (a) integrating the definite integral between limits of 0 and 0.5 and (b) substituting into the function found by use of the indefinite integral. Compare the results of these two methods.

51. Find the volume within the piece of tubing in an oil distribution line shown in Fig. 9-12. All cross sections are circular.

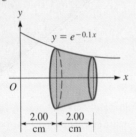

$y = e^{-0.1x}$

2.00 cm   2.00 cm

Fig. 9-12

52. A metal plate has the shape shown in Fig. 9-13. Find the x-coordinate of the centroid of the plate.

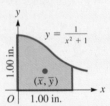

$y = \dfrac{1}{x^2 + 1}$

1.00 in.

$(\bar{x}, \bar{y})$

1.00 in.

Fig. 9-13

Writing Exercise

53. The side of a cutting blade designed using CAD (computer-assisted design) can be described as the region bounded by $y = 4/(1 + e^x)$, $x = 2.8$, and the axes. Write two or three paragraphs explaining how this area may be found by *algebraically* changing the form of the appropriate integral in either one of two ways. (The evaluation of the integral requires only the use of basic forms of this chapter.)

PRACTICE TEST

In Problems 1–6, evaluate the given integrals.

1. $\displaystyle\int (\sec x - \sec^3 x \tan x)\, dx$ **2.** $\displaystyle\int \sin^3 x\, dx$

3. $\displaystyle\int \tan^3 2x\, dx$ **4.** $\displaystyle\int \cos^2 4\theta\, d\theta$

5. $\displaystyle\int_4^7 \frac{\ln(x-3)}{x-3}\, dx$ **6.** $\displaystyle\int \frac{4e^{\tan 2x}}{\cos^2 2x}\, dx$

7. The electric current in a certain circuit is given by $i = \displaystyle\int \frac{6t+1}{4t^2+9}\, dt$, where t is the time. Integrate and find the resulting function if $i = 0$ for $t = 0$.

8. Find the first-quadrant area bounded by $y = \dfrac{1}{\sqrt{16 - x^2}}$ and $x = 3$.

10 METHODS OF INTEGRATION

For many functions that cannot be integrated directly by the basic forms, there are methods for changing them into forms that can be integrated. In Chapter 9 we saw that completing the square and trigonometric identities can be used for this purpose. In this chapter we shall develop several general methods of integration. We also will demonstrate the use of a table of integrals.

Even when using a table, we often find it necessary to change the form of an integral so that it fits a form in the table. Also, with tables and all of these methods, as well as many others that could be developed, there are many integrals that cannot be integrated exactly. This is one of the reasons that it is necessary to have approximate methods of integration such as the trapezoidal rule and Simpson's rule.

At the end of the chapter, we show how limits are used to evaluate definite integrals that have a limit of infinity or a discontinuous integrand.

When you have completed studying the methods of this chapter, you will have a better idea of which integrals can be integrated in an exact form and how to proceed with the integration.

In designing an industrial robot, the distance each joint moves must be carefully analyzed. In Section 10-3 we see how we may find the distance moved by a joint of a particular link.

10-1 INTEGRATION BY PARTS

In this section and the two sections that follow, we will develop three general methods of transforming integrals into forms that can be integrated by the basic formulas that were derived in Chapters 5 and 9. In this section the method of *integration by parts* is discussed. It can be used for many integrals in which the integrand can be expressed as the product of one function and the differential of another function. This is shown on the following page.

Since the derivative of a product of functions is found by use of the formula

$$\frac{d(uv)}{dx} = u\frac{dv}{dx} + v\frac{du}{dx}$$

the differential of a product of functions is given by $d(uv) = u\,dv + v\,du$. Integrating both sides of this equation, we have $uv = \int u\,dv + \int v\,du$. Solving for $\int u\,dv$, we obtain

$$\int u\,dv = uv - \int v\,du \qquad\qquad (10\text{-}1)$$

Integration by use of Eq. (10-1) is called **integration by parts.**

EXAMPLE 1   Integrate: $\int x \sin x\,dx$.

This integral does not fit any of the previous forms we have discussed, since neither x nor $\sin x$ can be made a factor of a proper du. However, by choosing $u = x$ and $dv = \sin x\,dx$, integration by parts may be used. Thus,

$$u = x \qquad dv = \sin x\,dx$$

By finding the differential of u and integrating dv, we find du and v. This gives us

$$du = dx \qquad v = -\cos x + C_1$$

Now, substituting in Eq. (10-1), we have

$$\int \underset{u}{}\ \underset{dv}{} \quad = \underset{u}{}\quad \underset{v}{} \quad - \int \quad \underset{v}{} \quad \underset{du}{}$$

$$\int (x)(\sin x\,dx) = (x)(-\cos x + C_1) - \int (-\cos x + C_1)(dx)$$

$$= -x \cos x + C_1 x + \int \cos x\,dx - \int C_1\,dx$$

$$= -x \cos x + C_1 x + \sin x - C_1 x + C$$

$$= -x \cos x + \sin x + C$$

Other choices of u and dv may be made, but they are not useful. For example, if we choose $u = \sin x$ and $dv = x\,dx$, then $du = \cos x\,dx$ and $v = \frac{1}{2}x^2 + C_2$. This makes $\int v\,du = \int (\frac{1}{2}x^2 + C_2)(\cos x\,dx)$, which is more complex than the integrand of the original problem.

We also note that the constant C_1 that was introduced when we integrated dv does not appear in the final result. This constant will always cancel out, and there-

NOTE▶   fore *we will not show any constant of integration when finding v.* ----------■

As in Example 1, there is often more than one choice as to the part of the integrand that is selected to be u and the part that is selected to be dv. There are no set rules that may be stated for the best choice of u and dv, but two guidelines may be stated.

CAUTION ▶

Guidelines for Choosing *u* and *dv*

1. *The quantity u is normally chosen such that du/dx is of simpler form than u.*

2. *The differential dv is normally chosen such that ∫ dv is easily integrated.*

Working examples, and thereby gaining experience in methods of integration, is the best way to determine when this method should be used and how to use it.

■EXAMPLE 2   Integrate: $\int x \sqrt{1 - x}\, dx$.

We see that this form does not fit the general power rule, for $x\, dx$ is not a factor of the differential of $1 - x$. By choosing $u = x$ and $dv = \sqrt{1 - x}\, dx$, we have $du/dx = 1$, and v can readily be determined. Thus,

$$u = x \qquad dv = \sqrt{1 - x}\, dx = (1 - x)^{1/2}\, dx$$

$$du = dx \qquad v = -\frac{2}{3}(1 - x)^{3/2}$$

Substituting in Eq. (10-1), we have

$$\int \overset{u}{\downarrow} \quad \overset{dv}{\downarrow} \qquad = \overset{u}{\downarrow} \qquad \overset{v}{\downarrow} \qquad - \int \qquad \overset{v}{\downarrow} \qquad \overset{du}{\downarrow}$$

$$\int x[(1 - x)^{1/2}\, dx] = x\left[-\frac{2}{3}(1 - x)^{3/2}\right] - \int \left[-\frac{2}{3}(1 - x)^{3/2}\right] dx$$

At this point, we see that we can complete the integration. Thus,

$$\int x(1 - x)^{1/2}\, dx = -\frac{2x}{3}(1 - x)^{3/2} + \frac{2}{3}\int (1 - x)^{3/2}\, dx$$

$$= -\frac{2x}{3}(1 - x)^{3/2} + \frac{2}{3}\left(-\frac{2}{5}\right)(1 - x)^{5/2} + C$$

$$= -\frac{2}{3}(1 - x)^{3/2}\left[x + \frac{2}{5}(1 - x)\right] + C$$

$$= -\frac{2}{15}(1 - x)^{3/2}(2 + 3x) + C$$

■EXAMPLE 3   Integrate: $\int \sqrt{x}\, \ln x\, dx$.

For this integral we have

$$u = \ln x \qquad dv = x^{1/2}\, dx$$

$$du = \frac{1}{x}\, dx \qquad v = \frac{2}{3}x^{3/2}$$

$$\int \sqrt{x}\, \ln x\, dx = \frac{2}{3}x^{3/2} \ln x - \frac{2}{3}\int x^{1/2}\, dx$$

$$= \frac{2}{3}x^{3/2} \ln x - \frac{4}{9}x^{3/2} + C$$

EXAMPLE 4   Integrate: $\int \sin^{-1} x \, dx$.
We write

$$u = \sin^{-1} x \qquad dv = dx$$

$$du = \frac{dx}{\sqrt{1 - x^2}} \qquad v = x$$

$$\int \sin^{-1} x \, dx = x \sin^{-1} x - \int \frac{x \, dx}{\sqrt{1 - x^2}} = x \sin^{-1} x + \frac{1}{2} \int \frac{-2x \, dx}{\sqrt{1 - x^2}}$$

$$= x \sin^{-1} x + \sqrt{1 - x^2} + C$$

EXAMPLE 5   In a certain electric circuit, the current i (in A) is given by the equation $i = te^{-t}$, where t is the time (in s). Find the charge q to pass a point in the circuit between $t = 0$ s and $t = 1.0$ s.

Since $i = \dfrac{dq}{dt}$, we have $\dfrac{dq}{dt} = te^{-t}$. Thus, with $q = \displaystyle\int_0^{1.0} te^{-t} \, dt$, we are to solve for q.

For this integral, $u = t \qquad dv = e^{-t} dt$
$$du = dt \qquad v = -e^{-t}$$

$$q = \int_0^{1.0} te^{-t} \, dt = -te^{-t} \Big|_0^{1.0} + \int_0^{1.0} e^{-t} \, dt = -te^{-t} - e^{-t} \Big|_0^{1.0}$$

$$= -e^{-1.0} - e^{-1.0} + 1.0 = 0.26 \text{ C}$$

Therefore, 0.26 C passes a given point in the circuit.

EXAMPLE 6   Integrate: $\int e^x \sin x \, dx$.
Let $u = \sin x$, $dv = e^x dx$, $du = \cos x \, dx$, $v = e^x$.

$$\int e^x \sin x \, dx = e^x \sin x - \int e^x \cos x \, dx$$

At first glance, it appears that we have made no progress in applying the method of integration by parts. We note, however, that when we integrated $\int e^x \sin x \, dx$, **NOTE ▶** part of the result was a term of $\int e^x \cos x \, dx$. This implies that *if $\int e^x$ **cos** $x \, dx$ were integrated, a term of $\int e^x$ **sin** $x \, dx$ might result.* Thus, the method of integration by parts is now applied to the integral $\int e^x \cos x \, dx$:

$$u = \cos x \qquad dv = e^x dx \qquad du = -\sin x \, dx \qquad v = e^x$$

And so $\int e^x \cos x \, dx = e^x \cos x + \int e^x \sin x \, dx$. Substituting this expression into the expression for $\int e^x \sin x \, dx$, we obtain

$$\int e^x \sin x \, dx = e^x \sin x - \left(e^x \cos x + \int e^x \sin x \, dx \right)$$

$$= e^x \sin x - e^x \cos x - \int e^x \sin x \, dx$$

$$2 \int e^x \sin x \, dx = e^x (\sin x - \cos x) + 2C$$

$$\int e^x \sin x \, dx = \frac{e^x}{2} (\sin x - \cos x) + C$$

Thus, by combining integrals of like form, we obtain the desired result.

EXERCISES *10-1*

In Exercises 1–16, integrate each of the given functions.

1. $\int \theta \cos \theta \, d\theta$

• **2.** $\int x \sin 2x \, dx$

• **3.** $\int x e^{2x} \, dx$

4. $\int 3x e^x \, dx$

• **5.** $\int x \sec^2 x \, dx$

6. $\int_0^{\pi/4} x \sec x \tan x \, dx$

• **7.** $\int 2 \tan^{-1} x \, dx$

8. $\int \ln s \, ds$

9. $\int_{-3}^0 \frac{4t \, dt}{\sqrt{1 - t}}$

10. $\int x \sqrt{x + 1} \, dx$

11. $\int x \ln x \, dx$

• **12.** $\int x^2 \ln 4x \, dx$

13. $\int 2\phi^2 \sin \phi \cos \phi \, d\phi$

14. $\int_0^1 r^2 e^{2r} \, dr$

• **15.** $\int_0^{\pi/2} e^x \cos x \, dx$

16. $\int e^{-x} \sin 2x \, dx$

In Exercises 17–28, solve the given problems by integration.

17. Find the area of the region bounded by $y = xe^{-x}$, $y = 0$, and $x = 2$. See Fig. 10-1.

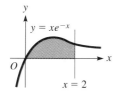

$y = xe^{-x}$

$x = 2$

Fig. 10-1

• **18.** Find the area of the region bounded by $y = 2(\ln x)/x^2$, $y = 0$, and $x = 3$.

19. Find the volume generated by revolving the region bounded by $y = \tan^2 x$, $y = 0$, and $x = 0.5$ about the y-axis.

(W) 20. Explain why a specific choice of u and dv is necessary in Exercise 3 but that there is a choice in Exercise 15.

21. Find the x-coordinate of the centroid of the region bounded by $y = \cos x$ and $y = 0$ for $0 \le x \le \pi/2$.

22. Find the moment of inertia with respect to its axis of the volume generated by revolving the region bounded by $y = e^x$, $x = 1$, and the coordinate axes about the y-axis.

23. The power p (in W) in a certain electric circuit is given by $p = \sqrt{\sin^{-1} t}$ for $0 \le t \le 1.0$ s. Find the root-mean-square value of the power.

24. Under a certain load distribution, the rate of change of the deflection y of a beam is given by $dy/dx = kx^3 \sqrt{1 + x^2}$, where k is a constant. Find the equation that describes the deflection if $y = 0$ for $x = 0$.

25. Computer simulation shows that the velocity v (in ft/s) of a test car is $v = t^3/\sqrt{t^2 + 1}$ from $t = 0$ to $t = 8.0$ s. Find the expression for the distance traveled by the car in t seconds.

26. The nose cone of a rocket has the shape of the volume that is generated by revolving the region bounded by $y = \ln x$, $y = 0$, and $x = 9.5$ about the x-axis. Find the volume (in m^3) of the nose cone. See Fig. 10-2.

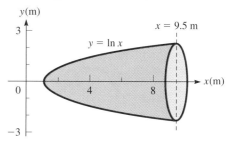

Fig. 10-2

27. The current in a given circuit is given by $i = e^{-2t} \cos t$. Find an expression for the amount of charge that passes a given point in the circuit as a function of the time, if $q_0 = 0$.

28. In finding the average length $\bar{x}$ (in nm) of a certain type of large molecule, we use the equation $\bar{x} = \lim_{b \to \infty} [0.1 \int_0^b x^3 e^{-x^2/8} \, dx]$.

Evaluate the integral and then use a calculator to show that $\bar{x} \to 3.2$ nm as $b \to \infty$.

10-2 INTEGRATION BY SUBSTITUTION

Another method for integrating some expressions is to put them into a standard form by changing the variable of integration by means of an appropriate substitution. There are many different substitutions that can be made, depending on the expression to be integrated. In this section one type of substitution is considered, and in the following section we will develop another.

Certain integrals containing a root of a linear expression may be integrated by substituting a new variable for the radical expression. If an expression of the form $(ax + b)^{p/q}$ appears in an integral that does not fit a standard form, a substitution that may be tried to put it in a standard form is $u = (ax + b)^{1/q}$. Consider the following examples.

■EXAMPLE 1 Integrate: $\int x\sqrt{1 - x}\,dx$.

Since the differential of $1 - x$ is $-dx$ and does not contain a factor of x, we cannot use the power rule. Therefore, let

$$u = \sqrt{1 - x}$$

CAUTION ▶ When we substitute this into the integral, we must *also substitute for x and for dx*. Therefore, solving for x and then finding dx, we have

$$u^2 = 1 - x$$
$$x = 1 - u^2$$
$$dx = -2u\,du$$

Now substituting for x, $\sqrt{1 - x}$, and dx in the integral, we have

$$\int x\sqrt{1 - x}\,dx = \int (1 - u^2)(u)(-2u\,du) = -2\int (u^2 - u^4)\,du$$

$$= -\frac{2}{3}u^3 + \frac{2}{5}u^5 + C$$

We have now integrated the expression, but it is in terms of a variable different from that in the original integral. Therefore, we must put it in terms of x.

$$-\frac{2}{3}u^3 + \frac{2}{5}u^5 + C = -\frac{2}{3}(1 - x)^{3/2} + \frac{2}{5}(1 - x)^{5/2} + C$$

This result may be written in another form which is sometimes more convenient.

$$-\frac{2}{3}(1 - x)^{3/2} + \frac{2}{5}(1 - x)^{5/2} + C = (1 - x)^{3/2}\left[-\frac{2}{3} + \frac{2}{5}(1 - x)\right] + C$$

$$= -\frac{2}{15}(1 - x)^{3/2}(2 + 3x) + C$$

Thus,

$$\int x\sqrt{1 - x}\,dx = -\frac{2}{3}(1 - x)^{3/2} + \frac{2}{5}(1 - x)^{5/2} + C$$

$$= -2(1 - x)^{3/2}\left[\frac{1}{3} - \frac{1}{5}(1 - x)\right] + C$$

$$= -\frac{2}{15}(1 - x)^{3/2}(2 + 3x) + C$$

See Example 2 of Section 10-1. We see that some integrals can be evaluated by more than one method.

■

EXAMPLE 2 Integrate: $\displaystyle\int \frac{2x\,dx}{(x+3)^{2/3}}$.

Here we let

$$u = (x+3)^{1/3}$$

Now solving for x and then finding dx, we have

$$u^3 = x + 3$$
$$x = u^3 - 3$$
$$dx = 3u^2\,du$$
$$\int \frac{2x\,dx}{(x+3)^{2/3}} = \int \frac{2(u^3-3)(3u^2\,du)}{u^2}$$
$$= 6\int (u^3 - 3)\,du$$
$$= 6\left(\frac{1}{4}u^4 - 3u\right) + C$$

Now substituting back in terms of x, we have

$$6\left(\frac{1}{4}u^4 - 3u\right) + C = 6\left[\frac{1}{4}(x+3)^{4/3} - 3(x+3)^{1/3}\right] + C$$

Now by writing the coefficient 3 of $(1+x)^{1/3}$ as 12/4 and then factoring the common factors of 1/4 and $(x+3)^{1/3}$ from the terms within the brackets, we can complete the solution.

$$6\left[\frac{1}{4}(x+3)^{4/3} - 3(x+3)^{1/3}\right] + C = \frac{6}{4}(x+3)^{1/3}(x+3-12) + C$$
$$= \frac{3}{2}(x+3)^{1/3}(x-9) + C$$

or

$$\int \frac{2x\,dx}{(x+3)^{2/3}} = \frac{3}{2}(x+3)^{1/3}(x-9) + C$$

If we are to evaluate a definite integral and use a method of substitution for the integration, there are two ways of handling the limits of integration:

1. Integrate with the new variable, substitute back, and then use the given limits of integration.
2. Integrate with the new variable and then change the limits of integration to the corresponding values of the new variable.

In the following example we will use the second method. It is generally easier to use, and eliminates the step of substituting back.

EXAMPLE 3   Evaluate: $\displaystyle\int_0^4 \frac{4x\,dx}{\sqrt{2x+1}}$.

We let

$$u = \sqrt{2x+1}$$

which means that

$$u^2 = 2x + 1$$

$$x = \frac{1}{2}(u^2 - 1)$$

$$dx = \frac{1}{2}(2u\,du) = u\,du$$

Also, when $x = 0$ (the lower limit), $u = \sqrt{2(0)+1} = 1$, and when $x = 4$ (the upper limit), $u = \sqrt{2(4)+1} = 3$. Thus,

$$\int_0^4 \frac{4x\,dx}{\sqrt{2x+1}} = \int_1^3 \frac{4(\frac{1}{2})(u^2-1)(u\,du)}{u}$$

$$= 2\int_1^3 (u^2 - 1)\,du = \frac{2}{3}u^3 - 2u \,\Big|_1^3 = \frac{2}{3}(27-1) - 2(3-1)$$

$$= \frac{52}{3} - 4 = \frac{40}{3}$$

EXAMPLE 4   Find the volume generated by rotating the first-quadrant region bounded by $y = \sqrt{x+4}$ and $x = 5$ about the y-axis. See Fig. 10-3.

From the figure we see that by using shells, the volume is given by

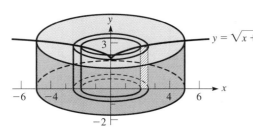

$$V = 2\pi\int_0^5 xy\,dx = 2\pi\int_0^5 x\sqrt{x+4}\,dx$$

Here we let

$$u = \sqrt{x+4}, \quad \text{or} \quad u^2 = x + 4$$
$$x = u^2 - 4$$
$$dx = 2u\,du$$

Fig. 10-3

For $x = 0$, $u = \sqrt{0+4} = 2$, and for $x = 5$, $u = \sqrt{5+4} = 3$. Therefore,

$$V = 2\pi\int_2^3 (u^2-4)(u)(2u\,du) = 4\pi\int_2^3 (u^4 - 4u^2)\,du$$

$$= 4\pi\left(\frac{1}{5}u^5 - \frac{4}{3}u^3\right)\Big|_2^3 = 4\pi\left[\frac{1}{5}(243-32) - \frac{4}{3}(27-8)\right]$$

$$= 4\pi\left(\frac{211}{5} - \frac{76}{3}\right)$$

$$= \frac{1012\pi}{15} = 212$$

EXERCISES $10\text{-}2$

In Exercises 1–16, integrate each of the given functions.

1. $\int x\sqrt{x+1}\,dx$

2. $\int x\sqrt{x+3}\,dx$

3. $\int x\sqrt{2x+1}\,dx$

4. $\int x\sqrt{3-x}\,dx$

5. $\int \dfrac{x\,dx}{\sqrt{x+3}}$

6. $\int \dfrac{x\,dx}{\sqrt{2x+7}}$

7. $\int \dfrac{x^2\,dx}{\sqrt{x-2}}$

8. $\int_{3}^{11} \dfrac{3x\,dx}{\sqrt{2x+3}}$

9. $\int_{0}^{3} x^2\sqrt{1+x}\,dx$

10. $\int \dfrac{3x^2\,dx}{\sqrt{1-x}}$

11. $\int x\sqrt[3]{x-1}\,dx$

12. $\int x\sqrt[3]{8-x}\,dx$

13. $\int \dfrac{x\,dx}{\sqrt[4]{2x+3}}$

14. $\int x(x-4)^{2/3}\,dx$

15. $\int_{0}^{7} x(x+1)^{2/3}\,dx$

16. $\int_{0}^{2} \dfrac{x^2\,dx}{(4x+1)^{5/2}}$

In Exercises 17–24, solve the given problems by integration.

17. Find the area bounded by $y = x^3\sqrt{x-1}$, $x = 2$, and $y = 0$. See Fig. 10-4.

18. Find the first-quadrant area bounded by $y = x\sqrt[3]{x+8}$ and $x = 8$.

19. Find the volume generated by revolving the first-quadrant region bounded by $y = x\sqrt[3]{x-1}$ and $x = 9$ about the y-axis.

20. Find the volume generated by revolving the region of Exercise 19 about the x-axis.

21. The force F (in lb) required to move a particular lever mechanism is a function of the displacement s (in ft) according to $F = 4s\sqrt{4s+3}$. Find the work done if the displacement changes from zero to 2.50 ft.

22. The electric current i (in μA) in a particular microprocessor circuit varies with the time t (in ns) according to the equation $i = t^3\sqrt{6t+1}$ for the first 3 ns. Find the charge that passes through the circuit during the first 2.50 ns.

23. Evaluate the integral in Example 1 by using the substitution $v = 1 - x$.

24. Evaluate the integral in Example 2 by using the substitution $v = x + 3$.

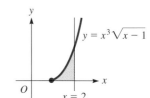

Fig. 10-4

$10\text{-}3$ INTEGRATION BY TRIGONOMETRIC SUBSTITUTION

We have seen that trigonometric relations provide a means of transforming trigonometric integrals into forms that can be integrated. As we will show in this section, they can also be useful for integrating certain types of algebraic integrals. Substitutions based on Eqs. (9-14), (9-15), and (9-16), which are referenced in the margin on the next page, prove to be particularly useful for integrating expressions that involve radicals. The following examples illustrate the method.

EXAMPLE 1   Integrate: $\int \dfrac{dx}{x^2 \sqrt{1 - x^2}}$.

If we let $x = \sin \theta$, the radical becomes $\sqrt{1 - \sin^2 \theta} = \cos \theta$. Therefore, by making this substitution, the integral can be transformed into a trigonometric integral. We must be careful to **replace all factors of the integral by proper expressions in terms of θ.** With $x = \sin \theta$, by finding the differential of x, we have $dx = \cos \theta \, d\theta$.

CAUTION ▶

For reference, Eqs. (9-14), (9-15), and (9-16) are
$\cos^2 x + \sin^2 x = 1$
$1 + \tan^2 x = \sec^2 x$
$1 + \cot^2 x = \csc^2 x$

Now, by substituting $x = \sin \theta$ and $dx = \cos \theta \, d\theta$ into the integral, we have

$$\int \frac{dx}{x^2 \sqrt{1 - x^2}} = \int \frac{\cos \theta \, d\theta}{\sin^2 \theta \sqrt{1 - \sin^2 \theta}} \qquad \text{substituting}$$

$$= \int \frac{\cos \theta \, d\theta}{\sin^2 \theta \cos \theta} = \int \csc^2 \theta \, d\theta \qquad \text{using trigonometric relations}$$

This last integral can be integrated by using Eq. (9-7). This leads to

$$\int \csc^2 \theta \, d\theta = -\cot \theta + C$$

CAUTION ▶

We have now performed the integration, but *the answer we now have is in terms of θ, and we must express the result in terms of x.* Making a right triangle with an angle θ such that $\sin \theta = x/1$ (see Fig. 10-5), we may express any of the trigonometric functions in terms of x. (This is the method used with inverse trigonometric functions.) Thus,

$$\cot \theta = \frac{\sqrt{1 - x^2}}{x}$$

Therefore, the result of the integration becomes

$$\int \frac{dx}{x^2 \sqrt{1 - x^2}} = -\cot \theta + C = -\frac{\sqrt{1 - x^2}}{x} + C \qquad \text{---------} \blacksquare$$

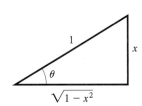

Fig. 10-5

EXAMPLE 2   Integrate: $\int \dfrac{dx}{\sqrt{x^2 + 4}}$.

If we let $x = 2 \tan \theta$, the radical in this integral becomes

$$\sqrt{x^2 + 4} = \sqrt{4 \tan^2 \theta + 4} = 2\sqrt{\tan^2 \theta + 1} = 2\sqrt{\sec^2 \theta} = 2 \sec \theta$$

Therefore, with $x = 2 \tan \theta$ and $dx = 2 \sec^2 \theta \, d\theta$, we have

$$\int \frac{dx}{\sqrt{x^2 + 4}} = \int \frac{2 \sec^2 \theta \, d\theta}{\sqrt{4 \tan^2 \theta + 4}} = \int \frac{2 \sec^2 \theta \, d\theta}{2 \sec \theta} \qquad \text{substituting}$$

$$= \int \sec \theta \, d\theta = \ln |\sec \theta + \tan \theta| + C \qquad \text{using Eq. (9-12)}$$

$$= \ln \left| \frac{\sqrt{x^2 + 4}}{2} + \frac{x}{2} \right| + C = \ln \left| \frac{\sqrt{x^2 + 4} + x}{2} \right| + C \qquad \text{see Fig. 10-6}$$

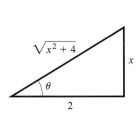

Fig. 10-6

This answer is acceptable, but by using the properties of logarithms, we have

$$\ln \left| \frac{\sqrt{x^2 + 4} + x}{2} \right| + C = \ln |\sqrt{x^2 + 4} + x| + (C - \ln 2)$$

$$= \ln |\sqrt{x^2 + 4} + x| + C'$$

Combining constants, as we did in writing $C' = C - \ln 2$, is a common practice in integration problems. $\qquad \text{---------} \blacksquare$

EXAMPLE 3 Integrate: $\displaystyle\int \frac{2\,dx}{x\sqrt{x^2 - 9}}$.

If we let $x = 3 \sec \theta$, the radical in this integral becomes

$$\sqrt{x^2 - 9} = \sqrt{9 \sec^2 \theta - 9} = 3\sqrt{\sec^2 \theta - 1} = 3\sqrt{\tan^2 \theta} = 3 \tan \theta$$

Therefore, with $x = 3 \sec \theta$ and $dx = 3 \sec \theta \tan \theta\, d\theta$, we have

$$\int \frac{2\,dx}{x\sqrt{x^2 - 9}} = 2\int \frac{3 \sec \theta \tan \theta\, d\theta}{3 \sec \theta \sqrt{9 \sec^2 \theta - 9}} = 2\int \frac{\tan \theta\, d\theta}{3 \tan \theta}$$

$$= \frac{2}{3}\int d\theta = \frac{2}{3}\theta + C = \frac{2}{3}\sec^{-1}\frac{x}{3} + C$$

It is not necessary to refer to a triangle to express the result in terms of x. The solution is found by solving $x = 3 \sec \theta$ for θ, as indicated. ∎

These examples show that by making the proper substitution, we can integrate algebraic functions by using the equivalent trigonometric forms. In summary, for the indicated radical form, the following trigonometric substitutions are used:

BASIC TRIGONOMETRIC SUBSTITUTIONS

For $\sqrt{a^2 - x^2}$	use	$x = a \sin \theta$
For $\sqrt{a^2 + x^2}$	use	$x = a \tan \theta$
For $\sqrt{x^2 - a^2}$	use	$x = a \sec \theta$

(10-2)

See the chapter introduction.

For reference, Eq. (6-31) is

$$s = \int_a^b \sqrt{1 + \left(\frac{dy}{dx}\right)^2}\,dx$$

EXAMPLE 4 The joint between two links of a robot arm moves back and forth along the curve defined by $y = 3.0 \ln x$ from $x = 1.0$ in. to $x = 4.0$ in. Find the distance the joint moves in one cycle.

The required distance is found by use of Eq. (6-31). Therefore, we find the derivative as $dy/dx = 3.0/x$, which means the total distance s moved in one cycle is

$$s = 2\int_{1.0}^{4.0} \sqrt{\left[1 + \left(\frac{3.0}{x}\right)^2\right]}\,dx = 2\int_{1.0}^{4.0} \frac{\sqrt{x^2 + 9.0}}{x}\,dx$$

To integrate, we make the substitution $x = 3.0 \tan \theta$ and $dx = 3.0 \sec^2 \theta\, d\theta$. Thus,

$$\int \frac{\sqrt{x^2 + 9.0}}{x}\,dx = \int \frac{\sqrt{9.0 \tan^2 \theta + 9.0}}{3.0 \tan \theta}(3.0 \sec^2 \theta\, d\theta) = 3.0\int \frac{\sec \theta}{\tan \theta}(1 + \tan^2 \theta)\, d\theta$$

$$= 3.0\left(\int \frac{\sec \theta}{\tan \theta}\,d\theta + \int \tan \theta \sec \theta\, d\theta\right) = 3.0\left(\int \csc \theta\, d\theta + \int \tan \theta \sec \theta\, d\theta\right) \quad \text{see Fig. 10-7}$$

$$= 3.0(\ln|\csc \theta - \cot \theta| + \sec \theta) + C = 3.0\left[\ln\left|\frac{\sqrt{x^2 + 9.0}}{x} - \frac{3.0}{x}\right| + \frac{\sqrt{x^2 + 9.0}}{3.0}\right] + C$$

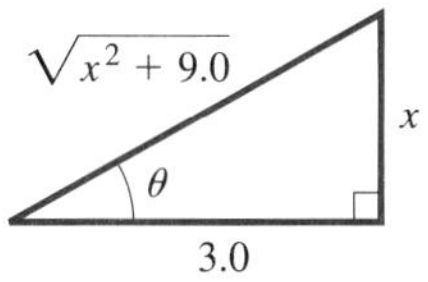

Limits have not been included, due to the change in variables. Evaluating, we have

$$s = 2\int_{1.0}^{4.0} \frac{\sqrt{x^2 + 9.0}}{x}\,dx = 6.0\left[\ln\left|\frac{\sqrt{x^2 + 9.0} - 3.0}{x}\right| + \frac{\sqrt{x^2 + 9.0}}{3.0}\right]\Bigg|_{1.0}^{4.0}$$

$$= 6.0\left[\left(\ln 0.50 - \ln\frac{\sqrt{10.0} - 3.0}{1.0}\right) + \frac{5.0}{3.0} - \frac{\sqrt{10.0}}{3.0}\right] = 10.4 \text{ in.} \quad ∎$$

Fig. 10-7

EXERCISES 10-3

In Exercises 1–16, integrate each of the given functions.

· 1. $\displaystyle\int \frac{\sqrt{1-x^2}}{x^2}\,dx$

· 2. $\displaystyle\int_0^4 \frac{dt}{(t^2+9)^{3/2}}$

· 3. $\displaystyle\int \frac{2\,dx}{\sqrt{x^2-4}}$

4. $\displaystyle\int \frac{\sqrt{x^2-25}}{x}\,dx$

5. $\displaystyle\int \frac{6\,dz}{z^2\sqrt{z^2+9}}$

6. $\displaystyle\int \frac{3\,dx}{x\sqrt{4-x^2}}$

· 7. $\displaystyle\int \frac{4\,dx}{(4-x^2)^{3/2}}$

8. $\displaystyle\int \frac{6p^3\,dp}{\sqrt{9+p^2}}$

9. $\displaystyle\int_0^{0.5} \frac{x^3\,dx}{\sqrt{1-x^2}}$

· 10. $\displaystyle\int_4^5 \frac{\sqrt{x^2-16}}{x^2}\,dx$

· 11. $\displaystyle\int \frac{5\,dx}{\sqrt{x^2+2x+2}}$

12. $\displaystyle\int \frac{dx}{\sqrt{x^2+2x}}$

13. $\displaystyle\int_{2.5}^3 \frac{dy}{y\sqrt{4y^2-9}}$

14. $\displaystyle\int \sqrt{16-x^2}\,dx$

15. $\displaystyle\int \frac{2\,dx}{\sqrt{e^{2x}-1}}$

16. $\displaystyle\int \frac{12\sec^2 u\,du}{(4-\tan^2 u)^{3/2}}$

In Exercises 17–24, solve the given problems by integration.

· 17. Find the area of a circle of radius 1.

18. Find the area bounded by
$y = \dfrac{1}{x^2\sqrt{x^2-1}}$, $x = \sqrt{2}$,
$x = \sqrt{5}$, and $y = 0$. See
Fig. 10-8.

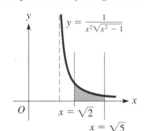

Fig. 10-8

(W) 19. The integral $\displaystyle\int \frac{x}{\sqrt{x^2+4}}\,dx$ can be integrated in more than one way. Explain what methods can be used and which is simpler.

20. Find the moment of inertia of a sphere of radius a with respect to its axis in terms of its mass.

· 21. Find the volume generated by revolving the region bounded by
$y = \dfrac{\sqrt{x^2-16}}{x^2}$, $y = 0$, and $x = 5$ about the y-axis.

22. The vertical cross section of a highway culvert is defined by the region within the ellipse $1.00x^2 + 9.00y^2 = 9.00$, where dimensions are in meters. Find the area of the cross section of the culvert.

23. If an electric charge Q is distributed along a straight wire of length $2a$, the electric potential V at a point P, which is at a distance b from the center of the wire, is $V = kQ\displaystyle\int_{-a}^{a} \frac{dx}{\sqrt{b^2+x^2}}$.
Here, k is a constant, and x is the distance along the wire. Evaluate the integral.

24. An electric insulating ring for a machine part can be described as the volume generated by revolving the region bounded by $y = x^2\sqrt{x^2-4}$, $y = 0$, and $x = 2.5$ cm about the y-axis. Find the volume (in cm^3) of material in the ring.

10-4   INTEGRATION BY PARTIAL FRACTIONS: NONREPEATED LINEAR FACTORS

In Chapter 9 we saw how *reversing differential forms* allows us to integrate various types of functions. Then, in the previous three sections of this chapter we have seen how the appropriate use of a *derivative of a product* of functions, *substituting a new variable for a radical expression,* and *trigonometric identities* are used to put integrals into a form that can be integrated. Now in this section and the next section, we show an *algebraic method* by which integrands that are fractions can be changed into a form that can be integrated.

In algebra we combine fractions into a single fraction by means of addition. However, if we wish to integrate an expression that contains a rational function, in which both the numerator and the denominator are polynomials, it is often advantageous to reverse the operation of addition and express the rational function as the sum of simpler fractions.

EXAMPLE 1 In attempting to integrate $\int \dfrac{7-x}{x^2+x-2}\,dx$, we find that it does not fit any of the standard forms in this chapter. However, we can show that

$$\frac{7-x}{x^2+x-2} = \frac{2}{x-1} - \frac{3}{x+2}$$

This means that

$$\int \frac{7-x}{x^2+x-2}\,dx = \int \frac{2\,dx}{x-1} - \int \frac{3\,dx}{x+2}$$
$$= 2\ln|x-1| - 3\ln|x+2| + C$$

We see that by writing the fraction in the original integrand as the sum of the simpler fractions, each resulting integrand can be integrated. ■

In Example 1 we saw that the integral is readily determined once the rational function $(7-x)/(x^2+x-2)$ is replaced by the simpler fractions. In this section and the next, we describe how certain rational functions can be expressed in terms of simpler fractions and thereby be integrated. *This technique is called the* **method of partial fractions.**

NOTE ▶

In order to express the rational function $f(x)/g(x)$ in terms of simpler partial fractions, *the degree of the numerator $f(x)$ must be less than that of the denominator $g(x)$*. If this is not the case, we divide numerator by denominator until the remainder is of the proper form. Then the denominator $g(x)$ is factored into a product of linear and quadratic factors. The method of determining the partial fractions depends on the factors that are obtained. In advanced algebra, the form of the partial fractions is shown (but we shall not show the proof here).

There are four cases for the types of factors of the denominator. They are (1) *nonrepeated linear factors,* (2) *repeated linear factors,* (3) *nonrepeated quadratic factors,* and (4) *repeated quadratic factors.* In this section, we consider the case of nonrepeated linear factors, and the other cases are discussed in the next section.

Nonrepeated Linear Factors

For the case of nonrepeated linear factors, we use the fact that *corresponding to each linear factor ax + b, occurring once in the denominator, there will be a partial fraction of the form*

$$\frac{A}{ax+b}$$

where A is a constant to be determined. The following examples illustrate the method.

The method of partial fractions essentially reverses the process of combining fractions over a common denominator. By determining that

$$\frac{7 - x}{x^2 + x - 2} = \frac{2}{x - 1} - \frac{3}{x + 2}$$

reverses the process by which

$$\frac{2}{x - 1} - \frac{3}{x + 2} = \frac{2(x + 2) - 3(x - 1)}{(x - 1)(x + 2)}$$
$$= \frac{7 - x}{x^2 + x - 2}$$

■EXAMPLE 2 Integrate: $\int \frac{7 - x}{x^2 + x - 2} dx$. (This is the same integral as in Example 1. Here we will see how the partial fractions are found.)

First, we note that the degree of the numerator is 1 (the highest power term is x) and that of the denominator is 2 (the highest power term is x^2). Since the degree of the denominator is higher, we may proceed to factoring it. Thus,

$$\frac{7 - x}{x^2 + x - 2} = \frac{7 - x}{(x - 1)(x + 2)}$$

There are two linear factors, $(x - 1)$ and $(x + 2)$, in the denominator, and they are different. This means that there are two partial fractions. Therefore, we write

$$\frac{7 - x}{(x - 1)(x + 2)} = \frac{A}{x - 1} + \frac{B}{x + 2} \qquad (1)$$

We are to determine constants A and B so that Eq. (1) is an identity. In finding A and B, we clear Eq. (1) of fractions by multiplying both sides by $(x - 1)(x + 2)$.

$$7 - x = A(x + 2) + B(x - 1) \qquad (2)$$

Equation (2) is also an identity, which means that there are two ways of determining the values of A and B.

NOTE▶ *Solution by substitution:* Since Eq. (2) is an identity, ***it is true for any value of x.*** Thus, in turn we pick $x = -2$ and $x = 1$, for each of these values makes a factor on the right equal to zero, and the values of B and A are easily found. Therefore,

$$\text{for } x = -2: \quad 7 - (-2) = A(-2 + 2) + B(-2 - 1)$$
$$9 = -3B, \quad B = -3$$
$$\text{for } x = 1: \quad 7 - 1 = A(1 + 2) + B(1 - 1)$$
$$6 = 3A, \quad A = 2$$

NOTE▶ *Solution by equating coefficients:* Since Eq. (2) is an identity, another way of finding the constants A and B is to ***equate coefficients of like powers of x from each side.*** Thus, writing Eq. (2) as

$$7 - x = (2A - B) + (A + B)x$$

we have

$2A - B = 7$
$\underline{A + B = -1}$
$3A = 6$
$A = 2$
$2(2) - B = 7$
$-B = 3$
$B = -3$

$$2A - B = 7 \quad \text{(equating constants: } x^0 \text{ terms)}$$
$$A + B = -1 \quad \text{(equating coefficients of } x)$$

Now, using the values $A = 2$ and $B = -3$ (as found at the left), we have

$$\frac{7 - x}{(x - 1)(x + 2)} = \frac{2}{x - 1} - \frac{3}{x + 2}$$

Therefore, the integral is found as in Example 1.

$$\int \frac{7 - x}{x^2 + x - 2} dx = \int \frac{7 - x}{(x - 1)(x + 2)} dx = \int \frac{2\,dx}{x - 1} - \int \frac{3\,dx}{x + 2}$$
$$= 2 \ln|x - 1| - 3 \ln|x + 2| + C$$

Using the properties of logarithms, we may write this as

$$\int \frac{7 - x}{x^2 + x - 2} dx = \ln \left| \frac{(x - 1)^2}{(x + 2)^3} \right| + C$$

■

EXAMPLE 3 Integrate: $\displaystyle\int \frac{6x^2 - 14x - 11}{(x + 1)(x - 2)(2x + 1)}\,dx$.

The denominator is factored and is of degree 3 (when multiplied out, the highest power term of x is x^3). This means we have three nonrepeated linear factors.

$$\frac{6x^2 - 14x - 11}{(x + 1)(x - 2)(2x + 1)} = \frac{A}{x + 1} + \frac{B}{x - 2} + \frac{C}{2x + 1}$$

Multiplying through by $(x + 1)(x - 2)(2x + 1)$, we have

$$6x^2 - 14x - 11 = A(x - 2)(2x + 1) + B(x + 1)(2x + 1) + C(x + 1)(x - 2)$$

We now substitute the values of 2, $-\frac{1}{2}$, and -1 for x. Again, these are chosen because they make factors of the coefficients of A, B, or C equal to zero, although any values may be chosen. Therefore,

for $x = 2$: $\qquad 6(4) - 14(2) - 11 = A(0)(5) + B(3)(5) + C(3)(0), \qquad\qquad B = -1$

for $x = -\frac{1}{2}$: $\quad 6\left(\frac{1}{4}\right) - 14\left(-\frac{1}{2}\right) - 11 = A\left(-\frac{5}{2}\right)(0) + B\left(\frac{1}{2}\right)(0) + C\left(\frac{1}{2}\right)\left(-\frac{5}{2}\right), \qquad C = 2$

for $x = -1$: $\qquad 6(1) - 14(-1) - 11 = A(-3)(-1) + B(0)(-1) + C(0)(-3), \qquad\qquad A = 3$

Therefore,

$$\int \frac{6x^2 - 14x - 11}{(x + 1)(x - 2)(2x + 1)}\,dx = \int \frac{3\,dx}{x + 1} - \int \frac{dx}{x - 2} + \int \frac{2\,dx}{2x + 1}$$

$$= 3\ln|x + 1| - \ln|x - 2| + \ln|2x + 1| + C_1 = \ln\left|\frac{(2x + 1)(x + 1)^3}{x - 2}\right| + C_1$$

Here we have let the constant of integration be C_1 since we used C as the numerator of the third partial fraction. ∎

EXAMPLE 4 Integrate: $\displaystyle\int \frac{2x^4 - x^3 - 9x^2 + x - 12}{x^3 - x^2 - 6x}\,dx$.

Since the numerator is of a higher degree than the denominator, we must first divide the numerator by the denominator. This gives

$$\frac{2x^4 - x^3 - 9x^2 + x - 12}{x^3 - x^2 - 6x} = 2x + 1 + \frac{4x^2 + 7x - 12}{x^3 - x^2 - 6x}$$

We must now express this rational fraction in terms of its partial fractions.

$$\frac{4x^2 + 7x - 12}{x^3 - x^2 - 6x} = \frac{4x^2 + 7x - 12}{x(x + 2)(x - 3)} = \frac{A}{x} + \frac{B}{x + 2} + \frac{C}{x - 3}$$

Clearing fractions, we have

$$4x^2 + 7x - 12 = A(x + 2)(x - 3) + Bx(x - 3) + Cx(x + 2)$$

Now, using values of x of -2, 3, and 0 for substitution, we obtain the values of $B = -1$, $C = 3$, and $A = 2$, respectively. Therefore,

$$\int \frac{2x^4 - x^3 - 9x^2 + x - 12}{x^3 - x^2 - 6x}\,dx = \int \left(2x + 1 + \frac{2}{x} - \frac{1}{x + 2} + \frac{3}{x - 3}\right)dx$$

$$= x^2 + x + 2\ln|x| - \ln|x + 2| + 3\ln|x - 3| + C_1$$

$$= x^2 + x + \ln\left|\frac{x^2(x - 3)^3}{x + 2}\right| + C_1 \qquad ∎$$

In Example 2 we showed two ways of finding the values of A and B. The method of substitution is generally easier to use with linear factors, and we used it in Examples 3 and 4. However, as we will see in the next section, the method of equating coefficients can be very useful for other cases with partial fractions.

EXERCISES 10-4

In Exercises 1–16, integrate each of the given functions.

1. $\displaystyle\int \frac{x + 3}{(x + 1)(x + 2)} \, dx$

2. $\displaystyle\int \frac{x + 2}{x(x + 1)} \, dx$

3. $\displaystyle\int \frac{dx}{x^2 - 4}$

4. $\displaystyle\int \frac{p - 9}{2p^2 - 3p + 1} \, dp$

5. $\displaystyle\int \frac{x^2 + 3}{x^2 + 3x} \, dx$

6. $\displaystyle\int \frac{x^3}{x^2 + 3x + 2} \, dx$

7. $\displaystyle\int_0^1 \frac{2t + 4}{3t^2 + 5t + 2} \, dt$

8. $\displaystyle\int_1^3 \frac{x - 1}{4x^2 + x} \, dx$

9. $\displaystyle\int \frac{4x^2 - 10}{x(x + 1)(x - 5)} \, dx$

10. $\displaystyle\int \frac{4x^2 + 21x + 6}{(x + 2)(x - 3)(x + 4)} \, dx$

11. $\displaystyle\int \frac{6x^2 - 2x - 1}{4x^3 - x} \, dx$

12. $\displaystyle\int_2^3 \frac{dR}{R^3 - R}$

13. $\displaystyle\int_1^2 \frac{x^3 + 7x^2 + 9x + 2}{x(x^2 + 3x + 2)} \, dx$

14. $\displaystyle\int \frac{2x^3 + x - 1}{x^3 + x^2 - 4x - 4} \, dx$

15. $\displaystyle\int \frac{dV}{(V^2 - 4)(V^2 - 9)}$

16. $\displaystyle\int \frac{5x^3 - 2x^2 - 15x + 24}{x^4 - 2x^3 - 11x^2 + 12x} \, dx$

In Exercises 17–24, solve the given problems by integration.

17. Find the area bounded by $y = (x - 16)/(x^2 - 5x - 14)$, $y = 0$, $x = 2$, and $x = 4$.

18. Find the first-quadrant area bounded by $y = 1/(x^3 + 3x^2 + 2x)$, $x = 1$, and $x = 3$.

19. Find the volume generated if the region of Exercise 18 is revolved about the *y*-axis.

20. Find the *x*-coordinate of the centroid of the region bounded by $y(x^2 - 1) = 1$, $y = 0$, $x = 2$, and $x = 4$.

21. The general expression for the slope of a curve is $(3x + 5)/(x^2 + 5x)$. Find the equation of the curve if it passes through $(1, 0)$.

22. The current i (in A) as a function of the time t (in s) in a certain electric circuit is given by $i = (4t + 3)/(2t^2 + 3t + 1)$. Find the total charge that passes a given point in the circuit during the first second.

23. The force F (in N) applied by a stamping machine in making a certain computer part is $F = 4x/(x^2 + 3x + 2)$, where x is the distance (in cm) through which the force acts. Find the work done by the force from $x = 0$ to $x = 0.500$ cm.

24. Under specified conditions, the time t (in min) required to form x grams of a substance during a chemcial reaction is given by $t = \int dx/[(4 - x)(2 - x)]$. Find the equation relating t and x if $x = 0$ g when $t = 0$ min.

$10\text{-}5$ INTEGRATION BY PARTIAL FRACTIONS: OTHER CASES

In the previous section, we introduced the method of partial fractions and considered the case of nonrepeated linear factors. In this section, we develop the use of partial fractions for the cases of repeated linear factors and nonrepeated quadratic factors. We also briefly discuss the case of repeated quadratic factors.

Repeated Linear Factors

For the case of repeated linear factors, we use the fact that *corresponding to each linear factor $ax + b$ that occurs n times in the denominator there will be n partial fractions*

$$\frac{A_1}{ax + b} + \frac{A_2}{(ax + b)^2} + \cdots + \frac{A_n}{(ax + b)^n}$$

where $A_1, A_2, \ldots, A_n$ are constants to be determined.

EXAMPLE 1 Integrate: $\displaystyle\int \frac{dx}{x(x+3)^2}$.

Here we see that the denominator has a factor of x and two factors of $x + 3$. For the factor of x, we use a partial fraction as in the previous section, for it is a nonrepeated factor. For the factor $x + 3$, we need two partial fractions, one with a denominator of $x + 3$ and the other with a denominator of $(x + 3)^2$. Thus, we write

NOTE ▶

$$\frac{1}{x(x+3)^2} = \boxed{\frac{A}{x} + \frac{B}{x+3} + \frac{C}{(x+3)^2}}$$

Multiplying each side by $x(x+3)^2$, we have

$$1 = A(x+3)^2 + Bx(x+3) + Cx \qquad (1)$$

Using the values of x of -3 and 0, we have

for $x = -3$: $\quad 1 = A(0^2) + B(-3)(0) + (-3)C, \qquad C = -\dfrac{1}{3}$

for $x = 0$: $\quad 1 = A(3^2) + B(0)(3) + C(0), \qquad A = \dfrac{1}{9}$

Since there are no other values that make a factor in Eq. (1) equal to zero, we must either choose some other value of x or equate coefficients of some power of x in Eq. (1). We will let $x = 1$.

Since Eq. (1) is an identity, we may choose any value of x. With $x = 1$, we have

$$1 = A(4^2) + B(1)(4) + C(1)$$
$$1 = 16A + 4B + C$$

Using the known values of A and C, we have

$$1 = 16\left(\frac{1}{9}\right) + 4B - \frac{1}{3}, \qquad B = -\frac{1}{9}$$

This means that

$$\frac{1}{x(x+3)^2} = \frac{\frac{1}{9}}{x} + \frac{-\frac{1}{9}}{x+3} + \frac{-\frac{1}{3}}{(x+3)^2}$$

or

$$\int \frac{dx}{x(x+3)^2} = \frac{1}{9}\int \frac{dx}{x} - \frac{1}{9}\int \frac{dx}{x+3} - \frac{1}{3}\int \frac{dx}{(x+3)^2}$$

$$= \frac{1}{9}\ln|x| - \frac{1}{9}\ln|x+3| - \frac{1}{3}\left(\frac{1}{-1}\right)(x+3)^{-1} + C_1$$

$$= \frac{1}{9}\ln\left|\frac{x}{x+3}\right| + \frac{1}{3(x+3)} + C_1$$

We have used the properties of logarithms in order to write the final form of the answer. ------------ ∎

EXAMPLE 2   Integrate: $\displaystyle\int \frac{3x^3 + 15x^2 + 21x + 15}{(x-1)(x+2)^3}\,dx$.

First, we set up the partial fractions as

$$\frac{3x^3 + 15x^2 + 21x + 15}{(x-1)(x+2)^3} = \frac{A}{x-1} + \frac{B}{x+2} + \frac{C}{(x+2)^2} + \frac{D}{(x+2)^3}$$

Next we clear fractions.

$$3x^3 + 15x^2 + 21x + 15 = A(x+2)^3 + B(x-1)(x+2)^2$$
$$+ C(x-1)(x+2) + D(x-1) \qquad (1)$$

for $x = 1$: $3 + 15 + 21 + 15 = 27A$, $54 = 27A$, $A = 2$

for $x = -2$: $3(-8) + 15(4) + 21(-2) + 15 = -3D$, $9 = -3D$, $D = -3$

To find B and C, we equate coefficients of powers of x. Therefore, we write Eq. (1) as

$$3x^3 + 15x^2 + 21x + 15 = (A+B)x^3 + (6A + 3B + C)x^2$$
$$+ (12A + C + D)x + (8A - 4B - 2C - D)$$

coefficients of x^3: $3 = A + B$, $3 = 2 + B$, $B = 1$

coefficients of x^2: $15 = 6A + 3B + C$, $15 = 12 + 3 + C$, $C = 0$

$$\frac{3x^3 + 15x^2 + 21x + 15}{(x-1)(x+2)^3} = \frac{2}{x-1} + \frac{1}{x+2} + \frac{0}{(x+2)^2} + \frac{-3}{(x+2)^3}$$

$$\int \frac{3x^3 + 15x^2 + 21x + 15}{(x-1)(x+2)^3}\,dx = 2\int \frac{dx}{x-1} + \int \frac{dx}{x+2} - 3\int \frac{dx}{(x+2)^3}$$

$$= 2\ln|x-1| + \ln|x+2| - 3\left(\frac{1}{-2}\right)(x+2)^{-2} + C_1$$

$$= \ln|(x-1)^2(x+2)| + \frac{3}{2(x+2)^2} + C_1 \qquad \blacksquare$$

If there is one repeated factor in the denominator and it is the only factor present in the denominator, a substitution is easier and more convenient than using partial fractions. This is illustrated in the following example.

EXAMPLE 3   Integrate: $\displaystyle\int \frac{x\,dx}{(x-2)^3}$.

This could be integrated by first setting up the appropriate partial fractions. However, the solution is more easily found by using the substitution $u = x - 2$. Using this, we have

$$u = x - 2 \qquad x = u + 2 \qquad dx = du$$

$$\int \frac{x\,dx}{(x-2)^3} = \int \frac{(u+2)(du)}{u^3} = \int \frac{du}{u^2} + 2\int \frac{du}{u^3}$$

$$= \int u^{-2}\,du + 2\int u^{-3}\,du = \frac{1}{-u} + \frac{2}{-2}u^{-2} + C$$

$$= -\frac{1}{u} - \frac{1}{u^2} + C = -\frac{u+1}{u^2} + C$$

$$= -\frac{x-2+1}{(x-2)^2} + C = \frac{1-x}{(x-2)^2} + C \qquad \blacksquare$$

Nonrepeated Quadratic Factors

For the case of nonrepeated quadratic factors, we use the fact that *corresponding to each irreducible quadratic factor $ax^2 + bx + c$ that occurs once in the denominator there is a partial fraction of the form*

$$\frac{Ax + B}{ax^2 + bx + c}$$

where A and B are constants to be determined. (Here, an *irreducible quadratic factor* is one that cannot be further factored into linear factors involving only real numbers.) This case is illustrated in the following examples.

EXAMPLE 4   Integrate: $\int \frac{4x + 4}{x^3 + 4x}\, dx.$

In setting up the partial fractions, we note that the denominator factors as $x^3 + 4x = x(x^2 + 4)$. Here the factor $x^2 + 4$ cannot be further factored. This means we have

NOTE ▶

$$\frac{4x + 4}{x^3 + 4x} = \frac{4x + 4}{x(x^2 + 4)} = \boxed{\frac{A}{x} + \frac{Bx + C}{x^2 + 4}}$$

Clearing fractions, we have

$$4x + 4 = A(x^2 + 4) + Bx^2 + Cx$$
$$= (A + B)x^2 + Cx + 4A$$

Equating coefficients of powers of x gives us

$$\text{for } x^2: \quad 0 = A + B$$
$$\text{for } x: \quad 4 = C$$
$$\text{for constants:} \quad 4 = 4A, \quad A = 1$$

Therefore, we easily find that $B = -1$ from the first equation. This means that

$$\frac{4x + 4}{x^3 + 4x} = \frac{1}{x} + \frac{-x + 4}{x^2 + 4}$$

and

$$\int \frac{4x + 4}{x^3 + 4x}\, dx = \int \frac{1}{x}\, dx + \int \frac{-x + 4}{x^2 + 4}\, dx$$
$$= \int \frac{1}{x}\, dx - \int \frac{x\, dx}{x^2 + 4} + \int \frac{4\, dx}{x^2 + 4}$$
$$= \ln |x| - \frac{1}{2} \ln |x^2 + 4| + 2 \tan^{-1} \frac{x}{2} + C_1$$

We could use the properties of logarithms to combine the first two terms of the result. ∎

EXAMPLE 5   Integrate: $\displaystyle\int \frac{x^3 + 3x^2 + 2x + 4}{x^2(x^2 + 2x + 2)}\, dx$.

In the denominator we have a repeated linear factor, x^2, and a quadratic factor. Therefore,

$$\frac{x^3 + 3x^2 + 2x + 4}{x^2(x^2 + 2x + 2)} = \frac{A}{x} + \frac{B}{x^2} + \frac{Cx + D}{x^2 + 2x + 2}$$

$$x^3 + 3x^2 + 2x + 4 = Ax(x^2 + 2x + 2) + B(x^2 + 2x + 2) + Cx^3 + Dx^2$$

$$= (A + C)x^3 + (2A + B + D)x^2 + (2A + 2B)x + 2B$$

Equating coefficients, we find that

for constants:	$2B = 4,$	$B = 2$	
for x:	$2A + 2B = 2,$	$A + B = 1,$	$A = -1$
for x^2:	$2A + B + D = 3,$	$-2 + 2 + D = 3,$	$D = 3$
for x^3:	$A + C = 1,$	$-1 + C = 1,$	$C = 2$

$$\frac{x^3 + 3x^2 + 2x + 4}{x^2(x^2 + 2x + 2)} = -\frac{1}{x} + \frac{2}{x^2} + \frac{2x + 3}{x^2 + 2x + 2}$$

$$\int \frac{x^3 + 3x^2 + 2x + 4}{x^2(x^2 + 2x + 2)}\, dx = -\int \frac{dx}{x} + 2\int \frac{dx}{x^2} + \int \frac{2x + 3}{x^2 + 2x + 2}\, dx$$

$$= -\ln|x| - 2\left(\frac{1}{x}\right) + \int \frac{2x + 2 + 1}{x^2 + 2x + 2}\, dx$$

$$= -\ln|x| - \frac{2}{x} + \int \frac{2x + 2}{x^2 + 2x + 2}\, dx + \int \frac{dx}{(x^2 + 2x + 1) + 1}$$

$$= -\ln|x| - \frac{2}{x} + \ln|x^2 + 2x + 2| + \tan^{-1}(x + 1) + C_1$$

CAUTION ▶   *Note the manner in which the integral with the quadratic denominator was handled for the purpose of integration.* First, the numerator, $2x + 3$, was written in the form $(2x + 2) + 1$ so that we could fit the logarithmic form with the $2x + 2$. Then, we completed the square in the denominator of the final integral so that it then fit an inverse tangent form.

Repeated Quadratic Factors

Finally, considering the case of repeated quadratic factors, we use the fact that *corresponding to each irreducible quadratic factor $ax^2 + bx + c$ that occurs n times in the denominator there will be n partial fractions*

$$\frac{A_1x + B_1}{ax^2 + bx + c} + \frac{A_2x + B_2}{(ax^2 + bx + c)^2} + \cdots + \frac{A_nx + B_n}{(ax^2 + bx + c)^n}$$

where $A_1, A_2, \ldots, A_n, B_1, B_2, \ldots, B_n$ are constants to be determined. The procedures that lead to the solution are the same as those for the other cases. Exercises 15 and 16 in the following set are solved by using these partial fractions for repeated quadratic factors.

— EXERCISES 10-5 —

In Exercises 1–16, integrate each of the given functions.

1. $\displaystyle\int \frac{x-8}{x^3-4x^2+4x}\,dx$

2. $\displaystyle\int \frac{dT}{T^3-T^2}$

3. $\displaystyle\int \frac{2\,dx}{x^2(x^2-1)}$

4. $\displaystyle\int_1^3 \frac{3x^3+8x^2+10x+2}{x(x+1)^3}\,dx$

5. $\displaystyle\int_1^2 \frac{2s\,ds}{(s-3)^3}$

6. $\displaystyle\int \frac{x\,dx}{(x+2)^4}$

7. $\displaystyle\int \frac{x^3-2x^2-7x+28}{(x+1)^2(x-3)^2}\,dx$

8. $\displaystyle\int \frac{4\,dx}{(x+1)^2(x-1)^2}$

9. $\displaystyle\int_0^2 \frac{x^2+x+5}{(x+1)(x^2+4)}\,dx$

10. $\displaystyle\int \frac{v^2+v-1}{(v^2+1)(v-2)}\,dv$

11. $\displaystyle\int \frac{5x^2+8x+16}{x^2(x^2+4x+8)}\,dx$

12. $\displaystyle\int \frac{2x^2+x+3}{(x^2+2)(x-1)}\,dx$

13. $\displaystyle\int \frac{10x^3+40x^2+22x+7}{(4x^2+1)(x^2+6x+10)}\,dx$

14. $\displaystyle\int_3^4 \frac{5x^3-4x}{x^4-16}\,dx$

15. $\displaystyle\int \frac{2r^3}{(r^2+1)^2}\,dr$

16. $\displaystyle\int \frac{-x^3+x^2+x+3}{(x+1)(x^2+1)^2}\,dx$

In Exercises 17–24, solve the given problems by integration.

17. Find the area bounded by $y = \dfrac{x-3}{x^3+x^2}$, $y = 0$, and $x = 1$.

18. Find the first quadrant area bounded by $y = \dfrac{3x^2+2x+9}{(x^2+9)(x+1)}$ and $x = 2$.

19. Find the volume generated by revolving the first-quadrant region bounded by $y = 4/(x^4+6x^2+5)$ and $x = 2$ about the y-axis.

20. Find the volume generated by revolving the first-quadrant region bounded by $y = x/(x+3)^2$ and $x = 3$ about the x-axis.

21. Under certain conditions, the velocity v (in m/s) of an object moving along a straight line as a function of the time t (in s) is given by $v = \dfrac{t^2+14t+27}{(2t+1)(t+5)^2}$. Find the distance traveled by the object during the first 2.00 s.

22. By a computer analysis, the electric current i (in A) in a certain circuit is given by $i = \dfrac{0.0010(7t^2+16t+48)}{(t+4)(t^2+16)}$, where t is the time (in s). Find the total charge that passes a point in the circuit in the first 0.250 s.

23. Find the x-coordinate of the centroid of the first-quadrant region bounded by $y = 4/(x^3+x)$, $x = 1$, and $x = 2$.

24. The slope of a curve is given by $\dfrac{dy}{dx} = \dfrac{29x^2+36}{4x^4+9x^2}$. Find the equation of the curve if it passes through $(1, 5)$.

10-6 INTEGRATION BY USE OF TABLES

In this chapter we have introduced certain methods of reducing integrals to basic forms. Often this transformation and integration requires a number of steps to be performed, and therefore integrals are tabulated for reference. The integrals found in tables have been derived by using the methods introduced thus far, as well as many other methods that can be used. Therefore, an understanding of the basic forms and some of the basic methods is very useful in finding integrals from tables. Such an understanding forms a basis for proper recognition of the forms that are used in the tables, as well as the types of results that may be expected. Therefore,

CAUTION ▶ *the use of the tables depends on proper recognition of the form and the variables and constants of the integral.* The following examples illustrate the use of the table of integrals found in Appendix E. More extensive tables are available in other sources.

EXAMPLE 1 Integrate: $\displaystyle\int \frac{x\,dx}{\sqrt{2+3x}}$.

We first note that this integral fits the form of Formula 6 of Appendix E, with $u = x$, $a = 2$, and $b = 3$. Therefore,

For reference, Formula 6 is

$$\int \frac{u\,du}{\sqrt{a+bu}} = -\frac{2(2a-bu)\sqrt{a+bu}}{3b^2}.$$

$$\int \frac{x\,dx}{\sqrt{2+3x}} = -\frac{2(4-3x)\sqrt{2+3x}}{27} + C$$

For reference, Formula 18 is

$$\int \frac{\sqrt{a^2 - u^2}}{u}\, du =$$

$$\sqrt{a^2 - u^2} - a \ln\left(\frac{a + \sqrt{a^2 - u^2}}{u}\right)$$

EXAMPLE 2 Integrate: $\int \frac{\sqrt{4 - 9x^2}}{x}\, dx$.

This fits the form of Formula 18, with proper identification of constants; $u = 3x$, $du = 3\, dx$, $a = 2$. Hence,

$$\int \frac{\sqrt{4 - 9x^2}}{x}\, dx = \int \frac{\sqrt{4 - 9x^2}}{3x}\, 3\, dx$$

$$= \sqrt{4 - 9x^2} - 2 \ln\left(\frac{2 + \sqrt{4 - 9x^2}}{3x}\right) + C$$

For reference, Formula 37 is

$$\int \sec^n u\, du =$$

$$\frac{\sec^{n-2} u \tan u}{n - 1} + \frac{n - 2}{n - 1}\int \sec^{n-2} u\, du$$

EXAMPLE 3 Integrate: $\int 5 \sec^3 2x\, dx$.

This fits the form of Formula 37; $n = 3$, $u = 2x$, $du = 2\, dx$. And so,

$$\int 5 \sec^3 2x\, dx = 5\left(\frac{1}{2}\right)\int \sec^3 2x(2\, dx)$$

$$= \frac{5}{2}\frac{\sec 2x \tan 2x}{2} + \frac{5}{2}\left(\frac{1}{2}\right)\int \sec 2x(2\, dx)$$

For reference, Eq. (9-12) is

$$\int \sec u\, du = \ln |\sec u + \tan u| + C.$$

To complete this integral, we must use the basic form of Eq. (9-12). Thus, we complete it by

$$\int 5 \sec^3 2x\, dx = \frac{5 \sec 2x \tan 2x}{4} + \frac{5}{4} \ln|\sec 2x + \tan 2x| + C$$

EXAMPLE 4 Find the area bounded by $y = x^2 \ln 2x$, $y = 0$, and $x = e$.
From Fig. 10-9, we see that the area is

$$A = \int_{0.5}^{e} x^2 \ln 2x\, dx$$

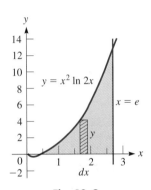

Fig. 10-9

This integral fits the form of Formula 46 if $u = 2x$. Thus, we have

$$A = \frac{1}{8}\int_{0.5}^{e} (2x)^2 \ln 2x(2\, dx) = \frac{1}{8}(2x)^3\left[\frac{\ln 2x}{3} - \frac{1}{9}\right]\Big|_{0.5}^{e}$$

$$= e^3\left(\frac{\ln 2e}{3} - \frac{1}{9}\right) - \frac{1}{8}\left(\frac{\ln 1}{3} - \frac{1}{9}\right) = e^3\left(\frac{3 \ln 2e - 1}{9}\right) + \frac{1}{72}$$

$$= 9.118$$

NOTE▶ The proper identification of u and du is the key step in the use of tables. Therefore, for the integrals in the following example the proper u and du, along with the appropriate formula from the table, are identified, but the integrations are not performed.

EXAMPLE 5 (a) $\int x \sqrt{1 - x^4}\, dx$ $\qquad u = x^2, \qquad du = 2x\, dx$ $\qquad$ Formula 15

(b) $\int \frac{(4x^6 - 9)^{3/2}}{x}\, dx$ $\qquad u = 2x^3, \qquad du = 6x^2\, dx$ $\qquad$ Formula 22
$\qquad\qquad\qquad\qquad\qquad$ introduce a factor of x^2 into numerator and denominator

(c) $\int x^3 \sin x^2\, dx$ $\qquad u = x^2, \qquad du = 2x\, dx$ $\qquad$ Formula 47

Following is a brief summary of the approach to integration we have used to obtain the exact result. Also, definite integrals may be approximated by methods such as the trapezoidal rule or Simpson's rule.

Basic Approach to Integrating a Function

1. *Write the integral such that it fits an integral form. Either a basic form as developed in Chapter 9 or a form from a table of integrals may be used.*

2. *Use a method of transforming the integral such that an integral form may be used. Appropriate methods are covered in this chapter or in other sources.*

EXERCISES *10-6*

In the following exercises, integrate each function by using the table in Appendix E. In Exercise 32, explain your solution.

1. $\displaystyle\int \frac{3x\,dx}{2 + 5x}$

2. $\displaystyle\int \frac{4x\,dx}{(1 + x)^2}$

3. $\displaystyle\int_2^7 4x\sqrt{2 + x}\,dx$

4. $\displaystyle\int \frac{dx}{x^2 - 4}$

5. $\displaystyle\int \frac{dy}{(y^2 + 4)^{3/2}}$

6. $\displaystyle\int_0^{\pi/3} \sin^3 x\,dx$

7. $\displaystyle\int \sin 2x \sin 3x\,dx$

8. $\displaystyle\int 6 \sin^{-1} 3x\,dx$

9. $\displaystyle\int \frac{\sqrt{4x^2 - 9}}{x}\,dx$

10. $\displaystyle\int \frac{(9x^2 + 16)^{3/2}}{x}\,dx$

11. $\displaystyle\int \cos^5 4x\,dx$

12. $\displaystyle\int 0.2 \tan^2 2\phi\,d\phi$

13. $\displaystyle\int 6r \tan^{-1} r^2\,dr$

14. $\displaystyle\int 5xe^{4x}\,dx$

15. $\displaystyle\int_1^2 (4 - x^2)^{3/2}\,dx$

16. $\displaystyle\int \frac{3\,dx}{9 - 16x^2}$

17. $\displaystyle\int \frac{dx}{x\sqrt{4x^2 + 1}}$

18. $\displaystyle\int \frac{\sqrt{4 + x^2}}{x}\,dx$

19. $\displaystyle\int \frac{8\,dx}{x\sqrt{1 - 4x^2}}$

20. $\displaystyle\int \frac{dx}{x(1 + 4x)^2}$

21. $\displaystyle\int_0^{\pi/12} \sin \theta \cos 5\theta\,d\theta$

22. $\displaystyle\int_0^2 x^2 e^{3x}\,dx$

23. $\displaystyle\int x^5 \cos x^3\,dx$

24. $\displaystyle\int 5 \sin^3 t \cos^2 t\,dt$

25. $\displaystyle\int \frac{2x\,dx}{(1 - x^4)^{3/2}}$

26. $\displaystyle\int \frac{dx}{x(1 - 4x)}$

27. $\displaystyle\int_1^3 \frac{\sqrt{3 + 5x^2}\,dx}{x}$

28. $\displaystyle\int_0^1 \frac{\sqrt{9 - 4x^2}}{x}\,dx$

29. $\displaystyle\int x^3 \ln x^2\,dx$

30. $\displaystyle\int \frac{1.2u\,du}{u^2 \sqrt{u^4 - 9}}$

31. $\displaystyle\int \frac{9x^2\,dx}{(x^6 - 1)^{3/2}}$

W 32. $\displaystyle\int x^7 \sqrt{x^4 + 4}\,dx$

33. Find the length of arc of the curve $y = x^2$ from $x = 0$ to $x = 1$.

34. Find the moment of inertia with respect to its axis of the volume generated by revolving the region bounded by $y = 3 \ln x$, $x = e$, and the x-axis about the y-axis.

35. Find the area of an ellipse with a major axis $2a$ and a minor axis $2b$.

36. The voltage across a 5.0-μF capacitor in an electric circuit is zero. What is the voltage after 5.00 μs if a current i (in mA) as a function of the time t (in s) given by $i = \tan^{-1} 2t$ charges the capacitor?

37. Find the force (in lb) on the region bounded by $x = 1/\sqrt{1 + y}$, $y = 0$, $y = 3$, and the y-axis, if the surface of the water is at the upper edge of the area.

38. If 6.00 g of a chemical are placed in water, the time t (in min) it takes to dissolve half of the chemical is given by
$$t = 560 \int_3^6 \frac{dx}{x(x + 4)},$$ where x is the amount of undissolved chemical at any time. Evaluate t.

39. The dome of a sports arena is the surface generated by revolving $y = 20.0 \cos 0.0196x$ ($0 \le x \le 80.0$ m) about the y-axis. Find the volume within the dome.

40. If an electric charge Q is distributed along a wire of length $2a$, the force F exerted on an electric charge q placed at point P is
$$F = kqQ \int \frac{b\,dx}{(b^2 + x^2)^{3/2}}.$$ Integrate to find F as a function of x.

10-7 IMPROPER INTEGRALS

In defining the definite integral $\int_a^b f(x)\,dx$ in Chapter 5, we assumed that both limits of integration are finite; we also assumed that the integrand $f(x)$ is continuous for $a \le x \le b$. We now extend the definition of a definite integral to **improper integrals,** *for which one or both limits of integration are infinite, or which have infinite discontinuous integrands.*

First, we discuss the type of improper integral for which one or both limits of integration are infinite. Consider the area in the following example.

■**EXAMPLE 1**   Find the area bounded by $y = 1/x^3$, $y = 0$, $x = 1$, and $x = b$, where $b > 1$. See Fig. 10-10.

From the figure we see that the area is given by

$$A = \int_1^b \frac{1}{x^3}\,dx = -\frac{1}{2x^2}\Big|_1^b = -\frac{1}{2b^2} + \frac{1}{2}$$
$$= \frac{1}{2}\left(1 - \frac{1}{b^2}\right)$$

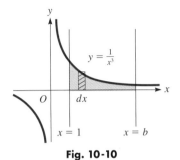

Fig. 10-10

If we now let b increase without bound, we see that the area increases, with a limiting value of 1/2. This we can show as

$$A = \lim_{b \to +\infty} \int_1^b \frac{1}{x^3}\,dx = \lim_{b \to +\infty} \frac{1}{2}\left(1 - \frac{1}{b^2}\right) = \frac{1}{2}$$

Representing this limit by $\int_1^{+\infty} \frac{1}{x^3}\,dx$, we see that we have a definite integral with an infinite limit. ■

Based on reasoning similar to that in Example 1, we have the following *definition for improper integrals for which one or both limits of integration are infinite.*

$\int_a^{+\infty} f(x)\,dx = \lim\limits_{b \to +\infty} \int_a^b f(x)\,dx$	**(10-3)**
$\int_{-\infty}^b f(x)\,dx = \lim\limits_{a \to -\infty} \int_a^b f(x)\,dx$	**(10-4)**
$\int_{-\infty}^{+\infty} f(x)\,dx = \lim\limits_{a \to -\infty} \int_a^0 f(x)\,dx + \lim\limits_{b \to +\infty} \int_0^b f(x)\,dx$	**(10-5)**

NOTE ▶   The improper integrals in Eqs. (10-3), (10-4), and (10-5) have the indicated values *only if the limits exist.* In this case, *if the limits exist, the improper integrals are said to be* **convergent.** On the other hand, *if the limits do not exist, the improper integrals are said to be* **divergent.** In using Eq. (10-3), we assume that $f(x)$ is continuous for $x \ge a$. Also, for Eq. (10-4) we assume that $f(x)$ is continuous for $x \le b$, and for Eq. (10-5) we assume that $f(x)$ is continuous for all x.

■**EXAMPLE 2**   Evaluate $\int_{-\infty}^0 e^x\,dx$ if it exists.

Using Eq. (10-4), we have

$$\int_{-\infty}^0 e^x\,dx = \lim_{a \to -\infty} \int_a^0 e^x\,dx = \lim_{a \to -\infty} e^x\Big|_a^0$$
$$= \lim_{a \to -\infty}(1 - e^a) = 1$$

Here the integral converges, since the limit exists. ■

EXAMPLE 3 A rocket is launched vertically with a velocity v (in mi/h), as a function of time t (in h), of $v = 30,000/(t + 1)$. Show that the rocket will eventually be as far from the earth as may be required.

The distance s from the earth is given by

$$s = \int_0^{+\infty} \frac{30,000 \, dt}{t + 1} = \lim_{b \to +\infty} \int_0^b \frac{30,000 \, dt}{t + 1}$$

$$= 30,000 \lim_{b \to +\infty} \ln|t + 1| \Big|_0^b = 30,000 \lim_{b \to +\infty} \ln|b + 1| = +\infty$$

Since the integral diverges and the limit is infinite, the distance is unbounded.

EXAMPLE 4 Evaluate $\displaystyle\int_{-\infty}^{+\infty} \frac{dx}{x^2 + 4x + 5}$ if it exists.

Using Eq. (10-5), we have

$$\int_{-\infty}^{+\infty} \frac{dx}{x^2 + 4x + 5} = \lim_{a \to -\infty} \int_a^0 \frac{dx}{x^2 + 4x + 5} + \lim_{b \to +\infty} \int_0^b \frac{dx}{x^2 + 4x + 5}$$

$$= \lim_{a \to -\infty} \int_a^0 \frac{dx}{(x + 2)^2 + 1} + \lim_{b \to +\infty} \int_0^b \frac{dx}{(x + 2)^2 + 1}$$

$$= \lim_{a \to -\infty} \tan^{-1}(x + 2) \Big|_a^0 + \lim_{b \to +\infty} \tan^{-1}(x + 2) \Big|_0^b$$

$$= \tan^{-1} 2 - \left(-\frac{\pi}{2}\right) + \frac{\pi}{2} - \tan^{-1} 2 = \pi$$

Here, $\tan^{-1}(x + 2) \to -\pi/2$ when $x \to -\infty$ and $\tan^{-1}(x + 2) \to \pi/2$ when $x \to +\infty$.

NOTE ▶ *An improper integral may diverge without becoming infinite.* See Exercise 31 at the end of this section. Also, the reason that it is necessary to split the improper integral for which both limits are infinite [Eq. (10-5)] is illustrated in Exercise 32.

We next consider improper integrals with discontinuous integrands. To illustrate this type of improper integral, consider the area in the following example.

EXAMPLE 5 Find the first quadrant area bounded by $y = 1/\sqrt{x}$ and $x = 4$.

We set up the area (see Fig. 10-11) as

$$A = \int_0^4 \frac{1}{\sqrt{x}} \, dx$$

However, we note that the integrand $1/\sqrt{x}$ is discontinuous at $x = 0$, which is the lower limit. Therefore, we set up the integral for the area as

$$A = \lim_{h \to 0} \int_h^4 \frac{1}{\sqrt{x}} \, dx$$

where $0 < h < 4$. We can now find the limiting value of the area it if exists. Thus

$$A = \lim_{h \to 0} \int_h^4 x^{-1/2} \, dx = \lim_{h \to 0} 2\sqrt{x} \Big|_h^4 = \lim_{h \to 0} (4 - 2\sqrt{h}) = 4$$

Since the limit exists, the limiting value of the area is 4.

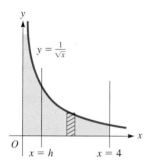

Fig. 10-11

Based on reasoning similar to that of Example 5, we have the following definitions for improper integrals that have infinite discontinuous integrands.

If $f(x)$ is continuous for $a < x \leq b$ (but is discontinuous for $x = a$),

$$\int_a^b f(x)\,dx = \lim_{h \to 0} \int_{a+h}^b f(x)\,dx \qquad \textbf{(10-6)}$$

If $f(x)$ is continuous for $a \leq x < b$ (but is discontinuous for $x = b$),

$$\int_a^b f(x)\,dx = \lim_{h \to 0} \int_a^{b-h} f(x)\,dx \qquad \textbf{(10-7)}$$

If $f(x)$ is continuous for $a \leq x \leq b$, except at c where $a < c < b$,

$$\int_a^b f(x)\,dx = \lim_{h \to 0} \int_a^{c-h} f(x)\,dx + \lim_{h' \to 0} \int_{c+h'}^b f(x)\,dx \qquad \textbf{(10-8)}$$

Here we assume $h > 0$ and $h' > 0$. These integrals are convergent if the limits exist.

CAUTION ▶ This means that the integral in Eq. (10-8) is convergent *only if **both** limits exist.*

▌EXAMPLE 6 Evaluate $\displaystyle\int_0^1 \frac{dx}{(x-1)^{1/3}}$ if it is convergent.

Since the integrand is discontinuous at the upper limit, we use Eq. (10-7).

$$\int_0^1 \frac{dx}{(x-1)^{1/3}} = \lim_{h \to 0} \int_0^{1-h} \frac{dx}{(x-1)^{1/3}}$$

$$= \lim_{h \to 0} \left[\frac{3}{2}(x-1)^{2/3} \right]\Bigg|_0^{1-h}$$

$$= \lim_{h \to 0} \left[\frac{3}{2}(-h)^{2/3} - \frac{3}{2}(-1)^{2/3} \right] = -\frac{3}{2} \quad\text{---------}\ ▮$$

▌EXAMPLE 7 Evaluate $\displaystyle\int_0^2 \frac{dx}{(x-1)^2}$ if it is convergent.

Here we see that the integrand is discontinuous as $x = 1$, which is between the limits of integration. Therefore, we use Eq. (10-8).

$$\int_0^2 \frac{dx}{(x-1)^2} = \lim_{h \to 0} \int_0^{1-h} \frac{dx}{(x-1)^2} + \lim_{h' \to 0} \int_{1+h'}^2 \frac{dx}{(x-1)^2}$$

$$= \lim_{h \to 0} \left[-\frac{1}{x-1} \right]_0^{1-h} + \lim_{h' \to 0} \left[-\frac{1}{x-1} \right]\Bigg|_{h'+1}^2$$

$$= \lim_{h \to 0} \left(\frac{1}{h} - 1 \right) + \lim_{h' \to 0} \left(-1 + \frac{1}{h'} \right)$$

Neither limit exists. The divergence of *either* would mean the integral diverges.

If we were to integrate this improper integral as an ordinary integral, we obtain

> Note that we would obtain an incorrect result by not treating it as an improper integral.

$$\int_0^2 \frac{dx}{(x-1)^2} = -\frac{1}{x-1}\Bigg|_0^2 = -1 - 1 = -2$$

This negative result cannot be valid, because the integrand is never negative. This shows the need to be careful to handle improper integrals correctly. ----------▮

--- EXERCISES *10-7* ---

In Exercises 1–24, determine whether the given improper integral is convergent or divergent. If it is convergent, evaluate it.

1. $\displaystyle\int_{1}^{+\infty} \frac{dx}{(x+2)^2}$ **2.** $\displaystyle\int_{2}^{+\infty} \frac{dx}{3x+4}$ **3.** $\displaystyle\int_{-\infty}^{0} \frac{dx}{\sqrt{1-3x}}$

4. $\displaystyle\int_{0}^{+\infty} \frac{dx}{(1+x)^3}$ **5.** $\displaystyle\int_{1}^{+\infty} \frac{x\,dx}{1+x^2}$ **6.** $\displaystyle\int_{3}^{+\infty} \frac{x\,dx}{\sqrt{x^2-4}}$

7. $\displaystyle\int_{0}^{+\infty} xe^{-x^2}\,dx$ **8.** $\displaystyle\int_{-\infty}^{-1} \frac{-x\,dx}{(1+x^2)^2}$ **9.** $\displaystyle\int_{-\infty}^{0} \frac{2\,dx}{(1-x)^3}$

10. $\displaystyle\int_{0}^{+\infty} \frac{x\,dx}{\sqrt{x+1}}$ **11.** $\displaystyle\int_{-\infty}^{+\infty} e^{-x}\,dx$ **12.** $\displaystyle\int_{-\infty}^{+\infty} \frac{dx}{x^2+4}$

13. $\displaystyle\int_{0}^{9} \frac{dx}{\sqrt{9-x}}$ **14.** $\displaystyle\int_{0}^{1} \frac{dx}{x}$ **15.** $\displaystyle\int_{\pi/4}^{\pi/3} \tan 2x\,dx$

16. $\displaystyle\int_{0}^{3} \frac{dx}{\sqrt{9-x^2}}$ **17.** $\displaystyle\int_{0}^{3} \frac{dx}{(x-2)^2}$ **18.** $\displaystyle\int_{-1}^{1} \frac{dx}{x^{1/3}}$

19. $\displaystyle\int_{0}^{1} \frac{2\,dx}{1-x^2}$ **20.** $\displaystyle\int_{2}^{3} \frac{dx}{\sqrt{x-2}}$ **21.** $\displaystyle\int_{1}^{2} \frac{dx}{\sqrt{x^2-1}}$

22. $\displaystyle\int_{1}^{3} \frac{3\,dx}{3x-x^2}$ **23.** $\displaystyle\int_{1}^{3} \frac{dx}{(x-2)^{1/3}}$ **24.** $\displaystyle\int_{-\pi/2}^{\pi/2} \sec x\,dx$

In Exercises 25–32, solve the given problems by evaluating the appropriate improper integrals (if convergent).

25. (a) Find the area bounded by $y = 1/x$, $y = 0$, and $x = 1$ (to the right of $x = 1$) See Fig. 10-12. (b) Find the volume generated if the region in part (a) is revolved about the x-axis.

26. Compare the area bounded by $y = 1/\sqrt{x}$, $y = 0$, and $x = 4$ (to the right of $x = 4$) with the area of Example 5.

27. Find the area bounded by $y = x/\sqrt{x^2-1}$, $x = 1$, $x = 2$, and $y = 0$.

28. Find the volume generated by revolving the region bounded by $y = 1/(x-2)$, $x = 0$, $y = 0$, and $x = 2$ about the x-axis.

29. The gravitational force F (in lb) of the earth on a satellite (the weight of the satellite) is given by $F = 10^{11}/(x+3960)^2$, where x is the vertical distance (in mi) from the surface of the earth to the satellite. How much work is done in moving the satellite from the earth's surface to an infinite distance from the earth?

30. The electric field intensity E around an infinitely long straight wire carrying an electric current i is given by

$$E = kiam \int_{-\infty}^{+\infty} \frac{dx}{(a^2+x^2)^{3/2}}$$

where k, i, a, and m are constants. Evaluate the integral.

31. To show that an improper integral may diverge without becoming infinite, evaluate $\int_{0}^{+\infty} \cos x\,dx$.

(W) 32. To show why it is necessary to split the integral $\int_{-\infty}^{+\infty} f(x)\,dx$ into two separate integrals, as in Eq. (10-5), evaluate
(a) $\displaystyle\lim_{b\to+\infty} \int_{-b}^{b} x\,dx$ and
(b) $\int_{-\infty}^{+\infty} x\,dx$ and compare results.

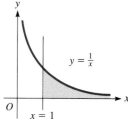

$y = \frac{1}{x}$

Fig. 10-12

CHAPTER EQUATIONS

Integration by parts	$\displaystyle\int u\,dv = uv - \int v\,du$	(10-1)
Trigonometric substitutions	For $\sqrt{a^2-x^2}$ use $x = a\sin\theta$	
	For $\sqrt{a^2+x^2}$ use $x = a\tan\theta$	(10-2)
	For $\sqrt{x^2-a^2}$ use $x = a\sec\theta$	
Improper integrals	$\displaystyle\int_{a}^{+\infty} f(x)\,dx = \lim_{b\to+\infty} \int_{a}^{b} f(x)\,dx$	(10-3)
	$\displaystyle\int_{-\infty}^{b} f(x)\,dx = \lim_{a\to-\infty} \int_{a}^{b} f(x)\,dx$	(10-4)
	$\displaystyle\int_{-\infty}^{+\infty} f(x)\,dx = \lim_{a\to-\infty} \int_{a}^{0} f(x)\,dx + \lim_{b\to+\infty} \int_{0}^{b} f(x)\,dx$	(10-5)
	$\displaystyle\int_{a}^{b} f(x)\,dx = \lim_{h\to0} \int_{a+h}^{b} f(x)\,dx$	(10-6)
	$\displaystyle\int_{a}^{b} f(x)\,dx = \lim_{h\to0} \int_{a}^{b-h} f(x)\,dx$	(10-7)
	$\displaystyle\int_{a}^{b} f(x)\,dx = \lim_{h\to0} \int_{a}^{c-h} f(x)\,dx + \lim_{h'\to0} \int_{c+h'}^{b} f(x)\,dx$	(10-8)

REVIEW EXERCISES

In Exercises 1–24, integrate each of the given functions without the use of a table of integrals.

1. $\displaystyle\int x \csc^2 2x \, dx$

2. $\displaystyle\int x \tan^{-1} x \, dx$

3. $\displaystyle\int x\sqrt{x-4}\, dx$

4. $\displaystyle\int \frac{2x\, dx}{\sqrt{2x+5}}$

5. $\displaystyle\int \frac{dx}{\sqrt{4x^2-9}}$

6. $\displaystyle\int \frac{x^2\, dx}{\sqrt{9-x^2}}$

7. $\displaystyle\int \frac{x+25}{x^2-25}\, dx$

8. $\displaystyle\int \frac{2x^2+6x+1}{2x^3-x^2-x}\, dx$

9. $\displaystyle\int \frac{x^2-2x+3}{(x-1)^3}\, dx$

10. $\displaystyle\int \frac{2x^2+3x+18}{x^3+9x}\, dx$

11. $\displaystyle\int_{1}^{+\infty} \frac{dx}{2x+7}$

12. $\displaystyle\int_{-1}^{1} \frac{dx}{\sqrt{x+1}}$

13. $\displaystyle\int \frac{x\, dx}{(x+3)^{2/3}}$

14. $\displaystyle\int x \tan^2 x \, dx$

15. $\displaystyle\int \frac{2\, dx}{(4x^2+1)^{3/2}}$

16. $\displaystyle\int \frac{3x^2-6x-2}{3x^3+x^2}\, dx$

17. $\displaystyle\int \frac{x^2+3}{x^4+3x^2+2}\, dx$

18. $\displaystyle\int x^3\sqrt{4-x^2}\, dx$

19. $\displaystyle\int_{0}^{2} \ln(x+2)\, dx$

20. $\displaystyle\int_{-2}^{1} \frac{x^2\, dx}{\sqrt{2-x}}$

21. $\displaystyle\int \frac{6(2-x^2)}{(x^2-1)(x^2-4)}\, dx$

22. $\displaystyle\int \frac{-x^3+9x^2-24x+36}{x^2(x-3)^2}\, dx$

23. $\displaystyle\int_{0}^{8} \frac{x\, dx}{(8-x)^{1/3}}$

24. $\displaystyle\int_{1}^{+\infty} \frac{dx}{x^2(x^2+1)}$

In Exercises 25–36, solve the given problems by integration.

25. Find the area bounded by $y = 1/\sqrt{x-2}$, $x = 2$, $x = 6$, and $y = 0$.

26. Find the area bounded by $y = x/(1+x)^2$, $y = 0$, and $x = 4$.

27. Find the area inside the circle $x^2 + y^2 = 25$ and to the right of the line $x = 3$.

28. Find the area bounded by $y = x\sqrt{x+4}$, $y = 0$, and $x = 5$.

29. Find the volume generated by revolving the region bounded by $y = xe^x$, $y = 0$, and $x = 2$ about the y-axis.

30. Find the volume generated by revolving about the y-axis the region bounded by $y = x + \sqrt{x+1}$, $x = 3$, and the axes.

31. The force F (in N) applied by a stamping machine in punching out a hole 0.500 cm deep is given by $F = 1000x/(2x+1)^2$, where x is the depth (in cm) of the hole at any instant. Find the work done by the machine in punching out the hole.

32. In electrical theory, the integral $W = \displaystyle\int_{0}^{\infty} Ri^2\, dt$ occurs. If $i = Ie^{-Rt/L}$, evaluate the integral. R, I, and L are constants.

33. Find the centroid of the region bounded by $y = \ln x$, $x = 2$, and the x-axis.

34. The nose cone of a space vehicle is to be covered with a heat shield. The cone is designed such that a cross section x feet from the tip and perpendicular to its axis is a circle of radius $1.5x^{2/3}$ feet. Find the surface area of the heat shield if the nose cone is 16.0 ft long. See Fig. 10-13.

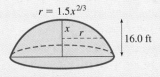

Fig. 10-13

Fig. 10-14

35. A window has the shape of a semiellipse, as shown in Fig. 10-14. What is the area of the window?

36. Find the radius of gyration with respect to the y-axis of the first-quadrant region bounded by $y = \sqrt{x+1}$ and $x = 3$.

Writing Exercise

37. A student was writing a computer program to find the area bounded by $y = x/(4-x^2)$, $y = 0$, and $x = 1$. The student also noted that this area could be found by integration using more than one method. Write one or two paragraphs explaining how the resulting integral can be evaluated by a method in this chapter and one developed before this chapter.

PRACTICE TEST

In Problems 1–5, evaluate the given integrals.

1. $\displaystyle\int \frac{dx}{x^2\sqrt{4-x^2}}$

2. $\displaystyle\int xe^{-2x}\, dx$

3. $\displaystyle\int \frac{x^3+5x^2+x+2}{x^4+x^2}\, dx$

4. $\displaystyle\int_{0}^{3} \frac{dx}{\sqrt{9-x^2}}$

5. $\displaystyle\int_{2}^{3} x\sqrt{x-2}\, dx$

6. Find the area bounded by $y = x/(x^2-3x+2)$, $y = 0$, $x = 3$ and $x = 4$.

11

INTRODUCTION TO PARTIAL DERIVATIVES AND DOUBLE INTEGRALS

To this point we have been dealing with functions that have a single independent variable. There are, however, numerous applications in many fields, including science, engineering, and technology, in which functions with more than one independent variable are used. Although a number of these functions involve three or more independent variables, many involve only two, and we shall concern ourselves primarily with these.

In the first two sections of this chapter we shall establish the meaning of a function of two independent variables, and then discuss how the graph of this type of function is shown. The remaining sections of the chapter develop some of the basic concepts of the calculus of functions of two variables and show some of the applications in technology.

In Section 11-1 we see how the design of a box to hold a product involves a function of two variables.

11-1 FUNCTIONS OF TWO VARIABLES

Many familiar formulas express one variable in terms of two or more other variables. The following example illustrates one from geometry.

Fig. 11-1

EXAMPLE 1 The total surface area of a right circular cylinder is a function of the radius and the height of the cylinder. That is, the area will change if either or both of these change. The formula for the total surface area is

$$A = 2\pi r^2 + 2\pi rh$$

We say that A is a function of r and h. See Fig. 11-1.

There are numerous applications of functions of two variables. For example, the voltage in a simple electric circuit depends on the resistance and current. The moment of a force depends on the magnitude of the force and the distance of the force from the axis. The pressure of a gas depends on its temperature and volume. The temperature at a point on a plate depends on the coordinates of the point.

We define a function of two variables as follows: *If z is uniquely determined for given values of x and y, then z is a function of x and y.* The notation used is similar to that used for one independent variable. It is $z = f(x, y)$, where both x and y are independent variables. Therefore, it follows that $f(a, b)$ means "*the value of the function when x = a and y = b.*"

EXAMPLE 2   If $f(x, y) = 3x^2 + 2xy - y^3$, find $f(-1, 2)$. Substituting, we have

$$f(-1, 2) = 3(-1)^2 + 2(-1)(2) - (2)^3$$
$$= 3 - 4 - 8 = -9$$

EXAMPLE 3   For a certain electric circuit, the current i (in A), in terms of the voltage E and resistance R (in Ω) is given by

$$i = \frac{E}{R + 0.25}$$

The formula in Example 3 is based on Ohm's law, which states that the current is proportional to the voltage for a constant resistance. The ohm (Ω) is named for the German physicist Georg Ohm (1787–1854).

Find the current for $E = 1.50$ V and $R = 1.20$ Ω and for $E = 1.60$ V and $R = 1.05$ Ω.

Substituting the first values, we have

$$i = \frac{1.50}{1.20 + 0.25} = 1.03 \text{ A}$$

For the second pair of values we have

$$i = \frac{1.60}{1.05 + 0.25} = 1.23 \text{ A}$$

For this circuit, the current generally changes if either or both of E and R change.

EXAMPLE 4   If $f(x, y) = 2xy^2 - y$, find $f(x, 2x) - f(x, x^2)$.
We note that in each evaluation the x factor remains as x, but that we are to substitute $2x$ for y and subtract the function for which x^2 is substituted for y.

$$f(x, 2x) - f(x, x^2) = [2x(2x)^2 - (2x)] - [2x(x^2)^2 - x^2]$$
$$= [8x^3 - 2x] - [2x^5 - x^2]$$
$$= 8x^3 - 2x - 2x^5 + x^2$$
$$= -2x^5 + 8x^3 + x^2 - 2x$$

This type of difference of functions is important in Section 11-5.

Restricting values of the function to real numbers means that certain restrictions may be placed on the independent variables. *Values of either x or y or both that lead to division by zero or to imaginary values for z are not permissible.*

EXAMPLE 5   If $f(x, y) = \dfrac{3xy}{(x - y)(x + 3)}$,

all values of x and y are permissible except those for which $x = y$ and $x = -3$. Each of these would indicate division by zero.
If $f(x, y) = \sqrt{4 - x^2 - y^2}$, neither x nor y may be greater than 2 in absolute value, and the sum of their squares may not exceed 4. Otherwise, imaginary values of the function would result.

One of the primary difficulties students have with functions is setting them up from stated conditions. Although many examples of this have been encountered in our previous work, an example of setting up a function of two variables should prove helpful. Although no general rules can be given for this procedure, a careful analysis of the statement should lead to the desired function.

SOLVING A WORD PROBLEM

See the chapter introduction.

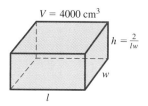

$V = 4000$ cm^3

$h = \dfrac{2}{lw}$

w

l

Fig. 11-2

▌EXAMPLE 6   The open rectangular bottom box for a jigsaw puzzle is designed to have a volume of 4000 cm^3. The cost of cardboard for this type of box is c cents/cm^2. Express the cost of the cardboard needed to make the box as a function of the length and width of the box.

The cost in question depends on the surface area of the sides of the box. An "open" box is one that has no top. Thus, the surface area of the box is

$$S = lw + 2lh + 2hw$$

where l is the length, w the width, and h the height of the box (see Fig. 11-2). However, this expression contains three independent variables. Using the condition that the volume of the box is 4000 cm^3, we have $lwh = 4000$. Since we wish to have only l and w, we solve this expression for h and find that $h = 4000/lw$. Substituting for h in the expression for the surface area, we have

$$S = lw + 2l\left(\frac{4000}{lw}\right) + 2\left(\frac{4000}{lw}\right)w$$

$$= lw + \frac{8000}{w} + \frac{8000}{l}$$

Since the cost C is given by $C = cS$, we have the function for the cost as

$$C = c\left(lw + \frac{8000}{w} + \frac{8000}{l}\right)$$

EXERCISES *11-1*

In Exercises 1–8, determine the indicated functions.

1. Express the volume of a right circular cylinder as a function of the radius and the height.

2. Express the volume of a right circular cone as a function of the radius of the base and the height.

3. Express the area of a triangle as a function of the base and the altitude.

4. Express the length of a diagonal of a rectangle as a function of the length and the width.

5. A cylindrical can is to be made to contain a volume V. Express the total surface area (including the top) of the can as a function of V and the radius of the can.

6. The angle between two forces $\mathbf{F}_1$ and $\mathbf{F}_2$ is 30°. Express the magnitude of the resultant $\mathbf{R}$ in terms of F_1 and F_2. See Fig. 11-3.

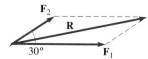

$\mathbf{F}_2$

$\mathbf{R}$

$30°$ $\mathbf{F}_1$

Fig. 11-3

7. A right circular cylinder is to be inscribed in a sphere of radius r. Express the volume of the cylinder as a function of the height h of the cylinder and r. See Fig. 11-4.

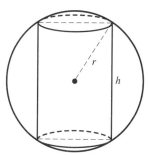

r

h

Fig. 11-4

8. A computing-leasing firm charges a monthly fee F based on the length of time a corporation has used the service, plus $100 for every hour the computer system is used during the month. Express the total monthly charge T as a function of F and the number of hours h the computer is used.

In Exercises 9–24, evaluate the given functions.

9. $f(x, y) = 2x - 6y$; find $f(0, -4)$.

10. $F(x, y) = x^2 - 5y + y^2$; find $F(2, -2)$.

• 11. $g(r, s) = r - 2rs - r^2 s$; find $g(-2, 1)$.

12. $f(r, \theta) = 2r(r \tan \theta - \sin 2\theta)$; find $f(3, \pi/4)$.

13. $Y(y, t) = \dfrac{2 - 3y}{t - 1} + 2y^2 t$; find $Y(y, 2)$.

• 14. $f(r, t) = r^3 - 3r^2 t + 3rt^2 - t^3$; find $f(3, t)$.

15. $X(x, t) = -6xt + xt^2 - t^3$; find $X(x, -t)$.

16. $g(y, z) = 2yz^2 - 6y^2 z - y^2 z^2$; find $g(y, 2y)$.

♦ 17. $H(p, q) = p - \dfrac{p - 2q^2 - 5q}{p + q}$; find $H(p, q + k)$.

18. $g(x, z) = z \tan^{-1}(x^2 + xz)$; find $g(-x, z)$.

19. $f(x, y) = x^2 - 2xy - 4x$; find $f(x + h, y + k) - f(x, y)$.

20. $g(y, z) = 4yz - z^3 + 4y$; find $g(y + 1, z + 2) - g(y, z)$.

21. $f(x, y) = xy + x^2 - y^2$; find $f(x, x) - f(x, 0)$.

• 22. $f(x, y) = 4x^2 - xy - 2y$; find $f(x, x^2) - f(x, 1)$.

23. $g(y, z) = 3y^3 - y^2 z + 5z^2$; find $g(3z^2, z) - g(z, z)$.

24. $X(x, t) = 2x - \dfrac{t^2 - 2x^2}{x}$; find $X(2t, t) - X(2t^2, t)$.

In Exercises 25–28, determine which values of x and y, if any, are not permissible. In Exercise 27, explain your answer.

25. $f(x, y) = \dfrac{\sqrt{y}}{2x}$

26. $f(x, y) = \dfrac{x^2 - 4y^2}{x^2 + 9}$

Ⓦ 27. $f(x, y) = \sqrt{x^2 + y^2 - x^2 y - y^3}$

28. $f(x, y) = \dfrac{1}{xy - y}$

In Exercises 29–40, solve the given problems.

• 29. The voltage v across a resistor R in an electric circuit is given by $v = iR$, where i is the current. What is the voltage if the current is 3 A and the resistance is 6 Ω?

30. The centripetal acceleration a of an object moving in a circular path is $a = v^2/R$, where v is the velocity of the object and R is the radius of the circle. What is the centripetal acceleration of an object moving at 6 ft/s in a circular path of radius 4 ft?

31. The pressure p (in Pa) of a gas as a function of its volume V and temperature T is $p = nRT/V$. If $n = 3.00$ mol and $R = 8.31$ J/mol · K, find p for $T = 300$ K and $V = 50.0$ m^3.

32. The power P (in W) supplied to a resistance R in an electric circuit by a current i is $P = i^2 R$. Find the power supplied by a current of 4.0 A to a resistance of 2.4 Ω.

33. The atmospheric temperature T near ground level in a certain region is $T = ax^2 + by^2$, where a and b are constants. What type of curve is each isotherm (along which the temperature is constant) in this region?

34. The pressure p exerted by a force F on an area A is $p = F/A$. If a given force is doubled on an area that is 2/3 of a given area, what is the ratio of the initial pressure to the final pressure?

35. For a certain electric circuit, the current i (in A) in terms of the voltage E and resistance R (in Ω) is given by $i = \dfrac{E}{R + 0.25}$. Find the current for $E = 1.50$ V and $R = 1.20$ Ω and for $E = 1.60$ V and $R = 1.05$ Ω.

36. The reciprocal of the image distance q from a lens as a function of the object distance p and the focal length f of the lens is

$$\frac{1}{q} = \frac{1}{f} - \frac{1}{p}$$

Find the image distance of an object 20 cm from a lens whose focal length is 5 cm.

37. A rectangular solar cell panel has a perimeter p and a width w. Express the area A of the panel in terms of p and w and evaluate the area for $p = 250$ cm and $w = 55$ cm. See Fig. 11-5.

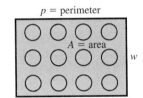

p = perimeter

A = area

w

Fig. 11-5

38. A gasoline storage tank is in the shape of a right circular cylinder with a hemisphere at each end, as shown in Fig. 11-6. Express the volume V of the tank in terms of r and h and then evaluate the volume for $r = 3.75$ ft and $h = 12.5$ ft.

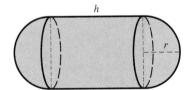

h

r

Fig. 11-6

39. The crushing load L of a pillar varies as the fourth power of its radius r and inversely as the square of its length l. Express L as a function of r and l for a pillar 20 ft tall and 1 ft in diameter that is crushed by a load of 20 tons.

40. The resonant frequency f (in Hz) of an electric circuit containing an inductance L and capacitance C is inversely proportional to the square root of the product of the inductance and the capacitance. If the resonant frequency of a circuit containing a 4-H inductor and a 64-μF capacitor is 10 Hz, express f as a function of L and C.

11-2 CURVES AND SURFACES IN THREE DIMENSIONS

We will now undertake a brief description of the graphical representation of a function of two variables. We shall show first a method of representation in the rectangular coordinate system in two dimensions. The following example illustrates the method.

■EXAMPLE 1   In order to represent $z = 2x^2 + y^2$, we will assume various values of z and sketch the resulting equation in the xy-plane. For example, if $z = 2$ we have

$$2x^2 + y^2 = 2$$

We recognize this as an ellipse with its major axis along the y-axis and vertices at $(0, \sqrt{2})$ and $(0, -\sqrt{2})$. The ends of the minor axis are at $(1, 0)$ and $(-1, 0)$. However, the ellipse $2x^2 + y^2 = 2$ represents the function $z = 2x^2 + y^2$ only for the value of $z = 2$. If $z = 4$, we have $2x^2 + y^2 = 4$, which is another ellipse. In fact, for all positive values of z, an ellipse is the resulting curve. Negative values of z are not possible, since neither x^2 nor y^2 may be negative. Figure 11-7 shows the ellipses that are obtained by using the indicated values of z.

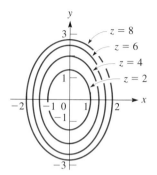

Fig. 11-7

The method of representation illustrated in Example 1 is useful if only a few specific values of z are to be used, or at least if the various curves do not intersect in such a way that they cannot be distinguished. If a general representation of z as a function of x and y is desired, it is necessary to use three coordinate axes, one each for x, y, and z. The most widely applicable system of this kind is to place a third coordinate axis at right angles to each of the x- and y-axes. In this way we employ three dimensions for the representation.

The three mutually perpendicular coordinate axes, the x-axis, the y-axis, and the z-axis are the basis of the **rectangular coordinate system in three dimensions.** Together they form three mutually perpendicular planes in space, the xy-plane, the yz-plane, and the xz-plane. To every point in space of the coordinate system is associated the set of numbers (x, y, z). The point at which the axes meet is the *origin.* The positive directions of the axes are indicated in Fig. 11-8. *That part of space in which all values of the coordinates are positive is called the first* **octant.** Numbers are not assigned to the other octants.

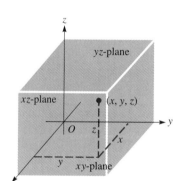

Fig. 11-8

■EXAMPLE 2   Represent the point $(2, 4, 3)$ in rectangular coordinates.

We first note that a certain distortion is necessary to represent values of x reasonably, since the x-axis "comes out" of the plane of the page. Units $\sqrt{2}/2$ ($= 0.7$) as long as those used on the other axes give a good representation. With this in mind, we draw a line 4 units long from the point $(2, 0, 0)$ on the x-axis in the xy-plane. This locates the point $(2, 4, 0)$. From this point a line 3 units long is drawn vertically upward. This locates the desired point, $(2, 4, 3)$. The point may be located by starting from $(0, 4, 0)$, proceeding 2 units *parallel* to the x-axis to $(2, 4, 0)$ and then proceeding vertically 3 units to $(2, 4, 3)$. It may also be located by starting from $(0, 0, 3)$ (see Fig. 11-9).

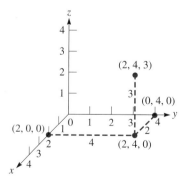

Fig. 11-9

We now show certain basic techniques by which three-dimensional figures may be drawn. We start by showing the general equation of a plane.

In Chapter 2 we showed that the graph of the equation $Ax + By + C = 0$ in two dimensions is a straight line. By the following example, we will verify that *the graph of the equation*

EQUATION OF A PLANE

$$Ax + By + Cz + D = 0$$

(11-1)

is a **plane** *in three dimensions.*

■EXAMPLE 3   Show that the graph of $2x + 3y + z - 6 = 0$ in three dimensions is a plane.

NOTE ▶   *If we let any of the three variables take on a specific value, we obtain a linear equation in the other two variables.* For example, the point $(\frac{1}{2}, 1, 2)$ satisfies the equation and therefore lies on the graph of the equation. For $x = \frac{1}{2}$, we have

$$3y + z - 5 = 0$$

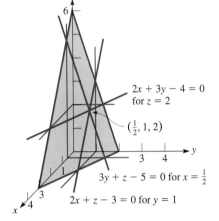

$2x + 3y - 4 = 0$
for $z = 2$

$(\frac{1}{2}, 1, 2)$

$3y + z - 5 = 0$ for $x = \frac{1}{2}$

$2x + z - 3 = 0$ for $y = 1$

Fig. 11-10

which is the equation of a straight line. This means that all pairs of values of y and z that satisfy this equation, along with $x = \frac{1}{2}$, satisfy the given equation. Thus, for $x = \frac{1}{2}$, the straight line $3y + z - 5 = 0$ lies on the graph of the equation.

For $z = 2$, we have

$$2x + 3y - 4 = 0$$

which is also a straight line. By similar reasoning this line lies on the graph of the equation. Since two lines through a point define a plane, these lines through $(\frac{1}{2}, 1, 2)$ define a plane. This plane is the graph of the equation (see Fig. 11-10).

For any point on the graph, there is a straight line on the graph that is parallel to one of the coordinate planes. Therefore, there are intersecting straight lines through the point. Thus, the graph is a plane. A similar analysis can be made for any equation of the same form.   ------------■

Since we know that the graph of an equation of the form of Eq. (11-1) is a plane, its graph can be found by determining its three intercepts, and the plane can then be represented by drawing in the lines between these intercepts. If the plane passes through the origin, by letting two of the variables in turn be zero, two straight lines that define the plane are found.

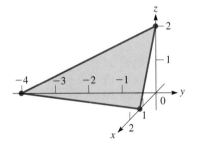

Fig. 11-11

■EXAMPLE 4   Sketch the graph of $3x - y + 2z - 4 = 0$.

The intercepts of the graph of an equation are those points where it crosses the respective axes. Thus, by letting two of the variables at a time equal zero, we obtain the intercepts. For the given equation the intercepts are $(\frac{4}{3}, 0, 0)$, $(0, -4, 0)$, and $(0, 0, 2)$. These points are located (see Fig. 11-11), and lines are drawn between them to represent the plane.   ------------■

The graph of an equation in three variables, which is essentially equivalent to a function with two independent variables, is a **surface** *in space. This is seen for the plane and will be verified for other equations in the examples that follow.*

TRACES

*The intersection of two surfaces is a **curve** in space.* This has been seen in Examples 3 and 4, since the intersections of the given planes and the coordinate planes are lines (which in the general sense are curves). *We define the **traces** of a surface to be the curves of intersection of the surface and the coordinate planes.* The traces of a plane are those lines drawn between the intercepts to represent the plane. Many surfaces may be sketched by finding their traces and intercepts.

■EXAMPLE 5   Find the intercepts and traces, and sketch the graph of the equation $z = 4 - x^2 - y^2$.

The intercepts of the graph of the equation are $(2, 0, 0)$, $(-2, 0, 0)$, $(0, 2, 0)$, $(0, -2, 0)$, and $(0, 0, 4)$.

Since the traces of a surface lie within the coordinate planes, for each trace one of the variables is zero. Thus, by ***letting each variable in turn be zero, we find the trace of the surface in the plane of the other two variables.*** Therefore, the traces of this surface are

in the yz-plane: $z = 4 - y^2$ (a parabola)

in the xz-plane: $z = 4 - x^2$ (a parabola)

in the xy-plane: $x^2 + y^2 = 4$ (a circle)

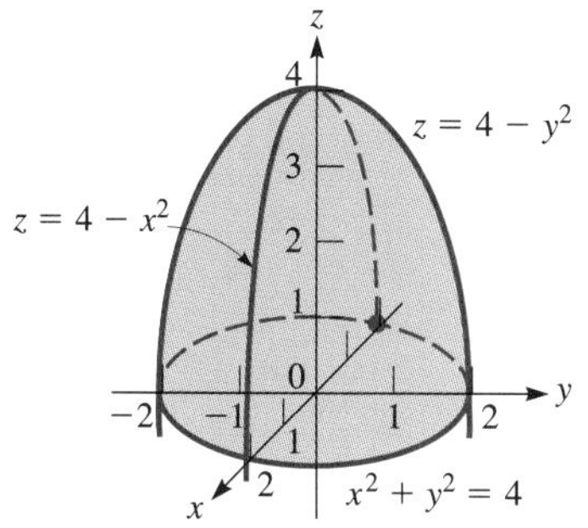

Using the intercepts and sketching the traces, we obtain the surface represented by the equation as shown in Fig. 11-12. This figure is called a **circular paraboloid.**■

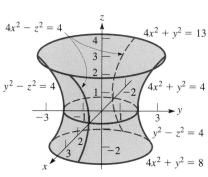

Fig. 11-12

There are numerous techniques for analyzing the equation of a surface in order to obtain its graph. Another which we shall discuss here, which is closely associated with a trace, is that of a **section.** *By assuming a specific value of one of the variables, we obtain an equation in two variables, the graph of which lies in a plane parallel to the coordinate plane of the two variables.* The following example illustrates sketching a surface by use of intercepts, traces, and sections.

SECTION

■EXAMPLE 6   Sketch the graph of $4x^2 + y^2 - z^2 = 4$.

The intercepts are $(1, 0, 0)$, $(-1, 0, 0)$, $(0, 2, 0)$, and $(0, -2, 0)$. We note that there are no intercepts on the z-axis, for this would necessitate $z^2 = -4$.

The traces are

in the yz-plane: $y^2 - z^2 = 4$ (a hyperbola)

in the xz-plane: $4x^2 - z^2 = 4$ (a hyperbola)

in the xy-plane: $4x^2 + y^2 = 4$ (an ellipse)

The surface is reasonably defined by these curves, but by assuming suitable values of z we may indicate its shape better. For example, if $z = 3$, we have $4x^2 + y^2 = 13$, which is an ellipse. In using it we must remember that it is valid for $z = 3$ and therefore should be drawn 3 units above the xy-plane. If $z = -2$, we have $4x^2 + y^2 = 8$, which is also an ellipse. Thus, we have the following sections:

for $z = 3$: $4x^2 + y^2 = 13$ (an ellipse)

for $z = -2$: $4x^2 + y^2 = 8$ (an ellipse)

Other sections could be found, but these are sufficient to obtain a good sketch of the graph (see Fig. 11-13). The figure is called an **elliptic hyperboloid.**■

Fig. 11-13

Having developed the rectangular coordinate system in three dimensions, we can compare the graph of a function using two dimensions and three dimensions. The next example shows the surface for the function of Example 1.

EXAMPLE 7 Sketch the graph of $z = 2x^2 + y^2$.
The only intercept is $(0, 0, 0)$. The traces are

in the yz-plane: $z = y^2$ (a parabola)

in the xz-plane: $z = 2x^2$ (a parabola)

in the xy-plane: the origin

The trace in the xy-plane is only the point at the origin, since $2x^2 + y^2 = 0$ may be written as $y^2 = -2x^2$, which is true only for $x = 0$ and $y = 0$.

To get a better graph we should use some positive values for z. As we noted in Example 1, negative values of z cannot be used. Since we used $z = 2$, $z = 4$, $z = 6$, and $z = 8$ in Example 1, we shall use these values here. Therefore,

for $z = 2$: $2x^2 + y^2 = 2$ for $z = 4$: $2x^2 + y^2 = 4$

for $z = 6$: $2x^2 + y^2 = 6$ for $z = 8$: $2x^2 + y^2 = 8$

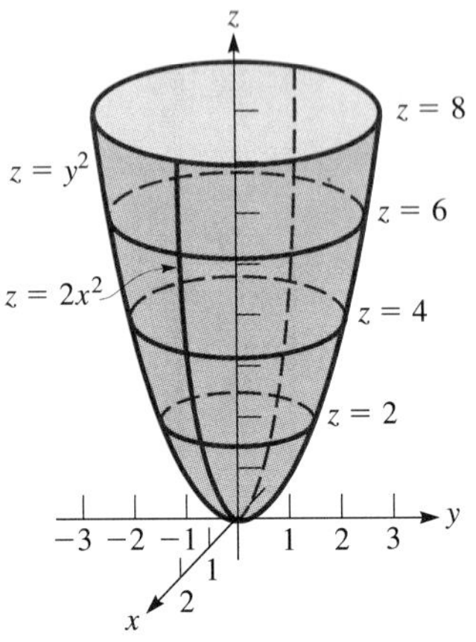

Each of these sections is an ellipse. The surface, called an **elliptic paraboloid,** is shown in Fig. 11-14. Carefully, note that the elliptical sections shown in Fig. 11-14 are equivalent to the ellipses shown in Fig. 11-7.

NOTE ▶

Fig. 11-14

Example 7 illustrates how *topographic maps* may be drawn. These maps represent three-dimensional terrain in two dimensions. For example, if Fig. 11-14 represents an excavation in the surface of the earth, then Fig. 11-7 represents the curves of constant elevation, or *contours*, with equally spaced elevations measured from the bottom of the excavation.

An equation with only two variables may represent a surface in space. Since only two variables are included in the equation, the surface is independent of the other variable. Another interpretation is that all sections, for all values of the variable not included, are the same. That is, *all sections parallel to the coordinate plane of the included variables are the same as the trace in that plane.*

NOTE ▶

EXAMPLE 8 Sketch the graph of $x + y = 2$ in the rectangular coordinate system in three dimensions and in two dimensions.

Since z does not appear in the equation, we can consider the equation to be $x + y + 0z = 2$. Therefore, we see that *for any value of z the section is the straight line $x + y = 2$.* Thus, the graph is a plane as shown in Fig. 11-15(a). The graph as a straight line in two dimensions is shown in Fig. 11-15(b).

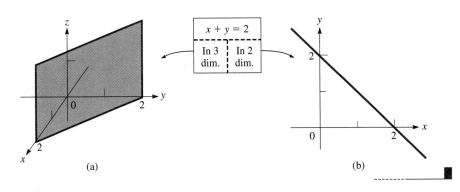

Fig. 11-15   (a)   (b)

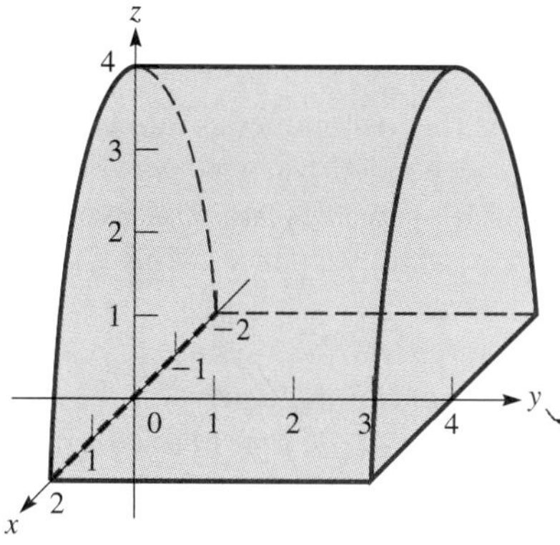

Fig. 11-16

EXAMPLE 9   The graph of the equation $z = 4 - x^2$ in three dimensions is a surface whose sections, for all values of y, are given by the parabola $z = 4 - x^2$. The surface is shown in Fig. 11-16.

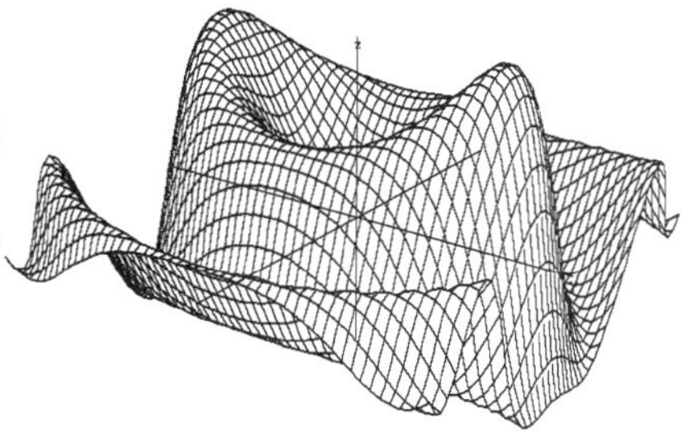

Fig. 11-17

The surface in Example 9 is known as a **cylindrical surface.** In general, *a cylindrical surface is one that can be generated by a line moving parallel to a fixed line while passing through a plane curve.*

It must be realized that most of the figures extend beyond the ranges indicated by the traces and sections. However, these traces and sections are convenient for representing and visualizing these surfaces.

There are various computer programs (and some graphing calculators) that can be used to display three-dimensional surfaces. They generally use sections in planes that are perpendicular to both the x- and y-axes. Such computer-drawn surfaces are of great value for all types of surfaces, especially those of a complex nature.

Figure 11-17 shows the graph of $z = \dfrac{\sin(2x^2 + y^2)}{x^2 + 1}$ drawn by a computer program.

— EXERCISES *11-2* —

In Exercises 1–4, use the method of Example 1 and draw graphs of the indicated equations for the given values of z.

1. $z = x^2 + y^2, z = 1, z = 4, z = 9$

2. $z = x - 2y, z = -2, z = 2, z = 5$

3. $z = y - x^2, z = 0, z = 2, z = 4$

4. $z = x^2 - 4y^2, z = -4, z = 1, z = 4$

In Exercises 5–24, sketch the graphs of the given equations in the rectangular coordinate system in three dimensions.

5. $x + y + 2z - 4 = 0$ **6.** $2x - y - z + 6 = 0$

7. $4x - 2y + z - 8 = 0$ **8.** $3x + 3y - 2z - 6 = 0$

9. $z = y - 2x - 2$ **10.** $z = x - 4y$

11. $x + 2y = 4$ **12.** $2x - 3z = 6$

13. $x^2 + y^2 + z^2 = 4$ **14.** $2x^2 + 2y^2 + z^2 = 8$

15. $z = 4 - 4x^2 - y^2$ **16.** $z = x^2 + y^2$

17. $z = 2x^2 + y^2 + 2$ **18.** $x^2 + y^2 - 4z^2 = 4$

19. $x^2 - y^2 - z^2 = 9$ **20.** $z^2 = 9x^2 + 4y^2$

21. $x^2 + y^2 = 16$ **22.** $4z = x^2$

23. $y^2 + 9z^2 = 9$ **24.** $xy = 2$

In Exercises 25–32, sketch the indicated curves and surfaces.

25. Curves that represent a constant temperature are called *isotherms.* The temperature at a point (x, y) of a flat plate is t (°C), where $t = 4x - y^2$. In two dimensions draw the isotherms for $t = -4, 0, 8$.

26. At a point (x, y) in the xy-plane, the electric potential V (in volts) is given by $V = y^2 - x^2$. Draw the lines of equal potential for $V = -9, 0, 9$.

27. An electric charge is so distributed that the electric potential at all points on an imaginary surface is the same. Such a surface is called an *equipotential surface.* Sketch the graph of the equipotential surface whose equation is $2x^2 + 2y^2 + 3z^2 = 6$.

28. The surface of a small hill can be roughly approximated by the equation $z(2x^2 + y^2 + 100) = 1500$, where the units are meters. Draw the surface of the hill and the contours for $z = 3$ m, $z = 6$ m, $z = 9$ m, $z = 12$ m, and $z = 15$ m.

29. The pressure p (in kPa), volume V (in m³), and temperature T (in K) for a certain gas are related by the equation $p = T/2V$. Sketch the p-V-T surface by using the z-axis for p, the x-axis for V, and the y-axis for T. Use units of 100 K for T and 10 m³ for V. Sections must be used for this surface, *a thermodynamic surface,* since none of the variables may equal zero.

30. Sketch the line in space defined by the intersection of the planes $x + 2y + 3z - 6 = 0$ and $2x + y + z - 4 = 0$.

31. Sketch the graph of $x^2 + y^2 - 2y = 0$ in three dimensions and in two dimensions.

32. Sketch the curve in space defined by the intersection of the surfaces $x^2 + (z - 1)^2 = 1$ and $z = 4 - x^2 - y^2$.

11-3 PARTIAL DERIVATIVES

In Chapter 3, when we showed that the derivative is the instantaneous rate of change of one variable with respect to another, only one independent variable was involved. To extend the derivative to functions of two (or more) variables, we find the derivative of the function with respect to one of the independent variables, while the other is held constant.

If $z = f(x, y)$ and y is held constant, z becomes a function of x alone. The derivative of this function with respect to x is termed the **partial derivative** *of z with respect to x. Similarly, if x is held constant, the derivative of the function with respect to y is the* **partial derivative** *of z with respect to y.*

For the function $z = f(x, y)$, the notations used for the partial derivative of z with respect to x include

The symbol ∂ was introduced by the German mathematician Carl Jacobi (1804–1851).

$$\frac{\partial z}{\partial x} \qquad \frac{\partial f}{\partial x} \qquad f_x \qquad \frac{\partial}{\partial x} f(x, y) \qquad f_x(x, y)$$

Similarly, $\partial z/\partial y$ denotes the partial derivative of z with respect to y. In speaking, this is often shortened to "the partial of z with respect to y."

EXAMPLE 1 If $z = 4x^2 + xy - y^2$, find $\partial z/\partial x$ and $\partial z/\partial y$.

To find the partial derivative of z with respect to x, we treat y as a constant.

$$z = 4x^2 + xy - y^2 \qquad \text{treat as constant}$$

$$\frac{\partial z}{\partial x} = 8x + y$$

To find the partial derivative of z with respect to y, we treat x as a constant.

$$z = 4x^2 + xy - y^2 \qquad \text{treat as constant}$$

$$\frac{\partial z}{\partial y} = x - 2y$$

EXAMPLE 2 If $z = \dfrac{x \ln y}{x^2 + 1}$, find $\partial z/\partial x$ and $\partial z/\partial y$.

$$\frac{\partial z}{\partial x} = \frac{(x^2 + 1)(\ln y) - (x \ln y)(2x)}{(x^2 + 1)^2} = \frac{(1 - x^2)\ln y}{(1 + x^2)^2}$$

$$\frac{\partial z}{\partial y} = \left(\frac{x}{x^2 + 1}\right)\left(\frac{1}{y}\right) = \frac{x}{y(x^2 + 1)}$$

We note that in finding $\partial z/\partial x$ it is necessary to use the quotient rule, since x appears in both numerator and denominator. However, when finding $\partial z/\partial y$, the only derivative needed is that of $\ln y$.

EXAMPLE 3 For the function $f(x, y) = x^2 y\sqrt{2 + xy^2}$, find $f_y(2, 1)$.

The notation $f_y(2, 1)$ means the partial derivative of f with respect to y, evaluated for $x = 2$ and $y = 1$. Thus, first finding $f_y(x, y)$, we have

$$f(x, y) = x^2 y(2 + xy^2)^{1/2}$$

$$f_y(x, y) = x^2 y\left(\frac{1}{2}\right)(2 + xy^2)^{-1/2}(2xy) + (2 + xy^2)^{1/2}(x^2)$$

$$= \frac{x^3 y^2}{(2 + xy^2)^{1/2}} + x^2(2 + xy^2)^{1/2}$$

$$= \frac{x^3 y^2 + x^2(2 + xy^2)}{(2 + xy^2)^{1/2}} = \frac{2x^2 + 2x^3 y^2}{(2 + xy^2)^{1/2}}$$

$$f_y(2, 1) = \frac{2(4) + 2(8)(1)}{(2 + 2)^{1/2}} = 12$$

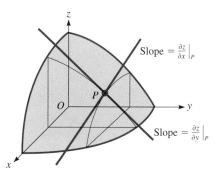

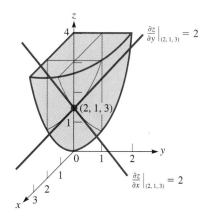

Fig. 11-18

To determine the geometric interpretation of a partial derivative, assume that $z = f(x, y)$ is the surface shown in Fig. 11-18. Choosing a point P on the surface, we then draw a plane through P parallel to the xz-plane. On this plane through P, the value of y is constant. The intersection of this plane and the surface is the curve as indicated. *The partial derivative of z with respect to x represents the slope of a line tangent to this curve.* When the values of the coordinates of point P are substituted into the expression for this partial derivative, it gives the slope of the tangent line at that point. In the same way, the partial derivative of z with respect to y, evaluated at P, gives the slope of the line tangent to the curve that is found from the intersection of the surface and the plane parallel to the yz-plane through P.

EXAMPLE 4 Find the slope of a line tangent to the surface $2z = x^2 + 2y^2$ and parallel to the xz-plane at the point $(2, 1, 3)$. Also, find the slope of a line tangent to this surface and parallel to the yz-plane at the same point.

Finding the partial derivative with respect to x, we have

$$2\frac{\partial z}{\partial x} = 2x \quad \text{or} \quad \frac{\partial z}{\partial x} = x$$

This derivative, evaluated at the point $(2, 1, 3)$, will give us the slope of the line tangent that is also parallel to the xz-plane. Therefore, the first required slope is

$$\left.\frac{\partial z}{\partial x}\right|_{(2,1,3)} = 2$$

The partial derivative of z with respect to y, evaluated at $(2, 1, 3)$, will give us the second required slope. Thus,

$$\frac{\partial z}{\partial y} = 2y, \quad \left.\frac{\partial z}{\partial y}\right|_{(2,1,3)} = 2$$

Therefore, both slopes are 2. See Fig. 11-19.

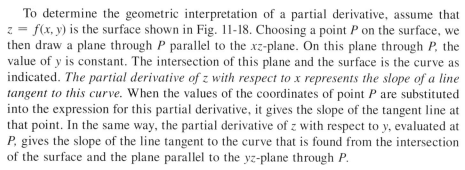

Fig. 11-19

The general interpretation of the partial derivative follows that of a derivative of a function with one independent variable. *The partial derivative $f_x(x_0, y_0)$ is the instantaneous rate of change of the function $f(x, y)$ with respect to x, with y held constant at the value of y_0.* This holds regardless of what the variables represent.

Applications of partial derivatives are found in many fields of technology. We show here an application from electricity, and others are found in the exercises.

■**EXAMPLE 5** An electric circuit in a television transmitter has parallel resistances r and R. The current through r can be found from

$$i = \frac{IR}{r + R}$$

where I is the total current for the two branches. Assuming that I is constant at 85.4 mA, find $\partial i/\partial r$ and evaluate it for $R = 0.150\ \Omega$ and $r = 0.032\ \Omega$.

Substituting for I and finding the partial derivative, we have the following:

$$i = \frac{85.4R}{r + R} = 85.4R(r + R)^{-1}$$

$$\frac{\partial i}{\partial r} = (-1)(85.4R)(r + R)^{-2}(1) = \frac{-85.4R}{(r + R)^2}$$

$$\frac{\partial i}{\partial r}\bigg|_{\substack{r=0.032 \\ R=0.150}} = \frac{-85.4(0.150)}{(0.032 + 0.150)^2} = -387\ \text{mA}/\Omega$$

This result tells us that the current is decreasing at the rate of 387 mA per ohm of change in the smaller resistor at the instant when $r = 0.032\ \Omega$, for a fixed value of $R = 0.150\ \Omega$. ■

Since the partial derivatives $\partial f/\partial x$ and $\partial f/\partial y$ are functions of x and y, we can take partial derivatives of each of them. This gives rise to **partial derivatives of higher order,** in a manner similar to the higher derivatives of a function of one independent variable. *The possible* **second-order partial derivatives** *of a function $f(x, y)$ are*

$$\frac{\partial^2 f}{\partial x^2} = \frac{\partial}{\partial x}\left(\frac{\partial f}{\partial x}\right) \qquad \frac{\partial^2 f}{\partial y^2} = \frac{\partial}{\partial y}\left(\frac{\partial f}{\partial y}\right)$$

$$\frac{\partial^2 f}{\partial x\, \partial y} = \frac{\partial}{\partial x}\left(\frac{\partial f}{\partial y}\right) \qquad \frac{\partial^2 f}{\partial y\, \partial x} = \frac{\partial}{\partial y}\left(\frac{\partial f}{\partial x}\right)$$

■**EXAMPLE 6** Find the second-order partial derivatives of $z = x^3y^2 - 3xy^3$. First, we find $\partial z/\partial x$ and $\partial z/\partial y$:

$$\frac{\partial z}{\partial x} = 3x^2y^2 - 3y^3 \qquad \frac{\partial z}{\partial y} = 2x^3y - 9xy^2$$

Therefore, we have the following second-order partial derivatives:

$$\frac{\partial^2 z}{\partial x^2} = \frac{\partial}{\partial x}\left(\frac{\partial z}{\partial x}\right) = 6xy^2 \qquad\qquad \frac{\partial^2 z}{\partial y^2} = \frac{\partial}{\partial y}\left(\frac{\partial z}{\partial y}\right) = 2x^3 - 18xy$$

$$\frac{\partial^2 z}{\partial x\, \partial y} = \frac{\partial}{\partial x}\left(\frac{\partial z}{\partial y}\right) = 6x^2y - 9y^2 \qquad \frac{\partial^2 z}{\partial y\, \partial x} = \frac{\partial}{\partial y}\left(\frac{\partial z}{\partial x}\right) = 6x^2y - 9y^2$$ ■

Television was invented in the 1920s and first used commercially in the 1940s.

In Example 6 we note that

$$\frac{\partial^2 z}{\partial x\, \partial y} = \frac{\partial^2 z}{\partial y\, \partial x} \qquad\qquad \textbf{(11-2)}$$

In general, this is true if the function and partial derivatives are continuous.

EXAMPLE 7   For $f(x, y) = \tan^{-1}\dfrac{y}{x^2}$, show that $\dfrac{\partial^2 f}{\partial x\, \partial y} = \dfrac{\partial^2 f}{\partial y\, \partial x}$.

Finding $\partial f/\partial x$ and $\partial f/\partial y$, we have

$$\frac{\partial f}{\partial x} = \frac{1}{1 + \left(\dfrac{y}{x^2}\right)^2}\left(\frac{-2y}{x^3}\right) = \frac{x^4}{x^4 + y^2}\left(-\frac{2y}{x^3}\right) = \frac{-2xy}{x^4 + y^2}.$$

$$\frac{\partial f}{\partial y} = \frac{1}{1 + \left(\dfrac{y}{x^2}\right)^2}\left(\frac{1}{x^2}\right) = \frac{x^4}{x^4 + y^2}\left(\frac{1}{x^2}\right) = \frac{x^2}{x^4 + y^2}$$

Now, finding $\partial^2 f/\partial x\, \partial y$ and $\partial^2 f/\partial y\, \partial x$, we have

$$\frac{\partial^2 f}{\partial x\, \partial y} = \frac{(x^4 + y^2)(2x) - x^2(4x^3)}{(x^4 + y^2)^2} = \frac{-2x^5 + 2xy^2}{(x^4 + y^2)^2}$$

$$\frac{\partial^2 f}{\partial y\, \partial x} = \frac{(x^4 + y^2)(-2x) - (-2xy)(2y)}{(x^4 + y^2)^2} = \frac{-2x^5 + 2xy^2}{(x^4 + y^2)^2}$$

We see that they are equal.　　　　　　　　　　　　　　■

EXERCISES *11-3*

In Exercises 1–24, find the partial derivative of the dependent variable or function with respect to each of the independent variables.

1. $z = 5x + 4x^2 y$

2. $z = 3x^2 y^3 - 3x + 4y$

3. $z = \dfrac{x^2}{y} - 2xy$

4. $f(x, y) = \dfrac{x + y}{x - y}$

5. $f(x, y) = xe^{2y}$

6. $z = 3y \cos 2x$

7. $f(x, y) = \dfrac{2 + \cos x}{1 - \sec 3y}$

8. $f(x, y) = \dfrac{\tan^{-1} y}{2 + x^2}$

9. $\phi = r\sqrt{1 + 2rs}$

10. $w = uv^2\sqrt{1 - u}$

11. $z = (x^2 + xy^3)^4$

12. $f(x, y) = (2xy - x^2)^5$

13. $z = \sin xy$

14. $y = \tan^{-1}\left(\dfrac{x}{t}\right)$

15. $y = \ln(r^2 + s)$

16. $u = e^{x + 2y}$

17. $f(x, y) = \dfrac{2\sin^3 2x}{1 - 3y}$

18. $f(x, y) = \dfrac{3x + \ln y}{x^2 + y^2}$

19. $z = \dfrac{\sin^{-1} xy}{3 + x^2}$

20. $z = \dfrac{\sqrt{1 - \tan xy}}{xy + y^2}$

21. $z = \sin x + \cos xy - \cos y$

22. $t = 2re^{rs^2} - \tan(2r + s)$

23. $f(x, y) = e^x \cos xy + e^{-2x}\tan y$

24. $u = \ln\dfrac{y^2}{x - y} + e^{-x}(\sin y - \cos 2y)$

In Exercises 25–28, evaluate the indicated partial derivatives at the given points.

25. $z = 3xy - x^2$, $\left.\dfrac{\partial z}{\partial x}\right|_{(1, -2, -7)}$

26. $z = x^2 \cos 4y$, $\left.\dfrac{\partial z}{\partial y}\right|_{(2, \frac{\pi}{2}, 4)}$

27. $z = x\sqrt{x^2 - y^2}$, $\left.\dfrac{\partial z}{\partial x}\right|_{(5, 3, 20)}$

28. $z = e^y \ln xy$, $\left.\dfrac{\partial z}{\partial y}\right|_{(e, 1, e)}$

In Exercises 29–32, find all second partial derivatives.

29. $z = 2xy^3 - 3x^2 y$

30. $F(x, y) = y \ln(x + 2y)$

31. $z = \dfrac{x}{y} + e^x \sin y$

32. $f(x, y) = \dfrac{2 + \cos y}{1 + x^2}$

In Exercises 33–48, solve the given problems.

33. Find the slope of a line tangent to the surface $z = 9 - x^2 - y^2$ and parallel to the *yz*-plane that passes through (1, 2, 4). Repeat the instructions for the line through (2, 2, 1). Draw an appropriate figure.

34. A metal plate in the shape of a circular segment of radius *r* expands by being heated. Express the width *w* (straight dimension) as a function of *r* and the height *h*. Then find both $\partial w / \partial r$ and $\partial w / \partial h$.

35. Two resistors R_1 and R_2, placed in parallel, have a combined resistance R_T given by $\dfrac{1}{R_T} = \dfrac{1}{R_1} + \dfrac{1}{R_2}$. Find $\dfrac{\partial R_T}{\partial R_1}$.

Ⓦ 36. Find $\partial z / \partial y$ for the function $z = 4x^2 - 8$. Explain your result. Draw an appropriate figure.

37. A metallic machine part contracts while cooling. It is in the shape of a hemisphere attached to a cylinder, as shown in Fig. 11-20. Find the rate of change of volume with respect to *r* when $r = 2.65$ cm and $h = 4.20$ cm.

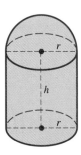

Fig. 11-20

38. Two masses *M* and *m* are attached as shown in Fig. 11-21. If $M > m$, the downward acceleration *a* of mass *M* is given by $a = \dfrac{M - m}{M + m} g$, where *g* is the acceleration due to gravity. Show that $M \dfrac{\partial a}{\partial M} + m \dfrac{\partial a}{\partial m} = 0$.

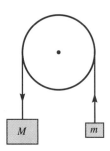

Fig. 11-21

39. In quality testing, a rectangular sheet of vinyl is stretched. Set up the length of the diagonal *d* of the sheet as a function of the sides *x* and *y*. Find the rate of change of *d* with respect to *x* for $x = 6.50$ ft if *y* remains constant at 4.75 ft.

40. If an observer and a source of sound are moving toward or away from each other, the observed frequency of sound is different from that emitted. This is known as the *Doppler effect*. The equation relating the frequency f_0 the observer hears and the frequency f_s emitted by the source (a constant) is $f_0 = f_s \left(\dfrac{v + v_0}{v - v_s} \right)$, where *v* is the velocity of sound in air (a constant), v_0 is the velocity of the observer, and v_s is the velocity of the source. Show that $f_s \dfrac{\partial f_0}{\partial v_s} = f_0 \dfrac{\partial f_0}{\partial v_0}$.

41. The *mutual conductance* (in $1/\Omega$) of a certain electronic device is defined as $g_m = \partial i_b / \partial e_c$. Under certain circumstances, the current i_b (in μA) is given by $i_b = 50(e_b + 5e_c)^{1.5}$. Find g_m when $e_b = 200$ V and $e_c = -20$ V.

42. The *amplification factor* of the electronic device of Exercise 41 is defined as $\mu = -\partial e_b / \partial e_c$. For the device of Exercise 41, under the given conditions, find the amplification factor.

43. The *coefficient of linear expansion* α of a wire whose length *L* is a function of the tension and temperature *T* is given by $\alpha = \dfrac{1}{L} \left(\dfrac{\partial L}{\partial T} \right)$. If *L* is a function of *T* and the tension *F* given by $L = L_0 + k_1 F + k_2 T + k_3 FT^2$, where k_1, k_2, and k_3 are constants, find the expression for α.

44. The fundamental frequency of vibration *f* of a string varies directly as the square root of the tension *T* and inversely as the length *L*. If a string 60 cm long is under a tension of 65 N and has a fundamental frequency of 30 Hz, find the partial derivative of *f* with respect to *T* and evaluate it for the given values.

45. An *isothermal process* is one during which the temperature does not change. If the volume *V*, pressure *p*, and temperature *T* of an ideal gas are related by the equation $pV = nRT$, where *n* and *R* are constants, find the expression for $\partial p / \partial V$, which is the rate of change of pressure with respect to volume for an isothermal process.

46. For the ideal gas of Exercise 45, show that $\left(\dfrac{\partial V}{\partial T} \right) \left(\dfrac{\partial T}{\partial p} \right) \left(\dfrac{\partial p}{\partial V} \right) = -1$.

47. The temperature *u* in a metal bar depends on the distance *x* from one end and the time *t*. Show that $u(x, t) = 5e^{-t} \sin 4x$ satisfies the *one-dimensional heat conduction equation* $\dfrac{\partial u}{\partial t} = k \dfrac{\partial^2 u}{\partial x^2}$, where *k* is called the *diffusivity*. In this case $k = 1/16$.

48. The displacement *y* at any point in a taut, flexible string depends on the distance *x* from one end of the string and the time *t*. Show that $y(x, t) = 2 \sin 2x \cos 4t$ satisfies the *wave equation* $\dfrac{\partial^2 y}{\partial t^2} = a^2 \dfrac{\partial^2 y}{\partial x^2}$ with $a = 2$.

$11\text{-}4$ CERTAIN APPLICATIONS OF PARTIAL DERIVATIVES

See Appendix A-2 for an application of partial derivatives in finding the equation of a curve to fit a set of data points.

We noted some important areas of application of partial derivatives in the last section. However, just as we found that we can apply the derivative to problems such as related rates, maximum and minimum problems, and error calculations by the use of differentials, we can also solve similar problems by the use of partial derivatives. In this section we shall show how to calculate the differential of a function of two variables and use it in certain applied situations. Then we will discuss applied maximum and minimum problems for functions of two variables.

In Section 4-8 the differential of a function of one independent variable was defined to be $dy = f'(x)\, dx$. In a similar manner, *we define the* **total differential** *of a function $z = f(x, y)$ to be*

$$dz = \frac{\partial z}{\partial x}\, dx + \frac{\partial z}{\partial y}\, dy \tag{11-3}$$

It will be noted that this definition is consistent with our original definition of the differential, in that if y is constant, $dz = (\partial z/\partial x)\, dx$. Also, is x is constant, we have $dz = (\partial z/\partial y)\, dy$. Therefore, each term measures the differential of z with respect to one of the independent variables, assuming the other is constant.

EXAMPLE 1   Find the expression for the total differential of the function $z = x^3 y + x^2 y^2 - 3xy^3$.

$$\frac{\partial z}{\partial x} = 3x^2 y + 2xy^2 - 3y^3$$

$$\frac{\partial z}{\partial y} = x^3 + 2x^2 y - 9xy^2$$

Therefore,

$$dz = (3x^2 y + 2xy^2 - 3y^3)\, dx + (x^3 + 2x^2 y - 9xy^2)\, dy$$

EXAMPLE 2   Find the total differential of $z = e^{-x} \cos 2y - \ln xy$.
First finding $\partial z/\partial x$ and $\partial z/\partial y$, we have

$$\frac{\partial z}{\partial x} = (-e^{-x}) \cos 2y - \frac{y}{xy} = -e^{-x} \cos 2y - \frac{1}{x}$$

$$\frac{\partial z}{\partial y} = e^{-x}(-\sin 2y)(2) - \frac{x}{xy} = -2e^{-x} \sin 2y - \frac{1}{y}$$

Therefore,

$$dz = -\left(e^{-x} \cos 2y + \frac{1}{x}\right) dx - \left(2e^{-x} \sin 2y + \frac{1}{y}\right) dy$$

We found that we could apply the differential of a function of one variable to finding changes in the value of the function for small changes in the independent variable. This led to applied problems in error calculations. We may also apply the total differential of a function of two variables in the same manner. The following examples illustrate the method.

■**EXAMPLE 3**   The power P (in W) delivered to a resistor of resistance R (in Ω) in an electric circuit is given by $P = i^2R$, where i is the current (in A). Find the approximate change in power if the current changes from 4.0 A to 4.1 A and the resistance changes from 22.0 Ω to 22.4 Ω.

By finding dP and evaluating it for $i = 4.0$ A, $di = 0.1$ A, $R = 22.0$ Ω, and $dR = 0.4$ Ω, we will have the desired result.

$$dP = \frac{\partial P}{\partial i}\, di + \frac{\partial P}{\partial R}\, dR$$
$$= (2iR)\, di + (i^2)\, dR$$

Evaluating, we have

$$dP = [2(4.0)\,(22.0)]\,(0.1) + (4.0^2)\,(0.4)$$
$$= 17.6 + 6.4 = 24.0 \text{ W}$$

Therefore, the approximate change in the power is 24 W.   ■

■**EXAMPLE 4**   A conical funnel, with a very small opening, was to be made such that the diameter across the top was 10.0 cm and the depth was 6.0 cm. However, the diameter was made only 9.8 cm and the depth 5.8 cm. What was the resulting percentage error in the volume of liquid the funnel will hold? See Fig. 11-22.

The volume of a cone as a function of its radius r and height h is

$$V = \frac{1}{3}\,\pi r^2 h$$

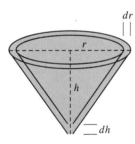

Fig. 11-22

Finding dV, we have

$$dV = \left(\frac{2}{3}\,\pi r h\right) dr + \left(\frac{1}{3}\,\pi r^2\right) dh$$

Using the values $r = 5.0$ cm, $dr = -0.1$ cm, $h = 6.0$ cm, and $dh = -0.2$ cm, we have

$$dV = \frac{2}{3}\,\pi(5.0)\,(6.0)\,(-0.1) + \frac{1}{3}\,\pi(25)\,(-0.2)$$
$$= -2.0\pi - \frac{1}{3}(5.0\pi) = -11.5 \text{ cm}^3$$

Calculating the intended volume V, we have

$$V = \frac{1}{3}\,\pi(25)\,(6) = 50\pi = 157 \text{ cm}^3$$

The percentage error in the volume is

$$100\left(\frac{dV}{V}\right) = 100\left(\frac{-11.5}{157}\right) = -7.3\%$$ ■

We now consider the problem of finding the relative maximum value or the relative minimum value of a function of two variables. Considering the geometric interpretation of the relative maximum value of a function of two variables, we observe the surface in Fig. 11-23(a). Point M is a relative maximum point, which means the **NOTE ▶** value of M is greater than at any other point very near M *in any direction*. It does not mean the value of the function is greater at M than at any point *anywhere* on the surface.

At the relative maximum point M, a line tangent to the surface, *in any direction*, must be horizontal. This means that *at a **relative maximum** point*

$$\frac{\partial z}{\partial x} = 0 \quad \text{and} \quad \frac{\partial z}{\partial y} = 0 \qquad \qquad \textbf{(11-4)}$$

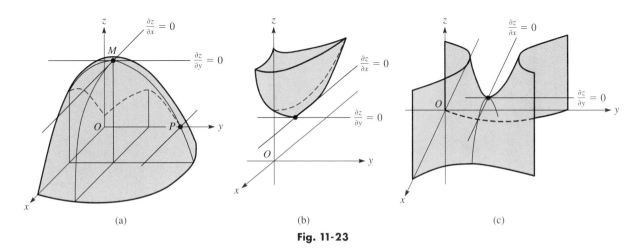

Fig. 11-23

CAUTION ▶ From Fig. 11-23(b) we see that Eqs. (11-4) also hold at a relative minimum point. It should be noted that *it is not sufficient that just one of the partial derivatives in Eqs. (11-4) equal zero* to have a relative maximum or relative minimum point. This is seen by examining point P in Fig. 11-23(a), where $\partial z/\partial x = 0$, but obviously $\partial z/\partial y \neq 0$.

It is possible that Eqs. (11-4) are satisfied but that there is neither a relative maximum nor a relative minimum. See Fig. 11-23(c), where a **saddle point** is illustrated. Therefore, Eqs. (11-4) do not provide a complete analysis. Although the proof is beyond the scope of this book, we now state the necessary conditions, which are analogous to the second-derivative test for functions of one independent variable.

If, for a function $f(x, y)$ at a point (a, b),

$$\frac{\partial f}{\partial x} = 0 \quad \text{and} \quad \frac{\partial f}{\partial y} = 0$$

and also

$$D = \frac{\partial^2 f}{\partial x^2} \frac{\partial^2 f}{\partial y^2} - \left(\frac{\partial^2 f}{\partial x \, \partial y}\right)^2$$

*then there is a **relative maximum** at (a, b) if $D > 0$ and $\partial^2 f/\partial x^2 < 0$, or a **relative minimum** at (a, b) if $D > 0$ and $\partial^2 f/\partial x^2 > 0$. If $D < 0$, there is neither a maximum nor a minimum, and if $D = 0$, the test fails (no conclusion can be drawn).*

■EXAMPLE 5   Examine the function $z = x^2 + 2xy + 2y^2 - 4x$ for relative maximum and minimum points.

First, finding the partial derivatives, we have

$$\frac{\partial z}{\partial x} = 2x + 2y - 4 \quad \text{and} \quad \frac{\partial z}{\partial y} = 2x + 4y$$

Setting these expressions equal to zero, we obtain the equations

$$2x + 2y - 4 = 0$$
$$2x + 4y = 0$$

Since *these equations must hold at the same point,* we solve them simultaneously. This gives us $x = 4$ and $y = -2$.

Now finding the second derivatives, we have

$$\frac{\partial^2 z}{\partial x^2} = 2 \qquad \frac{\partial^2 z}{\partial y^2} = 4 \qquad \frac{\partial^2 z}{\partial x\,\partial y} = 2$$

This means that

$$D = 2(4) - 2^2 = 4$$

Since $D > 0$ and $\partial^2 z/\partial x^2 > 0$, there is a relative minimum point at $(4, -2, -8)$.

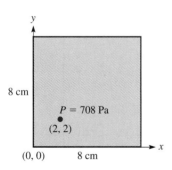

(0, 0)   8 cm

Fig. 11-24

■EXAMPLE 6   A computer analysis shows the pressure P (in Pa), created by a press at each point of a square plate 8 cm on a side, is $P = 6xy - x^3 - y^3 + 700$. Find the greatest pressure at any point, if distances are measured as shown in Fig. 11-24.

Finding the partial derivatives of P, we have

$$\frac{\partial P}{\partial x} = 6y - 3x^2 \qquad \frac{\partial P}{\partial y} = 6x - 3y^2$$

Setting each of these equal to zero, we have

$$6y - 3x^2 = 0 \quad \text{or} \quad 2y - x^2 = 0$$
$$6x - 3y^2 = 0 \quad \text{or} \quad 2x - y^2 = 0$$

Solving the first equation for y and substituting into the second equation gives us

$$y = \frac{1}{2}x^2$$
$$2x - \left(\frac{1}{2}x^2\right)^2 = 0$$
$$8x - x^4 = x(8 - x^3) = 0$$
$$x = 0, 2$$

For $x = 0$, $y = 0$, and $x = 2$, $y = 2$. We now find the second derivatives.

$$\frac{\partial^2 P}{\partial x^2} = -6x \qquad \frac{\partial^2 P}{\partial y^2} = -6y \qquad \frac{\partial^2 P}{\partial x\,\partial y} = 6$$

At $(0, 0)$, $D = -36$, which means $(0, 0)$, a corner of the plate, is neither a relative maximum nor a relative minimum. For $y = 0$, $\partial P/\partial x = -3x^2$, which means the pressure is decreasing near $(0, 0)$. At $(2, 2)$ $D = (-12)(-12) - 6^2 = 108$, and $\partial^2 P/\partial x^2 = -12$. Thus, there is a relative maximum value of $P = 708$ Pa at $(2, 2)$.

EXAMPLE 7   A rectangular box is to be made of material such that the base costs twice as much per square inch as the sides and top. Find the dimensions of the box if it is to hold 324 in.^{3}.

Letting the length, width, and height be l, w, and h, respectively, we may write the equations for the volume and cost in terms of these quantities. The equation for the volume is $324 = lwh$. Since the cost per square inch was not actually given, let us assume that the cost for the sides and top is a cents/in.^{2}. This means the cost of the base is $2a$ cents/in.^{2}. Therefore, the expression for the total cost C is

$$C = 2a(lw) + a(lw + 2wh + 2lh)$$
$$= a(3lw + 2wh + 2lh) \tag{1}$$

Solving $lwh = 324$ for h, and substituting into Eq. (1), we express C as a function of l and w.

$$C = a\left(3lw + \frac{648}{l} + \frac{648}{w}\right) \tag{2}$$

Taking the partial derivatives of C with respect to l and w, we may find the values of l and w that give the minimum cost. From these values we then can find h:

$$\frac{\partial C}{\partial w} = a\left(3l - \frac{648}{w^2}\right)$$
$$\frac{\partial C}{\partial l} = a\left(3w - \frac{648}{l^2}\right) \tag{3}$$

Setting each of these expressions equal to zero, we obtain the following equations:

$$l = \frac{216}{w^2} \qquad w = \frac{216}{l^2}$$

Substituting the expression for l into the solution for w, we have

$$w = \frac{216}{\left(\dfrac{216}{w^2}\right)^2} = \frac{w^4}{216}$$

$$w^3 = 216, \qquad w = 6 \text{ in.}$$

Substituting, we find $l = 6$ in. and $h = 9$ in. The l and w in the denominators of Eq. (2) show the maximum cost is unbounded. Thus, the solution represents a minimum. Also, the second-derivative test will show the solution to be a minimum. ∎

— EXERCISES *11-4* —

In Exercises 1–12, find the total differential of the given functions.

1. $z = 2x^2 - y^2 + 3x$

2. $z = xy - 3yx^2 + y^3$

3. $z = xe^y - y^2$

4. $z = x^3 - y \ln x$

5. $z = x(y - 2x^2)^5$

6. $z = y^2\sqrt{1 - 2xy}$

7. $z = \sin xy - y \cos x$

8. $z = x \tan 2y - x^3$

9. $z = \dfrac{x - 3y^2}{1 - \sin y}$

10. $z = \dfrac{\sec x^2}{x - y}$

11. $z = y \tan^{-1}\dfrac{y}{x^2}$

12. $z = e^{2xy} \ln(x^2 + y^2)$

In Exercises 13–20, examine the given functions for relative maximum and minimum points.

13. $z = x^2 + y^2 - 2x - 6y + 10$

14. $z = x^2 + xy + y^2 + 3x - 3y + 1$

• **15.** $z = x^2 + xy + y^2 - 3y + 2$

16. $z = x^4 - 4x + y^2$

17. $z = 2y - xy + x^3$

• **18.** $z = xy + \dfrac{1}{x} + \dfrac{8}{y}$

• **19.** $z = x^2 + 4 - 4x \cos y$ $(0 \le x < 2\pi, 0 \le y < 2\pi)$

20. $z = x - y \ln x$

In Exercises 21–34, solve the given problems.

21. The centripetal acceleration of an object moving in a circular path is given by $\alpha = r\omega^2$, where r is the radius of the circle and ω is the angular velocity. Find the approximate change in the centripetal acceleration if the radius changes from 10.0 cm to 10.3 cm and the angular velocity changes from 400 rad/min to 406 rad/min.

22. A voltage V is across a resistance R (in Ω). The current i (in A) is given by $i = V/R$. Find the approximate change in current if the voltage changes from 220 V to 225 V and the resistance changes from 20.0 Ω to 21.0 Ω.

23. The index of refraction n of a medium is defined as $n = \sin i / \sin r$, where i is the angle of incidence of light on the surface of the medium and r is the angle of refraction. In a particular experiment, n was calculated from values of $i = 45°$ and $r = 30°$. Then an error was discovered, and each angle was 1° too high. What was the approximate percentage error in the calculated value of n?

24. The electric current i (in A) in a circuit containing a variable resistor R (in Ω) and an inductance of 10.0 H is a function of R and the time t (in s) given by $i = 10.0(1 - e^{-0.100Rt})$. Find the approximate change in the current as R changes from 2.00 Ω to 2.04 Ω and t changes from 10.0 s to 11.0 s.

25. Two sides and the included angle of a triangular machine part were measured to be 2.30 cm, 2.10 cm, and 90.0°, respectively. The length of the third side was calculated from these values. It was then discovered that each of the sides should have been 0.10 cm more. What was the percentage error in the calculation of the third side?

26. If the radius and height of a right circular cylindrical container were measured to be 3.00 ft and 5.00 ft and then the radius was found to be 3.10 ft and the height 5.20 ft, what was the error in the original value of the volume?

• **27.** If the sum of the length l, width w, and height h, of a rectangular crate is 3 m, find the dimensions such that the volume is a maximum.

28. Find the dimensions of a closed rectangular box (top included) of the largest volume if the total surface area is 64.0 ft^2.

29. An open (no top) rectangular cargo container is to have a volume of 32.0 m^3. Find the dimensions that require the least material in building the container.

Ⓦ **30.** Find the point of the plane $2x + 3y - z = 12$ that is closest to the origin. Explain your solution.

31. The pressure P (in Pa) at each point of a circular plastic sheet of radius 5 cm is given by $P = x^4 + y^2 - 4x + 20$. What is the minimum pressure, and where is it located, if distances are measured from the center of the sheet?

• **32.** A flat plate is heated such that the temperature T (in °C) at any point (x, y) is $T = x^2 + 2y^2 - x$. Find the temperature at the coldest point.

33. A plate is cooled such that the temperature T (in °C) at any point (x, y) is $T = 2xy - 5x^2 - 2y^2 + 4x + 4y - 4$. Find the temperature at the warmest point.

34. A long metal strip 9 in. wide is to be made into a trough by bending up equal edges of length x, making equal angles θ with the bottom. Find the width of the bottom and the angle between the bottom and the edges if the cross-sectional area A of the trough is a maximum. See Fig. 11-25.

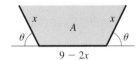

Fig. 11-25

If x, y, and z are functions of time t and we divide each term of Eq. (11-3) by the differential of t, dt, we obtain an equation for the **total derivative** *of z with respect to t. Therefore, the total derivative is*

$$\frac{dz}{dt} = \frac{\partial z}{\partial x}\frac{dx}{dt} + \frac{\partial z}{\partial y}\frac{dy}{dt} \qquad (11\text{-}5)$$

In Exercises 35–40, solve the given problems by use of Eq. (11-5).

35. Find dz/dt for $z = x^2 - xy + y^2$, $x = 1 + t^2$, $y = 1 - t^2$.

36. Find dz/dt for $z = \ln(x + y^2)$, $x = \sqrt{1 + t}$, $y = 1 + \sqrt{t}$.

37. Evaluate dz/dt for $t = \pi$, if $z = x^2 + y^2$, $x = \cos t$, $y = \sin t$.

38. Evaluate dz/dt for $t = 2$ if $z = \sin(xy)$, $x = t^2 - 2t$, $y = t^3 - 5t$.

39. The voltage V of a battery slowly drops, and the resistance R (in Ω) slowly increases with time as the battery wears down. If the current I (in A) in the circuit is given by $I = V/R$, evaluate the time rate of change of the current for $dR/dt = 0.50$ Ω/s, $dV/dt = -0.010$ V/s, when $R = 600$ Ω and $I = 0.040$ A.

40. If the radius r and height h of a right circular cylinder are increasing at 0.050 ft/s and 0.20 ft/s, respectively, find the rate at which the volume V is increasing when $r = 6.00$ ft and $h = 3.00$ ft.

11-5 DOUBLE INTEGRALS

We now turn our attention to integration in the case of a function of two variables. The analysis has similarities to that of partial differentiation, in that an operation is performed while holding one of the independent variables constant.

If $z = f(x, y)$ and we wish to integrate with respect to x and y, we first consider either x or y constant and integrate with respect to the other. After this integral is evaluated, we then integrate with respect to the variable first held constant. We shall now define this type of integral and then give an appropriate geometric interpretation.

If $z = f(x, y)$ the **double integral** *of the function over x and y is defined as*

$$\int_a^b \left[\int_{g(x)}^{G(x)} f(x, y)\, dy \right] dx$$

NOTE *It will be noted that the limits on the inner integral are functions of x and those on the outer integral are explicit values of x.* In performing the integration, **x is held constant while the inner integral is found and evaluated.** This results in a function of x only. This function is then integrated and evaluated.

It is customary not to include the brackets in stating a double integral. Therefore, we write

$$\int_a^b \left[\int_{g(x)}^{G(x)} f(x, y)\, dy \right] dx = \int_a^b \int_{g(x)}^{G(x)} f(x, y)\, dy\, dx \qquad \textbf{(11-6)}$$

The following example illustrates the use of Eq. (11-6).

EXAMPLE 1   Evaluate: $\displaystyle\int_0^1 \int_{x^2}^x xy\, dy\, dx$.

First, we integrate the inner integral with y as the variable and x as a constant.

$$\int_{x^2}^x \overset{\text{treat as constant}}{xy}\, dy = \left(x\frac{y^2}{2} \right)\Bigg|_{x^2}^x = x\left(\frac{x^2}{2} - \frac{x^4}{2} \right) = \frac{1}{2}(x^3 - x^5)$$

This means

$$\int_0^1 \int_{x^2}^x xy\, dy\, dx = \int_0^1 \frac{1}{2}(x^3 - x^5)\, dx$$

$$= \frac{1}{2}\left(\frac{x^4}{4} - \frac{x^6}{6} \right)\Bigg|_0^1$$

$$= \frac{1}{2}\left(\frac{1}{4} - \frac{1}{6} \right) - \frac{1}{2}(0) = \frac{1}{24} \qquad\blacksquare$$

It is also common to have an integral that is evaluated by integrating over x first and then over y. The next example on the following page illustrates such a case.

■EXAMPLE 2   Evaluate: $\int_1^4 \int_0^{4y} \sqrt{2x + y}\, dx\, dy$.

CAUTION ▶   Since the inner differential is dx, *the inner limits must be functions of y (which may be constant)*. Thus, *we first integrate with x as the variable and y as a constant.*

$$\int_1^4 \int_0^{4y} \sqrt{2x + y}\, dx\, dy = \int_1^4 \left[\frac{1}{2} \int_0^{4y} (2x + y)^{1/2} 2\, dx \right] dy$$

$$= \int_1^4 \left[\frac{1}{3}(2x + y)^{3/2} \right]_0^{4y} dy = \frac{1}{3} \int_1^4 [(9y)^{3/2} - (y)^{3/2}]\, dy$$

$$= \frac{1}{3} \int_1^4 (27y^{3/2} - y^{3/2})\, dy = \frac{26}{3} \int_1^4 y^{3/2}\, dy$$

$$= \frac{26}{3} \cdot \frac{2}{5}\, y^{5/2} \Big|_1^4 = \frac{52}{15}(32 - 1) \doteq \frac{1612}{15}$$ ■

■EXAMPLE 3   Evaluate: $\int_0^{\pi/2} \int_0^{\sin y} e^{2x} \cos y\, dx\, dy$.

We note here that the inner differential is dx. As in Example 2, this means that we first integrate with x as the variable and y as a constant. The final integration is performed with y as the variable.

$$\int_0^{\pi/2} \int_0^{\sin y} e^{2x} \cos y\, dx\, dy = \int_0^{\pi/2} \left[\frac{1}{2} e^{2x} \cos y \right]_0^{\sin y} dy$$

$$= \frac{1}{2} \int_0^{\pi/2} (e^{2\sin y} \cos y - \cos y)\, dy$$

$$= \frac{1}{2} \left[\frac{1}{2} e^{2\sin y} - \sin y \right]_0^{\pi/2} = \frac{1}{2}\left(\frac{1}{2}e^2 - 1 \right) - \frac{1}{2}\left(\frac{1}{2} - 0 \right)$$

$$= \frac{1}{4}e^2 - \frac{1}{2} - \frac{1}{4} = \frac{1}{4}(e^2 - 3) = 1.097$$ ■

For the geometric interpretation of a double integral, consider the surface shown in Fig. 11-26(a). An **element of volume** (dimensions of dx, dy, and z) extends from the xy-plane to the surface. With x a constant, sum (integrate) these elements of volume from the left boundary, $y = g(x)$, to the right boundary, $y = G(x)$. Now the volume of the vertical slice is a function of x, as shown in Fig. 11-26(b). By summing (integrating) the volumes of these slices from $x = a$ ($x = 0$ in the figure) to $x = b$, we have the complete volume as shown in Fig. 11-26(c).

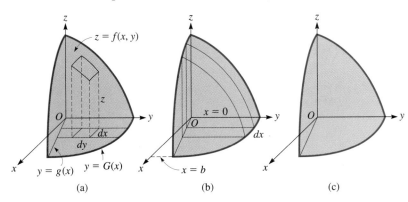

Fig. 11-26
(a)   (b)   (c)

Thus, *we may interpret a double integral as the* **volume under a surface,** in the same way as the integral was interpreted as the area of a plane figure. We may find the volume of a more general figure, as this volume is not necessarily a volume of revolution, as discussed in Section 6-3.

EXAMPLE 4   Find the volume that is in the first octant and under the plane $x + 2y + 4z - 8 = 0$. See Fig. 11-27.

This figure is a tetrahedron, for which $V = \frac{1}{3}Bh$. Assuming the base is in the xy-plane, $B = \frac{1}{2}(4)(8) = 16$, and $h = 2$. Therefore, $V = \frac{1}{3}(16)(2) = \frac{32}{3}$ cubic units. We shall use this value to check the one we find by double integration.

To find $z = f(x, y)$, we solve the given equation for z. Thus,

$$z = \frac{8 - x - 2y}{4}$$

CAUTION▶

Next, we must find the limits on y and x. Choosing to integrate over y first, we see that y goes from $y = 0$ to $y = (8 - x)/2$. ***This last limit is the trace of the surface in the xy-plane.*** Next, we note that x goes from $x = 0$ to $x = 8$. Therefore, we set up and evaluate the integral:

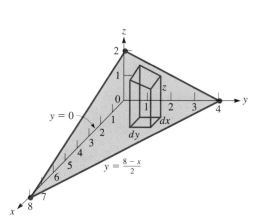

Fig. 11-27

$$V = \int_0^8 \int_0^{(8-x)/2} \left(\frac{8 - x - 2y}{4} \right) dy\, dx$$

$$= \int_0^8 \left[\frac{1}{4} \left(8y - xy - y^2 \right) \Big|_0^{(8-x)/2} \right] dx$$

$$= \frac{1}{4} \int_0^8 \left[8\left(\frac{8 - x}{2} \right) - x\left(\frac{8 - x}{2} \right) - \left(\frac{8 - x}{2} \right)^2 \right] dx$$

$$= \frac{1}{4} \int_0^8 \left(32 - 4x - 4x + \frac{x^2}{2} - 16 + 4x - \frac{x^2}{4} \right) dx$$

$$= \frac{1}{4} \int_0^8 \left(16 - 4x + \frac{x^2}{4} \right) dx$$

$$= \frac{1}{4} \left(16x - 2x^2 + \frac{x^3}{12} \right) \Big|_0^8 = \frac{1}{4} \left(128 - 128 + \frac{512}{12} \right)$$

$$= \frac{1}{4} \left(\frac{128}{3} \right) = \frac{32}{3} \text{ cubic units}$$

We see that the values obtained by the two different methods agree. ∎

EXAMPLE 5   Find the volume above the xy-plane, below the surface $z = xy$, and enclosed by the cylinder $y = x^2$ and the plane $y = x$.

Constructing the figure, shown in Fig. 11-28, we now note that $z = xy$ is the desired function of x and y. Integrating over y first, the limits on y are $y = x^2$ to $y = x$. The corresponding limits on x are $x = 0$ to $x = 1$. Therefore, the double integral to be evaluated is

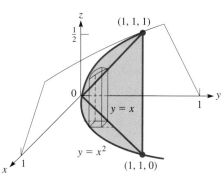

Fig. 11-28

$$V = \int_0^1 \int_{x^2}^x xy\, dy\, dx$$

This integral has already been evaluated in Example 1 of this section, and we can now see the geometric interpretation of that integral. Using the result from Example 1, we see that the required volume is 1/24 cubic unit. ∎

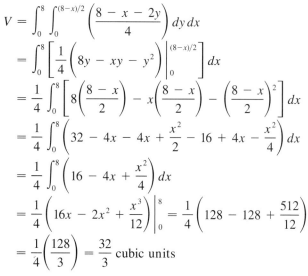

EXAMPLE 6   Find the volume that is in the first octant under the surface $z = 4 - x^2 - y^2$ and that is between the cylinder $x^2 = 3y$ and the plane $y = 1$. See Fig. 11-29.

Setting up the integration such that we integrate over x first, we have

$$V = \int_0^1 \int_0^{\sqrt{3y}} (4 - x^2 - y^2)\, dx\, dy$$

$$= \int_0^1 \left[4x - \frac{x^3}{3} - y^2 x \right]_0^{\sqrt{3y}} dy$$

$$= \int_0^1 (4\sqrt{3y} - \sqrt{3}y^{3/2} - \sqrt{3}y^{5/2})\, dy$$

$$= \sqrt{3} \left[4\left(\frac{2}{3}\right) y^{3/2} - \frac{2}{5} y^{5/2} - \frac{2}{7} y^{7/2} \right] \Big|_0^1$$

$$= \sqrt{3} \left(\frac{8}{3} - \frac{2}{5} - \frac{2}{7} \right) = \frac{208\sqrt{3}}{105} = 3.431 \text{ cubic units}$$

If we integrate over y first, the integral is

$$V = \int_0^{\sqrt{3}} \int_{x^2/3}^1 (4 - x^2 - y^2)\, dy\, dx$$

Integrating and evaluating this integral, we arrive at the same result.

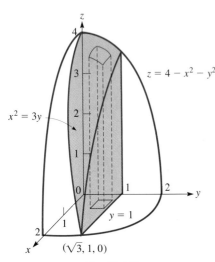

$z = 4 - x^2 - y^2$

$x^2 = 3y$

$y = 1$

$(\sqrt{3}, 1, 0)$

Fig. 11-29

─── **EXERCISES** *11-5* ───

In Exercises 1–16, evaluate the given double integrals.

1. $\displaystyle\int_2^4 \int_0^1 xy^2\, dx\, dy$

2. $\displaystyle\int_0^2 \int_0^1 \frac{y}{(xy+1)^2}\, dx\, dy$

3. $\displaystyle\int_0^1 \int_0^x 2y\, dy\, dx$

4. $\displaystyle\int_0^2 \int_1^x 2x\, dy\, dx$

5. $\displaystyle\int_1^2 \int_0^{y^2} xy^2\, dx\, dy$

6. $\displaystyle\int_0^4 \int_1^{\sqrt{y}} (x - y)\, dx\, dy$

7. $\displaystyle\int_0^1 \int_0^{\sqrt{1-x^2}} y\, dy\, dx$

8. $\displaystyle\int_4^9 \int_0^x \sqrt{x - y}\, dy\, dx$

9. $\displaystyle\int_0^{\pi/6} \int_{\pi/3}^y \sin x\, dx\, dy$

10. $\displaystyle\int_0^{\sqrt{3}} \int_{x^2/3}^1 (4 - x^2 - y^2)\, dy\, dx$

11. $\displaystyle\int_1^e \int_1^y \frac{1}{x}\, dx\, dy$

12. $\displaystyle\int_{-1}^1 \int_1^{e^x} \frac{1}{xy}\, dy\, dx$

13. $\displaystyle\int_1^2 \int_0^x yx^3 e^{xy^2}\, dy\, dx$

14. $\displaystyle\int_0^{\pi/6} \int_0^1 y \sin x\, dy\, dx$

15. $\displaystyle\int_0^{\ln 3} \int_0^x e^{2x+3y}\, dy\, dx$

16. $\displaystyle\int_0^{1/2} \int_y^{y^2} \frac{dx\, dy}{\sqrt{y^2 - x^2}}$

In Exercises 17–26, find the indicated volumes by double integration.

17. The first-octant volume under the plane
$x + y + z - 4 = 0$

18. The first-octant volume under the surface $z = y^2$ and bounded by the planes $x = 2$ and $y = 3$

19. The volume above the xy-plane and under the surface $z = 4 - x^2 - y^2$

20. The volume above the xy-plane, below the surface $z = x^2 + y^2$, and inside the cylinder $x^2 + y^2 = 4$

21. The first-octant volume bounded by the xy-plane, the planes $x = y$, $y = 2$, and $z = 2 + x^2 + y^2$

22. The volume bounded by the planes $x + 3y + 2z - 6 = 0$, $2x = y$, $x = 0$, and $z = 0$

23. The first-octant volume under the plane $z = x + y$ and inside the cylinder $x^2 + y^2 = 9$

24. The volume above the xy-plane and bounded by the cylinders $x = y^2$, $y = 8x^2$, and $z = x^2 + 1$. Integrate over y first and then check by integrating over x first.

25. A wedge is to be made in the shape shown in Fig. 11-30 (all vertical cross sections are equal right triangles). By double integration, find the volume of the wedge.

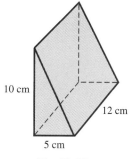

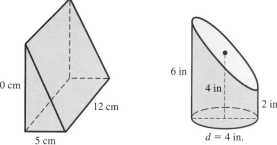

Fig. 11-30

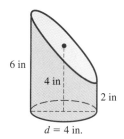

26. A circular piece of pipe is cut as shown in Fig. 11-31. Find the volume within the pipe. (*Hint:* Place the *z*-axis along the axis of the pipe, with the base of the pipe in the *xy*-plane.)

6 in
4 in
2 in
d = 4 in.

Fig. 11-31

In Exercises 27 and 28, draw the appropriate figure.

27. Draw an appropriate figure indicating a volume that is found from the integral

$$\int_0^{1/2} \int_{x^2}^1 (4 - x - 2y)\, dy\, dx$$

W 28. Repeat Exercise 27 for the integral

$$\int_1^2 \int_0^{2-y} \sqrt{1 + x^2 + y^2}\, dx\, dy. \quad \text{Explain.}$$

11-6 CENTROIDS AND MOMENTS OF INERTIA BY DOUBLE INTEGRATION

Among the important applications of double integration are those of finding centroids and moments of inertia of plane regions. In this section we shall show how double integration is used in these cases.

Considering the region shown in Fig. 11-32, we see that *we may determine the area of this region from the double integral*

$$A = \int_a^b \int_{g(x)}^{G(x)} dy\, dx \tag{11-7}$$

This is the case since, by summing (integrating) the elements of area *dy dx* over *y* while holding *x* constant, we find the area of the vertical strip. Then by summing (integrating) from *x = a* to *x = b* we find the total area.

In Section 6-4, we showed that the moment of the mass of an element of the region with respect to the *y*-axis is *kx dA*, where *k* is the mass per unit area, *x* is the moment arm, and *dA* is the element of area. Recalling that the *x*-coordinate of the centroid is the moment with respect to the *y*-axis, divided by the mass of the region, *we have*

$$\overline{x} = \frac{\displaystyle\int_a^b \int_{g(x)}^{G(x)} x\, dy\, dx}{A} \tag{11-8}$$

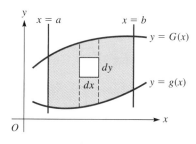

Fig. 11-32

as the equation by which we may find the x-coordinate of the centroid by double integration. (The factor *k* that would appear in the numerator and the denominator cancels out.)

Also, since the moment of the mass of the element of the region with respect to the *x*-axis is *ky dA, the y-coordinate of the centroid is found from*

$$\overline{y} = \frac{\displaystyle\int_a^b \int_{g(x)}^{G(x)} y\, dy\, dx}{A} \tag{11-9}$$

For many problems, use of Eqs. (11-8) and (11-9) is more convenient than the equivalent method in Section 6-4. One advantage is that the integration is often simplified. The following example illustrates the use of Eqs. (11-8) and (11-9).

EXAMPLE 1 Find the centroid of the region bounded by the lines $x = 0$, $y = 2x$, and $y = 3 - x$ (see Fig. 11-33).

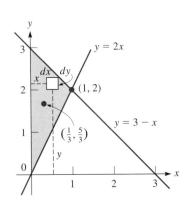

Fig. 11-33

$$A = \int_0^1 \int_{2x}^{3-x} dy\, dx = \int_0^1 y \Big|_{2x}^{3-x} dx$$

$$= \int_0^1 (3 - 3x)\, dx = 3x - \frac{3x^2}{2} \Big|_0^1 = \frac{3}{2}$$

$$A\overline{x} = \int_0^1 \int_{2x}^{3-x} x\, dy\, dx = \int_0^1 y \Big|_{2x}^{3-x} x\, dx$$

$$= \int_0^1 (3x - 3x^2)\, dx = \frac{3x^2}{2} - x^3 \Big|_0^1 = \frac{1}{2}$$

$$A\overline{y} = \int_0^1 \int_{2x}^{3-x} y\, dy\, dx = \int_0^1 \frac{y^2}{2} \Big|_{2x}^{3-x} dx$$

$$= \frac{1}{2} \int_0^1 (9 - 6x - 3x^2)\, dx$$

$$= \frac{1}{2}(9x - 3x^2 - x^3) \Big|_0^1 = \frac{5}{2}$$

Therefore, $\overline{x} = \frac{1}{2} \cdot \frac{2}{3} = \frac{1}{3}$, and $\overline{y} = \frac{5}{2} \cdot \frac{2}{3} = \frac{5}{3}$. ▬

In many problems, integrating over x first is definitely preferable. The following example illustrates this type of problem.

EXAMPLE 2 Find the centroid of the region bounded by $x = 2y - y^2$ and the y-axis (see Fig. 11-34).

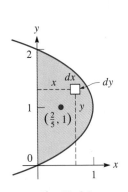

Fig. 11-34

$$A = \int_0^2 \int_0^{2y-y^2} dx\, dy = \int_0^2 (2y - y^2)\, dy = \left[y^2 - \frac{y^3}{3} \right] \Big|_0^2 = \frac{4}{3}$$

$$A\overline{x} = \int_0^2 \int_0^{2y-y^2} x\, dx\, dy = \int_0^2 \frac{x^2}{2} \Big|_0^{2y-y^2} dy$$

$$= \frac{1}{2} \int_0^2 (4y^2 - 4y^3 + y^4)\, dy$$

$$= \frac{1}{2} \left(\frac{4}{3} y^3 - y^4 + \frac{y^5}{5} \right) \Big|_0^2 = \frac{8}{15}$$

$$A\overline{y} = \int_0^2 \int_0^{2y-y^2} y\, dx\, dy = \int_0^2 (2y^2 - y^3)\, dy$$

$$= \left(\frac{2}{3} y^3 - \frac{1}{4} y^4 \right) \Big|_0^2 = \frac{4}{3}$$

Therefore, $\overline{x} = \frac{8}{15} \cdot \frac{3}{4} = \frac{2}{5}$, and $\overline{y} = \frac{4}{3} \cdot \frac{3}{4} = 1$. ▬

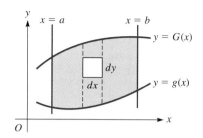

Fig. 11-35

From Section 6-5, we recall that the **moment of inertia** of the mass of an element of a region with respect to the y-axis is $I_y = kx^2\,dA$. Therefore, using double integrals to find the moments of inertia of the region shown in Fig. 11-35 (which is the same as Fig. 11-32), we have (assuming $k = 1$)

and

$$I_y = \int_a^b \int_{g(x)}^{G(x)} x^2\,dy\,dx \tag{11-10}$$

$$I_x = \int_a^b \int_{g(x)}^{G(x)} y^2\,dy\,dx \tag{11-11}$$

Also, from Section 6-5, the **radius of gyration** of the region with respect to an axis was defined by $R^2 = I/A$. The following examples illustrate the use of double integrals to find the moment of inertia and radius of gyration of a region.

EXAMPLE 3   Find I_y and I_x for the region bounded by $y = 2x$ and $y = x^2$. See Fig. 11-36.

$$I_y = \int_0^2 \int_{x^2}^{2x} x^2\,dy\,dx = \int_0^2 y \Big|_{x^2}^{2x} x^2\,dx$$

$$= \int_0^2 (2x^3 - x^4)\,dx = \frac{x^4}{2} - \frac{x^5}{5} \Big|_0^2 = \frac{8}{5}$$

$$I_x = \int_0^2 \int_{x^2}^{2x} y^2\,dy\,dx = \int_0^2 \frac{y^3}{3} \Big|_{x^2}^{2x} dx$$

$$= \frac{1}{3} \int_0^2 (8x^3 - x^6)\,dx$$

$$= \frac{1}{3} \left(2x^4 - \frac{x^7}{7} \right) \Big|_0^2 = \frac{32}{7}$$

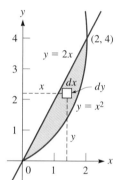

Fig. 11-36

EXAMPLE 4   Find the radius of gyration with respect to the y-axis for the region bounded by $y = x^3$, $x = 2$, and $y = 0$ (see Fig. 11-37).

Since integration is more convenient over y first than over x first, we set up our integrals in this way.

$$A = \int_0^2 \int_0^{x^3} dy\,dx = \int_0^2 y \Big|_0^{x^3} dx = \int_0^2 x^3\,dx = \frac{x^4}{4} \Big|_0^2 = 4$$

$$I_y = \int_0^2 \int_0^{x^3} x^2\,dy\,dx = \int_0^2 y \Big|_0^{x^3} x^2\,dx = \int_0^2 x^5\,dx = \frac{x^6}{6} \Big|_0^2 = \frac{32}{3}$$

$$R_y^2 = \frac{32}{3} \cdot \frac{1}{4} = \frac{8}{3}$$

$$R_y = \frac{2}{3} \sqrt{6} = 1.63$$

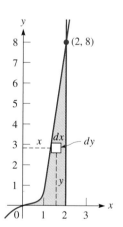

Fig. 11-37

EXERCISES *11-6*

In Exercises 1–4, use double integrals to find the coordinates of the centroid of the region bounded by the given curves.

1. $y = 2 - x$, $y = 0$, $x = 0$ **2.** $y = x^2$, $y = 4$, $x = 0$

3. $y = 9 - x^2$, $y = 0$, $x = 0$ **4.** $y = 2x$, $x = 1$, $x = 2$, $y = 0$

In Exercises 5–20, use double integrals to solve the given problems.

5. Find I_y and I_x for the region of Exercise 1.

6. Find I_y and I_x for the region of Exercise 2.

7. Find R_y for the region of Exercise 3.

8. Find R_x for the region of Exercise 4.

9. Find the coordinates of the centroid and R_y for the region bounded by $y = 1 - x^2$, $x = 1$, and $y = 1$.

10. Find R_y for the region bounded by $y = 3 + 2x - x^2$ and $y = 3 - x$.

11. Find the coordinates of the centroid of a right triangular region with legs a and b.

12. Find the coordinates of the centroid of a quarter circular region of radius a.

13. Find the moment of inertia of a rectangular region with sides a and b with respect to side a. Express the result in terms of the mass of the area.

14. Find the moment of inertia of an isosceles right triangular region with respect to one of the legs a. Express the result in terms of the mass of the area.

15. Find the radius of gyration of a square region of side 4 with respect to one of the sides.

16. Find the radius of gyration of a region in the shape of a rhombus of side 2, if the angle between adjacent sides is 45°, with respect to one of the sides.

17. A flat metal plate is in the shape of an isosceles trapezoid, as shown in Fig. 11-38. Where is the centroid of the plate?

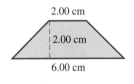

2.00 cm

2.00 cm

6.00 cm

Fig. 11-38

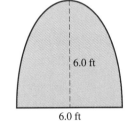

6.0 ft

6.0 ft

Fig. 11-39

18. An aquarium view window has the shape of a parabolic section, as shown in Fig. 11-39. Locate the centroid of the window.

19. Find the moment of inertia of a uniform flat square plate with respect to a diagonal.

20. Find the centroid of the infinite region in the second quadrant bounded by the axes and $y = e^x$.

CHAPTER EQUATIONS

Equation of a plane	$Ax + By + Cz + D = 0$	**(11-1)**
Partial derivatives	$\dfrac{\partial^2 z}{\partial x\, \partial y} = \dfrac{\partial^2 z}{\partial y\, \partial x}$	**(11-2)**
Total differential	$dz = \dfrac{\partial z}{\partial x}\, dx + \dfrac{\partial z}{\partial y}\, dy$	**(11-3)**
Relative maxima and minima	$\dfrac{\partial z}{\partial x} = 0 \quad \text{and} \quad \dfrac{\partial z}{\partial y} = 0$	**(11-4)**
Double integral	$\displaystyle\int_a^b \left[\int_{g(x)}^{G(x)} f(x, y)\, dy \right] dx = \int_a^b \int_{g(x)}^{G(x)} f(x, y)\, dy\, dx$	**(11-6)**
Area	$A = \displaystyle\int_a^b \int_{g(x)}^{G(x)} dy\, dx$	**(11-7)**

Coordinates of centroid	$$\overline{x} = \dfrac{\displaystyle\int_a^b \int_{g(x)}^{G(x)} x\,dy\,dx}{A}$$	(11-8)
	$$\overline{y} = \dfrac{\displaystyle\int_a^b \int_{g(x)}^{G(x)} y\,dy\,dx}{A}$$	(11-9)
Moment of inertia	$$I_y = \int_a^b \int_{g(x)}^{G(x)} x^2\,dy\,dx$$	(11-10)
	$$I_x = \int_a^b \int_{g(x)}^{G(x)} y^2\,dy\,dx$$	(11-11)

REVIEW EXERCISES

In Exercises 1–10, find the partial derivatives of the given functions with respect to each of the independent variables.

1. $z = 5x^3y^2 - 2xy^4$

2. $z = 2x\sqrt{y} - x^2y$

3. $z = \sqrt{x^2 - 3y^2}$

4. $u = \dfrac{r}{(r - 3s)^2}$

5. $z = \dfrac{2x - 3y}{x^2y + 1}$

6. $z = x(y^2 + xy + 2)^4$

7. $u = y \ln \sin(x^2 + 2y)$

8. $q = p \ln(r + 1) - \dfrac{rp}{r + 1}$

9. $z = \sin^{-1} \sqrt{x + y}$

10. $z = ye^{xy} \sin(2x - y)$

In Exercises 11 and 12, find all of the second partial derivatives of the given functions.

11. $z = 3x^2y - y^3 + 2xy$

12. $z = x\sqrt{2y + 1} + y^2(x - 2)^3$

In Exercises 13–20, evaluate each of the given double integrals.

13. $\displaystyle\int_0^2 \int_1^2 (3y + 2xy)\,dx\,dy$

14. $\displaystyle\int_2^7 \int_0^1 x\sqrt{2 + x^2y}\,dx\,dy$

15. $\displaystyle\int_0^3 \int_1^x (x + 2y)\,dy\,dx$

16. $\displaystyle\int_1^2 \int_0^{\pi/4} r \sec^2 \theta\,d\theta\,dr$

17. $\displaystyle\int_0^1 \int_0^{2x} x^2 e^{xy}\,dy\,dx$

18. $\displaystyle\int_{\pi/4}^{\pi/2} \int_1^{\sqrt{\cos\theta}} r \sin \theta\,dr\,d\theta$

19. $\displaystyle\int_1^e \int_1^x \dfrac{\ln y}{xy}\,dy\,dx$

20. $\displaystyle\int_1^3 \int_0^x \dfrac{2}{x^2 + y^2}\,dy\,dx$

In Exercises 21–24, use the function $z = \sqrt{x^2 + 4y^2}$.

21. Sketch the surface representing the function.

22. Find the equation of a line tangent to the surface of the function at $(2, 1, 2\sqrt{2})$ that is parallel to the yz-plane.

23. Find the approximate change in the function as x changes from 2.00 to 2.04 and y changes from 1.00 to 1.06.

24. Find the expression for the total differential of the function.

In Exercises 25–28, use the function $z = e^{x+y}$.

25. Find the equation of the line tangent to the surface of the function at $(1, 1, e^2)$ that is parallel to the xz-plane.

26. Sketch the surface representing the function.

27. Find the volume in the first octant under the surface of the function and inside the planes $x = 1$ and $y = x$.

28. Find the approximate change in the function as x changes from 1.00 to 1.01 and y changes from 1.00 to 1.02.

In Exercises 29–48, solve the given problems.

29. In a simple series electric circuit, with two resistors r and R connected across a voltage source E, the voltage v across r is $v = rE/(r + R)$. Assuming E to be constant, find $\partial v/\partial r$ and $\partial v/\partial R$.

30. For a gas, the volume expansivity is defined as $\beta = \dfrac{1}{V}\dfrac{\partial V}{\partial T}$, where V is the volume and T is the temperature of the gas. If V as a function of T and the pressure p is given by $V = a + bT/p - c/T^2$, where a, b, and c are constants, find β.

31. In the theory dealing with transistors, the current gain α of a transistor is defined as $\alpha = \partial i_c / \partial i_e$, where i_c is the collector current and i_e is the emitter current. If i_c is a function of i_e and the collector voltage v_c given by $i_c = i_e(1 - e^{-2v_c})$, find α if v_c is 2 V.

32. Young's modulus, which measures the ratio of the stress to strain in a stretched wire, is defined as $Y = \dfrac{L}{A} \dfrac{\partial F}{\partial L}$, where L is the length of the wire, A is its cross-sectional area, and F is the tension in the wire. Find Y (in Pa) if $F = -0.0100 \, T/L^2$ for $L = 1.10$ m, $T = 300$ K, and $A = 1.00 \times 10^{-6}$ m^2.

33. The period T of the pendulum as a function of its length l and the acceleration due to gravity g is given by $T = 2\pi\sqrt{l/g}$. Show that $\partial T/\partial l = T/2l$.

34. The volume of a right circular cone of radius r and height h is given by $V = \frac{1}{3}\pi r^2 h$. Show that $\partial V/\partial r = 2V/r$.

35. The image distance q from a lens as a function of the object distance p and the focal distance f of the lens is given by $q = pf/(p - f)$. Find the approximate change in q if f changes from 20.0 cm to 21.0 cm and p changes from 100 cm to 105 cm.

36. The impedance Z for an alternating current circuit is given by $Z = \sqrt{R^2 + X^2}$, where R is the resistance and X is the reactance. If R is measured to be 6 Ω and X to be 8 Ω and each has a possible error of 2%, what is the maximum possible error in the impedance?

37. A flat plate 5 in. square is heated such that the temperature T as a function of distances x and y from intersecting edges is $T = 2xy - y^2 - 2x^2 + 3y + 2$. Find the hottest point.

38. A closed rectangular box is made to contain 1000 in.^{3}. What dimensions should it have to make its surface area a minimum?

39. Find the volume in the first octant below the plane $x + y + z - 6 = 0$ and inside the cylinder $y = 4 - x^2$.

40. Find the first-octant volume bounded by $x + z = 4$, $x = 1$, and $y = 1$.

41. Find the first-octant volume bounded by $x^2 + y^2 = 16$ and $x + z = 8$.

42. Find the first-octant volume below the surface $z = 4 - y^2$ and inside the plane $x + y = 2$.

43. Find, by double integration, the coordinates of the centroid of the region bounded by $y = 2$, $x + y = 4$, $x = 0$, and $y = 0$.

44. Find, by double integration, the coordinates of the centroid of the region bounded by $y = x$ and $y = 4x - x^2$.

45. Find, by double integration, R_y for the region of Exercise 43.

46. Find, by double integration, R_x for the region of Exercise 44.

47. The power P (in W) delivered to an electric resistor R (in Ω) is given by $P = V^2/R$, where V is the voltage across the resistor. If V is increasing at the rate of 1.0 V/min and R is increasing at the rate of 2.0 Ω/min, find the time rate of change of power when $V = 60$ V and $R = 20$ Ω (see Exercise 35 on page 356).

48. The kinetic energy E of a moving object is given by $E = \frac{1}{2}mv^2$, where m is the mass of the object and v is its velocity. If m is increasing at the rate of 0.0100 kg/s and v is increasing at the rate of 5.00 m/s^2, find the time rate of change of E (in J/s) when $m = 5.00$ kg and $v = 100$ m/s (see Exercise 35 on page 356).

Writing Exercise

49. An architectural design student determined the area of a patio could be described by the double integral $\displaystyle\int_0^4 \int_0^{\sqrt{y}} f(x, y) \, dx \, dy$.

Write the integration with the order of integration interchanged. Write one or two paragraphs to explain your method of interchanging the order of integration.

PRACTICE TEST

1. Sketch the surface representing the function $z = 4 - x^2 - 4y^2$.

2. Evaluate: $\displaystyle\int_0^2 \int_{x^2}^{2x} (x^3 + 4y) \, dy \, dx$.

3. Find the total differential of the function $z = x^3 - x^2y + 3y^2$.

4. Find the volume of the solid in the first octant bounded by the coordinate planes and the cylinders $x^2 + y^2 = 9$ and $y^2 + z^2 = 9$.

5. An open (no top) rectangular storage container is to be constructed to have a volume of 12 m^3. The cost of the material to be used is \$4/m^2 for the bottom, \$3/m^2 for two of the opposite sides, and \$2/m^2 for the other two opposite sides. Find the dimensions that will minimize the cost of constructing the container.

6. By double integration, find the y-coordinate of the centroid of the region bounded by $y = \sin x$, $y = 0$, $x = 0$, and $x = \pi$.

To this point we have graphed plane curves in one coordinate system. This system, the rectangular coordinate system, is probably the most useful and widely applicable system. However, other coordinate systems prove to be better adapted for certain types of curves. One of these coordinate systems, the *polar coordinate system,* is widely used when certain types of applications are involved.

The rectangular coordinate system is based on distances from two fixed lines, the axes. The polar coordinate system is based on the distance from a fixed point, and the angle measured from a fixed line. Actually, this type of description is frequently used in locating geographic points. For example, when we say that the center of a low pressure area is 100 mi to the northeast, we are using this method of describing the location.

Also, as we extended the use of rectangular coordinates to three dimensions by the use of the *z*-axis, we can also extend the polar coordinates system to three dimensions in the same way. In doing so we are using *cylindrical coordinates.* These will also be briefly introduced in this chapter.

Great advances in our knowledge of the universe have been made through the use of the Hubble space telescope (shown above) since the mid-1990s. In Section 12-2 we illustrate the use of polar coordinates in the study of planetary motion.

The Swiss mathematician Jakob Bernoulli (1654–1705) was among the first to make significant use of polar coordinates.

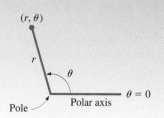

Fig. 12-1

12-1 POLAR COORDINATES

Instead of designating a point by its *x*- and *y*-coordinates, we can specify its location by its radius vector and the angle the radius vector makes with the *x*-axis. Therefore, the *r* and θ that are used in the definitions of the trigonometric functions can also be used as the coordinates of points in the plane. The important aspect of choosing coordinates is that, for each set of values, there must be only one point which corresponds to this set. We can see that this condition is satisfied by the use of *r* and θ as coordinates. *In* **polar coordinates,** *the origin is called the* **pole,** *and the half-line for which the angle is zero (equivalent to the positive x-axis) is called the* **polar axis.** The coordinates of a point are designated as (r, θ). We shall use radians when measuring the value of θ. See Fig. 12-1.

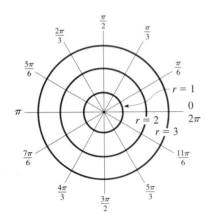

When using polar coordinates, we generally label the lines for some of the values of θ; namely, those for $\theta = 0$ (the polar axis), $\theta = \pi/2$ (equivalent to the positive y-axis), $\theta = \pi$ (equivalent to the negative x-axis), $\theta = 3\pi/2$ (equivalent to the negative y-axis), and possibly others. In Fig. 12-2, these lines and those for multiples of $\pi/6$ are shown. Also, the circles for $r = 1$, $r = 2$, and $r = 3$ are shown in this figure.

Fig. 12-2

■EXAMPLE 1 (a) If $r = 2$ and $\theta = \pi/6$, we have the point as shown in Fig. 12-3. The coordinates (r, θ) of this point are written as $(2, \pi/6)$ when polar coordinates are used. This point corresponds to $(\sqrt{3}, 1)$ in rectangular coordinates.

(b) In Fig. 12-3, the polar coordinate point $(1, 3\pi/4)$ is also shown. It is equivalent to the point $(-\sqrt{2}/2, \sqrt{2}/2)$ in rectangular coordinates.

(c) In Fig. 12-3, the polar coordinate point $(2, 5)$ is also shown. It is equivalent approximately to the point $(0.6, -1.9)$ in rectangular coordinates. Remember, the 5 is an angle in radian measure. --------‎‐‐___■

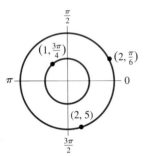

Fig. 12-3

NOTE ▶

One difference between rectangular coordinates and polar coordinates is that, for each point in the plane, there are limitless possibilities for the polar coordinates of that point. For example, the point $(2, \frac{\pi}{6})$ can also be represented by $(2, \frac{13\pi}{6})$ since the angles $\frac{\pi}{6}$ and $\frac{13\pi}{6}$ are coterminal. We also remove one restriction on r that we imposed in the definition of the trigonometric functions. That is, r is allowed to take on positive and negative values. If r is negative, θ is located as before, but **the point is found r units from the pole but on the opposite side** from that on which it is positive.

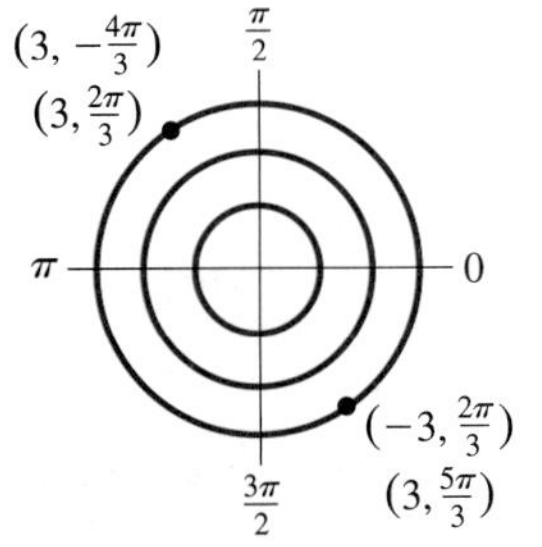

■EXAMPLE 2 The coordinates $(3, 2\pi/3)$ and $(3, -4\pi/3)$ represent the same point. However, the point $(-3, 2\pi/3)$ is on the opposite side of the pole, three units from the pole. Another possible set of coordinates for the point $(-3, 2\pi/3)$ is $(3, 5\pi/3)$. See Fig. 12-4. --------___■

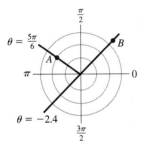

Fig. 12-4

When plotting a point in polar coordinates, it is generally easier to *first locate the terminal side of θ and then measure r along this terminal side*. This is illustrated in the following example.

■EXAMPLE 3 Plot the points $A(2, 5\pi/6)$ and $B(-3.2, -2.4)$ in the polar coordinate system.

To locate A we determine the terminal side of $\theta = 5\pi/6$ and then determine $r = 2$. See Fig. 12-5.

To locate B we find the terminal side of $\theta = -2.4$, measuring clockwise from the polar axis (and recalling that $\pi = 3.14 = 180°$). Then we locate $r = -3.2$ on the opposite side of the pole. See Fig. 12-5.

We will find that points with negative values of r occur frequently when plotting curves in polar coordinates. --------___■

Fig. 12-5

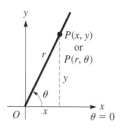

Fig. 12-6

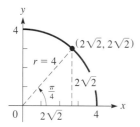

Fig. 12-7

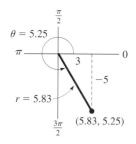

Fig. 12-8

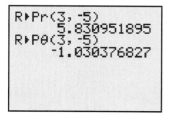

Fig. 12-9

The cyclotron was invented in 1931 at the University of California. It was the first accelerator to deflect particles into circular paths.

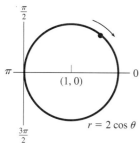

Fig. 12-10

Polar and Rectangular Coordinates

The relationships between the polar coordinates of a point and the rectangular coordinates of the same point come from the definitions of the trigonometric functions. Those most commonly used are (see Fig. 12-6):

$$x = r \cos \theta \qquad y = r \sin \theta \tag{12-1}$$

$$\tan \theta = \frac{y}{x} \qquad r = \sqrt{x^2 + y^2} \tag{12-2}$$

The following examples show the use of Eqs. (12-1) and (12-2) in changing coordinates in one system to coordinates in the other system. Also, these equations are used to transform equations from one system to the other.

EXAMPLE 4 Using Eqs. (12-1), we can transform the polar coordinates of $(4, \pi/4)$ into the rectangular coordinates $(2\sqrt{2}, 2\sqrt{2})$, since

$$x = 4 \cos \frac{\pi}{4} = 4\left(\frac{\sqrt{2}}{2}\right) = 2\sqrt{2} \quad \text{and} \quad y = 4 \sin \frac{\pi}{4} = 4\left(\frac{\sqrt{2}}{2}\right) = 2\sqrt{2}$$

See Fig. 12-7. ▪

EXAMPLE 5 Using Eqs. (12-2), we can transform the rectangular coordinates $(3, -5)$ into polar coordinates.

$$\tan \theta = -\frac{5}{3}, \qquad \theta = 5.25 \quad (\text{or} -1.03)$$

$$r = \sqrt{3^2 + (-5)^2} = 5.83$$

We know that θ is a fourth-quadrant angle since x is positive and y is negative. Therefore, the point $(3, -5)$ in rectangular coordinates can be expressed as the point $(5.83, 5.25)$ in polar coordinates (see Fig. 12-8). Other polar coordinates for the point are also possible. ▪

Calculators are programmed to make conversions between rectangular coordinates and polar coordinates. The manual should be consulted to determine how any particular model is used for these conversions. For a calculator that uses the *angle* feature, the display for the conversions of Example 5 is shown in Fig. 12-9.

EXAMPLE 6 If a protron (a positively charged particle) enters a magnetic field at right angles to the field, it follows a circular path, a fact used in the design of nuclear particle accelerators. If the path of a protron is described by the rectangular equation $x^2 + y^2 = 2x$ (measurements in meters), find the polar equation.

We change this rectangular equation into a polar equation by using the relations $r^2 = x^2 + y^2$ and $x = r \cos \theta$ as follows:

$$x^2 + y^2 = 2x \qquad \text{rectangular equation}$$
$$r^2 = 2r \cos \theta \qquad \text{substitute}$$
$$r = 2 \cos \theta \qquad \text{divide by } r$$

This is the polar equation of the circle. See Fig. 12-10. ▪

EXAMPLE 7   Find the rectangular equation of the *rose* $r = 4 \sin 2\theta$.

Using the trigonometric identity $\sin 2\theta = 2 \sin \theta \cos \theta$ and Eqs. (12-1) and (12-2) leads to the solution.

$$r = 4 \sin 2\theta \qquad \text{polar equation}$$

$$= 4(2 \sin \theta \cos \theta) = 8 \sin \theta \cos \theta \qquad \text{using identity}$$

$$\sqrt{x^2 + y^2} = 8 \left(\frac{y}{r} \right) \left(\frac{x}{r} \right) = \frac{8xy}{r^2} = \frac{8xy}{x^2 + y^2} \qquad \text{using Eqs. (12-1) and (12-2)}$$

$$x^2 + y^2 = \frac{64x^2y^2}{(x^2 + y^2)^2} \qquad \text{squaring both sides}$$

$$(x^2 + y^2)^3 = 64x^2y^2 \qquad \text{simplifying}$$

Plotting the graph of this equation from the rectangular equation would be complicated. However, as we will see in the next section, plotting this graph in polar coordinates is quite simple. The curve is shown in Fig. 12-11.

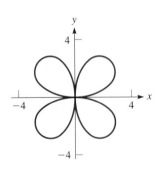

Fig. 12-11

EXERCISES *12-1*

In Exercises 1–12, plot the given polar coordinate points on polar coordinate paper.

1. $\left(3, \dfrac{\pi}{6} \right)$ **2.** $(2, \pi)$ **3.** $\left(\dfrac{5}{2}, -\dfrac{2\pi}{5} \right)$

4. $\left(5, -\dfrac{\pi}{3} \right)$ **5.** $\left(-2, \dfrac{7\pi}{6} \right)$ **6.** $\left(-5, \dfrac{\pi}{4} \right)$

7. $\left(-3, -\dfrac{5\pi}{4} \right)$ **8.** $\left(-4, -\dfrac{5\pi}{3} \right)$ **9.** $\left(0.5, -\dfrac{8\pi}{3} \right)$

10. $(2.2, -6\pi)$ **11.** $(2, 2)$ **12.** $(-1, -1)$

In Exercises 13–16, find a set of polar coordinates for each of the points for which the rectangular coordinates are given.

13. $(\sqrt{3}, 1)$ **14.** $(-1, -1)$

15. $\left(-\dfrac{\sqrt{3}}{2}, -\dfrac{1}{2} \right)$ **16.** $(-5, 4)$

In Exercises 17–20, find the rectangular coordinates for each of the points for which the polar coordinates are given.

17. $\left(8, \dfrac{4\pi}{3} \right)$ **18.** $(-4, -\pi)$

19. $(3.0, -0.40)$ **20.** $(-1.0, 1.0)$

In Exercises 21–28, find the polar equation of each of the given rectangular equations.

21. $x = 3$ **22.** $y = x$

23. $x + 2y = 3$ **24.** $x^2 + y^2 = 0.81$

25. $x^2 + (y - 2)^2 = 4$ **26.** $x^2 - y^2 = 0.01$

27. $x^2 + 4y^2 = 4$ **28.** $y^2 = 4x$

In Exercises 29–40, find the rectangular equation of each of the given polar equations. In Exercises 29–36, identify the curve that is represented by the equation.

29. $r = \sin \theta$ **30.** $r = 4 \cos \theta$

31. $r \cos \theta = 4$ **32.** $r \sin \theta = -2$

33. $r = \dfrac{2}{\cos \theta - 3 \sin \theta}$ **34.** $r = e^{r \cos \theta} \csc \theta$

35. $r = 4 \cos \theta + 2 \sin \theta$ **36.** $r \sin(\theta + \pi/6) = 3$

37. $r = 2(1 + \cos \theta)$ **38.** $r = 1 - \sin \theta$

39. $r^2 = \sin 2\theta$ **40.** $r^2 = 16 \cos 2\theta$

In Exercises 41–44, find the required equations.

41. Under certain conditions, the *x*- and *y*-components of a magnetic field *B* are given by the equations

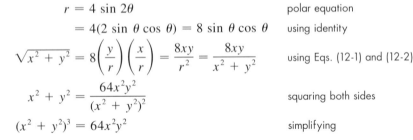

$$B_x = \frac{-ky}{x^2 + y^2} \quad \text{and} \quad B_y = \frac{kx}{x^2 + y^2}$$

Write these equations in terms of polar coordinates.

42. In designing a domed roof for a building, an architect uses the equation $x^2 + \dfrac{y^2}{k^2} = 1$, where *k* is a constant. Write this equation in polar form.

Ⓦ 43. The perimeter of a certain type of machine part can be described by the equation $r = a \sin \theta + b \cos \theta$, $a > 0$, $b > 0$. Explain why all such machine parts are circular.

44. The polar equation of the path of a weather satellite of the earth is $r = \dfrac{4800}{1 + 0.14 \cos \theta}$, where *r* is measured in miles. Find the rectangular equation of the path of this satellite. The path is an ellipse, with the earth at one of the foci.

12-2 CURVES IN POLAR COORDINATES

The basic method for finding a curve in polar coordinates is the same as in rectangular coordinates. We assume values of the independent variable, in this case θ, and then find the corresponding values of the dependent variable r. These points are then plotted and joined, thereby forming the curve that represents the function in polar coordinates.

Before using the basic method, it is useful to point out that there are certain basic curves that can be sketched directly from the equation. This is done by noting the meaning of each of the polar coordinate variables, r and θ. This is illustrated in the following example.

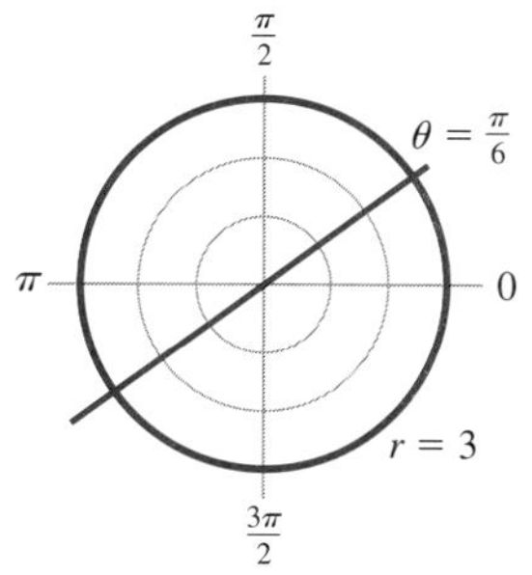

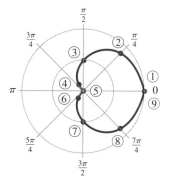

Fig. 12-12

EXAMPLE 1   **(a)** The graph of the polar equation $r = 3$ is a circle of radius 3, with center at the pole. This can be seen to be the case, since $r = 3$ for all possible values of θ. It is not necessary to find specific points for this circle, which is shown in Fig. 12-12.

(b) The graph of $\theta = \pi/6$ is a straight line through the pole. It represents all points for which $\theta = \pi/6$ for all possible values of r, positive or negative. This line is also shown in Fig. 12-12.

EXAMPLE 2   Plot the graph of $r = 1 + \cos \theta$.
We find the following values of r corresponding to the chosen values of θ.

θ	0	$\frac{\pi}{4}$	$\frac{\pi}{2}$	$\frac{3\pi}{4}$	π	$\frac{5\pi}{4}$	$\frac{3\pi}{2}$	$\frac{7\pi}{4}$	2π
r	2	1.7	1	0.3	0	0.3	1	1.7	2
Point number	1	2	3	4	5	6	7	8	9

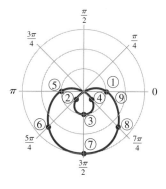

Fig. 12-13

We now see that the points start repeating, and it is unnecessary to find additional points. The curve is called a **cardioid** and is shown in Fig. 12-13.

EXAMPLE 3   Plot the graph of $r = 1 - 2 \sin \theta$.
Choosing values of θ, and then finding the corresponding values of r, we find the following table of values.

θ	0	$\frac{\pi}{4}$	$\frac{\pi}{2}$	$\frac{3\pi}{4}$	π	$\frac{5\pi}{4}$	$\frac{3\pi}{2}$	$\frac{7\pi}{4}$	2π
r	1	-0.4	-1	-0.4	1	2.4	3	2.4	1
Point number	1	2	3	4	5	6	7	8	9

Fig. 12-14

Particular care should be taken in plotting the points for which r is negative. This curve is known as a **limaçon** and is shown in Fig. 12-14.

■**EXAMPLE 4**   A cam is shaped such that the edge of the upper "half" is represented by the equation $r = 2.0 + \cos \theta$ and the lower "half" by the equation $r = \dfrac{3.0}{2.0 - \cos \theta}$, where measurements are in inches. Plot the curve that represents the shape of the cam.

We get the points for the edge of the cam by using values of θ from 0 to π for the upper "half" and from π to 2π for the lower "half." The table of values follows.

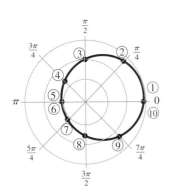

Fig. 12-15

$r = 2.0 + \cos \theta$

θ	0	$\frac{\pi}{4}$	$\frac{\pi}{2}$	$\frac{3\pi}{4}$	π
r	3.0	2.7	2.0	1.3	1.0
Point number	1	2	3	4	5

$r = \dfrac{3.0}{2.0 - \cos \theta}$

θ	π	$\frac{5\pi}{4}$	$\frac{3\pi}{2}$	$\frac{7\pi}{4}$	2π
r	1.0	1.1	1.5	2.3	3.0
Point number	6	7	8	9	10

The upper "half" is part of a limaçon, and the lower "half" is a semiellipse. The cam is shown in Fig. 12-15.   ■

■**EXAMPLE 5**   Plot the graph of $r = 2 \cos 2\theta$.

In finding values of r we must be careful first to multiply the values of θ by 2 before finding the cosine of the angle. Also, for this reason, we take values of θ as multiples of $\pi/12$, so as to get enough useful points. The table of values follows.

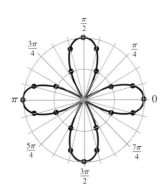

Fig. 12-16

θ	0	$\frac{\pi}{12}$	$\frac{\pi}{6}$	$\frac{\pi}{4}$	$\frac{\pi}{3}$	$\frac{5\pi}{12}$	$\frac{\pi}{2}$
r	2	1.7	1	0	-1	-1.7	-2

θ	$\frac{7\pi}{12}$	$\frac{2\pi}{3}$	$\frac{3\pi}{4}$	$\frac{5\pi}{6}$	$\frac{11\pi}{12}$	π
r	-1.7	-1	0	1	1.7	2

For values of θ starting with π, the values of θ repeat. We have a four-leaf **rose,** as shown in Fig. 12-16.   ■

■**EXAMPLE 6**   Plot the graph of $r^2 = 9 \cos 2\theta$.

Choosing the indicated values of θ, we get the values of r shown in the following table of values.

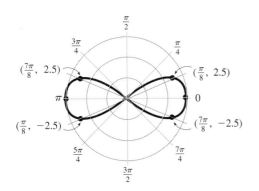

Fig. 12-17

θ	0	$\frac{\pi}{8}$	$\frac{\pi}{4}$	$\cdots$	$\frac{3\pi}{4}$	$\frac{7\pi}{8}$	π
r	± 3	± 2.5	0		0	± 2.5	± 3

There are no values of r corresponding to values of θ in the range $\pi/4 < \theta < 3\pi/4$, since twice these angles are in the second and third quadrants and the cosine is negative for such angles. The value of r^2 cannot be negative. Also, the values of r repeat for $\theta > \pi$. The figure is called a **lemniscate** and is shown in Fig. 12-17.   ■

See the chapter introduction.

EXAMPLE 7 The planet Mercury travels around the Sun in an orbit that is represented by the polar equation.

$$r = \frac{3.44 \times 10^7}{1 - 0.206 \cos \theta}$$

where r is measured in miles. View the graph of this curve on a graphing calculator.

Using the *mode* feature, a polar equation is displayed using the polar graph option or the parametric graph option, depending on the calculator. (Review the manual for the calculator.)

With the polar graph option, the function is entered directly as it is shown. The values for the viewing window are determined by settings for x, y, and the angles θ that will be used.

With the parametric graph option, to graph $r = f(\theta)$, we note that $x = r \cos \theta$ and $y = r \sin \theta$. This tells us that

$$x = f(\theta)\cos \theta$$
$$y = f(\theta)\sin \theta$$

Therefore, for the polar equation representing the polar equation for Mercury's orbit, by using

$$x = \left(\frac{3.44 \times 10^7}{1 - 0.206 \cos \theta}\right)\cos \theta \quad \text{and} \quad y = \left(\frac{3.44 \times 10^7}{1 - 0.206 \cos \theta}\right)\sin \theta$$

the graph can be displayed. From Fig. 12-18 we see that the orbit appears to be an ellipse. Actually, it is an ellipse with the sun at a focus, which in this case is at the pole.

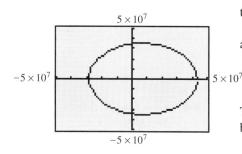

Fig. 12-18

— **EXERCISES** $12\text{-}2$ —

In Exercises 1–28, plot the curves of the given polar equations in polar coordinates.

1. $r = 4$　　　　**2.** $r = 2$　　　　**3.** $\theta = 3\pi/4$

4. $\theta = -1.5$　　**5.** $r = 4 \sec \theta$　　**6.** $r = 4 \csc \theta$

7. $r = 2 \sin \theta$　　**8.** $r = 3 \cos \theta$

9. $r = 1 - \cos \theta$ (cardioid)

10. $r = \sin \theta - 1$ (cardioid)

11. $r = 2 - \cos \theta$ (limaçon)

12. $r = 2 + 3 \sin \theta$ (limaçon)

13. $r = 4 \sin 2\theta$ (rose)　　**14.** $r = 2 \sin 3\theta$ (rose)

15. $r^2 = 4 \sin 2\theta$ (lemniscate)

16. $r^2 = 2 \sin \theta$

17. $r = 1.2^\theta$ (spiral)　　**18.** $r = 1.5^{-\theta}$ (spiral)

19. $r = 4|\sin 3\theta|$　　**20.** $r = 2 \sin \theta \tan \theta$ (cissoid)

21. $r = \dfrac{1}{2 - \cos \theta}$ (ellipse)

22. $r = \dfrac{1}{1 - \cos \theta}$ (parabola)

23. $r = \dfrac{6}{1 - 2 \cos \theta}$ (hyperbola)

24. $r = \dfrac{6}{3 - 2 \sin \theta}$ (ellipse)

25. $r = 4 \cos \frac{1}{2}\theta$　　　**26.** $r = 2 + \cos 3\theta$

27. $r = 2(1 - \sin(\theta - \pi/4))$　　**28.** $r = 4 \tan \theta$

In Exercises 29–32, view the curves of the given polar equations on a graphing calculator.

29. $r = \theta$ $(-20 \le \theta \le 20)$　　**30.** $r = 0.5^{\sin \theta}$

31. $r = 2 \sec \theta + 1$　　**32.** $r = 2 \cos(\cos 2\theta)$

In Exercises 33–36, sketch the indicated graphs.

33. An architect designs a patio shaped such that it can be described as the region within the polar curve $r = 4.0 - \sin \theta$, where measurements are in meters. Sketch the curve that represents the perimeter of the patio.

34. A missile is fired at an airplane and is always directed toward the airplane. The missile is traveling at twice the speed of the airplane. An equation that describes the distance r between the missile and the airplane is $r = \dfrac{70 \sin \theta}{(1 - \cos \theta)^2}$, where θ is the angle between their directions at all times. See Fig. 12-19. This is a *relative pursuit curve*. Sketch the graph of this equation for $\pi/4 \le \theta \le \pi$.

W **35.** Describe the type of conic section that is defined by the polar equation $r = a \sec^2 \frac{1}{2}\theta$. Use a graphing calculator to display the graph for $a = 1$.

36. Sketch the graph of the rectangular equation
$$4(x^6 + 3x^4y^2 + 3x^2y^4 + y^6 - x^4 - 2x^2y^2 - y^4) + y^2 = 0$$
(*Hint:* The equation can be written as
$4(x^2 + y^2)^3 - 4(x^2 + y^2)^2 + y^2 = 0$.) Transform this equation to polar coordinates and then sketch the curve.

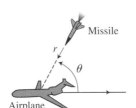

Fig. 12-19

Missile

Airplane

12-3 APPLICATIONS OF DIFFERENTIATION AND INTEGRATION IN POLAR COORDINATES

Polar coordinates are used extensively in applications and theoretical discussions in advanced mechanics, electricity, and certain other fields. In addition, a number of applications of differentiation and integration that we have previously discussed may be solved in polar coordinates. In this section we shall discuss an application of differentiation and one of integration. Certain other applications are shown in the exercises.

The application of differentiation is that of velocity. In Section 4-3 we were able to determine the resultant velocity of an object if we knew its x- and y-components. Here we shall consider the components of velocity in the direction of increasing r and increasing θ. These components are denoted by v_r and v_θ.

*The **radial component** of velocity v_r, and the **transverse component** of velocity v_θ, are given by*

$$v_r = \frac{dr}{dt} \quad \text{and} \quad v_\theta = r\frac{d\theta}{dt} \tag{12-3}$$

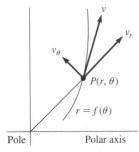

Fig. 12-20

See Fig. 12-20. To show that these are the proper relations we recall that $x = r \cos \theta$ and $y = r \sin \theta$. Taking derivatives with respect to time, treating both r and θ as functions of time, we have

$$\frac{dx}{dt} = \cos \theta \frac{dr}{dt} - r \sin \theta \frac{d\theta}{dt}$$

$$\frac{dy}{dt} = \sin \theta \frac{dr}{dt} + r \cos \theta \frac{d\theta}{dt}$$

Since $dx/dt = v_x$ and $dy/dt = v_y$, and using Eqs. (12-3), we have

$$v_x = v_r \cos \theta - v_\theta \sin \theta$$

$$v_y = v_r \sin \theta + v_\theta \cos \theta$$

Squaring both sides of each of these equations and adding, we have

$$v_x^2 = v_r^2 \cos^2 \theta - 2v_r v_\theta \cos \theta \sin \theta + v_\theta^2 \sin^2 \theta$$
$$v_y^2 = v_r^2 \sin^2 \theta + 2v_r v_\theta \cos \theta \sin \theta + v_\theta^2 \cos^2 \theta$$

$$\boxed{v_x^2 + v_y^2 = v_r^2 + v_\theta^2 = v^2} \qquad (12\text{-}4)$$

This analysis, along with the fact that dr/dt is the velocity in the r-direction for constant θ and $r(d\theta/dt)$ is the velocity in the θ-direction for constant r, verifies that these are the correct relations. The following examples illustrate the use of Eqs. (12-3) and (12-4).

EXAMPLE 1   Find the radial and transverse components of velocity of an object moving such that $r = t^2 - 2$ and $\theta = \frac{1}{4}t^2$, when $t = 2$. From these values calculate the magnitude of the velocity when $t = 2$.

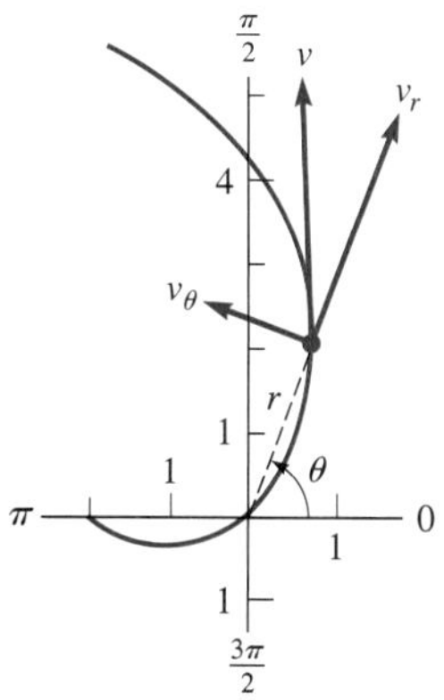

$$v_r = \frac{dr}{dt} = 2t \qquad v_\theta = r\frac{d\theta}{dt} = (t^2 - 2)\left(\frac{t}{2}\right) = \frac{1}{2}(t^3 - 2t)$$

Evaluating for $t = 2$, we have

$$v_r\big|_{t=2} = 4$$
$$v_\theta\big|_{t=2} = \frac{1}{2}(8 - 4) = 2$$

Thus,

$$v = \sqrt{16 + 4} = \sqrt{20} = 4.47$$

See Fig. 12-21.

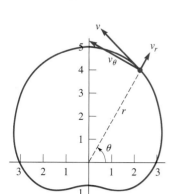

Fig. 12-21

EXAMPLE 2   A particle is moving along the curve $r = 3 + 2\sin\theta$ with constant angular velocity, $\omega = d\theta/dt = 2.00$ rad/s (see Fig. 12-22). Find the magnitude of the velocity (in cm/s) when $\theta = 1.05$.

Taking derivatives of $r = 3 + 2\sin\theta$ with respect to t, we have

$$\frac{dr}{dt} = 2\cos\theta\frac{d\theta}{dt}$$

Therefore,

$$v_r = \frac{dr}{dt} = 2\cos\theta\frac{d\theta}{dt}$$
$$v_\theta = r\frac{d\theta}{dt} = (3 + 2\sin\theta)\frac{d\theta}{dt}$$

Evaluating, we have

$$v_r\big|_{\theta=1.05} = 2(\cos 1.05)(2.00) = 1.99 \text{ cm/s}$$
$$v_\theta\big|_{\theta=1.05} = (3 + 2\sin 1.05)(2.00) = 9.47 \text{ cm/s}$$
$$v = \sqrt{1.99^2 + 9.47^2} = 9.68 \text{ cm/s}$$

Fig. 12-22

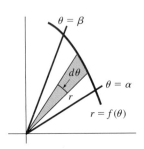

Fig. 12-23

The application of integration in polar coordinates we shall consider is the area bounded by polar curves. We recall Eq. (7-28), which states that the area of a circular sector is $A = \frac{1}{2}\theta r^2$. In Fig. 12-23 consider the element of area with radius r and central angle $d\theta$, as shown. If these elements are summed (integrated) from $\theta = \alpha$ to $\theta = \beta$, the sum will represent the area shown. Thus, *the area is found by*

$$A = \frac{1}{2}\int_{\alpha}^{\beta} r^2\, d\theta \qquad (12\text{-}5)$$

The following examples illustrate the use of Eq. (12-5).

■EXAMPLE 3 Find the area enclosed by one loop of the lemniscate $r^2 = \cos 2\theta$.

Using Eq. (12-5) we substitute $r^2 = \cos 2\theta$. To determine the limits α and β, we note the symmetry of the curve (see Fig. 12-24). By choosing $\alpha = 0$ and β to be the value where r is first zero, we will find the area of the upper half of the loop. Doubling this will give the desired area. Therefore, by setting $r^2 = \cos 2\theta = 0$, we find $2\theta = \pi/2$, or $\theta = \pi/4$. This means $\beta = \pi/4$. This means the integral for the area is

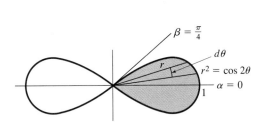

Fig. 12-24

$$A = 2\left[\frac{1}{2}\int_0^{\pi/4} \cos 2\theta\, d\theta\right]$$

$$= \frac{1}{2}\int_0^{\pi/4} \cos 2\theta (2\, d\theta)$$

$$= \frac{1}{2}\sin 2\theta\bigg|_0^{\pi/4}$$

$$= \frac{1}{2}(1 - 0) = \frac{1}{2} \qquad\qquad \blacksquare$$

■EXAMPLE 4 Find the area enclosed by the curve $r = 2 - \cos \theta$.

Using Eq. (12-5) we substitute $r^2 = (2 - \cos \theta)^2 = 4 - 4\cos \theta + \cos^2 \theta$. To find the limits α and β, we note the sketch of the curve in Fig. 12-25. Since it is symmetric to the polar axis, we may let $\alpha = 0$ and $\beta = \pi$ if we also multiply by 2. (Limits of $\alpha = 0$ and $\beta = 2\pi$ would also be proper.) Thus,

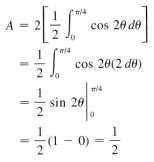

Fig. 12-25

$$A = 2\left[\frac{1}{2}\int_0^{\pi}(4 - 4\cos \theta + \cos^2 \theta)\, d\theta\right]$$

By using Formula 32 of Appendix E and the identity $\sin 2\theta = 2\sin \theta \cos \theta$, we have

$$A = \left[4\theta - 4\sin \theta + \frac{\theta}{2} + \frac{1}{4}\sin 2\theta\right]_0^{\pi}$$

$$= \left(4\pi + \frac{\pi}{2}\right) = \frac{9\pi}{2} \qquad\qquad \blacksquare$$

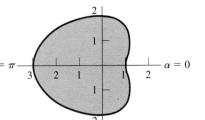

For reference, Formula 32 is
$$\int \cos^2 u\, du = \frac{u}{2} + \frac{1}{2}\sin u \cos u$$

Using Eq. (12-5) we can find the area between two polar curves. The following example illustrates how this is done.

EXAMPLE 5   Find the area outside the cardioid $r = 1 + \sin\theta$ and inside the circle $r = 3\sin\theta$.

After drawing the figure, we see that the required area is the shaded portion of Fig. 12-26. We find the limits of integration by solving the equations simultaneously. By substituting $r = 3\sin\theta$ into the equation of the cardioid, we have

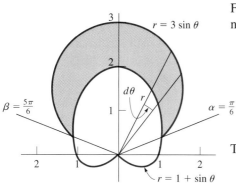

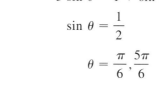

$$3\sin\theta = 1 + \sin\theta$$

$$\sin\theta = \frac{1}{2}$$

$$\theta = \frac{\pi}{6}, \frac{5\pi}{6}$$

Fig. 12-26

This means the points $(3/2, \pi/6)$ and $(3/2, 5\pi/6)$ are common to each curve.

Therefore, the area inside the circle between $\theta = \pi/6$ and $\theta = 5\pi/6$ is

$$A_1 = \frac{1}{2}\int_{\pi/6}^{5\pi/6}(3\sin\theta)^2\,d\theta$$

The area inside the cardioid between $\theta = \pi/6$ and $\theta = 5\pi/6$ is

$$A_2 = \frac{1}{2}\int_{\pi/6}^{5\pi/6}(1+\sin\theta)^2\,d\theta$$

The desired area is the difference of A_1 and A_2. Thus,

$$A = \frac{1}{2}\int_{\pi/6}^{5\pi/6}[(3\sin\theta)^2 - (1+\sin\theta)^2]\,d\theta$$

$$= \frac{1}{2}\int_{\pi/6}^{5\pi/6}(8\sin^2\theta - 2\sin\theta - 1)\,d\theta$$

$$= \frac{1}{2}\int_{\pi/6}^{5\pi/6}(4 - 4\cos 2\theta - 2\sin\theta - 1)\,d\theta$$

$$= \frac{1}{2}(3\theta - 2\sin 2\theta + 2\cos\theta)\,\Big|_{\pi/6}^{5\pi/6}$$

$$= \frac{1}{2}\left[3\left(\frac{5\pi}{6} - \frac{\pi}{6}\right) - 2\left(\sin\frac{5\pi}{3} - \sin\frac{\pi}{3}\right) + 2\left(\cos\frac{5\pi}{6} - \cos\frac{\pi}{6}\right)\right]$$

$$= \frac{1}{2}\left[2\pi - 2\left(-\frac{\sqrt{3}}{2} - \frac{\sqrt{3}}{2}\right) + 2\left(-\frac{\sqrt{3}}{2} - \frac{\sqrt{3}}{2}\right)\right]$$

$$= \pi$$

Advantage could have been taken of the symmetry by integrating from $\pi/6$ to $\pi/2$ and doubling the result.

EXERCISES $12\text{-}3$

In Exercises 1–4, r and θ are given as functions of t. Find the magnitudes of the velocities by finding v_r and v_θ for the given values of t.

1. $r = 3t$, $\theta = t^3$, $t = 1$

• 2. $r = 4t$, $\theta = \sqrt{t + 1}$, $t = 3$

3. $r = e^{0.1t}$, $\theta = \cos 2t$, $t = 2$

4. $r = \sin \dfrac{\pi t}{6}$, $\theta = \dfrac{1}{2t + 1}$, $t = 2$

In Exercises 5–12, find the magnitude of the velocity of an object moving along the given curves with a constant angular velocity $\omega = 2.00$ rad/s for the indicated values of θ and r (in ft).

5. $r = 2$, $\theta = \pi/3$ 6. $r = 4\theta$, $\theta = 2$

• 7. $r = \sin \theta$, $\theta = \pi/6$ 8. $r = \cos 2\theta$, $\theta = \pi/4$

9. $r = 4 - \cos \theta$, $\theta = \pi/2$

• 10. $r = 5 + 2 \sin 2\theta$, $\theta = \pi/6$

11. $r = \dfrac{1}{2 - \sin \theta}$, $\theta = \pi/6$

12. $r = 1 + \cos^2 \theta$, $\theta = \pi/4$

In Exercises 13–28, find the area bounded by the given curves.

13. $\theta = \frac{1}{6}\pi$, $\theta = \frac{2}{3}\pi$, $r = 2$ 14. $\theta = 0$, $\theta = \frac{4}{3}\pi$, $r = 3$

15. $\theta = 0$, $\theta = \frac{1}{4}\pi$, $r = \sec \theta$

16. $\theta = \frac{1}{6}\pi$, $\theta = \frac{1}{2}\pi$, $r \sin \theta = 2$

17. $r = 2 \cos \theta$ 18. $\theta = 0$, $\theta = \frac{1}{3}\pi$, $r = 4 \sin \theta$

• 19. $\theta = 0$, $\theta = \frac{1}{2}\pi$, $r = e^{2\theta}$ 20. $r = 1 - \cos \theta$

21. $r = 2 \sin 3\theta$ • 22. $r = 3 \cos 2\theta$

23. $r = 2 + \cos \theta$ 24. $r^2 = 4 \sin 2\theta$

• 25. Inside $r = 1 + \sin \theta$, outside $r = 1$

26. Inside $r = \sin \theta$, outside $r = 1 + \cos \theta$

• 27. Area common to $r = 2 \sin \theta$ and $r = 2 \cos \theta$

• 28. Inside $r = 5 \cos \theta$, outside $r = 2 + \cos \theta$

In Exercises 29–36, solve the given problems.

29. The outer edge of a rotating disk moves in a counterclockwise direction along the circle $r = 2.00 \sin \theta$. For $\theta = 1.05$ and $v = 10.0$ ft/s, find v_r and v_θ.

30. An object moves along the parabolic path $r = 4.00/(1 + \cos \theta)$ with a constant speed of 4.00 m/s. Find v_r and v_θ when $\theta = 1.57$.

31. A cam is in the shape of the curve $r = 2.00(3 - \cos \theta)$. Find the area (in cm^2) of one face of the cam.

32. Change $(x^2 + y^2)^2 - (x^2 + y^2) = 2xy$ into a polar coordinate equation and find the area bounded by $\theta = 0$, $\theta = \pi/4$, and the curve.

33. Using $x = r \cos \theta$ and $y = r \sin \theta$, take derivatives of each with respect to θ and treat r as a function of θ. From these results show that $\dfrac{dy}{dx} = \dfrac{r' \tan \theta + r}{r' - r \tan \theta}$ represents the slope of a line tangent to a polar curve. Here $r' = dr/d\theta$.

34. Find the slope of a line tangent to $r = \cos \theta$ at $\theta = \pi/6$. (See Exercise 33.)

35. The length of arc along a polar curve is given by $s = \displaystyle\int_\alpha^\beta \sqrt{r^2 + (r')^2}\, d\theta$, where $r' = dr/d\theta$. Using this equation find the total perimeter of the curve of $r = \cos^2(\theta/2)$.

36. Find the circumference of the circle $r = \sin \theta + \cos \theta$. (See Exercise 35.)

$12\text{-}4$ CYLINDRICAL COORDINATES

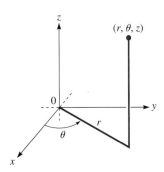

Fig. 12-27

Another set of coordinates that can be used to display a three-dimensional figure are **cylindrical coordinates,** *which are polar coordinates and the z-axis combined.* In using cylindrical coordinates, every point in space is designated by the coordinates (r, θ, z), as shown in Fig. 12-27. The equations relating the rectangular coordinates (x, y, z) and the cylindrical coordinates (r, θ, z) of a point are

$$x = r \cos \theta \qquad y = r \sin \theta \qquad z = z$$
$$r^2 = x^2 + y^2 \qquad \tan \theta = \frac{y}{x}$$

(12-6)

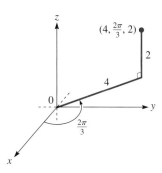

Fig. 12-28

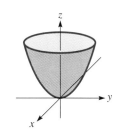

Fig. 12-29

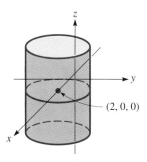

Fig. 12-30

■**EXAMPLE 1**   **(a)** Plot the point with cylindrical coordinates $(4, 2\pi/3, 2)$ and find its corresponding rectangular coordinates.

The point is shown in Fig. 12-28. From Eqs. (12-6) we have

$$x = 4 \cos \frac{2\pi}{3} = 4\left(-\frac{1}{2}\right) = -2 \qquad y = 4 \sin \frac{2\pi}{3} = 4\left(\frac{\sqrt{3}}{2}\right) = 2\sqrt{3} \qquad z = 2$$

Therefore, the rectangular coordinates of the point are $(-2, 2\sqrt{3}, 2)$.

(b) Find the cylindrical coordinates of the point with rectangular coordinates $(2, -2, 6)$.

Using Eqs. (12-6) we have

$$r = \sqrt{2^2 + (-2)^2} = 2\sqrt{2} \qquad z = 6$$

$$\tan \theta = \frac{-2}{2} = -1 \quad \text{(quadrant IV)} \qquad \theta = 2\pi - \frac{\pi}{4} = \frac{7\pi}{4}$$

The cylindrical coordinates are $(2\sqrt{2}, 7\pi/4, 6)$. As with polar coordinates, the value of θ can be $7\pi/4 + 2n\pi$, where n is an integer.   ▪

■**EXAMPLE 2**   Find the equation for the surface $4x^2 + 4y^2 = z$ in cylindrical coordinates and sketch the surface.

Factoring the 4 from the terms on the left, we have $4(x^2 + y^2) = z$. We now use the fact that $x^2 + y^2 = r^2$ to write this equation in cylindrical coordinates as $4r^2 = z$. From this equation we see that $z \ge 0$ and that as z increases we have circular sections of increasing radius. Therefore, it is a **circular paraboloid.** See Fig. 12-29.   ▪

■**EXAMPLE 3**   Find the equation for the surface $r = 4 \cos \theta$ in rectangular coordinates and sketch the surface.

If we multiply each side of the equation by r, we obtain $r^2 = 4r \cos \theta$. Now from Eqs. (12-6) we have $r^2 = x^2 + y^2$ and $r \cos \theta = x$, which means that $x^2 + y^2 = 4x$, or

$$x^2 - 4x + 4 + y^2 = 4$$
$$(x - 2)^2 + y^2 = 4$$

We recognize this as a cylinder that has as its trace in the xy-plane a circle with center at $(2, 0, 0)$, as shown in Fig. 12-30.   ▪

We can see from Examples 2 and 3 that certain equations take on a simpler form when written in cylindrical coordinates. The equation of a right circular cylinder, with the center of its trace in the xy-plane at the origin, is simply $r = a$, where a is the radius of the cylinder. This is the reason that these coordinates have been named "cylindrical" coordinates.

Volumes in Cylindrical Coordinates

We can use cylindrical coordinates to find the volume of a three-dimensional figure in a manner very similar to that used with rectangular coordinates. We recall that, when using double integration, the element of volume in rectangular coordinates is $z \, dx \, dy$. Since the z-coordinate is the same in cylindrical coordinates, we again use z for the height of the element. However, we must be careful to get the proper element of area in terms of dr and $d\theta$.

Noting the element of area shown in Fig. 12-31, we can see that as $\Delta\theta$ becomes smaller, the element of area more nearly approximates a rectangle with sides of Δr and $r\Delta\theta$. It can be shown that in the limit as $\Delta r \to 0$ and $\Delta\theta \to 0$, the element of area is $r\,dr\,d\theta$. Therefore, we see that *the volume above the xy-plane, bounded by* $z = f(r,\theta)$, $r_1 = f_1(\theta)$, $r_2 = f_2(\theta)$, $\theta = \alpha$, *and* $\theta = \beta$ *is*

CAUTION ▶

$$V = \int_{\alpha}^{\beta} \int_{r_1}^{r_2} f(r,\theta) r\,dr\,d\theta \qquad \textbf{(12-7)}$$

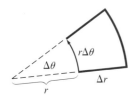

Fig. 12-31

The following example illustrates finding a volume by use of Eq. (12-7).

■EXAMPLE 4 Find the volume of the solid cut from the sphere $x^2 + y^2 + z^2 = 9$ by the cylinder $x^2 + y^2 - 3y = 0$.

First, using Eqs. (12-6) we change the rectangular equations to polar equations. The equation of the sphere becomes $r^2 + z^2 = 9$, and the equation of the cylinder becomes $r^2 - 3r\sin\theta = 0$, or $r = 3\sin\theta$. Because of the symmetry, the volume in the first octant is one-fourth of the total volume. Therefore, in Fig. 12-32, one-fourth of the volume in the first octant is shown. We see that $z = f(r,\theta) = \sqrt{9 - r^2}$, the limits on r are 0 and $3\sin\theta$, and the limits on θ are 0 and $\pi/2$. Therefore, the volume is found as

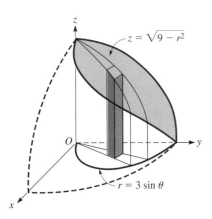

Fig. 12-32

$$V = 4\int_0^{\pi/2} \int_0^{3\sin\theta} \sqrt{9 - r^2}\, r\,dr\,d\theta$$

$$= \frac{4}{3}\int_0^{\pi/2} \left[-(9 - r^2)^{3/2}\right]\Big|_0^{3\sin\theta} d\theta$$

$$= \frac{4(27)}{3}\int_0^{\pi/2} (1 - \cos^3\theta)\,d\theta = 6(3\pi - 4)$$

In evaluating the first integral, we factored the common factor of 27 and also used the identity $\sin^2\theta + \cos^2\theta = 1$. The final integration is performed by the method of Chapter 9 or by use of Formula 33 in Appendix E. ■

EXERCISES *12-4*

In Exercises 1 and 2, find the rectangular coordinates of the points for which the cylindrical coordinates are given.

1. (a) $(2, \pi/2, 4)$ (b) $(3, -\pi/4, 5)$

2. (a) $(4, \pi/3, 2)$ (b) $(3, 5\pi/6, 2)$

In Exercises 3 and 4, find the cylindrical coordinates of the points for which the rectangular coordinates are given.

3. (a) $(6, 8, 5)$ (b) $(4, 4, -3)$

4. (a) $(8, 15, -6)$ (b) $(\sqrt{3}, -2, 1)$

In Exercises 5 and 6, describe the surface for which the cylindrical equation is given.

5. (a) $r = 2$ (b) $\theta = 2$ **6.** (a) $\theta = -2$ (b) $z = -2$

In Exercises 7 and 8, the given equations are in rectangular coordinates. Write them in cylindrical coordinates and describe the surface.

7. (a) $x = 2$ (b) $z = 2$ **8.** (a) $y = x$ (b) $x^2 + y^2 = 4$

In Exercises 9–14, the given equations are in rectangular coordinates. Write them in cylindrical coordinates and sketch the curves.

9. $x^2 + y^2 = 16$ **10.** $x^2 + y^2 = 4z$

11. $x^2 + y^2 + 4z^2 = 4$ **12.** $x = 2$

13. $9z = 4x^2 + 4y^2$ **14.** $z^2 + 9 = x^2 + y^2$

In Exercises 15–20, find the equation in rectangular coordinates of the surfaces whose equations are given in cylindrical coordinates and sketch the curves.

15. $r^2 = 4z$ **16.** $r = 4 \cos \theta$

17. $r^2 + 4z^2 = 16$ **18.** $r = z$

19. $r \sin \theta = 3$ **20.** $r = \dfrac{1}{\sin \theta - \cos \theta}$

In Exercises 21–24, find the volumes bounded by the indicated surfaces.

21. $z = 1 - r^2, z = 0$ **22.** $z = r, r = 2 \cos \theta, z = 0$

23. A grain storage silo can be described as the volume within the cylinder $x^2 + y^2 = 16$ and under a segment of the hemisphere $z = \sqrt{25 - x^2 - y^2}$. Find the volume (in m^3) of the silo.

W 24. The double integral $\displaystyle\int_0^{\pi/2} \int_0^{2\cos\theta} r^3 \, dr \, d\theta$ describes the volume of a machine part. Describe the part.

CHAPTER EQUATIONS

Polar and rectangular coordinates	$x = r \cos \theta \qquad y = r \sin \theta$	(12-1)
	$\tan \theta = \dfrac{y}{x} \qquad r = \sqrt{x^2 + y^2}$	(12-2)
Radial and transverse components of velocity	$v_r = \dfrac{dr}{dt} \quad \text{and} \quad v_\theta = r \dfrac{d\theta}{dt}$	(12-3)
	$v_x^2 + v_y^2 = v_r^2 + v_\theta^2 = v^2$	(12-4)
Area in polar coordinates	$A = \dfrac{1}{2} \displaystyle\int_\alpha^\beta r^2 \, d\theta$	(12-5)
Cylindrical coordinates	$x = r \cos \theta \qquad y = r \sin \theta \qquad z = z$	
	$r^2 = x^2 + y^2 \qquad \tan \theta = \dfrac{y}{x}$	(12-6)
Volume in cylindrical coordinates	$V = \displaystyle\int_\alpha^\beta \int_{r_1}^{r_2} f(r, \theta) r \, dr \, d\theta$	(12-7)

REVIEW EXERCISES

In Exercises 1–4, find the polar equation of each of the given rectangular equations.

1. $y = 2x$ **2.** $2xy = 1$

3. $x^2 - y^2 = 16$ **4.** $x^2 + y^2 = 7 - 6y$

In Exercises 5–8, find the rectangular equation of each of the given polar equations.

5. $r = 2 \sin 2\theta$ **6.** $r^2 = \sin \theta$

7. $r = \dfrac{4}{2 - \cos \theta}$ **8.** $r = \dfrac{2}{1 - \sin \theta}$

In Exercises 9–20, plot the curves of the given polar equations.

9. $r = 4(1 + \sin \theta)$ **10.** $r = 1 + 2 \cos \theta$

11. $r = 4 \cos 3\theta$ **12.** $r = -3 \sin \theta$

13. $r = \cot \theta$ **14.** $r = \dfrac{1}{2(\sin \theta - 1)}$

15. $r = 2 \sin\left(\dfrac{\theta}{2}\right)$ **16.** $r = \theta$

17. $r = 2 + \cos 3\theta$ **18.** $r = 3 - \cos 2\theta$

19. $r^2 = \theta$ **20.** $r^2 = 4 \cos \theta$

In Exercises 21–24, view the graphs of the given polar equations on a graphing calculator.

21. $r = 2 + 3 \csc \theta$ **22.** $r = 2 - 3^{\sin \theta}$

23. $r = \dfrac{\sin \theta}{1 + \cos \theta}$ **24.** $r = 1 + 3 \sin(\cos 3\theta)$

In Exercises 25–32, find the area bounded by the curves of the given equations.

25. $r = 4 \cos \theta, \theta = 0, \theta = \dfrac{\pi}{4}$

26. $r = 2 \csc \theta, \theta = \dfrac{\pi}{4}, \theta = \dfrac{\pi}{2}$

27. $r = 3 - \sin\theta$ 28. $r = \sqrt{\sin\theta}$

29. Inside $r = 6$, to the right of $r\cos\theta = 5$

30. Smaller loop of $r = 1 - 2\cos\theta$

31. Between $r = \theta$ and $\theta = \pi/2$ ($\theta \geq 0$)

32. Between $r = e^\theta$, $\theta = 0$, and $\theta = 2\pi$

In Exercises 33–44, solve the given problems.

33. Find the cylindrical coordinates of the points with the given rectangular coordinates: (a) $(5, 12, 3)$ (b) $(3, -3, 2)$.

34. Find the rectangular coordinates of the points with the given cylindrical coordinates: (a) $(8, \pi/6, 3)$ (b) $(5, -\pi/2, 4)$.

35. Using a graphing calculator, determine the number of points of intersection of the polar curves $r = 4|\cos 2\theta|$ and $r = 6\sin[\cos(\cos 3\theta)]$.

(W) 36. Describe the solid given by the inequality $1 \leq z \leq 2 - r^2$. Sketch the graph.

37. An object moves such that $r = t^3 - 6t$ and $\theta = t/(t+1)$, where t is the time. Find the magnitude of the velocity of the object when $t = 1$.

38. Express the equation of the planet Pluto (see Exercise 32 on page 51) in polar coordinates.

39. Under certain conditions the electric potential V is given by $V = \left(\dfrac{a}{r^2} - br\right)\cos\theta$, where a and b are constants. The polar components of the electric field are given by $E_r = -\dfrac{\partial V}{\partial r}$ and $E_\theta = -\dfrac{1}{r}\dfrac{\partial V}{\partial\theta}$. Find the expressions for E_r and E_θ in terms of r and θ.

40. The volume within a certain container can be described as being bounded by $z = 3.00$ and $z = r^2$. Find the volume (in cm^3) of the container.

41. The sound produced by a jet engine was measured at a distance of 100 m in all directions. The loudness of the sound d (in decibels) was found to be $d = 115 + 10\cos\theta$, where the $0°$ line for the angle θ is directed in front of the engine. Sketch the graph of d vs. θ in polar coordinates (use d as r).

42. From a fire tower, the terrain effectively seen is within a circle whose equation is described by $r = 10\cos\theta$. What is the observation area beyond the straight road shown in Fig. 12-33?

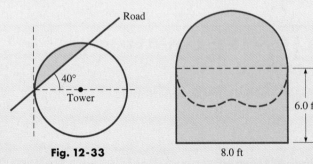

Fig. 12-33   **Fig. 12-34**

43. The top section of the perimeter of a window is part of a limaçon, and the bottom section is rectangular, as shown in Fig. 12-34. If the limaçon can be described by the equation $r = 4.0 + \sin\theta$, what is the area of the window?

44. The path of a certain plane is $r = 200(\sec\theta + \tan\theta)^{-5}/\cos\theta$, $0 < \theta < \pi/2$. Sketch the path and check it on a graphing calculator.

Writing Exercise

45. If a force varies inversely as the square of the distance from an attracting object (such as the sun exerts on earth), an object follows a path given by $r^{-1} = a + b\cos\theta$, where a and b are constants. Write this equation in rectangular coordinates and then write two or three paragraphs explaining why it represents one of the conic sections, depending on the values of a and b. It is through this kind of analysis that we know the paths of the planets and comets are conic sections.

PRACTICE TEST

1. Find the polar equation of the curve whose rectangular equation is $x^2 = 2x - y^2$.

2. Find the rectangular coordinates for the point $(2, 3\pi/4, 3)$ in cylindrical coordinates.

3. Plot the polar curve $r = 3 + \cos\theta$.

4. Find the rectangular equation for the polar equation $r = 4\cos\theta + 2\sin\theta$. Identify the curve.

5. Find the area bounded by one loop of the polar curve $r^2 = 4\sin 2\theta$.

6. Find the volume bounded by the cylindrical coordinate surfaces $4z = r^2$, $r = 2\sec\theta$, $z = 0$, $\theta = 0$, and $\theta = \pi/4$.

7. Find the magnitude of the velocity (in m/s) of an object moving along the polar curve $r = 2.0 + 3.0\cos 2\theta$ with a constant angular velocity $\omega = 3.0$ rad/s for $\theta = 0.26$.

13

EXPANSION OF FUNCTIONS IN SERIES

The values of the trigonometric functions can be determined exactly only for a few particular angles. The value of *e* can be approximated only in decimal form. The question arises as to how these values may be found, particularly if a specified degree of accuracy is necessary. In this chapter we show how a given function may be expressed in terms of a polynomial. Once this polynomial that approximates the function has been found, we shall be able to evaluate the function to any desired degree of accuracy. A number of other applications and uses of this polynomial are shown as well.

In the final two sections we show how certain types of functions can be represented by a series of sine and cosine terms. Such series are very useful in the study of functions and applications that are periodic in nature, such as mechanical vibrations and the currents and voltages in an ac electric circuit.

We begin by reviewing the meanings and properties of sequences and series.

Many types of electronic devices are used to control the current in a circuit. In Section 13-6 we see how series are used to analyze one such device.

13-1 INFINITE SERIES

In algebra certain sequences of numbers are introduced. These include **arithmetic sequences** and **geometric sequences** (also called *progressions*). In an arithmetic sequence each term after the first is obtained by adding a fixed number (the *common difference*) to the preceding term. In a geometric sequence each term after the first is obtained by multiplying the preceding term by a fixed number (the *common ratio*).

EXAMPLE 1   (a) An example of an arithmetic sequence is 2, 5, 8, 11,..., where the common difference is 3.

(b) An example of a geometric sequence is 2, 4, 8, 16,..., where the common ratio is 2.

There are many other ways of generating sequences. Squares of 1, 2, 3, 4,... form the sequence 1, 4, 9, 16,.... Also, the successive approximations $x_1, x_2, x_3, \ldots$ found by using Newton's method in solving a particular equation form a sequence.

In general, *a* **sequence** (*or* **infinite sequence**) *is an infinite succession of numbers. Each of the numbers is a* **term** *of the sequence.* Each term of the sequence is associated with a positive integer, although at times it is convenient to associate the first term with zero (or some specified positive integer). We shall use a_n to designate the term of the sequence corresponding to the integer n.

■ EXAMPLE 2 Find the first three terms of the sequence for which $a_n = 2n + 1$, $n = 1, 2, 3, \ldots$.

Substituting the values of n, we obtain the values

$$a_1 = 2(1) + 1 = 3 \qquad a_2 = 2(2) + 1 = 5 \qquad a_3 = 2(3) + 1 = 7, \ldots$$

Therefore, we have the sequence

$$3, 5, 7, \ldots$$

If we are given $a_n = 2n + 1$ for $n = 0, 1, 2, \ldots$, the sequence is

$$1, 3, 5, \ldots \qquad \qquad \blacksquare$$

The indicated sum of the terms of a sequence is called an **infinite series.** Thus, for the sequence

$$a_1, a_2, a_3, \ldots, a_n, \ldots$$

the associated infinite series is

$$a_1 + a_2 + a_3 + \cdots + a_n + \cdots$$

Using the summation sign Σ, the sum is indicated as

INFINITE SERIES

$$\sum_{n=1}^{\infty} a_n = a_1 + a_2 + a_3 + \cdots + a_n + \cdots \tag{13-1}$$

We must realize that an infinite series, as shown in Eq. (13-1), does not have a sum in the ordinary sense of the word, for it is not possible actually to carry out the addition of infinitely many terms. Therefore, we define the sum for an infinite series in terms of a limit.

For the infinite series of Eq. (13-1), we let S_n represent the sum of the first n terms. Therefore,

$$S_1 = a_1$$
$$S_2 = a_1 + a_2$$
$$S_3 = a_1 + a_2 + a_3$$
$$S_n = a_1 + a_2 + a_3 + \cdots + a_n$$

PARTIAL SUM

The numbers $S_1, S_2, S_3, \ldots, S_n, \ldots$ form a sequence. *Each term of this sequence is called a* **partial sum.** *We say that the infinite series, Eq. (13-1), is* **convergent** *and has the sum S given by*

$$S = \lim_{n \to \infty} S_n = \lim_{n \to \infty} \sum_{i=1}^{n} a_i \tag{13-2}$$

if this limit exists. If the limit does not exist, the series is **divergent.**

Many calculators can display sequences. In Fig. 13-1, a calculator display of the sequence for Example 2 is shown.

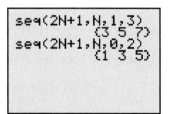

```
seq(2N+1,N,1,3)
          {3 5 7}
seq(2N+1,N,0,2)
          {1 3 5}
```

Fig. 13-1

EXAMPLE 3 For the infinite series

$$\sum_{n=0}^{\infty} \frac{1}{5^n} = \frac{1}{5^0} + \frac{1}{5^1} + \frac{1}{5^2} + \cdots + \frac{1}{5^n} + \cdots$$

the first six partial sums are

$$S_0 = 1 \qquad\qquad \text{first term}$$

$$S_1 = 1 + \frac{1}{5} = 1.2 \qquad\qquad \text{sum of first two terms}$$

$$S_2 = 1 + \frac{1}{5} + \frac{1}{25} = 1.24 \qquad \text{sum of first three terms}$$

$$S_3 = 1 + \frac{1}{5} + \frac{1}{25} + \frac{1}{125} = 1.248$$

$$S_4 = 1 + \frac{1}{5} + \frac{1}{25} + \frac{1}{125} + \frac{1}{625} = 1.2496$$

$$S_5 = 1 + \frac{1}{5} + \frac{1}{25} + \frac{1}{125} + \frac{1}{625} + \frac{1}{3125} = 1.24992$$

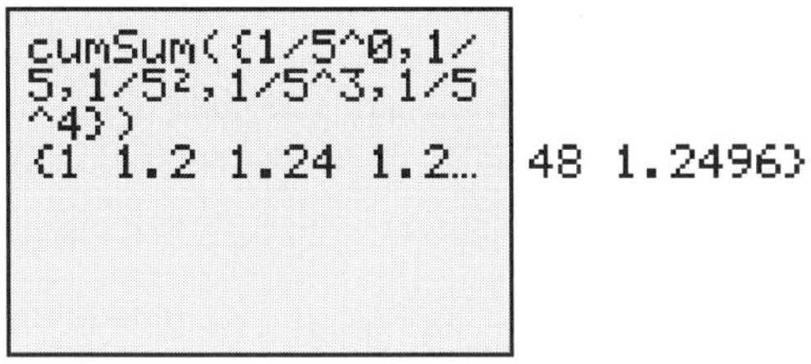

```
cumSum({1/5^0,1/
5,1/5²,1/5^3,1/5
^4})
{1 1.2 1.24 1.2…   48 1.2496}
```

Fig. 13-2

These values are found using the standard calculational features of a calculator or by the *cumulative sum* feature as shown in Fig. 13-2. Here it appears that the sequence of partial sums approaches the value 1.25. We therefore conclude that this infinite series converges and that its sum is approximately 1.25. (In Example 5 of this section, we show that this infinite series does in fact converge and have a sum of 1.25.)

EXAMPLE 4 (a) The infinite series

$$\sum_{n=1}^{\infty} 5^n = 5 + 5^2 + 5^3 + \cdots + 5^n + \cdots$$

is a divergent series. The first four partial sums are

$$S_1 = 5 \qquad S_2 = 30 \qquad S_3 = 155 \qquad S_4 = 780$$

Obviously, they are increasing without bound.
(b) The infinite series

$$\sum_{n=0}^{\infty} (-1)^n = 1 + (-1) + 1 + (-1) + \cdots + (-1)^n + \cdots$$

has as its first five partial sums

$$S_0 = 1 \qquad S_1 = 0 \qquad S_2 = 1 \qquad S_3 = 0 \qquad S_4 = 1$$

The values of these partial sums do not approach a limiting value, and therefore the series diverges.

See Appendix C for a graphing calculator program PARTSUMS. It evaluates the first *n* partial sums of an infinite series.

Since convergent series are those that have a value associated with them, they are the ones that are of primary use to us. However, generally it is not easy to determine whether a given series is convergent, and many types of tests have been developed for this purpose. These tests for convergence may be found in most textbooks that include the more advanced topics in calculus.

Subtracting rS_n from S_n gives

$S_n - rS_n = a_1 - a_1 r^n$

$S_n = \dfrac{a_1(1 - r^n)}{1 - r}$ $(r \neq 1)$

If $|r| < 1$, $\lim\limits_{n \to \infty} r^n = 0$

One important series for which we are able to determine the convergence, and its sum if convergent, is the geometric series. For this series the nth partial sum is

$$S_n = a_1 + a_1 r + a_1 r^2 + \cdots + a_1 r^{n-1}$$

where r is the fixed number by which we multiply a given term to get the next term. As shown at the left, if $|r| < 1$, the sum S of the infinite geometric series is

$$S = \lim_{n \to \infty} S_n = \frac{a_1}{1 - r} \tag{13-3}$$

CAUTION ▶

If $r = 1$, we see that the series is $a_1 + a_1 + a_1 + \cdots + a_1 + \cdots$ and is therefore divergent. If $r = -1$, the series is $a_1 - a_1 + a_1 - a_1 + \cdots$ and is also divergent. If $|r| > 1$, $\lim\limits_{n \to \infty} r^n$ is unbounded. Therefore, ***the geometric series is convergent only if $|r| < 1$*** *and has the value given by Eq. (13-3).*

■EXAMPLE 5   Show that the infinite series

$$\sum_{n=0}^{\infty} \frac{1}{5^n} = \frac{1}{5^0} + \frac{1}{5^1} + \frac{1}{5^2} + \cdots + \frac{1}{5^n} + \cdots$$

is convergent and find its sum. This is the same series as in Example 3.

We see that this is a geometric series with $r = \frac{1}{5}$. Since $|r| < 1$, the series is convergent. We find the sum to be

$$S = \frac{1}{1 - \frac{1}{5}} = \frac{1}{\frac{4}{5}} = \frac{5}{4} = 1.25 \qquad \text{using Eq. (13-3)}$$

This value agrees with the conclusion in Example 3. ■

━━━━━━━━━━━━━━━━━ **EXERCISES** *13-1* ━━━━━━━━━━━━━━━━━

In Exercises 1–4, give the first four terms of the sequences for which a_n is given.

1. $a_n = n^2$, $n = 1, 2, 3, \ldots$

2. $a_n = \dfrac{2}{3^n}$, $n = 1, 2, 3, \ldots$

3. $a_n = \dfrac{1}{n + 2}$, $n = 0, 1, 2, \ldots$

4. $a_n = \dfrac{n^2 + 1}{2n + 1}$, $n = 0, 1, 2, \ldots$

In Exercises 5–8, give (a) the first four terms of the sequence for which a_n is given and (b) the first four terms of the infinite series associated with the sequence.

5. $a_n = \left(-\dfrac{2}{5}\right)^n$, $n = 1, 2, 3, \ldots$

6. $a_n = \dfrac{1}{n} + \dfrac{1}{n + 1}$, $n = 1, 2, 3, \ldots$

7. $a_n = \cos\dfrac{n\pi}{2}$, $n = 0, 1, 2, \ldots$

8. $a_n = \dfrac{1}{n(n + 1)}$, $n = 2, 3, 4, \ldots$

In Exercises 9–12, find the nth term of the given infinite series for which $n = 1, 2, 3, \ldots$.

9. $\dfrac{1}{2} + \dfrac{1}{3} + \dfrac{1}{4} + \dfrac{1}{5} + \cdots$

10. $\dfrac{1}{2} + \dfrac{1}{4} + \dfrac{1}{8} + \dfrac{1}{16} + \cdots$

11. $\dfrac{1}{2 \times 3} + \dfrac{1}{3 \times 4} + \dfrac{1}{4 \times 5} + \dfrac{1}{5 \times 6} + \cdots$

12. $-\dfrac{2}{3} + \dfrac{4}{9} - \dfrac{8}{27} + \dfrac{16}{81} - \cdots$

In Exercises 13–20, find the first five partial sums of the given series and determine whether the series appears to be convergent or divergent. If it is convergent, find its approximate sum.

13. $1 + \dfrac{1}{8} + \dfrac{1}{27} + \dfrac{1}{64} + \dfrac{1}{125} + \cdots$

14. $1 + 2 + 5 + 10 + 17 + \cdots$

15. $1 + \dfrac{1}{2} + \dfrac{2}{3} + \dfrac{3}{4} + \dfrac{4}{5} + \cdots$

16. $\dfrac{1}{3} - \dfrac{1}{9} + \dfrac{1}{27} - \dfrac{1}{81} + \dfrac{1}{243} - \cdots$

17. $\displaystyle\sum_{n=0}^{\infty} \sqrt{n}$

18. $\displaystyle\sum_{n=1}^{\infty} \dfrac{2}{n(n+1)}$

19. $\displaystyle\sum_{n=1}^{\infty} \dfrac{2n+1}{n^2(n+1)^2}$

20. $\displaystyle\sum_{n=1}^{\infty} \dfrac{n}{2n+1}$

In Exercises 21–28, test each of the given geometric series for convergence or divergence. Find the sum of each series that is convergent.

21. $1 + 2 + 4 + \cdots + 2^n + \cdots$

22. $1 + \dfrac{1}{2} + \dfrac{1}{4} + \cdots + \dfrac{1}{2^n} + \cdots$

23. $1 - \dfrac{1}{3} + \dfrac{1}{9} - \cdots + \left(-\dfrac{1}{3}\right)^n + \cdots$

24. $1 - \dfrac{3}{2} + \dfrac{9}{4} - \cdots + \left(-\dfrac{3}{2}\right)^n + \cdots$

25. $10 + 9 + 8.1 + 7.29 + 6.561 + \cdots$

26. $4 + 1 + \dfrac{1}{4} + \dfrac{1}{16} + \dfrac{1}{64} + \cdots$

27. $512 - 64 + 8 - 1 + \dfrac{1}{8} - \cdots$

28. $16 + 12 + 9 + \dfrac{27}{4} + \dfrac{81}{16} + \cdots$

In Exercises 29–36, solve the given problems as indicated.

29. Using a calculator, take successive square roots of 2 and find at least 20 approximate values for the terms of the sequence $2^{1/2}$, $2^{1/4}$, $2^{1/8}$, $2^{1/16}$, ... From the values that are obtained, (a) what do you observe about the value of $\displaystyle\lim_{n\to\infty} 2^{1/2^n}$? (b) Determine whether the infinite series for this sequence converges or diverges.

(W) 30. Using a calculator, (a) take successive square roots of 0.01 and then (b) take successive square roots of 100. From these sequences of square roots, state any general conclusions that might be drawn.

31. The sum of the first n terms of a geometric sequence is given by

$$S_n = \dfrac{a_1(1 - r^n)}{1 - r} \quad (r \neq 1)$$

where a_1 is the first term and r is the common ratio. We can visualize the corresponding infinite series by graphing the function $f(x) = a_1(1 - r^x)/(1 - r)$ $(r \neq 1)$ (or using a calculator that can graph a sequence.) The graph represents the sequence of partial sums for values where $x = n$, since $f(n) = S_n$.

Use a graphing calculator to visualize the first five partial sums of the series

$$\dfrac{1}{2} + \dfrac{1}{4} + \dfrac{1}{8} + \cdots$$

What value does the infinite series approach? (*Remember:* Only points for which x is an integer have real meaning.)

32. Following Exercise 31, use a graphing calculator to show that the sum of the infinite series of Example 5 is 1.25. (*Be careful:* Because of the definition of the series, $x = 1$ corresponds to $n = 0$.)

33. The value V (in dollars) of a certain investment after n years can be expressed as

$$V = 100(1.05 + 1.05^2 + 1.05^3 + \cdots + 1.05^n)$$

(a) By finding partial sums, determine whether this series converges or diverges. (b) Following Exercise 31, use a graphing calculator to visualize the first ten partial sums.

34. If an electric discharge is passed through hydrogen gas, a spectrum of isolated parallel lines, called the Balmer series, is formed. See Fig. 13-3. The wavelengths λ (in nm) of the light for these lines is given by the formula

$$\dfrac{1}{\lambda} = 1.097 \times 10^{-2}\left(\dfrac{1}{2^2} - \dfrac{1}{n^2}\right) \quad (n = 3, 4, 5, \dots)$$

Find the wavelengths of the first three lines and the shortest wavelength of all the lines of the series.

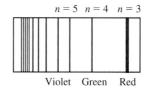

Fig. 13-3   Violet   Green   Red

35. Use geometric series to show that $\displaystyle\sum_{n=0}^{\infty} x^n = \dfrac{1}{1 - x}$ for $|x| < 1$.

36. Use geometric series to show that $\displaystyle\sum_{n=0}^{\infty} (-1)^n x^n = \dfrac{1}{1 + x}$ for $|x| < 1$.

$13\text{-}2$ MACLAURIN SERIES

In this section we develop a very important basic polynomial form of a function. Before developing the method using calculus, we shall review how this can be done for some functions algebraically.

EXAMPLE 1   By using long division (as started at the left), we have

$$\frac{2}{2-x} = 1 + \frac{1}{2}x + \frac{1}{4}x^2 + \cdots + \left(\frac{1}{2}x\right)^{n-1} + \cdots \tag{1}$$

$$
\begin{array}{r}
1 + \dfrac{x}{2} \\
2 - x \overline{\smash{\big)}\, 2 } \\
\underline{2 - x} \\
x \\
\underline{x - \dfrac{x^2}{2}} \\
\dfrac{x^2}{2}
\end{array}
$$

where n is the number of the term of the expression on the right. Since x represents a number, the right-hand side of Eq. (1) becomes a geometric series.

From Eq. (13-3) we know that the sum of a geometric series with first term a_1 and common ratio r is

$$S = \frac{a_1}{1 - r}$$

where $|r| < 1$ and the series converges.

If $x = 1$, the right-hand side of Eq. (1) is

$$1 + \frac{1}{2} + \frac{1}{4} + \cdots + \left(\frac{1}{2}\right)^{n-1} + \cdots$$

For this series, $r = \frac{1}{2}$ and $a_1 = 1$, which means that the series converges and $S = 2$. If $x = 3$, the right-hand side of Eq. (1) is

$$1 + \frac{3}{2} + \frac{9}{4} + \cdots + \left(\frac{3}{2}\right)^{n-1} + \cdots$$

which diverges since $r > 1$. Referring to the left side of Eq. (1), we see that it also equals 2 when $x = 1$. Thus, we see that the two sides agree for $x = 1$, but that the series diverges for $x = 3$. In fact, as long as $|x| < 2$, the series will converge to the value of the function on the left. From this we conclude that the series on the right properly represents the function on the left, as long as $|x| < 2$. ◼

From Example 1 we see that a rational function may be properly represented by a function of the form

POWER SERIES

$$\boxed{f(x) = a_0 + a_1 x + a_2 x^2 + \cdots + a_n x^n + \cdots} \tag{13-4}$$

Equation (13-4) is known as a **power-series expansion** *of the function f(x).* The problem now arises as to whether or not functions in general may be represented in this form. If such a representation were possible, it would provide a means of evaluating the transcendental functions for the purpose of making tables of values. Also, since a power-series expansion is in the form of a polynomial, it makes algebraic operations much simpler due to the properties of polynomials. A further study of calculus shows many other uses of power series.

In Example 1 we saw that the function could be represented by a power series as long as $|x| < 2$. That is, if we substitute any value of x in this interval into the series and also into the function, the series will converge to the value of the function. *This interval of values for which the series converges is called the* **interval of convergence.**

EXAMPLE 2   In Example 1 the interval of convergence for the series

$$1 + \frac{1}{2}x + \frac{1}{4}x^2 + \cdots + \left(\frac{1}{2}x\right)^{n-1} + \cdots$$

is $|x| < 2$. We saw that the series converges for $x = 1$, with $S = 2$, and that the value of the function is 2 for $x = 1$. This verifies that $x = 1$ is in the interval of convergence.

Also, we saw that the series diverges for $x = 3$, which verifies that $x = 3$ is not in the interval of convergence. --------------■

At this point we shall assume that unless otherwise noted the functions with which we shall be dealing may be properly represented by a power-series expansion (it takes more advanced methods to prove that this is generally possible), for appropriate intervals of convergence. We shall find that the methods of calculus are very useful in developing the method of general representation. Thus, writing a general power series, along with the first few derivatives, we have

$$f(x) = a_0 + a_1x + a_2x^2 + a_3x^3 + a_4x^4 + a_5x^5 + \cdots + a_nx^n + \cdots$$
$$f'(x) = a_1 + 2a_2x + 3a_3x^2 + 4a_4x^3 + 5a_5x^4 + \cdots + na_nx^{n-1} + \cdots$$
$$f''(x) = 2a_2 + 2(3)a_3x + 3(4)a_4x^2 + 4(5)a_5x^3 + \cdots + (n-1)na_nx^{n-2} + \cdots$$
$$f'''(x) = 2(3)a_3 + 2(3)(4)a_4x + 3(4)(5)a_5x^2 + \cdots + (n-2)(n-1)na_nx^{n-3} + \cdots$$
$$f^{iv}(x) = 2(3)(4)a_4 + 2(3)(4)(5)a_5x + \cdots + (n-3)(n-2)(n-1)na_nx^{n-4} + \cdots$$

NOTE ▶   Regardless of the values of the constants a_n for any power series, *if x = 0, the left and right sides must be equal,* and all the terms on the right are zero except the first. Thus, setting $x = 0$ in each of the above equations, we have

$$f(0) = a_0 \qquad f'(0) = a_1 \qquad f''(0) = 2a_2$$
$$f'''(0) = 2(3)a_3 \qquad f^{iv}(0) = 2(3)(4)a_4$$

Solving each of these for the constants a_n, we have

$$a_0 = f(0) \qquad a_1 = f'(0) \qquad a_2 = \frac{f''(0)}{2!} \qquad a_3 = \frac{f'''(0)}{3!} \qquad a_4 = \frac{f^{iv}(0)}{4!}$$

Substituting these into the expression for $f(x)$, we have

MACLAURIN SERIES

$$f(x) = f(0) + f'(0)x + \frac{f''(0)x^2}{2!} + \frac{f'''(0)x^3}{3!} + \cdots + \frac{f^n(0)x^n}{n!} + \cdots \qquad \textbf{(13-5)}$$

Named for the Scottish mathematician
Colin Maclaurin (1698–1746).

Equation (13-5) is known as the **Maclaurin series expansion** *of a function.* For a function to be represented by a Maclaurin expansion, the function and all of its derivatives must exist at $x = 0$. Here we have used the **factorial notation** $n!$, where

$$n! = n(n-1)(n-2)(\cdots)(1) \qquad \textbf{(13-6)}$$

For example $4! = (4)(3)(2)(1) = 24$. As we can see, this notation is useful for power-series expansions. It is also useful in other fields of mathematics.

As we mentioned earlier, one of the uses we will make of series expansions is that of determining the values of functions for particular values of x. If x is sufficiently small, successive terms become smaller and smaller, and the series will converge rapidly. This is considered in the sections that follow.

EXAMPLE 3   Find the first four terms of the Maclaurin series expansion of $f(x) = \dfrac{2}{2 - x}$.

This is written as

$$f(x) = \frac{2}{2 - x} \qquad f(0) = 1$$

find derivatives and evaluate each at $x = 0$

$$f'(x) = \frac{2}{(2 - x)^2} \qquad f'(0) = \frac{1}{2}$$

$$f''(x) = \frac{4}{(2 - x)^3} \qquad f''(0) = \frac{1}{2}$$

$$f'''(x) = \frac{12}{(2 - x)^4} \qquad f'''(0) = \frac{3}{4}$$

$$f(x) = 1 + \frac{1}{2} x + \frac{1}{2} \left(\frac{x^2}{2!} \right) + \frac{3}{4} \left(\frac{x^3}{3!} \right) + \cdots \qquad \text{using Eq. (13-5)}$$

or

$$\frac{2}{2 - x} = 1 + \frac{1}{2} x + \frac{1}{4} x^2 + \frac{1}{8} x^3 + \cdots$$

This result agrees with that obtained by direct division on page 388.

EXAMPLE 4   Find the first four terms of the Maclaurin series expansion of $f(x) = e^{-x}$.

We write

$$f(x) = e^{-x} \qquad f(0) = 1$$

find derivitives and evaluate each at $x = 0$

$$f'(x) = -e^{-x} \qquad f'(0) = -1$$

$$f''(x) = e^{-x} \qquad f''(0) = 1$$

$$f'''(x) = -e^{-x} \qquad f'''(0) = -1$$

$$f(x) = 1 + (-1)x + 1 \left(\frac{x^2}{2!} \right) + (-1) \left(\frac{x^3}{3!} \right) + \cdots \qquad \text{using Eq. (13-5)}$$

or

$$e^{-x} = 1 - x + \frac{x^2}{2!} - \frac{x^3}{3!} + \cdots$$

EXAMPLE 5   Find the first three nonzero terms of the Maclaurin series expansion of $f(x) = \sin 2x$.

We have

$$f(x) = \sin 2x \qquad f(0) = 0 \qquad f'''(x) = -8 \cos 2x \qquad f'''(0) = -8$$

$$f'(x) = 2 \cos 2x \qquad f'(0) = 2 \qquad f^{iv}(x) = 16 \sin 2x \qquad f^{iv}(0) = 0$$

$$f''(x) = -4 \sin 2x \qquad f''(0) = 0 \qquad f^{v}(x) = 32 \cos 2x \qquad f^{v}(0) = 32$$

$$f(x) = 0 + 2x + 0 + (-8) \frac{x^3}{3!} + 0 + 32 \frac{x^5}{5!} + \cdots$$

$$\sin 2x = 2x - \frac{4}{3} x^3 + \frac{4}{15} x^5 - \cdots$$

This series is called an **alternating series,** *since every other term is negative.*

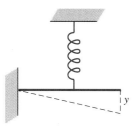

Fig. 13-4

EXAMPLE 6 A lever is attached to a spring as shown in Fig. 13-4. Frictional forces in the spring are just sufficient so that the lever does not oscillate after being depressed. Such motion is called *critically damped*. The displacement y of the lever as a function of the time t for one such case is $y = (1 + t)e^{-t}$. In order to study the motion for small values of t, a polynomial form of $y = f(t)$ is to be used. Find the first four nonzero terms of the expansion.

$$f(t) = (1 + t)e^{-t} \qquad\qquad\qquad f(0) = 1$$
$$f'(t) = (1 + t)e^{-t}(-1) + e^{-t} = -te^{-t} \qquad f'(0) = 0$$
$$f''(t) = te^{-t} - e^{-t} \qquad\qquad\qquad f''(0) = -1$$
$$f'''(t) = -te^{-t} + e^{-t} + e^{-t} = 2e^{-t} - te^{-t} \qquad f'''(0) = 2$$
$$f^{iv}(t) = -2e^{-t} + te^{-t} - e^{-t} = te^{-t} - 3e^{-t} \qquad f^{iv}(0) = -3$$

$$f(t) = 1 + 0 + (-1)\frac{t^2}{2!} + 2\frac{t^3}{3!} + (-3)\frac{t^4}{4!} + \cdots$$

or

$$(1 + t)e^{-t} = 1 - \frac{t^2}{2} + \frac{t^3}{3} - \frac{t^4}{8} + \cdots$$

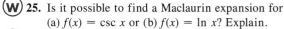

EXERCISES *13-2*

In Exercises 1–16, find the first three nonzero terms of the Maclaurin expansion of the given functions.

1. $f(x) = e^x$

2. $f(x) = \sin x$

3. $f(x) = \cos x$

4. $f(x) = \ln(1 + x)$

5. $f(x) = \sqrt{1 + x}$

6. $f(x) = \dfrac{1}{(1 - x)^{1/3}}$

7. $f(x) = e^{-2x}$

8. $f(x) = \dfrac{1}{2}(e^x + e^{-x})$

9. $f(x) = \cos 4\pi x$

10. $f(x) = e^x \sin x$

11. $f(x) = \dfrac{1}{1 - x}$

12. $f(x) = \dfrac{1}{(1 + x)^2}$

13. $f(x) = \ln(1 - 2x)$

14. $f(x) = (1 + x)^{3/2}$

15. $f(x) = \cos^2 x$

16. $f(x) = \ln(1 + 4x)$

In Exercises 17–24, find the first two nonzero terms of the Maclaurin expansion of the given functions.

17. $f(x) = \tan^{-1} x$

18. $f(x) = \cos x^2$

19. $f(x) = \tan x$

20. $f(x) = \sec x$

21. $f(x) = \ln \cos x$

22. $f(x) = xe^{\sin x}$

23. $f(x) = \sin^2 x$

24. $f(x) = e^{-x^2}$

In Exercises 25–32, solve the given problems.

(W) 25. Is it possible to find a Maclaurin expansion for (a) $f(x) = \csc x$ or (b) $f(x) = \ln x$? Explain.

(W) 26. Is it possible to find a Maclaurin expansion for (a) $f(x) = \sqrt{x}$ or (b) $f(x) = \sqrt{1 + x}$? Explain.

27. Find the first three nonzero terms of the Maclaurin expansion for (a) $f(x) = e^x$ and (b) $f(x) = e^{x^2}$. Compare these expansions.

28. By finding the Maclaurin expansion of $f(x) = (1 + x)^n$, derive the first four terms of the *binomial series*. Its interval of convergence is $|x| < 1$ for all values of n.

29. If $f(x) = e^{3x}$, compare the Maclaurin expansion with the linearization for $a = 0$.

30. If $f(x) = x^4 + 2x^2$, show that this function is obtained when a Maclaurin expansion is found.

31. The reliability R ($0 \le R \le 1$) of a certain computer system is $R = e^{-0.001t}$, where t is the time of operation (in min). Express $R = f(t)$ in polynomial form by using the first three terms of the Maclaurin expansion.

32. In the analysis of the optical paths of light from a narrow slit S to a point P as shown in Fig. 13-5, the law of cosines is used to obtain the equation

$$c^2 = a^2 + (a + b)^2 - 2a(a + b)\cos\frac{s}{a}$$

where s is part of the circular arc $\overarc{AB}$. By using two nonzero terms of the Maclaurin expansion of $\cos\frac{s}{a}$, simplify the right side of the equation. (In finding the expansion, let $x = \frac{s}{a}$ and then substitute back into the expansion.)

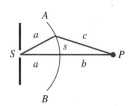

Fig. 13-5

13-3 CERTAIN OPERATIONS WITH SERIES

The series found in the first four exercises and Exercise 28 (the binomial series) of Section 13-2 are of particular importance. They are used to evaluate exponential functions, trigonometric functions, logarithms, powers, and roots, as well as develop other series. For reference, we give them here with their intervals of convergence.

$$e^x = 1 + x + \frac{x^2}{2!} + \frac{x^3}{3!} + \cdots \qquad \text{(all } x) \qquad (13\text{-}7)$$

$$\sin x = x - \frac{x^3}{3!} + \frac{x^5}{5!} - \cdots \qquad \text{(all } x) \qquad (13\text{-}8)$$

$$\cos x = 1 - \frac{x^2}{2!} + \frac{x^4}{4!} - \cdots \qquad \text{(all } x) \qquad (13\text{-}9)$$

$$\ln(1 + x) = x - \frac{x^2}{2} + \frac{x^3}{3} - \frac{x^4}{4} + \cdots \qquad (|x| < 1) \qquad (13\text{-}10)$$

$$(1 + x)^n = 1 + nx + \frac{n(n-1)}{2!} x^2 + \cdots \qquad (|x| < 1) \qquad (13\text{-}11)$$

See Appendix C for a graphing calculator program BINEXPAN. It gives the first K coefficients of $(AX + B)^N$.

In the next section we shall see how to use these series in finding values of functions. In this section we see how new series are developed by using the above basic series, and we also show other uses of series.

When we discussed functions in Chapter 1, we mentioned functions such as $f(2x)$ and $f(-x)$. *By using functional notation and the preceding series, we can find the series expansions of many other series without using direct expansion.* This can often save time in finding a desired series.

NOTE ▶

■ EXAMPLE 1 Find the Maclaurin expansion of e^{2x}.
From Eq. (13-7), we know the expansion of e^x. Hence,

$$f(x) = 1 + x + \frac{x^2}{2!} + \frac{x^3}{3!} + \cdots$$

Since $e^{2x} = f(2x)$, we have

$$f(2x) = 1 + (2x) + \frac{(2x)^2}{2!} + \frac{(2x)^3}{3!} + \cdots \qquad \text{in } f(x)\text{, replace } x \text{ by } 2x$$

$$e^{2x} = 1 + 2x + 2x^2 + \frac{4x^3}{3} + \cdots$$

■ EXAMPLE 2 Find the Maclaurin expansion of $\sin x^2$.
From Eq. (13-8), we know the expansion of $\sin x$. Therefore,

$$f(x) = x - \frac{x^3}{3!} + \frac{x^5}{5!} - \cdots$$

$$f(x^2) = (x^2) - \frac{(x^2)^3}{3!} + \frac{(x^2)^5}{5!} - \cdots \qquad \text{in } f(x)\text{, replace } x \text{ by } x^2$$

$$\sin x^2 = x^2 - \frac{x^6}{3!} + \frac{x^{10}}{5!} - \cdots$$

Direct expansion of this series is quite lengthy.

The basic algebraic operations may be applied to series in the same manner they are applied to polynomials. That is, we may add, subtract, multiply, or divide series in order to obtain other series. The interval of convergence for the resulting series is that which is common to those of the series being used. The multiplication of series is illustrated in the following example.

EXAMPLE 3   Multiply the series for e^x and $\cos x$ in order to obtain the series expansion for $e^x \cos x$.

Using the series expansion for e^x and $\cos x$ as shown in Eqs. (13-7) and (13-9), we have the following indicated multiplication:

$$e^x \cos x = \left(1 + x + \frac{x^2}{2!} + \frac{x^3}{3!} + \frac{x^4}{4!} + \cdots\right)\left(1 - \frac{x^2}{2!} + \frac{x^4}{4!} - \cdots\right)$$

By multiplying the series on the right, we have the following result, considering through the x^4 terms in the product.

$$1\left(1 - \frac{x^2}{2!} + \frac{x^4}{4!}\right) \quad x\left(1 - \frac{x^2}{2!}\right) \quad \frac{x^2}{2!}\left(1 - \frac{x^2}{2!}\right) \quad \left(\frac{x^3}{3!} + \frac{x^4}{4!}\right) \text{(1)}$$

$$e^x \cos x = 1 - \frac{x^2}{2} + \frac{x^4}{24} + x - \frac{x^3}{2} + \frac{x^2}{2} - \frac{x^4}{4} + \frac{x^3}{6} + \frac{x^4}{24} + \cdots$$

$$= 1 + x - \frac{1}{3}x^3 - \frac{1}{6}x^4 + \cdots$$

It is also possible to use the operations of differentiation and integration to obtain series expansions, although the proof of this is found in more advanced texts. Consider the following example.

EXAMPLE 4   Show that by differentiating the expansion for $\ln(1 + x)$ term by term, the result is the same as the expansion for $\dfrac{1}{1 + x}$.

The series for $\ln(1 + x)$ is shown in Eq. (13-10) as

$$\ln(1 + x) = x - \frac{x^2}{2} + \frac{x^3}{3} - \frac{x^4}{4} + \cdots$$

Differentiating, we have

$$\frac{1}{1 + x} = 1 - \frac{2x}{2} + \frac{3x^2}{3} - \frac{4x^3}{4} + \cdots$$

$$= 1 - x + x^2 - x^3 + \cdots$$

Using the binomial expansion for $\dfrac{1}{1 + x} = (1 + x)^{-1}$, we have

$$(1 + x)^{-1} = 1 + (-1)x + \frac{(-1)(-2)}{2!}x^2 + \frac{(-1)(-2)(-3)}{3!}x^3 + \cdots \qquad \text{using Eq. (13-11) with } n = -1$$

$$= 1 - x + x^2 - x^3 + \cdots$$

We see that the results are the same.

For reference, the standard forms of a complex number are as follows ($j = \sqrt{-1}$):

rectangular: $a + bj$
polar: $r(\cos\theta + j\sin\theta)$
exponential: $re^{j\theta}$
$r = \sqrt{a^2 + b^2}$, $\tan\theta = b/a$

Complex numbers were named by the German mathematician Karl Friedrich Gauss (1777–1855).

We can use algebraic operations on series to verify that the definition of the exponential form of a complex number, as shown in the reference at the left, is consistent with other definitions. The only assumption required here is that the Maclaurin expansions for e^x, $\sin x$, and $\cos x$ are also valid for complex numbers. This is shown in advanced calculus. Thus,

$$e^{j\theta} = 1 + j\theta + \frac{(j\theta)^2}{2!} + \frac{(j\theta)^3}{3!} + \cdots = 1 + j\theta - \frac{\theta^2}{2!} - j\frac{\theta^3}{3!} + \cdots \quad \textbf{(13-12)}$$

$$j\sin\theta = j\theta - j\frac{\theta^3}{3!} + \cdots \quad \textbf{(13-13)}$$

$$\cos\theta = 1 - \frac{\theta^2}{2!} + \cdots \quad \textbf{(13-14)}$$

When we add the terms of Eq. (13-13) to those of Eq. (13-14), the result is the series given in Eq. (13-12). Thus,

$$\boxed{e^{j\theta} = \cos\theta + j\sin\theta} \quad \textbf{(13-15)}$$

A comparison of Eq. (13-15) and the exponential form of a complex number indicates the reason for the choice of the definition of the exponential form.

An additional use of power series is now shown. Many integrals that occur in practice cannot be integrated by methods given in the preceding chapters. However, power series can be very useful in giving excellent approximations to some definite integrals.

EXAMPLE 5 Find the first-quadrant area bounded by $y = \sqrt{1 + x^3}$ and $x = 0.5$.

From Fig. 13-6, we see that the area is

$$A = \int_0^{0.5} \sqrt{1 + x^3}\, dx$$

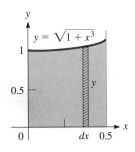

Fig. 13-6

This integral does not fit any form we have used. However, its value can be closely approximated by using the binomial expansion for $\sqrt{1 + x^3}$ and then integrating.

Using the binomial expansion to find the first three terms of the expansion for $\sqrt{1 + x^3}$, we have

$$\sqrt{1 + x^3} = (1 + x^3)^{0.5} = 1 + 0.5x^3 + \frac{0.5(-0.5)}{2}(x^3)^2 + \cdots$$

$$= 1 + 0.5x^3 - 0.125x^6 + \cdots$$

Substituting in the integral, we have

$$A = \int_0^{0.5} (1 + 0.5x^3 - 0.125x^6 + \cdots)\, dx$$

$$= x + \frac{0.5}{4}x^4 - \frac{0.125}{7}x^7 + \cdots \Big|_0^{0.5}$$

$$= 0.5 + 0.0078125 - 0.0001395 + \cdots = 0.507673 + \cdots$$

We can see that each of the terms omitted was very small. The result shown is correct to four decimal places, or $A = 0.5077$. Additional accuracy can be obtained by using more terms of the expansion.

EXAMPLE 6   Evaluate: $\int_0^{0.1} e^{-x^2}\,dx$.

We write

$$e^{-x^2} = 1 + (-x^2) + \frac{(-x^2)^2}{2!} + \cdots \qquad \text{using Eq. (13-7)}$$

Thus,

$$\int_0^{0.1} e^{-x^2}\,dx = \int_0^{0.1}\left(1 - x^2 + \frac{x^4}{2} - \cdots\right)dx \qquad \text{substitute}$$

$$= \left(x - \frac{x^3}{3} + \frac{x^5}{10} - \cdots\right)\Bigg|_0^{0.1} \qquad \text{integrate}$$

$$= 0.1 - \frac{0.001}{3} + \frac{0.00001}{10} = 0.0996677 \qquad \text{evaluate}$$

This answer is correct to the indicated accuracy.   ∎

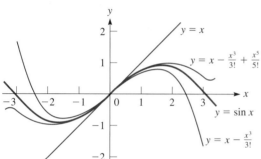

Fig. 13-7

The question of accuracy now arises. The integrals just evaluated indicate that the more terms used, the greater is the accuracy of the result. To graphically show the accuracy involved, Fig. 13-7 depicts the graphs of $y = \sin x$ and the graphs of

$$y = x \qquad y = x - \frac{x^3}{3!} \qquad y = x - \frac{x^3}{3!} + \frac{x^5}{5!}$$

which are the first three approximations of $y = \sin x$. We can see that each term added gives a better fit to the curve of $y = \sin x$. Also, this gives a graphical representation of the meaning of a series expansion.

We have just shown that the more terms included, the more accurate the result. For small values of x, a Maclaurin series gives good accuracy with a very few terms. In this case the series *converges* rapidly, as we mentioned earlier. For this reason, a Maclaurin series is of particular use for small values of x. For larger values of x, usually a function is expanded in a Taylor series (see Section 13-5). Of course, if we omit any term in a series, there is some error in the calculation.

EXERCISES *13-3*

In Exercises 1–8, find the first four nonzero terms of the Maclaurin expansions of the given functions by using Eqs. (13-7) to (13-11).

1. $f(x) = e^{3x}$ **2.** $f(x) = e^{-2x}$

3. $f(x) = \sin \frac{1}{2}x$ **4.** $f(x) = \sin x^4$

5. $f(x) = x \cos 4x$ **6.** $f(x) = \sqrt{1 - x^4}$

7. $f(x) = \ln(1 + x^2)$ **8.** $f(x) = x^2 \ln(1 - x)^2$

In Exercises 9–12, evaluate the given integrals by using three terms of the appropriate series.

9. $\int_0^1 \sin x^2\,dx$ **10.** $\int_0^{0.4} \sqrt[4]{1 - 2x^2}\,dx$

11. $\int_0^{0.2} \cos\sqrt{x}\,dx$ **12.** $\int_{0.1}^{0.2} \frac{\cos x - 1}{x}\,dx$

In Exercises 13–20, find the indicated series by the given operation.

13. Find the first four terms of the Maclaurin expansion of the function $f(x) = \dfrac{2}{1 - x^2}$ by adding the terms of the series for the functions $\dfrac{1}{1 - x}$ and $\dfrac{1}{1 + x}$.

14. Find the first four nonzero terms of the expansion of the function $f(x) = \frac{1}{2}(e^x - e^{-x})$ by subtracting the terms of the appropriate series. The result is the series for $\sinh x$. (See Exercise 47 of Section 8-1.)

●15. Find the first three terms of the expansion for $e^x \sin x$ by multiplying the proper expansions together, term by term.

16. Find the first three nonzero terms of the expansion for $f(x) = \tan x$ by dividing the series for $\sin x$ by that for $\cos x$.

● 17. Show that by differentiating term by term the expansion for $\sin x$, the result is the expansion for $\cos x$.

18. Show that by differentiating term by term the expansion for e^x, the result is also the expansion for e^x.

19. Show that by integrating term by term the expansion for $\cos x$, the result is the expansion for $\sin x$.

20. Show that by integrating term by term the expansion for $-1/(1 - x)$ (see Exercise 11 of Section 13-2), the result is the expansion for $\ln(1 - x)$.

In Exercises 21–28, solve the given problems.

21. Evaluate $\int_0^1 e^x \, dx$ directly and compare the result obtained by using four terms of the series for e^x and then integrating.

● 22. Evaluate $\lim\limits_{x \to 0} \dfrac{\sin x}{x}$ by using the series expansion for $\sin x$. Compare the result with Eq. (7-29).

● 23. Find the approximate value of the area bounded by $y = x^2 e^x$, $x = 0.2$, and the x-axis by using three terms of the appropriate Maclaurin series.

24. Find the approximate area bounded by $y = e^{-x^2}$, $x = -1$, $x = 1$, and $y = 0$ by using three terms of the appropriate series. See Fig. 13-8.

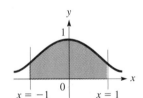

Fig. 13-8

25. The *Fresnel integral* $\int_0^x \cos t^2 \, dt$ is used in the analysis of beam displacements (and in optics). Evaluate this integral for $x = 0.2$ by using two terms of the appropriate series.

26. The dome of a sports arena is designed as the surface generated by revolving the curve of $y = 20.0 \cos 0.0196x$ ($0 \le x \le 80.0$ m) about the y-axis. Find the volume within the dome by using three terms of the appropriate series. (This is the same dome as in Exercise 39, page 331. Compare the results).

27. The impedance Z of an alternating current circuit, containing a resistance R, inductance I, and capacitance C, may be defined by the equation $Z = R + j(X_L - X_C)$, where X_L and X_C are known as the inductive reactance and capacitive reactance, respectively. Show, by use of the forms of a complex number, that the angle between Z and R is given by $\theta = \tan^{-1}(X_L - X_C)/R$.

28. The charge q on a capacitor in a certain electric circuit is given by $q = ce^{-at} \sin 6at$, where t is the time. By multiplication of series, find the first four nonzero terms of the expansion for q.

In Exercises 29–32, use a graphing calculator to display (a) the given function and (b) the first three series approximations of the function in the same display. Each display will be similar to that in Fig. 13-7 for the function $y = \sin x$ and its first three approximations. Be careful in choosing the appropriate window values.

29. $y = e^x$

30. $y = \cos x$

31. $y = \ln(1 + x)$ ($|x| < 1$)

32. $y = \sqrt{1 + x}$ ($|x| < 1$)

13-4 COMPUTATIONS BY USE OF SERIES EXPANSIONS

As we mentioned at the beginning of the previous section, power-series expansions can be used to compute numerical values of exponential functions, trigonometric functions, logarithms, powers, and roots. By including a sufficient number of terms in the expansion, we can calculate these values to any degree of accuracy that may be required.

It is through such calculations that tables of values can be made, and decimal approximations of numbers such as e and π can be found. Also, many of the values found on a calculator or a computer are calculated by using series expansions that have been programmed into the chip that is in the calculator or computer.

EXAMPLE 1   Calculate the value of $e^{0.1}$.

In order to evaluate $e^{0.1}$, we substitute 0.1 for x in the expansion for e^x. The more terms that are used, the more accurate a value we can obtain. The limit of the partial sums would be the actual value. However, since $e^{0.1}$ is irrational, we cannot express the exact value in decimal form.

Therefore, the value is found as follows:

$$e^x = 1 + x + \frac{x^2}{2!} + \cdots \qquad \text{Eq. (13-7)}$$

$$e^{0.1} = 1 + 0.1 + \frac{(0.1)^2}{2} + \cdots \qquad \text{substitute 0.1 for } x$$

$$= 1.105 \qquad \text{using 3 terms}$$

Using a calculator, we find that $e^{0.1} = 1.105170918$, which shows that our answer is valid to the accuracy shown. ◼

EXAMPLE 2   Calculate the value of sin 2°.

CAUTION ▶   In finding trigonometric values, we must be careful to *express the angle in radians.* Thus, the value of sin 2° is found as follows:

$$\sin x = x - \frac{x^3}{3!} + \cdots \qquad \text{Eq. (13-8)}$$

$$\sin 2° = \left(\frac{\pi}{90}\right) - \frac{(\pi/90)^3}{6} + \cdots \qquad 2° = \frac{\pi}{90} \text{ rad}$$

$$= 0.0348994963 \qquad \text{using 2 terms}$$

A calculator gives the value 0.0348994967. Here we note that the second term is much smaller than the first. In fact, a good approximation of 0.0349 can be found by using just one term. We now see that $\sin \theta \approx \theta$ for small values of θ. ◼

EXAMPLE 3   Calculate the value of cos 0.5429.

Since the angle is expressed in radians, we have

$$\cos 0.5429 = 1 - \frac{0.5429^2}{2} + \frac{0.5429^4}{4!} - \cdots \qquad \text{using Eq. (13-9)}$$

$$= 0.8562495 \qquad \text{using 3 terms}$$

A calculator shows that cos 0.5429 = 0.8562140824. Since the angle is not small, additional terms are needed to obtain this accuracy. With one more term, the value 0.8562139 is obtained. ◼

EXAMPLE 4   Calculate the value of ln 1.2.

$$\ln(1 + x) = x - \frac{x^2}{2} + \frac{x^3}{3} - \cdots \qquad \text{Eq. (13-10)}$$

$$\ln 1.2 = \ln(1 + 0.2)$$

$$= 0.2 - \frac{(0.2)^2}{2} + \frac{(0.2)^3}{3} - \cdots = 0.1827$$

To four significant digits, ln(1.2) = 0.1823. One more term is required to obtain this accuracy. ◼

We now illustrate the use of series in error calculations. We also discussed this as an application of differentials. A series solution allows as close a value of the calculated error as needed, whereas only one term can be found using differentials.

EXAMPLE 5 The velocity v of an object that has fallen h feet is $v = 8.00\sqrt{h}$. Find the approximate error in calculating the velocity of an object that has fallen 100.0 ft, with a possible error of 2.0 ft.

NOTE ▶ If we *let $v = 8.00\sqrt{100.0 + x}$, where x is the error in h,* we may express v as a Maclaurin expansion in x:

$$f(x) = 8.00(100.0 + x)^{1/2} \qquad f(0) = 80.0$$
$$f'(x) = 4.00(100.0 + x)^{-1/2} \qquad f'(0) = 0.400$$
$$f''(x) = -2.00(100.0 + x)^{-3/2} \qquad f''(0) = -0.00200$$

Therefore,

$$v = 8.00\sqrt{100.0 + x} = 80.0 + 0.400x - 0.00100x^2 + \cdots$$

Since the calculated value of v for $x = 0$ is 80.0, the error e in the value of v is

$$e = 0.400x - 0.00100x^2 + \cdots$$

Calculating, the error for $x = 2.0$ is

$$e = 0.400(2.0) - 0.00100(4.0) = 0.800 - 0.0040 = 0.796 \text{ ft/s}$$

The value 0.800 is that which is found using differentials. The additional terms are corrections to this term. The additional term in this case shows that the first term is a good approximation to the error. Although this problem can be done numerically, a series solution allows us to find the error for any value of x. ∎

SOLVING A WORD PROBLEM

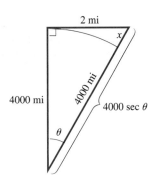

2 mi

4000 mi

4000 mi

4000 sec θ

x

θ

Fig. 13-9

EXAMPLE 6 From a point on the surface of the earth, a laser beam is aimed tangentially toward a vertical rod 2 mi distant. How far up on the rod does the beam touch? (Assume the earth is a perfect sphere of radius 4000 mi.)

From Fig. 13-9, we see that

$$x = 4000 \sec \theta - 4000$$

Finding the series for $\sec \theta$, we have

$$f(\theta) = \sec \theta \qquad\qquad f(0) = 1$$
$$f'(\theta) = \sec \theta \tan \theta \qquad\qquad f'(0) = 0$$
$$f''(\theta) = \sec^3 \theta + \sec \theta \tan^2 \theta \qquad f''(0) = 1$$

Thus, the first two nonzero terms are $\sec \theta = 1 + (\theta^2/2)$. Therefore,

$$x = 4000(\sec \theta - 1)$$
$$= 4000\left(1 + \frac{\theta^2}{2} - 1\right) = 2000\, \theta^2$$

The first two terms of the expansion for $\tan \theta$ are $\theta + \theta^3/3$, which means $\tan \theta = \theta$ if θ is very small ($\theta^3/3$ becomes negligible). From Fig. 13-9, $\tan \theta = 2/4000$, and therefore $\theta = 1/2000$. Therefore, we have

$$x = 2000\left(\frac{1}{2000}\right)^2 = \frac{1}{2000} = 0.0005 \text{ mi}$$

This means the 2-mi-long beam touches the rod only 2.6 ft above the surface! ∎

— EXERCISES *13-4* —

In Exercises 1–16, calculate the value of each of the given functions. Use the indicated number of terms of the appropriate series. Compare with the value found directly on a calculator.

1. $e^{0.2}$ (3)

2. 1.01^{-1} (4)

3. $\sin 0.1$ (2)

4. $\cos 0.05$ (2)

5. e (7)

6. $1/\sqrt{e}$ (5)

7. $\cos 3°$ (2)

8. $\sin 4°$ (2)

9. $\ln(1.4)$ (4)

10. $\ln(0.95)$ (4)

11. $\sin 0.3625$ (3)

12. $\cos 0.4072$ (3)

13. $\ln 0.9861$ (3)

14. $\ln 1.0534$ (3)

15. 1.032^6 (3)

16. 0.9982^8 (3)

In Exercises 17–20, calculate the value of each of the given functions. In Exercises 17 and 18, use the expansion for $\sqrt{1 + x}$, and in Exercises 19 and 20 use the expansion for $\sqrt[3]{1 + x}$. Use three terms of the appropriate series.

17. $\sqrt{1.1076}$

18. $\sqrt{0.7915}$

19. $\sqrt[3]{0.9628}$

20. $\sqrt[3]{1.1392}$

In Exercises 21–24, calculate the maximum error of the values indicated. If a series is alternating (every other term is negative), the maximum possible error in the calculated value is the value of the first term omitted.

21. The value found in Exercise 3

22. The value found in Exercise 2

23. The value found in Exercise 7

24. The value found in Exercise 9

In Exercises 25–32, solve the given problems by using series expansions.

25. We can evaluate π by use of $\frac{1}{4}\pi = \tan^{-1}\frac{1}{2} + \tan^{-1}\frac{1}{3}$ (see Exercise 66 of Section 7-5), along with the series for $\tan^{-1} x$. The first three terms are $\tan^{-1} x = x - \frac{1}{3}x^3 + \frac{1}{5}x^5$. Using these terms, expand $\tan^{-1}\frac{1}{2}$ and $\tan^{-1}\frac{1}{3}$ and approximate the value of π.

26. Use the fact that $\frac{1}{4}\pi = \tan^{-1}\frac{1}{7} + 2\tan^{-1}\frac{1}{3}$ to approximate the value of π. (See Exercise 25.)

27. The time t (in years) for an investment to increase by 10% when the interest rate is 6% is given by $t = \dfrac{\ln 1.1}{0.06}$. Evaluate this expression by using the first four terms of the appropriate series.

28. The period T of a pendulum of length L is given by

$$T = 2\pi \sqrt{\frac{L}{g}} \left(1 + \frac{1}{4}\sin^2\frac{\theta}{2} + \frac{9}{64}\sin^4\frac{\theta}{2} + \cdots \right)$$

where g is the acceleration due to gravity and θ is the maximum angular displacement. If $L = 1.000$ m and $g = 9.800$ m/s^2, calculate T for $\theta = 10.0°$ (a) if only one term (the 1) of the series is used and (b) if two terms of the indicated series are used. In the second term, substitute one term of the series for $\sin^2(\theta/2)$.

(W) **29.** The current in a circuit containing a resistance R, an inductance L, and a battery whose voltage is E is given by the equation $i = \dfrac{E}{R}(1 - e^{-Rt/L})$, where t is the time. Approximate this expression by using the first three terms of the appropriate exponential series. Under what conditions will this approximation be valid?

30. The image distance q from a certain lens as a function of the object distance p is given by $q = 20p/(p - 20)$. Find the first three nonzero terms of the expansion of the right side. From this expression, calculate q for $p = 2.00$ cm and compare it with the value found by substituting 2.00 in the original expression.

31. At what height above the shoreline of Lake Ontario must an observer be in order to see a point 15 km distant on the surface of the lake? (The radius of the earth is 6400 km.)

32. The efficiency E (in %) of an internal combustion engine in terms of its compression ratio e is given by $E = 100(1 - e^{-0.40})$. Determine the possible approximate error in the efficiency for a compression ratio measured to be 6.00 with a possible error of 0.50. (*Hint:* Set up a series for $(6 + x)^{-0.40}$.)

13-5 TAYLOR SERIES

To obtain accurate values of a function for values of x that are not close to zero, it is usually necessary to use many terms of a Maclaurin expansion. However, we can use another type of series, called a **Taylor series,** *which is a more general expansion than a Maclaurin expansion.* Also, functions for which a Maclaurin series may not be found may have a Taylor series.

The basic assumption in formulating a Taylor expansion is that a function may be expanded in a polynomial of the form

$$f(x) = c_0 + c_1(x - a) + c_2(x - a)^2 + \cdots \qquad \text{(13-16)}$$

Following the same line of reasoning as in deriving the Maclaurin expansion, we may find the constants $c_0, c_1, c_2, \ldots$. That is, derivatives of Eq. (13-16) are taken, and the function and its derivatives are evaluated at $x = a$. This leads to

$$f(x) = f(a) + f'(a)(x - a) + \frac{f''(a)(x - a)^2}{2!} + \cdots \qquad \text{(13-17)}$$

Named for the English mathematician Brook Taylor (1685–1737).

Equation (13-17) is the **Taylor series expansion** *of a function.* It converges rapidly for values of x that are close to a, and this is illustrated in Examples 3 and 4.

EXAMPLE 1   Expand $f(x) = e^x$ in a Taylor series with $a = 1$.

$$\begin{array}{lll} f(x) = e^x & f(1) = e & \text{find derivatives and evaluate each at } x = 1 \\ f'(x) = e^x & f'(1) = e & \\ f''(x) = e^x & f''(1) = e & \\ f'''(x) = e^x & f'''(1) = e & \end{array}$$

$$f(x) = e + e(x - 1) + e\frac{(x - 1)^2}{2!} + e\frac{(x - 1)^3}{3!} + \cdots \qquad \text{using Eq. (13-17)}$$

$$e^x = e\left[1 + (x - 1) + \frac{(x - 1)^2}{2} + \frac{(x - 1)^3}{6} + \cdots\right]$$

This series can be used in evaluating e^x for values of x near 1.

EXAMPLE 2   Expand $f(x) = \sqrt{x}$ in powers of $(x - 4)$.

Another way of stating this is to find the Taylor series for $f(x) = \sqrt{x}$, with $a = 4$. Thus,

$$\begin{array}{ll} f(x) = x^{1/2} & f(4) = 2 \qquad \text{find derivatives and} \\ & \qquad\qquad\qquad \text{evaluate each at } x = 4 \\ f'(x) = \dfrac{1}{2x^{1/2}} & f'(4) = \dfrac{1}{4} \\ f''(x) = -\dfrac{1}{4x^{3/2}} & f''(4) = -\dfrac{1}{32} \\ f'''(x) = \dfrac{3}{8x^{5/2}} & f'''(4) = \dfrac{3}{256} \end{array}$$

$$f(x) = 2 + \frac{1}{4}(x - 4) - \frac{1}{32}\frac{(x - 4)^2}{2!} + \frac{3}{256}\frac{(x - 4)^3}{3!} - \cdots \qquad \text{using Eq. (13-17)}$$

$$\sqrt{x} = 2 + \frac{(x - 4)}{4} - \frac{(x - 4)^2}{64} + \frac{(x - 4)^3}{512} - \cdots$$

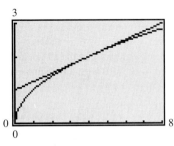

3

0

0   8

Fig. 13-10

This series would be used to evaluate square roots of numbers near 4.

In Fig. 13-10 we show a graphing calculator view of $y = \sqrt{x}$ and $y = 1 + x/4$, which are the first two terms of the Taylor series, as well as being the linearization of $f(x)$ at $x = 4$. Each passes through $(4, 2)$, and they have nearly equal values of y for values of x near 4.

In the last section we evaluated functions by using Maclaurin series. In the following examples we use Taylor series to evaluate functions.

EXAMPLE 3   By using Taylor series, evaluate $\sqrt{4.5}$.

Using the four terms of the series found in Example 2, we have

$$\sqrt{4.5} = 2 + \frac{(4.5 - 4)}{4} - \frac{(4.5 - 4)^2}{64} + \frac{(4.5 - 4)^3}{512} \qquad \text{substitute 4.5 for } x$$

$$= 2 + \frac{(0.5)}{4} - \frac{(0.5)^2}{64} + \frac{(0.5)^3}{512}$$

$$= 2.121337891$$

The value found directly on a calculator is 2.121320344. Therefore, the value found by these terms of the series expansion is correct to four decimal places.

CAUTION ▶

In Example 3 we saw that successive terms become small rapidly. If a value of x is chosen such that $x - a$ is larger, the successive terms may not become small rapidly, and many terms may be required. Therefore, *we should choose the value of a conveniently as close as possible to the x-values that will be used.* Also, we should note that a Maclaurin expansion for $\sqrt{x}$ cannot be used since the derivatives of $\sqrt{x}$ are not defined for $x = 0$.

EXAMPLE 4   Calculate the approximate value of sin 29° by using three terms of the appropriate Taylor expansion.

Since the value of sin 30° is known to be $\frac{1}{2}$, if we let $a = \frac{\pi}{6}$ (remember, we must use values expressed in radians), when we evaluate the expansion for $x = 29°$ (when expressed in radians) the quantity $(x - a)$ is $-\frac{\pi}{180}$ (equivalent to $-1°$). This means that its numerical values are small and become smaller when it is raised to higher powers. Therefore,

$$f(x) = \sin x \qquad f\left(\frac{\pi}{6}\right) = \frac{1}{2} \qquad \text{find derivatives and evaluate each at } x = \frac{\pi}{6}$$

$$f'(x) = \cos x \qquad f'\left(\frac{\pi}{6}\right) = \frac{\sqrt{3}}{2}$$

$$f''(x) = -\sin x \qquad f''\left(\frac{\pi}{6}\right) = -\frac{1}{2}$$

$$f(x) = \frac{1}{2} + \frac{\sqrt{3}}{2}\left(x - \frac{\pi}{6}\right) - \frac{1}{4}\left(x - \frac{\pi}{6}\right)^2 - \cdots \qquad \text{using Eq. (13-17)}$$

$$\sin x = \frac{1}{2} + \frac{\sqrt{3}}{2}\left(x - \frac{\pi}{6}\right) - \frac{1}{4}\left(x - \frac{\pi}{6}\right)^2 - \cdots \qquad f(x) = \sin x$$

$$\sin 29° = \sin\left(\frac{\pi}{6} - \frac{\pi}{180}\right) \qquad 29° = 30° - 1° = \frac{\pi}{6} - \frac{\pi}{180}$$

$$= \frac{1}{2} + \frac{\sqrt{3}}{2}\left(\frac{\pi}{6} - \frac{\pi}{180} - \frac{\pi}{6}\right) - \frac{1}{4}\left(\frac{\pi}{6} - \frac{\pi}{180} - \frac{\pi}{6}\right)^2 - \cdots \qquad \text{substitute } \frac{\pi}{6} - \frac{\pi}{180} \text{ for } x$$

$$= \frac{1}{2} + \frac{\sqrt{3}}{2}\left(-\frac{\pi}{180}\right) - \frac{1}{4}\left(-\frac{\pi}{180}\right)^2 - \cdots$$

$$= 0.4848088509$$

The value found directly on a calculator is 0.4848096202.

EXERCISES *13-5*

In Exercises 1–8, evaluate the given functions by using the series developed in the examples of this section.

1. $e^{1.2}$ **2.** $e^{0.7}$

3. $\sqrt{4.2}$ **4.** $\sqrt{3.5}$

5. $\sin 31°$ **6.** $\sin 28°$

7. $\sin 29.53°$ **8.** $\sqrt{3.8527}$

In Exercises 9–16, find the first three nonzero terms of the Taylor expansion for the given function and given value of a.

9. e^{-x} $(a = 2)$ **10.** $\cos x$ $(a = \frac{\pi}{4})$

11. $\sin x$ $(a = \frac{\pi}{3})$ **12.** $\ln x$ $(a = 3)$

13. $\sqrt[3]{x}$ $(a = 8)$ **14.** $\dfrac{1}{x}$ $(a = 2)$

15. $\tan x$ $(a = \frac{\pi}{4})$ **16.** $\ln \sin x$ $(a = \frac{\pi}{2})$

In Exercises 17–24, evaluate the given functions by using three terms of the appropriate Taylor series.

17. e^{π} **18.** $\ln 3.1$

19. $\sqrt{9.3}$ **20.** 2.056^{-1}

21. $\sqrt[3]{8.3}$ **22.** $\tan 46°$

23. $\sin 61°$ **24.** $\cos 42°$

In Exercises 25–28, solve the given problems.

25. By completing the steps indicated before Eq. (13-17) in the text, complete the derivation of Eq. (13-17).

26. Find the first three terms of the Taylor expansion of $f(x) = \ln x$ with $a = 1$. Compare this Taylor expansion with the linearization $L(x)$ of $f(x)$ with $a = 1$. Compare the graphs of $f(x)$, $L(x)$, and the Taylor expansion on a graphing calculator.

27. Calculate $\sin 31°$ by using three terms of the Maclaurin expansion for $\sin x$. Also calculate $\sin 31°$ by using three terms of the Taylor expansion in Example 4. Compare the accuracy of the values obtained with that found directly on a calculator.

28. In the analysis of the electric potential of an electric charge distributed along a straight wire of length L, the expression $\ln \dfrac{x + L}{x}$ is used. Find three terms of the Taylor expansion of this expression in powers of $(x - L)$.

(W) *In Exercises 29–32, use a graphing calculator to display (a) the function in the indicated exercise of this set and (b) the first two terms of the Taylor series found for that exercise in the same display. Describe how closely the graph in part (b) fits the graph in part (a). Use the given values of x for Xmin and Xmax.*

29. Exercise 11 ($\sin x$), $x = 0$ to $x = 2$

30. Exercise 13 ($\sqrt[3]{x}$), $x = 0$ to $x = 16$

31. Exercise 14 ($1/x$), $x = 0$ to $x = 4$

32. Exercise 15 ($\tan x$), $x = 0$ to $x = 1.5$

13-6 INTRODUCTION TO FOURIER SERIES

Many problems encountered in the various fields of science and technology involve functions that are periodic. *A periodic function is one for which $F(x + P) = F(x)$, where P is the period.* We noted that the trigonometric functions are periodic when we discussed their graphs in Chapter 7. Illustrations of applied problems that involve periodic functions are alternating-current voltages and mechanical oscillations.

Therefore, in this section we use a series made of terms of sines and cosines. This allows us to represent complicated periodic functions in terms of the simpler sines and cosines. It also provides us a good approximation over a greater interval than Maclaurin and Taylor series, which give good approximations with a few terms only near a specific value. Illustrations of applications of this type of series are given in Example 3 and in the exercises.

We shall assume that a function $f(x)$ may be represented by the series of sines and cosines as indicated:

$$f(x) = a_0 + a_1 \cos x + a_2 \cos 2x + \cdots + a_n \cos nx + \cdots$$
$$+ b_1 \sin x + b_2 \sin 2x + \cdots + b_n \sin nx + \cdots \qquad \text{(13-18)}$$

Named for the French mathematician and physicist Jean Baptiste Joseph Fourier (1768–1830).

Since all the sines and cosines indicated in this expansion have a period of 2π (the period of any given term may be less than 2π, but all do repeat every 2π units—for example, $\sin 2x$ has a period of π, but it also repeats every 2π), the series expansion indicated in Eq. (13-18) will also have a period of 2π. *This series is called a* **Fourier series.**

The principal problem to be solved is that of finding the coefficients a_n and b_n. Derivatives proved to be useful in finding the coefficients for a Maclaurin expansion. We use the properties of certain integrals to find the coefficients of a Fourier series. To utilize these properties, we multiply all terms of Eq. (13-18) by $\cos mx$ and then evaluate from $-\pi$ to π (in this way we take advantage of the period 2π). Thus, we have

$$\int_{-\pi}^{\pi} f(x)\cos mx\, dx = \int_{-\pi}^{\pi} (a_0 + a_1 \cos x + a_2 \cos 2x + \cdots)(\cos mx)\, dx$$
$$+ \int_{-\pi}^{\pi} (b_1 \sin x + b_2 \sin 2x + \cdots)(\cos mx)\, dx \qquad \text{(13-19)}$$

Using the methods of integration of Chapter 10, we now find the values of the coefficients a_n and b_n. For the coefficients a_n we find that the values differ depending on whether or not $n = m$. Therefore, first considering the case for which $n \neq m$, we have

$$\int_{-\pi}^{\pi} a_0 \cos mx\, dx = \frac{a_0}{m} \sin mx \Big|_{-\pi}^{\pi}$$

$$= \frac{a_0}{m}(0 - 0) = 0 \qquad \text{(13-20)}$$

$$\int_{-\pi}^{\pi} a_n \cos nx \cos mx\, dx$$

$$= a_n \left(\frac{\sin(n - m)x}{2(n - m)} + \frac{\sin(n + m)x}{2(n + m)} \right) \Big|_{-\pi}^{\pi} = 0 \qquad (n \neq m) \qquad \text{(13-21)}$$

These values are all equal to zero since the sine of any multiple of π is zero.

Now, considering the case for which $n = m$, we have

$$\int_{-\pi}^{\pi} a_n \cos nx \cos nx\, dx = \int_{-\pi}^{\pi} a_n \cos^2 nx\, dx$$

$$= \left(\frac{a_n x}{2} + \frac{a_n}{2n} \sin nx \cos nx \right) \Big|_{-\pi}^{\pi}$$

$$= \frac{a_n x}{2} \Big|_{-\pi}^{\pi} = \pi a_n \qquad \text{(13-22)}$$

On the next page, we continue by finding the values of the coefficient b_n in Eq. (13-19).

Now, finding the values of the coefficient b_n, we have

$$\int_{-\pi}^{\pi} b_n \sin nx \cos mx \, dx = b_n \left(-\frac{\cos(n-m)x}{2(n-m)} - \frac{\cos(n+m)x}{2(n+m)} \right) \Bigg|_{-\pi}^{\pi}$$

$$= b_n \left(-\frac{\cos(n-m)\pi}{2(n-m)} - \frac{\cos(n+m)\pi}{2(n+m)} \right.$$

$$\left. + \frac{\cos(n-m)(-\pi)}{2(n-m)} + \frac{\cos(n+m)(-\pi)}{2(n+m)} \right)$$

$$= 0 \ [\text{since } \cos \theta = \cos(-\theta)] \qquad (n \neq m) \qquad (13\text{-}23)$$

$$\int_{-\pi}^{\pi} b_n \sin nx \cos nx \, dx = \frac{b_n}{2n} \sin^2 nx \Big|_{-\pi}^{\pi} = 0 \qquad (13\text{-}24)$$

These integrals are seen to be zero, except for the one specific case of $\int_{-\pi}^{\pi} a_n \cos nx \cos mx \, dx$ when $n = m$, for which the result is indicated in Eq. (13-22). Using these results in Eq. (13-19), we have

$$\int_{-\pi}^{\pi} f(x) \cos nx \, dx = a_n \int_{-\pi}^{\pi} \cos^2 nx \, dx = \pi a_n$$

$$a_n = \frac{1}{\pi} \int_{-\pi}^{\pi} f(x) \cos nx \, dx \qquad (13\text{-}25)$$

This equation allows us to find the coefficients a_n, except a_0. We find the term a_0 by direct integration of Eq. (13-18) from $-\pi$ to π. When we perform this integration, all the sine and cosine terms integrate to zero, thereby giving the result

$$\int_{-\pi}^{\pi} f(x) \, dx = \int_{-\pi}^{\pi} a_0 \, dx = a_0 x \Big|_{-\pi}^{\pi} = 2\pi a_0$$

$$a_0 = \frac{1}{2\pi} \int_{-\pi}^{\pi} f(x) \, dx \qquad (13\text{-}26)$$

By multiplying all terms of Eq. (13-18) by $\sin mx$ and then integrating from $-\pi$ to π, we find the coefficients b_n. We obtain the result

$$b_n = \frac{1}{\pi} \int_{-\pi}^{\pi} f(x) \sin nx \, dx \qquad (13\text{-}27)$$

We can restate our equations for the Fourier series of a function $f(x)$:

$$f(x) = a_0 + a_1 \cos x + a_2 \cos 2x + \cdots + a_n \cos nx + \cdots$$
$$+ b_1 \sin x + b_2 \sin 2x + \cdots + b_n \sin nx + \cdots \qquad (13\text{-}18)$$

where the coefficients are found by

$$a_0 = \frac{1}{2\pi} \int_{-\pi}^{\pi} f(x) \, dx \qquad (13\text{-}26)$$

$$a_n = \frac{1}{\pi} \int_{-\pi}^{\pi} f(x) \cos nx \, dx \qquad (13\text{-}25)$$

$$b_n = \frac{1}{\pi} \int_{-\pi}^{\pi} f(x) \sin nx \, dx \qquad (13\text{-}27)$$

EXAMPLE 1   Find the Fourier series for the square wave function

$$f(x) = \begin{cases} -1 & -\pi \le x < 0 \\ 1 & 0 \le x < \pi \end{cases}$$

(Many of the functions we shall expand in Fourier series are discontinuous (not continuous) like this one. See Section 3-1 for a discussion of continuity.)

CAUTION ▶   Since $f(x)$ is defined differently for the intervals of x indicated, **it requires two integrals for each coefficient:**

using Eq. (13-26) $a_0 = \dfrac{1}{2\pi}\displaystyle\int_{-\pi}^{0}(-1)\,dx + \dfrac{1}{2\pi}\int_{0}^{\pi}(1)\,dx = -\dfrac{x}{2\pi}\Big|_{-\pi}^{0} + \dfrac{x}{2\pi}\Big|_{0}^{\pi} = -\dfrac{1}{2} + \dfrac{1}{2} = 0$

using Eq. (13-25) $a_n = \dfrac{1}{\pi}\displaystyle\int_{-\pi}^{0}(-1)\cos nx\,dx + \dfrac{1}{\pi}\int_{0}^{\pi}(1)\cos nx\,dx = -\dfrac{1}{n\pi}\sin nx\Big|_{-\pi}^{0} + \dfrac{1}{n\pi}\sin nx\Big|_{0}^{\pi} = 0 + 0 = 0$

for all values of n, since $\sin n\pi = 0$;

using Eq. (13-27)
with $n = 1$ $b_1 = \dfrac{1}{\pi}\displaystyle\int_{-\pi}^{0}(-1)\sin x\,dx + \dfrac{1}{\pi}\int_{0}^{\pi}(1)\sin x\,dx = \dfrac{1}{\pi}\cos x\Big|_{-\pi}^{0} - \dfrac{1}{\pi}\cos x\Big|_{0}^{\pi}$

$\qquad = \dfrac{1}{\pi}(1 + 1) - \dfrac{1}{\pi}(-1 - 1) = \dfrac{4}{\pi}$

using Eq. (13-27)
with $n = 2$ $b_2 = \dfrac{1}{\pi}\displaystyle\int_{-\pi}^{0}(-1)\sin 2x\,dx + \dfrac{1}{\pi}\int_{0}^{\pi}(1)\sin 2x\,dx = \dfrac{1}{2\pi}\cos 2x\Big|_{-\pi}^{0} - \dfrac{1}{2\pi}\cos 2x\Big|_{0}^{\pi}$

$\qquad = \dfrac{1}{2\pi}(1 - 1) - \dfrac{1}{2\pi}(1 - 1) = 0$

using Eq. (13-27)
with $n = 3$ $b_3 = \dfrac{1}{\pi}\displaystyle\int_{-\pi}^{0}(-1)\sin 3x\,dx + \dfrac{1}{\pi}\int_{0}^{\pi}(1)\sin 3x\,dx = \dfrac{1}{3\pi}\cos 3x\Big|_{-\pi}^{0} - \dfrac{1}{3\pi}\cos 3x\Big|_{0}^{\pi}$

$\qquad = \dfrac{1}{3\pi}(1 + 1) - \dfrac{1}{3\pi}(-1 - 1) = \dfrac{4}{3\pi}$

In general, if n is even, $b_n = 0$, and if n is odd, then $b_n = 4/n\pi$. Therefore,

$$f(x) = \frac{4}{\pi}\sin x + \frac{4}{3\pi}\sin 3x + \frac{4}{5\pi}\sin 5x + \cdots = \frac{4}{\pi}\left(\sin x + \frac{1}{3}\sin 3x + \frac{1}{5}\sin 5x + \cdots\right)$$

A graph of the function as defined, and the curve found by using the first three terms of the Fourier series, are shown in Fig. 13-11.

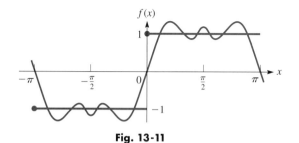

Fig. 13-11

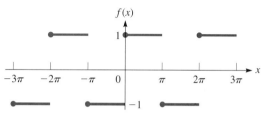

Fig. 13-12

Since functions found by Fourier series have a period of 2π, they can represent functions with this period. If the function $f(x)$ were defined to be periodic with period 2π, with the same definitions as originally indicated, we would graph the function as shown in Fig. 13-12. The Fourier series representation would follow it as in Fig. 13-11. If more terms were used, the fit would be closer.

EXAMPLE 2 Find the Fourier series for the function

$$f(x) = \begin{cases} 1 & -\pi \le x < 0 \\ x & 0 \le x < \pi \end{cases}$$

For the periodic function, let $f(x + 2\pi) = f(x)$ for all x.

A graph of three periods of this function is shown in Fig. 13-13.

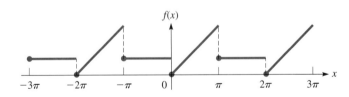

Fig. 13-13

Now, finding the coefficients, we have

$$a_0 = \frac{1}{2\pi} \int_{-\pi}^{0} dx + \frac{1}{2\pi} \int_{0}^{\pi} x\,dx = \frac{x}{2\pi}\bigg|_{-\pi}^{0} + \frac{x^2}{4\pi}\bigg|_{0}^{\pi} \qquad \text{using Eq. (13-26)}$$

$$= \frac{1}{2} + \frac{\pi}{4} = \frac{2 + \pi}{4}$$

$$a_1 = \frac{1}{\pi} \int_{-\pi}^{0} \cos x\,dx + \frac{1}{\pi} \int_{0}^{\pi} x \cos x\,dx \qquad \begin{array}{l}\text{using Eq. (13-25)}\\ \text{with } n = 1\end{array}$$

$$= \frac{1}{\pi} \sin x\bigg|_{-\pi}^{0} + \frac{1}{\pi} (\cos x + x \sin x)\bigg|_{0}^{\pi} = -\frac{2}{\pi}$$

$$a_2 = \frac{1}{\pi} \int_{-\pi}^{0} \cos 2x\,dx + \frac{1}{\pi} \int_{0}^{\pi} x \cos 2x\,dx \qquad \begin{array}{l}\text{using Eq. (13-25)}\\ \text{with } n = 2\end{array}$$

$$= \frac{1}{2\pi} \sin 2x\bigg|_{-\pi}^{0} + \frac{1}{4\pi} (\cos 2x + 2x \sin 2x)\bigg|_{0}^{\pi} = 0$$

$$a_3 = \frac{1}{\pi} \int_{-\pi}^{0} \cos 3x\,dx + \frac{1}{\pi} \int_{0}^{\pi} x \cos 3x\,dx \qquad \begin{array}{l}\text{using Eq. (13-25)}\\ \text{with } n = 3\end{array}$$

$$= \frac{1}{3\pi} \sin 3x\bigg|_{-\pi}^{0} + \frac{1}{9\pi} (\cos 3x + 3x \sin 3x)\bigg|_{0}^{\pi} = -\frac{2}{9\pi}$$

$$b_1 = \frac{1}{\pi} \int_{-\pi}^{0} \sin x\,dx + \frac{1}{\pi} \int_{0}^{\pi} x \sin x\,dx \qquad \begin{array}{l}\text{using Eq. (13-27)}\\ \text{with } n = 1\end{array}$$

$$= -\frac{1}{\pi} \cos x\bigg|_{-\pi}^{0} + \frac{1}{\pi} (\sin x - x \cos x)\bigg|_{0}^{\pi} = \frac{\pi - 2}{\pi}$$

$$b_2 = \frac{1}{\pi} \int_{-\pi}^{0} \sin 2x\,dx + \frac{1}{\pi} \int_{0}^{\pi} x \sin 2x\,dx \qquad \begin{array}{l}\text{using Eq. (13-27)}\\ \text{with } n = 2\end{array}$$

$$= -\frac{\cos 2x}{2\pi}\bigg|_{-\pi}^{0} + \frac{\sin 2x - 2x \cos 2x}{4\pi}\bigg|_{0}^{\pi} = -\frac{1}{2}$$

Therefore, the first few terms of the Fourier series are

$$f(x) = \frac{2 + \pi}{4} - \frac{2}{\pi} \cos x - \frac{2}{9\pi} \cos 3x - \cdots + \left(\frac{\pi - 2}{\pi}\right) \sin x - \frac{1}{2} \sin 2x + \cdots$$

See the chapter introduction.

EXAMPLE 3 Certain electronic devices allow an electric current to pass through in only one direction. When an alternating current is applied to the circuit, the current exists for only half the cycle. Figure 13-14 is a representation of such a current as a function of time. This type of electronic device is called a *half-wave rectifier*. Derive the Fourier series for a rectified wave for which half is defined by $f(t) = \sin t$ $(0 \le t \le \pi)$ and for which the other half is defined by $f(t) = 0$.

In finding the Fourier coefficients, we first find a_0 as

$$a_0 = \frac{1}{2\pi} \int_0^\pi \sin t \, dt = \frac{1}{2\pi} (-\cos t)\Big|_0^\pi = \frac{1}{2\pi} (1 + 1) = \frac{1}{\pi}$$

In the previous example we evaluated each of the coefficients individually. Here we show how to set up a general expression for a_n and another for b_n. Once we have determined these, we can substitute values of n in the formula to obtain the individual coefficients:

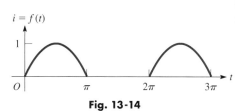

$i = f(t)$

Fig. 13-14

$$a_n = \frac{1}{\pi} \int_0^\pi \sin t \cos nt \, dt = -\frac{1}{2\pi} \left[\frac{\cos(1 - n)t}{1 - n} + \frac{\cos(1 + n)t}{1 + n} \right] \Bigg|_0^\pi$$

$$= -\frac{1}{2\pi} \left[\frac{\cos(1 - n)\pi}{1 - n} + \frac{\cos(1 + n)\pi}{1 + n} - \frac{1}{1 - n} - \frac{1}{1 + n} \right]$$

See Formula 40 in the table of integrals in Appendix E. It is valid for all values of n except $n = 1$. Now we write

$$a_1 = \frac{1}{\pi} \int_0^\pi \sin t \cos t \, dt = \frac{1}{2\pi} \sin^2 t \Big|_0^\pi = 0$$

$$a_2 = -\frac{1}{2\pi} \left(\frac{-1}{-1} + \frac{-1}{3} - \frac{1}{-1} - \frac{1}{3} \right) = -\frac{2}{3\pi}$$

$$a_3 = -\frac{1}{2\pi} \left(\frac{1}{-2} + \frac{1}{4} - \frac{1}{-2} - \frac{1}{4} \right) = 0$$

$$a_4 = -\frac{1}{2\pi} \left(\frac{-1}{-3} + \frac{-1}{5} - \frac{1}{-3} - \frac{1}{5} \right) = -\frac{2}{15\pi}$$

$$b_n = \frac{1}{\pi} \int_0^\pi \sin t \sin nt \, dt = \frac{1}{2\pi} \left[\frac{\sin(1 - n)t}{1 - n} - \frac{\sin(1 + n)t}{1 + n} \right] \Bigg|_0^\pi$$

$$= \frac{1}{2\pi} \left[\frac{\sin(1 - n)\pi}{1 - n} - \frac{\sin(1 + n)\pi}{1 + n} \right]$$

See Formula 39 in Appendix E. It is valid for all values of n except $n = 1$. Therefore, we have

$$b_1 = \frac{1}{\pi} \int_0^\pi \sin t \sin t \, dt = \frac{1}{\pi} \int_0^\pi \sin^2 t \, dt = \frac{1}{2\pi} (t - \sin t \cos t)\Big|_0^\pi = \frac{1}{2}$$

We see that $b_n = 0$ if $n > 1$, since each is evaluated in terms of the sine of a multiple of π.

Therefore, the Fourier series for the rectified wave is

$$f(t) = \frac{1}{\pi} + \frac{1}{2} \sin t - \frac{2}{\pi} \left(\frac{1}{3} \cos 2t + \frac{1}{15} \cos 4t + \cdots \right)$$

The graph of these terms of the Fourier series is shown in the graphing calculator display in Fig. 13-15.

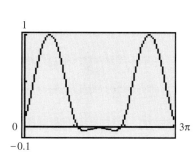

Fig. 13-15

All the types of periodic functions included in this section (as well as many others) may actually be seen on an oscilloscope when the proper signal is sent into it. In this way the oscilloscope may be used to analyze the periodic nature of such phenomena as sound waves and electric currents.

EXERCISES 13-6

In Exercises 1–10, find at least three nonzero terms (including a_0 and at least two cosine terms and two sine terms if they are not all zero) of the Fourier series for the given functions and sketch at least three periods of the function.

1. $f(x) = \begin{cases} 1 & -\pi \le x < 0 \\ 0 & 0 \le x < \pi \end{cases}$

2. $f(x) = \begin{cases} 0 & -\pi \le x < 0 \\ 1 & 0 \le x < \pi \end{cases}$

3. $f(x) = \begin{cases} 1 & -\pi \le x < 0 \\ 2 & 0 \le x < \pi \end{cases}$

4. $f(x) = \begin{cases} 0 & -\pi \le x < 0, \dfrac{\pi}{2} < x < \pi \\ 1 & 0 \le x \le \dfrac{\pi}{2} \end{cases}$

5. $f(x) = \begin{cases} 0 & -\pi \le x < 0 \\ x & 0 \le x < \pi \end{cases}$

6. $f(x) = x \quad -\pi \le x < \pi$

7. $f(x) = \begin{cases} -1 & -\pi \le x < 0 \\ 0 & 0 \le x < \dfrac{\pi}{2} \\ 1 & \dfrac{\pi}{2} \le x < \pi \end{cases}$

8. $f(x) = x^2 \quad -\pi \le x < \pi$

9. $f(x) = \begin{cases} -x & -\pi \le x < 0 \\ x & 0 \le x < \pi \end{cases}$

10. $f(x) = \begin{cases} 0 & -\pi \le x < 0 \\ x^2 & 0 \le x < \pi \end{cases}$

In Exercises 11–14, use a graphing calculator to display the terms of the Fourier series given in the indicated example or answer for the indicated exercise. Compare with the sketch of the function. For each calculator display use Xmin $= -8$ and Xmax $= 8$.

11. Example 1   **12.** Example 2

13. Exercise 5   **14.** Exercise 9

In Exercises 15 and 16, solve the given problems.

15. Find the Fourier expansion of the electronic device known as a *full-wave rectifier*. This is found by using as the function for the current $f(t) = -\sin t$ for $-\pi \le t \le 0$ and $f(t) = \sin t$ for $0 < t \le \pi$. The graph of this function is shown in Fig. 13-16. The portion of the curve to the left of the origin is dashed because from a physical point of view we can give no significance to this part of the wave, although mathematically we can derive the proper form of the Fourier expansion by using it.

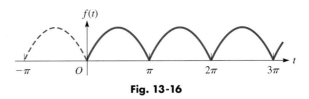

Fig. 13-16

16. The loudness L (in decibels) of a certain siren as a function of time t (in s) can be described by the function

$$\begin{aligned} L &= 0 & -\pi \le t < 0 \\ L &= 100t & 0 \le t < \pi/2 \\ L &= 100(\pi - t) & \pi/2 \le t < \pi \end{aligned}$$

with a period of 2π seconds (where only positive values of t have physical significance). Find a_0, the first nonzero cosine term, and the first two nonzero sine terms of the Fourier expansion for the loudness of the siren. See Fig. 13-17.

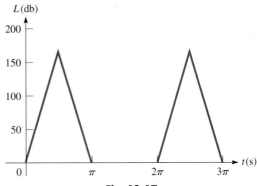

Fig. 13-17

13-7 MORE ABOUT FOURIER SERIES

When finding the Fourier expansion of some functions, it may turn out that all the sine terms evaluate to be zero or that all the cosine terms evaluate to be zero. In fact, in Example 1 on page 405 we see that all of the cosine terms were zero and that the expansion contained only sine terms. We now show how to quickly determine if an expansion will contain only sine terms, or only cosine terms.

Even Functions and Odd Functions

In Chapter 2 (page 39) we showed that when $-x$ replaces x in a function $f(x)$ and the function does not change, the curve of the function is symmetric to the y-axis. *Such a function is called an* **even function.**

EXAMPLE 1   We can show that the function $y = \cos x$ is an even function by using the Maclaurin expansions for $\cos x$ and $\cos(-x)$. These are

$$\cos x = 1 - \frac{x^2}{2} + \frac{x^4}{24} - \cdots$$

$$\cos(-x) = 1 - \frac{(-x)^2}{2} + \frac{(-x)^4}{24} \cdots = 1 - \frac{x^2}{2} + \frac{x^4}{24} - \cdots$$

Since the expansions are the same, $\cos x$ is an even function.   ∎

NOTE ▶ Since $\cos x$ is an even function and all of its terms are even functions, it follows that *an* **even function** *will have a Fourier series that contains only* **cosine** *terms (and possibly a constant term).*

EXAMPLE 2   The Fourier series for the function

$$f(x) = \begin{cases} 0 & -\pi \le x < -\pi/2,\ \pi/2 \le x < \pi \\ 1 & -\pi/2 \le x < \pi/2 \end{cases}$$

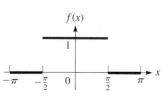

Fig. 13-18

is $f(x) = \dfrac{1}{2} + \dfrac{2}{\pi}\left(\cos x - \dfrac{1}{3} \cos 3x + \dfrac{1}{5} \cos 5x - \cdots \right).$ We   see   that $f(x) = f(-x)$, which means it is an even function. We also see that its Fourier series expansion contains only cosine terms (and a constant). Thus, *when finding the Fourier series we do not have to find any sine terms.* The graph of $f(x)$ in Fig. 13-18 shows its symmetry to the y-axis.   ∎

Again referring to Chapter 2 (page 39), we recall that if $-x$ replaces x and $-y$ replaces y at the same time, and the function does not change, then the function is symmetric to the origin. *Such a function is called an* **odd function.**

EXAMPLE 3   We can show that the function $y = \sin x$ is an odd function by using the Maclaurin expansions for $\sin x$ and $-\sin(-x)$ (the $-$ sign before $\sin(-x)$ is equivalent to making y negative). These are

$$\sin x = x - \frac{x^3}{6} + \frac{x^5}{120} - \cdots$$

$$-\sin(-x) = -\left((-x) - \frac{(-x)^3}{6} + \frac{(-x)^5}{120} - \cdots \right) = x - \frac{x^3}{6} + \frac{x^5}{120} - \cdots$$

Since $\sin x = -\sin(-x)$, $\sin x$ is an odd function.   ∎

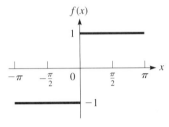

Fig. 13-19

Since sin x is an odd function and all its terms are odd functions, it follows that an *odd function* *will have a Fourier series that contains only **sine** terms (and no constant term).*

EXAMPLE 4 As we showed in Example 1 on page 405, the Fourier series for the function

$$f(x) = \begin{cases} -1 & -\pi \leq x < 0 \\ 1 & 0 \leq x < \pi \end{cases}$$

is $f(x) = \dfrac{4}{\pi}\left(\sin x + \dfrac{1}{3}\sin 3x + \dfrac{1}{5}\sin 5x + \cdots\right)$. We can see that $f(-x) = -f(-x)$, which means $f(x)$ is an odd function. We also see that its Fourier series expansion contains only sine terms. Therefore, *when finding the Fourier series, we do not have to find any cosine terms.* The graph of $f(x)$ in Fig. 13-19 shows its symmetry to the origin.

If a constant k is added to a function $f_1(x)$, the resulting function $f(x)$ is

$$f(x) = k + f_1(x)$$

NOTE ▶ Therefore, if we know the Fourier series expansion for $f_1(x)$, *the Fourier series expansion of $f(x)$ is found by adding k to the Fourier series expansion of $f_1(x)$.*

EXAMPLE 5 The values of the function

$$f(x) = \begin{cases} 1 & -\pi \leq x < -\pi/2,\ \pi/2 \leq x < \pi \\ 2 & -\pi/2 \leq x < \pi/2 \end{cases}$$

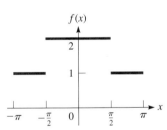

Fig. 13-20

are all 1 greater than those of the function of Example 2. Therefore, denoting the function of Example 2 as $f_1(x)$, we have $f(x) = 1 + f_1(x)$. This means that the Fourier series for $f(x)$ is

$$f(x) = 1 + \left(\frac{1}{2} + \frac{2}{\pi}\left(\cos x - \frac{1}{3}\cos 3x + \frac{1}{5}\cos 5x - \cdots\right)\right)$$
$$= \frac{3}{2} + \frac{2}{\pi}\left(\cos x - \frac{1}{3}\cos 3x + \frac{1}{5}\cos 5x - \cdots\right)$$

In Fig. 13-20, we see that the graph of $f(x)$ is shifted up vertically by 1 unit from the graph of $f_1(x)$ in Fig. 13-18. This is equivalent to a vertical translation of axes. We also note that $f(x)$ is an even function.

EXAMPLE 6 The values of the function

$$f(x) = \begin{cases} -\frac{3}{2} & -\pi \leq x < 0 \\ \frac{1}{2} & 0 \leq x < \pi \end{cases}$$

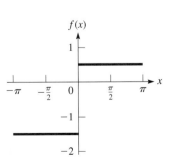

Fig. 13-21

are all $\frac{1}{2}$ less than those of the function of Example 4. Therefore, denoting the function of Example 4 as $f_1(x)$, we have $f(x) = -\frac{1}{2} + f_1(x)$. This means that the Fourier series for $f(x)$ is

$$f(x) = -\frac{1}{2} + \frac{4}{\pi}\left(\sin x - \frac{1}{3}\sin 3x + \frac{1}{5}\sin 5x - \cdots\right)$$

In Fig. 13-21, we see that the graph of $f(x)$ is shifted vertically down by $\frac{1}{2}$ unit from the graph of $f_1(x)$ in Fig. 13-19. Although $f(x)$ is not an odd function, it would be an odd function if its origin were translated to $(0, -\frac{1}{2})$.

Fourier Series with Period 2L

The standard form of a Fourier series we have considered to this point is defined over the interval from $x = -\pi$ to $x = \pi$. At times it is preferable to have a series that is defined over a different interval.

Noting that

$$\sin \frac{n\pi}{L}(x + 2L) = \sin n\left(\frac{\pi x}{L} + 2\pi\right) = \sin \frac{n\pi x}{L}$$

we see that $\sin(n\pi x/L)$ has a period of $2L$. Thus, by using $\sin(n\pi x/L)$ and $\cos(n\pi x/L)$ and the same method of derivation, the following equations are found for the coefficients for the Fourier series for the interval from $x = -L$ to $x = L$.

$$a_0 = \frac{1}{2L} \int_{-L}^{L} f(x)\, dx \tag{13-28}$$

$$a_n = \frac{1}{L} \int_{-L}^{L} f(x)\cos \frac{n\pi x}{L}\, dx \tag{13-29}$$

$$b_n = \frac{1}{L} \int_{-L}^{L} f(x)\sin \frac{n\pi x}{L}\, dx \tag{13-30}$$

EXAMPLE 7   Find the Fourier series for the function

$$f(x) = \begin{cases} 0 & -4 \le x < 0 \\ 2 & 0 \le x < 4 \end{cases}$$

and for which the period is 8. See Fig. 13-22.

Since the period is 8, $L = 4$. Next we note that $f(x) = 1 + f_1(x)$, where $f_1(x)$ is an odd function (from the definition of $f(x)$ and from Fig. 13-22, we can see the symmetry to the point $(0, 1)$). Therefore, *the constant is 1, and there are no cosine terms* in the Fourier series for $f(x)$. Now, finding the sine terms, we have

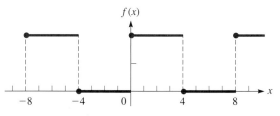

$$f(x)$$

Fig. 13-22

$$b_n = \frac{1}{4}\int_{-4}^{0} 0 \sin \frac{n\pi x}{4}\, dx + \frac{1}{4}\int_{0}^{4} 2 \sin \frac{n\pi x}{4}\, dx \qquad \text{using Eq. (13-30)}$$

$$= \frac{1}{2}\left(\frac{4}{n\pi}\right)\int_{0}^{4} \sin \frac{n\pi x}{4}\left(\frac{n\pi\, dx}{4}\right) = -\frac{2}{n\pi}\cos \frac{n\pi x}{4}\Big|_{0}^{4}$$

$$= -\frac{2}{n\pi}(\cos n\pi - \cos 0) = \frac{2}{n\pi}(1 - \cos n\pi)$$

$$b_1 = \frac{2}{\pi}[1 - (-1)] = \frac{4}{\pi} \qquad b_2 = \frac{2}{2\pi}(1 - 1) = 0$$

$$b_3 = \frac{2}{3\pi}[1 - (-1)] = \frac{4}{3\pi} \qquad b_4 = \frac{2}{4\pi}(1 - 1) = 0$$

Therefore, the Fourier series is

$$f(x) = 1 + \frac{4}{\pi}\sin \frac{\pi x}{4} + \frac{4}{3\pi}\sin \frac{3\pi x}{4} + \cdots$$

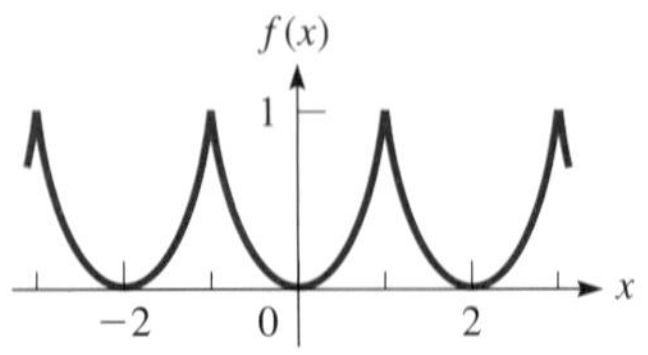

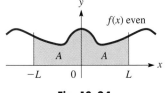

f(x)

Fig. 13-23

EXAMPLE 8 Find the Fourier series for the function

$$f(x) = x^2 \quad -1 \le x < 1$$

for which the period is 2. See Fig. 13-23.

Since the period is 2, $L = 1$. Next we note that $f(x) = f(-x)$, which means it is an even function. Therefore, there are no sine terms in the Fourier series. Finding the constant and the cosine terms, we have

$$a_0 = \frac{1}{2(1)} \int_{-1}^{1} x^2 \, dx = \frac{1}{6} x^3 \Big|_{-1}^{1} = \frac{1}{6}(1+1) = \frac{1}{3}$$

$$a_n = \frac{1}{1} \int_{-1}^{1} x^2 \cos \frac{n\pi x}{1} \, dx = \int_{-1}^{1} x^2 \cos n\pi x \, dx \qquad \text{integrating by parts: } u = x^2, \; du = 2x \, dx, \\ dv = \cos n\pi x \, dx, \; v = (1/n\pi)\sin n\pi x$$

$$= x^2 \left(\frac{1}{n\pi} \sin n\pi x \right) \Big|_{-1}^{1} - \frac{2}{n\pi} \int_{-1}^{1} x \sin n\pi x \, dx \qquad \text{integrating by parts: } u = x, \; du = dx, \\ dv = \sin n\pi x \, dx, \; v = (-1/n\pi)\cos n\pi x$$

$$\sin n\pi = 0 \qquad = \frac{1}{n\pi} \sin n\pi - \frac{1}{n\pi} \sin(-n\pi) - \frac{2}{n\pi} \left[x \left(-\frac{1}{n\pi} \cos n\pi x \right) \Big|_{-1}^{1} - \left(-\frac{1}{n\pi} \int_{-1}^{1} \cos n\pi x \, dx \right) \right]$$

$$\sin n\pi = 0 \qquad = \frac{2}{n^2\pi^2} [\cos n\pi + \cos(-n\pi)] + \frac{1}{n^2\pi^2} \sin n\pi x \Big|_{-1}^{1} = \frac{2}{n^2\pi^2}(2 \cos n\pi) = \frac{4}{n^2\pi^2} \cos n\pi$$

$$a_1 = \frac{4}{\pi^2} \cos \pi = -\frac{4}{\pi^2} \qquad a_2 = \frac{4}{4\pi^2} \cos 2\pi = \frac{4}{4\pi^2} \qquad a_3 = \frac{4}{9\pi^2} \cos 3\pi = -\frac{4}{9\pi^2}$$

Therefore, the Fourier series is

$$f(x) = \frac{1}{3} - \frac{4}{\pi^2} \left(\cos \pi x - \frac{1}{4} \cos 2\pi x + \frac{1}{9} \cos 3\pi x - \cdots \right) \qquad \blacksquare$$

Half-Range Expansions

We have seen that the Fourier series expansion for an even function contains only cosine terms (and possibly a constant), and the expansion of an odd function contains only sine terms. It is also possible to specify a function to be even or odd, such that the expansion will contain only cosine terms or only sine terms.

Considering the symmetry of an even function, the area under the curve from $-L$ to 0 is the same as the area under the curve from 0 to L (see Fig. 13-24). This means the value of the integral from $-L$ to 0 equals the value of the integral from 0 to L. Therefore, the value of the integral from $-L$ to L equals twice the value of the integral from 0 to L, or

Fig. 13-24

y

f(x) even

A *A*

$-L$ 0 L

$$\int_{-L}^{L} f(x) \, dx = 2 \int_{0}^{L} f(x) \, dx \qquad f(x) \text{ even}$$

Therefore, to obtain the Fourier coefficients for an expression from $-L$ to L for an even fucntion, we can multiply the coefficients obtained using Eqs. (13-28) and (13-29) from 0 to L by 2. Similar reasoning shows that the Fourier coefficients for an expansion from $-L$ to L for an odd function may be found by multiplying the coefficients obtained using Eq. (13-30) from 0 to L by 2.

A **half-range Fourier cosine series** *is a series that contains only cosine terms*, and *a* **half-range Fourier sine series** *is a series that contains only sine terms*. To find the half-range expansion for a function $f(x)$, it is defined for interval 0 to L (*half* of the interval from $-L$ to L) and then specified as odd or even, thereby clearly defining the function in the interval from $-L$ to 0. This means that *the Fourier coefficients for a half-range cosine series are given by*

$$a_0 = \frac{1}{L}\int_0^L f(x)\,dx \quad \text{and} \quad a_n = \frac{2}{L}\int_0^L f(x)\cos\frac{n\pi x}{L}\,dx \qquad (n = 1, 2, \ldots)$$
(13-31)

Similarly, *the Fourier coefficients for a half-range sine series are given by*

$$b_n = \frac{2}{L}\int_0^L f(x)\sin\frac{n\pi x}{L}\,dx \qquad (n = 1, 2, \ldots)$$
(13-32)

EXAMPLE 9   Find $f(x) = x$ in a half-range cosine series for $0 \le x < 2$.

Since we are to have a cosine series, we extend the function to be an even function with its graph as shown in Fig. 13-25. The portion in color between $x = 0$ and $x = L$ shows the given function as defined, and portions in black show the extension that makes it an even function. Now, by use of Eqs. (13-31) we find the Fourier expansion coefficients, with $L = 2$.

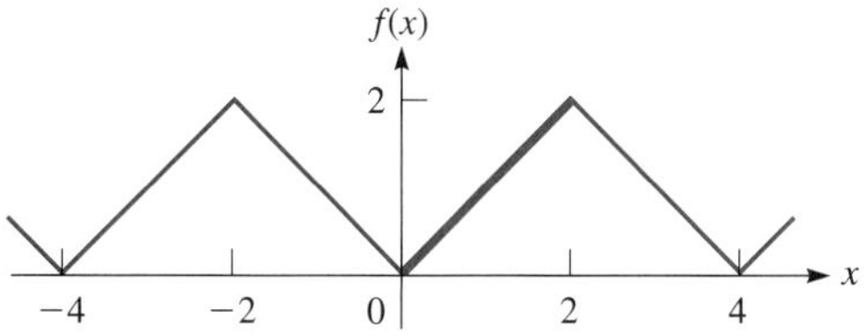

$$a_0 = \frac{1}{2}\int_0^2 x\,dx = \frac{1}{4}x^2\Big|_0^2 = 1$$

$$a_n = \frac{2}{2}\int_0^2 x\cos\frac{n\pi x}{2}\,dx = x\left(\frac{2}{n\pi}\sin\frac{n\pi x}{2}\right) - \left(\frac{-4}{n^2\pi^2}\cos\frac{n\pi x}{2}\right)\Big|_0^2$$

$$= \frac{4}{n^2\pi^2}(\cos n\pi - 1) \qquad (n \neq 0)$$

In n is even, $\cos n\pi - 1 = 0$. Therefore, we evaluate a_n for odd values of n, and find the expansion is

$$f(x) = 1 - \frac{8}{\pi^2}\left(\cos\frac{\pi x}{2} + \frac{1}{9}\cos\frac{3\pi x}{2} + \frac{1}{25}\cos\frac{5\pi x}{2} + \cdots\right)$$

f(x)

2

−4   −2   0   2   4   x

Fig. 13-25

EXAMPLE 10   Expand $f(x) = x$ in a half-range sine series of $0 \le x < 2$.

Since we are to have a sine series, we extend the function to be an odd function with its graph as shown in Fig. 13-26. Again, the portion in color shows the given function as defined, and portions in black show the extension that makes it an odd function. By using Eq. (13-32) we find the Fourier expansion coefficients, with $L = 2$.

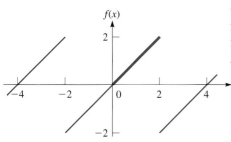

f(x)

2

−4   −2   0   2   4   x

−2

Fig. 13-26

$$b_n = \frac{2}{2}\int_0^2 x\sin\frac{n\pi x}{2}\,dx$$

$$= x\left(\frac{-2}{n\pi}\cos\frac{n\pi x}{2}\right) - \left(\frac{-4}{n^2\pi^2}\sin\frac{n\pi x}{2}\right)\Big|_0^2 = -\frac{4}{n\pi}\cos n\pi$$

$$f(x) = \frac{4}{\pi}\left(\sin\frac{\pi x}{2} - \frac{1}{2}\sin\pi x + \frac{1}{3}\sin\frac{3\pi x}{2} - \cdots\right)$$

─── **EXERCISES** *13-7* ───

In Exercises 1–8, determine whether the given function is even, odd, or neither. One period is defined for each function.

1. $f(x) = \begin{cases} 5 & -3 \leq x < 0 \\ 0 & 0 \leq x < 3 \end{cases}$ **2.** $f(x) = \begin{cases} -1 & -2 \leq x < 0 \\ 1 & 0 \leq x < 2 \end{cases}$

3. $f(x) = \begin{cases} 2 & -1 \leq x < 1 \\ 0 & -2 \leq x < -1, 1 \leq x < 2 \end{cases}$

4. $f(x) = \begin{cases} 0 & -2 \leq x < 0, 1 \leq x < 2 \\ 1 & 0 \leq x < 1 \end{cases}$

5. $f(x) = |x| \quad -4 \leq x < 4$ **6.** $f(x) = \begin{cases} 0 & -1 \leq x < 0 \\ e^x & 0 \leq x < 1 \end{cases}$

7. $f(x) = -x \cos 3x \quad -3 \leq x < 3$

8. $f(x) = x \sin 2x \cos x \quad -4 \leq x < 4$

In Exercises 9–12, write the Fourier series for each function by comparing it to an appropriate function given in an example of this section. Do not use any of the formulas for a_0, a_n, or b_n.

9. $f(x) = \begin{cases} 2 & -\pi \leq x < -\pi/2, \pi/2 \leq x < \pi \\ 3 & -\pi/2 \leq x < \pi/2 \end{cases}$

10. $f(x) = \begin{cases} -\frac{1}{2} & -\pi \leq x < 0 \\ \frac{3}{2} & 0 \leq x < \pi \end{cases}$

11. $f(x) = \begin{cases} -2 & -4 \leq x < 0 \\ 0 & 0 \leq x < 4 \end{cases}$

12. $f(x) = \begin{cases} -\frac{1}{3} & -\pi \leq x < -\pi/2, \pi/2 \leq x < \pi \\ \frac{2}{3} & -\pi/2 \leq x < \pi/2 \end{cases}$

In Exercises 13–18, find at least three nonzero terms (including a_0 and at least two cosine terms and two sine terms if they are not all zero) of the Fourier series for the function from the indicated exercise of this section. Sketch at least three periods of the function.

13. Exercise 1   **14.** Exercise 2   **15.** Exercise 3

16. Exercise 4   **17.** Exercise 5   **18.** Exercise 6

In Exercises 19–24, solve the given problems.

19. Expand $f(x) = 1$ in a half-range sine series for $0 \leq x < 4$.

20. Expand $f(x) = 1$ $(0 \leq x < 2)$, $f(x) = 0$ $(2 \leq x < 4)$ in a half-range cosine series for $0 \leq x < 4$.

21. Expand $f(x) = x^2$ in a half-range cosine series for $0 \leq x < 2$.

22. Expand $f(x) = x^2$ in a half-range sine series for $0 \leq x < 2$.

23. Each pulse of a pulsating force F of a pressing machine is 8 N. The force lasts for 1 s, followed by a 3-s pause. Thus, it can be represented by $F = 0$ for $-2 \leq t < 0$ and $1 \leq t < 2$, and $F = 8$ for $0 \leq t < 1$, with a period of 4 s (only positive values of t have physical significance). Find the Fourier series for the force.

24. A pulsating electric current i (in mA) with a period of 2 s can be described by $i = e^{-t}$ for $-1 \leq t < 1$ s for one period (only positive values of t have physical significance). Find the Fourier series that represents this current.

CHAPTER EQUATIONS

Infinite series

$$\sum_{n=1}^{\infty} a_n = a_1 + a_2 + a_3 + \cdots + a_n + \cdots \tag{13-1}$$

Sum of series

$$S = \lim_{n \to \infty} S_n = \lim_{n \to \infty} \sum_{i=1}^{n} a_i \tag{13-2}$$

Sum of geometric series

$$S = \lim_{n \to \infty} S_n = \frac{a_1}{1 - r} \tag{13-3}$$

Power series

$$f(x) = a_0 + a_1 x + a_2 x^2 + \cdots + a_n x^n + \cdots \tag{13-4}$$

Maclaurin series

$$f(x) = f(0) + f'(0)x + \frac{f''(0)x^2}{2!} + \frac{f'''(0)x^3}{3!} + \cdots + \frac{f^{(n)}(0)x^n}{n!} + \cdots \tag{13-5}$$

Factorial

$$n! = n(n-1)(n-2)(\cdots)(1) \tag{13-6}$$

Special series
$$e^x = 1 + x + \frac{x^2}{2!} + \frac{x^3}{3!} + \cdots \qquad \text{(all } x) \qquad \text{(13-7)}$$

$$\sin x = x - \frac{x^3}{3!} + \frac{x^5}{5!} - \cdots \qquad \text{(all } x) \qquad \text{(13-8)}$$

$$\cos x = 1 - \frac{x^2}{2!} + \frac{x^4}{4!} - \cdots \qquad \text{(all } x) \qquad \text{(13-9)}$$

$$\ln(1 + x) = x - \frac{x^2}{2} + \frac{x^3}{3} - \frac{x^4}{4} + \cdots \qquad (|x| < 1) \qquad \text{(13-10)}$$

$$(1 + x)^n = 1 + nx + \frac{n(n-1)}{2!} x^2 + \cdots \qquad (|x| < 1) \qquad \text{(13-11)}$$

Taylor series
$$f(x) = f(a) + f'(a)(x - a) + \frac{f''(a)(x - a)^2}{2!} + \cdots \qquad \text{(13-17)}$$

Fourier series
$$f(x) = a_0 + a_1 \cos x + a_2 \cos 2x + \cdots + a_n \cos nx + \cdots$$
$$+ \, b_1 \sin x + b_2 \sin 2x + \cdots + b_n \sin nx + \cdots \qquad \text{(13-18)}$$

Period = 2π
$$a_0 = \frac{1}{2\pi} \int_{-\pi}^{\pi} f(x)\, dx \qquad \text{(13-26)}$$

$$a_n = \frac{1}{\pi} \int_{-\pi}^{\pi} f(x) \cos nx\, dx \qquad \text{(13-25)}$$

$$b_n = \frac{1}{\pi} \int_{-\pi}^{\pi} f(x) \sin nx\, dx \qquad \text{(13-27)}$$

Period = $2L$
$$a_0 = \frac{1}{2L} \int_{-L}^{L} f(x)\, dx \qquad \text{(13-28)}$$

$$a_n = \frac{1}{L} \int_{-L}^{L} f(x) \cos \frac{n\pi x}{L}\, dx \qquad \text{(13-29)}$$

$$b_n = \frac{1}{L} \int_{-L}^{L} f(x) \sin \frac{n\pi x}{L}\, dx \qquad \text{(13-30)}$$

Half-range expansions
$$a_0 = \frac{1}{L} \int_0^L f(x)\, dx \quad \text{and} \quad a_n = \frac{2}{L} \int_0^L f(x) \cos \frac{n\pi x}{L}\, dx \quad (n = 1, 2, \ldots) \qquad \text{(13-31)}$$

$$b_n = \frac{2}{L} \int_0^L f(x) \sin \frac{n\pi x}{L}\, dx \quad (n = 1, 2, \ldots) \qquad \text{(13-32)}$$

REVIEW EXERCISES

In Exercises 1–8, find the first three nonzero terms of the Maclaurin expansion of the given functions.

1. $f(x) = \dfrac{1}{1 + e^x}$

2. $f(x) = e^{\cos x}$

3. $f(x) = \sin 2x^2$

4. $f(x) = \dfrac{1}{(1 - x)^2}$

5. $f(x) = (x + 1)^{1/3}$

6. $f(x) = \dfrac{x^2}{1 + x^2}$

7. $f(x) = \sin^{-1} x$

8. $f(x) = \dfrac{1}{1 - \sin x}$

In Exercises 9–20, calculate the value of each of the given functions. Use three terms of the appropriate series.

9. $e^{-0.2}$

10. $\ln(1.10)$

11. $\sqrt[3]{1.3}$

12. $\sin 3.5°$

13. 1.086^{-1}

14. 0.9839^{10}

15. $\ln 0.8172$

16. $\cos 0.1376$

17. $\tan 43.62°$

18. $\sqrt[4]{260}$

19. $\sqrt{148}$

20. $\cos 47°$

In Exercises 21 and 22, evaluate the given integrals by using three terms of the appropriate series.

21. $\displaystyle\int_{0.1}^{0.2} \frac{\cos x}{\sqrt{x}}\, dx$

22. $\displaystyle\int_{0}^{0.1} \sqrt[3]{1 + x^2}\, dx$

In Exercises 23 and 24, find the first three terms of the Taylor expansion for the given function and value of a.

23. $\cos x$ $(a = \pi/3)$ **24.** $\ln \cos x$ $(a = \pi/4)$

In Exercises 25–28, write the Fourier series for each function by comparing it to an appropriate function in an example of either Section 13-6 or 13-7. One period is given for each function. Do not use any formulas for a_0, a_n, or b_n.

25. $f(x) = \begin{cases} 0 & -\pi \le x < 0 \\ x - 1 & 0 \le x < \pi \end{cases}$

26. $f(x) = x^2 - 1$ $-1 \le x < 1$

27. $f(x) = \begin{cases} \pi - 1 & -4 \le x < 0 \\ \pi + 1 & 0 \le x < 4 \end{cases}$

28. $f(x) = \begin{cases} 1 & -\pi \le x < 0 \\ 1 + \sin x & 0 \le x < \pi \end{cases}$

In Exercises 29–32, find at least three nonzero terms (including a_0 and at least two cosine terms and two sine terms if they are not all zero) of the Fourier series for the given function. One period is given for each function.

29. $f(x) = \begin{cases} 0 & -\pi \le x < -\pi/2, \pi/2 \le x < \pi \\ 1 & -\pi/2 \le x < \pi/2 \end{cases}$ (See Example 2, page 409)

30. $f(x) = \begin{cases} -x & -\pi \le x < 0 \\ 0 & 0 \le x < \pi \end{cases}$

31. $f(x) = x$ $-2 \le x < 2$

32. $f(x) = \begin{cases} -2 & -3 \le x < 0 \\ 2 & 0 \le x < 3 \end{cases}$

In Exercises 33–56, solve the given problems.

33. Find the sum of the series $64 + 48 + 36 + 27 + \cdots$.

34. Find the first five partial sums of the series $\sum_{n=1}^{\infty} \dfrac{n}{3n + 1}$ and determine whether it appears to be convergent or divergent.

35. If h is small, show that $\sin(x + h) - \sin(x - h) = 2h \cos x$.

36. Find the first three nonzero terms of the Maclaurin expansion of the function $\sin x + x \cos x$ by differentiating the expansion term by term for $x \sin x$.

37. Using the properties of logarithms and Eq. (13-10), find four terms of the Maclaurin expansion of $\ln(1 + x)^4$.

38. By multiplication of series, show that the first two terms of the Maclaurin series for $2 \sin x \cos x$ are the same as those of the series for $\sin 2x$.

39. Find the first four nonzero terms of the expansion for $\cos^2 x$ by using the identity $\cos^2 x = \frac{1}{2}(1 + \cos 2x)$ and the series for $\cos x$.

40. Evaluate the integral $\int_0^1 x \sin x \, dx$ (a) by methods of Chapter 10 and (b) by using three terms of the series for $\sin x$. Compare results.

41. Find the first three terms of the Maclaurin expansion for $\sec x$ by finding the reciprocal of the series for $\cos x$.

42. By simplifying the sum of the squares of the Maclaurin series for $\sin x$ and $\cos x$, verify (to this extent) that $\sin^2 x + \cos^2 x = 1$.

43. Expand $f(x) = \sin x$ in a half-range cosine series for $0 \le x < \pi$.

44. Expand $f(x) = e^x$ in a half-range sine series for $0 \le x < \pi$.

45. Find the approximate area between the curve of $y = \dfrac{x - \sin x}{x^2}$ and the x-axis between $x = 0.1$ and $x = 0.2$.

46. Find the approximate value of the moment of inertia with respect to its axis of the solid generated by revolving the smaller region bounded by $y = \sin x$, $x = 0.3$, and the x-axis about the y-axis. Use two terms of the appropriate series.

47. Find three terms of the Maclaurin series for $\tan^{-1} x$ by integrating the series for $1/(1 + x^2)$, term by term.

48. The electric current i (in mA) in an electric circuit as a function of the time t (in ms) is $i = 0.5 \sin 0.5t - 0.2 \sin 0.4t$. Find the first three terms of the Maclaurin series for the current.

49. The number N of radioactive nuclei in a radioactive sample is $N = N_0 e^{-\lambda t}$. Here t is the time, N_0 is the number at $t = 0$, and λ is the *decay constant*. By using four terms of the appropriate series, express the right side of this equation as a polynomial.

50. The length of Lake Erie is a great circle arc of 390 km. If the lake is assumed to be flat, use series to find the error in calculating the distance from the center of the lake to the center of the earth. The radius of the earth is 6400 km.

51. The electric potential V at a distance x along a certain surface is given by $V = \ln \dfrac{1 + x}{1 - x}$. Find the first four terms of the Maclaurin series for V.

52. If a mass M is hung from a spring of mass m, the ratio of the masses is $m/M = k\omega \tan k\omega$, where k is a constant and ω is a measure of the frequency of vibration. By using two terms of the appropriate series, express m/M as a polynomial in terms of ω.

53. In the study of electromagnetic radiation, the expression $\dfrac{N_0}{1 - e^{-k/T}}$ is used. Here T is the thermodynamic temperature, and N_0 and k are constants. Show that this expression can be written as $N_0(1 + e^{-k/T} + e^{-2k/T} + \cdots)$. (*Hint:* Let $x = e^{-k/T}$.)

54. In the analysis of reflection from a spherical mirror, it is necessary to express the x-coordinate on the surface shown in Fig. 13-27 in terms of the y-coordinate and the radius R. Using the equation of the semicircle shown, solve for x (note that $x \le R$). Then express the result as a series. (Note that the first approximation gives a parabolic surface.)

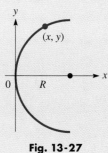

Fig. 13-27

55. A certain electric current is pulsating so that the current as a function of time is given by $f(t) = 0$ if $-\pi \leq t < 0$ and $\pi/2 < t < \pi$. If $0 < t < \pi/2$, $f(t) = \sin t$. Find the Fourier expansion for this pulsating current and sketch three periods.

56. The force F applied to a spring system as a function of the time t is given by $F = t/\pi$ if $0 \leq t \leq \pi$ and $F = 0$ if $\pi < t < 2\pi$. If the period of the force is 2π, find the first few terms of the Fourier series that represents the force.

Writing Exercise

57. A computer science class is assigned to write a program to make a table of values of the sine, cosine, and tangent of an angle in degrees to the nearest $0.1°$. Write a few paragraphs explaining how these values may be found using the known values for $0°$, $30°$, and $45°$, trigonometric relations in Chapter 7, and series from this chapter, without using many terms of any series.

PRACTICE TEST

1. By direct expansion, find the first four nonzero terms of the Maclaurin expansion for $f(x) = (1 + e^x)^2$.

2. Find the first three nonzero terms of the Taylor expansion for $f(x) = \cos x$, with $a = \pi/3$.

3. Evaluate $\ln 0.96$ by using four terms of the expansion for $\ln(1 + x)$.

4. Find the first three nonzero terms of the expansion for
$$f(x) = \frac{1}{\sqrt{1 - 2x}}$$
by using the binomial series.

5. Evaluate $\int_0^1 x \cos x \, dx$ by using three terms of the appropriate series.

6. An electric current is pulsating such that it is a function of the time with a period of 2π. If $f(t) = 2$ for $0 \leq t < \pi$ and $f(t) = 0$ for the other half-cycle, find the first three nonzero terms of the Fourier series for this current.

7. $f(x) = x^2 + 2$ for $-2 \leq x < 2$ (period $= 4$). Is $f(x)$ an even function, an odd function, or neither? Expressing the Fourier series as $F(x)$, what is the Fourier series for the function $g(x) = x^2 - 1$ for $-2 \leq x < 2$ (period $= 4$)? (Do *not* use integration to derive specific terms for either series.)

14 FIRST-ORDER DIFFERENTIAL EQUATIONS

In Section 14-5 we show how differential equations are used in the study of historical events by using the method of carbon dating.

A great many problems that arise in science, technology, and engineering involve rates of change. Since a derivative denotes a rate of change, equations that contain derivatives are of considerable importance in nearly all areas of application. These equations are called *differential equations*.

Among the areas that use the methods of differential equations for solution are those involving velocities, chemical reactions, interest calculations, changes in pressure and temperature, population growth, forces on beams and structures, nuclear energy, and many others. Also, they are used extensively in the study of currents and voltages in electric circuits.

This chapter gives an introduction to differential equations, and in it we develop a few methods of solving some basic types of differential equations. Actually, we already solved a few simple differential equations in earlier chapters when we started a solution with the expression for the slope of a tangent line or the velocity of an object.

14-1 SOLUTIONS OF DIFFERENTIAL EQUATIONS

A **differential equation** *is an equation that contains derivatives or differentials.* In this section we introduce the basic meaning of the solution of a differential equation. In the sections that follow, we consider certain methods of finding such solutions and show some applications.

Most of the differential equations we shall consider are those that contain first and second derivatives, although we will consider some with higher derivatives. *An equation that contains only first derivatives is called a* **first-order** *differential equation. An equation that contains second derivatives, and possibly first derivatives, is called a* **second-order** *differential equation. In general, the* **order** *of a differential equation is that of the highest derivative in the equation, and the* **degree** *is the highest power of that derivative.*

■EXAMPLE 1 (a) The equation $dy/dx + x = y$ is a first-order differential equation since it contains only a first derivative.

(b) The equations

$$\overbrace{\frac{d^2y}{dx^2}}^{\text{order}} + y = 3x^2 \quad \text{and} \quad \frac{d^2y}{dx^2} + 2\frac{dy}{dx} = x$$

are second-order equations since each contains a second derivative and no higher derivatives. The dy/dx in the second equation does not affect the order. ■

■EXAMPLE 2 The equation

$$\frac{d^2y}{dx^2} + \left(\frac{dy}{dx}\right)^4 - y = 6$$

is a differential equation of the second order and the first degree; that is, the highest derivative that appears is the second, and it is raised to the first power. Since the second derivative appears, the fourth power of the first derivative does not affect the degree. ■

In our discussion of differential equations, we shall restrict our attention to equations of the first degree.

A **solution** *of a differential equation is a relation between the variables that satisfies the differential equation; that is, when this relation is substituted into the differential equation, an algebraic identity results. A solution containing a number of independent arbitrary constants equal to the order of the differential equation is called the* **general solution** *of the equation. When specific values are given to at least one of these constants, the solution is called a* **particular solution.**

GENERAL SOLUTION
PARTICULAR SOLUTION

■EXAMPLE 3 Any coefficients that are not specified numerically after like terms have been combined are independent arbitrary contants. In the expression $c_1x + c_2 + c_3x$, there are only two arbitrary constants since the x-terms may be combined; $c_2 + c_4x$ is an equivalent expression with $c_4 = c_1 + c_3$. ■

■EXAMPLE 4 $y = c_1e^{-x} + c_2e^{2x}$ is the general solution of the differential equation

$$\frac{d^2y}{dx^2} - \frac{dy}{dx} = 2y$$

The order of this differential equation is 2, and there are two independent arbitrary constants in the solution. The equation $y = 4e^{-x}$ is a particular solution. It can be derived from the general solution by letting $c_1 = 4$ and $c_2 = 0$. Each of these solutions can be shown to satisfy the differential equation by taking two derivatives and substituting. ■

To solve a differential equation, we have to find some method of transforming the equation so that each term may be integrated. Some of these methods will be considered after this section. The purpose here is to show that a given equation is a solution of a differential equation *by taking the required derivatives and to show that an identity results after substitution.*

NOTE ▶

EXAMPLE 5   Show that $y = c_1 \sin x + c_2 \cos x$ is the general solution of the differential equation $y'' + y = 0$.

The function and its first two derivatives are

$$y = c_1 \sin x + c_2 \cos x$$

$$y' = c_1 \cos x - c_2 \sin x$$

$$y'' = -c_1 \sin x - c_2 \cos x$$

Substituting these into the differential equation, we have

$$\underset{\downarrow}{y''} \quad + \quad \underset{\downarrow}{y} \quad = 0$$

$$(-c_1 \sin x - c_2 \cos x) + (c_1 \sin x + c_2 \cos x) = 0 \qquad \text{or} \qquad 0 = 0$$

Therefore, this is the general solution, since there are two independent arbitrary constants and the order of the differential equation is 2.

EXAMPLE 6   Show that $y = cx + x^2$ is a solution of the differential equation $xy' - y = x^2$.

Taking one derivative of the function and substituting into the differential equation, we have

$$y = cx + x^2$$

$$xy' - y = x^2$$

$$y' = c + 2x$$

$$x(c + 2x) - (cx + x^2) = x^2 \qquad \text{or} \qquad x^2 = x^2$$

EXERCISES 14-1

In Exercises 1–4, determine whether the given equation is the general solution or a particular solution of the given differential equation.

1. $\dfrac{dy}{dx} + 2xy = 0, \quad y = e^{-x^2}$

2. $y' \ln x - \dfrac{y}{x} = 0, \quad y = c \ln x$

3. $y'' + 3y' - 4y = 3e^x, \quad y = c_1 e^x + c_2 e^{-4x} + \frac{3}{5} x e^x$

4. $\dfrac{d^2y}{dx^2} + 4y = 0, \quad y = c_1 \sin 2x + 3 \cos 2x$

In Exercises 5–8, show that each function $y = f(x)$ is a solution of the given differential equation.

5. $\dfrac{dy}{dx} - y = 1; \quad y = e^x - 1, \quad y = 5e^x - 1$

6. $\dfrac{dy}{dx} = 2xy^2; \quad y = -\dfrac{1}{x^2}, \quad y = -\dfrac{1}{x^2 + c}$

7. $y'' + 4y = 0; \quad y = 3 \cos 2x, \quad y = c_1 \sin 2x + c_2 \cos 2x$

8. $y'' = 2y'; \quad y = 3e^{2x}, \quad y = 2e^{2x} - 5$

In Exercises 9–28, show that the given equation is a solution of the given differential equation.

9. $\dfrac{dy}{dx} = 2x, \quad y = x^2 + 1$

10. $\dfrac{dy}{dx} = 1 - 3x^2, \quad y = 2 + x - x^3$

11. $\dfrac{dy}{dx} - 3 = 2x, \quad y = x^2 + 3x$

12. $xy' = 2y, \quad y = cx^2$

13. $y' + 2y = 2x, \quad y = ce^{-2x} + x - \frac{1}{2}$

14. $y' - 3x^2 = 1, \quad y = x^3 + x + c$

15. $y'' + 9y = 4 \cos x, \quad 2y = \cos x$

16. $y'' - 4y' + 4y = e^{2x}, \quad y = e^{2x}\left(c_1 + c_2 x + \dfrac{x^2}{2} \right)$

17. $x^2y' + y^2 = 0, \quad xy = cx + cy$

18. $xy' - 3y = x^2, \quad y = cx^3 - x^2$

19. $x\dfrac{d^2y}{dx^2} + \dfrac{dy}{dx} = 0, \quad y = c_1 \ln x + c_2$

20. $y'' + 4y = 10e^x, \quad y = c_1 \sin 2x + c_2 \cos 2x + 2e^x$

21. $y' + y = 2 \cos x, \quad y = \sin x + \cos x - e^{-x}$

22. $(x + y) - xy' = 0, \quad y = x \ln x - cx$

23. $y'' + y' = 6 \sin 2x, \quad y = e^{-x} - \frac{3}{5}\cos 2x - \frac{6}{5}\sin 2x$

24. $xy'' + y' = 16x^3, \quad y = x^4 + c_1 + c_2 \ln x$

25. $\cos x\dfrac{dy}{dx} + \sin x = 1 - y, \quad y = \dfrac{x + c}{\sec x + \tan x}$

26. $2xyy' + x^2 = y^2, \quad x^2 + y^2 = cx$

27. $(y')^2 + xy' = y, \quad y = cx + c^2$

28. $x^4(y')^2 - xy' = y, \quad y = c^2 + \dfrac{c}{x}$

14-2 SEPARATION OF VARIABLES

We shall now solve differential equations of the first order and first degree. Of the many methods for solving such equations, a few are presented in this and the next two sections. The first of these is *the method of* **separation of variables.**

A differential equation of the first order and first degree contains the first derivative to the first power. That is, it may be written as $dy/dx = f(x, y)$. This type of equation is more commonly expressed in its differential form,

$$M(x, y)\, dx + N(x, y)\, dy = 0 \qquad (14\text{-}1)$$

where $M(x, y)$ and $N(x, y)$ may represent constants, functions of either x or y, or functions of x and y.

To solve an equation of the form of Eq. (14-1), we must integrate. However, if $M(x, y)$ contains y, the first term cannot be integrated. Also, if $N(x, y)$ contains x, the second term cannot be integrated. If it is possible to rewrite Eq. (14-1) as

$$A(x)\, dx + B(y)\, dy = 0 \qquad (14\text{-}2)$$

NOTE ▶

where $A(x)$ does not contain y and $B(y)$ does not contain x, then *we may find the solution by integrating each term and adding the constant of integration.* (In rewriting Eq. (14-1), if division is used, the solution is not valid for values that make the divisor zero.) Many differential equations can be solved in this way.

■EXAMPLE 1 Solve the differential equation $dx - 4xy^3dy = 0$.

We can write this equation as

$$(1)\ dx + (-4xy^3)\, dy = 0$$

which means that $M(x, y) = 1$ and $N(x, y) = -4xy^3$.

CAUTION ▶

We must remove the x from the coefficient of dy **without introducing y into the coefficient of dx.** We do this by dividing each term by x, which gives us

$$dx/x - 4y^3dy = 0$$

It is now possible to integrate each term. Performing this integration, we have

$$\ln|x| - y^4 = c$$

The constant of integration c becomes the arbitrary constant of the solution. ∎

In Example 1 we showed the integration of dx/x as $\ln|x|$, which follows our discussion in Section 9-2. We know $\ln|x| = \ln x$ if $x > 0$ and $\ln|x| = \ln(-x)$ if $x < 0$. **NOTE ▶** Since we know the values being used when we find a particular solution, *we generally will not use the absolute value notation when integrating logarithmic forms.* We would show the integration of dx/x as $\ln x$, with the understanding that we know $x > 0$. When using negative values of x, we would express it as $\ln(-x)$.

■EXAMPLE 2 Solve the differential equation $xy\,dx + (x^2 + 1)\,dy = 0$.

In order to integrate each term, it is necessary to divide each term by $y(x^2 + 1)$. When this is done, we have

$$\frac{x\,dx}{x^2 + 1} + \frac{dy}{y} = 0$$

Integrating, we obtain the solution

$$\frac{1}{2}\ln(x^2 + 1) + \ln y = c$$

It is possible to make use of the properties of logarithms to make the form of this solution neater. If we write the constant of integration as $\ln c_1$, rather than c, we have $\frac{1}{2}\ln(x^2 + 1) + \ln y = \ln c_1$. Multiplying through by 2 and using the property of logarithms given by Eq. (8-9), we have $\ln(x^2 + 1) + \ln y^2 = \ln c_1^2$. Next, using the property of logarithms given by Eq. (8-7), we then have $\ln(x^2 + 1)y^2 = \ln c_1^2$, which means

For reference, Eq. (8-9) is
$\log_b x^n = n\log_b x$
and Eq. (8-7) is
$\log_b xy = \log_b x + \log_b y$.

$$(x^2 + 1)y^2 = c_1^2$$

CAUTION ▶
NOTE ▶
This form of the solution is more compact and generally would be preferred. However, *any expression that represents a constant may be chosen as the constant of integration* and leads to a correct solution. In checking answers, we must remember that a different choice of constant will lead to a different form of the solution. Thus, two different-appearing answers may both be correct. *Often there is more than one reasonable choice of a constant, and different forms of the solution may be expected.* ---------■

■EXAMPLE 3 Solve the differential equation $\dfrac{d\theta}{dt} = \dfrac{\theta}{t^2 + 4}$.

The solution proceeds as follows:

$$\frac{d\theta}{\theta} = \frac{dt}{t^2 + 4} \qquad \text{separate variables by multiplying by } dt \text{ and dividing by } \theta$$

$$\ln\theta = \frac{1}{2}\tan^{-1}\frac{t}{2} + \frac{c}{2} \qquad \text{integrate}$$

$$2\ln\theta = \tan^{-1}\frac{t}{2} + c$$

$$\ln\theta^2 = \tan^{-1}\frac{t}{2} + c$$

Note the different forms of the result using $c/2$ as the constant of integration. These forms would differ somewhat had we chosen c as the constant.

The choice of $\ln c$ as the constant of integration (on the left) is also reasonable. It would lead to the result $2\ln c\theta = \tan^{-1}(t/2)$. ---------■

In order to separate the variables of a differential equation that contains exponential functions, it may be necessary to use the properties of exponents. Also, when trigonometric functions are involved, the basic trigonometric identities may be needed.

■EXAMPLE 4 Solve the differential equation $2e^{3x} \sin y \, dx + e^x \csc y \, dy = 0$.

In the dx-term we want only a function of x, which means that we must divide by $\sin y$. Also, the dy-term indicates that we must divide by e^x. Thus,

$$\frac{2e^{3x} \sin y \, dx}{e^x \sin y} + \frac{e^x \csc y \, dy}{e^x \sin y} = 0 \qquad \text{divide by } e^x \sin y$$

$$2e^{2x} \, dx + \csc^2 y \, dy = 0 \qquad \text{variables separated}$$

$$e^{2x}(2 \, dx) + \csc^2 y \, dy = 0 \qquad \text{form for integrating}$$

$$e^{2x} - \cot y = c \qquad \text{integrate} \qquad \blacksquare$$

Finding Particular Solutions

In order to find a particular solution of a differential equation, we must have information that allows us to evaluate the constant of integration. The following examples show how particular solutions of differential equations are found. Also, we will show graphically the difference between the general solution and the particular solution.

■EXAMPLE 5 Solve the differential equation $(x^2 + 1)^2 dy + 4x \, dx = 0$, subject to the condition that $x = 1$ when $y = 3$.

Separating variables, we have

$$dy + \frac{4x \, dx}{(x^2 + 1)^2} = 0 \qquad \text{dividing by } (x^2 + 1)^2$$

$$y - \frac{2}{x^2 + 1} = c \qquad \text{integrating}$$

$$y = \frac{2}{x^2 + 1} + c \qquad \text{general solution}$$

Since a specific set of values is given, we can evaluate the constant of integration and thereby get a particular solution. Using the values $x = 1$ and $y = 3$, we have

$$3 = \frac{2}{1 + 1} + c, \qquad c = 2 \qquad \text{evaluate } c$$

which gives us

$$y = \frac{2}{x^2 + 1} + 2 \qquad \text{particular solution}$$

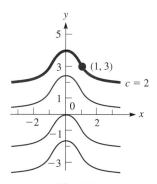

Fig. 14-1

The general solution defines a *family* of curves, one member of the family for each value of c that may be considered. A few of these curves are shown in Fig. 14-1. When c is specified as in the particular solution, we have the specific (darker) curve shown in Fig. 14-1. $\blacksquare$

See Section 16-1 for numerical methods of solving differential equations that can be written in the form $dy/dx = f(x, y)$.

EXAMPLE 6 Find the particular solution in Example 2 if the function is subject to the condition that $x = 0$ when $y = e$.

Using the solution $\frac{1}{2}\ln(x^2 + 1) + \ln y = c$, we have

$$\frac{1}{2}\ln(0 + 1) + \ln e = c, \qquad \frac{1}{2}\ln 1 + 1 = c, \qquad \text{or} \qquad c = 1$$

The particular solution is then

$$\frac{1}{2}\ln(x^2 + 1) + \ln y = 1 \qquad \text{substitute } c = 1$$

$$\ln(x^2 + 1) + 2\ln y = 2$$

$$\ln y^2(x^2 + 1) = 2 \qquad \text{using properties of logarithms}$$

$$y^2(x^2 + 1) = e^2 \qquad \text{exponential form}$$

Using the general solution $(x^2 + 1)y^2 = c_1^2$, we have

$$(0 + 1)e^2 = c_1^2, \qquad c_1^2 = e^2$$

$$y^2(x^2 + 1) = e^2$$

NOTE ▶ which is precisely the same solution as above. We see, therefore, that *the choice of the form of the constant does not affect the final result and that the constant is truly arbitrary.*

EXERCISES *14-2*

In Exercises 1–28, solve the given differential equations. Explain your method of solution for Exercise 13.

1. $2x\,dx + dy = 0$

2. $y^2\,dy + x^3\,dx = 0$

3. $y^2\,dx + dy = 0$

4. $y\,dx + x\,dy = 0$

5. $\dfrac{dV}{dP} = -\dfrac{V}{P^2}$

6. $\dfrac{2\,dy}{dx} = \dfrac{y(x + 1)}{x}$

7. $x^2 + (x^3 + 5)y' = 0$

8. $xyy' + \sqrt{1 + y^2} = 0$

9. $dy + \ln xy\,dx = (4x + \ln y)\,dx$

10. $r\sqrt{1 - \theta^2}\,\dfrac{dr}{d\theta} = \theta + 4$

11. $e^{x^2}\,dy = x\sqrt{1 - y}\,dx$

12. $\sqrt{1 + 4x^2}\,dy = y^3x\,dx$

(W) 13. $e^{x+y}\,dx + dy = 0$

14. $e^{2x}\,dy + e^x\,dx = 0$

15. $y' - y = 4$

16. $ds - s^2\,dt = 9\,dt$

17. $x\,\dfrac{dy}{dx} = y^2 + y^2\ln x$

18. $(yx^2 + y)\,\dfrac{dy}{dx} = \tan^{-1} x$

19. $y\tan x\,dx + \cos^2 x\,dy = 0$

20. $\sin x\sec y\,dx = dy$

21. $yx^2\,dx = y\,dx - x^2\,dy$

22. $e^{\cos\theta}\tan\theta\,d\theta + \sec\theta\,dy = 0$

23. $y\sqrt{1 - x^2}\,dy + 2\,dx = 0$

24. $(x^3 + x^2)\,dx + (x + 1)y\,dy = 0$

25. $2\ln t\,dt + t\,di = 0$

26. $2y(x^3 + 1)\,dy + 3x^2(y^2 - 1)\,dx = 0$

27. $y^2e^x + (e^x + 1)\dfrac{dy}{dx} = 0$

28. $y + 1 + \sec x(\sin x + 1)\dfrac{dy}{dx} = 0$

In Exercises 29–36, find the particular solution of the given differential equation for the indicated values.

29. $\dfrac{dy}{dx} + yx^2 = 0$; $\quad x = 0$ when $y = 1$

30. $\dfrac{dy}{dx} + 2y = 6$; $\quad x = 0$ when $y = 1$

31. $(xy^2 + x)\dfrac{dy}{dx} = \ln x$; $\quad x = 1$ when $y = 0$

32. $\dfrac{ds}{dt} = \sec s$; $\quad t = 0$ when $s = 0$

33. $y' = (1 - y)\cos x$; $\quad x = \pi/6$ when $y = 0$

34. $x\,dy = y\ln y\,dx$; $\quad x = 2$ when $y = e$

35. $y^2e^x\,dx + e^{-x}\,dy = y^2\,dx$; $\quad x = 0$ when $y = 2$

36. $2y\cos y\,dy - \sin y\,dy = y\sin y\,dx$; $\quad x = 0$ when $y = \pi/2$

14-3 INTEGRATING COMBINATIONS

NOTE ▶

Many differential equations cannot be solved by the method of separation of variables. Many other methods have been developed for solving such equations. One of these methods is based on the fact that *certain combinations of basic differentials can be integrated together as a unit.* The following differentials suggest some of these combinations that may occur:

$$d(xy) = x\,dy + y\,dx \tag{14-3}$$

$$d(x^2 + y^2) = 2(x\,dx + y\,dy) \tag{14-4}$$

$$d\left(\frac{y}{x}\right) = \frac{x\,dy - y\,dx}{x^2} \tag{14-5}$$

$$d\left(\frac{x}{y}\right) = \frac{y\,dx - x\,dy}{y^2} \tag{14-6}$$

Equation (14-3) suggests that if the combination $x\,dy + y\,dx$ occurs in a differential equation, we should look for a function of xy as a solution. Equation (14-4) suggests that if the combination $x\,dx + y\,dy$ occurs, we should look for a function of $x^2 + y^2$. Equations (14-5) and (14-6) suggest that if either of the combinations $x\,dy - y\,dx$ or $y\,dx - x\,dy$ occurs, we should look for a function of y/x or x/y.

■EXAMPLE 1   Solve the differential equation $x\,dy + y\,dx + xy\,dy = 0$.
By dividing through by xy, we have

$$\frac{x\,dy + y\,dx}{xy} + dy = 0$$

The left term is the differential of xy divided by xy. Thus, it integrates to $\ln xy$.

$$\frac{d(xy)}{xy} + dy = 0$$

for which the solution is

$$\ln xy + y = c \qquad \text{------------} ■$$

■EXAMPLE 2   Solve the differential equation $y\,dx - x\,dy + x\,dx = 0$.
The combination of $y\,dx - x\,dy$ suggests that this equation might make use of either Eq. (14-5) or (14-6). This would require dividing through by x^2 or y^2. If we divide by y^2, the last term cannot be integrated, but ***division by x^2 still allows integration of the last term.*** Performing this division, we obtain

CAUTION ▶

$$\frac{y\,dx - x\,dy}{x^2} + \frac{dx}{x} = 0$$

This left combination is the negative of Eq. (14-5). Thus, we have

$$-d\left(\frac{y}{x}\right) + \frac{dx}{x} = 0$$

for which the solution is $-\dfrac{y}{x} + \ln x = c$. $\qquad \text{------------} ■$

EXAMPLE 3   Solve the differential equation $(x^2 + y^2 + x)\,dx + y\,dy = 0$.
Regrouping the terms of this equation, we have

$$(x^2 + y^2)\,dx + (x\,dx + y\,dy) = 0 \qquad \text{divide each term by } x^2 + y^2$$

$$dx + \frac{x\,dx + y\,dy}{x^2 + y^2} = 0$$

The right term now can be put in the form of du/u (with $u = x^2 + y^2$) by multiplying each of the terms of the numerator by 2. This leads to

$$dx + \left(\frac{1}{2}\right)\overset{d(x^2 + y^2)}{\frac{2x\,dx + 2y\,dy}{x^2 + y^2}} = 0$$

$$x + \frac{1}{2}\ln(x^2 + y^2) = \frac{c}{2} \qquad \text{or} \qquad 2x + \ln(x^2 + y^2) = c$$

EXAMPLE 4   Find the particular solution of the differential equation

$$(x^3 + xy^2 + 2y)\,dx + (y^3 + x^2y + 2x)\,dy = 0$$

which satisfies the condition that $x = 1$ when $y = 0$.
Regrouping the terms of the equation, we have

$$x(x^2 + y^2)\,dx + y(x^2 + y^2)\,dy + 2(y\,dx + x\,dy) = 0$$

Factoring $x^2 + y^2$ from each of the first two terms gives

$$(x^2 + y^2)(x\,dx + y\,dy) + 2(y\,dx + x\,dy) = 0$$

$$\frac{1}{2}(x^2 + y^2)\overset{d(x^2 + y^2)}{(2x\,dx + 2y\,dy)} + 2\overset{d(xy)}{(y\,dx + x\,dy)} = 0$$

$$\frac{1}{2}\left(\frac{1}{2}\right)(x^2 + y^2)^2 + 2xy + \frac{c}{4} = 0 \qquad \text{integrating}$$

$$(x^2 + y^2)^2 + 8xy + c = 0$$

Using the given condition gives $(1 + 0)^2 + 0 + c = 0$, or $c = -1$. The particular solution is then

$$(x^2 + y^2)^2 + 8xy = 1$$

NOTE▶   The use of integrating combinations depends on proper recognition of the forms. It may take two or three arrangements to find the combination that leads to the solution. Of course, many equations cannot be arranged so as to give integrable combinations in all terms.

── **EXERCISES** *14-3* ──

In Exercises 1–16, solve the given differential equations.

1. $x\,dy + y\,dx + x\,dx = 0$

2. $(2y + x)\,dy + y\,dx = 0$

3. $y\,dx - x\,dy + x^3\,dx = 2\,dx$

4. $x\,dy - y\,dx + y^2\,dx = 0$

5. $A^3\,dr + A^2r\,dA + r\,dA = A\,dr$

6. $\sec(xy)\,dx + (x\,dy + y\,dx) = 0$

7. $x^3y^4(x\,dy + y\,dx) = 3\,dy$

8. $x\,dy + y\,dx + 4xy^3\,dy = 0$

9. $\sqrt{x^2 + y^2}\,dx - 2y\,dy = 2x\,dx$

10. $R\,dR + (R^2 + T^2 + T)\,dT = 0$

11. $\tan(x^2 + y^2)\,dy + x\,dx + y\,dy = 0$

12. $(x^2 + y^3)^2\,dy + 2x\,dx + 3y^2\,dy = 0$

(W) **13.** $y\,dy + (y^2 - x^2)\,dx = x\,dx$ (Explain your solution.)

14. $e^{x+y}(dx + dy) + 4x\,dx = 0$

15. $10x\,dy + 5y\,dx + 3y\,dy = 0$

16. $2(u\,dv + v\,du)\ln uv + 3u^3v\,du = 0$

In Exercises 17–20, find the particular solutions to the given differential equations that satisfy the given conditions.

17. $2(x\,dy + y\,dx) + 3x^2\,dx = 0$; $x = 1$ when $y = 2$

18. $t\,dt + s\,ds = 2(t^2 + s^2)\,dt$; $t = 1$ when $s = 0$

19. $y\,dx - x\,dy = y^3\,dx + y^2x\,dy$; $x = 2$ when $y = 4$

20. $e^{x/y}(x\,dy - y\,dx) = y^4\,dy$; $x = 0$ when $y = 2$

14-4 THE LINEAR DIFFERENTIAL EQUATION OF THE FIRST ORDER

There is one type of differential equation of the first order and first degree for which an integrable combination can always be found. *It is the* **linear differential equation** *of the first order and is of the form*

$$dy + Py\,dx = Q\,dx \qquad (14\text{-}7)$$

NOTE ▶   *where P and Q are functions of x only.* This type of equation occurs widely in applications.

If each side of Eq. (14-7) is multiplied by $e^{\int P\,dx}$, it becomes integrable, since the left side becomes of the form du with $u = ye^{\int P\,dx}$ and the right side is a function of x only. This is shown by finding the differential of $ye^{\int P\,dx}$. Thus,

$$d(ye^{\int P\,dx}) = e^{\int P\,dx}(dy + Py\,dx)$$

In finding the differential of $\int P\,dx$ we use the fact that, by definition, these are reverse processes. Thus, $d(\int P\,dx) = P\,dx$. Therefore, if each side is multiplied by $e^{\int P\,dx}$, the left side may be immediately integrated to $ye^{\int P\,dx}$, and the right-side integration may be indicated. The solution becomes

$$ye^{\int P\,dx} = \int Qe^{\int P\,dx}\,dx + c \qquad (14\text{-}8)$$

▌EXAMPLE 1   Solve the differential equation $dy + \left(\dfrac{2}{x}\right)y\,dx = 4x\,dx$.

This equation fits the form of Eq. (14-7) with $P = 2/x$ and $Q = 4x$. The first expression to find is $e^{\int P\,dx}$. In this case this is

$$e^{\int (2/x)dx} = e^{2\ln x} = e^{\ln x^2} = x^2 \qquad \text{see text comments following example}$$

The left side integrates to yx^2, while the right side becomes $\int 4x(x^2)\,dx$. Thus,

$$ye^{\int P\,dx} = \int Qe^{\int P\,dx}\,dx + c$$

$$y(x^2) = \int (4x)(x^2)\,dx + c \qquad \text{using Eq. (14-8)}$$

$$yx^2 = \int 4x^3\,dx + c = x^4 + c \qquad \text{integrating}$$

$$y = x^2 + cx^{-2} \qquad \text{divide by } x^2 \qquad \text{------------} ▐$$

NOTE ▶ As in Example 1, in finding the factor $e^{\int P\,dx}$ we often obtain an expression of the form $e^{\ln u}$. Using the properties of logarithms, we now show that $e^{\ln u} = u$.

$$\text{Let } y = e^{\ln u}$$
$$\ln y = \ln e^{\ln u} = \ln u(\ln e) = \ln u$$
$$y = u \qquad \text{or} \qquad e^{\ln u} = u$$

NOTE ▶ Also, in finding $e^{\int P\,dx}$, the constant of integration in the exponent $\int P\,dx$ can always be taken as zero, as we did in Example 1. To show why this is so, let $P = 2/x$ as in Example 1.

$$e^{\int (2/x)\,dx} = e^{\ln x^2 + c} = (e^{\ln x^2})(e^c) = x^2 e^c$$

The solution to the differential equation, as given in Eq. (14-8), is then

$$y(x^2)(e^c) = \int 4x(x^2)(e^c)\,dx + c_1 e^c$$

Regardless of the value of c, the factor e^c can be divided out. Therefore, it is convenient to let $c = 0$ and have $e^c = 1$.

■**EXAMPLE 2**   Solve the differential equation $x\,dy - 3y\,dx = x^3\,dx$.

Putting this equation in the form of Eq. (14-7) by dividing through by x gives $dy - (3/x)y\,dx = x^2\,dx$. Here $P = -3/x$, $Q = x^2$, and the factor $e^{\int P\,dx}$ becomes

$$e^{\int (-3/x)\,dx} = e^{-3\ln x} = e^{\ln x^{-3}} = x^{-3}$$

Therefore,

$$y\underbrace{e^{\int P\,dx}}_{} = \int Q\underbrace{e^{\int P\,dx}}_{}\,dx + c$$

$$yx^{-3} = \int x^2(x^{-3})\,dx + c \qquad \text{using Eq. (14-8)}$$

$$= \int x^{-1}\,dx + c = \ln x + c$$

$$y = x^3(\ln x + c)$$ ────────■

■**EXAMPLE 3**   Solve the differential equation $dy + y\,dx = x\,dx$.

Here, $P = 1$, $Q = x$, and $e^{\int P\,dx} = e^{\int (1)\,dx} = e^x$. Therefore,

$$ye^x = \int xe^x\,dx + c = e^x(x - 1) + c \qquad \text{using Eq. (14-8) and integrating by parts or tables}$$

$$y = x - 1 + ce^{-x}$$ ────────■

■**EXAMPLE 4**   Solve the differential equation $x^2\,dy + 2xy\,dx = \sin x\,dx$.

This equation is first written in the form of Eq. (14-7). This gives us

$$dy + \left(\frac{2}{x}\right)y\,dx = \frac{1}{x^2}\sin x\,dx$$

This shows us that $P = 2/x$ and $e^{\int P\,dx} = e^{\int (2/x)\,dx} = x^2$. Therefore,

$$yx^2 = \int \sin x\,dx + c = -\cos x + c \qquad \text{using Eq. (14-8)}$$

$$yx^2 + \cos x = c$$ ────────■

EXAMPLE 5   Solve the differential equation $\cos x \dfrac{dy}{dx} = 1 - y \sin x$.

Writing this in the form of Eq. (14-7), we have

$$dy + y \tan x \, dx = \sec x \, dx \qquad \text{dividing by } \cos x$$

Thus, with $P = \tan x$, we have

$$e^{\int P \, dx} = e^{\int \tan x \, dx} = e^{-\ln \cos x} = \sec x \qquad \text{see the first NOTE after Example 1}$$

$$y \sec x = \int \sec^2 x \, dx = \tan x + c \qquad \text{using Eq. (14-8)}$$

$$y = \sin x + c \cos x$$

EXAMPLE 6   For the differential equation $dy = (1 - 2y)x \, dx$, find the particular solution such that $x = 0$ when $y = 2$.

The solution proceeds as follows:

$$dy + 2xy \, dx = x \, dx \qquad \text{form of Eq. (14-7)}$$

$$e^{\int P \, dx} = e^{\int 2x \, dx} = e^{x^2} \qquad \text{find } e^{\int P \, dx}$$

$$ye^{x^2} = \int x e^{x^2} \, dx \qquad \text{using Eq. (14-8)}$$

$$= \frac{1}{2} e^{x^2} + c \qquad \text{general solution}$$

$$(2)(e^0) = \frac{1}{2}(e^0) + c, \qquad 2 = \frac{1}{2} + c, \qquad c = \frac{3}{2} \qquad x = 0, \ y = 2; \text{ evaluate } c$$

$$ye^{x^2} = \frac{1}{2} e^{x^2} + \frac{3}{2} \qquad \text{substitute } c = \frac{3}{2}$$

$$y = \frac{1}{2}(1 + 3e^{-x^2}) \qquad \text{particular solution}$$

EXERCISES *14-4*

In Exercises 1–24, solve the given differential equations.

1. $dy + y \, dx = e^{-x} \, dx$

2. $dy + 3y \, dx = e^{-3x} \, dx$

3. $dy + 2y \, dx = e^{-4x} \, dx$

4. $di + i \, dt = e^{-t} \cos t \, dt$

5. $\dfrac{dy}{dx} - 2y = 4$

6. $2 \dfrac{dy}{dx} = 5 - 6y$

7. $x \, dy - y \, dx = 3x \, dx$

8. $x \, dy + 3y \, dx = dx$

9. $2x \, dy + y \, dx = 8x^3 \, dx$

10. $3x \, dy - y \, dx = 9x \, dx$

11. $dr + r \cot \theta \, d\theta = d\theta$

12. $y' = x^2 y + 3x^2$

13. $\sin x \dfrac{dy}{dx} = 1 - y \cos x$

14. $\dfrac{dy}{dx} - \dfrac{y}{x} = \ln x$

15. $y' + y = 3$

16. $y' + 2y = \sin x$

17. $ds = (te^{4t} + 4s) \, dt$

18. $y' - 2y = 2e^{2x}$

19. $y' = x^3(1 - 4y)$

20. $y' + y \tan x = -\sin x$

21. $x \dfrac{dy}{dx} = y + (x^2 - 1)^2$

22. $dy = dt - \dfrac{y \, dt}{(1 + t^2)\tan^{-1} t}$

23. $x \, dy + (1 - 3x)y \, dx = 3x^2 e^{3x} \, dx$

24. $(1 + x^2) \, dy + xy \, dx = x \, dx$

(W) *In Exercises 25 and 26, solve the given differential equations. Explain how each can be solved using either of two different methods.*

25. $y' = 2(1 - y)$

26. $x \, dy = (2x - y) \, dx$

In Exercises 27–32, find the indicated particular solutions of the given differential equations.

27. $\dfrac{dy}{dx} + 2y = e^{-x}; \quad x = 0$ when $y = 1$

28. $dq - 4q \, du = 2 \, du; \quad q = 2$ when $u = 0$

29. $y' + 2y \cot x = 4 \cos x; \quad x = \pi/2$ when $y = 1/3$

30. $y' \sqrt{x} + \frac{1}{2}y = e^{\sqrt{x}}; \quad x = 1$ when $y = 3$

31. $(\sin x)y' + y = \tan x; \quad x = \pi/4$ when $y = 0$

32. $f(x) \, dy + 2yf'(x) \, dx = f(x)f'(x) \, dx; \quad f(x) = -1$ when $y = 3$

14-5 ELEMENTARY APPLICATIONS

The differential equations of the first order and first degree we have discussed thus far have numerous applications in geometry and the various fields of technology. In this section we illustrate some of these applications.

EXAMPLE 1 The slope of a given curve is given by the expression $6xy$. Find the equation of the curve if it passes through the point $(2, 1)$.

Since the slope is $6xy$, the differential equation for the curve is

$$\frac{dy}{dx} = 6xy$$

We now want to find the particular solution of this equation for which $x = 2$ when $y = 1$. The solution follows:

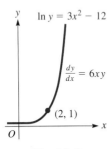

Fig. 14-2

$$\frac{dy}{y} = 6x\,dx \qquad \text{separate variables}$$

$$\ln y = 3x^2 + c \qquad \text{general solution}$$

$$\ln 1 = 3(2^2) + c \qquad \text{evaluate } c$$

$$0 = 12 + c, \qquad c = -12$$

$$\ln y = 3x^2 - 12 \qquad \text{particular solution}$$

The graph of this solution is shown in Fig. 14-2.

EXAMPLE 2 *A curve that intersects all members of a family of curves at right angles is called an* **orthogonal trajectory** *of the family.* Find the equations of the orthogonal trajectories of the parabolas $x^2 = cy$. As before, each value of c gives us a particular member of the family.

The derivative of the given equation is $dy/dx = 2x/c$. This equation contains the constant c, which depends on the point (x, y) on the parabola. *Eliminating this constant between the equations of the parabolas and the derivative,* we have

$$c = \frac{x^2}{y} \qquad \frac{dy}{dx} = \frac{2x}{c} = \frac{2x}{x^2/y} \qquad \text{or} \qquad \frac{dy}{dx} = \frac{2y}{x}$$

This equation gives a general expression for the slope of any of the members of the family. For a curve to be perpendicular, its slope must equal the negative reciprocal of this expression, or the slope of the orthogonal trajectories must be

$$\left.\frac{dy}{dx}\right|_{OT} = -\frac{x}{2y} \quad \longleftarrow \text{ this equation must not contain the constant } c$$

Solving this differential equation gives the family of orthogonal trajectories.

$$2y\,dy = -x\,dx$$

$$y^2 = -\frac{x^2}{2} + \frac{c}{2}$$

$$2y^2 + x^2 = c \qquad \text{orthogonal trajectories}$$

Thus the orthogonal trajectories are ellipses. Note in Fig. 14-3 that each parabola intersects each ellipse at right angles.

CAUTION ▶

Some calculators have a feature by which a family of curves can be graphed.

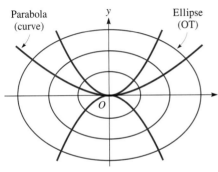

Fig. 14-3

SOLVING A WORD PROBLEM

See the chapter introduction.

Radioactivity was discovered in 1898 by the French physicist Henri Becquerel (1852–1908). Carbon dating was developed in 1947 by the U.S. chemist Willard Libby (1908–1980).

EXAMPLE 3 Radioactive elements decay at rates that are proportional to the amount of the element present. Carbon 14 decays such that one-half of an original amount decays into other forms in about 5730 years. By measuring the proportion of carbon 14 in remains at an ancient site, the approximate age of the remains can be determined. This method, called *carbon dating,* is used to determine the dates of prehistoric events.

The analysis of some wood at the site of the ancient city of Troy showed that the concentration of carbon 14 was 67.8% of the concentration that new wood would have. Determine the equation relating the amount of carbon 14 present with the time and then determine the age of the wood at the site of Troy.

Let N_0 be the original amount and N be the amount present at any time t (in years). The rate of decay can be expressed as a derivative. Therefore, since the rate of change is proportional to N, we have the equation

$$\frac{dN}{dt} = kN$$

Solving this differential equation, we have

$$\frac{dN}{N} = k\,dt \qquad \text{separate variables}$$

$$\ln N = kt + \ln c \qquad \text{general solution}$$

$$\ln N_0 = k(0) + \ln c \qquad N = N_0 \text{ for } t = 0$$

$$c = N_0 \qquad \text{solve for } c$$

$$\ln N = kt + \ln N_0 \qquad \text{substitute } N_0 \text{ for } c$$

$$\ln N - \ln N_0 = kt \qquad \text{use properties of logarithms}$$

$$\ln \frac{N}{N_0} = kt$$

$$N = N_0 e^{kt} \qquad \text{exponential form}$$

Now, using the condition that one-half of carbon 14 decays in 5730 years, we have $N = N_0/2$ when $t = 5730$ years. This gives

$$\frac{N_0}{2} = N_0 e^{5730k} \qquad \text{or} \qquad \frac{1}{2} = (e^k)^{5730}$$

CAUTION

$$\boxed{e^k = 0.5^{1/5730}}$$

Therefore, the equation relating N and t is

$$N = N_0(0.5)^{t/5730}$$

Since the present concentration is $N = 0.678N_0$, we can solve for t.

$$0.678N_0 = N_0(0.5)^{t/5730} \qquad \text{or} \qquad 0.678 = 0.5^{t/5730}$$

$$\ln 0.678 = \ln 0.5^{t/5730}, \qquad \ln 0.678 = \frac{t}{5730} \ln 0.5 \qquad \text{solving an exponential equation}$$

$$t = 5730 \frac{\ln 0.678}{\ln 0.5} = 3210 \text{ years}$$

Therefore, the wood at the Troy site is about 3210 years old.

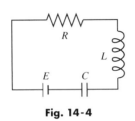

Fig. 14-4

EXAMPLE 4   The general equation relating the current *i*, voltage *E*, inductance *L*, capacitance *C*, and resistance *R* of a simple electric circuit (see Fig. 14-4) is

$$L\frac{di}{dt} + Ri + \frac{q}{C} = E \tag{14-9}$$

where *q* is the charge on the capacitor. Find the general expression for the current in a circuit containing an inductance, a resistance, and a voltage source if $i = 0$ when $t = 0$.

The differential equation for this circuit is

$$L\frac{di}{dt} + Ri = E$$

Using the method of the linear differential equation of the first order, we have the equation

$$di + \frac{R}{L}i\,dt = \frac{E}{L}dt$$

The factor $e^{\int P\,dt}$ is $e^{\int (R/L)dt} = e^{(R/L)t}$. This gives

$$ie^{(R/L)t} = \frac{E}{L}\int e^{(R/L)t}\,dt = \frac{E}{R}e^{(R/L)t} + c$$

Letting the current be zero for $t = 0$, we have $c = -E/R$. The result is

$$ie^{(R/L)t} = \frac{E}{R}e^{(R/L)t} - \frac{E}{R}$$

$$i = \frac{E}{R}(1 - e^{-(R/L)t})$$

(We can see that $i \to E/R$ as $t \to \infty$. In practice the exponential term becomes negligible very quickly.)

SOLVING A WORD PROBLEM

EXAMPLE 5   Fifty gallons of brine originally containing 20.0 lb of salt are in a tank into which 2.00 gal of water run each minute with the same amount of mixture running out each minute. How much salt is in the tank after 10.0 min?

Let x = the number of pounds of salt in the tank after *t* minutes. Each gallon of brine contains $x/50$ lb of salt, and in time dt, *2 dt gal of mixture leave the tank* **with $(x/50)(2\,dt)$ lb** *of salt*. The amount of salt that is leaving may also be written as $-dx$ (the minus sign is included to show that *x* is decreasing). Thus,

CAUTION ▶

$$-dx = \frac{2x\,dt}{50} \qquad \text{or} \qquad \frac{dx}{x} = -\frac{dt}{25}$$

This leads to $\ln x = -(t/25) + \ln c$. Using the fact that $x = 20$ lb when $t = 0$, we find that $\ln 20 = \ln c$, or $c = 20$. Therefore,

$$x = 20e^{-t/25}$$

is the general expression for the amount of salt in the tank at time *t*. Therefore, when $t = 10.0$ min, we have

$$x = 20e^{-10/25} = 20e^{-0.4} = 20(0.670) = 13.4 \text{ lb}$$

There are 13.4 lb of salt in the tank after 10.0 min. (Although the data were given with three significant digits, we did not use all significant digits in writing the equations that were used.)

EXAMPLE 6 An object moving through (or across) a resisting medium often experiences a retarding force that is approximately proportional to the velocity as well as the force that causes the motion. An example of this is a ball that falls due to the force of gravity and air resistance produces a retarding force. Applying Newton's laws of motion (from physics) to this ball leads to the equation

$$m\frac{dv}{dt} = F - kv \tag{14-10}$$

where m is the mass of the object, v is the velocity of the object, t is the time, F is the force causing the motion, and k ($k > 0$) is a constant. The quantity kv is the retarding force.

We assume that these conditions hold for a certain falling object whose mass is 1.0 slug (this is the unit of mass when force is expressed in pounds) and which experiences a force (its own weight) of 32.0 lb. The object starts from rest, and the air causes a retarding force that is numerically equal to 0.200 times the velocity. Substituting in Eq. (14-10), we have

$$\frac{dv}{dt} = 32 - 0.2v$$

Solving this differential equation, we have

$$\frac{dv}{32 - 0.2v} = dt \qquad \text{separate variables}$$

$$-5\ln(32 - 0.2v) = t - 5\ln c \qquad \text{integrate}$$

$$\ln(32 - 0.2v) = -\frac{t}{5} + \ln c \qquad \text{solve for } v$$

$$\ln\frac{32 - 0.2v}{c} = -\frac{t}{5}$$

$$32 - 0.2v = ce^{-t/5}$$

$$0.2v = 32 - ce^{-t/5}$$

$$v = 5(32 - ce^{-t/5}) \qquad \text{general solution}$$

Since the object started from rest, $v = 0$ when $t = 0$. Thus,

$$0 = 5(32 - c) \quad \text{or} \quad c = 32 \qquad \text{evaluate } c$$

$$v = 160(1 - e^{-t/5}) \qquad \text{particular solution}$$

Evaluating v for $t = 5.00$ s, we have

$$v = 160(1 - e^{-1.00}) = 160(1 - 0.368)$$

$$= 101 \text{ ft/s}$$

Therefore, after 5.00 s the velocity of the object is 101 ft/s. Without the air resistance, the velocity of the object would be about 160 ft/s after 5.00 s. (The data were given to three significant digits, but all significant digits were not used in writing the equations.)

EXERCISES *14-5*

In Exercises 1–4, find the equation of the curve for the given slope and point through which it passes. Use a graphing calculator to display the curve.

1. Slope given by $2x/y$; passes through $(2, 3)$
2. Slope given by $-y/(x + y)$; passes through $(-1, 3)$
3. Slope given by $y + x$; passes through $(0, 1)$
4. Slope given by $-2y + e^{-x}$; passes through $(0, 2)$

In Exercises 5–8, find the equation of the orthogonal trajectories of the curves for the given equations. Use a graphing calculator to display at least two members of the family of curves and at least two of the orthogonal trajectories.

5. The exponential curves $y = ce^x$
6. The cubic curves $y = cx^3$
7. The curves $y = c(\sec x + \tan x)$
8. The family of circles, all with centers at the origin

In Exercises 9–40, solve the given problems by solving the appropriate differential equation.

9. The isotope neon 23 decays such that half of an original amount disintegrates in 40.0 s. Find the relation between the amount present and the time and then find the percent remaining after 60.0 s.

10. Radium 226 decays such that 10% of an original amount disintegrates in 246 years. Find the half-life (the time for one-half of the original amount to disintegrate) of radium 226.

11. A possible health hazard in the home is radon gas. It is radioactive, and about 90.0% of an original amount disintegrates in 12.7 days. Find the half-life of radon gas. (The problem with radon is that it is a gas and is being continually produced by the radioactive decay of minute amounts of radioactive radium found in the soil and rocks of an area.)

12. A radioactive element leaks from a nuclear power plant at a constant rate r, and it decays at a rate that is proportional to the amount present. Find the relation between the amount N present in the environment in which it leaks and the time t, if $N = 0$ when $t = 0$.

13. The rate of change of the radial stress S on the walls of a pipe with respect to the distance r from the axis of the pipe is given by $r\dfrac{dS}{dr} = 2(a - S)$, where a is a constant. Solve for S as a function of r.

14. The velocity v of a meteor approaching the earth is given by $v\dfrac{dv}{dr} = -\dfrac{GM}{r^2}$, where r is the distance from the center of the earth, M is the mass of the earth, and G is a universal gravitational constant. If $v = 0$ for $r = r_0$, solve for v as a function of r.

15. Assume that the rate at which highway construction increases is directly proportional to the total mileage M of all highways already completed at time t (in years). Solve for M as a function of t if $M = 5250$ mi for a certain county when $t = 0$ and $M = 5460$ mi for $t = 2.00$ years.

16. The marginal profit function gives the change in the total profit P of a business due to a change in the business, such as adding new machinery or reducing the size of the sales staff. A company determines that the marginal profit dP/dx is $e^{-x^2} - 2Px$, where x is the amount invested in new machinery. Determine the total profit (in thousands of dollars) as a function of x, if $P = 0$ for $x = 0$.

17. According to Newton's law of cooling, the rate at which a body cools is proportional to the difference in temperature between it and the surrounding medium. Assuming Newton's law holds, how long will it take a cup of hot water, initially at 200°F, to cool to 100°F if the room temperature is 80.0°F, if it cools to 140°F in 5.0 min?

18. An object whose temperature is 100°C is placed in a medium whose temperature is 20°C. The temperature of the object falls to 50°C in 10 min. Express the temperature T of the object as a function of time t (in min). (Assume it cools according to Newton's law of cooling as stated in Exercise 17.)

19. If interest in a bank account is compounded continuously, the amount grows at a rate that is proportional to the amount present in the account. Interest that is compounded daily very closely approximates this situation. Determine the amount in an account after one year if $1000 is placed in the account and it pays 4% interest per year, compounded continuously.

20. The rate of change in the intensity I of light below the surface of the ocean with respect to the depth y is proportional to I. If the intensity at 15 ft is 50% of the intensity I_0 at the surface, at what depth is the intensity 15% of I_0?

21. If the current in an RL circuit with a voltage source E is zero when $t = 0$ (see Example 4), show that $\lim\limits_{t \to \infty} i = E/R$. See Fig. 14-5.

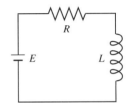

Fig. 14-5

22. If a circuit contains only an inductance and a resistance, with $L = 2.0$ H and $R = 30\ \Omega$, find the current i as a function of time t if $i = 0.020$ A when $t = 0$. See Fig. 14-6.

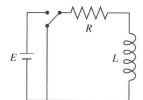

Fig. 14-6

23. An amplifier circuit contains a resistance R, an inductance L, and a voltage source $E \sin \omega t$. Express the current in the circuit as a function of the time t if the initial current is zero.

24. A radio transmitter circuit contains a resistance of $2.0\ \Omega$, a variable inductor of $100 - t$ henrys, and a voltage source of 4.0 V. Find the current i in the circuit as a function of the time t for $0 \le t \le 100$ s if the initial current is zero.

25. If a circuit contains only a resistance R and a capacitance C, find the expression relating the charge on the capacitor in terms of the time if $i = dq/dt$ and $q = q_0$ when $t = 0$. See Fig. 14-7.

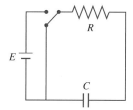

Fig. 14-7

26. A circuit contains a $4.0\text{-}\mu\text{F}$ capacitor, a $450\text{-}\Omega$ resistor, and a 20.0-mV voltage source. If the charge on the capacitor is 20.0 nC when $t = 0$, find the charge after 0.010 s. ($i = dq/dt$.)

27. One hundred gallons of brine originally containing 30 lb of salt are in a tank into which 5.0 gal of water run each minute. The same amount of mixture from the tank leaves each minute. How much salt is in the tank after 20 min?

28. Repeat Exercise 27 with the change that the water entering the tank contains 1.0 lb of salt per gallon.

29. An object falling under the influence of gravity has a variable acceleration given by $32 - v$, where v represents the velocity. If the object starts from rest, find an expression for the velocity in terms of the time. Also find the limiting value of the velocity (find $\lim\limits_{t \to \infty} v$).

30. In a ballistics test, a bullet is fired into a sandbag. The acceleration of the bullet within the sandbag is $-40\sqrt{v}$, where v is the velocity (in ft/s). When will the bullet stop if it enters the sandbag at 950 ft/s?

31. A boat with a mass of 10 slugs is being towed at 8.0 mi/h. The tow rope is then cut, and a motor that exerts a force of 20 lb on the boat is started. If the water exerts a retarding force that numerically equals twice the velocity, what is the velocity of the boat 3.0 min later?

32. A parachutist is falling at the rate of 196 ft/s when her parachute opens. If the air resists the fall with a force equal to $0.5v^2$, find the velocity as a function of time. The person and equipment have a combined mass of 5.00 slugs (weight is 160 lb).

33. For each cycle, a roller mechanism follows a path described by $y = 2x - x^2$, $y \ge 0$, such that $dx/dt = 6t - 3t^2$. Find x and y (in cm) in terms of the time t (in s) if x and y are zero for $t = 0$.

34. In studying the flow of water in a stream, it is found that an object follows the hyperbolic path $y(x + 1) = 10$ such that $(t + 1)\, dx = (x - 2)\, dt$. Find x and y (in ft) in terms of the time t (in s) if $x = 4$ ft and $y = 2$ ft for $t = 0$.

35. The rate of change of air pressure p (in lb/ft^2) with respect to height h (in ft) is approximately proportional to the pressure. If the pressure is 15.0 lb/in.^{2} when $h = 0$ and $p = 10.0$ lb/in.^{2} when $h = 9800$ ft, find the expression relating pressure and height.

36. Water flows from a vertical cylindrical storage tank through a hole of area A at the bottom of the tank. The rate of flow is $4.8\,A\sqrt{h}$, where h is the distance (in ft) from the surface of the water to the hole. If h changes from 9.0 ft to 8.0 ft in 16 min, how long will it take the tank to empty? See Fig. 14-8.

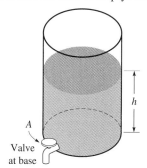

Fig. 14-8   Valve at base

37. Assume that the rate of depreciation of an object is proportional to its value at any time t. If a car costs $16,500 new and its value 3 years later is $9850, what is its value 11 years after it was purchased?

38. Assume that sugar dissolves at a rate proportional to the undissolved amount. If there are initially 525 g of sugar and 225 g remain after 4.00 min, how long does it take to dissolve 375 g?

39. Fresh air is being circulated into a room whose volume is 4000 ft^3. Under specified conditions the number of cubic feet x of carbon dioxide present at any time t (in min) is found by solving the differential equation $dx/dt = 1 - 0.25\ x$. Find x as a function of t if $x = 12$ ft^3 when $t = 0$.

40. The lines of equal potential in a field of force are all at right angles to the lines of force. In an electric field of force caused by charged particles, the lines of force are given by $x^2 + y^2 = cx$. Find the equation of the lines of equal potential. Use a graphing calculator to view a few members of the lines of force and those of equal potential.

CHAPTER EQUATIONS

Separation of variables	$M(x, y)\, dx + N(x, y)\, dy = 0$	**(14-1)**
	$A(x)\, dx + B(y)\, dy = 0$	**(14-2)**
Integrable combinations	$d(xy) = x\, dy + y\, dx$	**(14-3)**
	$d(x^2 + y^2) = 2(x\, dx + y\, dy)$	**(14-4)**
	$d\left(\dfrac{y}{x}\right) = \dfrac{x\, dy - y\, dx}{x^2}$	**(14-5)**
	$d\left(\dfrac{x}{y}\right) = \dfrac{y\, dx - x\, dy}{y^2}$	**(14-6)**
Linear differential equation of first order	$dy + Py\, dx = Q\, dx$	**(14-7)**
	$ye^{\int P\, dx} = \displaystyle\int Qe^{\int P\, dx}\, dx + c$	**(14-8)**
Electric circuit	$L\dfrac{di}{dt} + Ri + \dfrac{q}{C} = E$	**(14-9)**
Motion in resisting medium	$m\dfrac{dv}{dt} = F - kv$	**(14-10)**

REVIEW EXERCISES

In Exercises 1–20, find the general solution to the given differential equations.

1. $4xy^3\, dx + (x^2 + 1)\, dy = 0$ **2.** $\dfrac{dy}{dx} = e^{x-y}$

3. $\sin 2x\, dx + y \sin x\, dy = \sin x\, dx$

4. $x\, dy + y\, dx = y\, dy$

5. $(x + y)\, dx + (x + y^3)\, dy = 0$

6. $y \ln x\, dx = x\, dy$

7. $x\dfrac{dy}{dx} - 3y = x^2$ **8.** $dy - 2y\, dx = (x - 2)e^x\, dx$

9. $dy = 2y\, dx + y^2\, dx$ **10.** $x^2y\, dy = (1 + x)\csc y\, dx$

11. $y' + 4y = 2e^{-2x}$ **12.** $2xy\, dx = (2y - \ln y)\, dy$

13. $\sin x\dfrac{dy}{dx} + y \cos x + x = 0$

14. $y\, dy = (x^2 + y^2 - x)\, dx$

15. $(x^2 + 2x)\, dy = 2xy\, dx + 5y\, dx$

16. $\dfrac{dy}{dx} = \dfrac{1}{1 + e^x} - y$

17. $x\, dy - y\, dx = x^3y\, dx$

18. $x\, dy + y\, dx = x^3y^4\, dy$

19. $y\, dy = 2(x^2 + y^2)^2\, dx - x\, dx$

20. $y' = 2 + xe^{3x}$

In Exercises 21–28, find the indicated particular solution of the given differential equations.

21. $3y' = 2y \cot x;$ $x = \dfrac{\pi}{2}$ when $y = 2$

22. $x\, dy - y\, dx = y^3\, dy;$ $x = 1$ when $y = 3$

23. $y' = 4x - 2y;$ $x = 0$ when $y = -2$

24. $xy^2\, dx + e^x\, dy = 0;$ $x = 0$ when $y = 2$

25. $\dfrac{dy}{dx} = x - xy, \quad x = 2$ when $y = 0$

26. $\dfrac{dy}{dx} = 2 - y, \quad x = 0$ when $y = 1$

27. $xy' - 2y = x^3 \cos 4x, \quad x = \dfrac{\pi}{4}$ when $y = 0$

28. $2xy\,dy + dx = 2x^2 \sin x^2\,dx, \quad x = 1$ when $y = 0$

In Exercises 29–44, solve the given problems.

29. The time rate of change of volume of an evaporating substance is proportional to the surface area. Express the radius of an evaporating sphere of ice as a function of time. Let $r = r_0$ when $t = 0$. (*Hint:* Express both V and A in terms of the radius r.)

30. An insulated tank is filled with a solution containing radioactive cobalt. Due to the radioactivity, energy is released and the temperature T (in °F) of the solution rises with the time t (in h). The following equation expresses the relation between temperature and time for a specific case:

$$56{,}600 = 262(T - 70) + 20{,}200\dfrac{dT}{dt}$$

If the initial temperature is 70°F, what is the temperature 24 h later?

31. An object with a mass of 1.00 kg slides down a long inclined plane. The effective force of gravity is 4.00 N, and the motion is retarded by a force numerically equal to the velocity. If the object starts from rest, what is its velocity (in m/s) 4.00 s later?

32. A 192-lb object falls from rest under the influence of gravity. Find the equation for the velocity at any time t (in s) if the air resists the motion with a force numerically equal to twice the velocity.

33. Radioactive potassium 40 with a half-life of 1.28×10^9 years is used for dating rock samples. If a given rock sample has 75% of its original amount of potassium 40, how old is the rock?

34. When a gas undergoes an adiabatic change (no gain or loss of heat), the rate of change of pressure with respect to volume is directly proportional to the pressure and inversely proportional to the volume. Express the pressure in terms of the volume.

35. Under ideal conditions, the natural law of population change is that the population increases at a rate proportional to the population at any time. Under these conditions, project the population of the world in 2010 if it reached 5.0 billion in 1987 and 6.0 billion in 1999.

36. A spherical balloon is being blown up such that its volume V increases at a rate proportional to its surface area. Show that this leads to the differential equation $dV/dt = kV^{2/3}$ and solve for V as a function of t.

37. Find the orthogonal trajectories of the family of curves $y = cx^5$.

38. Find the equation of the curves for which their normals at all points are in the direction with the lines connecting the points and the origin.

39. If a circuit contains a resistance R, a capacitance C, and a source of voltage E, express the charge q on the capacitor as a function of time.

40. A 2-H inductor, a 40-Ω resistor, and a 20-V battery are connected in series. Find the current in the circuit as a function of time if the initial current is zero.

41. Air containing 20% oxygen passes into a 5.00-L container initially filled with 100% oxygen. A uniform mixture of the air and oxygen then passes from the container at the same rate. What volume of oxygen is in the container after 5.00 L of air have passed into it?

42. The gravitational acceleration of an object is inversely proportional to the square of its distance r from the center of the earth. Use the chain rule, Eq. (3-14), to show that the acceleration is $\dfrac{dv}{dt} = v\dfrac{dv}{dr}$, where $v = \dfrac{dr}{dt}$ is the velocity of the object. Then solve for v as a function r if $dv/dt = -g$ and $v = v_0$ for $r = R$, where R is the radius of the earth. Finally, show that a spacecraft must have a velocity of at least $v_0 = \sqrt{2gR}$ ($= 7$ mi/s) in order to escape from the earth's gravitation. (Note the expression for v^2 as $r \to \infty$.)

43. The angle ϕ between the radius vector and the curve $r = f(\theta)$ at (r, θ) is given by

$$\dfrac{dr}{d\theta} = \dfrac{r}{\tan \phi}$$

Find the polar equation of the orthogonal trajectories of the circles

$$r = a \sin \theta$$

See Fig. 14-9.

44. A differential equation of the form

$$M(x, y)\,dx + N(x, y)\,dy = 0$$

is termed *exact* if

$$\dfrac{\partial M}{\partial y} = \dfrac{\partial N}{\partial x}$$

Determine whether or not the equation

$$3x^2 \cos^2 y\,dx - x^3 \sin 2y\,dy = 0$$

is exact.

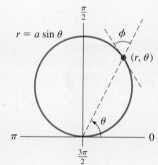

Fig. 14-9

Writing Exercise

45. Consider the differential equation for an ac electric circuit containing an inductor, a resistor, and a voltage source or an electric circuit containing a resistor, capacitor, and voltage source. Write two or three paragraphs explaining (a) that the method(s) of this chapter that can be used to solve either equation is (are) the same and (b) whether the method(s) used depend on the voltage source being constant or variable.

PRACTICE TEST

In Problems 1–4, find the general solution of each of the given differential equations.

1. $x\dfrac{dy}{dx} + 2y = 4$ **2.** $\csc y \, dx + e^x dy = 0$

3. $x \, dx + y \, dy = x^2 dx + y^2 dx$

4. $y' + y \tan x = 2x \cos x$

5. Find the particular solution of the differential equation

$(xy + y)\dfrac{dy}{dx} = 2$, if $y = 2$ when $x = 0$.

6. If interest in a bank account is compounded continuously, the amount grows at a rate that is proportional to the amount present. Derive the equation for the amount A in an account with continuous compounding in which the initial amount is A_0 and the interest rate is r as a function of the time t after A_0 is deposited.

7. Find the equation for the current as a function of the time (in s) in a circuit containing a 2-H inductance, an 8-Ω resistor, and a 6-V battery in series, if $i = 0$ when $t = 0$.

15

HIGHER-ORDER DIFFERENTIAL EQUATIONS

Until this point we have restricted our attention to differential equations of the first order and first degree. Another type of differential equation that is important in technical applications is the linear differential equation of the second order with constant coefficients. There are also some applications of the linear differential equation of order higher than two. Therefore, in this chapter we will consider higher-order equations, with our primary emphasis on second-order equations.

In the first section of this chapter we shall describe the general higher-order differential equation and the notation we will be using. In the sections that follow, we shall show how these types of equations are solved, again with emphasis on second-order equations. Finally, in Section 15-4, we shall consider some of the applications. These include electric circuits, simple harmonic motion, and deflection of beams.

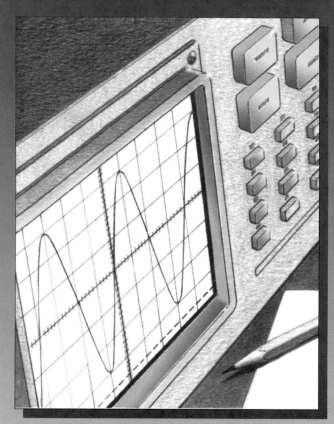

Second-order differential equations are very important in determining currents and voltages in electric circuits. In Section 15-4, we show how the equation relating current and time is found for a particular circuit.

15-1 HIGHER-ORDER HOMOGENEOUS EQUATIONS

The linear differential equation of the first order that we introduced in the previous chapter is a special case of the linear differential equation of degree n. On the following page we give the definition of the general linear differential equation and a special notation that we will use in writing it.

The general **linear differential equation of the** *n***th order** *is of the form*

$$a_0 \frac{d^n y}{dx^n} + a_1 \frac{d^{n-1} y}{dx^{n-1}} + \cdots + a_{n-1} \frac{dy}{dx} + a_n y = b \qquad (15\text{-}1)$$

where the a's and b are continuous functions of x (which may be constants). This equation is termed *linear* since *y* and all of its derivatives appear to the first power only.

For convenience of notation, the *n*th derivative with respect to the independent variable will be denoted by D^n. Here, *D* is called an **operator,** since it denotes the *operation* of differentiation. Using this notation with *x* as the independent variable, Eq. (15-1) becomes

$$a_0 D^n y + a_1 D^{n-1} y + \cdots + a_{n-1} D y + a_n y = b \qquad (15\text{-}2)$$

If b = 0, *the general linear equation is called* **homogeneous,** *and if b* ≠ 0, *it is called* **nonhomogeneous.** Both types of differential equations have important applications in technology.

■EXAMPLE 1   Using the operator form of Eq. (15-2), the differential equation

$$\frac{d^3 y}{dx^3} - 3 \frac{d^2 y}{dx^2} + 4 \frac{dy}{dx} - 2y = e^x \sec x$$

is written as

$$D^3 y - 3D^2 y + 4Dy - 2y = e^x \sec x$$

This equation is nonhomogeneous since $b = e^x \sec x$.   --------------■

Although the *a*'s may be functions of *x,* we shall restrict our attention to the cases in which they are constants. We shall, however, consider both homogeneous equations and nonhomogeneous equations. Also, as we noted in the chapter introduction, since second-order equations are the ones most commonly found in elementary applications, we shall devote most of our attention to them. The methods used to solve second-order equations may be applied to equations of higher order, and therefore we shall consider certain of these higher-order equations.

Second-Order Homogeneous Equations with Constant Coefficients

Using the operator notation, *a second-order, linear, homogeneous differential equation with constant coefficients is one of the form*

$$a_0 D^2 y + a_1 D y + a_2 y = 0 \qquad (15\text{-}3)$$

where the a's are constants. The following example indicates the kind of solution we should expect for this type of equation.

EXAMPLE 2   Solve the differential equation $D^2y - Dy - 2y = 0$.

First, we put this equation in the form $(D^2 - D - 2)y = 0$. This is another way of saying that we are to take the second derivative of y, subtract the first derivative, and finally subtract twice the function. This expression may now be factored as $(D - 2)(D + 1)y = 0$. (We shall not develop the algebra of the operator D. However, most such algebraic operations can be shown to be valid.) This formula tells us to find the first derivative of the function and add this to the function. Then twice this result is to be subtracted from the derivative of this result. If we let $z = (D + 1)y$, which is valid since $(D + 1)y$ is a function of x, we have $(D - 2)z = 0$. This equation is easily solved by separation of variables. Thus,

$$\frac{dz}{dx} - 2z = 0 \qquad \frac{dz}{z} - 2\,dx = 0 \qquad \ln z - 2x = \ln c_1$$

$$\ln \frac{z}{c_1} = 2x \qquad \text{or} \qquad z = c_1 e^{2x}$$

Replacing z by $(D + 1)y$, we have

$$(D + 1)y = c_1 e^{2x}$$

This is a linear equation of the first order. Then,

$$dy + y\,dx = c_1 e^{2x}\,dx$$

The factor $e^{\int P\,dx}$ is $e^{\int dx} = e^x$. And so,

$$ye^x = \int c_1 e^{3x}\,dx = \frac{c_1}{3}e^{3x} + c_2 \qquad \text{using Eq. (14-8)}$$

$$y = c_1' e^{2x} + c_2 e^{-x}$$

where $c_1' = \frac{1}{3}c_1$. This example indicates that solutions of the form e^{mx} result for this equation.

Based on the result of Example 2, assume that an equation of the form of Eq. (15-3) has a particular solution ce^{mx}. Substituting this into Eq. (15-3) gives

$$a_0 cm^2 e^{mx} + a_1 cme^{mx} + a_2 ce^{mx} = 0$$

Since the exponential function $e^{mx} > 0$ for all real x, this equation will be satisfied if m is a root of the equation

AUXILIARY EQUATION

$$a_0 m^2 + a_1 m + a_2 = 0 \qquad\qquad \text{(15-4)}$$

Equation (15-4) is called the **auxiliary equation** *of Eq. (15-3). Note that it may be formed directly by inspection from Eq. (15-3).*

There are two roots of the auxiliary Eq. (15-4), and there are two arbitrary constants in the solution of Eq. (15-3). These factors lead us to *the general solution of Eq. (15-3), which is*

GENERAL SOLUTION

$$y = c_1 e^{m_1 x} + c_2 e^{m_2 x} \qquad\qquad \text{(15-5)}$$

where m_1 and m_2 are the solutions of Eq. (15-4). We see that this is in agreement with the results of Example 2.

■**EXAMPLE 3**   Solve the differential equation $D^2y - 5\,Dy + 6y = 0$.

From this operator form of the differential equation, we write the auxiliary equation

$$m^2 - 5m + 6 = 0$$

Solving the auxiliary equation, we have

$$(m - 3)(m - 2) = 0$$
$$m_1 = 3 \qquad m_2 = 2$$

Now, using Eq. (15-5), we write the solution of the differential equation as

$$y = c_1 e^{3x} + c_2 e^{2x}$$

It makes no difference which constant is written with each exponential term.   ∎

■**EXAMPLE 4**   Solve the differential equation $y'' = 6y'$.

We first rewrite this equation using the D notation for derivatives. Also, we want to write it in the proper form of a homogeneous equation. This gives us

$$D^2y - 6\,Dy = 0$$

Proceeding with the solution, we have

$$m^2 - 6m = 0 \qquad \text{auxiliary equation}$$
$$m(m - 6) = 0 \qquad \text{solve for } m$$
$$m_1 = 0 \qquad m_2 = 6$$
$$y = c_1 e^{0x} + c_2 e^{6x} \qquad \text{using Eq. (15-5)}$$

Since $e^{0x} = 1$, we have

$$y = c_1 + c_2 e^{6x} \qquad \text{general solution}$$ ∎

THIRD-ORDER EQUATION

■**EXAMPLE 5**   Solve the differential equation $2\dfrac{d^3y}{dx^3} + \dfrac{d^2y}{dx^2} - 7\dfrac{dy}{dx} = 0$.

Although this is a third-order equation, the *method* of solution is the same as in the previous examples. Using the D-notation for derivatives, we have

$$2\,D^3y + D^2y - 7Dy = 0$$

This means the auxiliary equation is

$$2m^3 + m^2 - 7m = 0, \qquad m(2m^2 + m - 7) = 0$$

We can see that one root is $m = 0$. The quadratic factor is not factorable, but we can find the roots from it by using the quadratic formula. This gives

$$m = \frac{-1 \pm \sqrt{1 + 56}}{4} = \frac{-1 \pm \sqrt{57}}{4}$$

Since there are *three* roots, there are *three* arbitrary constants. Following the solution indicated by Eq. (15-5), we have

$$y = c_1 e^{0x} + c_2 e^{(-1+\sqrt{57})x/4} + c_3 e^{(-1-\sqrt{57})x/4}$$

Again, $e^{0x} = 1$. Also, factoring $e^{-x/4}$ from the second and third terms, we have

$$y = c_1 + e^{-x/4}(c_2 e^{x\sqrt{57}/4} + c_3 e^{-x\sqrt{57}/4})$$ ∎

EXAMPLE 6 Solve the differential equation $D^2y - 2\,Dy - 15y = 0$ and find the particular solution that satisfies the conditions $Dy = 2$ and $y = -1$, when $x = 0$. (It is necessary to give two conditions since there are two constants to evaluate.)

We have

$$m^2 - 2m - 15 = 0, \qquad (m - 5)(m + 3) = 0$$
$$m_1 = 5 \qquad m_2 = -3$$
$$y = c_1 e^{5x} + c_2 e^{-3x}$$

CAUTION ▶

This equation is the general solution. In order to evaluate the constants c_1 and c_2, *we use the given conditions to find two simultaneous equations in c_1 and c_2.* These are then solved to determine the particular solution. Thus,

$$y' = 5c_1 e^{5x} - 3c_2 e^{-3x}$$

Using the given conditions in the general solution and its derivative, we have

$$c_1 + c_2 = -1 \qquad y = -1 \text{ when } x = 0$$
$$5c_1 - 3c_2 = 2 \qquad Dy = 2 \text{ when } x = 0$$

The solution to this system of equations is $c_1 = -\frac{1}{8}$ and $c_2 = -\frac{7}{8}$. The particular solution becomes

$$y = -\frac{1}{8} e^{5x} - \frac{7}{8} e^{-3x} \qquad \text{or} \qquad 8y + e^{5x} + 7e^{-3x} = 0 \qquad ■$$

In all the examples and exercises of this section, all the roots of the auxiliary equation are different, and they do not include complex numbers. For repeated or complex roots, the solutions have a different form. They are the topic of the following section.

EXERCISES *15-1*

In Exercises 1–20, solve the given differential equations.

1. $\dfrac{d^2y}{dx^2} - \dfrac{dy}{dx} - 6y = 0$

2. $\dfrac{d^2y}{dx^2} + \dfrac{dy}{dx} = 0$

3. $3\dfrac{d^2y}{dx^2} + 4\dfrac{dy}{dx} + y = 0$

4. $\dfrac{d^2y}{dx^2} - 2\dfrac{dy}{dx} - 8y = 0$

5. $D^2y - 3\,Dy = 0$

6. $D^2y + 7\,Dy + 6y = 0$

7. $3\,D^2y + 12y = 20\,Dy$

8. $4\,D^2y + 12\,Dy = 7y$

9. $3y'' + 8y' - 3y = 0$

10. $8y'' + 6y' - 9y = 0$

11. $3y'' + 2y' - y = 0$

12. $2y'' - 7y' + 6y = 0$

13. $2\dfrac{d^2y}{dx^2} - 4\dfrac{dy}{dx} + y = 0$

14. $\dfrac{d^2y}{dx^2} + \dfrac{dy}{dx} - 5y = 0$

15. $4\,D^2y - 3\,Dy - 2y = 0$

16. $2\,D^2y - 3\,Dy - y = 0$

17. $y'' = 3y' + y$

18. $5y'' - y' = 3y$

19. $y'' + y' = 8y$

20. $8y'' = y' + y$

In Exercises 21–24, find the particular solutions of the given differential equations that satisfy the given conditions.

21. $D^2y - 4\,Dy - 21y = 0$; $Dy = 0$ and $y = 2$ when $x = 0$

22. $4\,D^2y - Dy = 0$; $Dy = 2$ and $y = 4$ when $x = 0$

23. $D^2y - Dy - 12y = 0$; $y = 0$ when $x = 0$, and $y = 1$ when $x = 1$

24. $2\,D^2y + 5\,Dy = 0$; $y = 0$ when $x = 0$, and $y = 2$ when $x = 1$

In Exercises 25–28, solve the given third- and fourth-order differential equations.

25. $y''' - 2y'' - 3y' = 0$

26. $D^3y - 6\,D^2y + 11\,Dy - 6y = 0$

27. $D^4y - 5\,D^2y + 4y = 0$

28. $D^4y - D^3y - 9\,D^2y + 9\,Dy = 0$

$15\text{-}2$ AUXILIARY EQUATION WITH REPEATED OR COMPLEX ROOTS

In solving higher-order homogeneous differential equations in the previous section, we purposely avoided repeated or complex roots of the auxiliary equation. In this section we develop the solutions for such equations. The following example indicates the type of solution that results from the case of repeated roots.

EXAMPLE 1   Solve the differential equation $D^2 y - 4\,Dy + 4y = 0$.

Using the method of Example 2 of the previous section, we have the following steps:

$$(D^2 - 4\,D + 4)y = 0, \qquad (D - 2)(D - 2)y = 0, \qquad (D - 2)z = 0$$

where $z = (D - 2)y$. The solution to $(D - 2)z = 0$ is found by separation of variables. And so

$$\frac{dz}{dx} - 2z = 0 \qquad \frac{dz}{z} - 2\,dx = 0$$

$$\ln z - 2x = \ln c_1 \qquad \text{or} \qquad z = c_1 e^{2x}$$

Substituting back, we have $(D - 2)y = c_1 e^{2x}$, which is a linear equation of the first order. Then

$$dy - 2y\,dx = c_1 e^{2x}\,dx \qquad e^{\int -2\,dx} = e^{-2x}$$

This leads to

$$ye^{-2x} = c_1 \int dx = c_1 x + c_2 \qquad \text{or} \qquad y = c_1 x e^{2x} + c_2 e^{2x}$$

This example indicates the type of solution that results when the auxiliary equation has repeated roots. If the method of the previous section were to be used, the solution of the above example would be $y = c_1 e^{2x} + c_2 e^{2x}$. This would not be the general solution, since both terms are similar, which means that there is only one independent constant. The constants can be combined to give a solution of the form $y = ce^{2x}$, where $c = c_1 + c_2$. This solution would contain only one constant for a second-order equation.   ∎

For reference, Eq. (15-3) is
$a_0 D^2 y + a_1\,Dy + a_2 y = 0$
and Eq. (15-4) is
$a_0 m^2 + a_1 m + a_2 = 0.$

Based on the above example, *the solution to Eq. (15-3) when the auxiliary Eq. (15-4) has repeated roots is*

SOLUTION WITH REPEATED ROOTS

$$\boxed{y = e^{mx}(c_1 + c_2 x)} \tag{15-6}$$

where m is the double root. (In Example 1, this double root is 2.)

EXAMPLE 2   Solve the differential equation $(D + 2)^2 y = 0$.

The auxiliary equation is $(m + 2)^2 = 0$, for which the solutions are $m = -2$, -2. Since we have repeated roots, the solution of the differential equation is

$$y = e^{-2x}(c_1 + c_2 x) \qquad \text{using Eq. (15-6)} \qquad ∎$$

EXAMPLE 3   Solve the differential equation $\dfrac{d^2y}{dx^2} - 10\,\dfrac{dy}{dx} + 25y = 0$.

The solution is as follows:

$$D^2y - 10\,Dy + 25y = 0 \qquad \text{using operator } D \text{ notation}$$
$$m^2 - 10m + 25 = 0 \qquad \text{auxiliary equation}$$
$$(m - 5)^2 = 0 \qquad \text{solve for } m$$
$$m = 5, 5 \qquad \text{double root}$$
$$y = e^{5x}(c_1 + c_2x) \qquad \text{using Eq. (15-6)} \quad \text{------}\blacksquare$$

When the auxiliary equation has complex roots, it can be solved by the method of the previous section and the solution can be put in a more useful form. For complex roots of the auxiliary equation $m = \alpha \pm j\beta$, the solution is of the form

$$y = c_1e^{(\alpha+j\beta)x} + c_2e^{(\alpha-j\beta)x} = e^{\alpha x}(c_1e^{j\beta x} + c_2e^{-j\beta x})$$

Using the exponential form of a complex number (see margin note on page 394), we have

$$y = e^{\alpha x}[c_1 \cos \beta x + jc_1 \sin \beta x + c_2 \cos(-\beta x) + jc_2 \sin(-\beta x)]$$
$$= e^{\alpha x}(c_1 \cos \beta x + c_2 \cos \beta x + jc_1 \sin \beta x - jc_2 \sin \beta x)$$
$$= e^{\alpha x}(c_3 \cos \beta x + c_4 \sin \beta x)$$

where $c_3 = c_1 + c_2$ and $c_4 = jc_1 - jc_2$.

Therefore, *if the auxiliary equation has complex roots of the form $\alpha \pm j\beta$,*

SOLUTION WITH COMPLEX ROOTS

$$\boxed{y = e^{\alpha x}(c_1 \sin \beta x + c_2 \cos \beta x)} \qquad (15\text{-}7)$$

is the solution to Eq. (15-3). The c_1 and c_2 here are not the same as those above. They are simply the two arbitrary constants of the solution.

EXAMPLE 4   Solve the differential equation $D^2y - Dy + y = 0$.
We have the following solution:

$$m^2 - m + 1 = 0 \qquad \text{auxiliary equation}$$
$$m = \frac{1 \pm j\sqrt{3}}{2} \qquad \text{complex roots}$$
$$\alpha = \frac{1}{2} \qquad \beta = \frac{\sqrt{3}}{2} \qquad \text{identify } \alpha \text{ and } \beta$$
$$y = e^{x/2}\left(c_1 \sin \frac{\sqrt{3}}{2}x + c_2 \cos \frac{\sqrt{3}}{2}x\right) \qquad \text{using Eq. (15-7)} \quad \text{------}\blacksquare$$

EXAMPLE 5   Solve the differential equation $D^3y + 4\,Dy = 0$.
This is a third-order equation, which means there are three arbitrary constants in the general solution.

$$m^3 + 4m = 0, \qquad m(m^2 + 4) = 0 \qquad \text{auxiliary equation}$$
$$m_1 = 0 \qquad m_2 = 2j \qquad m_3 = -2j \qquad \text{three roots, two of them complex}$$
$$\alpha = 0 \qquad \beta = 2 \qquad \text{identify } \alpha \text{ and } \beta \text{ for complex roots}$$
$$y = c_1e^{0x} + e^{0x}(c_2 \sin 2x + c_3 \cos 2x) \qquad \text{using Eqs. (15-5) and (15-7)}$$
$$= c_1 + c_2 \sin 2x + c_3 \cos 2x \qquad e^0 = 1 \quad \text{------}\blacksquare$$

EXAMPLE 6 Solve the differential equation $y'' - 2y' + 12y = 0$, if $y' = 2$ and $y = 1$ when $x = 0$.

$$D^2y - 2\,Dy + 12y = 0 \qquad \text{use operator } D \text{ notation}$$
$$m^2 - 2m + 12 = 0 \qquad \text{auxiliary equation}$$
$$m = \frac{2 \pm \sqrt{4 - 48}}{2} = 1 \pm j\sqrt{11} \qquad \text{complex roots: } \alpha = 1,\ \beta = \sqrt{11}$$
$$y = e^x(c_1 \cos \sqrt{11}x + c_2 \sin \sqrt{11}x) \qquad \text{general solution}$$

Using the condition that $y = 1$ when $x = 0$, we have

$$1 = e^0(c_1 \cos 0 + c_2 \sin 0) \qquad \text{or} \qquad c_1 = 1$$

Since $y' = 2$ when $x = 0$, we find the derivative and then evaluate c_2.

$$y' = e^x(c_1 \cos \sqrt{11}x + c_2 \sin \sqrt{11}x - \sqrt{11}c_1 \sin \sqrt{11}x + \sqrt{11}c_2 \cos \sqrt{11}x)$$
$$2 = e^0(\cos 0 + c_2 \sin 0 - \sqrt{11} \sin 0 + \sqrt{11}c_2 \cos 0) \qquad y' = 2 \text{ when } x = 0$$
$$2 = 1 + \sqrt{11}c_2, \qquad c_2 = \frac{1}{11}\sqrt{11} \qquad \text{solve for } c_2$$
$$y = e^x\left(\cos \sqrt{11}x + \frac{1}{11}\sqrt{11} \sin \sqrt{11}x\right) \qquad \text{particular solution} \quad \blacksquare$$

If the root of the auxiliary equation is repeated more than once—for example, a *triple root*—an additional term with another arbitrary constant and the next higher power of x is added to the solution for each additional root. Also, if a pair of complex roots is repeated, an additional term with a factor of x and another arbitrary constant is added for each root of the pair. These are illustrated in the following example.

ROOTS REPEATED MORE THAN ONCE

EXAMPLE 7 **(a)** For the differential equation $D^3y + 3\,D^2y + 3\,Dy + y = 0$, the auxiliary equation is

$$m^3 + 3m^2 + 3m + 1 = 0, \qquad (m + 1)^3 = 0$$

Each of the three roots is $m = -1$. The equation is a third-order equation, which means there are three arbitrary constants. Therefore, the general solution is

$$y = e^{-x}(c_1 + c_2x + c_3x^2)$$

(b) For the differential equation $D^4y + 8\,D^2y + 16y = 0$, the auxiliary equation

REPEATED COMPLEX ROOTS is

$$m^4 + 8m^2 + 16 = 0, \qquad (m^2 + 4)^2 = 0$$

With two factors of $m^2 + 4$, the roots are $2j$, $2j$, $-2j$, and $-2j$. The fourth-order equation, and four roots, indicate four arbitrary constants. Since $e^{0x} = 1$, the general solution is

$$y = (c_1 + c_2x)\sin 2x + (c_3 + c_4x)\cos 2x \qquad \blacksquare$$

Knowing the various types of possible solutions, it is possible to determine the differential equation if the solution is known. Consider the following example.

EXAMPLE 8 **(a)** A solution of $y = c_1e^x + c_2e^{2x}$ indicates an auxiliary equation with roots of $m_1 = 1$ and $m_2 = 2$. Thus, the auxiliary equation is $(m - 1)(m - 2) = 0$, and the simplest form of the differential equation is $D^2y - 3\,Dy + 2y = 0$.

(b) A solution of $y = e^{2x}(c_1 + c_2x)$ indicates repeated roots $m_1 = m_2 = 2$ of the auxiliary equation $(m - 2)^2 = 0$, and a differential equation of $D^2y - 4\,Dy + 4y = 0$.

$\blacksquare$

EXERCISES *15-2*

In Exercises 1–28, solve the given differential equations.

1. $\dfrac{d^2y}{dx^2} - 2\dfrac{dy}{dx} + y = 0$ **2.** $\dfrac{d^2y}{dx^2} - 6\dfrac{dy}{dx} + 9y = 0$

3. $D^2y + 12\,Dy + 36y = 0$ **4.** $16\,D^2y + 8\,Dy + y = 0$

5. $\dfrac{d^2y}{dx^2} + 9y = 0$ **6.** $\dfrac{d^2y}{dx^2} + y = 0$

7. $D^2y + Dy + 2y = 0$ **8.** $D^2y - 2\,Dy + 4y = 0$

9. $D^4y - y = 0$ **10.** $4\,D^2y = 12\,Dy - 9y$

11. $4\,D^2y + y = 0$ **12.** $9\,D^2y + 4y = 0$

13. $16y'' - 24y' + 9y = 0$ **14.** $9y'' - 24y' + 16y = 0$

15. $25y'' + 2y = 0$ **16.** $y'' - 4y' + 5y = 0$

17. $2\,D^2y + 5y = 4\,Dy$ **18.** $D^2y + 4\,Dy + 6y = 0$

19. $25y'' + 16y = 40y'$ **20.** $9y''' + 0.6y'' + 0.01y' = 0$

21. $2\,D^2y - 3\,Dy - y = 0$ **22.** $D^2y - 5\,Dy - 4y = 0$

23. $3\,D^2y + 12\,Dy = 2y$ **24.** $36\,D^2y = 25y$

25. $D^3y - 6\,D^2y + 12\,Dy - 8y = 0$

26. $D^4y - 2\,D^3y + 2\,D^2y - 2\,Dy + y = 0$

27. $D^4y + 2\,D^2y + y = 0$ **28.** $2\,D^4y - y = 0$

In Exercises 29–32, find the particular solutions of the given differential equations that satisfy the given conditions.

29. $y'' + 2y' + 10y = 0$; $y = 0$ when $x = 0$ and $y = e^{-\pi/6}$ when $x = \pi/6$

30. $9\,D^2y + 16y = 0$; $Dy = 0$ and $y = 2$ when $x = \pi/2$

31. $D^2y - 8\,Dy + 16y = 0$; $Dy = 2$ and $y = 4$ when $x = 0$

32. $D^4y + 3\,D^3y + 2\,D^2y = 0$; $y = 0$, $Dy = 4$, $D^2y = -8$, $D^3y = 16$ when $x = 0$

In Exercises 33–36, find the simplest form of the second-order homogeneous linear differential equation that has the given solution. In Exercises 34 and 35, explain how the equation is found.

33. $y = c_1e^{3x} + c_2e^{-3x}$ **W** **34.** $y = c_1e^{3x} + c_2xe^{3x}$

W **35.** $y = c_1\cos 3x + c_2\sin 3x$

36. $y = c_1e^{2x}\cos x + c_2e^{2x}\sin x$

15-3 SOLUTIONS OF NONHOMOGENEOUS EQUATIONS

We now consider the solution of a nonhomogeneous linear equation of the form

$$a_0D^2y + a_1Dy + a_2y = b \tag{15-8}$$

where b is a function of x or is a constant. When the solution is substituted into the left side, we must obtain b. Solutions found from the methods of Sections 15-1 and 15-2 give zero when substituted into the left side, but they do contain the arbitrary constants necessary in the solution. If we could find a particular solution that when substituted into the left side produced b, it could be added to the solution containing the arbitrary constants. Therefore, *the solution is of the form*

$$y = y_c + y_p \tag{15-9}$$

where y_c, called the **complementary solution,** *is obtained by solving the corresponding homogeneous equation and where y_p is the* **particular solution** *necessary to produce the expression b of Eq. (15-8). It should be noted that y_p satisfies the differential equation, but it has no arbitrary constants and therefore cannot be the general solution. The arbitrary constants are part of y_c.*

EXAMPLE 1   The differential equation $D^2y - Dy - 6y = e^x$ has the solution

$$y = c_1e^{3x} + c_2e^{-2x} - \tfrac{1}{6}e^x$$

where the complementary solution y_c and particular solution y_p are

$$y_c = c_1e^{3x} + c_2e^{-2x} \qquad y_p = -\tfrac{1}{6}e^x$$

The complementary solution y_c is obtained by solving the corresponding homogeneous equation $D^2y - Dy - 6y = 0$, and we shall discuss below the method of finding y_p. Again we note that y_c contains the arbitrary constants, y_p contains the expression needed to produce the e^x on the right, and therefore both are needed to have the complete general solution. ∎

By inspecting the form of b on the right side of the equation, we can find the form that the particular solution must have. Since a combination of the particular solution and its derivatives must form the function b,

CAUTION ▶   *y_p is an expression that contains all possible forms of b and its derivatives.*

The method that is used to find the exact form of y_p is called the **method of undetermined coefficients**.

EXAMPLE 2   (a) If the function b is $4x$, we choose the particular solution y_p to be of the form $y_p = A + Bx$. The Bx-term is included to account for the $4x$. Since the derivative of Bx is a constant, the A-term is included to account for any first derivative of the Bx-term that may be present. Since the derivative of A is zero, no other terms are needed to account for higher-derivative terms of the Bx-term.

NOTE ▶   (b) If the function b is e^{2x}, we choose the form of the particular solution to be $y_p = Ce^{2x}$. *Since all derivatives of Ce^{2x} are a constant times e^{2x}, no other forms appear in the derivatives, and no other forms are needed in y_p.*

(c) If the function b is $4x + e^{2x}$, we choose the form of the particular solution to be $y_p = A + Bx + Ce^{2x}$. ∎

EXAMPLE 3   (a) If b is of the form $x^2 + e^{-x}$, we choose the particular solution to be of the form $y_p = A + Bx + Cx^2 + Ee^{-x}$.

(b) If b is of the form $xe^{-2x} - 5$, we choose the form of the particular solution to be $y_p = Ae^{-2x} + Bxe^{-2x} + C$.

(c) If b is of the form $x \sin x$, we choose the form of the particular solution to be $y_p = A \sin x + B \cos x + Cx \sin x + Ex \cos x$. All these types of terms occur in the derivatives of $x \sin x$. ∎

EXAMPLE 4   (a) If b is of the form $e^x + xe^x$, we would then choose y_p to be of the form $y_p = Ae^x + Bxe^x$. These terms occur for xe^x and its derivatives. *Since the form of the e^x-term of b is already included in Ae^x, we do not include another e^x-term in y_p.*

NOTE ▶   (b) In the same way, if b is of the form $2x + 4x^2$, we choose the form of y_p to be $y_p = A + Bx + Cx^2$. *These are the only forms that occur in either $2x$ or $4x^2$ and their derivatives.* ∎

Once we have determined the form of y_p, we have to find the numerical values of the coefficients A, B, ... The *method of undetermined coefficients* is to

NOTE ▶   *substitute the chosen form of y_p into the differential equation and equate the coefficients of like terms.*

■**EXAMPLE 5**   Solve the differential equation $D^2y - Dy - 6y = e^x$.

In this case the solution of the auxiliary equation $m^2 - m - 6 = 0$ gives us the roots $m_1 = 3$ and $m_2 = -2$. Thus,

$$y_c = c_1 e^{3x} + c_2 e^{-2x}$$

The proper form of y_p is $y_p = Ae^x$. This means that $Dy_p = Ae^x$ and $D^2y_p = Ae^x$. Substituting y_p and its derivatives into the differential equation, we have

$$Ae^x - Ae^x - 6Ae^x = e^x$$

To produce equality, the coefficients of e^x must be the same on each side of the equation. Thus

$$-6A = 1 \qquad \text{or} \qquad A = -1/6$$

Therefore, $y_p = -\frac{1}{6}e^x$. This gives the complete solution $y = y_c + y_p$,

$$y = c_1 e^{3x} + c_2 e^{-2x} - \frac{1}{6}e^x \qquad \text{see Example 1}$$

This solution checks when substituted into the original differential equation.   ■

■**EXAMPLE 6**   Solve the differential equation $D^2y + 4y = x - 4e^{-x}$.

In this case we have $m^2 + 4 = 0$, which gives us $m_1 = 2j$ and $m_2 = -2j$. Therefore, $y_c = c_1 \sin 2x + c_2 \cos 2x$.

The proper form of the particular solution is $y_p = A + Bx + Ce^{-x}$. Finding two derivatives and then substituting into the differential equation gives

$$y_p = A + Bx + Ce^{-x} \qquad Dy_p = B - Ce^{-x} \qquad D^2y_p = Ce^{-x}$$

$$D^2y + 4y = x - 4e^{-x} \qquad \text{differential equation}$$

$$(Ce^{-x}) + 4(A + Bx + Ce^{-x}) = x - 4e^{-x} \qquad \text{substituting}$$

$$Ce^{-x} + 4A + 4Bx + 4Ce^{-x} = x - 4e^{-x}$$

$$(4A) + (4B)x + (5C)e^{-x} = 0 + (1)x + (-4)e^{-x} \qquad \text{note coefficients}$$

CAUTION ▶ *Equating the constants, and the coefficients* of x and e^{-x} on either side gives

$$4A = 0 \qquad 4B = 1 \qquad 5C = -4$$
$$A = 0 \qquad B = 1/4 \qquad C = -4/5$$

This means that the particular solution is

$$y_p = \frac{1}{4}x - \frac{4}{5}e^{-x}$$

In turn this tells us that the complete solution is

$$y = c_1 \sin 2x + c_2 \cos 2x + \frac{1}{4}x - \frac{4}{5}e^{-x}$$

Substitution into the original differential equation verifies this solution.   ■

█EXAMPLE 7   Solve the differential equation $D^3y - 3D^2y + 2Dy = 10 \sin x$.

$m^3 - 3m^2 + 2m = 0, \qquad m(m - 1)(m - 2) = 0 \qquad m_1 = 0 \qquad m_2 = 1 \qquad m_3 = 2$ auxiliary equation

$\qquad\qquad\qquad\qquad\qquad y_c = c_1 + c_2 e^x + c_3 e^{2x}$ complementary solution

We now find the particular solution:

$y_p = A \sin x + B \cos x$ particular solution form

$Dy_p = A \cos x - B \sin x \qquad D^2y_p = -A \sin x - B \cos x \qquad D^3y_p = -A \cos x + B \sin x$ find three derivatives

$(-A \cos x + B \sin x) - 3(-A \sin x - B \cos x) + 2(A \cos x - B \sin x) = 10 \sin x$ substitute into differential equation

$\qquad\qquad\qquad (3A - B)\sin x + (A + 3B) \cos x = 10 \sin x$

$\qquad\qquad\qquad\qquad 3A - B = 10 \qquad A + 3B = 0$ equate coefficients

The solution of this system is $A = 3$, $B = -1$.

$\qquad\qquad y_p = 3 \sin x - \cos x$ particular solution

$\qquad\qquad y = c_1 + c_2 e^x + c_3 e^{2x} + 3 \sin x - \cos x$ complete general solution

Check this solution in the differential equation.   ----------- █

█EXAMPLE 8   Find the particular solution of $y'' + 16y = 2e^{-x}$ if $Dy = -2$ and $y = 1$ when $x = 0$.

In this case we must not only find y_c and y_p, but we must also evaluate the constants of y_c from the given conditions. The solution is as follows:

$\qquad\qquad D^2y + 16y = 2e^{-x}$ operator D form

$\qquad\qquad m^2 + 16 = 0, \qquad m = \pm 4j$ auxiliary equation

$\qquad\qquad\qquad y_c = c_1 \sin 4x + c_2 \cos 4x$ complementary solution

$\qquad\qquad\qquad y_p = Ae^{-x}$ particular solution form

$\qquad\qquad\qquad Dy_p = -Ae^{-x} \qquad D^2y_p = Ae^{-x}$

$Ae^{-x} + 16Ae^{-x} = 2e^{-x}$ substituting

$\qquad\qquad 17Ae^{-x} = 2e^{-x}, \qquad A = \dfrac{2}{17}$ equate coefficients

$\qquad\qquad\qquad y_p = \dfrac{2}{17} e^{-x}$

$\qquad\qquad\qquad y = c_1 \sin 4x + c_2 \cos 4x + \dfrac{2}{17} e^{-x}$ complete general solution

We now evaluate c_1 and c_2 from the given conditions:

$Dy = 4c_1 \cos 4x - 4c_2 \sin 4x - \dfrac{2}{17} e^{-x}$

$1 = c_1(0) + c_2(1) + \dfrac{2}{17}(1), \qquad c_2 = \dfrac{15}{17}$ $y = 1$ when $x = 0$

$-2 = 4c_1(1) - 4c_2(0) - \dfrac{2}{17}(1), \qquad c_1 = -\dfrac{8}{17}$ $Dy = -2$ when $x = 0$

$y = -\dfrac{8}{17} \sin 4x + \dfrac{15}{17} \cos 4x + \dfrac{2}{17} e^{-x}$ required particular solution

This solution checks when substituted into the differential equation.   ----------- █

A Special Case

It may happen that a term of the proposed y_p is similar to a term of y_c. Since any term of y_c gives zero when substituted in the differential equation, so will that term of the proposed y_p. This means the proposed y_p must be modified. Therefore,

if a term of the proposed y_p is similar to a term of y_c, any term of the proposed y_p, included to account for the similar term of the function b, must be multiplied by the smallest possible integral power of x such that any resulting term of y_p is not similar to the term of y_c.

The following example shows that this is not as involved as it sounds.

> From Eq. (15-8), we note that the function b is the function on the right side of the differential equation.

■**EXAMPLE 9**   Solve the differential equation $D^2y - 2Dy + y = x + e^x$.
We find that the auxiliary equation and complementary solution are

$$m^2 - 2m + 1 = 0, \qquad (m-1)^2 = 0 \qquad m_1 = 1 \qquad m_2 = 1$$
$$y_c = e^x(c_1 + c_2x)$$

Based on the function b on the right side, the *proposed* form of y_p is

$$y_p = A + Bx + Ce^x \qquad \text{proposed form}$$

We now note that the term Ce^x is similar to the term c_1e^x of y_c. Therefore, we must multiply the term Ce^x by the smallest power of x such that it is not similar to any term of y_c. If we multiply by x, the term becomes similar to c_2xe^x. Therefore, we must multiply Ce^x by x^2 such that y_p is

> Note that the $A + Bx$ are not multiplied by x^2 since they are not included in y_p to account for the e^x on the right.

$$y_p = A + Bx + Cx^2e^x \qquad \text{correct modified form}$$

Using this form of y_p, we now complete the solution.

$$Dy_p = B + Cx^2e^x + 2Cxe^x \qquad D^2y_p = Cx^2e^x + 2Cxe^x + 2Ce^x + 2Cxe^x = 2Ce^x + 4Cxe^x + Cx^2e^x$$
$$(2Ce^x + 4Cxe^x + Cx^2e^x) - 2(B + Cx^2e^x + 2Cxe^x) + (A + Bx + Cx^2e^x) = x + e^x$$
$$(A - 2B) + Bx + 2Ce^x = x + e^x$$
$$A - 2B = 0 \qquad B = 1 \qquad 2C = 1 \qquad A = 2 \qquad B = 1 \qquad C = 1/2$$
$$y_p = 2 + x + \tfrac{1}{2}x^2e^x$$
$$y = e^x(c_1 + c_2x) + 2 + x + \tfrac{1}{2}x^2e^x$$

EXERCISES *15-3*

In Exercises 1–12, solve the given differential equations. The form of y_p is given.

1. $D^2y - Dy - 2y = 4$ (Let $y_p = A$.)

2. $D^2y - Dy - 6y = 4x$ (Let $y_p = A + Bx$.)

3. $D^2y - y = 2 + x^2$ (Let $y_p = A + Bx + Cx^2$.)

4. $D^2y + 4Dy + 3y = 2 + e^x$ (Let $y_p = A + Be^x$.)

5. $y'' - 3y' = 2e^x + xe^x$ (Let $y_p = Ae^x + Bxe^x$.)

6. $y'' + y' - 2y = 8 + 4x + 2xe^{2x}$
(Let $y_p = A + Bx + Ce^{2x} + Exe^{2x}$.)

7. $9D^2y - y = \sin x$ (Let $y_p = A\sin x + B\cos x$.)

8. $D^2y + 4y = \sin x + 4$
(Let $y_p = A + B\sin x + C\cos x$.)

9. $\dfrac{d^2y}{dx^2} - 2\dfrac{dy}{dx} + y = 2x + x^2 + \sin 3x$
(Let $y_p = A + Bx + Cx^2 + E\sin 3x + F\cos 3x$.)

10. $D^2y - y = e^{-x}$ (Let $y_p = Axe^{-x}$.)

11. $D^2y + 4y = -12\sin 2x$ (Let $y_p = Ax\sin 2x + Bx\cos 2x$.)

12. $y'' - 2y' + y = 3 + e^x$ (Let $y_p = A + Bx^2e^x$.)

In Exercises 13–28, solve the given differential equations. Explain your method of solution for Exercise 25.

13. $\dfrac{d^2y}{dx^2} - \dfrac{dy}{dx} - 30y = 10$

14. $2\dfrac{d^2y}{dx^2} + 11\dfrac{dy}{dx} - 6y = 8x$

15. $3\dfrac{d^2y}{dx^2} + 13\dfrac{dy}{dx} - 10y = 14e^{3x}$

16. $\dfrac{d^2y}{dx^2} + 4y = 2\sin 3x$

17. $D^2y - 4y = \sin x + 2\cos x$

18. $2\,D^2y + 5\,Dy - 3y = e^x + 4e^{2x}$

19. $D^2y + y = 4 + \sin 2x$

20. $D^2y - Dy + y = x + \sin x$

21. $D^2y + 5\,Dy + 4y = xe^x + 4$

22. $3\,D^2y + Dy - 2y = 4 + 2x + e^x$

23. $y''' - y' = \sin 2x$

24. $D^4y - y = x$

$\textbf{W}$ 25. $D^2y + y = \cos x$

26. $4y'' - 4y' + y = 4e^{x/2}$

27. $D^2y + 2\,Dy = 8x + e^{-2x}$

28. $D^3y - Dy = 4e^{-x} + 3e^{2x}$

In Exercises 29–32, find the particular solution of each differential equation for the given conditions.

29. $D^2y - Dy - 6y = 5 - e^x$; $Dy = 4$ and $y = 2$ when $x = 0$

30. $3y'' - 10y' + 3y = xe^{-2x}$; $y' = -\frac{9}{35}$ and $y = -\frac{13}{35}$ when $x = 0$

31. $y'' + y = x + \sin 2x$; $y' = 1$ and $y = 0$ when $x = \pi$

32. $D^2y - 2\,Dy + y = xe^{2x} - e^{2x}$; $Dy = 4$ and $y = -2$ when $x = 0$

15-4 APPLICATIONS OF HIGHER-ORDER EQUATIONS

We now show important applications of second-order differential equations to simple harmonic motion and simple electric circuits. Also, we will show an application of a fourth-order differential equation to the deflection of a beam.

SIMPLE HARMONIC MOTION

■**EXAMPLE 1** Simple harmonic motion may be defined as motion in a straight line for which the acceleration is proportional to the displacement and in the opposite direction. Examples of this type of motion are a weight on a spring, a simple pendulum, and an object bobbing in water. If x represents the displacement, d^2x/dt^2 is the acceleration.

Using the definition of simple harmonic motion, we have

$$\frac{d^2x}{dt^2} = -k^2x$$

(We chose k^2 for convenience of notation in the solution.) We write this equation in the form

$$D^2x + k^2x = 0 \qquad \text{here, } D = d/dt$$

The roots of the auxiliary equation are kj and $-kj$, and the solution is

$$x = c_1 \sin kt + c_2 \cos kt$$

This solution indicates an oscillating motion, which is known to be the case. If, for example, $k = 4$ and we know that $x = 2$ and $Dx = 0$ (which means the velocity is zero) for $t = 0$, we have

$$Dx = 4c_1 \cos 4t - 4c_2 \sin 4t$$
$$2 = c_1(0) + c_2(1) \qquad x = 2 \text{ for } t = 0$$
$$0 = 4c_1(1) - 4c_2(0) \qquad Dx = 0 \text{ for } t = 0$$

which gives $c_1 = 0$ and $c_2 = 2$. Therefore,

$$x = 2 \cos 4t$$

is the equation relating the displacement and time; Dx is the velocity, and D^2x is the acceleration. See Fig. 15-1. ┈┈┈┈┈ ■

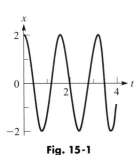

Fig. 15-1

EXAMPLE 2 In practice, an object moving with simple harmonic motion will in time cease to move due to unavoidable frictional forces. A "freely" oscillating object has a retarding force that is approximately proportional to the velocity. The differential equation for this case is $D^2x = -k^2x - b\,Dx$. This results from applying (from physics) Newton's second law of motion (see the margin note at the left). Again, using the operator $D^2x = d^2x/dt^2$, the term D^2x represents the acceleration of the object, the term $-k^2x$ is a measure of the restoring force (of the spring, for example), and the term $-b\,Dx$ represents the retarding (damping) force. This equation can be written as

$$D^2x + b\,Dx + k^2x = 0$$

The auxiliary equation is $m^2 + bm + k^2 = 0$, for which the roots are

$$m = \frac{-b \pm \sqrt{b^2 - 4k^2}}{2}$$

If $k = 3$ and $b = 4$, $m = -2 \pm j\sqrt{5}$, which means the solution is

$$x = e^{-2t}(c_1 \sin \sqrt{5}t + c_2 \cos \sqrt{5}t) \qquad (1)$$

Here, $4k^2 > b^2$, and this case is called **underdamped.** In this case the object oscillates as the amplitude becomes smaller.

If $k = 2$ and $b = 5$, $m = -1, -4$, which means the solution is

$$x = c_1e^{-t} + c_2e^{-4t} \qquad (2)$$

Here, $4k^2 < b^2$, and the case is called **overdamped.** It will be noted that the motion is not oscillatory, since no sine or cosine terms appear. In this case the object returns slowly to equilibrium without oscillating.

If $k = 2$ and $b = 4$, $m = -2, -2$, which means the solution is

$$x = e^{-2t}(c_1 + c_2t) \qquad (3)$$

Here, $4k^2 = b^2$, and the case is called **critically damped.** Again the motion is not oscillatory. In this case there is just enough damping to prevent any oscillations. The object returns to equilibrium in the minimum time.

See Fig. 15-2, in which Eqs. (1), (2), and (3) are represented in general. Of course, the actual values depend on c_1 and c_2, which in turn depend on the conditions imposed on the motion.

Newton's second law states that the net force acting on an object is equal to its mass times its acceleration. (This is one of Newton's best-known contributions to physics.)

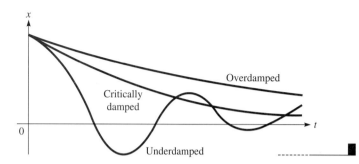

Fig. 15-2

SOLVING A WORD PROBLEM

EXAMPLE 3 In testing the characteristics of a particular type of spring, it is found that a weight of 16.0 lb stretches the spring 1.60 ft when the weight and spring are placed in a fluid that resists the motion with a force equal to twice the velocity. If the weight is brought to rest and then given a velocity of 12.0 ft/s, find the equation of motion. See Fig. 15-3.

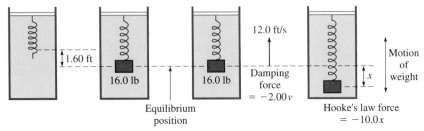

Fig. 15-3

In order to find the equation of motion, we use Newton's second law of motion (see Example 2). The weight (one force) at the end of the spring is offset by the equilibrium position force exerted by the spring, in accordance with Hooke's law (see Section 6-6). Therefore, the net force acting on the weight is the sum of the Hooke's law force due to the displacement from the equilibrium position and the resisting force. Using Newton's second law, we have

mass × acceleration = resisting force + Hooke's law force

$$m\,D^2x = -2.00\,Dx - kx$$

The mass of an object is its weight divided by the acceleration due to gravity. The weight is 16.0 lb, and the acceleration due to gravity is 32.0 ft/s². Thus, the mass m is

$$m = \frac{16.0\ \text{lb}}{32.0\ \text{ft/s}^2} = 0.500\ \text{slug}$$

where the slug is the unit of mass if the weight is in pounds.

The constant k for the Hooke's law force is found from the fact that the spring stretches 1.60 ft for a force of 16.0 lb. Thus, using Hooke's law,

$$16.0 = k(1.60), \qquad k = 10.0\ \text{lb/ft}$$

This means that the differential equation to be solved is

$$0.500\,D^2x + 2.00\,Dx + 10.0x = 0$$

or

$$1.00\,D^2x + 4.00\,Dx + 20.0x = 0$$

Solving this equation, we have

$$1.00m^2 + 4.00m + 20.0 = 0 \qquad \text{auxiliary equation}$$

$$m = \frac{-4.00 \pm \sqrt{16.0 - 4(20.0)(1.00)}}{2.00}$$

$$= -2.00 \pm 4.00j \qquad \text{complex roots}$$

$$x = e^{-2.00t}(c_1 \cos 4.00t + c_2 \sin 4.00t) \qquad \text{general solution}$$

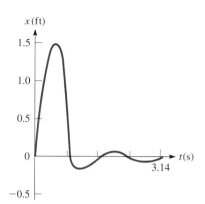

Fig. 15-4

Since the weight started from the equilibrium position with a velocity of 12.0 ft/s, we know that $x = 0$ and $Dx = 12.0$ for $t = 0$. Thus,

$$0 = e^0(c_1 + 0c_2) \quad \text{or} \quad c_1 = 0 \quad x = 0 \text{ for } t = 0$$

Thus, since $c_1 = 0$, we have

$$x = c_2 e^{-2.00t} \sin 4.00t$$

$$Dx = c_2 e^{-2.00t}(\cos 4.00t)(4.00) + c_2 \sin 4.00t(e^{-2.00t})(-2.00)$$

$$12.0 = c_2 e^0(1)(4.00) + c_2(0)(e^0)(-2.00) \quad Dx = 12.0 \text{ for } t = 0$$

$$c_2 = 3.00$$

This means that the equation of motion is

$$x = 3.00e^{-2.00t} \sin 4.00t$$

This motion is underdamped; the graph is shown in Fig. 15-4. ------

In Example 3 the force was given in pounds, and the mass was therefore expressed in slugs. If metric units are used, it is common to give the mass of the object in kilograms, and its weight is therefore in newtons. The weight, which is needed to determine the constant k in the Hooke's law force, is found by multiplying the mass, in kilograms, by the acceleration of gravity, 9.80 m/s^2.

It is possible to have an additional force acting on a weight such as the one in Example 3. For example, a vibratory force may be applied to the support of the spring. In such a case, called **forced vibrations,** this additional external force is added to the other net force. This means that the added force $F(t)$ becomes a nonzero function on the right side of the differential equation, and we must then solve a nonhomogeneous equation.

ELECTRIC CIRCUITS

■**EXAMPLE 4**   The impressed voltage in an electric circuit equals the sum of the voltages across the components of the circuit. For a circuit with a resistance R, an inductance L, a capacitance C, and a voltage source E (see Fig. 15-5), we have

$$L\frac{d^2q}{dt^2} + R\frac{dq}{dt} + \frac{q}{C} = E \tag{15-10}$$

By definition q represents the electric charge, $dq/dt = i$ is the current, and d^2q/dt^2 is the time rate of change of current. This equation may be written as

$$LD^2q + RDq + q/C = E$$

The auxiliary equation is $Lm^2 + Rm + 1/C = 0$. The roots are

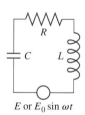

Fig. 15-5

$$m = \frac{-R \pm \sqrt{R^2 - 4L/C}}{2L} = -\frac{R}{2L} \pm \sqrt{\frac{R^2}{4L^2} - \frac{1}{LC}}$$

If we let $a = R/2L$ and $\omega = \sqrt{1/LC - R^2/4L^2}$, we have (assuming complex roots, which corresponds to realistic values of R, L, and C)

$$q_c = e^{-at}(c_1 \sin \omega t + c_2 \cos \omega t)$$

This indicates an oscillating charge, or an alternating current. However, the exponential term usually is such that the current dies out rapidly unless there is a source of voltage in the circuit. ------

If there is no source of voltage in the circuit of Example 4, we have a homogeneous differential equation to solve. If we have a constant voltage source, the particular solution is of the form $q_p = A$. If there is an alternating voltage source, the particular solution is of the form $q_p = A \sin \omega_1 t + B \cos \omega_1 t$, where ω_1 is the angular velocity of the source. After a very short time, the exponential factor in the complementary solution makes it negligible. For this reason it is referred to as the **transient** term, and *the particular solution is the* **steady-state** *solution*. Therefore, to find the steady-state solution, we need find only the particular solution.

■**EXAMPLE 5** An electric circuit is being analyzed on an oscilloscope. The circuit contains the elements $C = 400$ μF, $L = 1.00$ H, $R = 10.0$ Ω, and a voltage source of 500 sin 100t. See Fig. 15-6. Find the steady-state solution for the current.

This means the differential equation to be solved is

$$\frac{d^2q}{dt^2} + 10\frac{dq}{dt} + \frac{10^4}{4}q = 500 \sin 100t$$

See the chapter introduction.

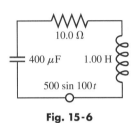

Fig. 15-6

Since we wish to find the steady-state solution, we must find q_p, from which we may find i_p by finding a derivative. The solution now follows:

$$q_p = A \sin 100t + B \cos 100t \qquad \text{particular solution form}$$

$$\frac{dq_p}{dt} = 100A \cos 100t - 100B \sin 100t$$

$$\frac{d^2q_p}{dt^2} = -10^4A \sin 100t - 10^4B \cos 100t$$

$$-10^4A \sin 100t - 10^4B \cos 100t + 10^3A \cos 100t - 10^3B \sin 100t \qquad \text{substitute into differential equation}$$

$$+ \frac{10^4}{4}A \sin 100t + \frac{10^4}{4}B \cos 100t = 500 \sin 100t$$

$$(-0.75 \times 10^4A - 10^3B)\sin 100t + (-0.75 \times 10^4B + 10^3A)\cos 100t = 500 \sin 100t$$

$$-7.5 \times 10^3A - 10^3B = 500 \qquad \text{equate coefficients of sin 100}t$$

$$10^3A - 7.5 \times 10^3B = 0 \qquad \text{equate coefficients of cos 100}t$$

In the 1880s it was decided that alternating current (favored by George Westinghouse) would be used to distribute electric power. Thomas Edison had argued for the use of direct current.

Solving these equations, we obtain

$$B = -8.73 \times 10^{-3} \qquad \text{and} \qquad A = -65.5 \times 10^{-3}$$

Therefore,

$$q_p = -65.5 \times 10^{-3} \sin 100t - 8.73 \times 10^{-3} \cos 100t$$

$$i_p = \frac{dq_p}{dt} = -6.55 \cos 100t + 0.87 \sin 100t$$

The use of inductance, the henry, is named for the U.S. physicist Joseph Henry (1797–1878).

which is the required solution. (We assumed three significant digits for the data but did not use all of them in most equations of the solution.) --------■

It should be noted that the complementary solutions of the mechanical and electric cases are of identical form. There is also an equivalent mechanical case to that of an impressed sinusoidal voltage source in the electric case. This arises in the case of forced vibrations, when an external force affecting the vibrations is applied to the system. Thus, we may have transient and steady-state solutions to mechanical and other nonelectric situations.

DEFLECTION OF BEAMS

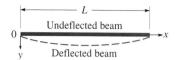

Fig. 15-7

In the study of the strength of materials and elasticity it is shown that the deflection y of a beam of length L satisfies the differential equation $EI\, d^4y/dx^4 = w(x)$, where EI is a measure of the stiffness of the beam and $w(x)$ is the weight distribution along the beam. See Fig. 15-7. Since this is a fourth-order equation, it is necessary to specify four conditions to obtain a solution. These conditions are determined by the way in which the ends, where $x = 0$ and where $x = L$, are held. For an end held in the specified manner, these conditions are: *clamped:* $y = 0$ and $y' = 0$: *hinged:* $y = 0$ and $y'' = 0$; *free:* $y'' = 0$ and $y''' = 0$ ($y' = 0$ indicates no change in alignment; $y'' = 0$ indicates no curvature; $y''' = 0$ indicates no shearing force). Since the conditions are given for specific positions, this kind of problem is called a *boundary value problem.* Consider the following example.

SOLVING A WORD PROBLEM

▌**EXAMPLE 6**   A uniform beam of length L is hinged at both ends and has a constant load distribution of w due to its own weight. Find the deflection y of the beam in terms of the distance x from one end.

Using the differential equation given above, we have $EI\, d^4y/dx^4 = w$. For convenience in the solution, let $k = w/EI$. Thus, the solution is as follows:

$$D^4y = k$$
$$m^4 = 0 \qquad m_1 = m_2 = m_3 = m_4 = 0$$

Since the four roots of the auxiliary equation are equal,

$$y_c = c_1 + c_2x + c_3x^2 + c_4x^3$$

The form of y_c indicates that we must multiply the *proposed* $y_p = A$ by x^4 so that it is not similar to any of the terms of y_c. This gives us $y_p = Ax^4$. Therefore,

$$y_p = Ax^4 \qquad Dy_p = 4Ax^3 \qquad D^2y_p = 12Ax^2 \qquad D^3y_p = 24Ax \qquad D^4y_p = 24A$$
$$24A = k, \qquad A = k/24$$

$$y = c_1 + c_2x + c_3x^2 + c_4x^3 + \frac{k}{24}x^4 \qquad \text{general solution}$$

We now find four known conditions to evaluate the four constants from the discussion above the example. For a beam hinged at both ends, at both $x = 0$ and $x = L$, $y = 0$ and $D^2y = 0$. We now find four derivatives and use these conditions.

$$Dy = c_2 + 2c_3x + 3c_4x^2 + \frac{k}{6}x^3 \qquad D^2y = 2c_3 + 6c_4x + \frac{k}{2}x^2 \qquad \text{find derivatives}$$

$$D^3y = 6c_4 + kx \qquad D^4y = k$$

At $x = 0$, $y = 0$: $c_1 = 0$;   At $x = 0$, $D^2y = 0$: $c_3 = 0$ use conditions to evaluate constants

At $x = L$, $y = 0$: $0 = c_2L + c_4L^3 + \dfrac{kL^4}{24}$

At $x = L$, $D^2y = 0$: $0 = 6c_4L + \dfrac{kL^2}{2}$

$$c_4 = -\frac{kL}{12} \qquad 0 = c_2L + \left(-\frac{kL}{12}\right)L^3 + \frac{kL^4}{24} \qquad c_2 = \frac{kL^3}{24}$$

$$y = \frac{kL^3}{24}x - \frac{kL}{12}x^3 + \frac{kx^4}{24} = \frac{k}{24}(L^3x - 2Lx^3 + x^4) \qquad \text{particular solution}$$

$$= \frac{w}{24EI}(L^3x - 2Lx^3 + x^4) \qquad\qquad k = w/EI \quad ▪$$

EXERCISES 15-4

1. When the angular displacement θ of a pendulum is small (less than about 6°), the pendulum moves with simple harmonic motion closely approximated by $D^2\theta + \dfrac{g}{l}\theta = 0$. Here, $D = d/dt$, g is the acceleration due to gravity, and l is the length of the pendulum. Find θ as a function of time (in s) if $g = 9.8$ m/s^2, $l = 1.0$ m, $\theta = 0.1$, and $D\theta = 0$ when $t = 0$. Sketch the curve.

2. A block of wood floating in oil is depressed from its equilibrium position such that its equation of motion is $D^2y + 8\,Dy + 3y = 0$, where y is the displacement (in in.) and $D = d/dt$. Find its displacement after 12 s if $y = 6.0$ in. and $Dy = 0$ when $t = 0$.

3. A car suspension is depressed from its equilibrium position such that its equation of motion is $D^2y + b\,Dy + 25y = 0$, where y is the displacement and $D = d/dt$. What must be the value of b if the motion is critically damped?

4. A mass of 0.820 kg stretches a given spring by 0.250 m. The mass is pulled down 0.150 m below the equilibrium position and released. Find the equation of motion of the mass if there is no damping.

5. A 4.00-lb weight stretches a certain spring 0.125 ft. With this weight attached, the spring is pulled 3.00 in. longer than its equilibrium length and released. Find the equation of the resulting motion, assuming no damping.

6. Find the solution for the spring of Exercise 5 if a damping force numerically equal to the velocity is present.

7. Find the solution for the spring of Exercise 5 if no damping is present but an external force of $4 \sin 2t$ is acting on the spring.

8. Find the solution for the spring of Exercise 5 if the damping force of Exercise 6 and the impressed force of Exercise 7 are both acting.

9. Find the equation relating the charge and the time in an electric circuit with the following elements: $L = 0.200$ H, $R = 8.00$ Ω, $C = 1.00$ μF, and $E = 0$. In this circuit, $q = 0$ and $i = 0.500$ A when $t = 0$.

10. For a given electric circuit, $L = 2$ mH, $R = 0$, $C = 50$ nF, and $E = 0$. Find the equation relating the charge and the time if $q = 10^5$ C and $i = 0$ when $t = 0$.

11. For a given circuit, $L = 0.100$ H, $R = 0$, $C = 100$ μF, and $E = 100$ V. Find the equation relating the charge and the time if $q = 0$ and $i = 0$ when $t = 0$.

12. Find the relation between the current and the time for the circuit of Exercise 11.

13. For a radio tuning circuit, $L = 0.500$ H, $R = 10.0$ Ω, $C = 200$ μF, and $E = 120 \sin 120\pi t$. Find the equation relating the charge and time.

14. Find the steady-state current for the circuit of Exercise 13.

15. In a given electric circuit $L = 8.00$ mH, $R = 0$, $C = 0.500$ μF, and $E = 20.0e^{-200t}$ mV. Find the relation between the current and the time if $q = 0$ and $i = 0$ for $t = 0$.

16. Find the current as a function of time for a circuit in which $L = 0.400$ H, $R = 60.0$ Ω, $C = 0.200$ μF, and $E = 0.800e^{-100t}$ V, if $q = 0$ and $i = 5.00$ mA for $t = 0$.

17. Find the steady-state current for a circuit with $L = 1.00$ H, $R = 5.00$ Ω, $C = 150$ μF, and $E = 120 \sin 100t$ V.

18. Find the steady-state solution for the current in an electric circuit containing the following elements: $C = 20.0$ μF, $L = 2.00$ H, $R = 20.0$ Ω, and $E = 200 \sin 10t$.

19. A *cantilever* beam is clamped at the end $x = 0$ and is free at the end $x = L$. Find the equation for the deflection y of the beam in terms of the distance x from one end if it has a constant load distribution of w due to its own weight. See Fig. 15-8.

Fig. 15-8

20. A beam 10 m in length is hinged at both ends and has a variable load distribution of $w = kEIx$, where $k = 7.2 \times 10^{-4}$/m and x is the distance from one end. Find the equation of the deflection y in terms of x.

CHAPTER EQUATIONS

General linear differential equation	$a_0\dfrac{d^n y}{dx^n} + a_1\dfrac{d^{n-1}y}{dx^{n-1}} + \cdots + a_{n-1}\dfrac{dy}{dx} + a_n y = b$	(15-1)
	$a_0 D^n y + a_1 D^{n-1} y + \cdots + a_{n-1}Dy + a_n y = b$	(15-2)
Homogeneous linear differential equation	$a_0 D^2 y + a_1 Dy + a_2 y = 0$	(15-3)
Auxiliary equation	$a_0 m^2 + a_1 m + a_2 = 0$	(15-4)
Distinct roots	$y = c_1 e^{m_1 x} + c_2 e^{m_2 x}$	(15-5)
Repeated roots	$y = e^{mx}(c_1 + c_2 x)$	(15-6)
Complex roots	$y = e^{\alpha x}(c_1 \sin \beta x + c_2 \cos \beta x)$	(15-7)
Nonhomogeneous linear differential equation	$a_0 D^2 y + a_1 Dy + a_2 y = b$	(15-8)
	$y = y_c + y_p$	(15-9)
Electric circuit	$L\dfrac{d^2 q}{dt^2} + R\dfrac{dq}{dt} + \dfrac{q}{C} = E$	(15-10)

REVIEW EXERCISES

In Exercises 1–20, find the general solution of each of the given differential equations.

1. $2D^2 y + Dy = 0$

2. $2D^2 y - 5Dy + 2y = 0$

3. $y'' + 2y' + y = 0$

4. $y'' + 2y' + 2y = 0$

5. $D^2 y + 2Dy + 6y = 0$

6. $4D^2 y - 4Dy + y = 0$

7. $\dfrac{d^2 y}{dx^2} = 0.08y - 0.2\dfrac{dy}{dx}$

8. $\dfrac{d^2 y}{dx^2} = 0.2\dfrac{dy}{dx} - 0.17y$

9. $2\dfrac{d^2 y}{dx^2} + \dfrac{dy}{dx} - 3y = 6$

10. $\dfrac{d^2 y}{dx^2} + 6\dfrac{dy}{dx} + 9y = 3x$

11. $y'' + y' - y = 2e^x$

12. $4D^3 y + 9Dy = xe^x$

13. $9D^2 y - 18Dy + 8y = 16 + 4x$

14. $y'' + y = 4\cos 2x$

15. $D^3 y - D^2 y + 9Dy - 9y = \sin x$

16. $y'' + y' = e^x + \cos 2x$

17. $y'' - 7y' - 8y = 2e^{-x}$

18. $3y'' - 6y' = 4 + xe^x$

19. $D^2 y + 25y = 50\cos 5x$

20. $D^2 y + 4y = 8x\sin 2x$

In Exercises 21–24, find the indicated particular solutions of the given differential equations.

21. $\dfrac{d^2 y}{dx^2} + \dfrac{dy}{dx} + 4y = 0$; $Dy = \sqrt{15}$, $y = 0$ when $x = 0$

22. $5y'' + 7y' - 6y = 0$; $y' = 10$, $y = 2$ when $x = 0$

23. $D^2 y + 4Dy + 4y = 4\cos x$; $Dy = 1$, $y = 0$ when $x = 0$

24. $y'' - 2y' + y = e^x + x$; $y = 0$, $y' = 0$ when $x = 0$

In Exercises 25–36, solve the given problems.

(W) 25. A certain spring stretches 0.50 m by a 40-N weight. With this weight suspended on it, the spring is stretched 0.50 m beyond the equilibrium position and released. Find the equation of the resulting motion if the medium in which the weight is suspended retards the motion with a force equal to 16 times the velocity. Classify the motion as underdamped, critically damped, or overdamped. Explain your choice. (Assume $g = 10$ m/s^2.)

26. Find the solution for the spring in Exercise 25 if there is no force retarding the motion.

27. A mass of 1.25 kg stretches a spring by 0.200 m. From the equilibrium position the mass is given a velocity of 4.50 m/s. Find the equation of motion if there is a resisting force that is numerically equal to four times the velocity.

28. A pendulum moves with simple harmonic motion according to the equation $D^2\theta + 0.2D\theta + (g/l)\theta = 0$ ($D = d/dt$). Find θ as a function of time t if $g = 9.8$ m/s^2, $l = 1.0$ m, $\theta = 0.10$, and $D\theta = 0$ when $t = 0$. Sketch the curve.

29. The end of a vibrating rod moves according to the equation $D^2y + 0.2\,Dy + 4000y = 0$, where y is the displacement and $D = d/dt$. Find y as a function of t if $y = 3.00$ cm and $Dy = -0.300$ cm/s when $t = 0$.

30. What is the expression for the steady-state displacement of x of an object that moves according to the equation $D^2x - 2Dx - 8x = 5 \sin 2t$?

31. A 0.5-H inductor, a 6-Ω resistor, and a 20-mF capacitor are connected in series with a generator for which $E = 24 \sin 10t$. Find the charge on the capacitor as a function of time if the initial charge and initial current are zero.

32. A 5.00-mH inductor and a 10.0-μF capacitor are connected in series with a voltage source of $0.200e^{-200t}$ V. Find the charge on the capacitor as a function of time if $q = 0$ and $i = 4.00$ mA when $t = 0$.

33. Find the equation for the current as a function of time if a resistance of 20 Ω, an inductor of 4 H, a capacitor of 100 μF, and a battery of 100 V are in series. The initial charge on the capacitor is 10 mC, and the initial current is zero.

34. If an electric circuit contains an inductance L, a capacitor with a capacitance C, and a sinusoidal source of voltage $E_0 \sin \omega t$, express the charge q on the capacitor as a function of the time. Assume $q = 0$, $i = 0$ when $t = 0$.

35. The approximate differential equation relating the displacement y of a beam at a horizontal distance x from one end is $EI\dfrac{d^2y}{dx^2} = M$, where E is the modulus of elasticity, I is the moment of inertia of the cross section of the beam perpendicular to its axis, and M is the bending moment at the cross section. If $M = 2000x - 40x^2$ for a particular beam of length L for which $y = 0$ when $x = 0$ and when $x = L$, express y in terms of x. Consider E and I as constants.

36. When a circular disk of mass m and radius r is suspended by a wire at the center of one of its flat faces and the disk is twisted through an angle θ, torsion in the wire tends to turn the disk back in the opposite direction. The differential equation for this case is $\dfrac{1}{2}mr^2\dfrac{d^2\theta}{dt^2} = -k\theta$, where k is a constant. Determine the equation of motion if $\theta = \theta_0$ and $d\theta/dt = \omega_0$ when $t = 0$. See Fig. 15-9.

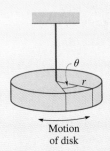

θ

r

Motion of disk

Fig. 15-9

Writing Exercise

37. The solutions for Example 3 on page 454 and Example 4 on page 455 are very similar. This means that the oscillation of the object on the spring is analogous to the motion of the electric charge in the *RLC* circuit. Write two or three paragraphs describing the mechanical and electrical quantities that are equivalent for these examples.

PRACTICE TEST

In Problems 1–4, find the general solution of each of the given differential equations.

1. $D^3y + 81\,Dy = 0$

2. $2\,D^2y - Dy = 2 \cos x$

3. $\dfrac{d^2y}{dx^2} - 4\dfrac{dy}{dx} + 4y = 3x$

4. $D^2y - 2\,Dy - 8y = 4e^{-2x}$

5. Find the particular solution of the differential equation $D^2y - 25y = 5x$, if $y = 3$ and $Dy = -5$ when $x = 0$.

6. A 16-lb weight stretches a certain spring 0.5 ft. With this weight attached, the spring is pulled 0.3 ft longer than its equilibrium length and released. Find the equation of the resulting motion, assuming no damping. (The acceleration due to gravity is 32 ft/s^2.)

CHAPTER 16

16 OTHER METHODS OF SOLVING DIFFERENTIAL EQUATIONS

In the previous two chapters we presented methods of solving certain types of differential equations. Other types of differential equations, many of which do not have exact solutions, have important technical and scientific applications. Numerous methods and techniques have been developed for solving these equations.

In the first section we present two methods of numerically solving differential equations. Such methods lend themselves well to calculators and computers. In the second section we present a method of finding a function that approximates the exact solution.

In the last two sections we present the method of *Laplace transforms*. Using this method, particular solutions are found by essentially algebraic means. Laplace transforms are of particular importance in solving equations related to electrical circuits and mechanical systems.

In Section 16-4 we show how Laplace transforms are used in solving differential equations for electric circuits that might be used in testing components used in various appliances such as radios.

$16\text{-}1$ NUMERICAL SOLUTIONS

Numerical methods of solving differential equations often require a great deal of time and computation to arrive at the desired solution. For this reason they lend themselves well to the use of computers and calculators, especially programmable calculators, by which the calculations can be done very rapidly. Also, since computers are used throughout industry, numerical methods are important in industry. The methods presented in this section were developed before the computer age and therefore can be used with any method of computation.

Euler's Method

To find an approximate solution to a differential equation of the form $dy/dx = f(x, y)$, that passes through a known point (x_0, y_0), we write the equation as $dy = f(x, y) \, dx$ and then approximate dy as $y_1 - y_0$, and replace dx with Δx. From Section 4-8, we recall that Δy closely approximates dy for a small dx and that $dx = \Delta x$. This gives us

$$y_1 = y_0 + f(x_0, y_0) \Delta x \qquad \text{and} \qquad x_1 = x_0 + \Delta x$$

Therefore, we now know another point (x_1, y_1) that is on (or very nearly on) the curve of the solution. We can now repeat this process using (x_1, y_1) as a known point to obtain a next point (x_2, y_2). Continuing this process, we can get a series of points that are approximately on the solution curve. The method is called ***Euler's method.***

■**EXAMPLE 1** For the differential equation $dy/dx = x + y$, use Euler's method to find the y-values of the solution for $x = 0$ to $x = 0.5$ with $\Delta x = 0.1$, if the curve of the solution passes through $(0, 1)$.

Using the method outline above, we have $x_0 = 0$, $y_0 = 1$, and

$$y_1 = 1 + (0 + 1)(0.1) = 1.1 \qquad \text{and} \qquad x_1 = 0 + 0.1 = 0.1$$

This tells us that the curve passes (or nearly passes) through the point $(0.1, 1.1)$. Assuming this point is correct, we use it to find the next point on the curve.

$$y_2 = 1.1 + (0.1 + 1.1)(0.1) = 1.22 \qquad \text{and} \qquad x_2 = 0.1 + 0.1 = 0.2$$

Therefore, the next approximate point is $(0.2, 1.22)$. Continuing this process we find a set of points that would approximately satisfy the function that is the solution of the differential equation. Tabulating results, we have the following table.

x	y	Correct value of y
0.0	1.0000	1.0000
0.1	1.1000	1.1103
0.2	1.2200	1.2428
0.3	1.3620	1.3997
0.4	1.5282	1.5836
0.5	1.7210	1.7974

the values shown have been rounded off, although more digits were carried in the calculations

In this case we are able to find the correct values since the equation can be written as $dy/dx - y = x$, and the solution is $y = 2e^x - x - 1$. Although numerical methods are generally used with equations that cannot be solved exactly, we chose this equation so that we could compare values obtained with known values.

We can see that as x increases, the error in y increases. More accurate values can be found by using smaller values of Δx. In Fig. 16-1 the solution curves using $\Delta x = 0.1$ and $\Delta x = 0.05$ are shown along with the correct values of y.

Euler's method is easy to use and understand, but it is less accurate than other methods. We will show one of the more accurate methods in the next example. ■

Named for the Swiss mathematician Leonhard Euler (1707–1783).

See Appendix C for a graphing calculator program EULRMETH. It gives numerical values of the solution of a differential equation using Euler's method.

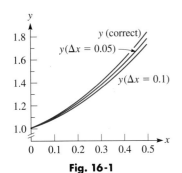

Fig. 16-1

Runge–Kutta Method

Named for the German mathematicians Carl Runge (1856–1927) and Martin Kutta (1867–1944).

For more accurate numerical solutions of a differential equation, the ***Runge–Kutta method*** is often used. Starting at a first point (x_0, y_0), the coordinates of the second point (x_1, y_1) are found by using a weighted average of the slopes calculated at the points where $x = x_0$, $x = x_0 + \frac{1}{2}\Delta x$, and $x = x_0 + \Delta x$. The formulas for y_1 and x_1 are

$$y_1 = y_0 + \frac{1}{6}H(J + 2K + 2L + M) \qquad \text{and} \qquad x_1 = x_0 + H \qquad \text{(for convenience, } H = \Delta x)$$

where

$$J = f(x_0, y_0)$$
$$K = f(x_0 + 0.5H, y_0 + 0.5HJ)$$
$$L = f(x_0 + 0.5H, y_0 + 0.5HK)$$
$$M = f(x_0 + H, y_0 + HL)$$

We have used uppercase letters to correspond to calculator use. Traditional sources normally use h for H and a lowercase letter (such as k) with subscripts for J, K, L, and M, and express 0.5 as 1/2.

As with Euler's method, once (x_1, y_1) is determined we use the formulas again to find (x_2, y_2) by replacing (x_0, y_0) with (x_1, y_1). The following example illustrates the use of the Runge–Kutta method.

See Appendix C for a graphing calculator program RUNGKUTT. It gives numerical values of the solution of a differential equation using the Runge–Kutta method.

■EXAMPLE 2 For the differential equation $dy/dx = x + \sin xy$, use the Runge–Kutta method to find y-values of the solution for $x = 0$ to $x = 0.5$ with $\Delta x = 0.1$, if the curve of the solution passes through $(0, 0)$.

Using the formulas and method outlined above, we have the following solution, with calculator notes to the right of the equations. Also, calculator symbols are used on the right sides of the equations to indicate the way in which they should be entered.

$$x_0 = 0$$
$$y_0 = 0$$
$$H = 0.1$$
$$J = X + \sin XY = 0 \qquad \text{store as } J$$
$$K = X + 0.5H + \sin[(X + 0.5H)(Y + 0.5HJ)] = 0.05 \qquad \text{store as } K$$
$$L = X + 0.5H + \sin[(X + 0.5H)(Y + 0.5HK)] = 0.050125 \qquad \text{store as } L$$
$$M = X + H + \sin[(X + H)(Y + HL)] = 0.10050125 \qquad \text{store as } M$$
$$y_1 = Y + (H/6)(J + 2K + 2L + M) = 0.0050125208 \qquad \text{store as } Y$$
$$x_1 = X + H = 0.1 \qquad \text{store as } X$$

We now use (x_1, y_1) as we just used (x_0, y_0) to get the next point (x_2, y_2). Following is a table indicating the values that are obtained on the calculator. The graph of these points is shown in Fig. 16-2.

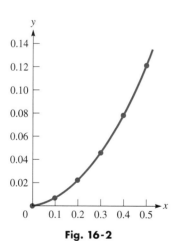

Fig. 16-2

x	y
0.0	0.0
0.1	0.0050125208
0.2	0.0202013395
0.3	0.0460278455
0.4	0.0832868181
0.5	0.1331460062

EXERCISES $16\text{-}1$

In Exercises 1–8, use Euler's method to find y-values of the solution for the given values of x and Δx, if the curve of the solution passes through the given point. Check the results against known values by solving the differential equations exactly. Plot the graphs of the solutions in Exercises 1–4.

1. $\dfrac{dy}{dx} = x + 1$; $x = 0$ to $x = 1$; $\Delta x = 0.2$; $(0, 1)$

2. $\dfrac{dy}{dx} = \sqrt{2x + 1}$; $x = 0$ to $x = 1.2$; $\Delta x = 0.3$; $(0, 2)$

3. $\dfrac{dy}{dx} = y(0.4x + 1)$; $x = -0.2$ to $x = 0.3$; $\Delta x = 0.1$; $(-0.2, 2)$

4. $\dfrac{dy}{dx} = y + e^x$; $x = 0$ to $x = 0.5$; $\Delta x = 0.1$; $(0, 0)$

5. The differential equation of Exercise 1 with $\Delta x = 0.1$

6. The differential equation of Exercise 2 with $\Delta x = 0.1$

7. The differential equation of Exercise 3 with $\Delta x = 0.05$

8. The differential equation of Exercise 4 with $\Delta x = 0.05$

In Exercises 9–14, use the Runge–Kutta method to find y-values of the solution for the given values of x and Δx, if the curve of the solution passes through the given point.

9. $\dfrac{dy}{dx} = xy + 1$; $x = 0$ to $x = 0.4$; $\Delta x = 0.1$; $(0, 0)$

10. $\dfrac{dy}{dx} = x^2 + y^2$; $x = 0$ to $x = 0.4$; $\Delta x = 0.1$; $(0, 1)$

11. $\dfrac{dy}{dx} = e^{xy}$; $x = 0$ to $x = 1$; $\Delta x = 0.2$; $(0, 0)$

12. $\dfrac{dy}{dx} = \sqrt{1 + xy}$; $x = 0$ to $x = 0.2$; $\Delta x = 0.05$; $(0, 1)$

13. $\dfrac{dy}{dx} = \cos(x + y)$; $x = 0$ to $x = 0.5$; $\Delta x = 0.1$; $(0, \pi/2)$

14. $\dfrac{dy}{dx} = y + \sin x$; $x = 0.5$ to $x = 1.0$; $\Delta x = 0.1$; $(0.5, 0)$

In Exercises 15 and 16, solve the given problems.

15. An electric circuit contains a 1-H inductor, a 2-Ω resistor, and a voltage source of $\sin t$. The resulting differential equation relating the current i and the time t is $di/dt + 2i = \sin t$. Find i after 0.5 s by Euler's method with $\Delta t = 0.1$ s if the initial current is zero. Solve the equation exactly and compare the values.

16. An object is being heated such that the rate of change of the temperature T (in °C) with respect to the time t (in min) is $dT/dt = \sqrt[3]{1 + t^3}$. Find T for $t = 5$ min by using the Runge–Kutta method with $\Delta t = 1$ min, if the initial temperature is 0°C.

$16\text{-}2$ A METHOD OF SUCCESSIVE APPROXIMATIONS

Another method of solving a differential equation to be discussed in this chapter is one that is based on successive approximations. In the previous section we saw how to develop numerical methods based essentially on successive approximations. In this section we shall demonstrate a method that leads to a function that approximates the solution.

The method applies to a differential equation of the form $y' = f(x, y)$. If both coordinates of one point of the solution are known, the value of the y-coordinate is substituted into the right side, thereby giving a function of x alone. This equation can then be integrated, yielding y as a function of x. This solution is referred to as the *first approximation*. The expression for this first approximation is then substituted into the right side of the original differential equation for y, and by integrating again, a *second approximation* is obtained. This process can be repeated as often as required, with each approximation being better than the one before it. The following examples illustrate the method.

EXAMPLE 1   Obtain the first three approximations to the solution of the differential equation $dy = (x + y)\, dx$, if the solution curve passes through $(0, 1)$.

By substituting $y = 1$ in the right side of the differential equation, we have

$$dy = (x + 1)\, dx$$

Integrating, we obtain the first approximation

$$y_1 = \frac{x^2}{2} + x + 1$$

where the constant of integration has been evaluated as 1 from the fact that the curve passes through $(0, 1)$. Substituting the expression for y_1 in the differential equation, we have

$$dy = \left(x + \frac{x^2}{2} + x + 1\right) dx = \left(\frac{x^2}{2} + 2x + 1\right) dx$$

$$y_2 = \frac{x^3}{6} + x^2 + x + 1$$

where the constant of integration again is 1 and y_2 is the second approximation. Substituting the expression for y_2 in the differential equation, we have

$$dy = \left(x + \frac{x^3}{6} + x^2 + x + 1\right) dx = \left(\frac{x^3}{6} + x^2 + 2x + 1\right) dx$$

$$y_3 = \frac{x^4}{24} + \frac{x^3}{3} + x^2 + x + 1$$

where, again, the constant of integration is 1 and y_3 is the third approximation

We note that this differential equation is the same as that of Example 1 of the previous section. Therefore, the exact solution is

$$y = 2e^x - x - 1$$

Figure 16-3 shows the graph of the exact solution and each of the three approximations. The accuracy of these approximations can be seen on the graph. ∎

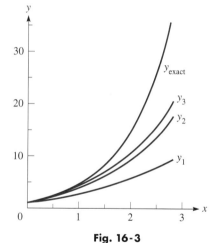

Fig. 16-3

EXAMPLE 2   Find the first two approximations of the solution of the differential equation $y' = 2 - xy^2$ if the solution passes through $(0, 1)$.

Writing the equation in differential form and substituting 1 for y, we have

$$dy = (2 - x)\, dx$$

which when integrated (the constant of integration is 1) gives

$$y_1 = 1 + 2x - \frac{x^2}{2}$$

Substituting this expression for y in the differential equation, we have

$$dy = \left[2 - x\left(1 + 2x - \frac{x^2}{2}\right)^2\right] dx$$

$$= \left(2 - x - 4x^2 - 3x^3 + 2x^4 - \frac{x^5}{4}\right) dx$$

$$y_2 = 1 + 2x - \frac{x^2}{2} - \frac{4x^3}{3} - \frac{3x^4}{4} + \frac{2x^5}{5} - \frac{x^6}{24}$$

where the constant of integration is 1 and y_2 is the desired approximation ∎

EXAMPLE 3   Find the third approximation to the solution of the differential equation $y' = y + \sin x$, if the solution curve passes through $(0, 1/2)$. Evaluate the function for $x = 1$.

Following the method as previously illustrated we have the following steps:

$$dy = \left(\frac{1}{2} + \sin x\right) dx$$

$$y_1 = \frac{x}{2} - \cos x + C_1$$

Since the curve passes through $(0, \frac{1}{2})$ we have

$$\frac{1}{2} = 0 - 1 + C_1 \quad \text{or} \quad C_1 = \frac{3}{2}$$

Thus

$$y_1 = \frac{3}{2} + \frac{x}{2} - \cos x$$

$$dy = \left(\frac{3}{2} + \frac{x}{2} - \cos x + \sin x\right) dx$$

$$y_2 = \frac{3x}{2} + \frac{x^2}{4} - \sin x - \cos x + C_2$$

Here $C_2 = \frac{3}{2}$, which means

$$y_1 = \frac{3}{2} + \frac{3x}{2} + \frac{x^2}{4} - \sin x - \cos x$$

$$dy = \left(\frac{3}{2} + \frac{3x}{2} + \frac{\cdot x^2}{4} - \cos x\right) dx$$

$$y_3 = \frac{3x}{2} + \frac{3x^2}{4} + \frac{x^3}{12} - \sin x + C_2$$

Here, $C_3 = \frac{1}{2}$, which means

$$y_3 = \frac{1}{2} + \frac{3x}{2} + \frac{3x^2}{4} + \frac{x^3}{12} - \sin x$$

Evaluating y_3 for $x = 1$, we have

$$y_3 = \frac{1}{2} + \frac{3}{2} + \frac{3}{4} + \frac{1}{12} - 0.84 = 1.99$$

The differential equation for this example can be solve exactly using methods developed in Chapter 14, and the solution is

$$y = e^x - \tfrac{1}{2}(\sin x + \cos x)$$

Evaluating this solution for $x = 1$, we find that $y = 2.03$. This shows that the third approximation is reasonably close for $x = 1$.

In finding the approximations, it should also be noted that the constant of integration was not the same for each of the approximations.   ------------- ∎

The method of this section has the disadvantage that it can be applied only if the resulting differentials can be integrated. This naturally limits the equations that can be solved in this manner. However, for those for which it is applicable, it often provides a reasonably quick and accurate value of y for a specified value of x.

EXERCISES 16-2

In Exercises 1–4, find the indicated approximation of the solution of each of the given differential equations for which the curve of the solution passes through the given point. Solve each exactly and evaluate the solution for the indicated value of x.

1. $y' = y^2$, y_2, $(0, 1)$, $x = 0.1$

2. $y' = x - y$, y_2, $(0, 1)$, $x = \frac{1}{2}$

3. $y' = 2x(1 + y)$, y_3, $(0, 0)$, $x = 1$

4. $y' = y + e^x$, y_3, $(0, 0)$, $x = 0.2$

In Exercises 5–8, find the indicated approximation of the solution of each of the given differential equations for which the curve of the solution passes through the given point.

5. $y' = x + y^2$, y_3, $(0, 0)$

6. $y' = 1 + xy^2$, y_2, $(0, 0)$

7. $y' = y - \cos 2x$, y_2, $(0, 1)$

8. $y' = x + y \cos x$, y_2, $(0, 1)$

In Exercises 9–12, solve the given problems.

9. Substitute the first few terms of the Maclaurin series for e^x in the expression $2e^x - x - 1$ and compare the results with the solution for Example 1.

10. Substitute the first few terms of the Maclaurin series for e^{x^2} in the expression for the exact solution to the differential equation of Exercise 3 and compare the results with the approximate solution.

11. A 1-slug object moves through a resisting medium such that the motion is resisted by a force equal to the cube of the velocity. If a 2-lb force causes the motion, the resulting differential equation is $dv/dt = 2 - v^3$. Find the velocity when $t = 0.3$ s if the object starts from rest. Use the second approximation of the solution.

(W) 12. Evaluate i for $t = 0.5$ s for the differential equation in Exercise 15 of Section 16-1 by finding i_2. Compare the results with those found in the indicated exercise. Discuss the accuracy of the different methods.

16-3 LAPLACE TRANSFORMS

Named for the French mathematician and astronomer Pierre Laplace (1749–1827).

The final method of solving differential equations that we shall discuss is by use of **Laplace transforms.** As we shall see, *Laplace transforms provide an **algebraic** method of obtaining a **particular** solution of a differential equation from stated initial conditions.* This is frequently what is desired in practice, and therefore Laplace transforms are often preferred for the solution of differential equations in engineering and electronics.

In this section we discuss the meaning of Laplace transforms and certain operations with them. The methods introduced in this section will then be applied to solving differential equations and certain technical applications in the next section. The treatment in this text is intended only as an introduction to the topic of Laplace transforms and their use in applications.

The Laplace transform of a function $f(t)$ is defined as the function $F(s)$ by the equation

$$F(s) = \int_0^\infty e^{-st}f(t)\,dt \tag{16-1}$$

By writing the transform as $F(s)$, we show that the result of integrating and evaluating is a function of s. To denote that we are dealing with "the Laplace transform of the function $f(t)$," the notation $\mathscr{L}(f)$ is used. Thus,

$$F(s) = \mathscr{L}(f) = \int_0^\infty e^{-st}f(t)\,dt \tag{16-2}$$

We shall see that both notations are quite useful.

We note that the integrals in Eqs. (16-1) and (16-2) are improper integrals, which were introduced in Section 10-7. Therefore, we find that it is necessary to evaluate the improper integral

$$\lim_{c \to \infty} \int_0^c e^{-st}f(t)\,dt$$

in order to find the Laplace transform of the function $f(t)$. In the following two examples we illustrate the use of Eq. (16-2).

■EXAMPLE 1   Find the Laplace transform of the function $f(t) = t$, $t > 0$.

By the definition of the Laplace transform

$$\mathscr{L}(f) = \mathscr{L}(t) = \int_0^\infty e^{-st}t\,dt$$

For reference, Formula 44 is

$$\int u e^{au}\,du = \frac{e^{au}(au - 1)}{a^2}$$

This may be integrated by parts or by Formula (44) in Appendix E. Using the formula, we have

$$\mathscr{L}(t) = \int_0^\infty te^{-st}\,dt = \lim_{c \to \infty} \int_0^c te^{-st}\,dt = \lim_{c \to \infty} \left. \frac{e^{-st}(-st - 1)}{s^2} \right|_0^c$$

$$= \lim_{c \to \infty} \left[\frac{e^{-sc}(-sc - 1)}{s^2} \right] + \frac{1}{s^2}$$

Now, for $s > 0$, as $c \to \infty$, $e^{-sc} \to 0$ and $sc \to \infty$. However, although we cannot prove it here, $e^{-sc} \to 0$ much faster than $sc \to \infty$. We can see that this is reasonable, for $ce^{-c} = 4.5 \times 10^{-4}$ for $c = 10$ and $ce^{-c} = 3.7 \times 10^{-42}$ for $c = 100$. Thus, the value at the upper limit approaches zero, which means the limit is zero. This means that

$$\mathscr{L}(t) = \frac{1}{s^2}$$

As we noted, this transform is defined for $s > 0$.

EXAMPLE 2 Find the Laplace transform of the function $f(t) = \cos at$.
By definition,

$$\mathscr{L}(f) = \mathscr{L}(\cos at) = \int_0^\infty e^{-st} \cos at \, dt$$

For reference, Formula 50 is
$$\int e^{au} \cos bu \, du = \frac{e^{au}(a \cos bu + b \sin bu)}{a^2 + b^2}$$

Using Formula (50) in Appendix E, we have

$$\mathscr{L}(\cos at) = \int_0^\infty e^{-st} \cos at \, dt = \lim_{c \to \infty} \int_0^c e^{-st} \cos at \, dt$$

$$= \lim_{c \to \infty} \frac{e^{-st}(-s \cos at + a \sin at)}{s^2 + a^2} \Bigg|_0^c$$

$$= \lim_{c \to \infty} \frac{e^{-sc}(-s \cos ac + a \sin ac)}{s^2 + a^2} - \left(-\frac{s}{s^2 + a^2} \right)$$

$$= 0 + \frac{s}{s^2 + a^2} = \frac{s}{s^2 + a^2} \qquad (s > 0)$$

Therefore, the Laplace transform of the function $\cos at$ is

$$\mathscr{L}(\cos at) = \frac{s}{s^2 + a^2}$$

In both examples the resulting transform was an algebraic function of s.

We now present a short table of Laplace transforms. They are sufficient for our work in this chapter. More complete tables are available in many references.

TABLE OF LAPLACE TRANSFORMS

	$f(t) = \mathscr{L}^{-1}(F)$	$\mathscr{L}(f) = F(s)$		$f(t) = \mathscr{L}^{-1}(F)$	$\mathscr{L}(f) = F(s)$
1.	1	$\dfrac{1}{s}$	**11.**	te^{-at}	$\dfrac{1}{(s+a)^2}$
2.	$\dfrac{t^{n-1}}{(n-1)!}$	$\dfrac{1}{s^n}(n=1,2,3,\ldots)$	**12.**	$t^{n-1}e^{-at}$	$\dfrac{(n-1)!}{(s+a)^n}$
3.	e^{-at}	$\dfrac{1}{s+a}$	**13.**	$e^{-at}(1-at)$	$\dfrac{s}{(s+a)^2}$
4.	$1 - e^{-at}$	$\dfrac{a}{s(s+a)}$	**14.**	$[(b-a)t+1]e^{-at}$	$\dfrac{s+b}{(s+a)^2}$
5.	$\cos at$	$\dfrac{s}{s^2+a^2}$	**15.**	$\sin at - at \cos at$	$\dfrac{2a^3}{(s^2+a^2)^2}$
6.	$\sin at$	$\dfrac{a}{s^2+a^2}$	**16.**	$t \sin at$	$\dfrac{2as}{(s^2+a^2)^2}$
7.	$1 - \cos at$	$\dfrac{a^2}{s(s^2+a^2)}$	**17.**	$\sin at + at \cos at$	$\dfrac{2as^2}{(s^2+a^2)^2}$
8.	$at - \sin at$	$\dfrac{a^3}{s^2(s^2+a^2)}$	**18.**	$t \cos at$	$\dfrac{s^2-a^2}{(s^2+a^2)^2}$
9.	$e^{-at} - e^{-bt}$	$\dfrac{b-a}{(s+a)(s+b)}$	**19.**	$e^{-at} \sin bt$	$\dfrac{b}{(s+a)^2+b^2}$
10.	$ae^{-at} - be^{-bt}$	$\dfrac{s(a-b)}{(s+a)(s+b)}$	**20.**	$e^{-at} \cos bt$	$\dfrac{s+a}{(s+a)^2+b^2}$

An important property of many important transforms is the **linearity property,**

$$\mathcal{L}[af(t) + bg(t)] = a\mathcal{L}(f) + b\mathcal{L}(g) \qquad (16\text{-}3)$$

We state this property here since it determines that the transform of a sum of functions is the sum of the transforms. This is of definite importance when dealing with a sum of functions. This property is a direct result of the definition of the Laplace transform.

Another Laplace transform important to the solution of a differential equation is the transform of the derivative of a function. Let us first find the Laplace transform of the first derivative of a function.

By definition,

$$\mathcal{L}(f') = \int_0^\infty e^{-st} f'(t)\, dt$$

To integrate by parts, let $u = e^{-st}$ and $dv = f'(t)\, dt$, so $du = -se^{-st}\, dt$ and $v = f(t)$ (the integral of the derivative of a function is the function). Therefore,

$$\mathcal{L}(f') = e^{-st} f(t)\Big|_0^\infty + s\int_0^\infty e^{-st} f(t)\, dt$$

$$= 0 - f(0) + s\mathcal{L}(f)$$

It is noted that the integral in the second term on the right is the Laplace transform of $f(t)$ by definition. Therefore, *the Laplace transform of the first derivative of a function is*

$$\mathcal{L}(f') = s\mathcal{L}(f) - f(0) \qquad (16\text{-}4)$$

Applying the same analysis, we may find *the Laplace transform of the second derivative of a function. It is*

$$\mathcal{L}(f'') = s^2\mathcal{L}(f) - sf(0) - f'(0) \qquad (16\text{-}5)$$

Here it is necessary to integrate by parts twice to derive the result. The transforms of higher derivatives are found in a similar manner.

Equations (16-4) and (16-5) allow us to express the transform of each derivative in terms of s and the transform itself. This is illustrated in the following example.

■EXAMPLE 3 Given that $f(0) = 0$ and $f'(0) = 1$, express the transform of $f''(t) - 2f'(t)$ in terms of s and the transform of $f(t)$.

By using the linearity property and the transforms of the derivatives, we have

$$\mathcal{L}[f''(t) - 2f'(t)] = \mathcal{L}(f'') - 2\mathcal{L}(f') \qquad \text{using Eq. (16-3)}$$

$$= [s^2\mathcal{L}(f) - sf(0) - f'(0)] - 2[s\mathcal{L}(f) - f(0)] \qquad \text{using Eqs. (16-5) and (16-4)}$$

$$= [s^2\mathcal{L}(f) - s(0) - 1] - 2[s\mathcal{L}(f) - 0] \qquad \text{substitute given values}$$

$$= (s^2 - 2s)\mathcal{L}(f) - 1$$

Inverse Transforms

If the Laplace transform of a function is known, it is then possible to find the function by finding the **inverse transform,**

$$\mathcal{L}^{-1}(F) = f(t) \qquad\qquad (16\text{-}6)$$

where $\mathcal{L}^{-1}$ denotes the inverse transform.

EXAMPLE 4   If $F(s) = \dfrac{s}{s^2 + a^2}$, from Transform (5) of the table we see that

$$\mathcal{L}^{-1}(F) = \mathcal{L}^{-1}\left(\frac{s}{s^2 + a^2}\right) = \cos\ at$$

$$f(t) = \cos\ at$$

EXAMPLE 5   If $(s^2 - 2s)\mathcal{L}(f) - 1 = 0$, then

$$\mathcal{L}(f) = \frac{1}{s^2 - 2s} \qquad \text{or} \qquad F(s) = \frac{1}{s(s - 2)}$$

Therefore, we have

$$
\begin{aligned}
f(t) = \mathcal{L}^{-1}(F) &= \mathcal{L}^{-1}\left[\frac{1}{s(s - 2)}\right] && \text{inverse transform} \\[2mm]
&= -\frac{1}{2}\,\mathcal{L}^{-1}\left[\frac{-2}{s(s - 2)}\right] && \text{fit form of Transform (4)} \\[2mm]
&= -\frac{1}{2}(1 - e^{2t}) && \text{use Transform (4)}
\end{aligned}
$$

The introduction of the factor -2 in Example 5 illustrates that it often takes some algebra to get $F(s)$ to match the proper form in the table. The following examples show the use of completing the square (see Section 2-3) and partial fractions (see Sections 10-4 and 10-5) can be used to assist in making $F(s)$ fit a form in the table.

COMPLETING THE SQUARE

EXAMPLE 6   If $F(s) = \dfrac{s + 5}{s^2 + 6s + 10}$, then

$$\mathcal{L}^{-1}(F) = \mathcal{L}^{-1}\left[\frac{s + 5}{s^2 + 6s + 10}\right]$$

It appears that this function does not fit any of the forms given. However,

$$s^2 + 6s + 10 = (s^2 + 6s + 9) + 1 = (s + 3)^2 + 1$$

By writing $F(s)$ as

$$F(s) = \frac{(s + 3) + 2}{(s + 3)^2 + 1} = \frac{s + 3}{(s + 3)^2 + 1} + \frac{2}{(s + 3)^2 + 1}$$

we can find the inverse of each term. Therefore,

$$\mathcal{L}^{-1}(F) = e^{-3t}\cos t + 2e^{-3t}\sin t \qquad \text{using Transforms (20) and (19)}$$

$$f(t) = e^{-3t}(\cos t + 2\sin t)$$

PARTIAL FRACTIONS   ∎**EXAMPLE 7**   If $F(s) = \dfrac{5s^2 - 17s + 32}{s^3 - 8s^2 + 16s}$, then

$$\mathscr{L}^{-1}(F) = \mathscr{L}^{-1}\left[\frac{5s^2 - 17s + 32}{s^3 - 8s^2 + 16s}\right]$$

To fit forms in the table, we will now use partial fractions.

$$\frac{5s^2 - 17s + 32}{s^3 - 8s^2 + 16s} = \frac{5s^2 - 17s + 32}{s(s - 4)^2} = \frac{A}{s} + \frac{B}{s - 4} + \frac{C}{(s - 4)^2} \qquad \begin{array}{l}\text{factor of } s,\\ \text{repeated factor } s - 4\end{array}$$

$$5s^2 - 17s + 32 = A(s - 4)^2 + Bs(s - 4) + Cs \qquad \text{multiply each side by } s(s - 4)^2$$

$s = 0$: $32 = 16A$, $A = 2$

$s = 4$: $5(4^2) - 17(4) + 32 = 4C$, $C = 11$

s^2 terms: $5 = A + B$, $5 = 2 + B$, $B = 3$

$$\mathscr{L}^{-1}(F) = \mathscr{L}^{-1}\left[\frac{2}{s} + \frac{3}{s - 4} + \frac{11}{(s - 4)^2}\right] \qquad \text{substitute in } F(s)$$

$$\mathscr{L}^{-1}(F) = f(t) = 2 + 3e^{4t} + 11te^{4t} \qquad \text{using Transforms (1), (3), (11)}$$

EXERCISES 16-3

In Exercises 1–4, verify the indicated transforms given in the table.

1. Transform 1   **2.** Transform 3

3. Transform 6   **4.** Transform 11

In Exercises 5–12, find the transforms of the given functions by use of the table.

5. $f(t) = e^{3t}$ **6.** $f(t) = 1 - \cos 2t$

7. $f(t) = 5t^3e^{-2t}$ **8.** $f(t) = 2e^{-3t}\sin 4t$

9. $f(t) = \cos 2t - \sin 2t$

10. $f(t) = 2t\sin 3t + e^{-3t}\cos t$

11. $f(t) = 3 + 2t\cos 3t$ **12.** $f(t) = t^3 - 3te^{-t}$

In Exercises 13–16, express the transforms of the given expressions in terms of s and $\mathscr{L}(f)$.

13. $y'' + y'$, $f(0) = 0$, $f'(0) = 0$

14. $y'' - 3y'$, $f(0) = 2$, $f'(0) = -1$

15. $2y'' - y' + y$, $f(0) = 1$, $f'(0) = 0$

16. $y'' - 3y' + 2y$, $f(0) = -1$, $f'(0) = 2$

In Exercises 17–28, find the inverse transforms of the given functions of s.

17. $F(s) = \dfrac{2}{s^3}$ **18.** $F(s) = \dfrac{3}{s^2 + 4}$

19. $F(s) = \dfrac{1}{2s + 6}$ **20.** $F(s) = \dfrac{3}{s^4 + 4s^2}$

21. $F(s) = \dfrac{1}{s^3 + 3s^2 + 3s + 1}$ **22.** $F(s) = \dfrac{s^2 - 1}{s^4 + 2s^2 + 1}$

23. $F(s) = \dfrac{s + 2}{(s^2 + 9)^2}$ **24.** $F(s) = \dfrac{s + 3}{s^2 + 4s + 13}$

25. $F(s) = \dfrac{4s^2 - 8}{(s + 1)(s - 2)(s - 3)}$ **26.** $F(s) = \dfrac{3s + 1}{(s - 1)(s^2 + 1)}$

27. $F(s) = \dfrac{2s + 3}{s^2 - 2s + 5}$

Ⓦ **28.** $F(s) = \dfrac{3s^4 + 3s^3 + 6s^2 + s + 1}{s^5 + s^3}$ (Explain your method of solution.)

$16\text{-}4$ SOLVING DIFFERENTIAL EQUATIONS BY LAPLACE TRANSFORMS

We will now show how certain differential equations can be solved by using Laplace transforms. *It must be remembered that these solutions are the **particular** solutions of the equations subject to the given conditions.* The necessary operations were developed in the preceding section. The following examples illustrate the method.

EXAMPLE 1   Solve the differential equation $2y' - y = 0$, if $y(0) = 1$. (Note that we are using y to denote the function.)

Taking transforms of each term in the equation, we have

$$\mathscr{L}(2y') - \mathscr{L}(y) = \mathscr{L}(0)$$
$$2\mathscr{L}(y') - \mathscr{L}(y) = 0$$

$\mathscr{L}(0) = 0$ by direct use of the definition of the transform. Now, using Eq. (16-4), $\mathscr{L}(y') = s\mathscr{L}(y) - y(0)$, we have

$$2[s\mathscr{L}(y) - 1] - \mathscr{L}(y) = 0 \qquad y(0) = 1$$

Solving for $\mathscr{L}(y)$, we obtain

$$2s\mathscr{L}(y) - \mathscr{L}(y) = 2$$
$$\mathscr{L}(y) = \frac{2}{2s - 1} = \frac{1}{s - \frac{1}{2}}$$

Finding the inverse transform, we have

$$y = e^{t/2} \qquad \text{using Transform (3)}$$

It is interesting to note that logarithms and Laplace transforms were both developed to simplify mathematical solutions. Logarithms were developed to "transform" the more complex arithmetical procedures into simpler procedures (see page 270). Laplace transforms were developed to "transform" differential equations into the simpler algebraic forms for solution.

NOTE ▶

The reader should check this solution with that obtained by methods developed earlier. Also, it should be noted that the solution was essentially an algebraic one. This points out the power and usefulness of Laplace transforms. *We are able to translate a differential equation into an algebraic form,* which can in turn be translated into the solution of the differential equation. Thus, we can solve a differential equation by using algebra and specific algebraic forms. ┄┄┄┄■

EXAMPLE 2   Solve the differential equation $y'' + 2y' + 2y = 0$, if $y(0) = 0$ and $y'(0) = 1$.

Using the same steps as outlined in Example 1, we have

$$\mathscr{L}(y'') + 2\mathscr{L}(y') + 2\mathscr{L}(y) = 0 \qquad \text{take transforms}$$
$$[s^2\mathscr{L}(y) - sy(0) - y'(0)] + 2[s\mathscr{L}(y) - y(0)] + 2\mathscr{L}(y) = 0 \qquad \text{using Eqs. (16-5) and (16-4)}$$
$$[s^2\mathscr{L}(y) - s(0) - 1] + 2[s\mathscr{L}(y) - 0] + 2\mathscr{L}(y) = 0 \qquad \text{substitute given values}$$
$$s^2\mathscr{L}(y) - 1 + 2s\mathscr{L}(y) + 2\mathscr{L}(y) = 0$$
$$(s^2 + 2s + 2)\mathscr{L}(y) = 1 \qquad \text{solve for } \mathscr{L}(y)$$
$$\mathscr{L}(y) = \frac{1}{s^2 + 2s + 2} = \frac{1}{(s + 1)^2 + 1} \qquad \text{take inverse transform}$$
$$y = e^{-t} \sin t \qquad \text{using Transform (19)} \qquad ■$$

EXAMPLE 3 Using Laplace transforms, solve the differential equation $y'' + y = \cos t$, if $y(0) = 1$ and $y'(0) = 2$.

$$\mathcal{L}(y'') + \mathcal{L}(y) = \mathcal{L}(\cos t) \qquad \text{take transforms}$$

$$[s^2\mathcal{L}(y) - s(1) - 2] + \mathcal{L}(y) = \frac{s}{s^2 + 1} \qquad \text{using Eq. (16-5) and Transform (5)}$$

$$(s^2 + 1)\mathcal{L}(y) = \frac{s}{s^2 + 1} + s + 2$$

$$\mathcal{L}(y) = \frac{s}{(s^2 + 1)^2} + \frac{s}{s^2 + 1} + \frac{2}{s^2 + 1}$$

$$y = \frac{t}{2}\sin t + \cos t + 2\sin t \qquad \text{using Transforms (16), (5), (6)}$$

SOLVING A WORD PROBLEM

EXAMPLE 4 A spring is stretched 1 ft by a weight of 16 lb (mass of 1/2 slug). The medium resists the motion of the object with a force of $4v$, where v is the velocity of motion. The differential equation describing the displacement y is

$$\frac{1}{2}\frac{d^2y}{dt^2} + 4\frac{dy}{dt} + 16y = 0 \qquad \text{see Example 3, page 454}$$

Find y as a function of time t, if $y(0) = 1$ and $dy/dt = 0$ for $t = 0$.

Clearing fractions and denoting derivatives by y'' and y', we have the following differential equation and solution.

$$y'' + 8y' + 32y = 0$$

$$\mathcal{L}(y'') + 8\mathcal{L}(y') + 32\mathcal{L}(y) = 0 \qquad \text{take transforms}$$

$$[s^2\mathcal{L}(y) - s(1) - 0] + 8[s\mathcal{L}(y) - 1] + 32\mathcal{L}(y) = 0 \qquad \text{substitute given values}$$

$$(s^2 + 8s + 32)\mathcal{L}(y) = s + 8 \qquad \text{solve for } \mathcal{L}(y)$$

$$\mathcal{L}(y) = \frac{s + 8}{(s + 4)^2 + 4^2} = \frac{s + 4}{(s + 4)^2 + 4^2} + \frac{4}{(s + 4)^2 + 4^2} \qquad \text{fit transform forms}$$

$$y = e^{-4t}\cos 4t + e^{-4t}\sin 4t = e^{-4t}(\cos 4t + \sin 4t) \qquad \text{take inverse transforms}$$

The graph of this solution is shown in Fig. 16-4.

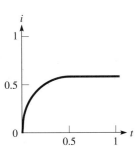

Fig. 16-4

EXAMPLE 5 The initial current in the circuit shown in Fig. 16-5 is zero. Find the current i as a function of the time t.

The differential equation for this circuit is

$$\frac{di}{dt} + 10i = 6 \qquad \text{using Eq. (15-10)}$$

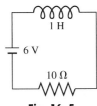

Fig. 16-5

Following the procedures outlined in the previous examples, the solution is found.

$$\mathcal{L}\left(\frac{di}{dt}\right) + 10\mathcal{L}(i) = \mathcal{L}(6) \qquad \text{take transforms}$$

$$[s\mathcal{L}(i) - 0] + 10\mathcal{L}(i) = \frac{6}{s} \qquad \text{substitute given values and find transform on right}$$

$$\mathcal{L}(i) = \frac{6}{s(s + 10)} \qquad \text{solve for } \mathcal{L}(i)$$

$$i = 0.6(1 - e^{-10t}) \qquad \text{take inverse transform}$$

Fig. 16-6

The graph of this solution is shown in Fig. 16-6.

See the chapter introduction.

FM radio was developed in the early 1930s.

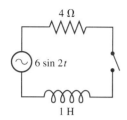

Fig. 16-7

$s = 0$: $12 = 4A + 4C$, $3 = A + C$
s terms: $0 = 4B + C$
s^2 terms: $0 = A + B$

EXAMPLE 6 An electric circuit in an FM radio transmitter contains a 1-H inductor and a 4-Ω resistor. It is being tested using a voltage source of $6 \sin 2t$. If the initial current is zero, find the current i as a function of the time t. See Fig. 16-7.

The solution is as follows:

$$(1)Di + 4i = 6 \sin 2t \qquad \text{differential equation, } D = d/dt$$

$$\mathcal{L}(Di) + 4\mathcal{L}(i) = 6\mathcal{L}(\sin 2t) \qquad \text{take transforms}$$

$$[s\mathcal{L}(i) - 0] + 4\mathcal{L}(i) = \frac{6(2)}{s^2 + 4} \qquad i(0) = 0$$

$$\mathcal{L}(i) = \frac{12}{(s + 4)(s^2 + 4)} = \frac{A}{s + 4} + \frac{Bs + C}{s^2 + 4} \qquad \text{use partial fractions}$$

$$12 = A(s^2 + 4) + B(s^2 + 4s) + C(s + 4)$$

The equations that give us the following values are shown in the margin at the left.

$$A = 0.6 \qquad B = -0.6 \qquad C = 2.4$$

$$\mathcal{L}(i) = 0.6\left(\frac{1}{s + 4}\right) - 0.6\left(\frac{s}{s^2 + 4}\right) + 1.2\left(\frac{2}{s^2 + 4}\right) \qquad \text{take inverse transforms}$$

$$i = 0.6e^{-4t} - 0.6 \cos 2t + 1.2 \sin 2t$$

This is checked by showing that $i(0) = 0$ and that it satisfies the original equation.

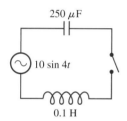

Fig. 16-8

$s = 0$: $400 = 16B + 200^2E$
s terms: $0 = 16A + 200^2C$
s^2 terms: $0 = B + E$
s^3 terms: $0 = A + C$

EXAMPLE 7 An electric circuit contains a 0.1-H inductor, a 250-μF capacitor, a voltage source of $10 \sin 4t$, and negligible resistance ($R = 0$). See Fig. 16-8. If the initial charge on the capacitor and initial current are zero, find the current as a function of time.

The solution is as follows:

$$0.1 D^2 q + \frac{1}{250 \times 10^{-6}} q = 10 \sin 4t \qquad \text{differential equation, } D = d/dt$$

$$D^2 q + 40{,}000q = 100 \sin 4t$$

$$\mathcal{L}(D^2 q) + 40{,}000\mathcal{L}(q) = 100\mathcal{L}(\sin 4t) \qquad \text{take transforms}$$

$$[s^2\mathcal{L}(q) - sq(0) - Dq(0)] + 40{,}000\mathcal{L}(q) = \frac{400}{s^2 + 16} \qquad q(0) = 0, D(q) = 0$$

$$\mathcal{L}(q) = \frac{400}{(s^2 + 200^2)(s^2 + 16)} = \frac{As + B}{s^2 + 200^2} + \frac{Cs + E}{s^2 + 16} \qquad \text{use partial fractions}$$

$$400 = (As + B)(s^2 + 16) + (Cs + E)(s^2 + 200^2)$$

The equations that give us the following values are shown in the margin at the left.

$$A = 0 \qquad B = -0.010 \qquad C = 0 \qquad E = 0.010$$

$$\mathcal{L}(q) = \frac{0.010}{s^2 + 16} - \frac{0.010}{s^2 + 200^2} = \frac{0.010}{4}\left(\frac{4}{s^2 + 16}\right) - \frac{0.010}{200}\left(\frac{200}{s^2 + 200^2}\right)$$

$$q = 0.0025 \sin 4t - 5.0 \times 10^{-5} \sin 200t \qquad \text{take inverse transforms}$$

$$i = 0.010 \cos 4t - 0.010 \cos 200t$$

EXERCISES *16-4*

In the following exercises, solve the given differential equations by using Laplace transforms, where the function is subject to the given conditions.

1. $y' + y = 0$, $y(0) = 1$

2. $y' - 2y = 0$, $y(0) = 2$

3. $2y' - 3y = 0$, $y(0) = -1$

4. $y' + 2y = 1$, $y(0) = 0$

5. $y' + 3y = e^{-3t}$, $y(0) = 1$

6. $y' + 2y = te^{-2t}$, $y(0) = 0$

7. $y'' + 4y = 0$, $y(0) = 0$, $y'(0) = 1$

8. $9y'' - 4y = 0$, $y(0) = 1$, $y'(0) = 0$

9. $y'' + 2y' = 0$, $y(0) = 0$, $y'(0) = 2$

10. $y'' + 2y' + y = 0$, $y(0) = 0$, $y'(0) = -2$

11. $y'' - 4y' + 5y = 0$, $y(0) = 1$, $y'(0) = 2$

12. $4y'' + 4y' + y = 0$, $y(0) = 1$, $y'(0) = 0$

13. $y'' + y = 1$, $y(0) = 1$, $y'(0) = 1$

14. $y'' + 4y = 2t$, $y(0) = 0$, $y'(0) = 0$

15. $y'' + 2y' + y = e^{-t}$, $y(0) = 1$, $y'(0) = 2$

16. $2y'' + 8y = 3 \sin 2t$, $y(0) = 0$, $y'(0) = 0$

17. $y'' - 4y = 10e^{3t}$, $y(0) = 5$, $y'(0) = 0$

18. $y'' - 2y' + y = e^{2t}$, $y(0) = 1$, $y'(0) = 3$

19. $y'' - y = 5 \sin 2t$, $y(0) = 0$, $y'(0) = 1$

20. $y'' + y' - 2y = \sin 3t$, $y(0) = 0$, $y'(0) = 0$

21. A constant force of 6 lb moves a 2-slug mass through a medium that resists the motion with a force equal to v, where v is the velocity. The differential equation relating the velocity and the time is $2\dfrac{dv}{dt} = 6 - v$. Find v as a function of t if the object starts from rest.

22. A pendulum moves with simple harmonic motion according to the differential equation $D^2\theta + 20\theta = 0$, where θ is the angular displacement and $D = d/dt$. Find θ as a function of t if $\theta = 0$ and $D\theta = 0.40$ rad/s when $t = 0$.

23. A 50-Ω resistor, a 4.0-μF capacitor, and a 40-V battery are connected in series. Find the charge on the capacitor as a function of the time t if the initial charge is zero.

24. A 2-H inductor, an 80-Ω resistor, and an 8-V battery are connected in series. Find the current in the circuit as a function of time if the initial current is zero.

25. A 10-H inductor, a 40-μF capacitor, and a voltage supply whose voltage is given by $100 \sin 50t$ are connected in series in an electric circuit. Find the current as a function of the time if the initial charge on the capacitor is zero and the initial current is zero.

26. A 20-mH inductor, a 40-Ω resistor, a 50-μF capacitor, and a voltage source of $100e^{-1000t}$ are connected in series in an electric circuit. Find the charge on the capacitor as a function of time t, if $q = 0$ and $i = 0$ when $t = 0$.

27. The weight on a spring undergoes forced vibrations according to the equation $D^2y + 9y = 18 \sin 3t$. Find its displacement y of the weight as a function of the time t, if $y = 0$ and $Dy = 0$ when $t = 0$.

28. A certain spring is stretched 1 m by a 2-kg (20-N) weight. The spring is stretched 0.5 m below the equilibrium position with the weight attached and then released. If the spring is in a medium that resists the motion with a force equal to $12v$, where v is the velocity, find the displacement y of the weight as a function of the time.

29. For the electric circuit shown in Fig. 16-9, find the current as a function of the time t if the initial current is zero.

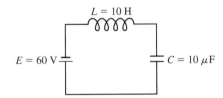

Fig. 16-9

30. For the electric circuit shown in Fig. 16-10, find the current as a function of the time t, if the initial charge on the capacitor is zero and the initial current is zero.

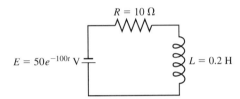

Fig. 16-10

31. For the electric circuit shown in Fig. 16-11, find the current as a function of the time t, if the initial charge on the capacitor is zero and the initial current is zero.

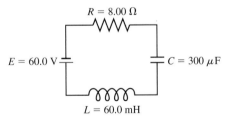

Fig. 16-11

(W) 32. For the beam in Example 6 on page 457, find the deflection y as a function of x, using Laplace transforms. The Laplace transform of the fourth derivative y^{iv} is given by

$$\mathcal{L}(f^{iv}) = s^4\mathcal{L}(f) - s^3f(0) - s^2f'(0) - sf''(0) - f'''(0)$$

Also, since $y'(0)$ and $y'''(0)$ are not given, but are constants, assume $y'(0) = a$, and $y'''(0) = b$. It is then possible to evaluate a and b to obtain the solution. Explain why choosing a and b in this way works.

CHAPTER EQUATIONS

Laplace transforms

$$F(s) = \int_0^\infty e^{-st} f(t)\, dt \tag{16-1}$$

$$F(s) = \mathcal{L}(f) = \int_0^\infty e^{-st} f(t)\, dt \tag{16-2}$$

$$\mathcal{L}[af(t) + bg(t)] = a\mathcal{L}(f) + b\mathcal{L}(g) \tag{16-3}$$

$$\mathcal{L}(f') = s\mathcal{L}(f) - f(0) \tag{16-4}$$

$$\mathcal{L}(f'') = s^2\mathcal{L}(f) - sf(0) - f'(0) \tag{16-5}$$

Inverse transform $\quad\mathcal{L}^{-1}(F) = f(t) \tag{16-6}$

REVIEW EXERCISES

In Exercises 1–8, solve the given differential equations by Euler's method for the specified values of x and Δx and for which the solution curve passes through the given point.

1. $y' = x^2 - y^2,\ 0 \le x \le 0.5,\ \Delta x = 0.1,\ (0, 1)$
2. $y' = \dfrac{1}{1 + x + y},\ 0 \le x \le 1,\ \Delta x = 0.2,\ (0, 0)$
3. $y' = \sqrt{x + y},\ 0 \le x \le 2,\ \Delta x = 0.2,\ (0, 1)$
4. $y' = x^2 + \sin y,\ 0 \le x \le 1,\ \Delta x = 0.1,\ (0, 1)$
5. $y' = \sin x + \tan y,\ 0 \le x \le 0.7,\ \Delta x = 0.1,\ (0, 0.5)$
6. $y' = e^{2x} + \sin xy,\ 0 \le x \le 1,\ \Delta x = 0.1,\ (0, 0)$
7. The equation of Exercise 1 with $\Delta x = 0.02$
8. The equation of Exercise 2 with $\Delta x = 0.05$

In Exercises 9–12, solve the given differential equations by the Runge–Kutta method for the specified values of x and Δx and for which the solution curve passes through the given point.

9. The equation of Exercise 1 for the given values
10. The equation of Exercise 2 for the given values
11. $y' = \sin x + \tan y,\ 0 \le x \le 0.4,\ \Delta x = 0.1,\ (0, 0.5)$
12. $y' = e^{2x} + \sin xy,\ 0 \le x \le 0.8,\ \Delta x = 0.2,\ (0, 0)$

In Exercises 13–16, find the indicated approximation of the solution of each of the given differential equations for which the solution curve passes through the given point.

13. $y' = y + x^2,\ y_3,\ (0, 1)$
14. $y' = \cos x - y,\ y_3,\ (0, 1)$
15. $y' = 1 + x^2 y,\ y_2,\ (1, 1)$
16. $y' = y^2 - \sin x,\ y_2,\ (0, 2)$

In Exercises 17–24, solve the given differential equations by use of Laplace transforms, where the function is subject to the given conditions.

17. $4y' - y = 0,\ y(0) = 1$
18. $2y' - y = 4,\ y(0) = 1$
19. $y' - 3y = e^t,\ y(0) = 0$
20. $y' + 2y = e^{-2t},\ y(0) = 2$
21. $y'' + y = 0,\ y(0) = 0,\ y'(0) = -4$
22. $y'' + 4y' + 5y = 0,\ y(0) = 1,\ y'(0) = 1$
23. $y'' + 9y = 3e^t,\ y(0) = 0,\ y'(0) = 0$
24. $y'' - 2y' + y = e^x + x,\ y(0) = 0,\ y'(0) = 1$

In Exercises 25–40, solve the given problems.

25. Find the second approximation to the solution of the differential equation $y' = 1 + ye^{-x}$ if the solution curve passes through the point $(0, 1)$. Evaluate y_2 for $x = 0.5$.
26. Repeat the instructions of Exercise 25 for $y' = 2x/(x^2 + e^y)$.
27. Solve $y' = 1 + ye^{-x}$ by Euler's method for $0 \le x \le 0.5$, using $\Delta x = 0.1$ if the solution curve passes through $(0, 1)$. Compare the value for $x = 0.5$ with that found in Exercise 25.
28. Find the solution for the equation of Exercise 27 by use of the Runge–Kutta method. Compare the value for $x = 0.5$ with those found in Exercises 25 and 27.
29. The differential equation relating the current and time for a certain electric circuit is $2\,di/dt + i = 12$. Solve this equation by use of Laplace transforms, given that the initial current is zero. Evaluate the current for $t = 0.300$ s.
30. A 6-H inductor and a 30-Ω resistor are connected in series with a voltage source of $10 \sin 20t$. Find the current as a function of time if the initial current is zero. Use Laplace transforms.

31. A 0.25-H inductor, a 4.0-Ω resistor, and a 100-μF capacitor are connected in series. If the initial charge on the capacitor is 400 μC and the initial current is zero, find the charge on the capacitor as a function of time. Use Laplace transforms.

32. An inductor of 0.5 H, a resistor of 6 Ω, and a capacitor of 200 μF are connected in series. If the initial charge on the capacitor is 10 mC and the initial current is zero, find the charge on the capacitor as a function of time after the switch is closed. Use Laplace transforms.

33. An 8.0-lb weight (1/4-slug mass) stretches a spring 6.0 in. An external force of cos 8t is applied to the spring. Express the displacement y of the object as a function of time if the initial displacement and velocity are zero. Use Laplace transforms.

34. A spring is stretched 1.00 m by a mass of 5.00 kg (assume the weight to be 50.0 N). Find the displacement y of the object as a function of time if $y(0) = 1$ m and $dy/dt = 0$ when $t = 0$. Use Laplace transforms.

35. Use Euler's method to find the current for $t = 0.300$ s for the equation of Exercise 29. Use $\Delta t = 0.05$.

36. Find the third approximation for i as a function of t for the equation of Exercise 29. Evaluate i for $t = 0.300$ s and compare this value with those found in Exercises 29 and 35.

37. Find the displacement y as a function of t for the spring and object of Exercise 33 with two conditions changed: The external force is removed, and the initial displacement is 9.0 in.

38. Find the solution of the spring system of Exercise 34 if it is in a medium that resists the motion with a force of 10v, where v is the velocity.

39. Assuming that only Transforms 1 through 6 in the table of transforms are known, derive the inverse Transform 9, given the form of $F(s)$.

40. Using the condition and method of Exercise 39, derive the inverse Transform 7.

Writing Exercise

41. Considering the differential equation for an *RLC* circuit, write one or two paragraphs describing the advantages and disadvantages of solving the equation by the methods of Chapter 15 and those of Laplace transforms.

PRACTICE TEST

In Problems 1–3, use the differential equation $dy/dx = (x + y)^2$, subject to the conditions that $y(0) = 1$.

1. Using Euler's method, find $y(0.2)$ with $\Delta x = 0.1$.

2. Using the Runge–Kutta method, find $y(0.2)$ with $\Delta x = 0.2$.

3. Using the method of successive approximations, find the first approximation y_1 and evaluate it for $x = 0.2$.

4. Find the Laplace transform of the function $f(t) = t^2 e^{2t}$.

5. Find the inverse Laplace transform of the function of s,

$$F(s) = \frac{10}{s^2 + 4s + 29}$$

6. Using Laplace transforms, solve the differential equation $D^2 y - Dy - 2y = 12$, if $y(0) = 0$ and $y'(0) = 0$.

7. An electric circuit contains an inductance of 2 H, a resistance of 8 Ω, and a battery of 6 V. Set up the differential equation for this circuit and solve it by use of Laplace transforms, given that the initial current is zero.

A P P E N D I X A — SUPPLEMENTARY TOPICS

A-1 ROTATION OF AXES

In Chapter 2 we discussed the circle, parabola, ellipse, and hyperbola and how these curves are represented by the second-degree equation

$$Ax^2 + Bxy + Cy^2 + Dx + Ey + F = 0 \qquad \text{(A1-1)}$$

Our discussion included the properties of the curves and their equations with center (vertex of a parabola) at the origin. However, except for the special case of the hyperbola $xy = c$, we did not cover what happens when the axes are rotated about the origin.

If a set of axes is rotated about the origin through an angle θ, as shown in Fig. A1-1, we say that there has been a **rotation of axes.** In this case each point P in the plane has two sets of coordinates, (x, y) in the original system and (x', y') in the rotated system.

If we now let r equal the distance from the origin O to point P and let ϕ be the angle between the x'-axis and the line OP, we have

$$x' = r \cos \phi \qquad y' = r \sin \phi \qquad \text{(A1-2)}$$
$$x = r \cos (\theta + \phi) \qquad y = r \sin (\theta + \phi) \qquad \text{(A1-3)}$$

Using the cosine and sine of the sum of two angles, we can write Eqs. (A1-3) as

$$x = r \cos \phi \cos \theta - r \sin \phi \sin \theta$$
$$y = r \cos \phi \sin \theta + r \sin \phi \cos \theta \qquad \text{(A1-4)}$$

Now, using Eqs. (A1-2), we have

$$x = x' \cos \theta - y' \sin \theta$$
$$y = x' \sin \theta + y' \cos \theta \qquad \text{(A1-5)}$$

In our derivation, we have used the special case when θ is acute and P is in the first quadrant of both sets of axes. When simplifying equations of curves using Eqs. (A1-5), we find that a rotation through a positive acute angle θ is sufficient. It can be shown, however, that Eqs. (A1-5) hold for any θ and position of P.

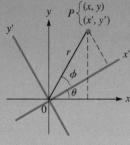

Fig. A1-1

479

▌EXAMPLE 1 Transform $x^2 - y^2 + 8 = 0$ by rotating the axes through 45°.

When $\theta = 45°$, the rotation Eqs. (A1-5) become

$$x = x' \cos 45° - y' \sin 45° = \frac{x'}{\sqrt{2}} - \frac{y'}{\sqrt{2}}$$

$$y = x' \sin 45° + y' \cos 45° = \frac{x'}{\sqrt{2}} + \frac{y'}{\sqrt{2}}$$

Substituting into the equation $x^2 - y^2 + 8 = 0$ gives

$$\left(\frac{x'}{\sqrt{2}} - \frac{y'}{\sqrt{2}}\right)^2 - \left(\frac{x'}{\sqrt{2}} + \frac{y'}{\sqrt{2}}\right)^2 + 8 = 0$$

$$\frac{1}{2}x'^2 - x'y' + \frac{1}{2}y'^2 - \frac{1}{2}x'^2 - x'y' - \frac{1}{2}y'^2 + 8 = 0$$

$$x'y' = 4$$

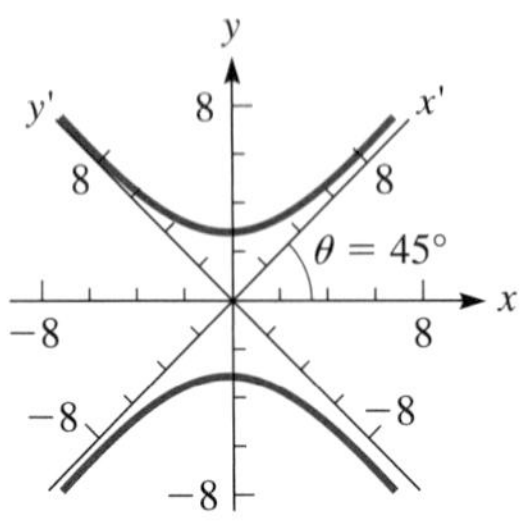

Fig. A1-2

The graph and both sets of axes are shown in Fig. A1-2. The original equation represents a hyperbola. We have the $xy = c$ form with rotation through 45°. ▌

When we showed the type of curve represented by the second-degree equation Eq. (A1-1) in Section 2-8, the standard forms of the parabola, ellipse, and hyperbola required that $B = 0$. This means that there is no xy-term in the equation. In Section 2-8, we considered only one case ($xy = c$) for which $B \neq 0$.

If we can remove the xy-term from a second-degree equation, the analysis of the graph is simplified. By a proper rotation of axes, we find that Eq. (A1-1) can be transformed into an equation that has no $x'y'$-term.

By substituting Eqs. (A1-5) into Eq. (A1-1) and then simplifying, we have

$$(A \cos^2 \theta + B \sin \theta \cos \theta + C \sin^2 \theta)x'^2 + [B \cos 2\theta - (A - C)\sin 2\theta]x'y'$$
$$+ (A \sin^2 \theta - B \sin \theta \cos \theta + C \cos^2 \theta)y'^2 + (D \cos \theta + E \sin \theta)x' + (E \cos \theta - D \sin \theta)y' + F = 0$$

If there is to be no $x'y'$-term, its coefficient must be zero. This means that $B \cos 2\theta - (A - C)\sin 2\theta = 0$, or

ANGLE OF ROTATION

$$\tan 2\theta = \frac{B}{A - C} \qquad (A \neq C) \tag{A1-6}$$

Equation (A1-6) gives the angle of rotation except when $A = C$. In this case the coefficient of the $x'y'$-term is $B \cos 2\theta$, which is zero if $2\theta = 90°$. Thus,

$$\theta = 45° \qquad (A = C) \tag{A1-7}$$

Consider the following example.

For reference: Eq. (7-4) is

$$\cos \theta = \frac{1}{\sec \theta}$$

Eq. (7-7) is

$$1 + \tan^2 \theta = \sec^2 \theta$$

Eq. (7-25) is

$$\sin \frac{\alpha}{2} = \pm \sqrt{\frac{1 - \cos \alpha}{2}}$$

Eq. (7-26) is

$$\cos \frac{\alpha}{2} = \pm \sqrt{\frac{1 + \cos \alpha}{2}}$$

EXAMPLE 2   By rotation of axes, transform $8x^2 + 4xy + 5y^2 = 9$ into a form without an xy-term. Identify and sketch the curve.

Here, $A = 8$, $B = 4$, and $C = 5$. Therefore, using Eq. (A1-6), we have

$$\tan 2\theta = \frac{4}{8 - 5} = \frac{4}{3}$$

Since $\tan 2\theta$ is positive, we may take 2θ as an acute angle, which means θ is also acute. For the transformation we need $\sin \theta$ and $\cos \theta$. We find these values by first finding the value of $\cos 2\theta$ and then using the half-angle formulas.

$$\cos 2\theta = \frac{1}{\sec 2\theta} = \frac{1}{\sqrt{1 + \tan^2 2\theta}} = \frac{1}{\sqrt{1 + (\frac{4}{3})^2}} = \frac{3}{5} \quad \begin{array}{l}\text{using Eqs. (7-4)}\\\text{and (7-7)}\end{array}$$

Now, using the half-angle formulas, Eqs. (7-25) and (7-26), we have

$$\sin \theta = \sqrt{\frac{1 - \cos 2\theta}{2}} = \sqrt{\frac{1 - \frac{3}{5}}{2}} = \frac{1}{\sqrt{5}}, \quad \cos \theta = \sqrt{\frac{1 + \cos 2\theta}{2}} = \sqrt{\frac{1 + \frac{3}{5}}{2}} = \frac{2}{\sqrt{5}}$$

Here, θ is about $26.6°$. Now substituting these values into Eqs. (A1-5), we have

$$x = x'\left(\frac{2}{\sqrt{5}}\right) - y'\left(\frac{1}{\sqrt{5}}\right) = \frac{2x' - y'}{\sqrt{5}}, \quad y = x'\left(\frac{1}{\sqrt{5}}\right) + y'\left(\frac{2}{\sqrt{5}}\right) = \frac{x' + 2y'}{\sqrt{5}}$$

Now, substituting into the equation $8x^2 + 4xy + 5y^2 = 9$ gives

$$8\left(\frac{2x' - y'}{\sqrt{5}}\right)^2 + 4\left(\frac{2x' - y'}{\sqrt{5}}\right)\left(\frac{x' + 2y'}{\sqrt{5}}\right) + 5\left(\frac{x' + 2y'}{\sqrt{5}}\right)^2 = 9$$

$$8(4x'^2 - 4x'y' + y'^2) + 4(2x'^2 + 3x'y' - 2y'^2) + 5(x'^2 + 4x'y' + 4y'^2) = 45$$

$$45x'^2 + 20y'^2 = 45$$

$$\frac{x'^2}{1} + \frac{y'^2}{\frac{9}{4}} = 1$$

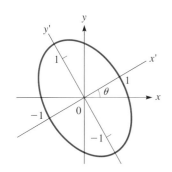

Fig. A1-3

This is an ellipse with semimajor axis of $3/2$ and semiminor axis of 1. See Fig. A1-3.   ∎

In Example 2, $\tan 2\theta$ was positive, and we made 2θ and θ positive. If, when using Eq. (A1-6), $\tan 2\theta$ is negative, we then make 2θ obtuse ($90° < 2\theta < 180°$). In this case $\cos 2\theta$ will be negative, but θ will be acute ($45° < \theta < 90°$).

In Section 2-7 we showed the use of translation of axes in writing an equation in standard form if $B = 0$. In this section we have seen how rotation of axes is used to eliminate the xy-term. It is possible that both a translation of axes and a rotation of axes are needed to write an equation in standard form.

In Section 2-8 we identified a conic section by inspecting the values of A and C when $B = 0$. If $B \neq 0$, these curves are identified as follows:

1. If $B^2 - 4AC = 0$, a parabola
2. If $B^2 - 4AC < 0$, an ellipse
3. If $B^2 - 4AC > 0$, a hyperbola

Special cases such as a point, parallel or intersecting lines, or no curve may result.

EXAMPLE 3   For the equation $16x^2 - 24xy + 9y^2 + 20x - 140y - 300 = 0$, identify the curve and simplify it to standard form. Sketch the graph and display it on a graphing calculator.

With $A = 16$, $B = -24$, and $C = 9$, using Eq. (A1-6), we have

$$\tan 2\theta = \frac{-24}{16 - 9} = -\frac{24}{7}$$

In this case $\tan 2\theta$ is negative, and we take 2θ to be an obtuse angle. We then find that $\cos 2\theta = -7/25$. In turn we find that $\sin \theta = 4/5$ and $\cos \theta = 3/5$. Here, θ is about $53.1°$. Using these values in Eqs. (A1-5), we find that

$$x = \frac{3x' - 4y'}{5} \qquad y = \frac{4x' + 3y'}{5}$$

Substituting these into the original equation and simplifying, we get

$$y'^2 - 4x' - 4y' - 12 = 0$$

This equation represents a parabola with its axis parallel to the x'-axis. The vertex is found by completing the square:

$$(y' - 2)^2 = 4(x' + 4)$$

The vertex is the point $(-4, 2)$ in the $x'y'$-rotated system. Therefore,

$$y''^2 = 4x''$$

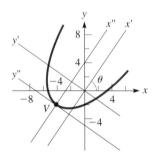

Fig. A1-4

is the equation in the $x''y''$-rotated and then translated system. The graph and the coordinate systems are shown in Fig. A1-4.

To display the curve on a graphing calculator, we solve for y by using the quadratic formula. Writing the equation as

$$9y^2 + (-24x - 140)y + (16x^2 + 20x - 300) = 0$$

Therefore, we see that in using the quadratic formula, $a = 9$, $b = -24x - 140$, and $c = 16x^2 + 20x - 300$. Now, solving for y, we have

$$y = \frac{24x + 140 \pm \sqrt{(-24x - 140)^2 - 4(9)(16x^2 + 20x - 300)}}{18}$$

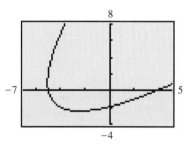

Fig. A1-5

We **enter both functions indicated by the ± sign** to get the display in Fig. A1-5.

EXERCISES A-1

In Exercises 1–4, transform the given equations by rotating the axes through the given angle. Identify and sketch each curve.

1. $x^2 - y^2 = 25$, $\theta = 45°$ **2.** $x^2 + y^2 = 16$, $\theta = 60°$

3. $8x^2 - 4xy + 5y^2 = 36$, $\theta = \tan^{-1} 2$

4. $2x^2 + 24xy - 5y^2 = 8$, $\theta = \tan^{-1} \frac{3}{4}$

In Exercises 5–10, transform each equation to a form without an xy-term by a rotation of axes. Identify and sketch each curve. Then display each curve on a graphing calculator.

5. $x^2 + 2xy + y^2 - 2x + 2y = 0$

6. $5x^2 - 6xy + 5y^2 = 32$

7. $3x^2 + 4xy = 4$

8. $9x^2 - 24xy + 16y^2 - 320x - 240y = 0$

9. $11x^2 - 6xy + 19y^2 = 20$

10. $x^2 + 4xy - 2y^2 = 6$

In Exercises 11 and 12, transform each equation to a form without an xy-term by a rotation of axes. Then transform the equation to a standard form by a translation of axes. Identify and sketch each curve. Then display each curve on a graphing calculator.

11. $16x^2 - 24xy + 9y^2 - 60x - 80y + 400 = 0$

12. $73x^2 - 72xy + 52y^2 + 100x - 200y + 100 = 0$

A-2 REGRESSION

Several methods of graphing functions have been developed in the earlier chapters. In this section we show how to start with a set of points and determine a function for which the graph passes through the given points. Our aim is to derive an equation that best "fits" given data obtained through experimentation or observation. Such equations can show a basic relationship between the variables and can be valuable in technology and industry for analyzing the properties of the quantities being measured.

We will show first a method of finding the equation of a straight line that passes through a set of data points, and in this way we *fit* the line to the points. In general, *the fitting of a curve to a set of points is called* **regression.** Fitting a straight line to a set of points is *linear regression,* and fitting some other type of curve to the points is *nonlinear regression.* Later in the section we consider nonlinear regression.

For a given set of several (at least 5 or 6) points representing pairs of data values, we cannot reasonably expect that the graph of any function will pass through all of the points *exactly.* Therefore, when we fit the graph of a function to such a set of points, we are finding the curve that best approximates passing through the points. It is possible that the curve that best fits the data will not actually pass directly through any of the points, although it should come reasonably close to most of them.

There are a number of different methods of determining the straight line that best fits the given data points. We shall employ the method that is most widely used: the **method of least squares.** *The basic principle of this method is that the sum of the squares of the deviations of all data points from the best line (in accordance with this method) has the least value possible. By* **deviation** *we mean the difference between the y-value of the line and the y-value for the point (of original data) for a particular value of x.*

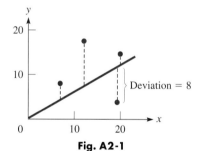

Fig. A2-1

EXAMPLE 1   In Fig. A2-1 the deviations of the points $(7, 7)$, $(12, 17)$, $(19, 3)$, and $(20, 14)$ are shown. The point $(19, 3)$ has a deviation of 8 from the indicated line in the figure. Thus, we square the value of this deviation to obtain 64. In order to find the equation of the straight line that best fits the given points, the method of least squares requires that the sum of all such squares be a minimum.

Therefore, in applying this method of least squares, it is necessary to use the equation of a straight line and the coordinates of the points of the data. The deviations of all of these data points are determined, and these values are then squared. It is then necessary to determine the constants for the slope m and the y-intercept b in the equation of a straight line $y = mx + b$ for which the sum of the squared values is a minimum. We now illustrate how this is done by using three data points and partial derivatives.

Given the data points (x_1, y_1), (x_2, y_2), and (x_3, y_3), find m and b in the equation

$$y = mx + b$$

for which the sum of the squares of the deviations is a minimum. See Fig. A2-2. The deviations are

$$d_1 = y_1 - (mx_1 + b)$$
$$d_2 = y_2 - (mx_2 + b)$$
$$d_3 = y_3 - (mx_3 + b)$$

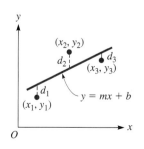

Fig. A2-2

Squaring the expression for each of the deviations and finding the sum S of these squares, we have

$$S = y_1^2 + y_2^2 + y_3^2 + m^2(x_1^2 + x_2^2 + x_3^2) + 3b^2 - 2m(x_1 y_1 + x_2 y_2 + x_3 y_3) - 2b(y_1 + y_2 + y_3) + 2mb(x_1 + x_2 + x_3)$$

We want S to be a minimum with respect to each of m and b. Therefore, this is a maximum-minimum problem of the type we studied in Section 11-4. We shall find the partial derivatives of S with respect to m and with respect to b, set each equal to zero, and solve the resulting equations simultaneously.

$$\frac{\partial S}{\partial m} = 2m(x_1^2 + x_2^2 + x_3^2) - 2(x_1 y_1 + x_2 y_2 + x_3 y_3) + 2b(x_1 + x_2 + x_3)$$

$$\frac{\partial S}{\partial b} = 6b - 2(y_1 + y_2 + y_3) + 2m(x_1 + x_2 + x_3)$$

Setting each of the partial derivatives equal to zero, we may obtain the equations

$$m(x_1^2 + x_2^2 + x_3^2) = (x_1 y_1 + x_2 y_2 + x_3 y_3) - b(x_1 + x_2 + x_3)$$
$$3b = (y_1 + y_2 + y_3) - m(x_1 + x_2 + x_3)$$

We are now going to generalize from these equations for n data points. Therefore, replacing 3 with n and replacing each of the sums of three terms with summations, such as $\Sigma x = x_1 + x_2 + x_3$, we have

$$m \Sigma x^2 = \Sigma xy - b \Sigma x \tag{1}$$

$$b = \frac{\Sigma y - m \Sigma x}{n} \tag{2}$$

Substituting the expression for b in Eq. (2) into Eq. (1), we have

$$m \Sigma x^2 = \Sigma xy - \left(\frac{\Sigma y - m \Sigma x}{n}\right) \Sigma x$$

$$mn \Sigma x^2 = n \Sigma xy - (\Sigma x)(\Sigma y) + m(\Sigma x)^2$$

$$m[n \Sigma x^2 - (\Sigma x)^2] = n \Sigma xy - (\Sigma x)(\Sigma y)$$

$$m = \frac{n \Sigma xy - \left(\Sigma x\right)\left(\Sigma y\right)}{n \Sigma x^2 - \left(\Sigma x\right)^2} \tag{A2-1}$$

and

$$b = \frac{\left(\Sigma x^2\right)\left(\Sigma y\right) - \left(\Sigma xy\right)\left(\Sigma x\right)}{n \Sigma x^2 - \left(\Sigma x\right)^2} \tag{A2-2}$$

Equation (A2-2) is found by substituting the expression for m in Eq. (A2-1) into Eq. (2) and simplifying.

Therefore, in order to obtain *the equation of the* **least-squares line**

$$y = mx + b \qquad \text{(A2-3)}$$

the slope *m* is found from Eq. (A2-1), and the *y*-intercept *b* is found from Eq. (A2-2).

■**EXAMPLE 2**   In a research project to determine the amount of a drug that remains in the bloodstream after a given dosage, the amounts *y* (in mg of drug/dL of blood) were recorded after *t* hours.

t (h)	1.0	2.0	4.0	8.0	10.0	12.0
y (mg/dL)	7.6	7.2	6.1	3.8	2.9	2.0

Find the least-squares line for these data, expressing *y* as a function of *t*. Sketch the graph of the line and data points.

The table and calculations are shown below.

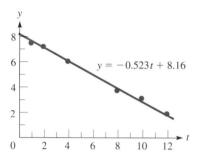

Fig. A2-3

t	*y*	*ty*	t^2
1.0	7.6	7.6	1.0
2.0	7.2	14.4	4.0
4.0	6.1	24.4	16.0
8.0	3.8	30.4	64.0
10.0	2.9	29.0	100
12.0	2.0	24.0	144
37.0	29.6	129.8	329

$$n = 6$$

$$m = \frac{6(129.8) - 37.0(29.6)}{6(329) - 37.0^2} = -0.523$$

$$b = \frac{(329)(29.6) - (129.8)(37.0)}{6(329) - 37.0^2} = 8.16$$

(a)

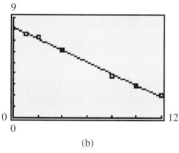

(b)

Fig. A2-4

The equation of the least-squares line is $y = -0.523t + 8.16$. The line and data points are shown in Fig. A2-3. This line is useful in determining the effectiveness of the drug. It can also be used to determine when additional medication may be administered.

Using the linear regression feature of a calculator, Fig. A2-4(a) shows the display of the coefficients of the line $y = ax + b$ (note that the slope is *a*). The value of *r* is called the *coefficient of correlation*. Figure A2-4(b) shows a calculator display of the points and the line. We see that they agree with Fig. A2-3.   ■

Nonlinear Regression

If the experimental points do not appear to be on a straight line but we recognize them as being approximately on some other type of curve, the method of least squares can be extended to use on these other curves. For example, if the points are apparently on a parabola, we could use the function $y = a + bx^2$. To use the above method, we shall extend the least-squares line to

$$y = m[f(x)] + b \qquad \text{(A2-4)}$$

Here, $f(x)$ must be calculated first, and then the problem can be treated as a least-squares line to find the values of *m* and *b*. Some of the functions $f(x)$ that may be considered for use are x^2, $1/x$, and 10^x.

EXAMPLE 3 In a physics experiment, the pressure p and volume V of a gas were measured at constant temperature. When the points were plotted, they were seen to approximate the hyperbola $y = c/x$. Find the least-squares approximation to the hyperbola $y = m(1/x) + b$ for the given data. See Fig. A2-5.

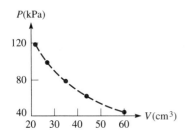

P(kPa)

Fig. A2-5

P (kPa)	V (cm^3)	$x\ (=V)$	$f(x) = \frac{1}{x}$	$y\ (=P)$	$(\frac{1}{x})y$	$(\frac{1}{x})^2$
120.0	21.0	21.0	0.0476190	120.0	5.7142857	0.0022676
99.2	25.0	25.0	0.0400000	99.2	3.9680000	0.0016000
81.3	31.8	31.8	0.0314465	81.3	2.5566038	0.0009889
60.6	41.1	41.1	0.0243309	60.6	1.4744526	0.0005920
42.7	60.1	60.1	0.0166389	42.7	0.7104825	0.0002769
			0.1600353	403.8	14.4238246	0.0057254

(*Calculator note:* The final digits for the values shown may vary depending on the calculator and how the values are used. Here, all individual values are shown with eight digits (rounded off), although more digits were used. The value of $1/x$ was found from the value of x, with the eight digits shown. However, the values of $(1/x)y$ and $(1/x)^2$ were found from the value of $1/x$, using the extra digits. The sums were found using the rounded-off values shown. However, since the data contain only three digits, any variation in the final digits for $1/x$, $(1/x)y$, or $(1/x)^2$ will not matter.)

$$m = \frac{5(14.4238246) - 0.1600353(403.8)}{5(0.0057254) - 0.1600353^2} = 2490$$

$$b = \frac{(0.0057254)(403.8) - (14.4238246)(0.1600353)}{5(0.0057254) - 0.1600353^2} = 1.2$$

The equation of the hyperbola $y = m(1/x) + b$ is

$$y = \frac{2490}{x} + 1.2$$

This hyperbola and the data points are shown in Fig. A2-6.

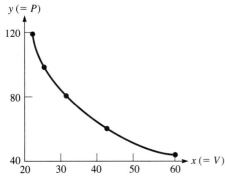

Fig. A2-6

As we noted and illustrated in Example 2, a graphing calculator can be used to determine the equation of a regression curve and to display its graph. A typical calculator can fit a number of different types of equations to data, and the regression equations available on a typical calculator are as follows:

Linear: $\quad y = ax + b$
Quadratic: $\quad y = ax^2 + bx + c$
Cubic: $\quad y = ax^3 + bx^2 + cx + d$
Quartic: $\quad y = ax^4 + bx^3 + cx^2 + dx + e$
Logarithmic: $\quad y = a + b \ln x$
Exponential: $\quad y = ab^x$
Power: $\quad y = ax^b$

Logistic: $\quad y = \dfrac{c}{1 - ae^{-bx}}$

Sinusoidal: $\quad y = a \sin(bx + c) + d$

— EXERCISES *A-2* —

In Exercises 1 and 2, find the equation of the least-squares line for the given data. Graph the line and data points on the same graph.

1.

x	4	6	8	10	12
y	1	4	5	8	9

2.

x	20	26	30	38	48	60
y	160	145	135	120	100	90

In Exercises 3–12, find the equation of the indicated least-squares curve. Sketch the curve and plot the data points on the same graph.

3. For the points in the following table, find the least-squares curve $y = m\sqrt{x} + b$.

x	0	4	8	12	16
y	1	9	11	14	15

4. For the points in the following table, find the least-squares curve $y = m(10^x) + b$.

x	0.00	0.200	0.500	0.950	1.325
y	6.00	6.60	8.20	14.0	26.0

5. In an electrical experiment, the following data were found for the values of current and voltage for a particular element of the circuit. Find the least-squares line for the voltage V as a function of the current i.

Current (mA)	15.0	10.8	9.30	3.55	4.60
Voltage (V)	3.00	4.10	5.60	8.00	10.50

6. In testing an air-conditioning system, the temperature T in a building was measured during the afternoon hours with the results shown in the table. Find the least-squares line for T as a function of the time t from noon.

t (h)	0.0	1.0	2.0	3.0	4.0	5.0
T (°C)	20.5	20.6	20.9	21.3	21.7	22.0

7. The pressure p was measured along an oil pipeline at different distances from a reference point, with results as shown. Find the least-squares line for p as a function of x. Check the values and line with a graphing calculator.

x (ft)	0	100	200	300	400
p (lb/in.²)	650	630	605	590	570

8. The heat loss L per hour through various thicknesses of a particular type of insulation was measured as shown in the table. Find the least-squares line for L as a function of t. Check the values and line with a graphing calculator.

t (in.)	3.0	4.0	5.0	6.0	7.0
L (Btu)	5900	4800	3900	3100	2450

9. The makers of a special blend of coffee found that the demand for the coffee depended on the price charged. The price P per pound and the monthly sales S are shown in the following table. Find the least-squares curve $P = m(1/S) + b$.

S (thousands)	240	305	420	480	560
P (dollars)	5.60	4.40	3.20	2.80	2.40

10. The resonant frequency f of an electric circuit containing a 4-μF capacitor was measured as a function of an inductance L in the circuit. The following data were found. Find the least-squares curve $f = m(1/\sqrt{L}) + b$.

L (H)	1.0	2.0	4.0	6.0	9.0
f (Hz)	490	360	250	200	170

11. The displacement y of an object at the end of a spring at given times t is shown in the following table. Find the least-squares curve $y = me^{-t} + b$.

t (s)	0.0	0.5	1.0	1.5	2.0	3.0
y (cm)	6.1	3.8	2.3	1.3	0.7	0.3

(W) 12. The average daily temperatures T (in °F) for each month in Minneapolis (National Weather Service records) are given in the following table.

t	J	F	M	A	M	J	J	A	S	O	N	D
T (°F)	11	18	29	46	57	68	73	71	61	50	33	19

Find the least-squares curve $T = m \cos[\frac{\pi}{6}(t - 0.5)] + b$. Assume the average temperature is for the 15th of each month. Then the values of t (in months) are 0.5, 1.5, ..., 11.5. (The fit is fairly good.) Compare the equation using the *sinusoidal regression* feature of a graphing calculator. What are the main reasons for the differences in the equations?

APPENDIX B

UNITS OF MEASUREMENT

Table B-1 Quantities and Their Associated Units

| Quantity | Quantity Symbol | U.S. Customary | | Metric (SI) | | |
		Name	Symbol	Name	Symbol	In Terms of Other SI Units
Length	s	foot	ft	**meter**	m	
Mass	m	slug		**kilogram**	kg	
Force	F	pound	lb	newton	N	$m \cdot kg/s^2$
Time	t	second	s	**second**	s	
Area	A		ft^2		m^2	
Volume	V		ft^3		m^3	
Capacity	V	gallon	gal	liter	L	$(1\ L = 1\ dm^3)$
Velocity	v		ft/s		m/s	
Acceleration	a		ft/s^2		m/s^2	
Density	d, ρ		lb/ft^3		kg/m^3	
Pressure	p		lb/ft^2	pascal	Pa	N/m^2
Energy, work	E, W		$ft \cdot lb$	joule	J	$N \cdot m$
Power	P	horsepower	hp	watt	W	J/s
Period	T		s		s	
Frequency	f		1/s	hertz	Hz	1/s
Angle	θ	radian	rad	radian	rad	
Electric current	I, i	ampere	A	**ampere**	A	
Electric charge	q	coulomb	C	coulomb	C	$A \cdot s$
Electric potential	V, E	volt	V	volt	V	$J/(A \cdot s)$
Capacitance	C	farad	F	farad	F	s/Ω
Inductance	L	henry	H	henry	H	$\Omega \cdot s$
Resistance	R	ohm	Ω	ohm	Ω	V/A
Thermodynamic temperature	T			**kelvin**	K	(temp. interval
Temperature	T	degrees Fahrenheit	°F	degrees Celsius	°C	$1\ °C = 1\ K)$
Quantity of heat	Q	British thermal unit	Btu	joule	J	
Amount of substance	n			**mole**	mol	
Luminous intensity	I	candlepower	cp	**candela**	cd	

Table B-2 Metric Prefixes

Prefix	Factor	Symbol	Prefix	Factor	Symbol
exa	10^{18}	E	deci	10^{-1}	d
peta	10^{15}	P	centi	10^{-2}	c
tera	10^{12}	T	milli	10^{-3}	m
giga	10^{9}	G	micro	10^{-6}	μ
mega	10^{6}	M	nano	10^{-9}	n
kilo	10^{3}	k	pico	10^{-12}	p
hecto	10^{2}	h	femto	10^{-15}	f
deca	10^{1}	da	atto	10^{-18}	a

Table B-3 Conversion Factors

1 in. = 2.54 cm (exact)	1 ft^3 = 28.32 L	1 lb = 453.6 g	1 Btu = 778.0 ft·lb
1 km = 0.6214 mi	1 L = 1.057 qt	1 kg = 2.205 lb	1 hp = 550 ft·lb/s (exact)
		1 lb = 4.448 N	1 hp = 746.0 W

Special Notes:

1. The SI base units are shown in boldface type.

2. The unit symbols shown above are those that are used in the text. Many of them were adopted with the adoption of the SI system. This means, for example, that we use s rather than sec for seconds and A rather than amp for amperes. Also, other units, such as volt, are not spelled out, a common practice in the past. When a given unit is used with both systems, we use the SI symbol for the unit.

3. The liter and degree Celsius are not actually SI units. However, they are recognized for use with the SI system due to their practical importance. Also, the symbol for liter has several variations. Presently, L is recognized for use in the United States and Canada, l is recognized by the International Committee of Weights and Measures, and ℓ is also recognized for use in several countries.

4. Other units of time, along with their symbols, which are recognized for use with the SI system and are used in this text, are minute, min; hour, h; day, d.

5. Many additional specialized units are used with the SI system. However, most of those that appear in this text are shown in the table. A few of the specialized units are noted when used in the text. One that is frequently used is that for revolution, r.

6. Other common U.S. units used in the text are inch, in.; yard, yd; mile, mi; ounce, oz.; ton; quart, qt; acre.

7. There are a number of units that were used with the metric system prior to the development of the SI system. However, many of these are not to be used with the SI system. Among those that were commonly used are the dyne, erg, and calorie.

Failure to Properly Change Units Can Lead to Major Problems

The failure to convert units caused the $125,000,000 Mars Climate Orbiter to fly too close to Mars and break up in the Martian atmosphere in September 1999. A spacecraft team submitted force data in pounds, but the mission controllers assumed the data were in newtons. The system for checking data did not note the change in units. As shown above, 1 lb = 4.448 N.

C THE GRAPHING CALCULATOR

C-1 INTRODUCTION

Until the 1970s the personal calculating device for engineers and scientists was the slide rule. The microprocessor chip became commercially available in 1971, and the scientific calculator became widely used during the 1970s. Then in the 1980s the graphing calculator was developed and is now used extensively.

The scientific calculator is still used by many when only accurate calculations are required. However, the graphing calculator can perform all the operations of a scientific calculator and numerous other operations. It also has the major advantage that entries can be seen in the viewing *window,* and this allows a visual check of entered data.

For these reasons, the graphing calculator is now used much more than the scientific calculator by engineers and scientists and in the more advanced mathematics courses. Therefore, we restrict our coverage of calculator use in this text to graphing calculators.

This appendix includes a brief discussion of graphing calculator features. We then show its use in the listing of the over 60 examples with sample screen displays throughout the text. A set of exercises that use the basic calculational and graphing features is also included. The final section of this appendix gives some graphing calculator programs that can be used to perform more extensive operations with fewer steps.

C-2 THE GRAPHING CALCULATOR

Since their introduction in the 1980s, many types and models of graphing calculators have been developed. All models can do all the calculational operations (and more) and display any of the graphs required in this text. Some models can also do symbolic operations. However, regardless of the model you are using, *to determine how the features of a particular model are used, refer to the manual for that model.*

In using a calculator, we must keep in mind that certain operations do not have defined results. If such an operation is attempted, the calculator will show an error display. Operations that can result in an error display include division by zero, square root of a negative number, logarithm of a negative number, and the inverse trigonometric function of a value outside the defined interval. Also, an error display may result if an improper sequence of keys is used.

Some calculators can display results involving imaginary numbers (those involving the square root of −1).

On a graphing calculator, negative numbers are entered by using the $(-)$ key, and subtraction is entered by using the $-$ key. Use of the $-$ key for the $(-)$ key will usually result in an error display.

Most of the keys are used to access two or three different features. These features may also have subroutines that may be used. To activate a second use, the *2nd* (or *shift*) key is used. The *alpha* key is used for entering alphabetical characters when using more than one literal symbol (*x* is usually available directly) or when entering a calculator program.

Graphing Calculator Features

We now list some of the more important features of a graphing calculator with an explanation of their use. (This listing uses the designations on a TI-83 calculator. Your calculator may use a different designation for some features.)

Feature	Use
MODE	Used to determine how numbers and graphs are displayed. A *menu* is displayed that shows settings that may be used. The *mode* can be set, for example, (1) for the number of decimal places displayed in numbers; (2) to display angles in degrees or in radians; (3) to display the graph of a function or of parametric equations; (4) to display a graph in rectangular or polar coordinates. See the *mode* feature for other possible settings.
Y=	Used to enter functions. These functions are to be graphed, or used in some other way.
WINDOW	Used to set the boundaries of the viewed portion of a graph. (On some calculators this is the *range* feature.)
GRAPH	Used to display the graph of a function. (On some models this feature puts the calculator in graphing mode.)
TRACE	Used to move the cursor from pixel to pixel along a displayed graph.
ZOOM	Used to adjust the viewing window, usually to magnify the view of a particular part of a graph.
STAT	Used to access various statistical features.
MATH	Used to access specific mathematical features. These include fractions, *n*th roots, maximum and minimum values, absolute values, and others.
TEST	Used to enter special symbols such as $>$ and $<$, and to enter special logic words such as *and* and *or*.
MATRX	Used to enter matrix values and perform matrix operations such as the evaluation of a determinant.
DRAW	Used to make certain types of drawings, such as shading in an area between curves or a tangent line to a curve.
PRGM	Used to enter and access calculator programs. A set of sample programs is given in the last section of this appendix.

There are many other features on a graphing calculator that are used to perform specific types of calculations and operations. Again, *to determine how the features of a particular model are used, refer to the manual for that model.*

Examples of the use of these features are found in the text examples in the listing that follows on the next page.

Examples of Calculator Use

When using a calculator, be sure to properly round off final answers. If you keep extra digits, you show a number with too great an accuracy, and it is incorrect to do so.

The following list shows the pages on which examples of the use of a graphing calculator are given in the text. The list is divided into calculator uses that are primarily calculational and those that are primarily graphical. For the graphical examples, the calculator *window* settings are shown outside of the display.

EXERCISES *C-2*

The following exercises provide an opportunity for practice in performing calculations and a few basic graphs on a graphing calculator. There are many additional exercises throughout the book that require these types of calculations. Many additional exercises that require graphing of various types of curves are found after graphing is introduced in Chapter 1.

In Exercises 1–92, perform the indicated calculations on a graphing calculator.

1. $47.08 + 8.94$

2. $654.1 + 407.7$

3. $4724 - 561.9$

4. $0.9365 - 8.077$

5. 0.0396×471

6. 26.31×0.9393

7. $76.7 \div 194$

8. $52{,}060 \div 75.09$

9. 3.76^2

10. 0.986^2

11. $\sqrt{0.2757}$

12. $\sqrt{60.36}$

13. $\dfrac{1}{0.0749}$

14. $\dfrac{1}{607.9}$

15. $(19.66)^{2.3}$

16. $(8.455)^{1.75}$

17. $\sin 47.3°$

18. $\sin 1.15$

19. $\cos 3.85$

20. $\cos 119.1°$

21. $\tan 306.8°$

22. $\tan 0.537$

23. $\sec 6.11$

24. $\csc 242.0°$

25. $\sin^{-1} 0.6607$ (in degrees)

26. $\cos^{-1}(-0.8311)$ (in radians)

27. $\tan^{-1}(-2.441)$ (in radians)

28. $\sin^{-1} 0.0737$ (in degrees)

29. log 3.857

30. log 0.9012

31. ln 808

32. ln 70.5

33. $10^{0.545}$

34. $10^{-0.0915}$

35. $e^{-5.17}$

36. $e^{1.672}$

37. $(4.38 + 9.07) \div 6.55$

38. $(382 + 964) \div 844$

39. $4.38 + (9.07 \div 6.55)$

40. $382 + (964 \div 844)$

41. $\dfrac{5.73 \times 10^{11}}{20.61 - 7.88}$

42. $\dfrac{7.09 \times 10^{23}}{284 + 839}$

43. $50.38\pi^2$

44. $\dfrac{5\pi}{14.6}$

45. $\sqrt{1.65^2 + 6.44^2}$

46. $\sqrt{0.735^2 + 0.409^2}$

47. $3(3.5)^4 - 4(3.5)^2$

48. $\dfrac{3(-1.86)}{(-1.86)^2 + 1}$

49. $29.4 \cos 72.5°$

50. $\dfrac{477}{\sin 58.7°}$

51. $\dfrac{4 + \sqrt{(-4)^2 - 4(3)(-9)}}{2(3)}$

52. $\dfrac{-5 - \sqrt{5^2 - 4(4)(-7)}}{2(4)}$

53. $\dfrac{0.176(180)}{\pi}$

54. $\dfrac{209.6\pi}{180}$

55. $\dfrac{1}{2}\left(\dfrac{51.4\pi}{180}\right)(7.06)^2$

56. $\dfrac{1}{2}\left(\dfrac{148.2\pi}{180}\right)(49.13)^2$

57. $\sin^{-1}\dfrac{27.3 \sin 36.5°}{46.8}$

58. $\dfrac{0.684 \sin 76.1°}{\sin 39.5°}$

59. $\sqrt{3924^2 + 1762^2 - 2(3924)(1762)\cos 106.2°}$

60. $\cos^{-1}\dfrac{8.09^2 + 4.91^2 - 9.81^2}{2(8.09)(4.91)}$

61. $\sqrt{5.81 \times 10^8} + \sqrt[3]{7.06 \times 10^{11}}$

62. $(6.074 \times 10^{-7})^{2/5} - (1.447 \times 10^{-5})^{4/9}$

63. $\dfrac{3}{2\sqrt{7} - \sqrt{6}}$

64. $\dfrac{7\sqrt{5}}{4\sqrt{5} - \sqrt{11}}$

65. $\tan^{-1}\dfrac{7.37}{5.06}$

66. $\tan^{-1}\dfrac{46.3}{-25.5}$

67. $2 + \dfrac{\log 12}{\log 7}$

68. $\dfrac{10^{0.4115}}{\pi}$

69. $\dfrac{26}{2}(-1.450 + 2.075)$

70. $\dfrac{4.55(1 - 1.08^{15})}{1 - 1.08}$

71. $\sin^2\left(\dfrac{\pi}{7}\right) + \cos^2\left(\dfrac{\pi}{7}\right)$

72. $\sec^2\left(\dfrac{2}{9}\pi\right) - \tan^2\left(\dfrac{2}{9}\pi\right)$

73. $\sin 31.6° \cos 58.4° + \sin 58.4° \cos 31.6°$

74. $\cos^2 296.7° - \sin^2 296.7°$

75. $\sqrt{(1.54 - 5.06)^2 + (-4.36 - 8.05)^2}$

76. $\sqrt{(7.03 - 2.94)^2 + (3.51 - 6.44)^2}$

77. $\dfrac{(4.001)^2 - 16}{4.001 - 4}$

78. $\dfrac{(2.001)^2 + 3(2.001) - 10}{2.001 - 2}$

79. $\dfrac{4\pi}{3}(8.01^3 - 8.00^3)$

80. $4\pi(76.3^2 - 76.0^2)$

81. $0.01\left(\dfrac{1}{2}\sqrt{2} + \sqrt{2.01} + \sqrt{2.02} + \dfrac{1}{2}\sqrt{2.03}\right)$

82. $0.2\left[\dfrac{1}{2}(3.5)^2 + 3.7^2 + 3.9^2 + \dfrac{1}{2}(4.1)^2\right]$

83. $\dfrac{e^{0.45} - e^{-0.45}}{e^{0.45} + e^{-0.45}}$

84. $\ln \sin 2e^{-0.055}$

85. $\ln\dfrac{2 - \sqrt{2}}{2 - \sqrt{3}}$

86. $\sqrt{\dfrac{9}{2} + \dfrac{9 \sin 0.2\pi}{8\pi}}$

87. $2 + \dfrac{0.3}{4} - \dfrac{(0.3)^2}{64} + \dfrac{(0.3)^2}{512}$

88. $\dfrac{1}{2} + \dfrac{\pi\sqrt{3}}{360} - \dfrac{1}{4}\left(\dfrac{\pi}{180}\right)^2$

89. $160(1 - e^{-1.50})$

90. $e^{-3.60}(\cos 1.20 + 2 \sin 1.20)$

91. $\dfrac{(10)(9)(8)(7)(6)}{5!}$

92. $\dfrac{20! - 15!}{20! + 15!}$

In Exercises 93–108, display the graphs of the given functions on a graphing calculator. Use window settings to properly display the curve.

93. $y = 2x$

94. $y = 6 - x$

95. $y = x^2$

96. $y = 4 - 2x^2$

97. $y = 0.5x^3$

98. $y = 2x^2 - x^4$

99. $y = \sqrt{4 - x}$

100. $y = \sqrt{25 - x^2}$

101. $y = \log x$

102. $y = 2^x$

103. $y = 2e^x$

104. $y = 3 \ln x$

105. $y = \sin 2x$

106. $y = 2 \cos x$

107. $y = \tan^{-1} x$

108. $y = \dfrac{6}{x}$

C-3 GRAPHING CALCULATOR PROGRAMS

Programs like those used on a computer can be stored in the memory of a graphing calculator. As mentioned earlier, such programs are used to perform more extensive operations with fewer steps, often significantly fewer steps. Generally, it is necessary only to enter certain data to obtain the required results.

Following are 14 programs that were written for use on a TI-83 graphing calculator. For other calculator models, it may be necessary to adapt the steps indicated to the format and designated operations of that model.

The chapter, program title, and page on which the program reference appears are shown. A brief description of each program is also given.

Chapter 2   SLOPEDIS   page 31

This program calculates the slope m of a line through two points (a, c) and (e, f) and the distance between the points. It also displays the line segment on a split screen. Set proper *window* values before running the program.

```
:Prompt A,C,E,F
:Horiz
:Output (2,1,"M=")
:Output (2,4,(F−C)/(E−A))
:Output (3,1,"D=")
:Output (3,4, √((E−A)²+(F−C)²))
:Line (A,C,E,F)
:Pause:Full
```

Chapter 2   GRAPHLIN   page 33

This program displays the graph of a line through two points, (a, c) and (d, e). The *window* values should be set appropriately before running the program. Default *window* is $(-9, 9)$ by $(-6, 6)$.

```
:Prompt A,C,D,E
:(E−C)/(D−A)→M:C−MA→B
:"MX+B"→Y₁
:−9→Xmin:9→Xmax:1→Xscl
:−6→Ymin:6→Ymax:1→Yscl
:DispGraph
:Pause:ClrHome
```

Chapter 2   GRAPHCON   page 63

This program displays the graph of a conic of the form $ax^2 + cy^2 + dx + ey + f = 0$. The *window* settings may have to be changed in the program. The default *window* is $(-9, 9)$ by $(-6, 6)$.

```
:Prompt A,C,D,E,F
:If C=0:Goto A
:(−E+√(E²−4C(AX²+DX+F)))/(2C)→Y₁
:(−E−√(E²−4C(AX²+DX+F)))/(2C)→Y₂
:Lbl B
:−9→Xmin:9→Xmax:1→Xscl
:−6→Ymin:6→Ymax:1→Yscl
:DispGraph
:Pause:ClrHome:Stop
:Lbl A:"(AX²+DX+F)/(−E)"→Y₁
:Goto B
```

Chapter 3   LIMFUNC   page 73

This program evaluates the limit of $f(x)$ as $x \to a^+$.

```
:Input "F(X)=",Y₁
:Input "A=",A
:1→D:1→N
:For(I,1,8):A+D→X
:Disp "FOR X=",X
:Disp "F(X)=",Y₁
:.1D→D
:Pause:End
```

Chapter 3   SECTOTAN   page 78

This program displays the graph of a function and four secant lines as the secant line approaches the tangent line at a given point. Input the function Y_1, A (the x-coordinate of the point of tangency), and H $(= \Delta x)$.

```
:ClrDraw
:Input "Y₁=",Y₁
:DispGraph:Pause
:Prompt A,H
:A→X:Y₁→B
:For(I,1,4):A+H→X
:(Y₁−B)/H→M
:DrawF M(X−A)+B
:Pause: .7H→H:
:End:ClrHome
```

Chapter 4   NEWTON   page 122

This program finds the nth approximation of a root of the equation $f(x) = 0$ using Newton's method. Input the function $f(x)$, its derivative dy/dx, n, and the estimate A.

```
:Input "Y=",Y₁
:Input "DY/DX=", Y₂
:Input "N=", N
:Input "A=",X
:For(I,2,N)
:Disp "ROOT",I:Disp "IS",X−Y₁/Y₂
:X−Y₁/Y₂→X
:Pause:End
```

Chapter 4 ADDVCTR page 125

This program adds two vectors, given their magnitudes and angles.

```
:Disp "ENTER MAGNITUDES"
:Input "A=",A:Input "B=",B
:Disp "ENTER ANGLES"
:Input "θA=",P:Input "θB=",Q
:Acos(P)+Bcos(Q)→R
:Asin(P)+Bsin(Q)→S
:tan⁻¹(S/R)→T
:Disp "R=", √(R²+S²)
:If R<0:Goto A:If S<0:Goto B
:Disp "θR=",T:Stop
:Lbl A:Disp "θR=",T+180:Stop
:Lbl B:Disp "θR=",T+360
```

Chapter 5 AREAUNCV page 167

This program approximates the area under a curve by summing the areas of n rectangles (inscribed for an increasing curve, circumscribed for a decreasing curve) from $x = a$ to $x = b$.

```
:Input "F(X)=",Y₁
:Prompt A,B,N
:(B−A)/N→D:0→R
:For (I,0,N−1)
:A+ID→X:R+DY₁→R
:End
:Disp "AREA=",R
```

Chapter 5 TRAPRULE page 175

This program evaluates the integral of $f(x)\,dx$ from $x = a$ to $x = b$ by using the trapezoidal rule with n intervals.

```
:Input "F(X)=",Y₁
:Prompt A,B,N
:(B−A)/N→D:A→X:Y₁→E
:B→X:Y₁→F:E+F→S
:For (I,1,N−1)
:A+ID→X:S+2Y₁→S:End
:Disp "APPROX VALUE OF":Disp "INTEGRAL IS"
:Disp SD/2
```

Chapter 7 SINECURV page 231

This program displays the graphs of $y_1 = \sin x$, $y_2 = 2 \sin x$, $y_3 = \sin 2x$, and $y_4 = 2 \sin(2x - \pi/3)$. There is no input.

```
:⁻2→Xmin:7→Xmax
:⁻2→Ymin:2→Ymax
:"sin(X)"→Y₁:"2sin(X)"→Y₂
:"sin(2X)"→Y₃:"2sin(2X−π/3)"→Y₄
:Disp "Y₁=sin(X)"
:Disp "Y₂=2sin(X)"
:Disp "Y₃=sin(2X)"
:Disp "Y₄=2sin(2X−π/3)"
:Pause:DispGraph:Pause:ClrHome
```

Chapter 13 PARTSUMS page 385

This program evaluates the first n partial sums of an infinite series.

```
:Input "GENERAL TERM=",Y₁
:Input "N=",J:0→S
:For (N,1,J)
:Disp "FOR N=",N
:Disp "SUM=",S+Y₁
:S+Y₁→S
:Pause:End
```

Chapter 13 BINEXPAN page 392

This program gives the first k coefficients of $(ax + b)^n$.

```
:Prompt K,A,B,N
:A^N→T
:Disp "COEFFICIENTS="
:For (C,1,K): Disp T
:T(N−C+1)B/(AC)→T
:Pause:End
```

Chapter 16 EULRMETH page 462

This program gives numerical values of y for the solution of the differential equation $dy/dx = f(x, y)$ from $x = a$ to $x = b$ with an initial y-value of y_0 and $dx = H$, using Euler's method.

```
:Input "F(X,Y)=",Y₁
:Prompt A,B,H
:Input "Y0=",Y
:abs(int((B−A)/H+.5))→N
:For (I,1,N+1):A+(I−1)H→X
:Disp "X=",X
:Disp "Y=",Y
:Y+Y₁H→Y
:Pause:End
```

Chapter 16 RUNGKUTT page 463

This program gives numerical values of y for the solution of the differential equation $dy/dx = f(x, y)$ from $x = a$ to $x = b$ with an initial y-value of y_0 and $dx = h$, using the Runge–Kutta method.

```
:Input "F(X,Y)=,",Y₁
:Prompt A,B,H
:Input "Y0=",C
:abs(int((B−A)/H+.5))→N:C→Y
:Disp "X=",A
:Disp "Y=",C:Pause
:For (I,1,N)
:A+(I−1)H→X:HY₁→J
:X+H/2→X:C+J/2→Y
:HY₁→K:C+K/2→Y
:HY₁→L:X+H/2→X
:C+L→Y:HY₁→M
:C+(J+2K+2L+M)/6→Y
:Y→C
:Disp "X=", X
:Disp "Y=", Y
:Pause:End
```

At the right is an explanation and example of Newton's method for solving equations. They were copied directly from *Essays on Several Curious and Useful Subjects in Speculative and Mix'd Mathematicks* by Thomas Simpson (of Simpson's rule). It was published in London in 1740.

See Exercise 20, page 123.

A new Method for the Solution of Equations in Numbers.

CASE I.

When only one Equation is given, and one Quantity (x) to be determined.

TAKE the Fluxion of the given Equation (be it what it will) suppofing, x, the unknown, to be the variable Quantity; and having divided the whole by $\dot{x}$, let the Quotient be reprefented by A. Eftimate the Value of x pretty near the Truth, fubftituting the fame in the Equation, as alfo in the Value of A, and let the Error, or refulting Number in the former, be divided by this numerical Value of A, and the Quotient be fubtracted from the faid former Value of x; and from thence will arife a new Value of that Quantity much nearer to he Truth than the former, wherewith proceeding as before, another new Value may be had, and fo another, &c. 'till we arrive to any Degree of Accuracy defired.

EXAMPLE I.

LET $300x - x^3 - 1000$ be given $= 0$; to find a Value of x. From $300\dot{x} - 3x^2\dot{x}$, the Fluxion of the given Equation, having expunged $\dot{x}$, (*Cafe* I.) there will be $300 - 3xx = A$: And, becaufe it appears by Infpection, that the Quantity $300x - x^3$, when x is $= 3$, will be lefs, and when $x = 4$, greater than 1000, I eftimate x at 3.5, and fubftitute inftead thereof, both in the Equation and in the Value of A, finding the Error in the former $= 7.125$, and the Value of the latter $= 263.25$: Wherefore, by taking $\frac{7.125}{263.25} = .027$ from 3.5 there will remain 3.473 for a new Value of x; with which proceeding as before, the next Error, and the next Value of A, will come out .00962518, and 263.815 refpectively; and from thence the third Value of $x = 3.47296351$; which is true, at leaft, to 7 or 8 Places.

The basic forms of Chapter 9 are not included. The constant of integration is omitted.

Forms containing $a + bu$ and $\sqrt{a + bu}$

1. $\displaystyle\int \frac{u\,du}{a + bu} = \frac{1}{b^2}[(a + bu) - a \ln(a + bu)]$

2. $\displaystyle\int \frac{du}{u(a + bu)} = -\frac{1}{a} \ln \frac{a + bu}{u}$

3. $\displaystyle\int \frac{u\,du}{(a + bu)^2} = \frac{1}{b^2}\left(\frac{a}{a + bu} + \ln(a + bu)\right)$

4. $\displaystyle\int \frac{du}{u(a + bu)^2} = \frac{1}{a(a + bu)} - \frac{1}{a^2} \ln \frac{a + bu}{u}$

5. $\displaystyle\int u\sqrt{a + bu}\,du = -\frac{2(2a - 3bu)(a + bu)^{3/2}}{15b^2}$

6. $\displaystyle\int \frac{u\,du}{\sqrt{a + bu}} = -\frac{2(2a - bu)\sqrt{a + bu}}{3b^2}$

7. $\displaystyle\int \frac{du}{u\sqrt{a + bu}} = \frac{1}{\sqrt{a}} \ln\left(\frac{\sqrt{a + bu} - \sqrt{a}}{\sqrt{a + bu} + \sqrt{a}}\right), \quad a > 0$

8. $\displaystyle\int \frac{\sqrt{a + bu}}{u}\,du = 2\sqrt{a + bu} + a\int \frac{du}{u\sqrt{a + bu}}$

Forms containing $\sqrt{u^2 \pm a^2}$ and $\sqrt{a^2 - u^2}$

9. $\displaystyle\int \frac{du}{u^2 - a^2} = \frac{1}{2a} \ln \frac{u - a}{u + a}$

10. $\displaystyle\int \frac{du}{\sqrt{u^2 \pm a^2}} = \ln(u + \sqrt{u^2 \pm a^2})$

11. $\displaystyle\int \frac{du}{u\sqrt{u^2 + a^2}} = -\frac{1}{a} \ln\left(\frac{a + \sqrt{u^2 + a^2}}{u}\right)$

12. $\displaystyle\int \frac{du}{u\sqrt{u^2 - a^2}} = \frac{1}{a} \sec^{-1}\frac{u}{a}$

13. $\displaystyle\int \frac{du}{u\sqrt{a^2 - u^2}} = -\frac{1}{a} \ln\left(\frac{a + \sqrt{a^2 - u^2}}{u}\right)$

14. $\displaystyle\int \sqrt{u^2 \pm a^2}\,du = \frac{u}{2}\sqrt{u^2 \pm a^2} \pm \frac{a^2}{2} \ln(u + \sqrt{u^2 \pm a^2})$

15. $\displaystyle\int \sqrt{a^2 - u^2}\,du = \frac{u}{2}\sqrt{a^2 - u^2} + \frac{a^2}{2} \sin^{-1}\frac{u}{a}$

16. $\displaystyle\int \frac{\sqrt{u^2 + a^2}}{u}\, du = \sqrt{u^2 + a^2} - a\, \ln\left(\frac{a + \sqrt{u^2 + a^2}}{u}\right)$

17. $\displaystyle\int \frac{\sqrt{u^2 - a^2}}{u}\, du = \sqrt{u^2 - a^2} - a\, \sec^{-1}\frac{u}{a}$

18. $\displaystyle\int \frac{\sqrt{a^2 - u^2}}{u}\, du = \sqrt{a^2 - u^2} - a\, \ln\left(\frac{a + \sqrt{a^2 - u^2}}{u}\right)$

19. $\displaystyle\int (u^2 \pm a^2)^{3/2}\, du = \frac{u}{4}(u^2 \pm a^2)^{3/2} \pm \frac{3a^2 u}{8}\sqrt{u^2 \pm a^2} + \frac{3a^4}{8}\ln(u + \sqrt{u^2 \pm a^2})$

20. $\displaystyle\int (a^2 - u^2)^{3/2}\, du = \frac{u}{4}(a^2 - u^2)^{3/2} + \frac{3a^2 u}{8}\sqrt{a^2 - u^2} + \frac{3a^4}{8}\sin^{-1}\frac{u}{a}$

21. $\displaystyle\int \frac{(u^2 + a^2)^{3/2}}{u}\, du = \frac{1}{3}(u^2 + a^2)^{3/2} + a^2\sqrt{u^2 + a^2} - a^3\ln\left(\frac{a + \sqrt{u^2 + a^2}}{u}\right)$

22. $\displaystyle\int \frac{(u^2 - a^2)^{3/2}}{u}\, du = \frac{1}{3}(u^2 - a^2)^{3/2} - a^2\sqrt{u^2 - a^2} + a^3\sec^{-1}\frac{u}{a}$

23. $\displaystyle\int \frac{(a^2 - u^2)^{3/2}}{u}\, du = \frac{1}{3}(a^2 - u^2)^{3/2} - a^2\sqrt{a^2 - u^2} + a^3\ln\left(\frac{a + \sqrt{a^2 - u^2}}{u}\right)$

24. $\displaystyle\int \frac{du}{(u^2 \pm a^2)^{3/2}} = \pm\, \frac{u}{a^2\sqrt{u^2 \pm a^2}}$

25. $\displaystyle\int \frac{du}{(a^2 - u^2)^{3/2}} = \frac{u}{a^2\sqrt{a^2 - u^2}}$

26. $\displaystyle\int \frac{du}{u(u^2 + a^2)^{3/2}} = \frac{1}{a^2\sqrt{u^2 + a^2}} - \frac{1}{a^3}\ln\left(\frac{a + \sqrt{u^2 + a^2}}{u}\right)$

27. $\displaystyle\int \frac{du}{u(u^2 - a^2)^{3/2}} = -\frac{1}{a^2\sqrt{u^2 - a^2}} - \frac{1}{a^3}\sec^{-1}\frac{u}{a}$

28. $\displaystyle\int \frac{du}{u(a^2 - u^2)^{3/2}} = \frac{1}{a^2\sqrt{a^2 - u^2}} - \frac{1}{a^3}\ln\left(\frac{a + \sqrt{a^2 - u^2}}{u}\right)$

Trigonometric forms

29. $\displaystyle\int \sin^2 u\, du = \frac{u}{2} - \frac{1}{2}\sin u \cos u$

30. $\displaystyle\int \sin^3 u\, du = -\cos u + \frac{1}{3}\cos^3 u$

31. $\displaystyle\int \sin^n u\, du = -\frac{1}{n}\sin^{n-1} u \cos u + \frac{n-1}{n}\int \sin^{n-2} u\, du$

32. $\displaystyle\int \cos^2 u\, du = \frac{u}{2} + \frac{1}{2}\sin u \cos u$

33. $\displaystyle\int \cos^3 u\, du = \sin u - \frac{1}{3}\sin^3 u$

34. $\displaystyle\int \cos^n u\, du = \frac{1}{n}\cos^{n-1} u \sin u + \frac{n-1}{n}\int \cos^{n-2} u\, du$

35. $\displaystyle\int \tan^n u\, du = \frac{\tan^{n-1} u}{n-1} - \int \tan^{n-2} u\, du$

36. $\displaystyle\int \cot^n u \; du = -\frac{\cot^{n-1}u}{n-1} - \int \cot^{n-2}u \; du$

37. $\displaystyle\int \sec^n u \; du = \frac{\sec^{n-2}u \tan u}{n-1} + \frac{n-2}{n-1} \int \sec^{n-2}u \; du$

38. $\displaystyle\int \csc^n u \; du = -\frac{\csc^{n-2}u \cot u}{n-1} + \frac{n-2}{n-1} \int \csc^{n-2}u \; du$

39. $\displaystyle\int \sin au \sin bu \; du = \frac{\sin(a-b)u}{2(a-b)} - \frac{\sin(a+b)u}{2(a+b)}$

40. $\displaystyle\int \sin au \cos bu \; du = -\frac{\cos(a-b)u}{2(a-b)} - \frac{\cos(a+b)u}{2(a+b)}$

41. $\displaystyle\int \cos au \cos bu \; du = \frac{\sin(a-b)u}{2(a-b)} + \frac{\sin(a+b)u}{2(a+b)}$

42. $\displaystyle\int \sin^m u \cos^n u \; du = \frac{\sin^{m+1}u \cos^{n-1}u}{m+n} + \frac{n-1}{m+n} \int \sin^m u \cos^{n-2}u \; du$

43. $\displaystyle\int \sin^m u \cos^n u \; du = -\frac{\sin^{m-1}u \cos^{n+1}u}{m+n} + \frac{m-1}{m+n} \int \sin^{m-2}u \cos^n u \; du$

Other forms

44. $\displaystyle\int u e^{au} du = \frac{e^{au}(au-1)}{a^2}$

45. $\displaystyle\int u^2 e^{au} du = \frac{e^{au}}{a^3}(a^2u^2 - 2au + 2)$

46. $\displaystyle\int u^n \ln u \; du = u^{n+1}\left(\frac{\ln u}{n+1} - \frac{1}{(n+1)^2}\right)$

47. $\displaystyle\int u \sin u \; du = \sin u - u \cos u$

48. $\displaystyle\int u \cos u \; du = \cos u + u \sin u$

49. $\displaystyle\int e^{au} \sin bu \; du = \frac{e^{au}(a \sin bu - b \cos bu)}{a^2 + b^2}$

50. $\displaystyle\int e^{au} \cos bu \; du = \frac{e^{au}(a \cos bu + b \sin bu)}{a^2 + b^2}$

51. $\displaystyle\int \sin^{-1}u \; du = u \sin^{-1}u + \sqrt{1 - u^2}$

52. $\displaystyle\int \tan^{-1}u \; du = u \tan^{-1}u - \frac{1}{2}\ln(1 + u^2)$

Since statements will vary for writing exercises **W**, answers here are in abbreviated form. Answers are not included for end-of-chapter writing exercises.

Exercises 1-1, page 6

1. (a) $A = \pi r^2$ (b) $A = \frac{1}{4}\pi d^2$ **3.** $V = \frac{1}{6}\pi d^3$ **5.** $A = 5l$

7. $A = s^2, s = \sqrt{A}$ **9.** $3, -1$ **11.** $11, 3.8$

13. $\frac{5}{2}, -\frac{1}{2}$ **15.** $\frac{1}{4}a + \frac{1}{2}a^2, 0$

17. $3s^2 + s + 6, 12s^2 - 2s + 6$ **19.** -8

21. 3 is an integer, rational real; $-\pi$ is irrational, real; $-\sqrt{-6}$ is imaginary; $\sqrt{7}/3$ is irrational, real

23. $3, \frac{7}{2}, \frac{6}{7}, \sqrt{3}, 3 - \sqrt{5}$ **25.** (a) $<$ (b) $>$ (c) $>$

27. (a) positive integer (b) negative integer (c) positive rational number less than 1

29. (a) yes (b) yes **31.** to right of origin

33. 10.4 m **35.** 75 ft, $2v + 0.2v^2$, 240 ft, 240 ft

Exercises 1-2, page 12

1. $\sqrt{x^2 - 2x + 5}$ **3.** $\dfrac{x - 1}{\sqrt{x^2 + 4}}$ **5.** $\sqrt[6]{x^6 + 1}$

7. $(4x - 5)^{7/2}$ **9.** $\dfrac{3x + 4}{\sqrt{2x + 1}}$ **11.** $\dfrac{1}{(x^2 + 1)^{3/2}}$

13. domain: all real numbers; range: all real numbers

15. domain: all real numbers except 0; range: all real numbers except 0

17. domain: all real numbers except 0; range: all real numbers $f(s) > 0$

19. domain: all real numbers $h \geq 0$; range: all real numbers $H(h) \geq 1$; $\sqrt{h}$ cannot be negative

21. all real numbers $y > 2$

23. all real numbers except $-4, 2,$ and 6 **25.** 2, not defined

27. $2, \dfrac{3}{4}$ **29.** $d = 80 + 55t$ **31.** $w = 5500 - 2t$

33. $m = 0.5h - 390$ **35.** $C = 5l + 250$

37. (a) $y = 3000 - 0.25x$ (b) 2900 L

39. $A = \dfrac{1}{16}p^2 + \dfrac{(60 - p)^2}{4\pi}$

41. $A = \pi(6 - x)^2$, domain $0 \leq x \leq 6$, range: $0 \leq A \leq 36\pi$

43. $s = \dfrac{300}{t}$, domain: $t > 0$, range: $s > 0$ (upper limits depend on truck)

45. Domain is all values of $C > 0$, with some upper limit depending on the circuit.

47. $m = \begin{cases} 0.5h - 390 \text{ for } h > 1000 \\ 110 \text{ for } 0 \leq h \leq 1000 \end{cases}$

Exercises 1-3, page 15

1. $(2, 1), (-1, 2), (-2, -3)$ **3.**

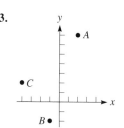

5. isosceles triangle **7.** rectangle **9.** $(5, 4)$

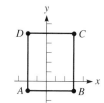 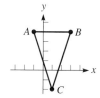

11. $(3, -2)$

13. on a line parallel to the y-axis, 1 unit to the right

15. on a line parallel to the x-axis, 3 units above

17. on a line bisecting the first and third quadrants

19. 0 **21.** to the right of the y-axis **23.** to the left of a line that is parallel to the y-axis, 1 unit to its left

25. first, third **27.** (a) 8 (b) 6

Exercises 1-4, page 21

1. **3.**

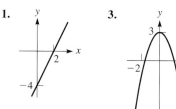

5. **7.**

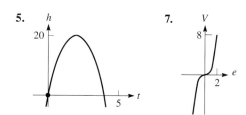

9.
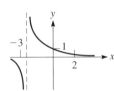

11.
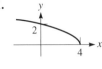

33. all real numbers $Y(y) > 3.46$ (approx.)
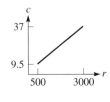

13.
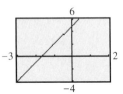

15.
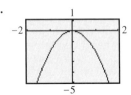

35. all real numbers
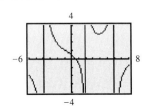

37.
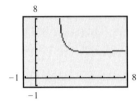

17.
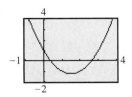

19.
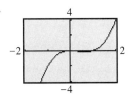

39.
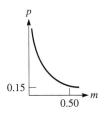

41. 24.4 cm, 29.4 cm
43. 18 s

21.
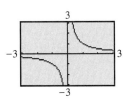

23.
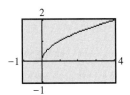

45. $A = 100w - w^2$
$30 \text{ m} \leq w \leq 70 \text{ m}$

47. 0.25 ft/min

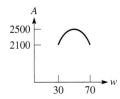

25. −6.4, 6.4

27. 1.4
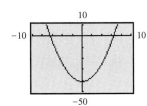

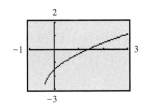

49.
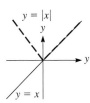

$y = x$ is same as
$y = |x|$ for $x \geq 0$.
$y = |x|$ is same as
$y = -x$ for $x < 0$.

29. all real numbers $y > 0$ or $y \leq -1$
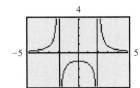

31. all real numbers $y \geq 0$ or $y \leq -4$
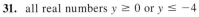
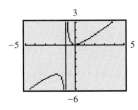

51.

53. yes

55. no

Review Exercises for Chapter 1, page 23

1. $A = 4\pi t^2$ **3.** $y = -\dfrac{10}{9}x + \dfrac{250}{9}$ **5.** 16, −47

7. 3, $\sqrt{1 - 4h}$ **9.** $3h^2 + 6hx - 2h$ **11.** −3

13. −3.67, 16.7 **15.** 0.16503, −0.21476

17. domain: all real numbers; range: all real numbers $f(x) \geq 1$

19. domain: all real numbers $t > -4$; range: all real numbers $g(t) > 0$

21. **23.** 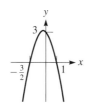 **25.**

27. 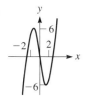 **29.** **31.**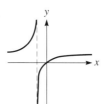

33. 0.4 **35.** 0.2, 5.8

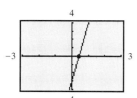 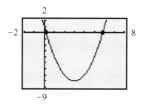

37. 1.4 **39.** −0.7, 0.7

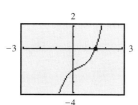 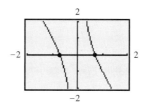

41. all real numbers $y \geq -6.25$

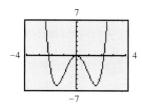

43. all real numbers $y \leq -2.83$ or $y \geq 2.83$

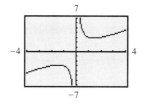

45. either a or b is positive, the other is negative

47. $(1, \sqrt{3})$ or $(1, -\sqrt{3})$ **49.** 13.4 **51.** 72.0°

53. 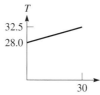 **55.**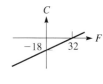

57. 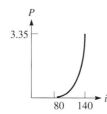 **59.**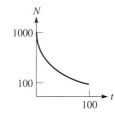

61. 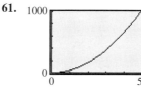 **63.**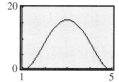

65. 3.5 h **67.** 33°C **69.** 1.03 ft **71.** 6.5 h

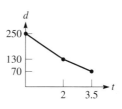

Exercises 2-1, page 31

1. $2\sqrt{29}$ **3.** 3 **5.** 55 **7.** $2\sqrt{53}$ **9.** 2.86

11. $\dfrac{5}{2}$ **13.** undefined **15.** $-\dfrac{3}{4}$ **17.** $-\dfrac{5}{9}$

19. 0.747 **21.** $\dfrac{1}{3}\sqrt{3}$ **23.** −1.084 **25.** 20.0°

27. 98.50° **29.** parallel **31.** perpendicular

33. $8, -2$ **35.** -3 **37.** two sides equal $2\sqrt{10}$

39. $m_1 = \dfrac{5}{12}, m_2 = \dfrac{4}{3}$ **41.** 10 **43.** $4\sqrt{10} + 4\sqrt{2} = 18.3$

45. $(1, 5)$ **47.** $(-2.8, 4.2)$

Exercises 2-2, page 36

1. $4x - y + 20 = 0$ **3.** $7x - 2y - 24 = 0$

5. $x - y + 2 = 0$ **7.** $y = -2.7$

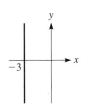   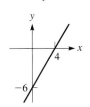

9. $x = -3$ **11.** $3x - 2y - 12 = 0$

    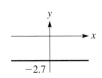

13. $x + 3y + 5 = 0$ **15.** $x + 4y - 9 = 0$

    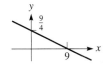

17. $0.4x + y + 1.7 = 0$ **19.** $3x + y - 18 = 0$

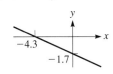   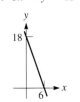

21. $y = 4x - 8; m = 4, (0, -8)$

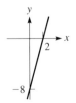

23. $y = -\dfrac{3}{5}x + 2; m = -\dfrac{3}{5}, (0, 2)$

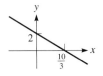

25. $y = \dfrac{3}{2}x - \dfrac{1}{2}; m = \dfrac{3}{2}, \left(0, -\dfrac{1}{2}\right)$

27. $y = 3.5x + 0.5; m = 3.5, (0, 0.5)$ **29.** -2

31. The slope of the first line is 3. A line perpendicular to it has a slope of $-\dfrac{1}{3}$. The slope of the second line is $-\dfrac{k}{3}$, so $k = 1$.

33. parallel   **35.** perpendicular   **37.** neither

39. perpendicular

41. $v = 12.2 + 5.16t$ **43.** $v = 0.607T + 331$

45. $5x + 6y = 1220$

47. $y = 150{,}000 - 0.80x$

49. $y = 10^{-5}(2.4 - 5.6x)$

51. $n = \dfrac{7}{6}t + 10$; at 6:30, $n = 10$; at 8:30, $n = 150$

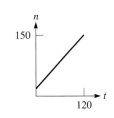

53.

55.

57.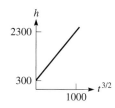

59.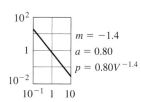

Exercises 2-3, page 41

1. $(2, 1)$, $r = 5$ **3.** $(-1, 0)$, $r = 2$ **5.** $x^2 + y^2 = 9$

7. $(x - 2)^2 + (y - 2)^2 = 16$, or
$x^2 + y^2 - 4x - 4y - 8 = 0$

9. $(x + 2)^2 + (y - 5)^2 = 5$, or
$x^2 + y^2 + 4x - 10y + 24 = 0$

11. $(x - 12)^2 + (y + 15)^2 = 324$, or
$x^2 + y^2 - 24x + 30y + 45 = 0$

13. $(x - 2)^2 + (y - 1)^2 = 8$, or
$x^2 + y^2 - 4x - 2y - 3 = 0$

15. $(x + 3)^2 + (y - 5)^2 = 25$, or
$x^2 + y^2 + 6x - 10y + 9 = 0$

17. $(x - 2)^2 + (y - 2)^2 = 4$, or
$x^2 + y^2 - 4x - 4y + 4 = 0$

19. $(x - 2)^2 + (y - 5)^2 = 25$, or
$x^2 + y^2 - 4x - 10y + 4 = 0$; and
$(x + 2)^2 + (y + 5)^2 = 25$, or
$x^2 + y^2 + 4x + 10y + 4 = 0$

21. $(0, 3)$, $r = 2$ **23.** $(-1, 5)$, $r = \dfrac{9}{2}$

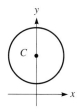

25. $(1, 0)$, $r = 3$ **27.** $(-2.1, 1.3)$, $r = 3.1$

29. $(0, 2)$, $r = \dfrac{5}{2}$ **31.** $(1, 2)$, $r = \dfrac{1}{2}\sqrt{22}$

33. symmetrical to both axes and origin

35. symmetrical to y-axis **37.** $(7, 0)$, $(-1, 0)$

39. $3x^2 + 3y^2 + 4x + 8y - 20 = 0$, circle

41. 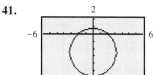 **43.** 2.82 in.

45. $x^2 + y^2 = 0.0100$

47. $(x - 500 \times 10^{-6})^2 + y^2 = 0.16 \times 10^{-6}$

Exercises 2-4, page 46

1. $F(1, 0)$, $x = -1$ **3.** $F(-1, 0)$, $x = 1$

5. $F(0, 2)$, $y = -2$ **7.** $F(0, -1)$, $y = 1$

 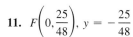

9. $F\left(\dfrac{5}{8}, 0\right)$, $x = -\dfrac{5}{8}$ **11.** $F\left(0, \dfrac{25}{48}\right)$, $y = -\dfrac{25}{48}$

13. $y^2 = 12x$ **15.** $x^2 = 16y$ **17.** $x^2 = 0.64y$

19. $x^2 = \dfrac{1}{8}y$ **21.** $y^2 - 2y - 12x + 37 = 0$

(3, 1)

23.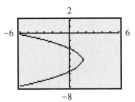

25.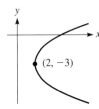
(2, −3)

27. $4p$ **29.** $x^2 = 14{,}700y$ **31.**
H
i

33. 57.6 m **35.** $y^2 = 1.2x$ **37.**
f
0.92
200 A

39. $y^2 = 8x$ or $x^2 = 8y$ with vertex midway between island and shore.

Exercises 2-5, page 51

1. $V(2, 0), V(-2, 0),$
$F(\sqrt{3}, 0), F(-\sqrt{3}, 0)$

3. $V(0, 6), V(0, -6),$
$F(0, \sqrt{11}), F(0, -\sqrt{11})$

5. $V(3, 0), V(-3, 0),$
$F(\sqrt{5}, 0), F(-\sqrt{5}, 0)$

7. $V(0, 7), V(0, -7),$
$F(0, \sqrt{45}), F(0, -\sqrt{45})$

9. $V(0, 4), V(0, -4),$
$F(0, \sqrt{14}), F(0, -\sqrt{14})$

11. $V(0.25, 0), V(-0.25, 0),$
$F(0.23, 0), F(-0.23, 0)$
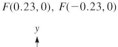

13. $\dfrac{x^2}{225} + \dfrac{y^2}{144} = 1$, or $144x^2 + 225y^2 = 32{,}400$

15. $\dfrac{y^2}{9} + \dfrac{x^2}{5} = 1$, or $9x^2 + 5y^2 = 45$

17. $\dfrac{x^2}{64} + \dfrac{15y^2}{144} = 1$, or $3x^2 + 20y^2 = 192$

19. $\dfrac{x^2}{5} + \dfrac{y^2}{20} = 1$, or $4x^2 + y^2 = 20$

21. $16x^2 + 25y^2 - 32x - 50y - 359 = 0$

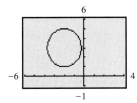

23.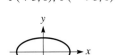

25.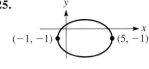
(−1, −1) (5, −1)

27. Write equation as $\dfrac{x^2}{1} + \dfrac{y^2}{1/k} = 1$. $\sqrt{\dfrac{1}{k}} > 1$, or $0 < k < 1$.

29. $2x^2 + 3y^2 - 8x - 4 = 2x^2 + 3(-y)^2 - 8x - 4$

31. $\dfrac{2}{3}\sqrt{2} = 0.943$ **33.** $7x^2 + 16y^2 = 112$

35. 27.5 m **37.** 13 ft **39.** 843 ft^3

Exercises 2-6, page 57

1. $V(5, 0), V(-5, 0),$
$F(13, 0), F(-13, 0)$

3. $V(0, 3), V(0, -3),$
$F(0, \sqrt{10}), F(0, -\sqrt{10})$

5. $V(1,0)$, $V(-1,0)$,
$F(\sqrt{5},0)$, $F(-\sqrt{5},0)$

7. $V(0,\sqrt{5})$, $V(0,-\sqrt{5})$,
$F(0,\sqrt{7})$, $F(0,-\sqrt{7})$

29. $x^2 - 2y^2 = 2$

31. $l^2 - x^2 = 2000^2$

9. $V(0,2)$, $V(0,-2)$,
$F(0,\sqrt{5})$, $F(0,-\sqrt{5})$

11. $V(0.4,0)$, $V(-0.4,0)$,
$F(0.9,0)$, $F(-0.9,0)$

33. $i = 6.00/R$

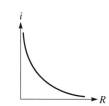

35.

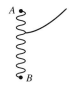

13. $\dfrac{x^2}{9} - \dfrac{y^2}{16} = 1$, or $16x^2 - 9y^2 = 144$

15. $\dfrac{y^2}{100} - \dfrac{x^2}{36} = 1$, or $9y^2 - 25x^2 = 900$

17. $\dfrac{x^2}{1} - \dfrac{y^2}{3} = 1$, or $3x^2 - y^2 = 3$

19. $\dfrac{x^2}{5} - \dfrac{y^2}{4} = 1$, or $4x^2 - 5y^2 = 20$

21.

Exercises 2-7, page 60

1. parabola, $(-1,2)$

3. hyperbola, $(1,2)$

5. ellipse, $(-1,0)$

7. parabola, $(-3,1)$

23. $9x^2 - 16y^2 - 108x + 64y + 116 = 0$

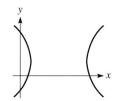

25.

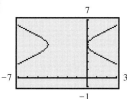

27.

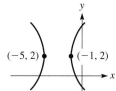

9. $(y-3)^2 = 16(x+1)$, or $y^2 - 6y - 16x - 7 = 0$

11. $y^2 = 24(x-6)$

13. $\dfrac{(x+2)^2}{25} + \dfrac{(y-2)^2}{16} = 1$, or
$16x^2 + 25y^2 + 64x - 100y - 236 = 0$

15. $\dfrac{(y-1)^2}{16} + \dfrac{(x+2)^2}{4} = 1$, or
$4x^2 + y^2 + 16x - 2y + 1 = 0$

17. $\dfrac{(y-2)^2}{1} - \dfrac{(x+1)^2}{3} = 1$, or
$x^2 - 3y^2 + 2x + 12y - 8 = 0$

19. $\dfrac{(x+1)^2}{9} - \dfrac{(y-1)^2}{16} = 1$, or
$16x^2 - 9y^2 + 32x + 18y - 137 = 0$

21. parabola, $(-1, -1)$

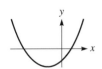

23. ellipse, $(-3, 0)$

25. hyperbola, $(0, 4)$

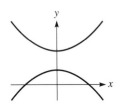

27. parabola, $(1, 0)$

29. hyperbola, $(-4, 5)$

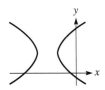

31. ellipse, $\left(\dfrac{2}{3}, -2\right)$

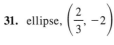

33. hyperbola, $(1, -8)$

35. circle, $\left(\dfrac{1}{3}, \dfrac{4}{3}\right)$

37. $x^2 - y^2 + 4x - 2y - 22 = 0$

39. $y^2 + 4x - 4 = 0$

41. $(x - 95)^2 = -\dfrac{95^2}{60}(y - 60)$

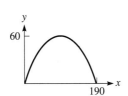

43. $\dfrac{x^2}{9.0} + \dfrac{y^2}{16} = 1, \dfrac{(x - 7.0)^2}{16} + \dfrac{y^2}{9.0} = 1$

Exercises 2-8, page 63

1. ellipse   **3.** hyperbola   **5.** circle   **7.** parabola

9. hyperbola   **11.** circle   **13.** none (st. line)

15. hyperbola   **17.** ellipse   **19.** ellipse

21. parabola; $V(-4, 0)$; $F(-4, 2)$

23. hyperbola; $C(1, -2)$;
     $V(1, -2 \pm \sqrt{2})$

25. ellipse; $C(5, 0)$;
     $V(5, \pm 2\sqrt{2})$

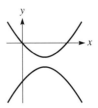

27. parabola; $V\left(-\dfrac{1}{2}, \dfrac{5}{2}\right)$; $F = \left(\dfrac{1}{2}, \dfrac{5}{2}\right)$

29. ellipse

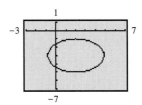

31. $y = \dfrac{-3x - 7 \pm \sqrt{60x + 139}}{9}$

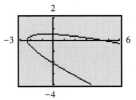

Enter both functions (from quadratic formula).

33. (a) circle (b) hyperbola (c) ellipse

35. a point at the origin   **37.** parabola

39. Circle if light beam is perpendicular to floor; otherwise, an ellipse.

Review Exercises for Chapter 2, page 65

1. $4x - y - 11 = 0$

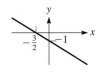

3. $2x + 3y + 3 = 0$

5. $x^2 + y^2 - 2x + 4y - 5 = 0$ **7.** $y^2 = 12x$

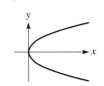

9. $9x^2 + 25y^2 = 900$ **11.** $144y^2 - 169x^2 = 24{,}336$

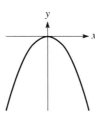

13. $(-3, 0)$, $r = 4$ **15.** $(0, -5)$, $y = 5$

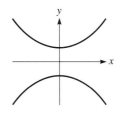

17. $V(0, 4)$, $V(0, -4)$, $F(0, \sqrt{15})$, $F(0, -\sqrt{15})$

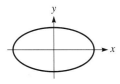

19. $V\left(\dfrac{1}{2\sqrt{2}}, 0\right)$, $V\left(-\dfrac{1}{2\sqrt{2}}, 0\right)$, $F\left(\dfrac{\sqrt{70}}{20}, 0\right)$, $F\left(-\dfrac{\sqrt{70}}{20}, 0\right)$

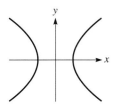

21. $V(4, -8)$, $F(4, -7)$ **23.** $(2, -1)$ **25.** 4   **27.** 2

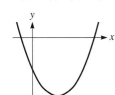

29. $y = \dfrac{6 \pm \sqrt{x^2 + 4x - 12}}{2}$ **31.**

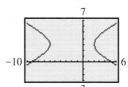

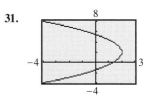

33. $m_1 = -\dfrac{12}{5}$, $m_2 = \dfrac{5}{12}$; $d_1^2 = 169$, $d_2^2 = 169$, $d_3^2 = 338$

35. 8   **37.** $x^2 - 6x - 8y + 1 = 0$ **39.** $R_T = R + 2.5$

41. $y = 25 - \dfrac{5}{3}x$

43. $y = 100.5T - 10{,}050$

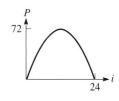

45. 11,000 ft^2 **47.** $y^2 = 32x$

49. $A = 300w - w^2$ **51.**

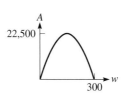

53. 7500 ft^2 **55.** 18 cm, 8 cm   **57.** 37.8 ft

59. $3y^2 - x^2 = 27$

Exercises 3-1, page 76

1. cont. all x **3.** not cont. $x = 0$ and $x = 1$, div. by zero

5. cont. $x \le 0$, $x > 2$; function not defined   **7.** cont. all x

9. not cont. $x = 1$, small change   **11.** cont. $x \le 2$

13. not cont. $x = 2$, small change   **15.** cont. all x

17.

x	2.900	2.990	2.999	3.001	3.010	3.100
$f(x)$	6.700	6.970	6.997	7.003	7.030	7.300

$\lim\limits_{x \to 3} f(x) = 7$

19.

x	0.900	0.990	0.999	1.001
$f(x)$	1.7100	1.9701	1.9970	2.0030

x	1.010	1.100
$f(x)$	2.0301	2.3100

$\lim\limits_{x \to 1} f(x) = 2$

21.

x	1.900	1.990	1.999	2.001
$f(x)$	-0.2516	-0.2502	-0.25002	-0.24998

x	2.010	2.100
$f(x)$	-0.2498	-0.2485

$\lim\limits_{x \to 2} f(x) = -0.25$

23.

x	10	100	1000
$f(x)$	0.4468	0.4044	0.4004

$\lim\limits_{x \to \infty} f(x) = 0.4$

25. 7   **27.** 1   **29.** 1   **31.** $-\dfrac{2}{3}$ **33.** 2   **35.** 2

37. Does not exist   **39.** 0   **41.** 3   **43.** 0

45.

x	-0.1	-0.01	-0.001	0.001
$f(x)$	-3.1	-3.01	-3.001	-2.999

x	0.01	0.1
$f(x)$	-2.99	-2.9

$\lim\limits_{x \to 0} f(x) = -3$

47.

x	10	100	1000
$f(x)$	2.1649	2.0106	2.0010

$\lim\limits_{x \to \infty} f(x) = 2$

49. 3 cm/s   **51.** 34.9° C, 0° C   **53.** e

55. no; div. by zero; $\lim\limits_{x \to 0^+} f(x) \neq \lim\limits_{x \to 0^-} f(x)$

Exercises 3-2, page 81

1. (slopes) 3.5, 3.9, 3.99   **3.** (slopes) $-2, -2.8,$
3.999; $m = 4$ $-2.98, -2.998; m = -3$

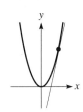

5. 4   **7.** -3 **9.** $m_{\tan} = 2x_1; 4, -2$

11. $m_{\tan} = 4x_1 + 5; -3, 7$ **13.** $m_{\tan} = 2x_1 + 4; -2, 8$

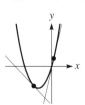   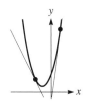

15. $m_{\tan} = 6 - 2x_1; 10, 0$ **17.** $m_{\tan} = 4x_1^3; 0, 0.5, 4$

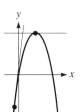

19. $m_{\tan} = 5x_1^4; 0, 0.31, 5$

21.   **23.**

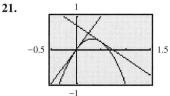

25. $\dfrac{\Delta y}{\Delta x} = 4.1, m_{\tan} = 4$ **27.** $\dfrac{\Delta y}{\Delta x} = -12.61, m_{\tan} = -12$

Exercises 3-3, page 85

1. 3   **3.** -2 **5.** $2x$ **7.** $10x$ **9.** $2x - 7$

11. $8 - 4x$ **13.** $3x^2 + 4$ **15.** $-\dfrac{1}{(x + 2)^2}$

17. $1 - \dfrac{4}{3x^2}$ **19.** $-\dfrac{4}{x^3}$ **21.** $4x^3 + 3x^2 + 2x + 1$

23. $4x^3 + \dfrac{2}{x^2}$ **25.** $6x - 2; -8$

27. $-\dfrac{33}{(3x + 2)^2}; -\dfrac{3}{11}$ **29.** $\dfrac{-2}{x^2}$, all real numbers except 0

31. $\dfrac{-6x}{(x^2 - 1)^2}$, all real numbers except -1 and 1

33. $\dfrac{1}{2\sqrt{x+1}}$ **35.** $\dfrac{1}{2\sqrt{x}}$

Exercises 3-4, page 89

1. $m = 4$ **3.** $m = -\dfrac{3}{4}$

  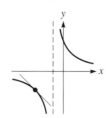

5. $4.00, 4.00, 4.00, 4.00, 4.00; \lim\limits_{t\to 3} v = 4$ ft/s

7. $5, 6.5, 7.7, 7.97, 7.997; \lim\limits_{t\to 2} v = 8$ ft/s   **9.** $4; 4$ ft/s

11. $6t - 4; 8$ ft/s   **13.** $3 + \dfrac{2}{5t^2}$ **15.** $6t - 6t^2$

17. $12t - 4$ **19.** $6t$ **21.** -2 **23.** $6w$

25. 460 W   **27.** -83.1 W/(m$^2\cdot$h)   **29.** πd^2

31. $24.2/\sqrt{\lambda}$

Exercises 3-5, page 94

1. $5x^4$ **3.** $-36x^8$ **5.** $4x^3$ **7.** $2x + 2$

9. $15r^2 - 2$ **11.** $8x^7 - 28x^6 - 1$

13. $-42x^6 + 15x^2$ **15.** $x^2 + x$ **17.** 16   **19.** 33

21. -4 **23.** -29

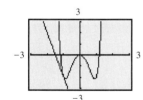 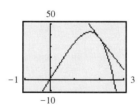

25. $30t^4 - 5$ **27.** $-6 - 6t^2$ **29.** 64   **31.** 45

33. 1   **35.** $(2, 4)$ **37.** $-\dfrac{1}{4}$ **39.** $3\pi r^2$

41. 84 W/A   **43.** $a(c_1 + 2c_2E + 3c_3E^2)$

45. -80.5 m/km   **47.** 391 mm^2

Exercises 3-6, page 98

1. $x^2(3) + (3x + 2)(2x) = 9x^2 + 4x$

3. $6x(6x - 5) + (3x^2 - 5x)(6) = 54x^2 - 60x$

5. $(3t + 2)(2) + (2t - 5)(3) = 12t - 11$

7. $(x^4 - 3x^2 + 3)(-6x^2) + (1 - 2x^3)(4x^3 - 6x)$
$= -14x^6 + 30x^4 + 4x^3 - 18x^2 - 6x$

9. $(2x - 7)(-2) + (5 - 2x)(2) = -8x + 24$

11. $(x^3 - 1)(4x - 1) + (2x^2 - x - 1)(3x^2)$
$= 10x^4 - 4x^3 - 3x^2 - 4x + 1$

13. $\dfrac{3}{(2x + 3)^2}$ **15.** $\dfrac{-2x}{(x^2 + 1)^2}$ **17.** $\dfrac{6x - 2x^2}{(3 - 2x)^2}$

19. $\dfrac{-6x^2 + 6x + 4}{(3x^2 + 2)^2}$ **21.** $\dfrac{-3x^2 - 16x - 26}{(x^2 + 4x + 2)^2}$

23. $\dfrac{-2x^3 + 2x^2 + 5x + 4}{x^3(x + 2)^2}$ **25.** -107 **27.** 75

29. 19   **31.** -5.64

33. (1) $\dfrac{-12x^3 + 45x^2 - 14x}{(3x - 7)^2}$ (2) $\dfrac{-12x^3 + 45x^2 - 14x}{(3x - 7)^2}$

35. 12   **37.** $1, -1$

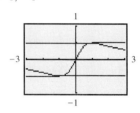

39. $8t^3 - 45t^2 - 14t - 8$ **41.** -0.07 V/Ω

43. $1.2°$C/h   **45.** $\dfrac{2R(R + 2r)}{3(R + r)^2}$ **47.** $\dfrac{E^2(R - r)}{(R + r)^3}$

Exercises 3-7, page 104

1. $\dfrac{1}{2x^{1/2}}$ **3.** $-\dfrac{6}{t^3}$ **5.** $-\dfrac{1}{x^{4/3}}$ **7.** $\dfrac{3}{2}x^{1/2} + \dfrac{1}{x^2}$

9. $10x(x^2 + 1)^4$ **11.** $-192x^2(7 - 4x^3)^7$

13. $\dfrac{2x^2}{(2x^3 - 3)^{2/3}}$ **15.** $\dfrac{24y}{(4 - y^2)^5}$ **17.** $\dfrac{24x^3}{(2x^4 - 5)^{1/4}}$

19. $\dfrac{-4x}{(1 - 8x^2)^{3/4}}$ **21.** $\dfrac{12x + 5}{(8x + 5)^{1/2}}$

23. $\dfrac{-1}{\sqrt{2R + 1}(4R + 1)^{3/2}}$ **25.** $\dfrac{3}{10}$ **27.** $\dfrac{5}{36}$

29. $\dfrac{x^3(0) - 1(3x^2)}{x^6} = -3x^{-4}$

31. $x = 0$ **33.** 1

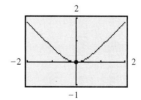

35. -1.35 cm/s   **37.** $\dfrac{-450,000}{V^{5/2}}$, -4.50 kPa/cm^3

39. -45.2 W/(m$^2 \cdot$ h)   **41.** $\dfrac{8a^3}{(4a^2 - \lambda^2)^{3/2}}$

43. $\dfrac{2(w + 1)}{(2w^2 + 4w + 4)^{1/2}}$

Exercises 3-8, page 108

1. $-\dfrac{3}{2}$ **3.** $\dfrac{6x + 1}{4}$ **5.** $\dfrac{x}{4y}$ **7.** $\dfrac{2x}{5y^4}$ **9.** $\dfrac{2x}{2y + 1}$

11. $\dfrac{-3y}{3x + 1}$ **13.** $\dfrac{-2x - y^3}{3xy^2 + 3}$

15. $\dfrac{3(y^2 + 1)(y^2 - 2x + 1)}{(y^2 + 1)^2 - 6x^2y}$ **17.** $\dfrac{4(2y - x)^3 - 2x}{8(2y - x)^3 - 1}$

19. $\dfrac{-3x(x^2 + 1)^2}{y(y^2 + 1)}$ **21.** 3   **23.** $-\dfrac{108}{157}$

25. 1   **27.** $-\dfrac{x}{y}$ **29.** $\dfrac{r - R + 1}{r + 1}$

31. $\dfrac{2C^2r(12CSr - 20Cr - 3L)}{3(C^2r^2 - L^2)}$

Exercises 3-9, page 111

1. $y' = 3x^2 + 2x, y'' = 6x + 2, y''' = 6, y^{(n)} = 0 \ (n \geq 4)$

3. $f'(x) = 3x^2 - 24x^3, f''(x) = 6x - 72x^2,$
    $f'''(x) = 6 - 144x, f^{(4)}(x) = -144, f^{(n)}(x) = 0 \ (n \geq 5)$

5. $y' = -8(1 - 2x)^3, y'' = 48(1 - 2x)^2,$
    $y''' = -192(1 - 2x), y^{(4)} = 384, y^{(n)} = 0 \ (n \geq 5)$

7. $f'(r) = (16r + 1)(4r + 1)^2, f''(r) = 24(8r + 1)(4r + 1),$
    $f'''(r) = 96(16r + 3), f^{iv}(r) = 1536, f^{(n)}(r) = 0 \ (n \geq 5)$

9. $84x^5 - 30x^4$ **11.** $-\dfrac{1}{4x^{3/2}}$ **13.** $-\dfrac{12}{(8x - 3)^{7/4}}$

15. $\dfrac{12}{(1 + 2p)^{5/2}}$ **17.** $600(2 - 5x)^2$

19. $30(27x^2 - 1)(3x^2 - 1)^3$ **21.** $\dfrac{4}{(1 - x)^3}$

23. $\dfrac{2}{(x + 1)^3}$ **25.** $-\dfrac{9}{y^3}$ **27.** $-\dfrac{6(x^2 - xy + y^2)}{(2y - x)^3}$

29. $\dfrac{9}{125}$ **31.** $-\dfrac{13}{384}$ **33.** -50 **35.** 48

37. -32.2 ft/s^2 **39.** $-\dfrac{1.60}{(2t + 1)^{3/2}}$

Review Exercises for Chapter 3, page 113

1. -4 **3.** $\dfrac{1}{4}$ **5.** 1   **7.** $\dfrac{7}{3}$ **9.** $\dfrac{2}{3}$ **11.** -2

13. 5   **15.** $-4x$ **17.** $-\dfrac{4}{x^3}$ **19.** $\dfrac{1}{2\sqrt{x + 5}}$

21. $14x^6 - 6x$ **23.** $\dfrac{2}{x^{1/2}} + \dfrac{3}{x^2}$ **25.** $\dfrac{3}{(1 - 5y)^2}$

---

27. $-12(2 - 3x)^3$ **29.** $\dfrac{9x}{(5 - 2x^2)^{7/4}}$

31. $\dfrac{1}{\sqrt{1 + \sqrt{1 + \sqrt{1 + 8x}}} \ \sqrt{1 + \sqrt{1 + 8x}} \ \sqrt{1 + 8x}}$

33. $\dfrac{-2x - 3}{2x^2(4x + 3)^{1/2}}$ **35.** $\dfrac{2x - 6(2x - 3y)^2}{1 - 9(2x - 3y)^2}$ **37.** $\dfrac{5}{48}$

39. $\dfrac{74}{5}$ **41.** $36x^2 - \dfrac{2}{x^3}$ **43.** $\dfrac{56}{(1 + 4x)^3}$

45. It appears to be 8, but using TRACE there is no value shown for $x = 2$.

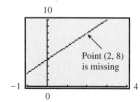
Point (2, 8) is missing

47. (a) 30 ft/s (b) 6 ft/s   **49.** -31

51. 5   **53.** $-k + k^2t - \dfrac{1}{2}k^3t^2$ **55.** $-\dfrac{2k}{r^3}$

57. $0.4(0.01t + 1)^2(0.04t + 1)$

59. $\dfrac{2R(R + 2r)}{3(R + r)^2}$ **61.** $-\dfrac{1}{4\pi\sqrt{C}(L + 2)^{3/2}}$

63. $\dfrac{-15}{(0.5t + 1)^2}$

65. $y' = \dfrac{w}{6EI}(3L^2x - 3Lx^2 + x^3)$

    $y'' = \dfrac{w}{2EI}(L - x)^2$

    $y''' = \dfrac{w}{EI}(x - L)$

    $y^{iv} = \dfrac{w}{EI}$

67. $p = 2w + \dfrac{150}{w}, \dfrac{dp}{dw} = 2 - \dfrac{150}{w^2}$

69. $A = 4x - x^3, \dfrac{dA}{dx} = 4 - 3x^2$

71. At $t = 5$ years, $dV/dt = -\$7500$/year (rate of appreciation is decreasing); $d^2V/dt^2 = \$1500$/year^2 (rate at which appreciation changes is increasing) (Machinery is depreciating, but depreciation is lessening.)

Exercises 4-1, page 119

1. $4x - y - 2 = 0$

3. $2y + x - 2 = 0$

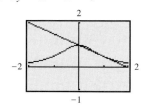

5. $x - 2y + 6 = 0$

7. $2x - 6y + 7 = 0$

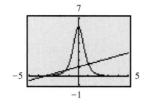

9. $\sqrt{3}x + 8y - 7 = 0$
$16x - 2\sqrt{3}y - 15\sqrt{3} = 0$

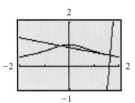

11. $2x - 12y + 37 = 0$
$72x + 12y + 37 = 0$

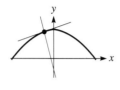

13. $y = 2x - 4$

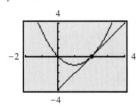

15. $y - 8 = -\frac{1}{24}(x - \frac{3}{2})$,
or $2x + 48y - 387 = 0$

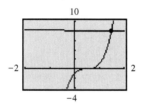

17. Take derivatives; evaluate at (a, b); show that product
$m_1 m_2 = -4a/b^2 = -1$.

19. $3x - 5y - 150 = 0$

21. $x + y - 6 = 0$

23. $x + 2y - 3 = 0$, $x = 0$, $x - 2y + 3 = 0$

Exercises 4-2, page 123

1. 3.4494897 **3.** −0.1804604 **5.** 0.5857864
7. 0.3488942 **9.** 2.5615528 **11.** −1.2360680
13. 0.9175433 **15.** 0.6180340
17. −1.8557725, 0.6783628, 3.1774097
19. Find the real root of $x^3 - 4 = 0$; 1.5874011
21. 29.4 m **23.** 5.05 ft

Exercises 4-3, page 129

1. 3.16, 341.6°

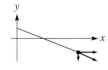

3. 8.07, 352.4°

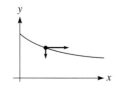

5. $a = 0$ **7.** 20.0, 3.7° **9.** 9.4 m/s, 302°
11. 1.3 ft/min², 288° **13.** 120 ft/s, 323°; 32 ft/s², 270°
15. 276 m/s, 43.5°; 2090 m/s, 16.7°
17. 22.1 m/s², 25.4°; 20.2 m/s², 8.5°
19. 21.2 mi/min, 296.6°
21. $x^2 + y^2 = 1.75^2$; $v_x = 28,800$ in./min,
$v_y = -27,100$ in./min
23. 370 m/s, 19°

Exercises 4-4, page 132

1. 0.0900 Ω/s **3.** 330 mi/h **5.** 4.1×10^{-6} m/s
7. $\dfrac{dB}{dt} = \dfrac{-3kr(dr/dt)}{[r^2 + (l/2)^2]^{5/2}}$ **9.** 0.15 mm²/month
11. −101 mm³/min **13.** −4.6 kPa/min
15. 3.18×10^6 mm³/s **17.** 0.48 m/min
19. 12.0 ft/s **21.** 820 mi/h **23.** 8.33 ft/s

Exercises 4-5, page 139

1. inc. $x > -1$, dec. $x < -1$
3. inc. $-2 < x < 2$, dec. $x < -2$, $x > 2$
5. min. $(-1, -1)$ **7.** min. $(-2, -16)$, max. $(2, 16)$
9. conc. up all x
11. conc. up $x < 0$, conc. down $x > 0$, infl. $(0, 0)$

13.

15.

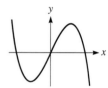

17. max. $(3, 18)$,
conc. down all x

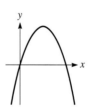

19. max. $(-2, 8)$, min. $(0, 0)$,
infl. $(-1, 4)$

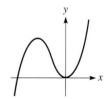

21. no max. or min.,
infl. $(-1, 1)$

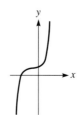

23. max. $(1, 16)$, min. $(3, 0)$,
infl. $(2, 8)$

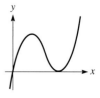

25. max. (1, 1),
inf1. (0, 0), $(\frac{2}{3}, \frac{16}{27})$

27. max. $(-1, 4)$, min. $(1, -4)$,
inf1. (0, 0)

5. int. $(-\sqrt[3]{2}, 0)$, min. (1, 3),
inf1. $(-\sqrt[3]{2}, 0)$, asym. $x = 0$

7. int. (1, 0), $(-1, 0)$,
asym. $x = 0$, $y = x$,
conc. up $x < 0$,
conc. down $x > 0$

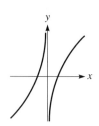

29. Where $y' > 0$, y inc.
$y' = 0$, y has a max. or min.
$y' < 0$, y dec.
$y'' > 0$, y conc. up
$y'' = 0$, y has infl.
$y'' < 0$, y conc. down

31. max. (200, 100)

33. max. (0, 75),
inf1. (1, 64), (3, 48)

9. int. (0, 0), max. $(-2, -4)$,
min. (0, 0), asym. $x = -1$

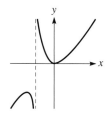

11. int. $(0, -1)$,
max. $(0, -1)$,
asym. $x = 1$,
$x = -1$, $y = 0$

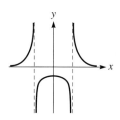

35. $V = 4x^3 - 40x^2 + 96x$,
max. (1.57, 67.6)

37.

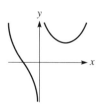

13. int. (1, 0), max. (2, 1),
inf1. $(3, \frac{8}{9})$,
asym. $x = 0$, $y = 0$

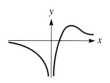

15. int. (0, 0), (1, 0), $(-1, 0)$,
max. $(\frac{1}{2}\sqrt{2}, \frac{1}{2})$,
min. $(-\frac{1}{2}\sqrt{2}, -\frac{1}{2})$

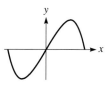

39.

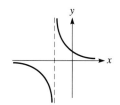

17. int. (0, 0), inf1. (0, 0)
asym. $x = -3$,
$x = 3$, $y = 0$

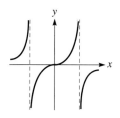

19. int. (0, 0), asym. $C_T = 6$,
inc. $C \geq 0$,
conc. down $C \geq 0$

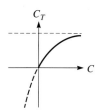

Exercises 4-6, page 143

1. inc. $x < 0$, dec. $x > 0$,
conc. up $x < 0$, $x > 0$,
asym. $x = 0$, $y = 0$

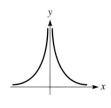

3. dec. $x < -1$, $x > -1$,
conc. up $x > -1$,
conc. down $x < -1$,
int. (0, 2),
asym. $x = -1$, $y = 0$

21. int. (0, 1), max. (0, 1)
inf1. (141, 0.82),
asym. $R = 0$

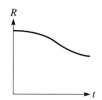

23. $A = 2\pi r^2 + \dfrac{40}{r}$,
min. $(1.47, 40.8)$,
asym. $r = 0$

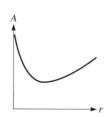

5. $x - 2y + 3 = 0$ **7.** $4.19, 72.6°$ **9.** 2.12

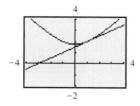

Exercises 4-7, page 148

1. 196 ft   **3.** $\dfrac{E}{2R}$ **5.** 35 m², \$8300   **7.** 1500 Ω

9. 12 in. by 12 in.   **11.** 5 mm, 5 mm   **13.** 1.1 h

15. 8.49 cm, 8.49 cm   **17.** 3.00 ft   **19.** 12

21. $0.58L$ **23.** 1.33 in.   **25.** 100 m

27. $w = 1.00$ ft, $d = 1.73$ ft   **29.** 8.0 mi from refinery

31. 59.2 m, 118 m

Exercises 4-8, page 154

1. $(5x^4 + 1)\,dx$ **3.** $\dfrac{-10\,dr}{r^6}$ **5.** $48t(3t^2 - 5)^3\,dt$

7. $\dfrac{-12x\,dx}{(3x^2 + 1)^2}$ **9.** $x(1 - x)^2(-5x + 2)\,dx$

11. $\dfrac{2\,dx}{(5x + 2)^2}$ **13.** 12.28, 12   **15.** 1.71275, 1.675

17. $-2.4, -2.4730881$ **19.** 0.6257, 0.6264903

21. $L(x) = 2x$ **23.** $L(x) = -2x - 3$

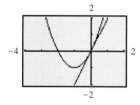 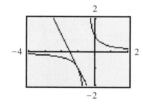

25. 0.0038 cm²   **27.** -31 nm   **29.** $\dfrac{dr}{r} = \dfrac{kd\lambda/(2\lambda^{1/2})}{k\lambda^{1/2}}$

31. $\dfrac{dA}{A} = \dfrac{2ds}{s}$ **33.** $L(x) = -\frac{1}{2}x + \frac{3}{2}$; 1.45

35. $L(V) = 1.73 - 0.13V$

Review Exercises for Chapter 4, page 155

1. $5x - y + 1 = 0$ **3.** $8x + 5y - 50 = 0$

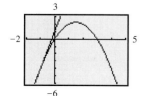 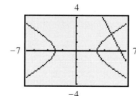

11. $2.00, 90.9°$ **13.** 0.7458983   **15.** 1.9111643

17. min. $(-2, -16)$,   **19.** int. $(0, 0)$, $(\pm 3\sqrt{3}, 0)$,
conc. up all x max. $(3, 54)$, min. $(-3, -54)$,
infl. $(0, 0)$

 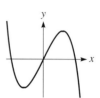

21. min. $(2, -48)$,   **23.** min. $(\pm\sqrt{2}, 2)$,
conc. up $x < 0$, $x > 0$ asym. $x = \pm 1$
conc. up $x < -1$, $x > 1$

25. $\left(12x^2 - \dfrac{1}{x^2}\right)dx$ **27.** $\dfrac{(1 - 4x)\,dx}{(1 - 3x)^{2/3}}$ **29.** 0.061

31. $L(x) = \dfrac{1}{3}(11x - 4)$ **33.** 1.85 m²   **35.** $\dfrac{R\,dR}{R^2 + X^2}$

37. $2x - y + 1 = 0$ **39.** 0.0 m, 6.527 m

41. 8.8 m/s, 336°   **43.** -7.44 cm/s   **45.** 22 m³/s

47.   **49.** -0.567 ft/min

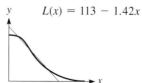

51. 38,000 m²/min   **53.** 5000 cm²

55. max. $(0, 100)$; infl. $(37, 63)$;   **57.** 710 mi/h
int. $(0, 100)$, $(89, 0)$

$L(x) = 113 - 1.42x$

59. 6 μF, 6 μF   **61.** 13.5 ft³   **63.** 2.2 in.

65. 0.318 ft/min   **67.** $r = 2.09$ cm
$h = 23.0$ cm

Exercises 5-1, page 161

1. 1   **3.** 3   **5.** 6   **7.** −1

9. $x^{5/2}$ **11.** $\frac{3}{2}t^4 + 2t$ **13.** $\frac{2}{3}x^3 - \frac{1}{2}x^2$ **15.** $\frac{4}{3}x^{3/2} + 3x$

17. $\dfrac{7}{5x^5}$ **19.** $2v^2 + 3\pi^2 v$ **21.** $\frac{1}{3}x^3 + 2x - \dfrac{1}{x}$

23. $(2x + 1)^6$ **25.** $(p^2 - 1)^4$ **27.** $\frac{1}{40}(2x^4 + 1)^5$

29. $(6x + 1)^{3/2}$ **31.** $\frac{1}{4}(3x + 1)^{4/3}$

Exercises 5-2, page 166

1. $x^2 + C$ **3.** $\frac{1}{8}x^8 + C$ **5.** $\frac{4}{5}x^{5/2} + C$

7. $-\dfrac{1}{3x^3} + C$ **9.** $\frac{1}{3}x^3 - \frac{1}{6}x^6 + C$

11. $3x^3 + \frac{1}{2}x^2 + 3x + C$ **13.** $\dfrac{1}{6}t^3 + \dfrac{2}{t} + C$

15. $\frac{2}{7}x^{7/2} - \frac{2}{5}x^{5/2} + C$ **17.** $6x^{1/3} + \frac{1}{9}x + C$

19. $s + \frac{4}{3}s^3 + \frac{4}{5}s^5 + C$ **21.** $\frac{1}{6}(x^2 - 1)^6 + C$

23. $\frac{1}{5}(x^4 + 3)^5 + C$ **25.** $\frac{1}{80}(2\theta^5 + 5)^8 + C$

27. $\frac{1}{12}(8x + 1)^{3/2} + C$ **29.** $\frac{1}{6}\sqrt{6x^2 + 1} + C$

31. $\sqrt{x^2 - 2x} + C$ **33.** $y = 2x^3 + 2$

35. $y = 5 - \frac{1}{18}(1 - x^3)^6$ **37.** $12y = 83 + (1 - 4x^2)^{3/2}$

39. $i = 2t^2 - 0.2t^3 + 2$ **41.** $f = \sqrt{0.01A + 1} - 1$

43. $y = 3x^2 + 2x - 3$

Exercises 5-3, page 171

1. 9, 12.15   **3.** 1.92, 2.28   **5.** 7.625, 8.208

7. 0.464, 0.5995   **9.** 1.92, 1.96   **11.** 13.5

13. $\frac{8}{3}$ **15.** 9   **17.** 0.8   **19.** 2

Exercises 5-4, page 174

1. 1   **3.** $\frac{254}{7}$ **5.** $6 + 2\sqrt{6} - 2\sqrt{3}$ **7.** 2.53

9. $-\frac{32}{3}$ **11.** $\frac{33}{20}\sqrt[3]{\frac{11}{5}} - \frac{3}{8}\sqrt[3]{\frac{1}{2}} - \frac{17}{5} = -1.552$ **13.** $\frac{4}{3}$

15. $-\frac{81}{4}$ **17.** 2   **19.** $\frac{1}{4}(20.5^{2/3} - 17.5^{2/3}) = 0.1875$

21. $\frac{88}{3249} = 0.0271$ **23.** 84   **25.** $\frac{364}{3}$ **27.** $\frac{3880}{9}$

29. $\frac{33}{784} = 0.0421$ **31.** $\frac{2}{3}(13\sqrt{13} - 27) = 13.25$

33. 64,000 ft·lb   **35.** 86.8 m²

Exercises 5-5, page 177

1. $\frac{11}{2} = 5.50$, $\frac{16}{3} = 5.33$ **3.** 7.661, $\frac{23}{3} = 7.667$

5. 0.2042   **7.** 18.98   **9.** 0.5205   **11.** 21.74

13. 45.36   **15.** 0.177k

Exercises 5-6, page 181

1. (a) 6 (b) 6   **3.** (a) 19.67 (b) 19.67   **5.** 0.2028

7. 19.27   **9.** 0.5114   **11.** 13.147   **13.** 44.63

15. 1.191 in.

Review Exercises for Chapter 5, page 182

1. $x^4 - \frac{1}{2}x^2 + C$ **3.** $\frac{2}{7}u^{7/2} + \frac{4}{3}u^{3/2} + C$ **5.** $\frac{19}{3}$

7. $\frac{16}{3}$ **9.** $3x - \dfrac{1}{x^2} + C$ **11.** 3

13. $\dfrac{1}{10(2 - 5u)^2} + C$ **15.** $-\frac{6}{7}(7 - 2x)^{7/4} + C$

17. $\frac{9}{8}(3\sqrt[3]{3} - 1)$ **19.** $-\frac{1}{30}(1 - 2x^3)^5 + C$

21. $-\dfrac{1}{2x - x^3} + C$ **23.** $\frac{3350}{3}$ **25.** $y = 3x - \frac{1}{3}x^3 + \frac{17}{3}$

27. (a) $x - x^2 + C_1$
(b) $-\frac{1}{4}(1 - 2x)^2 + C_2 = x - x^2 + C_2 - \frac{1}{4}$; $C_1 = C_2 - \frac{1}{4}$

29. 22   **31.** 0.842   **33.** 0.811   **35.** 13.6

37. 19.3016   **39.** 19.0356   **41.** 24.68 m²   **43.** 25.81 m²

45. $y = k(2L^3x - 6Lx^2 + \frac{2}{5}x^5)$

47. 14.9 m²

Exercises 6-1, page 189

1. 80 ft/s   **3.** $s = 8.00 - 0.25t$ **5.** 15 ft/s

7. 17,800 m   **9.** 76 ft/s   **11.** 256 ft

13. 0.345 nC   **15.** 0.017 C   **17.** 120 V

19. 4.65 mV   **21.** 970 rad   **23.** 66.7 A   **25.** $\dfrac{k}{x_1}$

27. $m = 1002 - 2\sqrt{t + 1}$, 2.51×10^5 min

Exercises 6-2, page 195

1. 2   **3.** $\frac{8}{3}$ **5.** $\frac{27}{8}$ **7.** $\frac{32}{3}$ **9.** $\frac{1}{6}$ **11.** $\frac{26}{3}$ **13.** 3

15. $\frac{15}{4}$ **17.** $\frac{7}{6}$ **19.** $\frac{256}{15}$ **21.** $\frac{343}{24}$ **23.** $\frac{65}{6}$ **25.** 1

27. $\frac{48}{5}$ **29.** 18.0 J   **31.** 80.8 km

33. 4 cm²   **35.** 0.683 m²

Exercises 6-3, page 200

1. $\frac{8}{3}\pi$ **3.** $\frac{8}{3}\pi$ **5.** $\frac{8}{3}\pi$ **7.** 72π **9.** $\frac{768}{7}\pi$ **11.** $\frac{348}{5}\pi$

13. $\frac{16}{3}\pi$ **15.** $\frac{128}{7}\pi$ **17.** $\frac{2}{5}\pi$ **19.** $\dfrac{10\pi}{3}\sqrt{5}$ **21.** $\frac{1296}{5}\pi$

23. $\frac{16}{3}\pi$ **25.** $\frac{8}{3}\pi$ **27.** $\frac{1}{3}\pi r^2 h$ **29.** 7.56 mm³

31. 18.3 cm³

Exercises 6-4, page 207

1. 2.9 cm   **3.** 1.2 cm   **5.** (−0.5 in., 0.5 in.)

7. (0.32 in., 0.23 in.)   **9.** $(0, \frac{6}{5})$ **11.** $(\frac{4}{3}, \frac{4}{3})$ **13.** $(\frac{3}{5}, \frac{12}{35})$

15. $(\frac{5}{3}, 4)$ **17.** $(\frac{7}{8}, 0)$ **19.** $(0, \frac{5}{6})$ **21.** $(\frac{2}{3}, 0)$

23. $(\frac{2}{3}b, \frac{1}{3}a)$. Place triangle with a vertex at origin, side b and right angle on x-axis. Equation of hypotenuse is $y = ax/b$. Use Eqs. (6-16) and (6-17).

25. 0.375 cm above center of base

27. 19.3 cm from larger base

Exercises 6-5, page 212

1. 68 g · cm², 2.9 cm   **3.** 2530 g · cm², 3.58 cm

5. $\frac{64}{15}k$ **7.** $\frac{2}{3}\sqrt{6}$ **9.** $\frac{1}{6}mb^2$ **11.** $\frac{4}{7}\sqrt{7}$ **13.** $\frac{8}{11}\sqrt{55}$

15. $\frac{64}{3}\pi k$ **17.** $\frac{2}{5}\sqrt{10}$ **19.** $\frac{3}{10}mr^2$ **21.** 0.324 g · cm²

23. 31.2 kg · cm²

Exercises 6-6, page 215

1. 8.0 lb · in.   **3.** 200 N · mm   **5.** 600 N · mm

7. 9.4×10^{-22} J   **9.** 99k J   **11.** 10,000 ft · lb

13. 1800 N · m   **15.** 3.00×10^5 ft · ton

17. 8.82×10^4 ft · lb   **19.** 1.26×10^6 N · m = 1.26 MJ

Exercises 6-7, page 217

1. 2340 lb   **3.** 28,100 lb   **5.** 20,800 lb   **7.** 5320 lb

9. 6500 N   **11.** 1570 lb

13. 3.92×10^4 N, 1.18×10^5 N, buoyant force   **15.** 11,700 lb

Exercises 6-8, page 220

1. $\frac{38}{3}$ **3.** 21   **5.** $4\pi\sqrt{2}$ **7.** $\frac{56}{3}\pi$ **9.** 2.7 A

11. 35.3%   **13.** 109 ft   **15.** $S = \pi r\sqrt{r^2 + h^2}$

Review Exercises for Chapter 6, page 222

1. 4.4 s   **3.** 4.2 s   **5.** 0.44 C   **7.** 55 V

9. $y = 20x + \frac{1}{120}x^3$ **11.** $\frac{2}{3}$ **13.** 18   **15.** $\frac{27}{4}$

17. $\frac{48}{5}\pi$ **19.** $\frac{512}{5}\pi$ **21.** $\frac{4}{3}\pi ab^2$ **23.** $(\frac{40}{21}, \frac{10}{3})$

25. $(\frac{14}{5}, 0)$ **27.** $\frac{8}{5}k$ **29.** 68.7 g · mm²   **31.** 8500 ft · lb

33. 1.8 m   **35.** 47 m³   **37.** 10,200 lb   **39.** 0.29 Ω

Exercises 7-1, page 231

1. 15.2°   **3.** 315°48′   **5.** $\frac{\pi}{12}, \frac{5\pi}{6}$ **7.** $\frac{5\pi}{12}, \frac{11\pi}{6}$

9. 72°, 270°   **11.** 10°, 315°   **13.** 0.401   **15.** 43.0°

17. −0.8290   **19.** 1.4663   **21.** 0.7071   **23.** 3.732

25. −1.732   **27.** −0.1161   **29.** 0.3141, 2.827

31. 2.932, 6.074   **33.** 0.8309, 5.452   **35.** 2.442, 3.841

37. 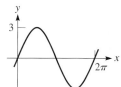   **39.**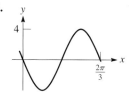

41.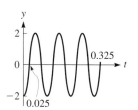

43.

45.

47.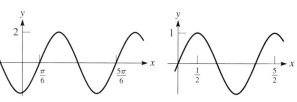

49. $y = 5 \sin(\frac{\pi}{8}x + \frac{\pi}{8})$. As shown: amplitude = 5; period = $\frac{2\pi}{b} = 16$, $b = \frac{\pi}{8}$; displacement = $-\frac{c}{b} = -1$; $c = \frac{\pi}{8}$

51. $y = -0.8 \cos 2x$. As shown: amplitude = $|-0.8| = 0.8$; period = $\frac{2\pi}{b} = \pi$, $b = 2$; displacement = $-\frac{c}{b} = 0$, $c = 0$

Exercises 7-2, page 237

[*Note:* "Answers" to trigonometric identities are intermediate steps of suggested reductions of the left member.]

1. $\frac{\cos x}{\sin x} \cdot \frac{1}{\cos x} = \frac{1}{\sin x}$ **3.** $\left(\frac{\sin y}{\cos y}\right)\left(\frac{1}{\sin y}\right) = \frac{1}{\cos y}$

5. $(1 - \sin^2 x) - \sin^2 x$

7. $\sin x\left(\frac{\sin x}{\cos x}\right) + \cos x = \frac{\sin^2 x + \cos^2 x}{\cos x} = \frac{1}{\cos x}$

9. $\tan^2 x + \tan x \cot x = \tan^2 x + 1$

11. $\frac{\sec\theta}{\frac{1}{\sec\theta}} - \frac{\tan\theta}{\frac{1}{\tan\theta}} = \sec^2\theta - \tan^2\theta$

13. $(\sin x \cos y + \cos x \sin y)(\sin x \cos y - \cos x \sin y)$
$= \sin^2 x \cos^2 y - \cos^2 x \sin^2 y$
$= \sin^2 x(1 - \sin^2 y) - (1 - \sin^2 x)\sin^2 y$

15. $\cos x \cos y + \sin x \sin y + \sin x \cos y + \cos x \sin y$
$= \cos x(\cos y + \sin y) + \sin x(\sin y + \cos y)$

17. $(\cos^2 x - \sin^2 x)(\cos^2 x + \sin^2 x) = (\cos^2 x - \sin^2 x)(1)$

19. $\frac{\sin 3x \cos x + \cos 3x \sin x}{\sin x \cos x} = \frac{\sin 4x}{\frac{1}{2}\sin 2x} = \frac{2 \sin 2x \cos 2x}{\frac{1}{2}\sin 2x}$

21. $\frac{1 - \cos\alpha}{2\sqrt{\frac{1}{2}(1 - \cos\alpha)}} = \sqrt{\frac{1 - \cos\alpha}{2}}$

23. $2\left(\dfrac{1 - \cos x}{2}\right) + \cos x$

25. $0 = \cos A \cos B \cos C + \sin A \sin B,$

$\cos C = -\dfrac{\sin A \sin B}{\cos A \cos B}$

27. $i_0 \sin(\omega t + \alpha) = i_0(\sin \omega t \cos \alpha + \cos \omega t \sin \alpha)$

29. $vi \sin \omega t \sin\left(\omega t - \dfrac{\pi}{2}\right)$

$= vi \sin \omega t\left(\sin \omega t \cos \dfrac{\pi}{2} - \cos \omega t \sin \dfrac{\pi}{2}\right)$

$= vi \sin \omega t[-(\cos \omega t)(1)] = -\dfrac{1}{2}vi(2 \sin \omega t \cos \omega t)$

31. $\sin \omega t = \pm\sqrt{\dfrac{1 - \cos 2\omega t}{2}}, \ \sin^2\omega t = \dfrac{1}{2}(1 - \cos 2\omega t)$

33. 25.7 ft   **35.** 34.73 m²

Exercises 7-3, page 242

1. $\cos(x + 2)$ **3.** $12x^2 \cos(2x^3 - 1)$ **5.** $-3 \sin \frac{1}{2}x$

7. $-6 \sin(3x - 1)$ **9.** $6\pi \sin 3\pi\theta \cos 3\pi\theta = 3\pi \sin 6\pi\theta$

11. $-45 \cos^2(5x + 2)\sin(5x + 2)$

13. $\sin 3x + 3x \cos 3x$ **15.** $9x^2 \cos 5x - 15x^3 \sin 5x$

17. $2x \cos x^2 \cos 2x - 2 \sin x^2 \sin 2x$

19. $\dfrac{2 \cos 4x}{\sqrt{1 + \sin 4x}}$ **21.** $\dfrac{3t \cos(3t - \pi/3) - \sin(3t - \pi/3)}{2t^2}$

23. $\dfrac{4x(1 - 3x)\sin x^2 - 6 \cos x^2}{(3x - 1)^2}$

25. $4 \sin 3x(3 \cos 3x \cos 2x - \sin 3x \sin 2x)$

27. $2 \cos 2t \cos(\sin 2t)$

29. $3 \sin^2 x \cos x + 2 \sin 2x$

31. $-\dfrac{\cos s}{\sin^2 s} + \dfrac{\sin s}{\cos^2 s}$ **33.** (a)

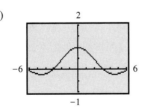

(b) See the table.

35. (a) 0.5403023, value of derivative
(b) 0.5402602, slope of secant line

37. Resulting curve is $y = \cos x$.

39. $\dfrac{2x - y \cos xy}{x \cos xy - 2 \sin 2y}$

41. $\dfrac{d \sin x}{dx} = \cos x, \dfrac{d^2 \sin x}{dx^2} = -\sin x,$

$\dfrac{d^3 \sin x}{dx^3} = -\cos x, \dfrac{d^4 \sin x}{dx^4} = \sin x$

43. 0.085   **45.** −2.36   **47.** 410 V/s

49. −199 cm/s   **51.** −38.5 km

Exercises 7-4, page 246

1. $5 \sec^2 5x$ **3.** $5 \csc^2(0.25\pi - \theta)$ **5.** $6 \sec 2x \tan 2x$

7. $\dfrac{3}{\sqrt{2x + 3}} \csc \sqrt{2x + 3} \cot \sqrt{2x + 3}$

9. $30 \tan 3x \sec^2 3x$ **11.** $-4 \cot^3 \dfrac{1}{2}x \csc^2 \dfrac{1}{2}x$

13. $2 \tan 4x\sqrt{\sec 4x}$ **15.** $-84 \csc^4 7x \cot 7x$

17. $0.5t^2 \sec^2 0.5t + 2t \tan 0.5t$

19. $-4 \csc x^2(2x \cos x \cot x^2 + \sin x)$

21. $-\dfrac{\csc x(x \cot x + 1)}{x^2}$

23. $\dfrac{2(-4 \sin 4x - 4 \sin 4x \cot 3x + 3 \cos 4x \csc^2 3x)}{(1 + \cot 3x)^2}$

25. $\sec^2 x(\tan^2 x - 1)$ **27.** $2 \cos 2\theta \sec^2(\sin 2\theta)$

29. $\dfrac{1 + 2 \sec^2 4x}{\sqrt{2x + \tan 4x}}$ **31.** $\dfrac{2 \cos 2x - \sec y}{x \sec y \tan y - 2}$

33. $24 \tan 3x \sec^2 3x \, dx$

35. $4 \sec 4x(\tan^2 4x + \sec^2 4x) \, dx$

37. (a) 3.4255188, value of derivative
(b) 3.4260524, slope of secant line

39. $2 \tan x \sec^2 x = 2 \sec x(\sec x \tan x)$ **41.** −12

43. $2 \sec^2 x - \sec x \tan x = \dfrac{2}{\cos^2 x} - \dfrac{\sin x}{\cos^2 x}$

45. −8.4 cm/s   **47.** 140 ft/s

Exercises 7-5, page 252

1. y is the angle whose tangent is x.

3. y is the angle whose cotangent is $3x$.

5. y is twice the angle whose sine is x.

7. y is 5 times the angle whose cosine is $2x - 1$.

9. $\dfrac{\pi}{3}$ **11.** $\dfrac{\pi}{4}$ **13.** $-\dfrac{\pi}{3}$ **15.** $\dfrac{\pi}{3}$

17. $-\dfrac{\pi}{4}$ **19.** $\dfrac{\pi}{4}$ **21.** 0   **23.** $\dfrac{1}{2}\sqrt{3}$

25. $\dfrac{1}{2}\sqrt{2}$ **27.** −1   **29.** $2/\sqrt{21}$ **31.** $2/\sqrt{3}$

33. −1.3090   **35.** −0.9838   **37.** 1.4413   **39.** 1.4503

41. −1.2389   **43.** −0.2239   **45.** $x = \dfrac{1}{3}\sin^{-1} y$

47. $x = 4 \tan y$ **49.** $x = \dfrac{1}{3}\sec^{-1}\left(\dfrac{y - 1}{3}\right)$

51. $x = 1 - \cos(1 - y)$ **53.** $\dfrac{x}{\sqrt{1 - x^2}}$ **55.** $\dfrac{1}{x}$

57. $\dfrac{3x}{\sqrt{9x^2 - 1}}$ **59.** $2x\sqrt{1 - x^2}$

61. $t = \dfrac{1}{2\omega}\cos^{-1}\dfrac{y}{A} - \dfrac{\phi}{\omega}$ **63.** $t = \dfrac{1}{\omega}\left(\sin^{-1}\dfrac{i}{I_m} - \alpha - \phi\right)$

65. $\sin\left(\sin^{-1}\dfrac{3}{5} + \sin^{-1}\dfrac{5}{13}\right) = \dfrac{3}{5}\cdot\dfrac{12}{13} + \dfrac{4}{5}\cdot\dfrac{5}{13} = \dfrac{56}{65}$

67. $\dfrac{\pi}{6} + \dfrac{\pi}{3} = \dfrac{\pi}{2}$ **69.** $\sin^{-1}\left(\dfrac{a}{c}\right)$

71. Let $y =$ height to top of pedestal; $\tan\alpha = \dfrac{151 + y}{d}$,

$\tan\beta = \dfrac{y}{d}$; $\tan\alpha = \dfrac{151 + d\tan\beta}{d}$

Exercises 7-6, page 256

1. $\dfrac{2x}{\sqrt{1 - x^4}}$ **3.** $\dfrac{18x^2}{\sqrt{1 - 9x^6}}$ **5.** $\dfrac{-1.8}{\sqrt{1 - 0.25s^2}}$

7. $\dfrac{1}{\sqrt{(x - 1)(2 - x)}}$ **9.** $\dfrac{1}{2\sqrt{x}(1 + x)}$ **11.** $-\dfrac{6}{x^2 + 1}$

13. $\dfrac{5x}{\sqrt{1 - x^2}} + 5\sin^{-1}x$ **15.** $\dfrac{0.8u}{1 + 4u^2} + 0.4\tan^{-1}2u$

17. $\dfrac{3\sqrt{1 - 4x^2}\sin^{-1}2x - 6x + 2}{\sqrt{1 - 4x^2}(\sin^{-1}2x)^2}$

19. $\dfrac{2(\cos^{-1}2x + \sin^{-1}2x)}{\sqrt{1 - 4x^2}(\cos^{-1}2x)^2}$ **21.** $\dfrac{-24(\cos^{-1}4x)^2}{\sqrt{1 - 16x^2}}$

23. $\dfrac{4\sin^{-1}(4x + 1)}{\sqrt{-4x^2 - 2x}}$ **25.** $\dfrac{-1}{t^2 + 1}$

27. $\dfrac{-2(2x + 1)^2}{(1 + 4x^2)^2}$ **29.** $\dfrac{18(4 - \cos^{-1}2x)^2}{\sqrt{1 - 4x^2}}$

31. $-\dfrac{x^2y^2 + 2y + 1}{2x}$

33. (a) 1.1547005, value of derivative
(b) 1.1547390, slope of secant line

35. $\dfrac{3(\sin^{-1}x)^2\,dx}{\sqrt{1 - x^2}}$ **37.** 0.41

39.

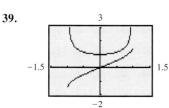

41. $\dfrac{-16x}{(1 + 4x^2)^2}$

43. Let $y = \sec^{-1}u$; solve for u;
take derivatives; substitute.

45. $\dfrac{E - A}{\omega m\sqrt{m^2E^2 - (A - E)^2}}$

47. $\theta = \tan^{-1}\dfrac{h}{x}$; $\dfrac{d\theta}{dx} = \dfrac{-h}{x^2 + h^2}$

Exercises 7-7, page 260

1. $d\sin x/dx = \cos x$ and $d\cos x/dx = -\sin x$, and
$\sin x = \cos x$ at points of intersection.

3. $\dfrac{1}{x^2 + 1}$ is always positive.

5.

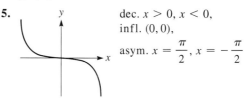

dec. $x > 0$, $x < 0$,
infl. $(0, 0)$,
asym. $x = \dfrac{\pi}{2}$, $x = -\dfrac{\pi}{2}$

7. $8\sqrt{2}x + 8y + 4\sqrt{2} - 5\pi\sqrt{2} = 0$ **9.** 1.9337538
11. 10 **13.** 0.58 ft/s, -1.7 ft/s^2 **15.** -0.072 lb/s
17. 98.96 in./s, 270° **19.** 3731 in./s^2, 0°
21. -0.073 rad/s **23.** 8.08 ft/s **25.** 0.020
27. 0.19 m **29.** $w = 9.24$ in., $d = 13.1$ in.
31. 14 ft

Review Exercises for Chapter 7, page 263

1. $-12\sin(4x - 1)$ **3.** $\dfrac{-0.2\sec^2\sqrt{3 - 2v}}{\sqrt{3 - 2v}}$

5. $-6\csc^2(3x + 2)\cot(3x + 2)$

7. $-24x\cos^3x^2\sin x^2$ **9.** $\dfrac{9}{9 + x^2}$ **11.** -1

13. $(-2\csc 4x)\sqrt{\csc 4x + \cot 4x}$

15. $8\cos 2x(1 + \sin 2x)^3$ **17.** $\dfrac{2x(1 + 4x^2)(\tan^{-1}2x) - 2x^2}{(1 + 4x^2)(\tan^{-1}2x)^2}$

19. $-\dfrac{2y^2\cos 2x + \sec^2x}{2y\sin 2x}$ **21.** $\cos^{-1}x$

23. $\dfrac{2y\sin 2x + \sin 2y}{\cos 2x - 2x\cos 2y}$

25. infl. $(\frac{1}{2}\pi, \frac{1}{2}\pi)$, $(\frac{3}{2}\pi, \frac{3}{2}\pi)$ **27.** $7.27x + y - 8.44 = 0$

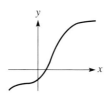

29. $2\sin x\cos x - 2\cos x\sin x = 0$ **31.** 0.5109734
33. $-0.064°$ C/day **35.** $-kE_0^2\cos\frac{1}{2}\theta\sin\frac{1}{2}\theta$
37. 45° **39.** 2π, 330° **41.** 0.568 rad/s
43. -0.0065 rad/s **45.** 7.07 in

47. $a = 16 \cos \theta(1 + \sin \theta)$

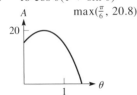

$\max(\frac{\pi}{6}, 20.8)$

Exercises 8-1, page 271

1. $4 = \log_4 256$ **3.** $16 = 8^{4/3}$ **5.** $1/3 = \log_8 2$

7. $16 = 0.5^{-4}$ **9.** $0 = \log_{12} 1$ **11.** $1/512 = 8^{-3}$

13. 343 **15.** 0.2 **17.**

19.

21. $\log_5 3$ **23.** $\log_6 x^{5/2}$ **25.** $\log_b 4y$ **27.** $\log_2(\frac{3}{5}x^2)$

29. 1.099 **31.** 1.921 **33.** $b_2 = b_1 10^{0.4(m_1 - m_2)}$

35. $N = N_0 e^{-kt}$

37.

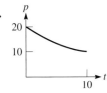

39.

41. 21.7 s   **43.** 2.3×10^4 **45.**

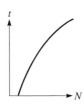

47. $[\frac{1}{2}(e^u + e^{-u})]^2 - [\frac{1}{2}(e^u - e^{-u})]^2$
    $= \frac{1}{4}e^{2u} + \frac{1}{2} + \frac{1}{4}e^{-2u} - \frac{1}{4}e^{2u} + \frac{1}{2} - \frac{1}{4}e^{-2u} = 1$

Exercises 8-2, page 275

1. $\dfrac{2 \log e}{x}$ **3.** $\dfrac{6 \log_5 e}{3x + 1}$ **5.** $\dfrac{-0.6}{1 - 3x}$

7. $\dfrac{4 \sec^2 2x}{\tan 2x} = 4 \sec 2x \csc 2x$ **9.** $\dfrac{1}{2x}$

11. $\dfrac{6(x + 1)}{x^2 + 2x}$ **13.** $\dfrac{6(t + 2)(t + \ln t^2)}{t}$

15. $\dfrac{3(2x + 1)\ln(2x + 1) - 6x}{(2x + 1)[\ln(2x + 1)]^2}$ **17.** $\dfrac{1}{x \ln x}$

19. $\dfrac{1}{x^2 + x}$ **21.** $\dfrac{\cos \ln x}{x}$ **23.** $6 \ln 2v + 3 \ln^2 2v$

25. $\dfrac{x \sec^2 x + \tan x}{x \tan x}$ **27.** $\dfrac{x + 4}{x(x + 2)}$

29. $\dfrac{\sqrt{x^2 + 1}}{x}$ **31.** $\dfrac{x + y - 2 \ln(x + y)}{x + y + 2 \ln(x + y)}$

33. 0.5 is value of derivative; 0.4999875 is slope of secant line.

35. (a)   (b) See the table.

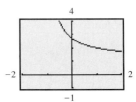

37. 1.15 **39.** $L(x) = 4x - \pi$ **41.** 2.73

43. $x^x(\ln x + 1)$. In Eq. (3-15) the exponent is constant.
    For x^x, both the base and the exponent are variables.

45. $\dfrac{10 \log e}{I} \dfrac{dI}{dt}$ **47.** 0.083 s^2/ft

Exercises 8-3, page 278

1. $(2 \ln 3)3^{2x}$ **3.** $(6 \ln 4)4^{6x}$ **5.** $\dfrac{e^{\sqrt{x}}}{2\sqrt{x}}$

7. $4e^{2t}(3e^t - 2)$ **9.** $e^{-x}(1 - x)$ **11.** $e^{\sin x}(x \cos x + 1)$

13. $8e^{-4s}$ **15.** $e^{-3x}(4 \cos 4x - 3 \sin 4x)$

17. $\dfrac{2e^{3x}(12x + 5)}{(4x + 3)^2}$ **19.** $\dfrac{2xe^{x^2}}{e^{x^2} + 4}$

21. $16e^{6x}(x \cos x^2 + 3 \sin x^2)$

23. $\dfrac{2(1 + 2te^{2t})}{t\sqrt{\ln 2t + e^{2t}}}$ **25.** $\dfrac{e^{xy}(xy + 1)}{1 - x^2 e^{xy} - \cos y}$

27. $\dfrac{3e^{2x}}{x} + 6e^{2x} \ln x$ **29.** $12e^{6x} \cot 2e^{6x}$

31. $\dfrac{4e^{2x}}{\sqrt{1 - e^{4x}}}$

33. (a) 2.7182818, value of derivative
    (b) 2.7184177, slope of secant line

35. -0.724 **37.** $\dfrac{2e^{4x}(4x + 7)\,dx}{(x + 2)^2}$

39.

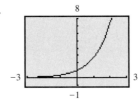

41. $(-xe^{-x} + e^{-x}) + (xe^{-x}) = e^{-x}$

43. $\dfrac{2e^{2x}(e^{2x} + 1) - 2e^{2x}(e^{2x} - 1)}{(e^{2x} + 1)^2} = \dfrac{4e^{2x}}{(e^{2x} + 1)^2}$

$= \dfrac{(e^{2x} + 1)^2 - (e^{2x} - 1)^2}{(e^{2x} + 1)^2}$

45. $-0.00164/\text{h}$ **47.** $e^{-66.7t}(999 \cos 226t - 295 \sin 226t)$

49. $\dfrac{d}{dx}\left[\dfrac{1}{2}(e^u - e^{-u})\right] = \dfrac{1}{2}(e^u + e^{-u})\dfrac{du}{dx}$

51. $2x \cosh 2x + \sinh 2x$

Exercises 8-4, page 282

1. int. $(0,0)$, max. $(0,0)$,
not defined for $\cos x < 0$,
asym. $x = -\frac{1}{2}\pi, \frac{1}{2}\pi, \ldots$

3. int. $(0,0)$, max. $\left(1, \dfrac{1}{e}\right)$,
infl. $\left(2, \dfrac{2}{e^2}\right)$, asym. $y = 0$

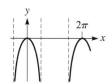

5. int. $(0,0)$, max. $(0,0)$,
infl. $(-1, -\ln 2)$, $(1, -\ln 2)$

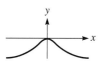

7. int. $(0, 4)$, max. $(0, 4)$,
infl. $\left(\dfrac{1}{2}\sqrt{2}, \dfrac{4}{e}\sqrt{e}\right)$, $\left(-\dfrac{1}{2}\sqrt{2}, \dfrac{4}{e}\sqrt{e}\right)$,
asym. $y = 0$

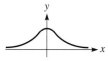

9. max. $(1, -1)$,
asym. $x = 0$

11. int. $(0,0)$, infl. $(0,0)$,
inc. all x

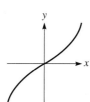

13. $y = x - 1$ **15.** $2\sqrt{2}x - 2y + 2\sqrt{2} - 3\pi\sqrt{2} = 0$

17. 1.3140968   **19.** -0.303 W/day

21. $\dfrac{p(-a + bT)}{T^2}$ **23.** $a = k^2 x$

25. int. $(0,0)$,
min. $(0,0)$, $(2\pi, 0), \ldots$,
asym. $x = -\dfrac{\pi}{2}, \dfrac{\pi}{2}, \dfrac{3\pi}{2}, \ldots$

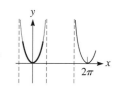

27. max. $(117.6°\text{W}, 50.2°\text{N})$, $(86.2°\text{W}, 47.7°\text{N})$
min. $(101.9°\text{W}, 41.2°\text{N})$, $(70.5°\text{W}, 43.0°\text{N})$

29. $v = -e^{-0.5t}(1.4 \cos 6t + 2.3 \sin 6t)$, -2.03 cm/s

31. $1/\sqrt{e} = 0.607$

Review Exercises for Chapter 8, page 283

1. $\dfrac{6x}{x^2 + 1}$ **3.** $2e^{2(x-3)}$ **5.** $-4t \cot t^2$

7. $\dfrac{2 \cos x \ln(3 + \sin x)}{3 + \sin x}$ **9.** $12e^{\sin 2\theta} \cos 2\theta$

11. $x^2(x + 3)e^x$ **13.** $\dfrac{2(1 + e^{-x})}{x - e^{-x}}$

15. $\dfrac{-\cos x(2e^{3x} \sin x + 3e^{3x} \cos x + 2 \sin x)}{(e^{3x} + 1)^2}$

17. $x(1 + 2 \ln x)$ **19.** $\dfrac{2e^{2x}(x^2 - x + 1)}{(x^2 + 1)^2}$

21. $3e^{-\theta} \sec 2\theta(2 \tan 2\theta - 1)$ **23.** $\dfrac{y(1 - 2x \ln y)}{x^2 - y}$

25. $1.73x + 3.00y - 0.48 = 0$

27. int. $(0, 0)$
asym. $x = -1$
conc. down $x > -1$

29. -0.7034674
31. 0.429
33. 4.55 mi/min

35.
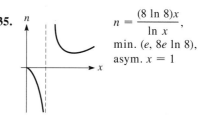
$n = \dfrac{(8 \ln 8)x}{\ln x}$,
min. $(e, 8e \ln 8)$,
asym. $x = 1$

37. 0.0007 ppm

39. $e^{-0.1t}(75.3 \cos 120\pi t - 301.6 \sin 120\pi t)$

41. 2.00 m wide, 1.82 m high

43. Find derivatives and substitute.

Exercises 9-1, page 287

1. $\frac{1}{5} \sin^5 x + C$ **3.** $-\frac{0.8}{3}(\cos \theta)^{3/2} + C$

5. $\frac{4}{3} \tan^3 x + C$ **7.** $\frac{1}{8}$ **9.** $\frac{1}{4}(\sin^{-1} x)^4 + C$

11. $\frac{1}{2}(\tan^{-1} 5x)^2 + C$ **13.** $\frac{1}{3}[\ln(x + 1)]^3 + C$

15. 0.179   **17.** $\frac{1}{4}(4 + e^x)^4 + C$

19. $\frac{1}{3}(1 - e^{-2r})^{3/2} + C$ **21.** $\frac{1}{10}(1 + \sec^2 x)^5 + C$

23. $\frac{1}{6}$ **25.** 1.102   **27.** $y = \frac{1}{3}(\ln x)^3 + 2$

29. $\frac{1}{3}mnv^2$ **31.** $q = (1 - e^{-t})^3$

Exercises 9-2, page 291

1. $\frac{1}{4} \ln|1 + 4x| + C$ **3.** $-\frac{1}{3} \ln|4 - 3x^2| + C$

5. $\frac{1}{3} \ln 4 = 0.462$ **7.** $-0.2 \ln|\cot 2\theta| + C$

9. $\ln 2 = 0.693$ **11.** $\ln|1 - e^{-x}| + C$

13. $\ln|x + e^x| + C$ **15.** $\frac{1}{4} \ln|1 + 4 \sec x| + C$

17. $\frac{1}{4} \ln 5 = 0.402$ **19.** $0.5 \ln|\ln r| + C$

21. $\ln|2x + \tan x| + C$ **23.** $-2\sqrt{1 - 2x} + C$

25. $\ln|x| - \frac{2}{x} + C$ **27.** $\frac{1}{3} \ln(\frac{5}{4}) = 0.0744$ **29.** 1.10

31. $\pi \ln 2 = 2.18$ **33.** $y = \ln \dfrac{3.5}{3 + \cos x} + 2$

35. 335 million   **37.** $i = \dfrac{E}{R}(1 - e^{-Rt/L})$ **39.** 1.41 m

Exercises 9-3, page 294

1. $e^{7x} + C$ **3.** $\frac{1}{2}e^{2x+5} + C$ **5.** 28.2

7. $2e^{x^3} + C$ **9.** $2(e^2 - e) = 9.34$ **11.** $2e^{2\sec\theta} + C$

13. $\frac{2}{3}(1 + e^y)^{3/2} + C$ **15.** $6 - \dfrac{3(e^6 - e^2)}{2} = -588.06$

17. $-\dfrac{4}{e^{\sqrt{x}}} + C$ **19.** $e^{\tan^{-1}x} + C$ **21.** $-\frac{1}{3}e^{\cos 3x} + C$

23. 0   **25.** $3e^2 - 3 = 19.2$ **27.** $\pi(e^4 - e) = 163$

29. $\frac{1}{8}(e^8 - 1) = 372$ **31.** $\ln b \int b^u \, du = b^u + C_1$

33. $q = EC(1 - e^{-t/RC})$ **35.** $s = -e^{-2t} - 0.6e^{-5t}$

Exercises 9-4, page 297

1. $\frac{1}{2} \sin 2x + C$ **3.** $0.1 \tan 3\theta + C$ **5.** $2 \sec \frac{1}{2}x + C$

7. 0.6365   **9.** $\frac{3}{2} \ln|\sec \phi^2 + \tan \phi^2| + C$

11. $\cos\left(\dfrac{1}{x}\right) + C$ **13.** $\frac{1}{2}\sqrt{3}$ **15.** $\frac{1}{5} \sec 5x + C$

17. $\frac{1}{2} \ln|\sec 2x + \tan 2x| + C$

19. $\frac{1}{2}(\ln|\sin 2x| + \sin 2x) + C$

21. $\csc x - \cot x - \ln|\csc x - \cot x| + \ln|\sin x| + C$

23. $\frac{1}{9}\pi + \frac{1}{3} \ln 2 = 0.580$ **25.** 0.693

27. $\pi\sqrt{3} = 5.44$ **29.** $\theta = 0.10 \cos 2.5t$

31. 0.7726 m

Exercises 9-5, page 301

1. $\frac{1}{3} \sin^3 x + C$ **3.** $-\frac{1}{2} \cos 2x + \frac{1}{6} \cos^3 2x + C$

5. $2 \sin 2\theta + C$

7. $\frac{1}{24}(64 - 43\sqrt{2}) = 0.1329$

9. $\frac{1}{2}x - \frac{1}{4} \sin 2x + C$

11. $\frac{1}{3}(9\phi + 4 \sin 3\phi + \sin 3\phi \cos 3\phi) + C$

13. $\frac{1}{2} \tan^2 x + \ln|\cos x| + C$ **15.** $\frac{3}{4}$

17. $\frac{1}{6} \tan^3 2x - \frac{1}{2} \tan 2x + x + C$

19. $\frac{1}{3} \sin^3 s + C$ **21.** $x - \frac{1}{2} \cos 2x + C$

23. $\frac{1}{4} \cot^4 x - \frac{1}{3} \cot^3 x + \frac{1}{2} \cot^2 x - \cot x + C$

25. $1 + \frac{1}{2} \ln 2 = 1.347$

27. $\frac{1}{5} \tan^5 x + \frac{2}{3} \tan^3 x + \tan x + C$

29. $\frac{1}{2}\pi^2 = 4.935$ **31.** $\sqrt{2} - 1 = 0.414$

33. $\int \sin x \cos x \, dx = \frac{1}{2} \sin^2 x + C_1 = -\frac{1}{2} \cos^2 x + C_2$;
$C_2 = C_1 + \frac{1}{2}$

35. $\frac{4}{3}$ **37.** $V = \sqrt{\dfrac{1}{1/60.0} \displaystyle\int_0^{1/60.0} (340 \sin 120\pi t)^2 \, dt} = 240$ V

39. $\dfrac{aA}{2} + \dfrac{A}{2b\pi} \sin ab\pi \cos 2bc\pi$

Exercises 9-6, page 305

1. $\sin^{-1} \frac{1}{2}x + C$ **3.** $\frac{1}{8} \tan^{-1} \frac{1}{8}x + C$ **5.** $\frac{1}{4} \sin^{-1} 4x + C$

7. 0.8634   **9.** $\frac{2}{5}\sqrt{5} \sin^{-1} \frac{1}{5}\sqrt{5} = 0.415$

11. $\frac{4}{9} \ln|9x^2 + 16| + C$ **13.** 2.356   **15.** $\sin^{-1} e^x + C$

17. $\tan^{-1}(x + 1) + C$ **19.** $4 \sin^{-1} \frac{1}{2}(x + 2) + C$

21. -0.714 **23.** $2 \sin^{-1}(\frac{1}{2}x) + \sqrt{4 - x^2} + C$

25. (a) inverse tangent, $\displaystyle\int \dfrac{du}{a^2 + u^2}$ where $u = 3x$,

$du = 3 \, dx$, $a = 2$; numerator cannot fit du of denominator. Positive $9x^2$ leads to inverse tangent

form.   (b) logarithmic, $\displaystyle\int \dfrac{du}{u}$ where $u = 4 + 9x$,

$du = 9 \, dx$ (c) general power, $\displaystyle\int u^{-1/2} \, du$ where

$u = 4 + 9x^2$, $du = 18x \, dx$

27. (a) general power, $\int u^{-1/2}\,du$ where $u = 4 - 9x^2$,

$du = -18x\,dx$; numerator can fit du of denominator. Square root becomes $-1/2$ power. Does not fit inverse sine form.   (b) inverse sine, $\int \dfrac{du}{\sqrt{a^2 - u^2}}$ where

$u = 3x$, $du = 3dx$, $a = 2$ (c) logarithmic, $\int \dfrac{du}{u}$

where $u = 4 - 9x$, $du = -9dx$

29. $\tan^{-1} 2 = 1.11$ **31.** $k\tan^{-1}\dfrac{x}{d} + C$

33. $\sin^{-1}\dfrac{x}{A} = \sqrt{\dfrac{k}{m}}\,t + \sin^{-1}\dfrac{x_0}{A}$ **35.** $0.22k$

Review Exercises for Chapter 9, page 307

1. $-\frac{1}{2}e^{-2x} + C$ **3.** $-\dfrac{1}{\ln 2x} + C$

5. $4\ln 2 = 2.773$ **7.** $\frac{2}{35}\tan^{-1}\frac{7}{5}x + C$ **9.** 0

11. $\frac{1}{2}\ln 2 = 0.3466$ **13.** $\frac{2}{3}\sin^3 t - \cos t + C$

15. $\tan^{-1} e^x + C$ **17.** $\frac{1}{9}\tan^3 3x + \frac{1}{3}\tan 3x + C$

19. $\dfrac{3}{\sqrt{e}} - 2 = -0.1804$ **21.** $\frac{3}{4}\tan^{-1}\dfrac{x^2}{2} + C$

23. $\frac{1}{4}\ln 3$ **25.** $2\ln|x + e^{2x}| + C$

27. $\sin^{-1}[\frac{1}{3}(x + 2)] + C$ **29.** $\dfrac{\pi}{4} = 0.7854$

31. $\frac{1}{2}\sin e^{2x} + C$ **33.** $3\sin 1 = 2.524$

35. $\frac{1}{2}x^2 - 2x + 3\ln|x + 2| + C$

37. $\frac{1}{3}(e^x + 1)^3 + C_1 = \frac{1}{3}e^{3x} + e^{2x} + e^x + C_2;\ C_2 = C_1 + \frac{1}{3}$

39. $\displaystyle\int \dfrac{1}{1 + \sin x}\,dx = \int \dfrac{1 - \sin x}{1 - \sin^2 x}\,dx = \int \dfrac{1 - \sin x}{\cos^2 x}\,dx$

$= \displaystyle\int \sec^2 x - \int \sec x \tan x\,dx = \tan x - \sec x + C$

41. $y = \frac{1}{3}\tan^3 x + \tan x$ **43.** $\ln 3 = 1.10$

45. $v = 64(1 - e^{-0.5t})$ **47.** $\sqrt{2}$ **49.** $\frac{2}{3}k$

51. 3.47 cm^3

Exercises 10-1, page 313

1. $\cos\theta + \theta\sin\theta + C$ **3.** $\frac{1}{2}xe^{2x} - \frac{1}{4}e^{2x} + C$

5. $x\tan x + \ln|\cos x| + C$

7. $2x\tan^{-1}x - 2\ln\sqrt{1 + x^2} + C$ **9.** $-\frac{32}{3}$

11. $\frac{1}{2}x^2\ln x - \frac{1}{4}x^2 + C$

13. $\frac{1}{2}\phi\sin 2\phi - \frac{1}{4}(2\phi^2 - 1)\cos 2\phi + C$

15. $\frac{1}{2}(e^{\pi/2} - 1) = 1.91$ **17.** $1 - \dfrac{3}{e^2} = 0.594$

19. 0.1104 **21.** $\frac{1}{2}\pi - 1 = 0.571$ **23.** 0.756

25. $s = \frac{1}{3}[(t^2 - 2)\sqrt{t^2 + 1} + 2]$

27. $q = \frac{1}{5}[e^{-2t}(\sin t - 2\cos t) + 2]$

Exercises 10-2, page 317

1. $\frac{2}{15}(3x - 2)(x + 1)^{3/2} + C$ **3.** $\frac{1}{15}(3x - 1)(2x + 1)^{3/2} + C$

5. $\frac{2}{3}(x - 6)(x + 3)^{1/2} + C$

7. $\frac{2}{15}(3x^2 + 8x + 32)(x - 2)^{1/2} + C$ **9.** $\frac{1696}{105}$

11. $\frac{3}{28}(4x + 3)(x - 1)^{4/3} + C$ **13.** $\frac{2}{7}(x - 2)(2x + 3)^{3/4} + C$

15. $\frac{3081}{40}$ **17.** $\frac{928}{315} = 2.946$ **19.** $\dfrac{30{,}024\pi}{35} = 2695$

21. $38.5 \text{ ft} \cdot \text{lb}$ **23.** $-\dfrac{2}{15}(1 - x)^{3/2}(2 + 3x) + C$

Exercises 10-3, page 320

1. $-\dfrac{\sqrt{1 - x^2}}{x} - \sin^{-1} x + C$

3. $2\ln|x + \sqrt{x^2 - 4}| + C$ **5.** $-\dfrac{2\sqrt{z^2 + 9}}{3z} + C$

7. $\dfrac{x}{\sqrt{4 - x^2}} + C$ **9.** $\dfrac{16 - 9\sqrt{3}}{24} = 0.017$

11. $5\ln|\sqrt{x^2 + 2x + 2} + x + 1| + C$

13. 0.03997 **15.** $2\sec^{-1} e^x + C$ **17.** π

19. Trig. substitution will work; power rule simpler

21. 2.68 **23.** $kQ\ln\dfrac{\sqrt{a^2 + b^2} + a}{\sqrt{a^2 + b^2} - a}$

Exercises 10-4, page 324

1. $\ln\left|\dfrac{(x + 1)^2}{x + 2}\right| + C$ **3.** $\frac{1}{4}\ln\left|\dfrac{x - 2}{x + 2}\right| + C$

5. $x + \ln\left|\dfrac{x}{(x + 3)^4}\right| + C$ **7.** 1.057

9. $\ln\left|\dfrac{x^2(x - 5)^3}{x + 1}\right| + C$ **11.** $\frac{1}{4}\ln\left|\dfrac{x^4(2x + 1)^3}{2x - 1}\right| + C$

13. $1 + \ln\dfrac{16}{3} = 2.674$ **15.** $\dfrac{1}{60}\ln\left|\dfrac{(V + 2)^3(V - 3)^2}{(V - 2)^3(V + 3)^2}\right| + C$

17. 1.322 **19.** $2\pi\ln\dfrac{6}{5} = 1.146$

21. $y = \ln\left|\dfrac{x(x + 5)^2}{36}\right|$ **23.** $0.1633 \text{ N} \cdot \text{cm}$

Exercises 10-5, page 329

1. $\dfrac{3}{x-2} + 2\ln\left|\dfrac{x-2}{x}\right| + C$ **3.** $\dfrac{2}{x} + \ln\left|\dfrac{x-1}{x+1}\right| + C$

5. $-\dfrac{5}{4}$ **7.** $-\dfrac{2}{x+1} - \dfrac{1}{x-3} + \ln|x+1| + C$

9. $\dfrac{1}{8}\pi + \ln 3 = 1.491$ **11.** $-\dfrac{2}{x} + \dfrac{3}{2}\tan^{-1}\dfrac{x+2}{2} + C$

13. $\dfrac{1}{4}\ln(4x^2 + 1) + \ln|x^2 + 6x + 10| + \tan^{-1}(x+3) + C$

15. $\ln(r^2 + 1) + \dfrac{1}{r^2+1} + C$ **17.** $2 + 4\ln\dfrac{2}{3} = 0.3781$

19. $\pi \ln\dfrac{25}{9} = 3.210$ **21.** 0.9190 m   **23.** 1.369

Exercises 10-6, page 331

1. $\dfrac{3}{25}[2 + 5x - 2\ln|2 + 5x|] + C$ **3.** $\dfrac{3544}{15} = 236.3$

5. $\dfrac{y}{4\sqrt{y^2+4}} + C$ **7.** $\dfrac{1}{2}\sin x - \dfrac{1}{10}\sin 5x + C$

9. $\sqrt{4x^2 - 9} - 3\sec^{-1}\left(\dfrac{2x}{3}\right) + C$

11. $\dfrac{1}{20}\cos^4 4x \sin 4x + \dfrac{1}{5}\sin 4x - \dfrac{1}{15}\sin^3 4x + C$

13. $3r^2 \tan^{-1} r^2 - \dfrac{3}{2}\ln(1 + r^4) + C$

15. $\dfrac{1}{4}(8\pi - 9\sqrt{3}) = 2.386$

17. $-\ln\left(\dfrac{1+\sqrt{4x^2+1}}{2x}\right) + C$

19. $-8\ln\left(\dfrac{1+\sqrt{1-4x^2}}{2x}\right) + C$ **21.** 0.0208

23. $\dfrac{1}{3}(\cos x^3 + x^3\sin x^3) + C$ **25.** $\dfrac{x^2}{\sqrt{1-x^4}} + C$

27. 4.892 **29.** $\dfrac{1}{4}x^4(\ln x^2 - \dfrac{1}{2}) + C$ **31.** $-\dfrac{3x^3}{\sqrt{x^6-1}} + C$

33. $\dfrac{1}{4}[2\sqrt{5} + \ln(2 + \sqrt{5})] = 1.479$

35. πab **37.** 208 lb

39. $187,000$ m^3

Exercises 10-7, page 335

1. $\dfrac{1}{3}$ **3.** divergent   **5.** divergent   **7.** $\dfrac{1}{2}$ **9.** 1

11. divergent   **13.** 6   **15.** divergent   **17.** divergent

19. divergent   **21.** $\ln(2 + \sqrt{3})$ **23.** 0

25. (a) divergent, (b) π **27.** $\sqrt{3}$ **29.** 2.53×10^7 mi $\cdot$ lb

31. $\sin x$ takes on all values from -1 to 1 as x increases

Review Exercises for Chapter 10, page 336

1. $-\dfrac{1}{2}x\cot 2x + \dfrac{1}{4}\ln|\sin 2x| + C$

3. $\dfrac{2}{15}(3x + 8)(x - 4)^{3/2} + C$ **5.** $\dfrac{1}{2}\ln|2x + \sqrt{4x^2 - 9}| + C$

7. $\ln\left|\dfrac{(x-5)^3}{(x+5)^2}\right| + C$ **9.** $\ln|x - 1| - \dfrac{1}{(x-1)^2} + C$

11. divergent   **13.** $\dfrac{3}{4}(x + 3)^{1/3}(x - 9) + C$

15. $\dfrac{2x}{\sqrt{4x^2 + 1}} + C$ **17.** $2\tan^{-1} x - \dfrac{1}{2}\sqrt{2}\tan^{-1}(\dfrac{1}{2}x\sqrt{2}) + C$

19. $6\ln 2 - 2 = 2.159$ **21.** $\ln\left|\dfrac{(x+1)(x+2)}{(x-1)(x-2)}\right| + C$

23. $\dfrac{144}{5}$ **25.** 4   **27.** 11.18   **29.** $4\pi(e^2 - 1) = 80.29$

31. 48.3 N $\cdot$ cm   **33.** $(1.65, 0.244)$ **35.** 1.10 m^2

Exercises 11-1, page 339

1. $V = \pi r^2 h$ **3.** $A = \dfrac{1}{2}bh$ **5.** $A = \dfrac{2V}{r} + 2\pi r^2$

7. $V = \dfrac{1}{4}\pi h(4r^2 - h^2)$ **9.** 24   **11.** -2

13. $2 - 3y + 4y^2$ **15.** $6xt + xt^2 + t^3$

17. $\dfrac{p^2 + pq + kp - p + 2q^2 + 4kq + 2k^2 + 5q + 5k}{p + q + k}$

19. $2hx - 2kx - 2hy + h^2 - 2hk - 4h$ **21.** 0

23. $81z^6 - 9z^5 - 2z^3$ **25.** $x \neq 0, y \geq 0$

27. $y \leq 1$ **29.** 18 V   **31.** 150 Pa

33. for a, b, and T with same sign: circle if $a = b$, ellipse if $a \neq b$:
for a and b of different signs, hyperbola

35. 1.03 A, 1.23 A   **37.** $A = \dfrac{pw - 2w^2}{2}$, 3850 cm^2

39. $L = \dfrac{(1.28 \times 10^5)r^4}{l^2}$

Exercises 11-2, page 345

1.
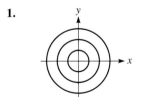

3.
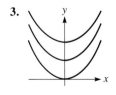

5.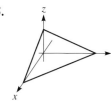

7.

9.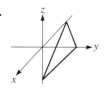

11.

13.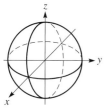

15.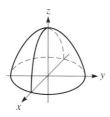

17.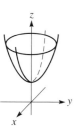

19.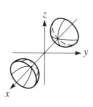

21.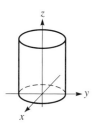

23.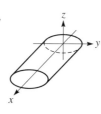

25.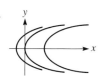

27.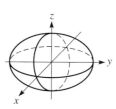

29.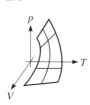

31.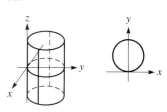

Exercises 11-3, page 349

1. $\dfrac{\partial z}{\partial x} = 5 + 8xy, \dfrac{\partial z}{\partial y} = 4x^2$

3. $\dfrac{\partial z}{\partial x} = \dfrac{2x}{y} - 2y, \dfrac{\partial z}{\partial y} = -\dfrac{x^2}{y^2} - 2x$

5. $\dfrac{\partial f}{\partial x} = e^{2y}, \dfrac{\partial f}{\partial y} = 2xe^{2y}$

7. $\dfrac{\partial f}{\partial x} = -\dfrac{\sin x}{1 - \sec 3y}, \dfrac{\partial f}{\partial y} = \dfrac{3(2 + \cos x)\sec 3y \tan 3y}{(1 - \sec 3y)^2}$

9. $\dfrac{\partial \phi}{\partial r} = \dfrac{1 + 3rs}{\sqrt{1 + 2rs}}, \dfrac{\partial \phi}{\partial s} = \dfrac{r^2}{\sqrt{1 + 2rs}}$

11. $\dfrac{\partial z}{\partial x} = 4(2x + y^3)(x^2 + xy^3)^3, \dfrac{\partial z}{\partial y} = 12xy^2(x^2 + xy^3)^3$

13. $\dfrac{\partial z}{\partial x} = y \cos xy, \dfrac{\partial z}{\partial y} = x \cos xy$

15. $\dfrac{\partial y}{\partial r} = \dfrac{2r}{r^2 + s}, \dfrac{\partial y}{\partial s} = \dfrac{1}{r^2 + s}$

17. $\dfrac{\partial f}{\partial x} = \dfrac{12 \sin^2 2x \cos 2x}{1 - 3y}, \dfrac{\partial f}{\partial y} = \dfrac{6 \sin^3 2x}{(1 - 3y)^2}$

19. $\dfrac{\partial z}{\partial x} = \dfrac{3y + x^2y - 2x\sqrt{1 - x^2y^2}\sin^{-1} xy}{(3 + x^2)^2\sqrt{1 - x^2y^2}},$

$\dfrac{\partial z}{\partial y} = \dfrac{x}{(3 + x^2)\sqrt{1 - x^2y^2}}$

21. $\dfrac{\partial z}{\partial x} = \cos x - y \sin xy, \dfrac{\partial z}{\partial y} = -x \sin xy + \sin y$

23. $\dfrac{\partial f}{\partial x} = e^x(\cos xy - y \sin xy) - 2e^{-2x}\tan y,$

$\dfrac{\partial f}{\partial y} = -xe^x \sin xy + e^{-2x}\sec^2 y$

25. -8 **27.** $\frac{41}{4}$

29. $\dfrac{\partial^2 z}{\partial x^2} = -6y, \dfrac{\partial^2 z}{\partial y^2} = 12xy, \dfrac{\partial^2 z}{\partial x\,\partial y} = 6y^2 - 6x$

31. $\dfrac{\partial^2 z}{\partial x^2} = e^x \sin y, \dfrac{\partial^2 z}{\partial y^2} = \dfrac{2x}{y^3} - e^x \sin y,$

$\dfrac{\partial^2 z}{\partial x\,\partial y} = \dfrac{\partial^2 z}{\partial y\,\partial x} = -\dfrac{1}{y^2} + e^x \cos y$

33. $-4, -4$

35. $\left(\dfrac{R_2}{R_1 + R_2}\right)^2$

37. 114 cm^2 **39.** 0.807   **41.** 3.75×10^{-3} 1/Ω

43. $\dfrac{k_2 + 2k_3 FT}{L_0 + k_1 F + k_2 T + k_3 FT^2}$ **45.** $-\dfrac{nRT}{V^2}$

47. $-5e^{-t}\sin 4x = \frac{1}{16}(-80e^{-t}\sin 4x)$

Exercises 11-4, page 355

1. $dz = (4x + 3)\,dx - 2y\,dy$ **3.** $dz = e^y\,dx + (xe^y - 2y)\,dy$

5. $dz = (y - 22x^2)(y - 2x^2)^4\,dx + 5x(y - 2x^2)^4\,dy$

7. $dz = (y\cos xy + y\sin x)\,dx + (x\cos xy - \cos x)\,dy$

9. $dz = \dfrac{dx}{1 - \sin y} + \dfrac{6y\sin y - 6y + x\cos y - 3y^2\cos y}{(1 - \sin y)^2}\,dy$

11. $dz = -\dfrac{2xy^2\,dx}{x^4 + y^2} + \left(\dfrac{yx^2}{x^4 + y^2} + \tan^{-1}\dfrac{y}{x^2}\right)dy$

13. min $(1, 3, 0)$ **15.** min $(-1, 2, -1)$

17. no max. or min.   **19.** min $(2, 0, 0)$ and $(-2, \pi, 0)$

21. 9.60×10^4 rad/min^2 **23.** 1.3%   **25.** 4.5%

27. $l = 1$ m, $w = 1$ m, $h = 1$ m

29. base 4 m by 4 m, height 2 m   **31.** $P_{min} = 17$ Pa at $(1, 0)$

33. $T_{max} = 0°$C at $(0.67, 1.33)$ **35.** $6xt - 6yt$

37. 0   **39.** -5.0×10^{-5} A/s

Exercises 11-5, page 360

1. $\dfrac{28}{3}$ **3.** $\dfrac{1}{3}$ **5.** $\dfrac{127}{14}$ **7.** $\dfrac{1}{3}$ **9.** $\dfrac{\pi - 6}{12}$

11. 1   **13.** 495.2   **15.** $\dfrac{74}{5}$ **17.** $\dfrac{32}{3}$ **19.** 8π

21. $\dfrac{28}{3}$ **23.** 18   **25.** 300 cm^3 **27.**

Exercises 11-6, page 364

1. $\left(\dfrac{2}{3}, \dfrac{2}{3}\right)$ **3.** $\left(\dfrac{9}{8}, \dfrac{18}{5}\right)$ **5.** $\dfrac{4}{3}, \dfrac{4}{3}$ **7.** 1.34

9. $\left(\dfrac{3}{4}, \dfrac{7}{10}\right)$, 0.775   **11.** $\left(\dfrac{a}{3}, \dfrac{b}{3}\right)$ **13.** $\dfrac{mb^2}{3}$ **15.** $\dfrac{4}{3}\sqrt{3}$

17. $\left(3, \dfrac{5}{6}\right)$ (origin at lower left corner)   **19.** $\dfrac{1}{12}a^4$

Review Exercises for Chapter 11, page 365

1. $\dfrac{\partial z}{\partial x} = 15x^2y^2 - 2y^4$, $\dfrac{\partial z}{\partial y} = 10x^3y - 8xy^3$

3. $\dfrac{\partial z}{\partial x} = \dfrac{x}{\sqrt{x^2 - 3y^2}}$, $\dfrac{\partial z}{\partial y} = \dfrac{-3y}{\sqrt{x^2 - 3y^2}}$

5. $\dfrac{\partial z}{\partial x} = \dfrac{2 - 2x^2y + 6xy^2}{(x^2y + 1)^2}$, $\dfrac{\partial z}{\partial y} = -\dfrac{3 + 2x^3}{(x^2y + 1)^2}$

7. $\dfrac{\partial u}{\partial x} = 2xy\cot(x^2 + 2y)$,

$\dfrac{\partial u}{\partial y} = 2y\cot(x^2 + 2y) + \ln\sin(x^2 + 2y)$

9. $\dfrac{\partial z}{\partial x} = \dfrac{\partial z}{\partial y} = \dfrac{1}{2\sqrt{(x + y)(1 - x - y)}}$

11. $\dfrac{\partial^2 z}{\partial x^2} = 6y$, $\dfrac{\partial^2 z}{\partial y^2} = -6y$, $\dfrac{\partial^2 z}{\partial x\,\partial y} = 6x + 2$

13. 12   **15.** $\dfrac{21}{2}$ **17.** $\dfrac{e^2 - 3}{4} = 1.097$ **19.** $\dfrac{1}{6}$

21.

23. 0.113

25. $z = e^2 x$

27. 1.48

29. $\dfrac{\partial v}{\partial r} = \dfrac{ER}{(r + R)^2}$, $\dfrac{\partial v}{\partial R} = \dfrac{-rE}{(r + R)^2}$ **31.** 0.982

33. $\dfrac{\pi}{\sqrt{gl}} = \dfrac{2\pi\sqrt{l/g}}{2l}$ **35.** 1.25 cm   **37.** $\left(\dfrac{3}{2}, 3\right)$

39. $\dfrac{292}{15}$ **41.** $32\left(\pi - \dfrac{2}{3}\right) = 79.20$ **43.** $\left(\dfrac{14}{9}, \dfrac{8}{9}\right)$

45. 1.83   **47.** -12 W/min

Exercises 12-1, page 370

1.   **3.** 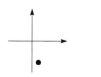 **5.**

7.   **9.** **11.**

13. $\left(2, \dfrac{\pi}{6}\right)$ **15.** $\left(1, \dfrac{7\pi}{6}\right)$ **17.** $(-4, -4\sqrt{3})$

19. $(2.76, -1.17)$ **21.** $r = 3 \sec \theta$

23. $r = \dfrac{3}{\cos \theta + 2 \sin \theta}$

25. $r = 4 \sin \theta$ **27.** $r^2 = \dfrac{4}{1 + 3 \sin^2 \theta}$

29. $x^2 + y^2 - y = 0$, circle **31.** $x = 4$, straight line

33. $x - 3y - 2 = 0$, straight line

35. $x^2 + y^2 - 4x - 2y = 0$, circle

37. $x^4 + y^4 - 4x^3 + 2x^2y^2 - 4xy^2 - 4y^2 = 0$

39. $(x^2 + y^2)^2 = 2xy$

41. $B_x = -\dfrac{k \sin \theta}{r}$, $B_y = \dfrac{k \cos \theta}{r}$

43. rectangular equation: $x^2 + y^2 - ax - by = 0$

Exercises 12-2, page 373

1.   **3.**

5.   **7.**   **9.**

11.   **13.**   **15.**

17.   **19.**

21.   **23.**

25. 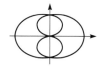 **27.**

29. 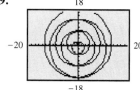 **31.**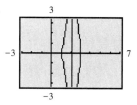

33.   **35.**

Exercises 12-3, page 378

1. 9.49 **3.** 1.85 **5.** 4.00 ft/s **7.** 2.00 ft/s
9. 8.25 ft/s **11.** 1.54 ft/s **13.** π **15.** $\frac{1}{2}$ **17.** π
19. $\frac{1}{8}(e^{2\pi} - 1)$ **21.** π **23.** $\frac{9}{2}\pi$ **25.** $\frac{1}{4}(8 + \pi)$
27. $\frac{1}{2}(\pi - 2)$ **29.** 4.98 ft/s, 8.67 ft/s **31.** 119 cm^2
33. Use indicated procedure. **35.** 4

Exercises 12-4, page 380

1. (a) $(0, 2, 4)$ (b) $(\frac{3}{2}\sqrt{2}, -\frac{3}{2}\sqrt{2}, 5)$
3. (a) $(10, \tan^{-1}(4/3), 5)$ (b) $(4\sqrt{2}, \frac{1}{4}\pi, -3)$
5. (a) cylinder, axis is z-axis, $r = 2$
   (b) plane $\theta = 2$ for all r and z
7. (a) $r = 2 \sec \theta$, plane 2 units in front of yz plane
   (b) $z = 2$, plane 2 units above xy plane

9. $r = 4$

11. $r^2 + 4z^2 = 4$

13. $9z = 4r^2$

15. $x^2 + y^2 = 4z$

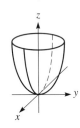

17. $x^2 + y^2 + 4z^2 = 16$

19. $y = 3$

21. $\frac{\pi}{2}$ **23.** 205 m^3

Review Exercises for Chapter 12, page 381

1. $\theta = \tan^{-1} 2 = 1.11$ **3.** $r^2 \cos 2\theta = 16$
5. $(x^2 + y^2)^3 = 16x^2y^2$ **7.** $3x^2 + 4y^2 - 8x - 16 = 0$

9.

11.

13.

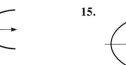

15.

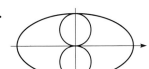

17.

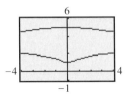

19.

21.

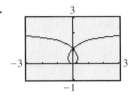

23.

25. $\pi + 2$ **27.** $\frac{19}{2}\pi$ **29.** 4.502   **31.** $\frac{1}{48}\pi^3$
33. (a) $(13, \tan^{-1} 2.4, 3)$ (b) $(3\sqrt{2}, \tan^{-1}(-1), 2)$ **35.** 4
37. $\frac{13}{4}$ **39.** $E_r = \left(\dfrac{2a}{r^3} + b\right)\cos\theta$, $E_\theta = \left(\dfrac{a}{r^3} - b\right)\sin\theta$

41.

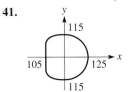

43. 81.9 ft^2

Exercises 13-1, page 386

1. 1, 4, 9, 16   **3.** $\frac{1}{2}, \frac{1}{3}, \frac{1}{4}, \frac{1}{5}$
5. (a) $-\frac{2}{5}, \frac{4}{25}, -\frac{8}{125}, \frac{16}{625}$ (b) $-\frac{2}{5} + \frac{4}{25} - \frac{8}{125} + \frac{16}{625} - \cdots$
7. (a) 1, 0, -1, 0   (b) $1 + 0 - 1 + 0 + \cdots$
9. $a_n = \dfrac{1}{n + 1}$ **11.** $a_n = \dfrac{1}{(n + 1)(n + 2)}$
13. 1, 1.125, 1.1620370, 1.1776620, 1.1856620; convergent; 1.2
15. 1, 1.5, 2.1666667, 2.9166667, 3.7166667; divergent
17. 0, 1, 2.4142, 4.1463, 6.1463; divergent
19. 0.75, 0.8888889, 0.9375000, 0.9600000, 0.9722222; convergent; 1   **21.** divergent
23. convergent, $S = \frac{3}{4}$ **25.** convergent, $S = 100$
27. convergent, $S = \frac{4096}{9}$ **29.** (a) 1   (b) diverges
31. 1   **33.** (a) diverges   (b) $y = 2100(1.05^x - 1)$

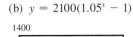
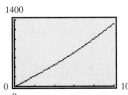

35. $r = x; S = \dfrac{x^0}{1 - x} = \dfrac{1}{1 - x}$

Exercises 13-2, page 391

1. $1 + x + \frac{1}{2}x^2 + \cdots$ **3.** $1 - \frac{1}{2}x^2 + \frac{1}{24}x^4 - \cdots$

5. $1 + \frac{1}{2}x - \frac{1}{8}x^2 + \cdots$ **7.** $1 - 2x + 2x^2 + \cdots$

9. $1 - 8\pi^2 x^2 + \frac{32}{3}\pi^4 x^4 - \cdots$ **11.** $1 + x + x^2 + \cdots$

13. $-2x - 2x^2 - \frac{8}{3}x^3 - \cdots$ **15.** $1 - x^2 + \frac{1}{3}x^4 - \cdots$

17. $x - \frac{1}{3}x^3 + \cdots$ **19.** $x + \frac{1}{3}x^3 + \cdots$

21. $-\frac{1}{2}x^2 - \frac{1}{12}x^4 - \cdots$ **23.** $x^2 - \frac{1}{3}x^4 + \cdots$

25. No, functions are not defined at $x = 0$.

27. $e^x = 1 + x + \dfrac{x^2}{2} + \cdots, \; e^{x^2} = 1 + x^2 + \dfrac{x^4}{2} + \cdots$

29. $f(x) = 1 + 3x + \frac{9}{2}x^2 + \cdots; L(x) = 1 + 3x$

31. $R = e^{-0.001t} = 1 - 0.001t + (5 \times 10^{-7})t^2 - \cdots$

Exercises 13-3, page 395

1. $1 + 3x + \frac{9}{2}x^2 + \frac{9}{2}x^3 + \cdots$

3. $\dfrac{x}{2} - \dfrac{x^3}{2^3 3!} + \dfrac{x^5}{2^5 5!} - \dfrac{x^7}{2^7 7!} + \cdots$

5. $x - 8x^3 + \frac{32}{3}x^5 - \frac{256}{45}x^7 + \cdots$

7. $x^2 - \frac{1}{2}x^4 + \frac{1}{3}x^6 - \frac{1}{4}x^8 + \cdots$ **9.** 0.3103

11. 0.1901 **13.** $2(1 + x^2 + x^4 + x^6 + \cdots)$

15. $x + x^2 + \frac{1}{3}x^3 + \cdots$

17. $\dfrac{d}{dx}(x - \frac{1}{6}x^3 + \frac{1}{120}x^5 - \cdots) = 1 - \frac{1}{2}x^2 + \frac{1}{24}x^4 - \cdots$

19. $\displaystyle\int \cos x \, dx = x - \dfrac{x^3}{3!} + \cdots$

21. $\displaystyle\int_0^1 e^x \, dx = 1.7182818,$

$\displaystyle\int_0^1 (1 + x + \frac{1}{2}x^2 + \frac{1}{6}x^3) \, dx = 1.7083333$

23. 0.003099 **25.** 0.199968

27. Use indicated method.

29. 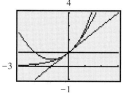 **31.**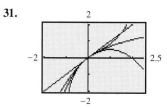

Exercises 13-4, page 399

1. 1.22, 1.2214028 **3.** 0.0998333, 0.0998334

5. 2.7180556, 2.7182818 **7.** 0.9986292, 0.9986295

9. 0.3349333, 0.3364722 **11.** 0.3546130, 0.3546129

13. $-0.0139975, -0.0139975$

15. 1.20736, 1.20803 **17.** 1.0523528

19. 0.9874462 **21.** 8.3×10^{-8} **23.** 3.1×10^{-7}

25. 3.146 **27.** 1.59 years

29. $i = \dfrac{E}{L}\left(t - \dfrac{Rt^2}{2L}\right);$ small values of t **31.** 18 m

Exercises 13-5, page 402

1. 3.32 **3.** 2.049 **5.** 0.51504 **7.** 0.49288

9. $e^{-2}\left[1 - (x - 2) + \dfrac{(x - 2)^2}{2!} - \cdots\right]$

11. $\dfrac{1}{2}\left[\sqrt{3} + \left(x - \dfrac{1}{3}\pi\right) - \dfrac{\sqrt{3}}{2!}\left(x - \dfrac{1}{3}\pi\right)^2 - \cdots\right]$

13. $2 + \frac{1}{12}(x - 8) - \frac{1}{288}(x - 8)^2 + \cdots$

15. $1 + 2(x - \frac{1}{4}\pi) + 2(x - \frac{1}{4}\pi)^2 + \cdots$ **17.** 23.1308

19. 3.0496 **21.** 2.0247 **23.** 0.87462

25. Use the indicated method.

27. 0.5150408, 0.5150388, 0.5150381

29. Graph of part (b) fits well near $x = \pi/3$. **31.** Graph of part (b) fits well near $x = 2$.

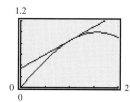 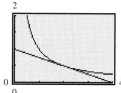

Exercises 13-6, page 408

1. $f(x) = \dfrac{1}{2} - \dfrac{2}{\pi}\sin x - \dfrac{2}{3\pi}\sin 3x - \cdots$

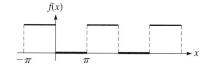

3. $f(x) = \dfrac{3}{2} + \dfrac{2}{\pi}\sin x + \dfrac{2}{3\pi}\sin 3x + \cdots$

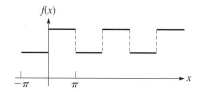

5. $f(x) = \dfrac{\pi}{4} - \dfrac{2}{\pi}\left(\cos x + \dfrac{1}{9}\cos 3x + \cdots\right)$
$+ \left(\sin x - \dfrac{1}{2}\sin 2x + \cdots\right)$

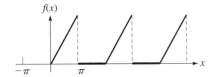

7. $f(x) = -\dfrac{1}{4} - \dfrac{1}{\pi}\cos x + \dfrac{1}{3\pi}\cos 3x - \cdots$
$+ \dfrac{3}{\pi}\sin x - \dfrac{1}{\pi}\sin 2x + \dfrac{1}{\pi}\sin 3x - \cdots$

9. $f(x) = \dfrac{\pi}{2} - \dfrac{4}{\pi}\cos x - \dfrac{4}{9\pi}\cos 3x - \cdots$

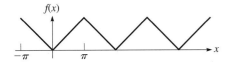

11.   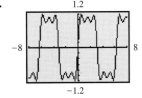   **13.**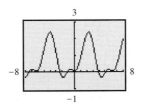

15. $f(t) = \dfrac{2}{\pi} - \dfrac{4}{3\pi}\cos 2t - \dfrac{4}{15\pi}\cos 4t - \cdots$

Exercises 13-7, page 414

1. Neither   **3.** Even   **5.** Even   **7.** Odd
9. $\dfrac{5}{2} + \dfrac{2}{\pi}\left(\cos x - \dfrac{1}{3}\cos 3x + \dfrac{1}{5}\cos 5x + \cdots\right)$
11. $-1 + \dfrac{4}{\pi}\sin\dfrac{\pi x}{4} + \dfrac{4}{3\pi}\sin\dfrac{3\pi x}{4} + \cdots$
13. $f(x) = \dfrac{5}{2} - \dfrac{10}{\pi}\left(\sin\dfrac{\pi x}{3} + \dfrac{1}{3}\sin \pi x - \cdots\right)$

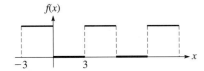

15. $f(x) = 1 + \dfrac{4}{\pi}\cos\dfrac{\pi x}{2} - \dfrac{4}{3\pi}\cos\dfrac{3\pi x}{2} + \cdots$

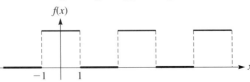

17. $f(x) = 2 - \dfrac{16}{\pi^2}\left(\cos\dfrac{\pi x}{4} + \dfrac{1}{9}\cos\dfrac{3\pi x}{4} + \cdots\right)$

19. $f(x) = \dfrac{4}{\pi}\left(\sin\dfrac{\pi x}{4} + \dfrac{1}{3}\sin\dfrac{3\pi x}{4} + \dfrac{1}{5}\sin\dfrac{5\pi x}{4} + \cdots\right)$

21. $f(x) = \dfrac{4}{3} - \dfrac{16}{\pi^2}\left(\cos\dfrac{\pi x}{2} - \dfrac{1}{4}\cos \pi x + \dfrac{1}{9}\cos\dfrac{3\pi x}{2} - \cdots\right)$

23. $f(t) = 2 + \dfrac{8}{\pi}\left(\cos\dfrac{\pi}{2}t - \dfrac{1}{3}\cos\dfrac{3\pi}{2}t + \cdots + \sin\dfrac{\pi}{2}t\right.$
$\left. + \sin \pi t + \dfrac{1}{3}\sin\dfrac{3\pi}{2}t + \cdots\right)$

Review Exercises for Chapter 13, page 415

1. $\dfrac{1}{2} - \dfrac{1}{4}x + \dfrac{1}{48}x^3 - \cdots$ **3.** $2x^2 - \dfrac{4}{3}x^6 + \dfrac{4}{15}x^{10} - \cdots$
5. $1 + \dfrac{1}{3}x - \dfrac{1}{9}x^2 + \cdots$ **7.** $x + \dfrac{1}{6}x^3 + \dfrac{3}{40}x^5 + \cdots$
9. 0.82 **11.** 1.09 **13.** 0.9214 **15.** -0.2015
17. 0.95299 **19.** 12.1655 **21.** 0.259
23. $\dfrac{1}{2} - \dfrac{1}{2}\sqrt{3}(x - \dfrac{1}{3}\pi) - \dfrac{1}{4}(x - \dfrac{1}{3}\pi)^2 + \cdots$
25. $f(x) = \dfrac{\pi - 2}{4} - \dfrac{2}{\pi}\left(\cos x + \dfrac{1}{9}\cos 3x + \cdots\right)$
$+ \left(\dfrac{\pi - 2}{\pi}\right)\sin x - \dfrac{1}{2}\sin 2x + \cdots$

27. $f(x) = \pi + \dfrac{4}{\pi}\left(\sin\dfrac{\pi x}{4} + \dfrac{1}{3}\sin\dfrac{3\pi x}{4} + \cdots\right)$
29. $f(x) = \dfrac{1}{2} + \dfrac{2}{\pi}\left(\cos x - \dfrac{1}{3}\cos 3x + \cdots\right)$

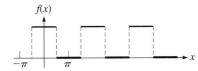

31. $f(x) = \dfrac{4}{\pi}\left(\sin\dfrac{\pi x}{2} - \dfrac{1}{2}\sin \pi x + \dfrac{1}{3}\sin\dfrac{3\pi x}{2} - \cdots\right)$

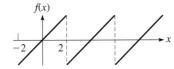

33. 256

35. $(x+h) - \dfrac{(x+h)^3}{3!} + \cdots - (x-h) + \dfrac{(x-h)^3}{3!} - \cdots$

$= 2h - \dfrac{2hx^2}{2!} + \cdots = 2h\left(1 - \dfrac{x^2}{2!} + \cdots\right)$

37. $4x - 2x^2 + \frac{4}{3}x^3 - x^4 + \cdots$

39. $1 - x^2 + \frac{1}{3}x^4 - \frac{2}{45}x^6 + \cdots$

41. $1 + \dfrac{x^2}{2} + \dfrac{5x^4}{24} + \cdots$

43. $f(x) = \dfrac{2}{\pi} - \dfrac{4}{\pi}\left(\dfrac{1}{3}\cos 2x + \dfrac{1}{15}\cos 4x + \dfrac{1}{35}\cos 6x + \cdots\right)$

45. 0.00249688   47. $x - \dfrac{x^3}{3} + \dfrac{x^5}{5} - \cdots$

49. $N_0\left(1 - \lambda t + \dfrac{\lambda^2 t^2}{2} - \dfrac{\lambda^3 t^3}{6} + \cdots\right)$

51. $2x + \frac{2}{3}x^3 + \frac{2}{5}x^5 + \frac{2}{7}x^7 + \cdots$

53. $N_0[1 + e^{-k/T} + (e^{-k/T})^2 + \cdots]$
$= N_0(1 + e^{-k/T} + e^{-2k/T} + \cdots)$

55. $f(t) = \dfrac{1}{2\pi} + \dfrac{1}{\pi}\left(\dfrac{1}{2}\cos t - \dfrac{1}{3}\cos 2t + \cdots\right)$

$+ \dfrac{1}{4}\sin t + \dfrac{2}{3\pi}\sin 2t + \cdots$

Exercises 14-1, page 420

1. particular solution   3. general solution

(The following "answers" are the unsimplified expressions obtained by substituting functions and derivatives.)

5. $e^x - (e^x - 1) = 1;\ 5e^x - (5e^x - 1) = 1$

7. $-12\cos 2x + 4(3\cos 2x) = 0;$
$(-4c_1\sin 2x - 4c_2\cos 2x) + 4(c_1\sin 2x + c_2\cos 2x) = 0$

9. $2x = 2x$ 11. $(2x+3) - 3 = 2x$

13. $(-2ce^{-2x} + 1) + 2(ce^{-2x} + x - \frac{1}{2}) = 2x$

15. $-\frac{1}{2}\cos x + \frac{9}{2}\cos x = 4\cos x$

17. $x^2\left[-\dfrac{c^2}{(x-c)^2}\right] + \left[\dfrac{cx}{(x-c)}\right]^2 = 0$

19. $x\left(-\dfrac{c_1}{x^2}\right) + \dfrac{c_1}{x} = 0$

21. $(\cos x - \sin x + e^{-x}) + (\sin x + \cos x - e^{-x})$
$= 2\cos x$

23. $(e^{-x} + \frac{12}{5}\cos 2x + \frac{24}{5}\sin 2x) +$
$(-e^{-x} + \frac{6}{5}\sin 2x - \frac{12}{5}\cos 2x) = 6\sin 2x$

25. $\cos x\left[\dfrac{(\sec x + \tan x) - (x+c)(\sec x \tan x + \sec^2 x)}{(\sec x + \tan x)^2}\right]$

$+ \sin x = 1 - \dfrac{x+c}{\sec x + \tan x}$

27. $c^2 + cx = cx + c^2$

Exercises 14-2, page 424

1. $y = c - x^2$ 3. $x - \dfrac{1}{y} = c$ 5. $\ln V = \dfrac{1}{P} + c$

7. $\ln(x^3 + 5) + 3y = c$ 9. $y = 2x^2 + x - x\ln x + c$

11. $4\sqrt{1-y} = e^{-x^2} + c$ 13. $e^x - e^{-y} = c$

15. $\ln(y+4) = x + c$ 17. $y(1 + \ln x)^2 + cy + 2 = 0$

19. $\tan^2 x + 2\ln y = c$

21. $x^2 + 1 + x\ln y + cx = 0$ 23. $y^2 + 4\sin^{-1} x = c$

25. $i = c - (\ln t)^2$ 27. $\ln(e^x + 1) - \dfrac{1}{y} = c$

29. $3\ln y + x^3 = 0$ 31. $\frac{1}{3}y^3 + y = \frac{1}{2}\ln^2 x$

33. $2\ln(1-y) = 1 - 2\sin x$ 35. $e^{2x} - \dfrac{2}{y} = 2(e^x - 1)$

Exercises 14-3, page 426

1. $2xy + x^2 = c$ 3. $x^3 - 2y = cx - 4$

5. $A^2 r - r = cA$ 7. $(xy)^4 = 12\ln y + c$

9. $2\sqrt{x^2 + y^2} = x + c$

11. $y = c - \frac{1}{2}\ln\sin(x^2 + y^2)$

13. $\ln(y^2 - x^2) + 2x = c$ 15. $5xy^2 + y^3 = c$

17. $2xy + x^3 = 5$ 19. $2x = 2xy^2 - 15y$

Exercises 14-4, page 429

1. $y = e^{-x}(x + c)$ 3. $y = -\frac{1}{2}e^{-4x} + ce^{-2x}$

5. $y = -2 + ce^{2x}$ 7. $y = x(3\ln x + c)$

9. $y = \dfrac{8}{7}x^3 + \dfrac{c}{\sqrt{x}}$ 11. $r = -\cot\theta + c\csc\theta$

13. $y = (x + c)\csc x$ 15. $y = 3 + ce^{-x}$

17. $2s = e^{4t}(t^2 + c)$ 19. $y = \frac{1}{4} + ce^{-x^4}$

21. $3y = x^4 - 6x^2 - 3 + cx$ 23. $xy = (x^3 + c)e^{3x}$

25. Can solve by separation of variables: $\dfrac{dy}{1 - y} = 2\,dx$.

Can also solve as linear differential equation of first order: $dy + 2y\,dx = 2\,dx$; $y = 1 + ce^{-2x}$

27. $y = e^{-x}$ 29. $y = \frac{4}{3}\sin x - \csc^2 x$

31. $y(\csc x - \cot x) = \ln\dfrac{(\sqrt{2} - 1)(\csc 2x - \cot 2x)}{\csc x - \cot x}$

Exercises 14-5, page 434

1. $y^2 = 2x^2 + 1$

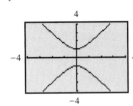

3. $y = 2e^x - x - 1$

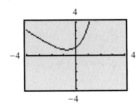

5. $y^2 = c - 2x$

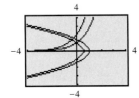

7. $y^2 = c - 2 \sin x$

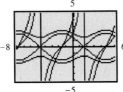

9. $N = N_0(0.5)^{t/40}$, 35.4%   **11.** 3.82 days

13. $S = a + \dfrac{c}{r^2}$ **15.** $5250e^{0.0196t}$ **17.** 13 min

19. \$1040.81   **21.** $\lim\limits_{t \to \infty} \dfrac{E}{R}(1 - e^{-Rt/L}) = \dfrac{E}{R}$

23. $i = \dfrac{E}{R^2 + \omega^2 L^2}(R \sin \omega t - \omega L \cos \omega t + \omega L e^{-Rt/L})$

25. $q = q_0 e^{-t/RC}$ **27.** 11 lb

29. $v = 32(1 - e^{-t})$, 32   **31.** 10 ft/s

33. $x = 3t^2 - t^3$, $y = 6t^2 - 2t^3 - 9t^4 + 6t^5 - t^6$

35. $p = 15(0.667)^{10^{-4}h}$ **37.** \$2490

39. $x = 4(1 + 2e^{-0.25t})$

Review Exercises for Chapter 14, page 436

1. $2 \ln(x^2 + 1) - \dfrac{1}{2y^2} = c$ **3.** $y^2 = 2x - 4 \sin x + c$

5. $2x^2 + 4xy + y^4 = c$ **7.** $y = cx^3 - x^2$

9. $y = c(y + 2)e^{2x}$ **11.** $y = e^{-2x} + ce^{-4x}$

13. $y = \frac{1}{2}(c - x^2) \csc x$ **15.** $y = \dfrac{cx^{5/2}}{(x + 2)^{1/2}}$

17. $y = cxe^{x^3/3}$ **19.** $(x^2 + y^2)(4x + c) + 1 = 0$

21. $y^3 = 8 \sin^2 x$ **23.** $y = 2x - 1 - e^{-2x}$

25. $x^2 + 2 \ln(1 - y) - 4 = 0$ **27.** $y = \frac{1}{4}x^2 \sin 4x$

29. $r = r_0 + kt$ **31.** 3.93 m/s   **33.** 5.31×10^8 years

35. 7.1 billion   **37.** $5y^2 + x^2 = c$

39. $q = c_1 e^{-t/RC} + EC$ **41.** 2.47 L

43. $r = c \cos \theta$

Exercises 15-1, page 443

1. $y = c_1 e^{3x} + c_2 e^{-2x}$ **3.** $y = c_1 e^{-x} + c_2 e^{-x/3}$

5. $y = c_1 + c_2 e^{3x}$ **7.** $y = c_1 e^{6x} + c_2 e^{2x/3}$

9. $y = c_1 e^{x/3} + c_2 e^{-3x}$ **11.** $y = c_1 e^{x/3} + c_2 e^{-x}$

13. $y = e^x(c_1 e^{x\sqrt{2}/2} + c_2 e^{-x\sqrt{2}/2})$

15. $y = e^{3x/8}(c_1 e^{x\sqrt{41}/8} + c_2 e^{-x\sqrt{41}/8})$

17. $y = e^{3x/2}(c_1 e^{x\sqrt{13}/2} + c_2 e^{-x\sqrt{13}/2})$

19. $y = e^{-x/2}(c_1 e^{x\sqrt{33}/2} + c_2 e^{-x\sqrt{33}/2})$

21. $y = \frac{1}{5}(3e^{7x} + 7e^{-3x})$ **23.** $y = \dfrac{e^3}{e^7 - 1}(e^{4x} - e^{-3x})$

25. $y = c_1 + c_2 e^{-x} + c_3 e^{3x}$

27. $y = c_1 e^x + c_2 e^{-x} + c_3 e^{2x} + c_4 e^{-2x}$

Exercises 15-2, page 447

1. $y = (c_1 + c_2 x)e^x$ **3.** $y = (c_1 + c_2 x)e^{-6x}$

5. $y = c_1 \sin 3x + c_2 \cos 3x$

7. $y = e^{-x/2}(c_1 \sin \frac{1}{2}\sqrt{7}x + c_2 \cos \frac{1}{2}\sqrt{7}x)$

9. $y = c_1 e^x + c_2 e^{-x} + c_3 \sin x + c_4 \cos x$

11. $y = c_1 \sin \frac{1}{2}x + c_2 \cos \frac{1}{2}x$ **13.** $y = (c_1 + c_2 x)e^{3x/4}$

15. $y = c_1 \sin \frac{1}{5}\sqrt{2}x + c_2 \cos \frac{1}{5}\sqrt{2}x$

17. $y = e^x(c_1 \cos \frac{1}{2}\sqrt{6}x + c_2 \sin \frac{1}{2}\sqrt{6}x)$

19. $y = (c_1 + c_2 x)e^{4x/5}$

21. $y = e^{3x/4}(c_1 e^{x\sqrt{17}/4} + c_2 e^{-x\sqrt{17}/4})$

23. $y = c_1 e^{x(-6+\sqrt{42})/3} + c_2 e^{x(-6-\sqrt{42})/3}$

25. $y = e^{2x}(c_1 + c_2 x + c_3 x^2)$

27. $y = (c_1 + c_2 x)\sin x + (c_3 + c_4 x)\cos x$

29. $y = e^{-x} \sin 3x$

31. $y = (4 - 14x)e^{4x}$

33. $D^2 y - 9y = 0$

35. $D^2 y + 9y = 0$. The sum of $\cos 3x$ and $\sin 3x$ with no exponential factor indicates imaginary roots with $\alpha = 0$ and $\beta = 3$.

Exercises 15-3, page 451

1. $y = c_1 e^{2x} + c_2 e^{-x} - 2$

3. $y = c_1 e^{-x} + c_2 e^x - 4 - x^2$

5. $y = c_1 + c_2 e^{3x} - \frac{3}{4}e^x - \frac{1}{2}xe^x$

7. $y = c_1 e^{x/3} + c_2 e^{-x/3} - \frac{1}{10} \sin x$

9. $y = (c_1 + c_2 x)e^x + 10 + 6x + x^2 - \frac{2}{25} \sin 3x + \frac{3}{50} \cos 3x$

11. $y = c_1 \sin 2x + c_2 \cos 2x + 3x \cos 2x$

13. $y = c_1 e^{-5x} + c_2 e^{6x} - \frac{1}{3}$

15. $y = c_1 e^{2x/3} + c_2 e^{-5x} + \frac{1}{4}e^{3x}$

17. $y = c_1 e^{2x} + c_2 e^{-2x} - \frac{1}{5} \sin x - \frac{2}{5} \cos x$

19. $y = c_1 \sin x + c_2 \cos x - \frac{1}{3} \sin 2x + 4$

21. $y = c_1 e^{-x} + c_2 e^{-4x} - \frac{7}{100} e^x + \frac{1}{10} x e^x + 1$

23. $y = c_1 + c_2 e^x + c_3 e^{-x} + \frac{1}{10} \cos 2x$

25. $y = c_1 \sin x + c_2 \cos x + \frac{1}{2} x \sin x$

27. $y = c_1 + c_2 e^{-2x} + 2x^2 - 2x - \frac{1}{2} x e^{-2x}$

29. $y = \frac{1}{6}(11e^{3x} + 5e^{-2x} + e^x - 5)$

31. $y = -\frac{2}{3} \sin x + \pi \cos x + x - \frac{1}{3} \sin 2x$

Exercises 15-4, page 458

1. $\theta = 0.1 \cos 3.1t$ **3.** 10

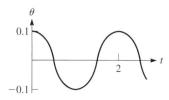

5. $y = 0.250 \cos 16.0t$

7. $y = 0.250 \cos 16.0t + 0.127 \sin 2.00t - 0.016 \sin 16.0t$

9. $q = 2.23 \times 10^{-4} e^{-20t} \sin 2240t$

11. $q = 0.01(1 - \cos 316t)$

13. $q = e^{-10t}(c_1 \sin 99.5t + c_2 \cos 99.5t)$
$- 1.81 \times 10^{-3} \sin 120\pi t$
$- 1.03 \times 10^{-4} \cos 120\pi t$

15. $i = 10^{-6}(2.00 \cos(1.58 \times 10^4 t)$
$+ 158 \sin(1.58 \times 10^4 t) - 2.00 e^{-200t})$

17. $i_p = 0.528 \sin 100t - 3.52 \cos 100t$

19. $y = \frac{w}{24EI}(6L^2 x^2 - 4Lx^3 + x^4)$

Review Exercises for Chapter 15, page 459

1. $y = c_1 + c_2 e^{-x/2}$ **3.** $y = (c_1 + c_2 x)e^{-x}$

5. $y = e^{-x}(c_1 \sin \sqrt{5}x + c_2 \cos \sqrt{5}x)$

7. $y = c_1 e^{0.2x} + c_2 e^{-0.4x}$ **9.** $y = c_1 e^x + c_2 e^{-3x/2} - 2$

11. $y = e^{-x/2}(c_1 e^{x\sqrt{5}/2} + c_2 e^{-x\sqrt{5}/2}) + 2e^x$

13. $y = c_1 e^{2x/3} + c_2 e^{4x/3} + \frac{1}{2}x + \frac{25}{8}$

15. $y = c_1 e^x + c_2 \sin 3x + c_3 \cos 3x - \frac{1}{16}(\sin x + \cos x)$

17. $y = c_1 e^{-x} + c_2 e^{8x} - \frac{2}{9} x e^{-x}$

19. $y = c_1 \sin 5x + c_2 \cos 5x + 5x \sin 5x$

21. $y = 2e^{-x/2} \sin(\frac{1}{2}\sqrt{15}x)$

23. $y = \frac{1}{25}[16 \sin x + 12 \cos x - 3e^{-2x}(4 + 5x)]$

25. $y = 0.25e^{-2t}(2 \cos 4t + \sin 4t)$, underdamped

27. $x = 0.661 e^{-1.60t} \sin 6.81t$ **29.** $y = 3.00 e^{-0.100t} \cos 63.2t$

31. $q = e^{-6t}(0.4 \cos 8t + 0.3 \sin 8t) - 0.4 \cos 10t$

33. $i = 0$ **35.** $y = \frac{10}{3EI}[100x^3 - x^4 + xL^2(L - 100)]$

Exercises 16-1, page 464

1.

x	0.0	0.2	0.4	0.6	0.8	1.0
y	1.00	1.20	1.44	1.72	2.04	2.40

$y = \frac{1}{2}x^2 + x + 1$

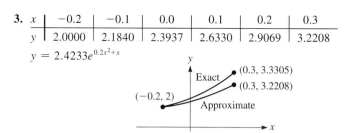

3.

x	-0.2	-0.1	0.0	0.1	0.2	0.3
y	2.0000	2.1840	2.3937	2.6330	2.9069	3.2208

$y = 2.4233 e^{0.2x^2 + x}$

5. (Not all values shown)

x	0.0	0.2	0.4	0.6	0.8	1.0
y	1.00	1.21	1.46	1.75	2.08	2.45

7. (Not all values shown)

x	-0.2	-0.1	0.0	0.1	0.2	0.3
y	2.000	2.1903	2.4079	2.6573	2.9436	3.2732

9.

x	0.0	0.1	0.2	0.3	0.4
y	0.0000	0.1003	0.2027	0.3092	0.4220

11.

x	0.0	0.2	0.4	0.6	0.8	1.0
y	0.0000	0.2027	0.4232	0.6884	1.0588	1.7722

13.

x	0.0	0.1	0.2	0.3	0.4	0.5
y	1.5708	1.5660	1.5521	1.5302	1.5011	1.4656

15. $i_{approx} = 0.0804$ A, $i_{exact} = 0.0898$ A

Exercises 16-2, page 467

1. $y_2 = 1 + x + x^2 + \frac{1}{3}x^3$; $y = \frac{1}{1 - x}$;
$y_2(0.1) = 1.1103$; $y(0.1) = 1.1111$

3. $y_3 = x^2 + \frac{1}{2}x^4 + \frac{1}{6}x^6$; $y = e^{x^2} - 1$; $y_3(1) = 1.67$; $y(1) = 1.72$

5. $y_3 = \frac{1}{2}x^2 + \frac{1}{20}x^5 + \frac{1}{160}x^8 + \frac{1}{4400}x^{11}$

7. $y_2 = \frac{3}{4} + x + \frac{1}{2}x^2 + \frac{1}{4}\cos 2x - \frac{1}{2}\sin 2x$

9. Maclaurin expansion: $y = 1 + x + x^2 + \frac{1}{3}x^3 + \frac{1}{12}x^4 + \cdots$
Example 1: $y_3 = 1 + x + x^2 + \frac{1}{3}x^3 + \frac{1}{24}x^4 + \cdots$

11. 0.58 ft/s

Exercises 16-3, page 472

1. $F(s) = \int_0^\infty e^{-st}\,dt = -\frac{1}{s}e^{-st}\Big|_0^\infty = \frac{1}{s}$

3. $F(s) = \int_0^\infty e^{-st}\sin at\,dt = \frac{e^{-st}(-s\sin at - a\cos at)}{s^2 + a^2}\Big|_0^\infty$
$= \frac{a}{s^2 + a^2}$

5. $\frac{1}{s-3}$ **7.** $\frac{30}{(s+2)^4}$ **9.** $\frac{s-2}{s^2+4}$

11. $\frac{3}{s} + \frac{2(s^2-9)}{(s^2+9)^2}$ **13.** $s^2\mathcal{L}(f) + s\mathcal{L}(f)$

15. $(2s^2 - s + 1)\mathcal{L}(f) - 2s + 1$ **17.** t^2 **19.** $\frac{1}{2}e^{-3t}$

21. $\frac{1}{2}t^2 e^{-t}$ **23.** $\frac{1}{54}(9t\sin 3t + 2\sin 3t - 6t\cos 3t)$

25. $-\frac{1}{3}e^{-t} - \frac{8}{3}e^{2t} + 7e^{3t}$ **27.** $\frac{1}{2}e^t(4\cos 2t + 5\sin 2t)$

Exercises 16-4, page 476

1. $y = e^{-t}$ **3.** $y = -e^{3t/2}$ **5.** $y = (1 + t)e^{-3t}$

7. $y = \frac{1}{2}\sin 2t$ **9.** $y = 1 - e^{-2t}$ **11.** $y = e^{2t}\cos t$

13. $y = 1 + \sin t$ **15.** $y = e^{-t}(\frac{1}{2}t^2 + 3t + 1)$

17. $y = 2e^{3t} + 3e^{-2t}$ **19.** $y = \frac{3}{2}e^t - \frac{3}{2}e^{-t} - \sin 2t$

21. $v = 6(1 - e^{-t/2})$ **23.** $q = 1.6 \times 10^{-4}(1 - e^{-5000t})$

25. $i = 5t\sin 50t$ **27.** $y = \sin 3t - 3t\cos 3t$

29. $i = 5.0e^{-50t} - 5.0e^{-100t}$ **31.** $i = 4.42e^{-66.7t}\sin 226t$

Review Exercises for Chapter 16, page 477

1.

x	0.0	0.1	0.2	0.3	0.4	0.5
y	1.0000	0.9000	0.8200	0.7568	0.7085	0.6743

3. (Not all values shown)

x	0.0	0.4	0.8	1.2	1.6	2.0
y	1.0000	1.4366	2.0115	2.7128	3.5330	4.4672

5. (Not all values shown)

x	0.0	0.1	0.2	0.4	0.6	0.7
y	0.5000	0.5546	0.6266	0.8359	1.1843	1.4864

7. (Not all values shown)

x	0.0	0.1	0.2	0.3	0.4	0.5
y	1.0000	0.9077	0.8329	0.7734	0.7282	0.6965

9.

x	0.0	0.1	0.2	0.3	0.4	0.5
y	1.0000	0.9094	0.8358	0.7772	0.7327	0.7018

11.

x	0.0	0.1	0.2	0.3	0.4
y	0.5000	0.5637	0.6476	0.7567	0.8996

13. $y_3 = 1 + x + \frac{1}{2}x^2 + \frac{1}{2}x^3 + \frac{1}{12}x^4 + \frac{1}{60}x^5$

15. $y_2 = -\frac{7}{36} + x - \frac{1}{9}x^3 + \frac{1}{4}x^4 + \frac{1}{18}x^6$

17. $y = e^{t/4}$ **19.** $y = \frac{1}{2}(e^{3t} - e^t)$ **21.** $y = -4\sin t$

23. $y = \frac{1}{10}(3e^t - \sin 3t - 3\cos 3t)$

25. $y_2 = \frac{7}{2} + x - (3 + x)e^{-x} + \frac{1}{2}e^{-2x}$; $y_2(0.5) = 2.0611$

27.

x	0.0	0.1	0.2	0.3	0.4	0.5
y	1.0000	1.2000	1.4086	1.6239	1.8442	2.0678

29. $i = 12(1 - e^{-t/2})$; $i(0.3) = 1.67$ A

31. $q = 10^{-4}e^{-8t}(4.0\cos 200t + 0.16\sin 200t)$

33. $y = 0.25t\sin 8t$

35.

t	0.00	0.05	0.10	0.15	0.20	0.25	0.30
i	0.0000	0.3000	0.5925	0.8777	1.1557	1.4269	1.6912

37. $y = 0.75\cos 8t$

39. $F(s) = \frac{b-a}{(s+a)(s+b)} = \frac{1}{s+a} - \frac{1}{s+b}$;
$L^{-1}(F) = e^{-at} - e^{-bt}$

Exercises A-1, page 482

1. hyperbola;
$2x'y' + 25 = 0$

3. ellipse;
$4x'^2 + 9y'^2 = 36$

5. parabola;
$x'^2 + \sqrt{2}y' = 0$

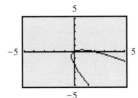

7. hyperbola;
$4x'^2 - y'^2 = 4$

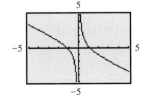

9. ellipse;
$x'^2 + 2y'^2 = 2$

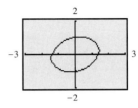

11. parabola;
$y'^2 - 4x' + 16 = 0$
$y''^2 = 4x''$

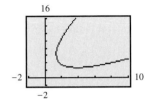

Exercises A-2, page 487

1. $y = 1.0x - 2.6$

3. $y = 3.5\sqrt{x} + 1.3$

5. $V = -0.590i + 11.3$

7. $p = -0.200x + 649$

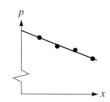

9. $P = \dfrac{1343}{S}$

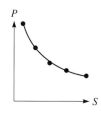

11. $y = 6.20e^{-t} - 0.05$

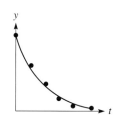

Exercises C-2, page 492

(Most answers have been rounded off to four significant digits.)

1. 56.02 **3.** 4162.1 **5.** 18.65 **7.** 0.3954

9. 14.14 **11.** 0.5251 **13.** 13.35 **15.** 944.6

17. 0.7349 **19.** −0.7594 **21.** −1.337

23. 1.015 **25.** 41.35° **27.** −1.182 **29.** 0.5862

31. 6.695 **33.** 3.508 **35.** 0.005685 **37.** 2.053

39. 5.765 **41.** 4.501×10^{10} **43.** 497.2

45. 6.648 **47.** 401.2 **49.** 8.841 **51.** 2.523

53. 10.08 **55.** 22.36 **57.** 20.3° **59.** 4729

61. 3.301×10^4 **63.** 1.056 **65.** 55.5°

67. 3.277 **69.** 8.125 **71.** 1.000 **73.** 1.000

75. 12.90 **77.** 8.001 **79.** 8.053 **81.** 0.04259

83. 0.4219 **85.** 0.7822 **87.** 2.0736465

89. 124.3 **91.** 252

93.

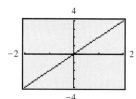

95.

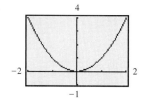

97.

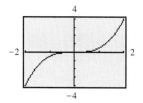

99.

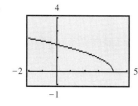

101.

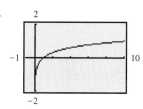

103.

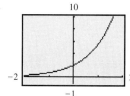

105.

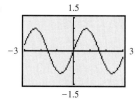

107.

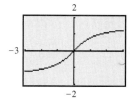

Chapter 1

1.
$$f(x) = 2x - x^2 + \frac{8}{x}$$

$$f(-4) = 2(-4) - (-4)^2 + \frac{8}{-4}$$
$$= -8 - 16 - 2$$
$$= -26$$

$$f(2.385) = 2(2.385) - (2.385)^2 + \frac{8}{2.385} = 2.436$$

2. $w = 2000 - 10t$

3. $f(x) = 4 - 2x$

$y = 4 - 2x$

$y = 4 - 2(-1) = 6$

$y = 4 - 2(0) = 4$

$y = 4 - 2(1) = 2$

$y = 4 - 2(2) = 0$

$y = 4 - 2(3) = -2$

$y = 4 - 2(4) = -4$

x	y
-1	6
0	4
1	2
2	0
3	-2
4	-4

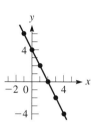

4. $y = 2x^2 - 3x - 3$

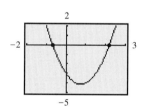

$x = -0.7$ and $x = 2.2$

5. $y = \sqrt{4 + 2x}$

$y = \sqrt{4 + 2(-2)} = 0$

$y = \sqrt{4 + 2(-1)} = 1.4$

$y = \sqrt{4 + 2(0)} = 2$

$y = \sqrt{4 + 2(1)} = 2.4$

$y = \sqrt{4 + 2(2)} = 2.8$

$y = \sqrt{4 + 2(4)} = 3.5$

x	y
-2	0
-1	1.4
0	2
1	2.4
2	2.8
4	3.5

6. On negative x-axis

7. $f(x) = \sqrt{6 - x}$
Domain: $x \leq 6$; x cannot be greater than 6 to have real values of $f(x)$.
Range: $f(x) \geq 0$; $\sqrt{6 - x}$ is the principal square root of $6 - x$ and cannot be negative.

8. Let r = radius of circular part

$$\overset{\text{square}}{} \quad \overset{\text{semicircle}}{}$$
$$A = (2r)(2r) + \tfrac{1}{2}(\pi r^2)$$
$$= 4r^2 + \tfrac{1}{2}\pi r^2$$

9. For graphing, use x for Q and y for P.
Graph $y = 0.00021x^2 + 0.013x - 1.2$ and find the positive zero.
From display, for $P = 1.2$ lb/in^2, $Q = 51$ gal/min

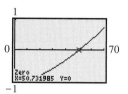

Chapter 2

1.
$$2(x^2 + x) = 1 - y^2$$
$$2x^2 + y^2 + 2x - 1 = 0$$
$A \neq C$ (same sign)
$B = 0$: ellipse

2. $4x - 2y + 5 = 0$
$$y = 2x + \tfrac{5}{2}$$
$$m = 2 \qquad b = \tfrac{5}{2}$$

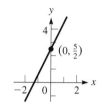

3. $x^2 = -12y$
$4p = -12, \qquad p = -3$
$V(0, 0) \qquad F(0, -3)$

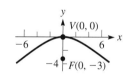

4. Center $(-1, 2)$; $h = -1$, $k = 2$
$$r = \sqrt{(2 + 1)^2 + (3 - 2)^2}$$
$$= \sqrt{10}$$
$$(x + 1)^2 + (y - 2)^2 = 10$$
or
$$x^2 + y^2 + 2x - 4y - 5 = 0$$

5. $m = \dfrac{-2 - 1}{2 + 4} = -\dfrac{1}{2}$
$$y - 1 = -\tfrac{1}{2}(x + 4)$$
$$2y - 2 = -x - 4$$
$$x + 2y + 2 = 0$$

6. $x^2 = 4py$
$(6.00)^2 = 4p(4.00)$
$p = 2.25$
Focus is 2.25 cm
from vertex.

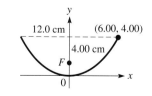

7. $a = 8 \qquad b = 4$

$$\frac{x^2}{64} + \frac{y^2}{16} = 1$$

Find y for $x = 4$ ft.

$$\frac{16}{64} + \frac{y^2}{16} = 1$$

$$y^2 = 12 \qquad y = 3.5 \text{ ft}$$

$$h = 10.0 + 3.5 = 13.5 \text{ ft}$$

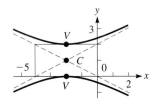

8. $4y^2 - x^2 - 4x - 8y - 4 = 0$

$$4(y^2 - 2y \qquad) - (x^2 + 4x \qquad) = 4$$

$$4(y^2 - 2y + 1) - (x^2 + 4x + 4) = 4 + 4 - 4$$

$$\frac{(y-1)^2}{1^2} - \frac{(x+2)^2}{2^2} = 1$$

$$C(-2, 1) \qquad V(-2, 0) \qquad V(-2, 2)$$

Chapter 3

1. $\displaystyle\lim_{x \to 1} \frac{x^2 - x}{x^2 - 1} = \lim_{x \to 1} \frac{x(x-1)}{(x+1)(x-1)} = \lim_{x \to 1} \frac{x}{x+1} = \frac{1}{2}$

2. $\displaystyle\lim_{x \to \infty} \frac{1 - 4x^2}{x + 2x^2} = \lim_{x \to \infty} \frac{\dfrac{1}{x^2} - 4}{\dfrac{1}{x} + 2} = -2$

3. $y = 3x^2 - \dfrac{4}{x^2}$

$$\frac{dy}{dx} = 6x + \frac{8}{x^3}$$

$$\left.\frac{dy}{dx}\right|_{x=2} = 6(2) + \frac{8}{2^3}$$

$$= 13$$

$$m_{\tan} = 13$$

```
nDeriv(3X²-4/X²,
X,2)
         13.0000005
```

4. $s = t\sqrt{10 - 2t}$

$$v = \frac{ds}{dt} = t(\tfrac{1}{2})(10 - 2t)^{-1/2}(-2) + (10 - 2t)^{1/2}(1)$$

$$= \frac{-t}{(10 - 2t)^{1/2}} + (10 - 2t)^{1/2} = \frac{10 - 3t}{(10 - 2t)^{1/2}}$$

$$v\big|_{t=4.00} = \frac{10 - 3(4.00)}{[10 - 2(4.00)]^{1/2}} = \frac{10 - 12.00}{2.00^{1/2}} = -1.41 \text{ cm/s}$$

5. $(1 + y^2)^3 - x^2 y = 7x$

$$3(1 + y^2)^2(2yy') - x^2 y' - y(2x) = 7$$

$$6y(1 + y^2)^2 y' - x^2 y' = 7 + 2xy$$

$$y' = \frac{7 + 2xy}{6y(1 + y^2)^2 - x^2}$$

6. $V = \dfrac{kq}{\sqrt{x^2 + b^2}} = kq(x^2 + b^2)^{-1/2}$

$$\frac{dV}{dx} = kq\left(-\frac{1}{2}\right)(x^2 + b^2)^{-3/2}(2x) = \frac{-kqx}{(x^2 + b^2)^{3/2}}$$

7. $y = \dfrac{2x}{3x + 2}$

$$\frac{dy}{dx} = \frac{(3x + 2)(2) - 2x(3)}{(3x + 2)^2} = \frac{4}{(3x + 2)^2} = 4(3x + 2)^{-2}$$

$$\frac{d^2 y}{dx^2} = -2(4)(3x + 2)^{-3}(3) = \frac{-24}{(3x + 2)^3}$$

8. $y = 5x - 2x^2$

$$y + \Delta y = 5(x + \Delta x) - 2(x + \Delta x)^2$$

$$\Delta y = 5\Delta x - 4x\Delta x - 2\Delta x^2$$

$$\frac{\Delta y}{\Delta x} = 5 - 4x - 2\Delta x$$

$$\lim_{\Delta x \to 0} \frac{\Delta y}{\Delta x} = 5 - 4x$$

Chapter 4

1. $y = x^4 - 3x^2$

$$\frac{dy}{dx} = 4x^3 - 6x$$

$$\left.\frac{dy}{dx}\right|_{x=1} = 4(1^3) - 6(1)$$

$$= -2$$

$$y - (-2) = -2(x - 1)$$

$$y = -2x$$

2. $y = 3x^2 - x$

$$y + \Delta y = 3(x + \Delta x)^2 - (x + \Delta x)$$

$$\Delta y = 6x\Delta x + 3\Delta x^2 - \Delta x$$

$$dy = (6x - 1)\,dx$$

For $x = 3$, $\Delta x = 0.1$.

$$\Delta y = 6(3)(0.1) + 3(0.1)^2 - (0.1)$$

$$= 1.73$$

$$dy = [6(3) - 1](0.1) = 1.7$$

$$\Delta y - dy = 0.03$$

3. $x = 3t^2$ $\qquad\qquad\qquad\qquad$ $y = 2t^3 - t^2$

$$v_x = \frac{dx}{dt} = 6t \qquad\qquad\qquad v_y = \frac{dy}{dt} = 6t^2 - 2t$$

$$a_x = \frac{dv_x}{dt} = \frac{d^2x}{dt^2} = 6 \qquad\quad a_y = \frac{dv_y}{dt} = \frac{d^2y}{dt^2} = 12t - 2$$

$$a_x\big|_{t=2} = 6 \qquad\qquad\qquad a_y\big|_{t=2} = 12(2) - 2 = 22$$

$$a\big|_{t=2} = \sqrt{6^2 + 22^2} = 22.8 \qquad \tan\theta = \tfrac{22}{6}, \qquad \theta = 74.7°$$

4. $P = \dfrac{144r}{(r + 0.6)^2}$

$$\frac{dP}{dr} = \frac{144[(r + 0.6)^2(1) - r(2)(r + 0.6)(1)]}{(r + 0.6)^4}$$

$$= \frac{144[(r + 0.6) - 2r]}{(r + 0.6)^3} = \frac{144(0.6 - r)}{(r + 0.6)^3}$$

$$\frac{dP}{dr} = 0; \qquad 0.6 - r = 0, \qquad r = 0.6 \ \Omega$$

$$\left(r < 0.6, \frac{dP}{dr} > 0; r > 0.6, \frac{dP}{dr} < 0 \right)$$

5. $x^2 - \sqrt{4x + 1} = 0;$ $\qquad f(x) = x^2 - \sqrt{4x + 1}$

$$f'(x) = 2x - \tfrac{1}{2}(4x + 1)^{-1/2}(4) = 2x - \frac{2}{(4x + 1)^{1/2}}$$

n	x_n	$f(x_n)$	$f'(x_n)$	$x_n - \dfrac{f(x_n)}{f'(x_n)}$
1	1.5	-0.3957513	2.2440711	1.6763542
2	1.6763542	0.0343001	2.6322118	1.6633233

$x_3 = 1.6633$

6. $y = \sqrt{2x + 4}, \ a = 6$

$$\frac{dy}{dx} = \frac{1}{2}(2x + 4)^{-1/2}(2) = \frac{1}{\sqrt{2x + 4}}$$

$$\frac{dy}{dx}\Big|_{x=6} = \frac{1}{\sqrt{2(6) + 4}} = \frac{1}{4}$$

$$f(6) = \sqrt{2(6) + 4} = 4$$

$$L(x) = 4 + \tfrac{1}{4}(x - 6) = \tfrac{1}{4}x + \tfrac{5}{2}$$

7. $y = x^3 + 6x^2$

$y' = 3x^2 + 12x = 3x(x + 4)$

$y'' = 6x + 12 = 6(x + 2)$

$\quad x < -4 \qquad y$ inc. $\qquad$ Max. $(-4, 32)$

$-4 < x < 0 \qquad y$ dec. $\qquad$ Min. $(0, 0)$

$\quad x > 0 \qquad y$ inc. $\qquad$ Infl. $(-2, 16)$

$\quad x < -2 \qquad y$ conc. down

$\quad x > -2 \qquad y$ conc. up

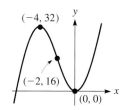

8. $y = \dfrac{4}{x^2} - x$

$$y' = -\frac{8}{x^3} - 1 = -\frac{8 + x^3}{x^3}$$

$$y'' = \frac{24}{x^4}$$

$\quad x < -2 \qquad y$ dec.

$-2 < x < 0 \qquad y$ inc.

$\quad x > 0 \qquad y$ dec., conc. up

$\quad x < 0 \qquad y$ conc. up

Min. $(-2, 3)$, no infl., int. $(\sqrt[3]{4}, 0)$, sym. none; as $x \to \pm\infty$, $y \to -x$, asym. $y = -x$, $x = 0$. Domain: all real x except 0; range: all real y.

9. Let V = volume of cube

$\qquad e$ = edge of cube

$$V = e^3 \qquad \frac{dV}{dt} = 3e^2\frac{de}{dt}$$

$$\frac{dV}{dt}\Big|_{e=4.00} = 3(4.00)^2(-0.50)$$

$$= -24 \text{ ft}^3/\text{s}$$

10.

$3x + 2y = 6000$

$$y = \frac{6000 - 3x}{2}$$

$$A = xy = x\left(\frac{6000 - 3x}{2}\right)$$

$$= 3000x - \tfrac{3}{2}x^2$$

$$\frac{dA}{dx} = 3000 - 3x$$

$3000 - 3x = 0, \ x = 1000$ m

$y = 1500$ m

$A_{max} = (1000)(1500) = 1.5 \times 10^6$ m^2

$$\left(x < 1000, \frac{dA}{dx} > 0; x > 1000, \frac{dA}{dx} < 0 \right)$$

Chapter 5

1. Power of x required for $2x$ is 2. Therefore, multiply by $1/2$. Antiderivative of $2x = \tfrac{1}{2}(2x^2) = x^2$. Power of $(1 - x)^4$ required is 5. Derivative of $(1 - x)^5$ is $5(1 - x)^4(-1)$. Writing $-(1 - x)^4$ as $\tfrac{1}{5}[5(1 - x)^4(-1)]$, the antiderivative of $-(1 - x)^4$ is $\tfrac{1}{5}(1 - x)^5$. Therefore, the antiderivative of $2x - (1 - x)^4$ is $x^2 + \tfrac{1}{5}(1 - x)^5$.

2. $\displaystyle\int x\sqrt{1 - 2x^2}\,dx = \int x(1 - 2x^2)^{1/2}\,dx$

$u = 1 - 2x^2 \qquad du = -4x\,dx \qquad n = \tfrac{1}{2} \qquad n + 1 = \tfrac{3}{2}$

$$\int x(1 - 2x^2)^{1/2}\,dx = -\tfrac{1}{4}\int (1 - 2x^2)^{1/2}(-4x\,dx)$$

$$= -\tfrac{1}{4}(\tfrac{2}{3})(1 - 2x^2)^{3/2} + C$$

$$= -\tfrac{1}{6}(1 - 2x^2)^{3/2} + C$$

3. $\dfrac{dy}{dx} = (6 - x)^4$, $dy = (6 - x)^4\,dx$

$$\int dy = \int (6 - x)^4\,dx$$

$$y = -\int (6 - x)^4(-dx) = -\frac{1}{5}(6 - x)^5 + C$$

$$2 = -\frac{1}{5}(6 - 5)^5 + C,\ C = \tfrac{11}{5}$$

$$y = -\frac{1}{5}(6 - x)^5 + \tfrac{11}{5}$$

4. $y = \dfrac{1}{x + 2}$ $n = 6$ $\Delta x = \dfrac{4 - 1}{6} = \dfrac{1}{2}$

$A = \frac{1}{2}(\frac{2}{7} + \frac{1}{4} + \frac{2}{9} + \frac{1}{5} + \frac{2}{11} + \frac{1}{6}) = 0.6532$

x	1	$\frac{3}{2}$	2	$\frac{5}{2}$	3	$\frac{7}{2}$	4
y	$\frac{1}{3}$	$\frac{2}{7}$	$\frac{1}{4}$	$\frac{2}{9}$	$\frac{1}{5}$	$\frac{2}{11}$	$\frac{1}{6}$

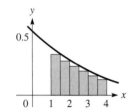

5. (See values for Problem 4.)

$$\int_1^4 \frac{dx}{x + 2} = \frac{1}{4}\left[\frac{1}{3} + 2\left(\frac{2}{7}\right) + 2\left(\frac{1}{4}\right) + 2\left(\frac{2}{9}\right)\right.$$
$$\left. + 2\left(\frac{1}{5}\right) + 2\left(\frac{2}{11}\right) + \frac{1}{6}\right]$$
$$= 0.6949$$

6. (See values for Problem 4.)

$$\int_1^4 \frac{dx}{x + 2} = \frac{1}{6}\left[\frac{1}{3} + 4\left(\frac{2}{7}\right) + 2\left(\frac{1}{4}\right) + 4\left(\frac{2}{9}\right)\right.$$
$$\left. + 2\left(\frac{1}{5}\right) + 4\left(\frac{2}{11}\right) + \frac{1}{6}\right] = 0.6932$$

7. $i = \displaystyle\int_1^3 \left(t^2 + \frac{1}{t^2}\right) dt = \frac{1}{3}t^3 - \frac{1}{t}\,\Big|_1^3$

$= \frac{1}{3}(27) - \frac{1}{3} - (\frac{1}{3} - 1) = 9.3$ A

Chapter 6

1. $A = \displaystyle\int_0^2 \frac{1}{4}x^2\,dx$

$= \dfrac{1}{12}x^3\Big|_0^2 = \dfrac{2}{3}$

2. $\bar{x} = \dfrac{\displaystyle\int_0^2 x(\frac{1}{4}x^2)\,dx}{\frac{2}{3}} = \dfrac{\frac{1}{4}\displaystyle\int_0^2 x^3\,dx}{\frac{2}{3}} = \dfrac{\frac{1}{16}x^4\big|_0^2}{\frac{2}{3}} = \dfrac{1}{\frac{2}{3}} = \dfrac{3}{2}$

$\bar{y} = \dfrac{\displaystyle\int_0^1 y(2 - 2\sqrt{y})\,dy}{\frac{2}{3}} = \dfrac{2\displaystyle\int_0^1 (y - y^{3/2})\,dy}{\frac{2}{3}}$

$= \dfrac{2(\frac{1}{2}y^2 - \frac{2}{5}y^{5/2})\big|_0^1}{\frac{2}{3}}$

$= \dfrac{1 - \frac{4}{5}}{\frac{2}{3}} = \dfrac{1}{5} \times \dfrac{3}{2} = \dfrac{3}{10}$

3. $V = \pi\displaystyle\int_0^2 \left(\frac{1}{4}x^2\right)^2 dx = \dfrac{\pi}{16}\displaystyle\int_0^2 x^4\,dx = \dfrac{\pi}{80}x^5\Big|_0^2$

$= \dfrac{32\pi}{80} = \dfrac{2\pi}{5}$

4. $V = \pi\displaystyle\int_0^3 9^2\,dx - \pi\displaystyle\int_0^3 (x^2)^2\,dx = 81\pi x\Big|_0^3 - \dfrac{\pi}{5}x^5\Big|_0^3$

$= 243\pi - \dfrac{243\pi}{5} = \dfrac{972\pi}{5}$

or $V = 2\pi\displaystyle\int_0^9 xy\,dy = 2\pi\displaystyle\int_0^9 y^{1/2}y\,dy = \dfrac{4\pi}{5}y^{5/2}\Big|_0^9$

$= \dfrac{4\pi}{5}(3^5 - 0) = \dfrac{972\pi}{5}$

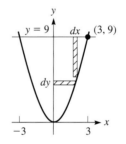

5. $I_y = k\displaystyle\int_0^3 x^2(9 - x^2)\,dx = k\displaystyle\int_0^3 (9x^2 - x^4)\,dx$

$= k\left(3x^3 - \dfrac{1}{5}x^5\right)\Big|_0^3 = k\left(81 - \dfrac{243}{5}\right) = \dfrac{162k}{5}$

6. $s = \displaystyle\int (60 - 4t)\,dt = 60t - 2t^2 + C$

$s = 10$ for $t = 0$, $10 = 60(0) - 2(0^2) + C$, $C = 10$
$s = 60t - 2t^2 + 10$

7. $F = kx$, $12 = k(2.0)$, $k = 6.0$ N/cm

$W = \displaystyle\int_{2.0}^{6.0} 6.0x\,dx = 3.0x^2\Big|_{2.0}^{6.0} = 3.0(36.0 - 4.0)$

$= 96$ N·cm

8. $F = 62.4 \int_{1.00}^{3.00} 6.00h\,dh = (62.4)(6.00)\frac{1}{2}h^2\Big|_{1.00}^{3.00}$

$\qquad = (62.4)(3.00)(9.00 - 1.00) = 1500 \text{ lb}$

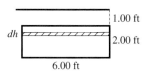

1.00 ft

2.00 ft

6.00 ft

Chapter 7

1. $y = \tan^3 2x + \tan^{-1} 2x$

$\dfrac{dy}{dx} = 3(\tan^2 2x)(\sec^2 2x)(2) + \dfrac{2}{1 + (2x)^2}$

$\qquad = 6 \tan^2 2x \sec^2 2x + \dfrac{2}{1 + 4x^2}$

2. $y = 2(3 + \cot 4x)^3$

$\dfrac{dy}{dx} = 2(3)(3 + \cot 4x)^2(-\csc^2 4x)(4)$

$\qquad = -24(3 + \cot 4x)^2 \csc^2 4x$

3. $y \sec 2x = \sin^{-1} 3y$

$y(\sec 2x \tan 2x)(2) + (\sec 2x)(y') = \dfrac{3y'}{\sqrt{1 - (3y)^2}}$

$\sqrt{1 - 9y^2} \sec 2x\,(2y \tan 2x + y') = 3y'$

$y' = \dfrac{2y\sqrt{1 - 9y^2} \sec 2x \tan 2x}{3 - \sqrt{1 - 9y^2} \sec 2x}$

4. $y = \dfrac{\cos^2(3x + 1)}{x}$

$dy = \dfrac{x\{2 \cos(3x + 1)[-\sin(3x + 1)(3)]\} - \cos^2(3x + 1)(1)}{x^2}\,dx$

$\quad = \dfrac{-6x \cos(3x + 1) \sin(3x + 1) - \cos^2(3x + 1)}{x^2}\,dx$

5. $y = x \cos^{-1} 2x$

$\dfrac{dy}{dx} = \dfrac{-2x}{\sqrt{1 - 4x^2}} + \cos^{-1} 2x$

$\dfrac{dy}{dx}\Big|_{x=0.1} = \dfrac{-0.2}{\sqrt{1 - 0.04}} + \cos^{-1} 0.2 = 1.165$

numerical derivative value: 1.165313676

6. $T = \dfrac{A}{1 + B \sin^2(\theta/2)}$

$\dfrac{dT}{dt} = -A(1 + B \sin^2(\theta/2))^{-2}\left[2B \sin\left(\dfrac{\theta}{2}\right)\cos\left(\dfrac{\theta}{2}\right)\left(\dfrac{1}{2}\right)\right]\dfrac{d\theta}{dt}$

$\quad = \dfrac{-AB \sin \theta}{2(1 + B \sin^2(\theta/2))^2}\dfrac{d\theta}{dt}$

7. $y = x - \cos x$

$y' = 1 + \sin x$

$1 + \sin x = 0, \qquad \sin x = -1, \qquad x = -\pi/2, 3\pi/2$

$y'' = \cos x \qquad \cos x = 0, x = -\pi/2, \pi/2, 3\pi/2$

y' is never negative, y increasing all x except $-\pi/2, 3\pi/2$

infl. $(-\pi/2, -\pi/2), (\pi/2, \pi/2), (3\pi/2, 3\pi/2)$

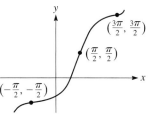

8. For $t = 8.0$ s, $x = 40$ ft.

$\theta = \tan^{-1} \dfrac{x}{250}$

$\dfrac{d\theta}{dt} = \dfrac{1}{1 + \dfrac{x^2}{250^2}}\dfrac{dx/dt}{250} = \dfrac{250^2}{250^2 + x^2}\dfrac{dx/dt}{250}$

$\dfrac{d\theta}{dt}\Big|_{t=8.0} = \dfrac{250}{250^2 + 40^2}(5.0) = 0.020 \text{ rad/s}$

250 ft

Chapter 8

1. $y = 2x^2 e^{x/2}$

$y' = 2x^2(e^{x/2})\left(\tfrac{1}{2}\right) + e^{x/2}(4x)$

$\quad = x(x + 4)e^{x/2}$

2. $y = \ln \cos(1 - x^2)$

$\dfrac{dy}{dx} = \dfrac{-[\sin(1 - x^2)](-2x)}{\cos(1 - x^2)}$

$\quad = 2x \tan(1 - x^2)$

3. $x \ln y = 2 \sin^{-1} x$

$x\left(\dfrac{y'}{y}\right) + \ln y = \dfrac{2}{\sqrt{1 - x^2}}$

$y' = \dfrac{2y}{x\sqrt{1 - x^2}} - \dfrac{y \ln y}{x}$

4. $y = \dfrac{3x}{2 - e^{-x}}$

$dy = \dfrac{(2 - e^{-x})(3) - 3x(-e^{-x})(-1)}{(2 - e^{-x})^2}$

$\quad = \dfrac{6 - 3e^{-x} - 3xe^{-x}}{(2 - e^{-x})^2}$

5. $y = \ln \dfrac{2x - 1}{1 + x^2}$

$\quad = \ln(2x - 1) - \ln(1 + x^2)$

$\dfrac{dy}{dx} = \dfrac{2}{2x - 1} - \dfrac{2x}{1 + x^2}$

$m_{\tan} = \dfrac{dy}{dx}\Big|_{x=2} = \dfrac{2}{4 - 1} - \dfrac{4}{1 + 4} = \dfrac{2}{3} - \dfrac{4}{5} = -\dfrac{2}{15}$

6. $i = 8e^{-t} \sin 10t$

$\dfrac{di}{dt} = 8[e^{-t} \cos 10t(10) + \sin 10t(-e^{-t})]$

$\quad = 8e^{-t}(10 \cos 10t - \sin 10t)$

7. $y = xe^x$

$y' = xe^x + e^x = e^x(x + 1)$

$y'' = xe^x + e^x + e^x = e^x(x + 2)$

$x < -1 \quad y$ dec.

$x > -1 \quad y$ inc.

$x < -2 \quad y$ conc. down

$x > -2 \quad y$ conc. up

Int. $(0, 0)$, min. $(-1, -e^{-1})$

infl. $(-2, -2e^{-2})$, asym. $y = 0$

8. $I = 100xe^{-x/10}$

$\dfrac{dI}{dx} = 100x(e^{-x/10})\left(-\dfrac{1}{10}\right) + 100e^{-x/10} = 100e^{-x/10}\left(-\dfrac{x}{10} + 1\right)$

$100e^{-x/10}\left(-\dfrac{x}{10} + 1\right) = 0, \quad x = 10$

$\dfrac{dI}{dx} > 0$ for $x < 10, \qquad \dfrac{dI}{dx} < 0$ for $x > 10$

Therefore, $x = 10$ items is a maximum.

Chapter 9

1. $\displaystyle\int (\sec x - \sec^3 x \tan x)\, dx$

$\quad = \displaystyle\int \sec x\, dx - \int \sec^2 x(\sec x \tan x)\, dx$

$\quad = \ln|\sec x + \tan x| - \dfrac{1}{3}\sec^3 x + C$

2. $\displaystyle\int \sin^3 x\, dx = \int \sin^2 x \sin x\, dx$

$\quad = \displaystyle\int (1 - \cos^2 x) \sin x\, dx$

$\quad = \displaystyle\int \sin x\, dx - \int \cos^2 x \sin x\, dx$

$\quad = -\cos x + \dfrac{1}{3}\cos^3 x + C$

3. $\displaystyle\int \tan^3 2x\, dx = \int \tan 2x(\tan^2 2x)\, dx$

$\quad = \displaystyle\int \tan 2x(\sec^2 2x - 1)\, dx$

$\quad = \dfrac{1}{2}\displaystyle\int \tan 2x \sec^2 2x(2\, dx)$

$\quad\quad - \dfrac{1}{2}\displaystyle\int \tan 2x(2\, dx)$

$\quad = \dfrac{1}{4}\tan^2 2x + \dfrac{1}{2}\ln|\cos 2x| + C$

4. $\displaystyle\int \cos^2 4\theta\, d\theta = \dfrac{1}{2}\int (1 + \cos 8\theta)\, d\theta$

$\quad = \dfrac{1}{2}\displaystyle\int d\theta + \dfrac{1}{16}\int \cos 8\theta(8\, d\theta)$

$\quad = \dfrac{1}{2}\theta + \dfrac{1}{16}\sin 8\theta + C$

5. $\displaystyle\int_4^7 \dfrac{\ln(x - 3)}{x - 3}\, dx = \dfrac{1}{2}[\ln(x - 3)]^2\Big|_4^7$

$\quad = \dfrac{1}{2}[(\ln 4)^2 - (\ln 1)^2] = 0.9609$

6. $\displaystyle\int \dfrac{4e^{\tan 2x}}{\cos^2 2x}\, dx = 2\int e^{\tan 2x}(2 \sec^2 2x\, dx)$

$\quad = 2e^{\tan 2x} + C$

7. $i = \displaystyle\int \dfrac{6t + 1}{4t^2 + 9}\, dt = \int \dfrac{6t\, dt}{4t^2 + 9} + \int \dfrac{dt}{4t^2 + 9}$

$\quad = \dfrac{6}{8}\displaystyle\int \dfrac{8t\, dt}{4t^2 + 9} + \dfrac{1}{2}\int \dfrac{2\, dt}{9 + (2t)^2}$

$\quad = \dfrac{3}{4}\ln(4t^2 + 9) + \dfrac{1}{2}\left(\dfrac{1}{3}\right)\tan^{-1}\dfrac{2t}{3} + C$

$i = 0$ for $t = 0$: $0 = \dfrac{3}{4}\ln 9 + \dfrac{1}{6}\tan^{-1} 0 + C$,

$C = -\dfrac{3}{4}\ln 9$

$i = \dfrac{3}{4}\ln(4t^2 + 9) + \dfrac{1}{6}\tan^{-1}\dfrac{2t}{3} - \dfrac{3}{4}\ln 9$

$\quad = \dfrac{3}{4}\ln\dfrac{4t^2 + 9}{9} + \dfrac{1}{6}\tan^{-1}\dfrac{2t}{3}$

8. $y = \dfrac{1}{\sqrt{16 - x^2}}$; $A = \displaystyle\int_0^3 y\, dx$

$\quad = \displaystyle\int_0^3 \dfrac{dx}{\sqrt{16 - x^2}}$

$\quad = \sin^{-1}\dfrac{x}{4}\Big|_0^3$

$\quad = \sin^{-1}\dfrac{3}{4} = 0.8481$

Chapter 10

1. Let $x = 2 \sin \theta$,

$\quad dx = 2 \cos \theta\, d\theta$.

$\displaystyle\int \dfrac{dx}{x^2\sqrt{4 - x^2}} = \int \dfrac{2 \cos \theta\, d\theta}{4 \sin^2 \theta\sqrt{4 - 4 \sin^2 \theta}}$

$\quad = \dfrac{1}{4}\displaystyle\int \dfrac{\cos \theta\, d\theta}{\sin^2 \theta\sqrt{\cos^2 \theta}} = \dfrac{1}{4}\int \csc^2 \theta\, d\theta$

$\quad = -\dfrac{1}{4}\cot \theta + C$

$\quad = -\dfrac{\sqrt{4 - x^2}}{4x} + C$

2. $\int xe^{-2x}\,dx$; $u = x$, $du = dx$, $dv = e^{-2x}\,dx$, $v = -\frac{1}{2}e^{-2x}$

$$\int xe^{-2x}\,dx = x(-\tfrac{1}{2}e^{-2x}) - \int (-\tfrac{1}{2}e^{-2x})\,dx$$

$$= -\tfrac{1}{2}xe^{-2x} - \tfrac{1}{4}e^{-2x} + C$$

3. $\int \dfrac{x^3 + 5x^2 + x + 2}{x^4 + x^2}\,dx$

$$\frac{x^3 + 5x^2 + x + 2}{x^4 + x^2} = \frac{x^3 + 5x^2 + x + 2}{x^2(x^2 + 1)}$$

$$= \frac{A}{x} + \frac{B}{x^2} + \frac{Cx + D}{x^2 + 1}$$

$$x^3 + 5x^2 + x + 2 = Ax(x^2 + 1) + B(x^2 + 1) + Cx^3 + Dx^2$$

x^3-terms: $1 = A + C$ x^2-terms: $5 = B + D$

x-terms: $1 = A$ $x = 0$: $2 = B$

$A = 1$, $B = 2$, $C = 0$, $D = 3$

$$\int \frac{x^3 + 5x^2 + x + 2}{x^4 + x^2}\,dx = \int \frac{dx}{x} + \int \frac{2dx}{x^2} + \int \frac{3dx}{x^2 + 1}$$

$$= \ln|x| - \frac{2}{x} + 3\tan^{-1}x + C$$

4. $\displaystyle\int_0^3 \frac{dx}{\sqrt{9 - x^2}} = \lim_{h \to 0} \int_0^{3-h} \frac{dx}{\sqrt{9 - x^2}} = \lim_{h \to 0} \sin^{-1}\left(\frac{x}{3}\right)\Big|_0^{3-h}$

$$= \lim_{h \to 0} \sin^{-1}\left(\frac{3 - h}{3}\right) = \sin^{-1} 1 = \frac{\pi}{2}$$

5. $\displaystyle\int_2^3 x\sqrt{x - 2}\,dx$

$u = \sqrt{x - 2}$, $u^2 = x - 2$, $x = u^2 + 2$, $dx = 2u\,du$

For $x = 2$, $u = 0$;   for $x = 3$, $u = 1$

$$\int_2^3 x\sqrt{x - 2}\,dx = \int_0^1 (u^2 + 2)(u)(2u\,du) = 2\int_0^1 (u^4 + 2u^2)\,du$$

$$= \tfrac{2}{5}u^5 + \tfrac{4}{3}u^3\big|_0^1 = \tfrac{2}{5} + \tfrac{4}{3} - 0 = \tfrac{26}{15}$$

6. $A = \displaystyle\int_3^4 \frac{x\,dx}{x^2 - 3x + 2}$

$$\frac{x}{x^2 - 3x + 2} = \frac{x}{(x - 1)(x - 2)} = \frac{A}{x - 1} + \frac{B}{x - 2}$$

$x = A(x - 2) + B(x - 1)$

$x = 1$: $1 = -A$, $A = -1$; $x = 2$: $B = 2$

$$A = \int_3^4 \frac{x\,dx}{x^2 - 3x + 2} = -\int_3^4 \frac{dx}{x - 1} + \int_3^4 \frac{2\,dx}{x - 2}$$

$$= -\ln|x - 1| + 2\ln|x - 2|\big|_3^4$$

$$= -\ln 3 + 2\ln 2 + \ln 2 - 2\ln 1 = 0.9808$$

Chapter 11

1. $z = 4 - x^2 - 4y^2$

Intercepts: $(2, 0, 0)$, $(-2, 0, 0)$, $(0, 1, 0)$, $(0, -1, 0)$, $(0, 0, 4)$

Traces: in yz-plane: $z = 4 - 4y^2$ (parabola)

in xz-plane: $z = 4 - x^2$ (parabola)

in xy-plane: $x^2 + 4y^2 = 4$ (ellipse)

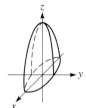

2. $\displaystyle\int_0^2 \int_{x^2}^{2x} (x^3 + 4y)\,dy\,dx = \int_0^2 (x^3y + 2y^2)\Big|_{x^2}^{2x}\,dx$

$$= \int_0^2 (2x^4 + 8x^2 - x^5 - 2x^4)\,dx = \int_0^2 (8x^2 - x^5)\,dx$$

$$= \tfrac{8}{3}x^3 - \tfrac{1}{6}x^6\big|_0^2 = \tfrac{64}{3} - \tfrac{64}{6} = \tfrac{32}{3}$$

3. $z = x^3 - x^2y + 3y^2$

$$\frac{\partial z}{\partial x} = 3x^2 - 2xy, \qquad \frac{\partial z}{\partial y} = -x^2 + 6y$$

$$dz = (3x^2 - 2xy)\,dx + (-x^2 + 6y)\,dy$$

4. $V = \displaystyle\int_0^3 \int_0^{\sqrt{9-y^2}} \sqrt{9 - y^2}\,dx\,dy$

$$= \int_0^3 x\sqrt{9 - y^2}\Big|_0^{\sqrt{9-y^2}}\,dy$$

$$= \int_0^3 (9 - y^2)\,dy$$

$$= 9y - \tfrac{1}{3}y^3\big|_0^3 = 18$$

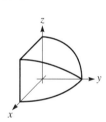

5. $xyz = 12$, $z = 12/xy$

$C = 4(xy) + 3(2yz) + 2(2xz)$

$$= 4xy + 6y\left(\frac{12}{xy}\right) + 4x\left(\frac{12}{xy}\right)$$

$$= 4xy + \frac{72}{x} + \frac{48}{y}$$

$$\frac{\partial C}{\partial x} = 4y - \frac{72}{x^2}, \qquad \frac{\partial C}{\partial y} = 4x - \frac{48}{y^2}$$

$$4y - \frac{72}{x^2} = 0, \qquad y = \frac{18}{x^2}, \qquad 4x - \frac{48}{y^2} = 0, \qquad x = \frac{12}{(18/x^2)^2}$$

$18^2x = 12x^4$, $x^3 = 27$, $x = 3$ m,

$y = 18/9 = 2$ m, $z = 12/6 = 2$ m

These are values for a minimum since C has unlimited maximum, due to x and y in denominators.

6. $\bar{y} = \dfrac{\displaystyle\int_0^\pi \int_0^{\sin x} y\, dy\, dx}{\displaystyle\int_0^\pi \int_0^{\sin x} dy\, dx}$

$= \dfrac{\displaystyle\int_0^\pi \frac{1}{2} y^2 \Big|_0^{\sin x} dx}{\displaystyle\int_0^\pi y \Big|_0^{\sin x} dx}$

$= \dfrac{\frac{1}{2}\displaystyle\int_0^\pi \sin^2 x\, dx}{\displaystyle\int_0^\pi \sin x\, dx} = \dfrac{\frac{1}{4}\displaystyle\int_0^\pi (1 - \cos 2x)\, dx}{-\cos x|_0^\pi} = \dfrac{\frac{1}{4}x - \frac{1}{8}\sin 2\,x|_0^\pi}{-(-1) - (-1)}$

$= \dfrac{\pi/4 - 0}{2} = \dfrac{\pi}{8} = 0.3927$

Chapter 12

1. $x^2 = 2x - y^2$
$x^2 + y^2 = 2x$
$r^2 = 2r\cos\theta$
$r = 2\cos\theta$

2. Cylindrical coordinates: $(2, 3\pi/4, 3)$
$x = r\cos\theta = 2\cos(3\pi/4) = 2(-\frac{\sqrt{2}}{2}) = -\sqrt{2}$
$y = r\sin\theta = 2\sin(3\pi/4) = 2(\frac{\sqrt{2}}{2}) = \sqrt{2}$
$z = z = 3$
Rectangular coordinates: $(-\sqrt{2}, \sqrt{2}, 3)$

3. $r = 3 + \cos\theta$

r	0	$\frac{\pi}{4}$	$\frac{\pi}{2}$	$\frac{3\pi}{4}$	π	$\frac{5\pi}{4}$	$\frac{3\pi}{2}$	$\frac{7\pi}{4}$	2π
θ	4.0	3.7	3.0	2.3	2.0	2.3	3.0	3.7	4.0

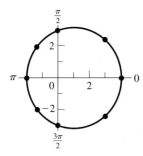

4. $r = 4\cos\theta + 2\sin\theta$
$r^2 = 4r\cos\theta + 2r\sin\theta$
$x^2 + y^2 = 4x + 2y$
$x^2 + y^2 - 4x - 2y = 0$,   circle

5. $A = \frac{1}{2}\int r^2\, d\theta$; $r^2 = 4\sin 2\theta$

$4\sin 2\theta = 0$, $\sin 2\theta = 0$, $\theta = 0, \pi/2$

$A = \dfrac{1}{2}\int_0^{\pi/2} 4\sin 2\theta\, d\theta = \int_0^{\pi/2} \sin 2\theta(2\, d\theta)$

$= -\cos 2\theta|_0^{\pi/2} = -(-1 - 1) = 2$

6. $V = \iint zr\, dr\, d\theta$; $z = r^2/4$, $r = 2\sec\theta$, $z = 0$, $\theta = 0$, $\theta = \pi/4$

$V = \displaystyle\int_0^{\pi/4} \int_0^{2\sec\theta} \frac{1}{4}r^2(r\, dr\, d\theta)$

$= \dfrac{1}{4}\displaystyle\int_0^{\pi/4} \int_0^{2\sec\theta} r^3\, dr\, d\theta$

$= \dfrac{1}{16}\displaystyle\int_0^{\pi/4} r^4 \Big|_0^{2\sec\theta} d\theta$

$= \displaystyle\int_0^{\pi/4} \sec^4\theta\, d\theta$

$= \displaystyle\int_0^{\pi/4} (\tan^2\theta + 1)\sec^2\theta\, d\theta$

$= \frac{1}{3}\tan^3\theta + \tan\theta|_0^{\pi/4} = \frac{1}{3} + 1 = \frac{4}{3}$

7. $r = 2.0 + 3.0\cos 2\theta$, $\omega = 3.0$ rad/s; $v_r = \dfrac{dr}{dt}$, $v_\theta = r\dfrac{d\theta}{dt}$

$v_r = -6.0\sin 2\theta\dfrac{d\theta}{dt}$

$v_r|_{\theta=0.26} = -6.0(\sin 0.52)(3.0) = -8.944$ m/s

$v_\theta = (2.0 + 3.0\cos 2\theta)\dfrac{d\theta}{dt}$

$v_\theta|_{\theta=0.26} = (2.0 + 3.0\cos 0.52)(3.0) = 13.81$ m/s

$v = \sqrt{8.944^2 + 13.81^2} = 16.5$ m/s

Chapter 13

1. $f(x) = (1 + e^x)^2$ $f(0) = (1 + 1)^2 = 4$
$f'(x) = 2(1 + e^x)(e^x)$ $f'(0) = 2(1 + 1)(1) = 4$
$\quad = 2e^x + 2e^{2x}$ $f''(0) = 2(1) + 4(1) = 6$
$f''(x) = 2e^x + 4e^{2x}$ $f'''(0) = 2(1) + 8(1) = 10$
$f'''(x) = 2e^x + 8e^{2x}$
$(1 + e^x)^2 = 4 + 4x + \frac{6}{2}x^2 + \frac{10}{6}x^3 + \cdots$
$\quad\quad = 4 + 4x + 3x^2 + \frac{5}{3}x^3 + \cdots$

2. $f(x) = \cos x$ $f\left(\dfrac{\pi}{3}\right) = \dfrac{1}{2}$

$f'(x) = -\sin x$ $f'\left(\dfrac{\pi}{3}\right) = -\dfrac{\sqrt{3}}{2}$

$f''(x) = -\cos x$ $f''\left(\dfrac{\pi}{3}\right) = -\dfrac{1}{2}$

$$\cos x = \frac{1}{2} - \frac{\sqrt{3}}{2}\left(x - \frac{\pi}{3}\right) - \frac{\frac{1}{2}\left(x - \frac{\pi}{3}\right)^2}{2} + \cdots$$

$$= \frac{1}{2}\left[1 - \sqrt{3}\left(x - \frac{\pi}{3}\right) - \frac{1}{2}\left(x - \frac{\pi}{3}\right)^2 + \cdots\right]$$

3. $\ln(1 + x) = x - \dfrac{x^2}{2} + \dfrac{x^3}{3} - \dfrac{x^4}{4} + \cdots$

$\ln 0.96 = \ln(1 - 0.04)$

$$= -0.04 - \frac{(-0.04)^2}{2} + \frac{(-0.04)^3}{3} - \frac{(-0.04)^4}{4}$$

$$= -0.040\,822\,0$$

4. $f(x) = \dfrac{1}{\sqrt{1 - 2x}} = (1 - 2x)^{-1/2}$

$(1 + x)^n = 1 + nx + \dfrac{n(n - 1)}{2}x^2 + \cdots$

$(1 - 2x)^{-1/2} = 1 + \left(-\dfrac{1}{2}\right)(-2x)$

$$+ \frac{-\frac{1}{2}\left(-\frac{3}{2}\right)}{2}(-2x)^2 + \cdots$$

$$= 1 + x + \tfrac{3}{2}x^2 + \cdots$$

5. $\displaystyle\int_0^1 x \cos x\, dx = \int_0^1 x\left(1 - \frac{x^2}{2} + \frac{x^4}{24}\right) dx$

$$= \int_0^1 \left(x - \frac{x^3}{2} + \frac{x^5}{24}\right) dx$$

$$= \frac{1}{2}x^2 - \frac{x^4}{8} + \frac{x^6}{144}\bigg|_0^1$$

$$= \tfrac{1}{2} - \tfrac{1}{8} + \tfrac{1}{144} - 0 = 0.3819$$

6. $f(t) = 0 \qquad -\pi \le t < 0$

$f(t) = 2 \qquad 0 \le t < \pi$

$a_0 = \dfrac{1}{2\pi}\displaystyle\int_0^\pi 2\, dt = \dfrac{1}{\pi}t\bigg|_0^\pi = 1$

$a_n = \dfrac{1}{\pi}\displaystyle\int_0^\pi 2\cos nx\, dx = \dfrac{2}{n\pi}\sin nx\bigg|_0^\pi$

$$= 0 \quad \text{for all } n$$

$b_n = \dfrac{1}{\pi}\displaystyle\int_0^\pi 2\sin nx\, dx = \dfrac{2}{n\pi}(-\cos nx)\bigg|_0^\pi$

$$= \frac{2}{n\pi}(1 - \cos n\pi)$$

$b_1 = \dfrac{2}{\pi}(1 + 1) = \dfrac{4}{\pi} \qquad b_2 = \dfrac{2}{2\pi}(1 - 1) = 0$

$b_3 = \dfrac{2}{3\pi}(1 + 1) = \dfrac{4}{3\pi}$

$f(t) = 1 + \dfrac{4}{\pi}\sin x + \dfrac{4}{3\pi}\sin 3x + \cdots$

7. $f(x) = x^2 + 2, f(-x) = (-x)^2 + 2 = x^2 + 2, -f(-x) \ne f(x)$

Since $f(x) = f(-x)$, $f(x)$ is an even function.

Since $f(x) \ne -f(-x)$, $f(x)$ is not an odd function.

$g(x) = x^2 - 1$, $g(x) = f(x) - 3$

Fourier series for $g(x)$ is $F(x) - 3$.

Chapter 14

1. $x\dfrac{dy}{dx} + 2y = 4$

$dy + \dfrac{2}{x}y\, dx = \dfrac{4}{x}dx$

$e^{\int \frac{2}{x}dx} = e^{2\ln x} = x^2$

$yx^2 = \displaystyle\int \frac{4}{x}x^2\, dx = \int 4x\, dx = 2x^2 + c$

$y = 2 + \dfrac{c}{x^2}$

2. $\csc y\, dx + e^x dy = 0$

$e^{-x}dx + \sin y\, dy = 0$

$-e^{-x} - \cos y + c = 0$

$e^{-x} + \cos y = c$

3. $x\, dx + y\, dy = x^2\, dx + y^2\, dx = (x^2 + y^2)\, dx$

$\dfrac{x\, dx + y\, dy}{x^2 + y^2} = dx$

$\tfrac{1}{2}\ln(x^2 + y^2) = x + \tfrac{1}{2}c$

$\ln(x^2 + y^2) = 2x + c$

4. $y' + y\tan x = 2x\cos x$

$e^{\int \tan x\, dx} = e^{-\ln\cos x} = e^{\ln(\cos x)^{-1}} = \sec x$

$y\sec x = \displaystyle\int 2x\cos x\sec x\, dx = \int 2x\, dx = x^2 + c$

$y = (x^2 + c)\cos x$

5. $(xy + y)\dfrac{dy}{dx} = 2$

$y(x + 1)\, dy = 2\, dx$

$y\, dy = \dfrac{2\, dx}{x + 1}$

$\tfrac{1}{2}y^2 = 2\ln(x + 1) + c$

$y = 2 \quad \text{when} \quad x = 0$

$\tfrac{1}{2}(4) = 2\ln(1) + c, \quad c = 2$

$\tfrac{1}{2}y^2 = 2\ln(x + 1) + 2$

$y^2 = 4\ln(x + 1) + 4$

6. $\dfrac{dA}{dt} = rA$

$\dfrac{dA}{A} = r\, dt$

$\ln A = rt + \ln c$

$\ln \dfrac{A}{c} = rt$

$A = ce^{rt}$

$A_0 = ce^0,$

$c = A_0$

$A = A_0 e^{rt}$

7. $L\dfrac{d^2q}{dt^2} + R\dfrac{dq}{dt} = E \qquad L\dfrac{di}{dt} + Ri = E$

$L = 2\ \text{H} \qquad R = 8\ \Omega \qquad E = 6\ \text{V}$

$2\dfrac{di}{dt} + 8i = 6 \qquad di + 4i\,dt = 3\,dt$

$e^{\int 4\,dt} = e^{4t} \qquad ie^{4t} = \int 3e^{4t}\,dt = \tfrac{3}{4}e^{4t} + c$

$i = 0$ for $t = 0 \qquad 0 = \tfrac{3}{4} + c \qquad c = -\tfrac{3}{4}$

$ie^{4t} = \tfrac{3}{4}e^{4t} - \tfrac{3}{4}$

$i = \tfrac{3}{4}(1 - e^{-4t}) = 0.75(1 - e^{-4t})$

Chapter 15

1. $D^3y + 81Dy = 0$

$m^3 + 81m = 0, \qquad m = 0,\ \pm 9j$

$y = c_1 + c_2 \sin 9x + c_3 \cos 9x$

2. $2\,D^2y - Dy = 2\cos x$

$2m^2 - m = 0 \qquad m = 0,\ \tfrac{1}{2}$

$y_c = c_1 + c_2 e^{x/2}$

$y_p = A \sin x + B \cos x$

$Dy_p = A \cos x - B \sin x$

$D^2y_p = -A \sin x - B \cos x$

$2(-A \sin x - B \cos x) -$

$\quad (A \cos x - B \sin x) = 2 \cos x$

$-2A + B = 0 \qquad -2B - A = 2 \qquad A = -\tfrac{2}{5} \qquad B = -\tfrac{4}{5}$

$y = c_1 + c_2 e^{x/2} - \tfrac{2}{5} \sin x - \tfrac{4}{5} \cos x$

3. $\dfrac{d^2y}{dx^2} - 4\dfrac{dy}{dx} + 4y = 3x$

$m^2 - 4m + 4 = 0 \qquad m = 2, 2$

$y_c = (c_1 + c_2 x)e^{2x}$

$y_p = A + Bx \qquad Dy_p = B \qquad D^2y_p = 0$

$0 - 4B + 4(A + Bx) = 3x$

$4A - 4B = 0 \qquad 4B = 3$

$B = \tfrac{3}{4} \qquad A = \tfrac{3}{4}$

$y = (c_1 + c_2 x)e^{2x} + \tfrac{3}{4} + \tfrac{3}{4}x$

4. $D^2y - 2\,Dy - 8y = 4e^{-2x}$

$m^2 - 2m - 8 = 0 \qquad (m + 2)(m - 4) = 0 \qquad m = -2, 4$

$y_c = c_1 e^{-2x} + c_2 e^{4x}$

$y_p = Axe^{-2x}$ (factor of x necessary due to first term of y_c)

$Dy_p = Ae^{-2x} - 2Axe^{-2x}$

$D^2y_p = -2Ae^{-2x} - 2Ae^{-2x} + 4Axe^{-2x} = -4Ae^{-2x} + 4Axe^{-2x}$

$(-4Ae^{-2x} + 4Axe^{-2x}) - 2(Ae^{-2x} - 2Axe^{-2x})$

$\quad - 8Axe^{-2x} = 4e^{-2x}$

$-6Ae^{-2x} = 4e^{-2x}, \qquad A = -\tfrac{2}{3}$

$y = c_1 e^{-2x} + c_2 e^{4x} - \tfrac{2}{3}xe^{-2x}$

5. $D^2y - 25y = 5x$

$m^2 - 25 = 0, \qquad m = \pm 5$

$y_c = c_1 e^{5x} + c_2 e^{-5x}$

$y_p = Ax + B, \qquad Dy_p = A, \qquad D^2y_p = 0$

$-25(Ax + B) = 5x, \qquad -25A = 5, \qquad A = -1/5, \qquad B = 0$

$y = c_1 e^{5x} + c_2 e^{-5x} - \tfrac{1}{5}x$

$Dy = 5c_1 e^{5x} - 5c_2 e^{-5x} - \tfrac{1}{5}$

$y = 3$ and $Dy = -5$ when $x = 0$

$3 = c_1 + c_2, \qquad -5 = 5c_1 - 5c_2 - \tfrac{1}{5}$

$25c_1 + 25c_2 = 75$

$25c_1 - 25c_2 = -24$

$\qquad 50c_1 = 51, \qquad c_1 = \tfrac{51}{50}, \qquad c_2 = \tfrac{99}{50}$

$y = \tfrac{1}{50}(51e^{5x} + 99e^{-5x} - 10x)$

6. $F = kx;\ 16 = k(0.5),\ k = 32\ \text{lb/ft}$

$m = \dfrac{16\ \text{lb}}{32\ \text{ft/s}^2} = 0.5\ \text{slug}$

$mD^2x = -kx, \qquad 0.5D^2x + 32x = 0$

$D^2x + 64x = 0$

$m^2 + 64 = 0; \qquad m = \pm 8j$

$x = c_1 \sin 8t + c_2 \cos 8t$

$Dx = 8c_1 \cos 8t - 8c_2 \sin 8t$

$x = 0.3\ \text{ft} \qquad Dx = 0$ for $t = 0$

$0.3 = c_1 \sin 0 + c_2 \cos 0, \qquad c_2 = 0.3$

$0 = 8c_1 \cos 0 - 8c_2 \sin 0, \qquad c_1 = 0$

$x = 0.3 \cos 8t$

Chapter 16

1. $\dfrac{dy}{dx} = (x + y)^2, \qquad y(0) = 1, \qquad \Delta x = 0.1$

$y_1 = 1 + (0 + 1)^2(0.1) = 1.1$

$y_2 = 1.1 + (0.1 + 1.1)^2(0.1) = 1.244$

2. $\dfrac{dy}{dx} = (x + y)^2, \qquad y(0) = 1, \qquad \Delta x = 0.2$

$x_0 = 0, \qquad y_0 = 1, \qquad H = 0.2$

$y_1 = y_0 + \tfrac{1}{6}H(J + 2K + 2L + M), \qquad \Delta x = H$

$J = f(x_0, y_0), \qquad K = f(x_0 + 0.5H, y_0 + 0.5HJ)$

$L = f(x_0 + 0.5H, y_0 = 0.5HK), \qquad M = f(x_0 + H, y_0 + HL)$

$J = (0 + 1)^2 = 1$

$K = \{[0 + 0.5(0.2)] + [1 + 0.5(0.2)(1)]\}^2 = 1.44$

$L = \{[0 + 0.5(0.2)] + [1 + 0.5(0.2)(1.44)]\}^2 = 1.547536$

$M = \{(0 + 0.2) + [1 + 0.2(1.547536)]\} = 2.278611987$

$y_1 = 1 + \tfrac{1}{6}(0.2)[1 + 2(1.44) + 2(1.547536) + 2.278611987]$

$\quad = 1.3085$

3. $\dfrac{dy}{dx} = (x + y)^2, \qquad y(0) = 1$

$dy = (x + 1)^2\, dx$

$y_1 = \frac{1}{3}(x + 1)^3 + c$

$1 = \frac{1}{3} + c, \qquad c = \frac{2}{3}$

$y_1 = \frac{1}{3}(x + 1)^3 + \frac{2}{3}$

$y_1(0.2) = \frac{1}{3}(1.2)^3 + \frac{2}{3} = 1.2427$

4. $f(t) = t^2 e^{2t}$

$\mathcal{L}(t^2 e^{2t}) = \dfrac{2!}{(s - 2)^3} = \dfrac{2}{(s - 2)^3}$

5. $F(s) = \dfrac{10}{s^2 + 4s + 29} = \dfrac{2(5)}{(s + 2)^2 + 5^2}$

$\mathcal{L}^{-1}(F) = f(t) = 2e^{-2t} \sin 5t$

6. $D^2 y - Dy - 2y = 12 \qquad y(0) = 0,\ y'(0) = 0$

$\mathcal{L}(y'') - \mathcal{L}(y') - 2\mathcal{L}(y) = \mathcal{L}(12)$

$s^2\mathcal{L}(y) - s(0) - 0 - [s\mathcal{L}(y) - 0] - 2\mathcal{L}(y) = \dfrac{12}{s}$

$\mathcal{L}(y)(s^2 - s - 2) = \dfrac{12}{s}$

$\mathcal{L}(y) = \dfrac{12}{s(s^2 - s - 2)} = \dfrac{12}{s(s + 1)(s - 2)}$

$= \dfrac{A}{s} + \dfrac{B}{s + 1} + \dfrac{C}{s - 2}$

$12 = A(s + 1)(s - 2) + Bs(s - 2) + Cs(s + 1)$

$s = 0: \quad 12 = -2A,\ A = -6$

$s = -1: \quad 12 = 3B,\ B = 4$

$s = 2: \quad 12 = 6C,\ C = 2$

$\mathcal{L}(y) = -\dfrac{6}{s} + \dfrac{4}{s + 1} + \dfrac{2}{s - 2}$

$y = -6 + 4e^{-t} + 2e^{2t}$

7. $L\dfrac{di}{dt} + Ri = E; \quad L = 2\text{ H}, \quad R = 8\ \Omega, \quad E = 6\text{ V}, \quad i(0) = 0$

$2\dfrac{di}{dt} + 8i = 6, \quad \dfrac{di}{dt} + 4i = 3$

$\mathcal{L}\!\left(\dfrac{di}{dt}\right) + \mathcal{L}(4i) = \mathcal{L}(3)$

$s\mathcal{L}(i) - i(0) + 4\mathcal{L}(i) = \dfrac{3}{s}, \quad \mathcal{L}(i)(s + 4) = \dfrac{3}{s}$

$\mathcal{L}(i) = \dfrac{3}{s(s + 4)} = \dfrac{3}{4}\left(\dfrac{4}{s(s + 4)}\right)$

$i = 0.75(1 - e^{-4t})$

The final review exercise in each chapter is a writing exercise. Each of these 16 exercises will require at least a paragraph to provide a good explanation of the problem presented. Also, there are over 110 other exercises throughout the text (at least five in each chapter and marked with Ⓦ before the exercise number or instructions) that require at least one complete sentence (up to a paragraph) to complete the solution.

Following is a listing of these writing exercises. The first number shown is the page number, and the numbers in parentheses are the exercise numbers. The * denotes the final review exercise of the chapter.

NOTES

NOTES

NOTES

NOTES

ALGEBRA

Exponents and Radicals

$a^m \cdot a^n = a^{m+n}$

$\dfrac{a^m}{a^n} = a^{m-n} \quad a \neq 0$

$(a^m)^n = a^{mn}$

$(ab)^n = a^n b^n$

$\left(\dfrac{a}{b}\right)^n = \dfrac{a^n}{b^n} \quad b \neq 0$

$a^0 = 1 \quad a \neq 0$

$a^{-n} = \dfrac{1}{a^n} \quad a \neq 0$

$a^{m/n} = \sqrt[n]{a^m} = (\sqrt[n]{a})^m$

$\sqrt{ab} = \sqrt{a}\,\sqrt{b}$

Special Products

$a(x + y) = ax + ay$

$(x + y)(x - y) = x^2 - y^2$

$(x + y)^2 = x^2 + 2xy + y^2$

$(x - y)^2 = x^2 - 2xy + y^2$

Quadratic Equation and Formula

$ax^2 + bx + c = 0$

$x = \dfrac{-b \pm \sqrt{b^2 - 4ac}}{2a}$

Properties of Logarithms

$\log_b x + \log_b y = \log_b xy$

$\log_b x - \log_b y = \log_b\left(\dfrac{x}{y}\right)$

$n \log_b x = \log_b (x^n)$

Complex Numbers

$\sqrt{-a} = j\sqrt{a} \quad (a > 0)$

$x + yj = r(\cos\theta + j\sin\theta)$

$\quad = re^{j\theta} = r\underline{/\theta}$

Variation

Direct variation: $y = kx$

Inverse variation: $y = k/x$

TRIGONOMETRY

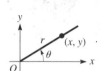

$\sin\theta = \dfrac{y}{r} \qquad \cos\theta = \dfrac{x}{r} \qquad \tan\theta = \dfrac{y}{x} \qquad \cot\theta = \dfrac{x}{y} \qquad \sec\theta = \dfrac{r}{x} \qquad \csc\theta = \dfrac{r}{y}$

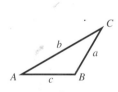

Law of Sines: $\dfrac{a}{\sin A} = \dfrac{b}{\sin B} = \dfrac{c}{\sin C}$

Law of Cosines: $a^2 = b^2 + c^2 - 2bc\cos A$

$\pi \text{ rad} = 180°$

Basic Identities

$\sin\theta = \dfrac{1}{\csc\theta} \qquad \cos\theta = \dfrac{1}{\sec\theta} \qquad \tan\theta = \dfrac{1}{\cot\theta} \qquad \tan\theta = \dfrac{\sin\theta}{\cos\theta} \qquad \cot\theta = \dfrac{\cos\theta}{\sin\theta}$

$\sin^2\theta + \cos^2\theta = 1 \qquad 1 + \tan^2\theta = \sec^2\theta \qquad 1 + \cot^2\theta = \csc^2\theta$

$\sin 2\alpha = 2\sin\alpha\cos\alpha \qquad \cos 2\alpha = \cos^2\alpha - \sin^2\alpha = 2\cos^2\alpha - 1 = 1 - 2\sin^2\alpha$

$\sin\dfrac{\alpha}{2} = \pm\sqrt{\dfrac{1 - \cos\alpha}{2}} \qquad \cos\dfrac{\alpha}{2} = \pm\sqrt{\dfrac{1 + \cos\alpha}{2}}$